DRYING '84

Edited by

Arun S. Mujumdar

McGill University, Montreal

HEMISPHERE PUBLISHING CORPORATION

Washington New York London

DISTRIBUTION OUTSIDE NORTH AMERICA

SPRINGER-VERLAG

Berlin Heidelberg New York Tokyo

DRYING '84

1 2 3 4 5 6 7 8 9 0 E B E B 8 9 8 7 6 5 4

Library of Congress Cataloging in Publication Data

Main entry under title:

Drying '84.

Includes bibliographical references and indexes.
1. Drying. I. Mujumdar, A. S.
TP363.D7924 1984 660.2′8426 84-689
ISBN 0-89116-298-4 Hemisphere Publishing Corporation

DISTRIBUTION OUTSIDE NORTH AMERICA:
ISBN 3-540-13429-8 Springer-Verlag Berlin

CONTENTS

PREFACE

It is indeed a great pleasure for me to present this third volume in the DRYING series. Like its predecessor volumes (DRYING '80, Vol. 1&2 and DRYING '82) this book presents a compendium of recent developments in drying theory and practice. With escalating energy costs it is hardly surprising that the interest in understanding and improving drying technology in general is rising rapidly. This is reflected in the appreciable increase in the number of scientific and technical publications on drying and allied topics e.g. flow, heat/mass transfer in porous media, gas/solid contacting processes etc. Judging from the strongly positive response to the earlier compendia on drying and the serial ADVANCES IN DRYING, it is my fervent hope that DRYING'84 will fill a widely recognized need and serve the drying community worldwide.

While the interest in drying technology is driven primarily (but not exclusively) by the shortage of energy little is said or done about the shortage of an even scarcer commodity viz. technical manpower resources. With limited technical resources it is more important now than ever to keep oneself abreast of latest developments and thinking in one's area of expertise. It is well known that the half-life of engineering education is now only about 5-6 years. In the case of drying R & D, a further complicating factor is its truly interdisciplinary and global nature. In compiling DRYING'84 and its companion volumes I have made special effort to obtain input from all parts of the world. Some of the work included in this volume is available in the English language for the first time. Also, all contributions were prepared specially for this book. I want to thank all contributors for their interest, cooperation, time and effort which will go a long way in serving their peers around the world for years to come.

I am grateful to the following individuals who went out of their way to make DRYING'84 what it is: Dr. L. Imre, Technical University of Budapest, Hungary; Dr. I. Filkova, Czech Technical University, Prague, Czechoslovakia; Dr. Z. Pakowski and Prof. S. Strumillo, Lodz Technical University, Poland; Dr. O.G. Martynenko, Heat and Mass Transfer Institute, Minsk, USSR. My special warm thanks and sincere appreciation goes to my Editorial Assistant, Purnima Mujumdar, whose untiring and prompt typing and retyping of numerous papers and formidable correspondence related to this book, made it possible for me to compile this book. Without her voluntary contribution this work would have not seen the light of the day.

Finally, I would like to record with pride that 1983-84 marks the 75th Anniversary of the Chemical Engineering Department at McGill University. I am happy to have this truly international book come out on this significant occasion, particularly so because this department has been so closely associated with the International Drying Symposia and various key reference works in drying.

McGill University
Montreal, Quebec,
Canada

Arun S. Mujumdar

SECTION I: INTERNATIONAL EFFORTS IN DRYING R & D

CANADIAN RESEARCH AND DEVELOPMENT IN DRYING - A SURVEY

A.S. Mujumdar[1] and G.S.V. Raghavan[2]

[1]Department of Chemical Engineering, McGill University, Montreal, Quebec, Canada H3A 2K6

[2]Department of Agricultural Engineering, Macdonald College, Ste. Anne de Bellevue, Quebec, Canada H9X 1C0

ABSTRACT

Results are presented of a survey on the drying R & D activity in Canadian academic institutions and government laboratories. It is observed that almost all activity is funded by government agencies and that a major fraction of the entire drying activity is stimulated by applications in agriculture and the forest product industries. On the basis of this survey, recommendations are made concerning additional areas for drying R & D in Canada, and improved communication between researchers, designers and users of drying equipment in Canada. In general, the emphasis appears to be on applied rather than on basic research. No attempt was made to survey product or service-oriented developmental work carried out by industry, principally because of the difficulty associated with accessibility of such information.

INTRODUCTION

The objective of this article is to present a concise survey of solids drying research and development activity in Canadian academic and research institutions. This survey is necessarily limited to R & D, the results of which, with few exceptions, are generally available. Product or process development work which is proprietory (and often in the nature of engineering service), is excluded because of the difficulty in compiling such information.

This survey report is based on recent papers and reports originating in Canadian institutions and, to a greater extent, on the response received to a short form questionnaire mailed to: (a) all chemical, agricultural and food engineering departments across Canada, (b) major Canadian government laboratories and industrial R & D centers, and (c) government agencies responsible for monitoring and funding of drying R & D in Canada. Additional mailings were made to individuals known to be active in solids drying or related research. It should be noted at the outset that this survey also covers topics that are closely related to drying (e.g. heat transfer from impinging jets to moving surfaces, an area directly pertinent to constant rate drying of continuous sheets).

The main objective of this effort has been to present to the ever-growing community of drying researchers, designers and users, the current status of drying R & D in Canada with a view to the establishment of collaborative links. Since drying is a truly interdisciplinary area, such a survey would hopefully help avoid "reinvention of the wheel" by engineers working independently in their own discipline (and often in their own country). In spite of the initiation of the biennial international symposia in drying and of the publication of a series of books by the first author, it is observed that further efforts are needed to improve interdisciplinary and international communication in the field of solids drying. It is hoped that the establishment of the new journal "DRYING TECHNOLOGY", an international journal, will help to alleviate the serious communication gaps which are clearly evident in the widely scattered drying literature.

CLASSIFICATION OF R & D AREAS

Several possible classification schemes were considered to present the survey results. It is well known that a fine classification of dryers leads to well over 100 dryer types. Since most returns could be readily categorized as follows, the survey results will be presented under these five headings:

(a) Drying of continuous sheets
(b) Drying of particulates
(c) Miscellaneous
(d) Drying of grains
(e) Drying in agricultural industry (other than grains)

It is instructive to make the following general observations at the outset. (Note that all statistics are based on entries received for this survey.)

1. Over 60% of drying R & D is in some way connected with agriculture. About 15% is related to the pulp and paper industry while the remainder is motivated by more general applications and by the need for fundamental understanding.
2. All drying R & D surveyed is apparently funded by one of the various government agencies or research institutes. No project was reported to be funded by Canadian industry. This is surprising in view of the energy-intensive nature

of drying. The pulp and paper industry alone spends over 500 million dollars per year for thermal drying of pulp and paper in Canada.

3. Only about 15% of the reported R & D is related to development of novel drying techniques. It is necessary to increase the component of innovation in drying R & D in Canada.
4. Drying research is confined to a few universities. It is further confined mainly to agricultural and chemical engineering disciplines although mechanical engineers can contribute very significantly to drying research and development. Electrical engineers could also contribute to specialized electrical drying methods, e.g. dielectric or microwave drying. Mining and mineral engineers can contribute to problems of drying of minerals, coal etc.

(a) Drying of Continuous Sheets

Contact drying using "steam cans" continues to be the industrial process for drying of newsprint and other grades of paper, even though the concept was patented some 160 years ago. Mujumdar (1) has reviewed the recent developments and trends in drying of paper and other materials in the form of thin continuous sheets, e.g. textiles, pulp web etc. Notable among the newer concepts in high speed drying of newsprint is the so-called Papridryer process (after the Pulp and Paper Research Institute of Canada) which combines the impingement drying process (already in common usage in drying tissue paper and in stenter drying of textiles) with the through drying process. Although successful trials on both pilot and mill scale have been reported, commercial exploitation has not yet been reported. Indeed, this process is based on very sound engineering concepts and with further development, promises successful commercial application within the next decade.

Douglas, Mujumdar and Crotogino of the Chemical Engineering Department of McGill University have examined transfer processes under single/multiple jets impinging on stationary and moving surfaces which may or may not be permeable. Both analytical and experimental studies have been made for laminar/turbulent jets impinging normally or obliquely. These have been partially reported in a series of publications (2 to 6) and theses. Effects of swirl in the jet, introduction of cross-flow (imposed or induced by neighboring jets), large temperature difference between the jet and the heat transfer surface, etc., have also been investigated. Mujumdar and Douglas (7) reviewed the flow and heat transfer under impinging jets and formed the basis for the long series of subsequent investigations in this important area.

The analytical studies in impinging jet heat transfer have consisted of numerical solutions of the full governing equations (continuity, momentum and energy) subject to appropriate boundry conditions. Both vorticity-stream function and velocity-pressure formulations have been used. For turbulent jets, the well-known two-equation (k-ε) model has been used for jets impinging on stationary surfaces and moving surfaces in the presence of imposed cross-flow. Current work in this area is examining better modelling of the wall region, a problem of considerable complexity. All computational work to date has been restricted to two-dimensional turbulent jets. The effect of wall suction has not been examined.

On the experimental side, several studies have been reported on the flow field development and wall static pressure distribution on single and multiple slot jets as well as on single round jets. Effects of partial confinement (also referred to as semi-confinement) and uniform wall suction have been studied. More definitive results are needed on the effect of nozzle exit turbulence levels on the jet development. For round jets, Obot et al. (8) have systematically examined the effect of nozzle configuration, which was varied from a long tube to an orifice punched in a thin rigid plate. Flow results are useful in explaining the heat transfer results. It is interesting to note that the effect of suction at the impingement surface is to suppress the turbulence level while steepening the velocity (and hence the temperature) profile at the surface, resulting in higher transfer rates.

Presence of a normal flux at the impingement surface (due to applied suction which causes the through drying of permeable mats supported on rotating rolls) is a unique boundary condition peculiar to the Papridryer process. Presence of a confining surface (parallel to the impingement surface) and motion of the surface are features not unique to the Papridryer process; however, no work in these complex flow systems other than that by Baines and Keffer (9) has been reported to date. They measured the surface shear stress on a rotating drum with an air jet impinging on it. The impinging jet group at McGill has successfully developed an experimental rig to permit measurement of the instantaneous heat transfer rate for a permeable rotating cylinder exposed to a confined impinging jet (10,11). The influence of oblique impingement and suction applied at the cylinder surface is currently being examined. Successful completion of these studies will allow accurate evaluation of impinging jet heat transfer rates under conditions closely simulating the Papridryer. It is noteworthy that such results are also useful in the design and operation of Yankee dryers, textile stenter dryers and cooling equipment.

Drying under impinging turbulent jets per se has not been studied at McGill. Mujumdar and Huang (12) have made numerical computations of coupled heat and mass transfer under laminar round impinging jets in the presence of wall suction. These results must be extended to turbulent jets. Jaussaud (13) extended the computer code to study the effects of oblique impingement and cross-flow on coupled heat and mass transfer under a slot jet. For practical usage it is necessary to study turbulent jets both experimentally and numerically in so far as coupled heat and mass transfer is concerned.

Through drying of paper is the subject of a current thesis project which uses the infrared absorption technique to monitor the instantaneous drying rates for uniform air flow through a carefully mounted sample sheet. A critical nozzle is employed to ensure unchanged airflow as the sample permeability varies with moisture content during drying. The experimental technique is similar to that used by Gummel (14). In a separate project, an attempt is also being made to model the transport of heat and moisture in paper.

Successful development of models and experimental results for impingement heat/mass transfer and through drying will permit combination of the two to simulate the Papridryer process. Even on the basis of rather simple assumptions, Mujumdar and Huang (12) have demonstrated the ability to simulate the reported performance of pilot and mill Papridryers. They conclude, however, that the successful predictions are possibly due to compensating errors rather than to precision of the model they developed.

"Sheet flutter" is one of several limitations of the conventional contact drying process. The sheet, while travelling in the open draw between succeeding steam cans, develops violent flutter as a result of aeromechanical causes not yet fully understood. This happens on typical, well-engineered newsprint machines at machine speeds of about 1000 m/minute. Mujumdar (15,16) has examined this problem in some detail with a view to identifying the aerodynamic and mechanical causes of sheet flutter. He also presented an elementary analytical model (the so-called travelling threadline model) which seems to adequately describe the sheet flutter phenomenon without accounting for the aerodynamics involved. The actual phenomenon is much more complex. Numerous remedies have been suggested in the trade and patent literature which has been reviewed by Mujumdar (17).

It is now recognized that superheated steam as a drying medium has some definite advantages over hot air. Loo and Mujumdar (18) have undertaken a technical feasibility study on superheated steam drying of paper and pulp in sheet form. It may be noted that flash drying of pulp in superheated steam is already a viable commercial process boasting extremely high energy efficiency. A process very close to drying in superheated steam is used commercially in India in textile dryers; here, the conventional hot air jets are replaced by superheated steam. The dryer exhaust steam is reheated and recycled.

Blackwell and MacCallum (19) have presented results of a cost-benefit analysis of commercially available hog fuel drying systems. They have covered economics of hog fuel drying, design and environmental constraints. The hog fuel dryers are (a) flue-gas dryers (rotary, cascade and flash types), and (b) steam dryers (hot air and shell-and-tube). They show that a hog fuel dryer is desirable as a retrofit on old boiler installations but not on new ones designed to burn wet hog fuel. Payback periods of 1.5 to 3.5 years have been estimated for different systems; the shortest periods are for flash dryers. More data are needed on drying characteristics of hog fuel.

(b) Drying of Particulates

Since drying of grains is considered in a separate section, this section will deal only with non-grain drying and related studies. Quite understandably, much of the reported work deals with gas-solid contactors developed over the past two or three decades, (e.g. fluidized beds and spouted beds). In particular, much work has been done with mechanically assisted fluidization (vibrated beds, agitated beds) and spouted beds including several variants thereof. Mujumdar et al. (20 to 25) have reported results of a series of investigations on the aerodynamics, heat transfer to immersed surfaces and drying in vibrated fluid beds (VFB). For sticky, agglomerating solids with wide size distribution, application of vibration aids fluidization (or, more correctly, pseudo-fluidization) at air flow rates which are small fractions of the minimum fluidization velocity. The fluid bed can therefore be designed for aeration rates dictated by the heat/mass transfer operation rather than by the requirement of fluidization.

Pakowski et al. (25) have reviewed world literature on VFB drying, while Karoly et al. (26) are currently working on a more general critical review of the world literature on vibrated beds of particulates. Much of the existing literature in this area is not available in the English language. In October 1982 at the first Latin American Heat and Mass Transfer Congress, Mujumdar presented a plenary lecture on VFB transfer operations. The text of this lecture will appear in an issue of the Latin American Journal of Heat and Mass Transfer in both English and Spanish (22).

For difficult-to-fluidize solids an alternative to vibration is agitation of the bed. Khalid and Mujumdar (21) examined the flow and drying characteristics in such a device using an on-line fast-response dew point meter to measure the transient drying rates. The results were not very favorable due to poor distribution of solids and severe channelling of the air due to stirring. The vibrated bed is a superior option to the agitated bed.

Mujumdar and Malhotra are currently examining heat transfer from multiple horizontal cylinders to a VFB of model particles - both dry and wet. With the help of high speed movie photography, two-dimentional VFB will be used to obtain a semi-quantitative picture of the solids distribution during vibration. Heat transfer data from immersed surfaces are needed for design of energy-efficient direct-indirect VFB dryers which are already in production. The vibrated bed consists of large sand particles (or glass beads); the paste is fed continuously (or batch-wise for very small pilot or lab-scale operation) while the dry powder is collected in a bag filter. The paste or slurry forms a thin coating on the inert solids, dries rapidly and is peeled, ground and entrained in the drying air. Aeration rate and air temperature can be controlled to obtain the desired residence time and final moisture content. This device is similar to the spouted bed dryer-grinder invented by Drs. Pallai, Nemeth et al. of the Hungarian

Academy of Sciences, Veszprem, Hungary (US Patent 4203 228/77, 1977).

Spouted beds, and modifications thereof, have been subjects of intense investigations at the University of British Columbia, Vancouver (Professors N. Epstein and J.R. Grace, Chemical Engineering Department), and more recently at McGill University. Although the work at UBC has not been directed specifically at drying, the results and fundamental understanding gained are nevertheless very relevant to the design and operation of the spouted bed as a particulate dryer, or as a dryer for pastes.

The research being carried out at McGill (27) is directed towards eventual application in grain drying. Variants of the two-dimensional spouted bed concept are being studied experimentally using various grains (wheat, oats, barley, corn, etc.). Current and recent studies have been confined to the aerodynamics of these novel designs. This work will be extended to include wall-to-bed heat transfer and convective (gas-particle) heat transfer. The eventual goal is to design and optimize a spouted bed dryer for grains (or other granular spoutable solids) using combined direct-indirect modes of operation for high energy efficiency. Mujumdar (28) recently presented a plenary lecture at the Symposium on Transport Processes in Particulate Systems, São Carlos, Brazil, summarizing recent developments in the spouted bed technology and also presenting a number of modifications that are worth pursuing in an effort to offset some of the disadvantages of the conventional tubular spouted bed first proposed by Mathur and Gishler of N.R.C., Ottawa, in 1955.

Swaminathan and Mujumdar (29) have presented results of their work on conventional tubular spouted beds, half-beds and full beds with a draft-tube. They used a flat bottom bed for fabrication simplicity in the field, and attempted to position the draft tube so as to minimize the dead zone on the bottom plate. This had limited success. One important conclusion of this study was that rough particles behave differently from smooth ones. In particular, half-beds of rough particles have better spoutability and lower peak pressure drop than full beds. An international collaborative effort now underway involves preparation of an exhaustive and critical review of world literature in the rapidly growing area of spouted bed drying technology. A.S. Mujumdar of McGill University has initiated collaboration with E. Pallai and J. Nemeth of the Hungarian Academy of Sciences, Veszprem, Hungary to prepare an article for Volume 4 of <u>Advances in Drying</u> (30). This unique effort will cover both the Western and the much less accessible but extensive East European studies on spouted beds.

Work on spouted beds at the University of British Columbia is concerned with the fundamental aspects; delineation of regimes, distribution of air between annulus and spout, pressure drops, fountain heights, solids circulation rates, etc., using half-column spout-fluid beds. Conflicting reports exist in the literature on spoutability of 50-500 μm particles. Chandanani and Epstein at UBC are examining the spouting of fine particles with respect to spouting regimes, usual spouting parameters (minimum spouting velocity, maximum spoutable bed depth, pressure profiles, solids circulation patterns, etc.) This will lead to a more careful categorization of spouted bed flow regimes than that which presently exists (31).

Spouted beds can also be used for simultaneous comminution (grinding) and drying. Epstein and Khoe are investigating a spouted bed coal grinder for a given degree of size reduction based on minimum pre-testing of the material to be ground. The investigation is aimed at determining the separate contributions to overall size reduction by the individual mechanisms involved; crushing, impaction and attrition.

(c) <u>Miscellaneous</u>

With the objective of improvement of product quality and reduction of transportation costs, Kybett and Vigrass (32) at the University of Regina are developing a thermal dewatering process which consists of heating lignite at temperatures above 150°C at a pressure above the saturated vapor pressure of water at that temperature. Water is thus removed in liquid phase.

Sridhar (33) at the Whiteshell Nuclear Research Establishment, AECL, has developed a new calciner for treatment of high level waste (HLLW) in nuclear waste management, and for manufacture of fuel-metal (U, Th) oxide powders in nuclear fuel fabrication. The objectives are to convert HLLW into solid products for further treatment and disposal, and to produce thoria and urania powders which would need minimal further treatment for use in nuclear fuel fabrication. The calciner is essentially a composite of the spray and drum dryer (calciner) with a slot-type atomizer (rather than the conventional round atomizers used in spray drying).

Movement of moisture and heat transfer in solids is not a subject exclusively restricted to drying of solids. Steward, at the University of New Brunswick, is interested in predicting the moisture and temperature profiles in soils exposed to various intensities of fire. Laboratory tests are underway to verify the model. At the fundamental level, such modelling work is also relevant to drying of solids, especially in high temperature drying.

Gauvin (34) at McGill University is actively pursuing the idea of using superheated steam for spray or atomized suspension drying of thermally stable materials, primarily of mineral origin. A techno-economic feasibility study (35) has shown that the concept is feasible but that experimental verfication is lacking. There appear to be some technological problems which need to be resolved. Use of steam plasmas for drying is being investigated experimentally. Clearly, such processes are feasible only where the cost of electricity is low.

Gauvin and Katta have developed a highly successful computer-aided design procedure for conventional spray dryers. With some developmental work, which needs to be carried out by industry, this code shows promise as a general design tool for spray dryers which are eminently difficult to scale-up from laboratory or even pilot plant scale to full scale.

(d) Drying of Grains

Recent research in this area focuses mainly on thin layer models, heat and mass transfer aspects, modelling, energy and efficiency aspects, low temperature drying, safety performance and control. Techniques used for enhancement of heat transfer, heat pump use, indirect drying methods, microwave techniques are also covered.

Muir of the Department of Agricultural Engineering at the University of Manitoba is pursuing research pertaining to heat amd moisture transfer in stored grain. His general objective is to develop research information that is needed to design and operate systems to dry, with near ambient temperature air, cereals and oilseed crops in storage. Optimum rates of flow of ambient or slightly heated air (<5°C in increase) and preferred methods of controlling stored crop ventilation systems are being determined using replicated experiments and mathematical models. The ecological interrelationships among the abiotic and biotic variables are being determined in slowly drying columns of seed. Further plans are underway to develop mathematical models in order to use them in computer simulations of the rates of drying and deterioration of the seeds in the experimental bins.

Becker and Douglas of the Department of Chemical Engineering are working on energy conservation, product quality preservation, design and operation. Diffusion properties of cereal grains are being studied to complete a data base for modelling. Results have been obtained for equilibrium moisture and diffusion properties of wheat, corn and other cereal grains. They are also interested in mathematical modelling, design optimization and the study of variability in drying properties.

Sokhansanj of the Department of Agricultural Engineering at the University of Saskatchewan is continually investigating thin layer drying aspects of cereal grains. His general objective is to develop a diffusion type model to predict single particle grain drying under low temperature conditions. The major results obtained are drying rates for rape seed, wheat and barley in low temperature conditions.

Otten, at the School of Engineering, University of Guelph, has performed research in thin layer drying aspects of agricultural products. In order to meet his general objectives of generating empirical data suitable for modelling of high and low temperature drying systems, he has designed and built thin layer drying apparatus. Ezeike and Otten (36) have performed some theoretical analyses of the tempering phase of a cyclic drying process. Hutchinson and Otten (37) have reported models using thin layer data for low temperature drying of soybeans and white beans.

Otten at the University of Guelph has looked into energy conservation aspects of grain drying. His general objective is to reduce energy consumption in commercial and on-farm grain drying through changes in dryer design and storage methods. In his project he has dealt with recycling of cooling and unsaturated exhaust air, low-temperature drying, solar-drying and combination drying of grain crops. Results of the studies include the redesign of some commercial dryers to recover cooling and exhaust air, the introduction of low temperature drying in the farm operation, and the establishment of management schemes. Investigations are also underway to produce suitable automatic control equipment and algorithms for high-temperature dryers. Meiering et al. (38), Brown et al. (39), Johnson et al. (40), Otten et al. (41) have investigated dryer performance and energy use in corn drying, grain corn quality affected by drying method, corn drying with solar assisted low temperature methods, and performance of commercial cross-flow grain dryers.

Sokhansanj of the University of Saskatchewan is actively involved in energy efficiency and quality aspects of grain dryers. His general objectives are to determine the energy efficiency of a direct recirculating dryer and to predict the quality of grain being dried with air of higher relative humidity. He is also working on aspects pertaining to low temperature drying. He predicts drying time, temperatures, and moisture in cereal grains stored in bins on a farm scale, and the predictions compare well with the field data.

Gnyp and St. Pierre of the Department of Chemical Engineering at the University of Windsor are evaluating emission and performance characteristics of cross-flow grain dryers with and without emission controls. Their general objective is to determine the feasibility of improving energy utilization and minimizing the dispersion of particulate matter over the area surrounding each installation. In order to achieve the objectives, they are making simultaneous measurements of total particulate loading in exhaust gases, particle size distributions of particulate matter, average volumetric flow rate of exhaust air, dry-wet bulb temperatures of drying air, cooling air, recirculating air, exhaust air and ambient air, inlet and outlet moisture contents of grain being dried, corn flow rates and specific energy consumption. The results (42,43) of this research indicate that the two kg/h particulate matter emission of uncontrollable dryers can be reduced to 0.8 kg/h with installation of filters. Although this increases capital cost of a unit by 40%, it improves energy utilization by 10%. Further research is planned in order to obtain field measurements of transient values of temperatures, gas humidities and grain moisture contents throughout an existing dryer operating under well documented conditions. On the subject of emission pertaining to commercial grain dryers, Otten of the University of Guelph has made a priority of obtaining emission data on commercial grain dryers and of developing design criteria to reduce emissions, preferably without resorting to emission control equipment. Some of the results are documented in Otten et al. (44).

Friesen, of Manitoba Agriculture, has not only put out a publication (45) categorizing the various aspects of hot-air grain dryers, but he is also involved in the investigation of ambient air drying techniques for grains and oil seeds. His general objectives are to determine equipment operational and design requirements for ambient air drying of crops, and to evaluate various alternative drying systems. His test program includes determination of outputs of grain aeration fans, distribution systems, and monitoring and trouble shooting of existing drying systems.

Mittal and Otten of the School of Engineering at the University of Guelph are concentrating their efforts on the modelling of low temperature aspects of corn drying (LTD) and on the validation of these models with field data. Their results indicate that the new management schemes using LTD are energy efficient compared to high temperature drying. Their microcomputer based system provides a 5 to 13% energy saving compared to the conventional systems. Further field testing is underway to develop more economical units. The detailed results of the studies are presented in Mittal and Otten (46) and Otten and Brown (47).

Limited studies are being conducted using nonconventional methods for grain drying. They are: desiccant use, microwave drying and heat pump application. Bilanski of the School of Engineering at the University of Guelph has tried to dry cereals and peanuts using desiccants. The philosophy here is to apply desiccating agents such as NaCl, $CaCl_2$, Silica gel and bentonite to wet grain mass for quick and safe removal of the grain moisture content. The desiccating agents are then dried using solar energy. Srivastava and Bilanski (48) report results of such techniques for drying maize. Sturton et al. (49) have used bentonite for drying cereal grains. Further work is essential in this area.

Jan of Canada Agricultural Research Station, Melfort, is comparing microwave sample drying techniques with other methods. His purpose is to evaluate the effectiveness and accuracy of microwave drying techniques as compared to Ohaus meter and oven drying methods. Microwave use in grain drying may be feasible. This area is not being covered by the researchers who answered the survey questionnaire; however, it is important to point out that there is great potential for development in this area.

Raghavan and Grant of the Department of Agricultural Engineering at McGill University and Kittler of Decron, Inc. have looked into the feasiblity of heatpump use for grain drying. Performance tests indicated that the heatpump use in grain drying proved to be more efficient than conventional methods. The main problem with the system is its capital cost. Further work is planned in this area in order to improve the coefficient of performance, optimization of blower capacity, better selection of refrigerant, and safety features.

Raghavan of the Department of Agricultural Engineering at McGill University is conducting research in the general area of enhancement of heat transfer for application to grain drying. To achieve this without product contamination, material to be dried or processed is immersed in an agitated bed of granular material. Several theoretical equations describing the technique have been developed through thesis work (Richard (50) and Tessier (51). Extensive experimental data have been generated for particulates of different types and shapes. An experimental dryer has been designed and built. This machine has been used to verify the theoretical models and also to dry corn. Experimental investigations point out that the technique works and results in energy savings. Detailed review, theoretical models and experimental data are presented in Richard and Raghavan (52,53,54,55) and Tessier and Raghavan (56). Sibley and Raghavan are presently working on the heat and mass transfer aspects of high temperature drying of grains in a particulate medium. This information will be useful for the design and control of the dryers. Pannu and Raghavan are concurrently attempting to come up with a better design for a particulate medium dryer. When the design is finalized, the machine will be built and used for obtaining experimental data both on heat and mass transfer aspects. The technique discussed here can be extended to processing, heat treating, etc..

(e) Drying in Agricultural Industry (Other than Grains)

Research efforts in this category should cover efforts in the university (Agr. Eng., Chem Eng., Food Eng., Food Science), government research labs, institutes and industry. Although effort was made to obtain information from these sources, response was quite limited. En-Zen Jan of Canada Agricultural Research Station at Melfort is working on the development of a hay tower system. His objectives are to develop a system capable of drying and storing of tough hay, and mechanically filling and unloading hay. Jan et al. (57) report on a prototype hay tower which was built for artificial drying, storage and handling of chopped hay. Many improvements are noted in this tower.

Jutras and Stratford of the Department of Agricultural Engineering at McGill University are working at developing a solar hay drying system for Quebec which would improve hay quality, energy savings and drying time. They are collaborating with Lawand of Brace Research Institute.

Sokhansanj of the University of Saskatchewan is studying energy conservation aspects in alfalfa dehydration. He is presently developing instruments to measure the humidity ratio in the exhaust stack of alfalfa dehydration. System control and optimization are also his interests.

It should be noted that there are very few researchers working on food drying, but there are many problems that need to be solved. LeMaguer and Toupin of the Food Science Department at the University of

Alberta are studying water and solute movement during water removal in plant materials. Their objectives are mathematical modelling using irreversible thermodynamics and diffusion models in complex cellular structures. They were able to measure transport parameters for the membranes, in order to understand single cell behavior under osmotic drying conditions. They are also working on complete tissue modelling. The detailed results are shown in Mazza and LeMaguer (58,59), Toupin and LeMaguer (60). LeMaguer, Mazza and Smyrl of the Food Science Department at the University of Alberta are actively involved in the determination of the effect of drying (freeze-drying, fluidized bed drying) on aroma retention. Their major findings are quantification of the effect of initial solids concentration, initial volatile concentration, sample thickness, freezing rate in freeze drying, critical moisture content controls lock-in phenomena for aroma in hot air drying of onion, and aroma losses during hot air drying (Smyrl and LeMaguer (61,62); LeMaguer and Mazza (63)).

Kok and Kwendakwema of the Department of Agricultural Engineering at McGill University are working towards the use of solar energy for the drying of food. They have designed, constructed and tested a solar food dryer in Zambia. They have successfully dried several vegetables and meat even during the unfavorable time of the year (cold season). Dried food quality was found to be better than that produced with traditional methods. They demonstrated the capabilities of the dryer in all operating modes and submodes; the flexibility of the design was adequate.

RECOMMENDATIONS

1. An interdisciplinary approach is necessary to study the subject of drying of solids.
2. Improved industry-university interaction will lead to evaluation of fruitful new approaches to design of dryers in view of the escalating energy costs. Greater emphasis should be placed on the influence of drying conditions on product quality.
3. Fundamental research in the field of heat, mass and momentum transfer can make significant contributions to the present, limited knowledge of the drying process. Although no short term breakthroughs can be expected through fundamental research in drying, a sustained effort in this area is needed at the academic institutions.
4. To improve communication between researchers, users and designers of dryers and ancillary equipment, a biennial informal meeting of interested individuals may be initiated. Such meetings will also become breeding grounds for interdisciplinary collaborative and interactive research projects. Also, organization of intensive short courses or workshops on Industrial Drying Technology would be very helpful to the industrial practitioner as well as to the researcher in academia to keep abreast of developments in other industries.

ACKNOWLEDGEMENTS

The assistance of Purnima Mujumdar and Rosemary Haraldsson in the preparation of this typescript is gratefully acknowledged. The authors also acknowledge the assistance of individuals who responded to the survey questionnaire which made the preparation of this report possible.

REFERENCES

1. Mujumdar, A.S., Recent developments in paper drying, IPPTA J. 19 (1): 72-76, 1982.
2. Saad, N.R., W.J.M. Douglas and A.S. Mujumdar, Prediction of heat transfer under an axisymmetric laminar impinging jet. Ind. Eng. Chem. Fund., 16: 148-156, 1977.
3. Huang, B., W.J.M. Douglas and A.S. Mujumdar, Heat transfer under a laminar swirling impinging jet. Heat Transfer 1978, Hemisphere, N.Y., 1978.
4. Mujumdar. A.S., Y.-K. Li and W.J.M. Douglas, Evaporation under an impinging jet: a numerical model. Can. J. Chem. Eng., 58: 448-453, 1980.
5. Huang, B., A.S. Mujumdar and W.J.M. Douglas, Flow characteristics of a swirling laminar impinging jet. Can. J. Chem. Eng., 59: 423, 1981.
6. Mujumdar, A.S. and P.G. Huang, A simulation model for combined impingement and through drying, IPPTA J., 19(1): 65-71, 1982.
7. Mujumdar, A.S. and W.J.M. Douglas, Impingement Heat Transfer - A Review, TAPPI Eng. Conf. New Orleans, USA, 1972.
8. Obot, N.T., A.S. Mujumdar and W.J.M. Douglas, Effect of semi-confinement on turbulent impinging jet heat transfer, 7th Int. Heat Transfer Conf., Munich FRG, 1982.
9. Baines, W.D. and J. Keffer, Drying '80, ed. A.S. Mujumdar, Vol. 1, Hemisphere, N.Y., 1980.
10. Huang, P.-G., A.S. Mujumdar and W.J.M. Douglas, Turbulent flow and heat transfer under a plane turbulent impinging jet including effects of cross-flow and surface motion, Hemisphere/McGraw-Hill, N.Y., 1982.
11. Saad, N.R., A.S. Mujumdar and W.J.M. Douglas, Heat transfer under multiple slot jets, Hemisphere/McGraw-Hill, N.Y., 1980.
12. Mujumdar, A.S. and P.G. Huang, A transient model for combined impingement and through drying, Drying '82, Hemisphere/McGraw-Hill, N.Y., 1982.
13. Jaussaud, J.-P., M.Eng. Thesis, McGill University, Montreal, 1981.
14. Gummel, P., Dr.-Ing. Thesis, Univ. Karlsruhe, F.R. Germany (1977). Also Drying '80, Vol. 2, ed. A.S. Mujumdar, Hemisphere, N.Y., 1980.
15. Mujumdar, A.S., Analytical modelling of sheet flutter, Svensk Paperstiding (Swedish Paper Journal), 79(6): 187-192, 1976.
16. Mujumdar, A.S., Sheet flutter: an examination of mechanical and aerodynamic causes. Austr. Pulp & Paper Assoc. J. 29(6): 475, 1976.
17. Mujumdar, A.S., Sheet Flutter - A Review, IPPTA J., 1974.

18. Loo, E. and A.S. Mujumdar, Combined impingement and through drying of permeable sheets using superheated steam, M.Eng. Project (in progress) 1983.
19. Blackwell,B. and C.MacCallum, Hog fuel drying concepts. Pulp and Paper Canada 84:1, 1983.
20. Mujumdar, A.S. and Z. Pakowski, Heat Transfer Augmentation due to Vibration of a Fluidized Bed, ASME-JSME Joint Thermal Eng. Conf., Hawaii, 1983.
21. Khalid, M. and A.S. Mujumdar, Pressure drop and drying characteristics of an agitated fluid bed, ibid, 1983.
22. Mujumdar, A.S. Keynote Lecture at Latin Am. Heat and Mass Transfer Conf., La Plant, Argentina: "Aerodynamics, Heat Transfer and Drying in Vibrated Fluid Beds", Nov. 1-4, 1982. Proceedings as an issue of the Latin Am. J. Heat & Mass Transfer, 1983.
23. Gupta, R. and A.S. Mujumdar, Aerodynamics of a vibrated fluid bed, Can. J. Chem. Eng., 58: 332-338, 1980.
24. Ringer, D. and A.S. Mujumdar, Immersed-surface heat transfer in vibrated fluid beds, Chem. Ing.-Tech., 54: 686, 1982.
25. Pakowski, Z., A.S. Mujumdar and H.C. Strumillo, Vibrated fluid bed drying, Advances in Drying, Vol. 3, Hemisphere/McGraw-Hill, N.Y., 1983.
26. Karoly, E., A.S. Mujumdar and D.U. Ringer, Vibrated Beds, Advances in Transport Processes, Vol. VII, ed. A.S. Mujumdar and R.A. Mashelkar, Wiley-Halsted. (to appear in 1985).
27. Anderson, K., G.S.V. Raghavan and A.S. Mujumdar, Drying '84, Hemisphere/McGraw-Hill, New York, 1984
28. Mujumdar, A.S., Keynote Lecture at Symp. Transport Processes in Particulate Systems, São Carlos, Brazil: "Recent Developments in Spouted Bed Technology" Oct. 20-22, 1982. Also, Drying '84, Hemisphere, N.Y., 1984.
29. Swaminathan, R. and A.S. Mujumdar, Some Aerodynamic Aspects of Spouted Beds, Drying '84, Hemisphere, N.Y., 1984.
30. Pallai, E., J. Nemeth and A.S. Mujumdar, Spouted Bed Drying - Principles and Applications, to appear in Advances in Drying, Vol. 4, Hemisphere/McGraw-Hill, N.Y., 1985.
31. Mathur, K.B. and N. Epstein, Spouted Beds, Academic Press, N.Y., 1974.
32. Kybett, B. and L. Vigrass, Thermal dewatering of lignite. 32nd Can. Chem. Eng. Conf., Vancouver, 1982.
33. Sridhar, T.S., Whiteshell roto-spray calciner U.S. Patent 4,334,953, 1982.
34. Gauvin, W.H., A novel approach to spray drying using plasmas of water vapor. J. Can. Soc. Chem. Eng. 59(6): 697-704, 1981.
35. Amelot, M.P. Spray drying with plasma superheated steam. M. Eng. Thesis, McGill Univ., 1983.
36. Ezeike, G.O.I. and L. Otten, Theoretical analysis of the tempering phase of a cyclic drying process. Trans ASAE 24 (6): 1590-1594, 1981.
37. Hutchinson, D. and L. Otten , Thin-layer drying of soybeans and white beans. CSAE Paper No. 82-104, Vancouver, 1982.
38. Meiering, A.G., T.B. Daynard, R. Brown and L. Otten, Drier performance and energy use in corn drying. Can.Agric.Eng. 19:49-54, 1977.
39. Brown, R., T. Daynard, G. Fulford, L. Otten and A. Meiering, Effect of drying method on grain corn quality. Am. Soc. Cereal Chem. 56(6): 529-532, 1979.
40. Johnson, P.D.A. and L. Otten, Solar-assisted low-temperature corn drying on Southern Ontario. Can. Agric. Eng. 22(1): 29-34, 1980.
41. Otten, L., R. Brown and K. Anderson, A study of a commercial cross-flow grain dryer. Can. Agric. Eng. 22(2): 163-170, 1980.
42. Gnyp, A.W., C.C. St. Pierre and D.S. Smith, Source sampling of "Ace" model 2165-D grain dryer equipped with Behlen "Rotovac" pollution controls. Ontario Ministry of the Environment ARB-TDA Rep.No. 60-80A, 72 p., 1980.
43. Gnyp, A.W., C.C. St. Pierre and D.S. Smith, Source sampling of "Ace" model 2165-D grain dryer equipped with wet grain precleaning system for Wellington Engineering Ltd. 29 Hickory St., Guelph, Ontario N1G 2X2, 103 p., 1980.
44. Otten. L., R. Brown, A.W. Gnyp, C.C. St. Pierre, and D.S. Smith, Emission and performance testing of a cross-flow grain dryer without emission control. Can. Agric. Eng. (In press) 1983.
45. Friesen, O. Heated air grain dryers. Agric. Can. Publ. 1700, 1980.
46. Mittal, G.S. and L. Otten, Evaluation of various fan and heater management schemes for low-temperature corn drying. Can. Agric. Eng 23(2): 97-100, 1981.
47. Otten, L. and R.B. Brown, Low-temperature and combination drying in Ontario. Can. Agric. Eng. 24(1): 51-55, 1982.
48. Srivastava, A.C. and W.K. Bilanski, Use of desiccants for drying maize. J. Agric. Mech. in Asia, Africa, Latin America 12(4): 49-54, 1981.
49. Sturton, S.L., W. Bilanski and D.R. Menzies, Drying of cereal grains with the desiccant bentonite. Can. Soc. Agric. Eng. 23(2): 101-103, 1981.
50. Richard, P. Heat transfer aspects of drying and processing by immersion in a particulate medium. M.Sc. Thesis. Department of Agric. Eng., McGill Univ., 1981.
51. Tessier, S. Heat transfer and drying in a solid medium rotating drum. Dept. Agric. Eng. McGill Univ. 1982
52. Richard, P. and G.S.V. Raghavan, Particle --article heat transfer applications to drying and processing - a review. Drying '80. Vol. 1. Developments in Drying 132-140, 1980.
53. Richard, P. and G.S.V. Raghavan, Heat transfer between flowing granular materials and immersed objects. Trans. ASAE 23(6): 1564-1568, 1572, 1980.
54. Richard, P. and G.S.V. Raghavan, A study of the heat transfer parameters for drying by immersion in a heated granular medium. Drying '80. Vol. II. 2nd Int. Symp. on Drying. p 272-281, 1980.
55. Richard, P. and G.S.V. Raghavan, Drying and processing by immersion in a heated particulate medium. Advances in Drying. Hemisphere Publ., N.Y., 1983.
56. Tessier, S. and G.S.V. Raghavan, Drying shelled corn in a continuous flow heated sand flighted drum. Energex '80 Proceedings, 1982.

57. Jan, E.Z., M. Feldman, J.A. Robertson and K.W. Lievers, Drying and storage of chopped hay with hay tower. CSAE Paper No. 80-220, 1980.

58. Mazza, G. and M. LeMaguer, Dehydration of onion: some theoretical and practical consideration. J. Food Technol. 15: 181-194, 1980.

59. Mazza, G. and M. LeMaguer, Volatiles retention during the dehydration of onions. LWT, 12: 333-337, 1979.

60. Toupin, C. and M. LeMaguer, Mass transfer in cellular foodstuffs in direct contact with aqueous freezants. 36th Annual Meeting Pacific Northwest Region of Am. Soc. Agric. Eng., Edmonton, 1981.

61. Smyrl, T.G. and M. LeMaguer, Effect of pH on volatile retention in freeze dried model solutions. J. Food Sci. 43: 1357, 1978.

62. Smyrl, T.G. and M. LeMaguer, Retention of sparingly soluble volatile components during the freeze drying of model solutions. J. Food Process Eng. 2 , 1978.

63. LeMaguer, M. and G. Mazza, The drying of onions: combined heat and mass transfer aspects in water and aroma losses. Proc. Annual Meeting of Can. Soc. Chem. Eng. Edmonton, 1980.

APPENDIX I

List of Academic and Other Institutions Included in Present Survey *

(a) Universities

1. University of Alberta
Dept. of Food Science
Edmonton, Alberta T6G 2P5
Investigators: M. LeMaguer, G. Mazza, T.G. Smyrl

2. University of British Columbia
Dept. of Chemical Engineering
Vancouver, B.C. V6T 1W5
Investigators: J.R. Grace, N. Epstein

3. University of Guelph
School of Engineering
Guelph, Ontario N1G 2W1
Investigators: L. Otten, W.K. Bilanski, G.S. Mittal

4. Macdonald College of McGill University
Dept. of Agricultural Engineering
Ste. Anne de Bellevue, Quebec H9X 1C0
Investigators: G.S.V. Raghavan, R. Kok

5. University of Manitoba
Dept. of Agricultural Engineering
Winnipeg, Manitoba R3T 2N2
Investigator: W.E. Muir

6. McGill University
Dept. of Chemical Engineering
3480 University St.
Montreal, Quebec H3A 2A7
Investigators: W.H. Gauvin, W.J.M. Douglas, A.S. Mujumdar, Z. Pakowski, R.H. Crotogino

7. Queen's University
Dept. of Chemical Engineering
Kingston, Ontario K7L 3N6
Investigators: H.A. Becker, P.L. Douglas

8. University of Regina
Dept. of Chemistry
Regina, Saskatchewan S4S 0A2
Investigator: B.D. Kybett

9. University of Saskatchewan
Department of Agricultrual Engineering
Saskatoon, Saskatchewan S7N 0W0
Investigator: S. Sokhansanj

10. University of Windsor
Department of Chemical Engineering
Windsor, Ontario N9B 3P4
Investigators: A.W. Gnyp, C.C. St. Pierre

(b) Others

1. Atomic Energy of Canada Ltd.
Nuclear Waste Management Div.
Pinawa, Manitoba R0E 1L0
Investigator: T.S. Sridhar

2. Canada Agriculture Research Station
Box 1240, Melfort
Saskatoon, Saskatchewan S0E 1A0
Investigator: E.Z. Jan

3. Manitoba Agriculture
Winnipeg, Manitoba R3C 0V8
Investigator: O. Friessen

4. Ontario Research Foundation
Mississagua, Ontario

5. Sandwell and Co. Ltd.
Vancouver, B.C.
Investigators: B. Blackwell, C. MacCallum

6. Pulp & Paper Research Institute of Canada
Pointe Claire, Quebec

APPENDIX II

International Drying Symposia and the Related Publications

1. Mujumdar, A.S. Founding Program Chairman, Biennial Int. Drying Symposia. Montreal (Aug. 1978), Montreal (July 1980), Birmingham, U.K. (Sept. 1982), Kyoto, Japan (July 1982), Cambridge, MA,USA (Aug. 1986), Prague, Czechoslovakia (Sept 1987), Canada (1989).

2. Mujumdar, A.S. ed. Proceedings of 1st Int. Drying Symposium (1978), (Science Press, Princeton, N.J., 1978); Drying '80 (vol. 1 and 2), Drying '82, Hemisphere/McGraw-Hill, N.Y. (1980, 82).

---

* The list is not intended to be all inclusive.

3. Mujumdar, A.S. ed. Advances in Drying, vol. 1 (1980), vol. 2 (1983), vol. 3 (1984), Hemisphere/McGraw-Hill, N.Y.,USA.

4. Mujumdar, A.S., ed. Handbook of Industrial Drying, Marcel Dekker, N.Y. USA. (1985).

DRYING RESEARCH AND DEVELOPMENT IN CZECHOSLOVAKIA

Iva Filková

ČVUT, Faculty of Mechanical Engineering
Praha, Czechoslovakia

ABSTRACT

This paper was prepared with the purpose of making information on the scope of drying research and development activities in Czechoslovakia generally available. The most important research centers are mentioned. A list of the companies manufacturing drying equipment is included. The list of references contains a selection of the papers published during the last several years in each research sub-field of drying.

A. RESEARCH ACTIVITIES

Drying represents a basic operation in a number of important industrial processes. Thus it stands to reason that research activities are particularly concentrated in selected industrial processes such as the drying of milk and other foodstuffs, the drying of grain, hops fodder and wood, the drying of textile and ceramic materials etc.

The National Research Institute for Machine Design(SVUSS) in Praha-Běchovice is a Research centre where a department devoted to drying R & D represents one of its important constituents. There are several other institutes also where attention is paid to drying research. Following is a partial list of such institutes:

The Wood Processing (SVUD) in Bratislava and Praha, the Research Institute of Food Industry (VUPP) in Praha and Bratislava, the Research Institute of Chemical Equipment in Brno and the Research Institute of Inorganic Chemistry in Ústí nad Labem.

Research in drying is also carried out at some universities, mainly at the Czech Technical University (ČVUT) -Faculty of Mechanical Engineering in Prague. The Slovak Technical University (SVSS) in Bratislava and the Forestry and Wood Processing College in Zvolen should be also mentioned in this connection.

Besides the centers listed above, each enterprise which manufactures dryers has its own research department where research and development activities are focussed to the dryers that they manufacture.

In the next part of this paper we shall provide a brief summary of the major drying topics which are the subject of research in Czechoslovakia.

1. Drying Theory. Here heat and mass transfer together with the phase change of moisture and the character of moisture bond in a material are investigated.

Heat and moisture transfer (1, 2, 3) in unsteady conditions has been described previously e.g. by Lykov. A general set of nonlinear equations is obtained as

$$c_1(x_i,\tau,u,v)\dot{u} = c_2(x_i,\tau,u,v)\nabla^2 u + c_3(x_i,\tau,u,v)\dot{v}$$

$$c_4(x_i,\tau,u,v)\dot{v} = c_5(x_i,\tau,u,v)\nabla^2 w + c_6(x_i,\tau,u,v)\nabla^2 v$$

Unknown functions u, v represent temperature and moisture in dried material and functions c_i, generally dependent on time and the unknown functions u, v, are proportional to the thermal and moisture conductivity. A research team has applied various mathematical methods to solve the equations numerically. The finite element method is used as it seems to be the most suitable one. The solution leads to matrices which are suitable for solution on a digital computer.

Solutions for planar and cylindrical geometry have been obtained. Agreement with experimental results has been very satisfactory. Solutions to more complex geometries have also been obtained recently.

The effort is concentrated on multilayer materials at the present time. Special attention is paid to the boundary conditions of the mass transfer equation expressed in potential form.

The Character of moisture bond (4. 5, 6, 7, 8, 9) in the material is usually described by a drying curve, thermogram or energogram of the drying process or by the use of sorption isotherms. One research group has developed an objective thermo-dynamic method which enables one to evaluate a moisture bond by interpreting the shape of the characteristic curve in a diagram of state functions. The curve can be obtained for any particular material using the following state functions of bond moisture: potential μ_w^{MR}, specific enthalpy h_w^{MR} and specific enthropy s_w^{MR}. The first two functions must be obtained experimentally at a constant temperature T when moisture values u change from zero to maximum. The enthropy has to be calculated. Each pair of the functions $\mu_w(u)$ and $h_w(u)$, for a particular moisture content u represents one point in the diagram of state functions. A plot of μ_w^{MR} versus h_w^{MR} forms the characteristic curve for the given material.

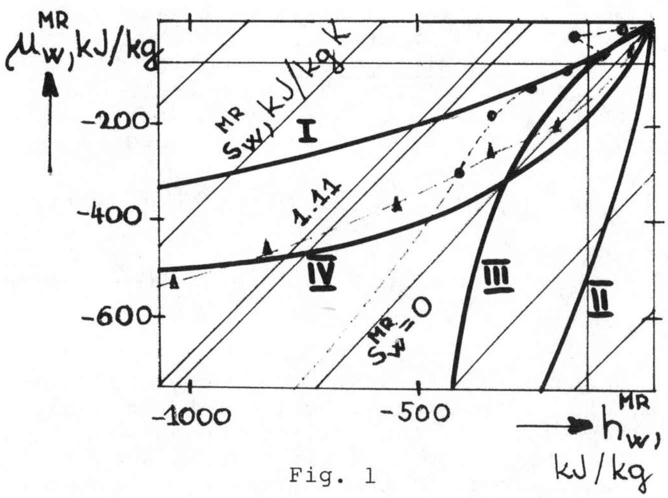

Fig. 1

The basic types of characteristic curves are shown schematically in Fig. 1. The interpretation is as follows:

type I..... stable, strongly hydrophilic material

type II.... stable, predominantly hydrophilic material

type III... stable material with negligible phase change

type IV.... unstable material, intensification of drying process may cause deterioration of product quality.

The '▲' points represent experimental data for potato flour and '●' points are the data for bentonite.

The development of the characteristic curve permits a new tool for an objective evaluation of the complex moisture bond in materials which helps select the most suitable drying method. The state functions of various materials are now being measured systematically so that a data bank could be established.

Further investigation is oriented to the multicomponent moisture bond in a material. A very promising field of future research seems to be the influence of moisture components on the change of the moisture bond character and the influence of liquid wettability on the drying rate even in the constant-rate period.

2. <u>Mathematical modelling</u> (10, 11, 12, 13, 14, 15) of convective drying is based on a set of general equations of mass, heat and momentum balance which describe a drying process in a volume element of a dryer. A general model of convective drying has been developed under the assumption that the material is a porous body consisting of small channels through which the drying medium passes.

The solution of a set of characteristic equations is arrived at by a finite difference method for each particular material. The dependence of physical and chemical properties on state variables is taken into account.

This model has already been applied to the calculation of drying rates of granular materials in a fixed, fluidized bed and moving bed, to the drying of a flat plate in a paralel flow, to the drying in a forced convection stream and to drying in gas-particle systems. The resulting process variables are the following: specific moisture and temperature of solids and gas, gas velocity and pressure drop. In Fig. 2 some experimental results such as a drying curve and local values of moisture u for a fixed bed of wheat, are compared with those which has been predicted by the model. The agreement is seen to be good.

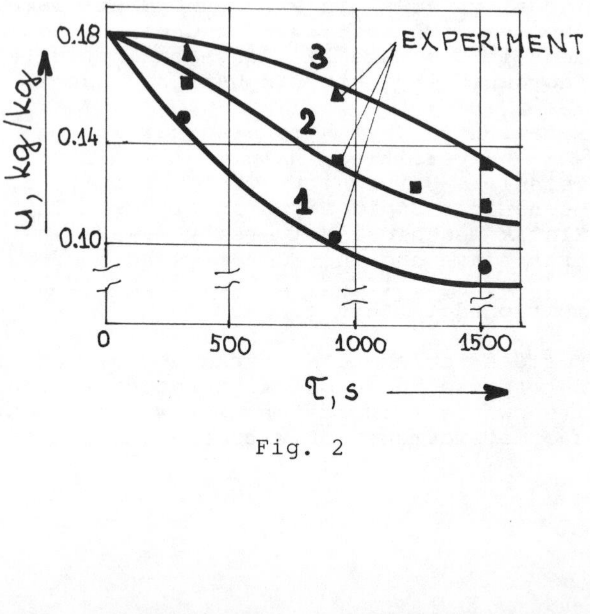

Fig. 2

The mathematical model mentioned above was also applied successfully to the drying of granular and powder materials in a vibro-fluidized bed, e.g. in the second stage of milk drying.

3. Impinging jet drying (16, 17, 18, 19, 20, 21) has been investigated extensively. Besides theoretical studies, its practical application in various industrial sectors has been studied. Special attention is paid to the application of impinging jet drying in the textile, wood processing and paper industries and in the metal-plate heat treatment where this method is widely used. The evaporation rate is very high but owing to high energy costs an optimum value needs to be found.

In Fig. 3 the dependence of specific energy consumption of a fan (n) on the evaporation rate (a) is demonstrated for an impinging jet system.

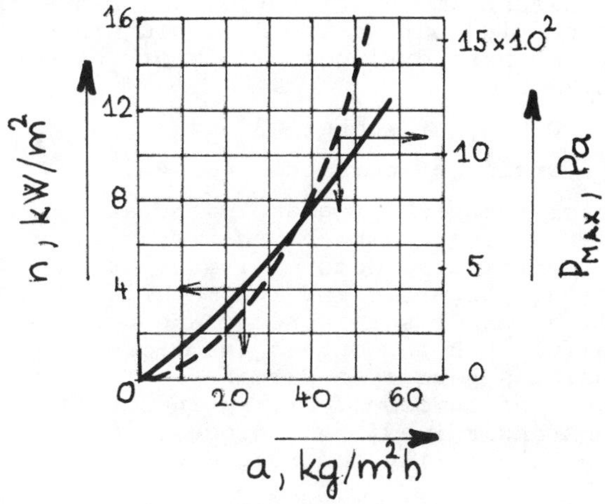

Fig. 3

An increase of maximum aerodynamic pressure (P_{max}) with n is also evident here. It is interesting to note that undesirable aerodynamic effects have been utilized for support of the dried material passing through the dryer. The material can be transported on an "air cushion" formed by the jets.

In Fig. 4 the dependence of the maximum length of a dryer (L) is shown for the drying of loose plate-metal, at the relative maximum sag (y/L_{max}). Two values of sag are considered: y=1m and 2m. The coefficient σ/g_M is a parameter, σ is the permissible tensile stress of the belt and g_M is the specific weight.

An example is also shown in Fig. 4. An aluminium belt (σ/g_M=100m) is to be dried, with the permissible sag y=1m. The resulting length of the dryer is L_{max}=28m.

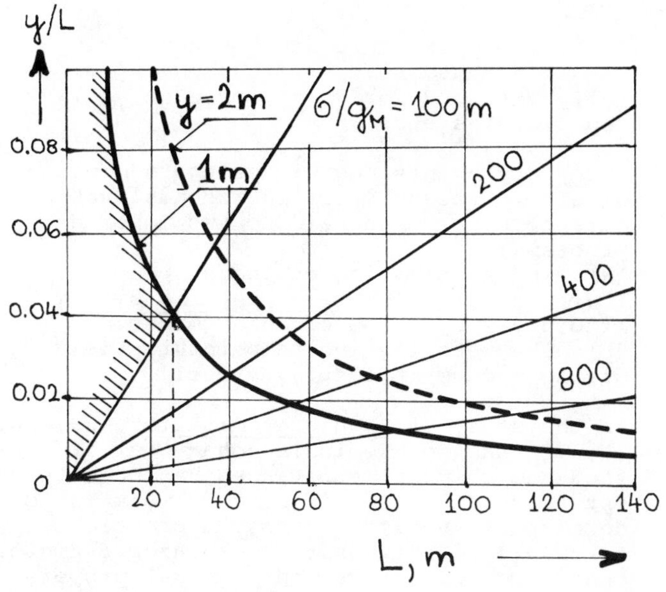

Fig. 4

However, for the large capacity modern units a longer drying chamber is required, up to 40m long. Aerodynamic suspension of the belt by jet streams makes it possible to construct such a dryer.

The research group involved in impinging jet drying has developed three methods by which the supporting effect of aerodynamic streams can be utilized. Thus, extensive experimental data for different materials and different jet systems have been obtained.

The intensity of evaporation from the material surface in the impact flow has been studied at the same time. The study has been carried out for evaporation of water and several organic solvents and solvent mixtures.

4. Drying in disperse systems has been intensively studied both in gas-droplet and gas-particle systems.

The basic theoretical corelations and the experimental studies relevant to many applications such as moving bed dryers, vibrofluidized bed dryers and fluidized bed dryers, spray dryers, flash dryers etc have been carried out. Pilot plant studies on fixed and vibro-fluid bed drying have been carried out.

Fluidized bed drying has been investigated mainly from the point of view of application to biological materials, which have a paste-like character because of their high moisture content. Recent research is focussed on the predrying operations (e.g. granulation), on the hydrodynamics of the fluidized bed, on temperature effects and drying kinetics.

Optimum bed moisture and residence time of particles have been studied and experimentally verified for yeast, caseine, activated slurries etc.

Spray drying has been investegated intensively in recent years in Czechoslovakia. Attention has been paid to two important problems:
(a) The atomization of the suspension, i.e. the formation of a disperse system liquid-gas by means of an atomizer.
(b) The velocity and temperature field of drying medium in a spray dryer.

In the study of liquid atomization, (29, 30, 31, 32) the Non-Newtonian character of many food and chemical substances has been taken into account. The aim of this work is to develop and verify the most appropriate correlations between the mean drop diameter, operating parameters and physical properties of a liquid, for both wheel and nozzle atomization. Such relations would enable one to achieve the required drop diameter, drop distribution and thus to influence the quality of the final product.

A spray is characterized by its size distribution curve and the mean volume-surface diameter (Sauter). The following functional relations have been investigated and tested experimentally:

$$D_{3,2} = f(N, R, M, b, \rho, \mu, \sigma)$$

where N - wheel rpm, R - wheel radius, b - width of vane, M - mass flow rate and ρ, μ, σ - physical properties of the fluid.

In the case of a pseudoplastic liquid, the dynamic viscosity is not constant, so it has been substituted by an apparent viscosity μ_a. This apparent viscosity was derived by means of the mean outlet liquid velocity on the wheel edge. Consistency coefficient K and flow index n were used to express μ_a as follows:

$$\mu_a = K\left[\left(\frac{\rho R \omega^2}{K}\right)^2 \cdot \left(\frac{V_k}{b}\right) \cdot \left(\frac{2n+1}{n}\right)\right]^{\frac{n-1}{2n+1}}$$

Here V_k is volumetric flow rate for one vane , and ω is angular velocity.

The apparent viscosity was employed rather than the dynamic in the empirical correlations chosen and the resulting droplet diameter $D_{3,2}$ was verified experimentally. Within the range of experimental conditions the following rearranged form of the Kremnev correlation yielded the best agreement with the experimental data.

$$D_{3,2} = 0.24\left(\frac{1}{N}\right)^{0.6} \cdot \left(\frac{1}{\rho}\right)^{0.3} \cdot \left(\frac{\mu_a M}{2R\rho}\right)^{0.2} \cdot \left(\frac{\sigma}{8b}\right)^{0.1}, \text{ m}$$

The effect of vane shape on the drop diameter was also studied. Similar research on nozzle atomization is proceeding.

The other area of spray drying being investigated is the motion of falling particles in the drying chamber and transport processes in the nozzle zone. For this purpose the velocity field of the drying gas flow must be measured accurately.

A method has been developed to calculate the kinematic characteristics of the spary e.g. velocity, particle trajectory and mass distribution along the chamber radius. This seems to provide explanation for some of the anomalous phenomena reported for spray dryers. For example , this explains the motion of droplets toward the upper part of the drying chamber where deposits of wet solids are often encountered in operation.

5. Miscellaneous

Significant amount of drying R & D has also been carried out in the following industrial sectors where drying is an important operation:

- Drying of lumber (35-42)
- Drying of ceramics (43,44).

A major research program in the area of lumber drying has been underway. It covers all important aspects of lumber drying, from physical properties of the material to the application of solar energy and wood waste as a fuel. Both contact and vacuum drying techniques have been examined. The complex process of lumder predrying and drying has been standardized. All process parameters are automatically controlled.

6. Energy savings and nonconventional heat sources:

Ways to save energy in drying processes have been looked at. This includes combination of granulation or disintegration processes with drying processes. The level and quality of the control system also affects the energy consumption in drying operations.

As alternative energy sources for drying the following sources are considered:

- Solar energy usage with several new types of solar collectors has been examined and tested; the thermal efficiencies are of the order of 42%. At current energy and capital costs , wide practical application is not foreseen for some time. Cheaper types of solar collectors may be used in fut the energy costs continue to soar.

- Waste heat from compressors at natural gas stations.

There are several compressor stations in Czechoslovakia as part of its natural gas supply system. They are each equipped with gas turbines with a capacity of 6 MW each. The exhaust gas can be utilized in heat exchangers for water heating up to 130 ^{0}C. This energy source could be employed for agricultural purposes, mainly for drying of fodder in the low temperature regime. First experiments have given very good results with a payout period of about 3.5 years.

At present attention is paid to other energy sources which promise to have application in drying process e.g.

- combustion of wood and domestic wastes
- biogas and heat pumps
- recuperation of waste heat from heat exchangers and drying equipment.

There is significant activity in drying research and its application, which has a long tradition in Czechoslovakia. The drying section was formed by the Czechoslovak Society of Science and Technology, that is the center of the numerous drying activities. Only some of them will be mentioned here: National Drying Conferences are organized every five years, with invited participents from other Eastern European countries. Regular national conferences on Drying of Dispersed Systems have been held every four years.

The Drying Section initiates or organizes postgraduate courses on drying theory and practice, courses on industrial drying, regular courses on drying in agriculture, wood, textile and ceramic industries.

It was decided in 1980 to form a new section on 'Drying' by the international chemical engineering congress (CHISA), which takes place every third year in Czechoslovakia. This section was first held in 1981. The next CHISA Congress will take place in 1984 in Prague. The 6th International Drying Symposium is scheduled to be held in Prague in 1987 in conjunction with CHISA'87.

B. EQUIPMENT MANUFACTURERS

There are several facilities in Czechoslovakia where drying equipment is manufactured. Following is only a representative list of the major supplies of drying equipment.

1. VZT Nové Mesto n.v. (50)

It is the biggest manufacturer of drying equipment in Czechoslovakia. The factory was founded about 80 years ago. The following types of dryers are manufactured here:

- spray dryers of milk and pharmaceutical products
- lumber dryers and predryers
- continuous tunnel dryers for agricultural products
- flash dryers with disintegration
- vibrofluidized bed dryers
- special types of tray and tunnel dryers.

The spray dryer VRA-4 equipped with wheel atomizer is the latest model. Its capacity is 1000kg H_2o/h, inlet air temperature is 180 ^{0}C. An additional heat exchanger enables to heat inlet air by means of outlet air. The VRA-4 can be also used as the 1st stage and followed by a vibrofluidized bed as the 2nd stage of drying.

Among other drying equipment manufactured in the VZT should be mentioned the flash dryer VDA with multistage disintergrator in which e.g. $CaCo_2$, waste slurries and poultry excrements can be dried.

The continuous dryer PCHB 750 for hops has lately been adapted for green crops drying. The waste heat of the natural gas supply system is utilized as fuel.

The latest types of tunnel dryers for lumber, TWA-1 and TWA-2, are manufactured as a single stage or a double stage equipment with longitudinal countercurrent flow. They are followed by a heat recuperator consisting of heat pipes.

Two types of pre-dryers, DWA 1 and 2, equipped with heat recuperation have been developed recently.

2. ELITEX Chrastava

The ELITEX enterprise manufactures machines for the textile industry, among them two dryers for drying textiles :

- drying and setting stenter type 4580.3, which can be used to dry light-weight woven and knitted fabrics of cotton, rayon, wool and linen. The maximum speed is 150m/min

- universal dryer type 4548.0, situated after the rotary screen printing machine. It is suitable for drying all kinds of textile materials and dyestuffs.

A new type of dryer 4580.4 with higher drying output, lower energy consumption was developed and is being tested now. The supporting effect of aerodynamic streams was utilized here. The dryer is controlled by microprocessors.

3. TMS Pardubice

The manufacture program of the TMS Pardubice contains complete plant equipment for the post-harvest grain treatment. Either the whole technological line or separate units can be delivered. One of them is the LSO hot-air dryer for drying grain, barley, corn and rape. It is an indirect - heat dryer manufactured in two designs:

- single-shaft unit with the capacity of 10,20,25 t/h, approx. 1000kg H_2O/h
- double-shaft unit with the capacity of 40,50,60 t/h, approx. 2000kg H_2O/H.

The drying output is automatically controlled in accordance to the required grain moisture at the outlet.

Baking ovens under licence from WINKLER Co. (Federal Republic of Germany) are also manufactured in different capacities, the effective area varying from 25 to 108m^2.

4. Kovofiniš Ledeč and Škoda Plzeň

They manufacture complete lines for metal-plate production and surface treatment. The production line consists of different units and one of them is a tunnel dryer where metal-plates are dried.

5. Škoda Ejpovice

The production of this enterprise is focussed on hot-air drum dryers of green crops and other agricultural materials under licence from PROMILL Co. (France). Three basic types have been manufactured lately:

- BS-6 with the capacity of 2200kg H_2O/h, fuel is either town gas or natural gas or heating oil. Together with this dryer pelletizing equipment for dry product is delivered.

- BS-18El, is a new type of green crops drum dryer with a capacity of 6500kg H_2O/h.

- BS-6S, is a drum dryer for animal excrements and slaughter-house wastes with a capacity of 1800kg H_2O/h; heating oil is used as fuel. The dryer is equipped with deodorization and catalytic combustion for odour removal.

- BS-6M, is a mobile dryer with the same capacity as BS-6, licence of TAARUP Co. (Denmark).

Two more types of drum dryers are designed for manufacture in 1983. Both of them demonstrate considerable energy savings.

There are a number of smaller enterprises where specific types of dryers are manufactured, like dryers for ceramics, leather, photographic materials, etc. These are not mentioned in the above list.

CLOSURE

An attempt is made to give the reader an overview of the diverse drying R & D carried out in Czechoslovakia. Further details can be found in the literature cited below. No pretense is made that this survey is exhaustive. References marked with (*) are in Czech and the author has provided the English translation of the paper titles for the benefit of the reader.

ACKNOWLEDGEMENTS

Thanks are due to Purnima Mujumdar for preparing the typescript. This article was edited by A.S. Mujumdar.

REFERENCES

1. Valchářová, J., Valchář, J.: The thermokinetics of the moisture transport in solids drying, DRYING'80, Vol. 2, Ed. A.S. Mujumdar, Hemisphere /McGraw-Hill, (53-81), 1980.

2. Research Report No. 80-09008, (in Czech), SVUSS Prague, 1980.

3. Research Report No. 82-09015, (in Czech), SVUSS Prague, 1982.

4. Research Report No. 79-09001, (in Czech) SVUSS Prague, 1979

5. Čermák, B.: Characterising moisture bond in materials with respect to drying using the state function diagram, DRYING'80, Vol. 2, Ed. A.S. Mujumdar, Hemisphere/McGraw-Hill (61-65), 1980.

6.* Čermák, B.: Catalogue of the thermodynamic state functions of moisture bond in materials, (in Czech), Technical Manual No. 5, SVUSS Prague, 1979.

7.* Čermák, B.: How to characterize moisture bond in materials by means of the state function diagram, (in Czech), ZTV 24, No. 2, Prague, 1981.

8.* Čermák, B.: Application of thermogravimetric methods for heat and mass transfer during the evaporation of moisture from a material, (in Russian), The International Seminar on the Experimental Methods in Heat and Mass Transfer, part 2, Minsk (Soviet Union), 1981.

9. Research Report No. 80-09128, (in Czech), SVUSS Prague, 1980.

10. Research Report No. 77-09008, (in Czech), SVUSS Prague, 1977.

11. Houška, K.: Method of calculation of convective dryers and its practical application, Congress CHISA, Prague, 1981.

12.* Viktorin, Z.: Mathematical and physical models of heat and mass transfer in convective drying, (in Russian), Teplomassoobmen-VI, Vol.X, Minsk(Soviet Union), 1980.

----

* Original papers in Czech language or as stated.

13.* Viktorin, Z., Houška, K.: Recommendations for pneumatic dryers calculation and selection, (in Czech), Technical Manual No. 6, SVUSS Prague, 1981.

14. Houška, K.: New method of calculation of the drying process and its application to dryer design, (in Czech), Průmysl potravin No. 11, Prague, 1979.

15.* Houška, K.: Computer application to calculation of drying processes and dryers, (in Czech), Proceeding of the 3rd Conference "Drying and Dryers in Textile Industry", Liberec, 1979.

16.* Korger, M.: An experimental study of heat and mass transfer in impact drying of a flat materials, (in Russian), Teplomassoobmen -VI, Vol. I, Minsk, (Soviet Union), 1980.

17.* Korger, M.: Cushioning effect of aerodynamic streams, (in Czech), Proceedings of the 6th Drying Conference, Trenčín, 1981.

18. Křížek, F.: Mass transfer in impact drying using a set of circular jets, Congress CHISA'81, Prague, 1981.

19.* Křížek, F.: Surface evaporation rate of liquid at various Schmidt numbers, (in Russian), IFZ, Vol. XXXIV, No. 3, (431-438), 1978.

20.* Korger, M.: Supporting effect of aerodynamic streams (in Czech), Proceedings of the 3rd Conference "Drying and Dryers in Textile Industry", Liberec, 1979.

21.* Křížek, F.: A new system of nozzles providing supporting effects, (in Czech), Proceedings of the 3rd Conference "Drying and Dryers in Textile Industry", Liberec, 1979.

22.* Hlavačka, V., Valchář, J., Viktorin, Z. : Technological and heat transfer processes in gas-particle systems, (in Czech), SNTL Prague, 1980.

23. Viktorin, Z.: Drying of granular material in gas-solid particles dispersed systems, Congress CHISA, Prague, 1981.

24.* Choc, M., Kolář, S. : Advances in theory and application of the pneumatic drying process, (in Czech), Proceedings of the 6th National Drying Conference, Trenčín, 1981.

25.* Viktorin, Z.: Estimation of hydrodynamic and thermokinetic properties of fixed beds, (in Czech), Manual of Project Engineer No. 03-1, SNTL Prague, 1978.

26.* Choc, M.: Drying of granular materials in a vibrofluidized layer, (in Czech), Proceedings of the conference "Drying and Dryers of granular materials", DT Pardubice, 1978.

27.* Viktorin, Z.: Future trends in drying and dryers of granular materials, (in Czech), Proceedings of the conference "Drying and Dryers of Granular Materials", DT Pardubice, 1978.

28.* Beran, Z., Smejkal, S., Staněk, J.: Drying of biological materials in fluidized beds, (in Czech), Proceedings of the 6th National Drying Conference, Trenčín, 1981.

29. Filková, I.; Dropsize distribution of Non-Newtonian Slurries, DRYING'80, Vol. I, Ed. A.S. Mujumdar (346-350), McGraw-Hill, 1980.

30. Filková, I., Weberschinke, J.: A generalized equation for predicting the dropsize of a pseudoplastic fluid in a spray dryer, Congress CHISA, Prague, 1981.

31. Weberschinke, J., Filková, I.: Apparent viscosity of Non-Newtonian droplet at the outlet of wheel atomizer, DRYING'82, Ed. A.S. Mujumdar (165-170), Hemisphere/McGraw-Hill, 1982.

32. Filková, I., Čedík, P.: Atomization in pneumatic nozzles - some effects of operating parameters on drop size, Proceedings of the third International Drying Symposium, Vol.1, (516-527), Birmingham, 1982.

33.* Kmeť, K.: Analysis of transport phenomena during atomization in a spray dryer, (in Czech), PhD thesis, ČVUT Prague, 1980.

34. Kolář, S.: Experimental study of the drying chamber aerodynamics in a pilot spray dryer, Congress CHISA, Prague, 1981.

35.* Viktorin, Z: Testing and evaluation of the dryers in wood industry, (in Czech), Proceedings of the Conference "Testing and evaluation of dryers", DT Prague, 1978.

36.* Viktorin, Z., Korger, M., Čermák, B.: New aspects of lumber drying in Czechoslovakia (in Russian), Proceedings of the symposium "Drying of Wood" Opatija (Yugoslavia), 1978.

37.* Viktorin, Z. and others: Evaluation of the lumber dryer newly designed and manufactured in Czechoslovakia, (in Russian), Proceedings of the symposium "Drying of Wood", Opatija

(Yugoslavia), 1978.

38.* Viktorin, Z. et.al.,: Future trends in drying of lumber in Czechoslovakia, (in Russian), Proceedings of the symposium "Drying of Wood", Opatija, (Yugoslavia), 1978.

39.* Viktorin, Z.: Determination of drying time in natural drying of lumber, (in Czech), Drevo, 34, No. 6 (162-167) 1979.

40.* Viktorin, Z., Čermák, B.: Moisture bond energy in a wood and its experimental determination, (in Czech), Drevársky Výzkum, XXV, Vol.1, ALFA Bratislava, 1981.

41.* Viktorin, Z., Drahoš, V.: Advances in drying process and dryers of lumber, Práce VUD, Vol. 26, SNTL Prague, 1982.

42.* Viktorin, Z.: Drying processes in a pile of lumber, Drevársky Výzkum, Vol. 4, ALFA Bratislava, 1982.

43.* Korger, M., Křížek, F.: The principles of drying in the ceramic industry, Textbook ČSVTS, DT Plzeň, 1978.

44.* Křížek, F.: Evaporation of a liquid from the surface of a ceramic material, Proceedings of the meeting "V. SILICHEM, DT Brno, 1981.

45.* Viktorin, Z.: Energy saving possibilities in lumber drying, Drevo, No. 35, 9, (251-255), 1980.

46.* Křížek, F.: Recuperation of waste heat by means of heat pipes, Textil, 9, 1980.

47. Strach, L.: Energy saving methods in drying, Congress CHISA, Prague, 1981.

48.* Viktorin, Z.: Solar collector for heating of air in a lumber dryer, Elektrotechnický obzor, 70, No. 7, 1981.

49.* Gondár, V., Viktorin, Z.: Low temperature dryer for food and agricultural purposes, Proceedings of the 6th National Drying Conference, Trenčín, 1981.

50.* Kmeť, K., Kolář, S., Polanský, A.: Development of spray dryers in the enterprise VZDUCHOTECHNIKA, Průmysl potravin, No. 1, Prague, 1980.

MECHANISM OF WATER SORPTION-DESORPTION IN POLYMERS

J.V. Alemán

J.L.G. Fierro

Instituto de Plásticos y Caucho, Juán de la Cierva 3
Madrid 6, Spain

ABSTRACT

Polymers contain water which must be reduced to a lower level for application. As an example of drying behaviour, polystyrene has been studied in a laboratory try drier at temperatures below the glass transition temperature (Tg = 373 K) as a function of the process variables (drying temperature, air relative humidity, air flow rate). This information is insufficient to furnish a quantitative interpretation of the computed effective diffusion coefficients (D_{eff}).
Sorption isotherms for epoxide prepolymers (Tg = 343 K) show that water is sorbed in increasing amounts as the temperature decreases, and the water vapor concentration increases, the sorption curve being linear below Tg and anomalous sigmoidal above Tg. The heat and entropy of mixing indicate that, at low concentrations, the water may be hydrogen bonded, while at moderate or high concentrations, clustering of the water takes place, with an increase in free volume and a decrease of Tg to 323 K. Deborah number shows that the relaxation of the macromolecular chains is to be taken into account.
Diffusion coefficients computed considering this fact, are some ten times smaller below Tg than above Tg, the behaviour at the beginning of the process being Fickean, followed by a non-Fickean zone characterized by a diffusion coefficient which decreases less rapidly as the water content of the polymer increases. Combined effect of temperature and sorption rate is described in terms of temperature shift factors a_T, with which the computed constant energy of activation for diffusion below Tg is of ca. 52.3 kJ/mol.

SYMBOLS

W_o = weight of polymer entirely dry (t=0)

W_1 = initial weight of polymer plus water

W_t = weight of polymer plus water at time t

W_∞ = final weight of polymer plus water (t = ∞)

$\Delta W_t = W_t - W_o$ = water content = X

$\Delta W_1 = W_1 - W_o$ = initial humidity = X_c

$\Delta W_\infty = W_\infty - W_o$ = equilibrium humidity = X_e

X % = $(\Delta W_t/W_t) \times 100$ = percent water

$\Delta Q = \Delta W_{t(n)} - \Delta W_{t(n-1)}$ = increment water content

V_S = drying rate = $\frac{d(\Delta W_t)}{A.dt}$

R.H.% = relative humidity (%)

$R.H._A$% = relative humidity (%) at room conditions

$R.H._B$% = relative humidity (%) in the preconditioning chamber

$R.H._{DC}$ % = relative humidity (%) in the drying chamber

P = water vapor pressure

P_o = water vapor saturation pressure

$w_1 = \frac{\text{grams (water)}}{\text{100 grams (dry polymer)}}$

C_o = water vapor concentration outside the slab

C_t = water vapor concentration inside the slab at time t
= $(\Delta W_t/W_o) \times 1{,}187$

C_∞ = equilibrium water concentration inside the slab at time t=∞
= $w_1 \times 1{,}187$

C_s = equilibrium concentration at the surface of the slab

ΔC = concentration difference
= driving force of the diffusion process

Subscripts

1 = initial diffusion stage

2 = decreasing diffusion stage

3 = increasing diffusion stage

I) THE DRYING PROCESS

Polymers in use normally have a certain amount of water which differs with polymer structure, processing conditions and with polymer storage history. For polymers to be useful, their water content generally needs to be reduced (1,2) to a level which depends upon the planned application. This drying process may be carried out to different limits according to the water-polymer interaction characteristics (3).
Water molecules have a pronounced tendency to hydrogen bond: a) with themselves to form agglomerates (pockets, etc.) and b) with polar groups of polymers to provide strong localized bonds which change with:
i) The sorption behavior: hydrophobic polymers absorb only small amounts of water and their behavior is thermodynamically near ideal (Henry's law), whereas the hydrophilic polymers absorb larger of water due to the strong interaction between their polar groups and water.
ii) The equilibrium diffusivity: the diffusivity of the hydrophobic polymers (polyolefins, etc.) is independent of the amount of water present. The not very hydrophilic polymers (polymethacrylates, etc.) have a diffusivity which decreases with increasing water concentration due to preferential water association. The hydrophilic polymers (polyamides, etc.) have a diffusivity which increases with increasing water content, because the adsorbed water swells the polymer, making it easier for water molecules to diffuse.

1) Materials
Polystyrene (PS) was chosen as a representative polymer to establish drying characteristics. The material used was Styron 634 from the Dow Chemical Company, in cylindrical shape (average radius 1.0 millimeter and length 3.0 millimeters), molecular weight of 120,000 and glass transition temperature (Tg) of ~373 K.

2) Apparatus
A laboratory-scale tray dryer was constructed (4) which allowed the use of a wide range of process variables: (the process temperature was always below Tg for the PS to be in its glassy state where most viscoelastic mechanisms of molecular motion are frozen in (3))temperature, air velocity, polymer size and shape, thickness of the drying bed, humidity in the drying chamber, and humidity in air preconditioning chamber.

3) Results
Results were plotted as percent humidity (X%) vs time (t), and vs drying rate (V_S) as follows:
Figure 1 shows the effect of the temperature on the percent water (X percent), and on the drying rate (V_S), at R.H.$_{DC}$ = 4.5 percent, R.H.$_B$ = 70 percent, V_V = 2.5 liters/s.
As may be observed,these curves have three main parts (2,5):
a) A zone of transient behavior (Zone I to the right of Zone II, not shown in Fig. 1), of some 10 min duration, during which

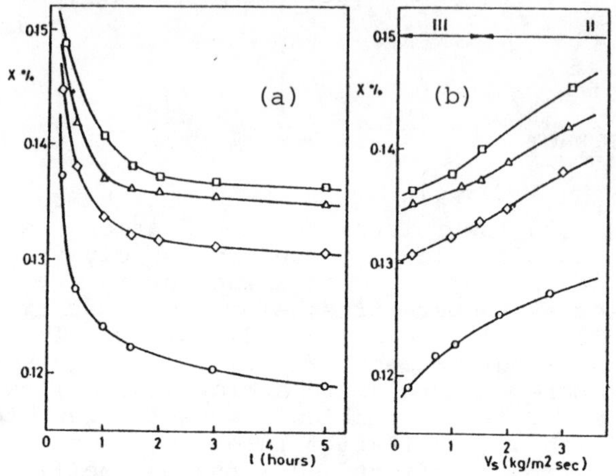

Fig.1. Effect of temperature on the percent water (X%), a) as a function of time (t), and b) as a function of drying rate (V_S), at air volumetric flow rate V_V = 2.5 liter/s, R.H.$_B$ = 70 percent, R.H.$_{DC}$ = 4.5 percent. □ ,333 K; △ ,353 K; ◇ ,363 K; ○ ,373 K. Courtesy Polym. Eng. Sc. 21 , 804 (1981).

the polymer temperature rise to the processing temperature takes place, and at the same time also the constant rate of evaporation of water from the surface of the solid polymer, through a typical process of heat transfer in which the main transfer mechanism is convection, since conduction and radiation are usually negligible.
b) A Zone II of decreasing drying rate, of length about 1.5 h, during which time the elimination of the water held by pores of all sizes takes place through a combined process of heat transfer from the air to the solid simultaneous with a mass transfer from the solid to the air. From these combined processes only it is possible to define a drying rate "averaged" between the diffusion in large pores (Stephan), the diffusion in pores of small diameter (Knudsen), and the simultaneous heat transfer process.
The drying rate in Zone II is non-linear (Fig.1) and becomes zero where the maximum hygroscopic humidity is reached.
c) A Zone III, of length three or more hours, in which hygroscopic (or associated) water is eliminated, and the drying rate decreases continuously, becoming zero when the equilibrium humidity (X_e) is

reached. In this work X_e is considered to be approximately equal to the humidity of the solid polymer at the exit of the dryer after more than five hours operation.
Drying in Zone III is a combination of diffusion of water vapor through the solid, and flow of the water through the capillaries, in a process ruled by a general differential equation for which only numerical solutions are possible (5,6,7). Because the flow of water in the capillaries is small, it is usually considered that the drying process in Zone III takes place only by diffusion of water vapor, and that (since the drying rate decreases almost linearly) (4) the "effective" diffusion coefficient (D) is constant and obeys Fick's law:

$$\frac{\partial X}{\partial t} = -D \frac{\partial^2 X}{\partial r^2} \qquad (1)$$

which gives for the diffusion coefficient in a sphere of volume equivalent to the cylindrical pellets used:

$$D = - \frac{r^2}{2.25\pi^2} \cdot \frac{1}{t} \cdot \ln\left(\frac{X - X_e}{X_c - X_e} \cdot \frac{\pi^2}{6}\right) \qquad (2)$$

Effect of temperature on the diffusion coefficients computed by means of equation 2 is described by the Arrhenius equation:

$$D(T) = D_o \cdot e^{-\Delta E/RT} = 3.6 \times 10^{-5} e^{-294/T} \qquad (3)$$

The effect of water concentration on the diffusion coefficient was then studied. These data were treated as in the previous case of the effect of temperature on the diffusion coeficient. Results are shown in Table 1 for both humidity of the drying chamber (R.H.$_{DC}$ percent) and percent humidity (X percent). In this later case some

Table 1. Effect of Water Content on the Diffusion Coefficient

R.H.$_{DC}$ (percent)	D.10^{-7} (m^2/h)	X percent (4h)
22	0.35	0.142
17	0.35	0.135
13	0.47	0.136
8	0.21	0.136
4.5	0.31	0.130

anomalies were observed, mostly due to the narrow interval of final humidities attainned (X percent = 0.13-0.14).
The diffusion coefficients thus obtained show a not very pronounced increase with water content, and obey (within the limits of variation of these data) the equation

$$\ln D(X\%) = \ln D_{X\%=0} + a\, X\% =$$

$$= - 19.30 + 4.0\ (X\%) \qquad (4)$$

The source of this behavior is provided by the non-hydrophilic, but unsymmetrical substitutions, in the hydrocarbon backbone of PS: the water interacts with the polymer chains producing reorientation and chain displacement (8), i.e., "loosening" of the structure; as a consequence, the adsorbed water becomes more and more mobile (local cooperative vibrations of only a few structural units are sufficient for the water molecules to diffuse) increasing the diffusion coefficient.
In order to ascertain in a less empirical way the effect of concentration on the diffusion coefficient of polymers, the study of the sorption behavior of water in epoxideprepolymers was carried out as follows:

II) SORPTION EQUILIBRIUM

1) Materials
The solid epoxide prepolymer used, prepared by the fusion process (9), was Araldite 6097 supplied by Ciba-Geigy AG (Swizerland), with glass transition temperature Tg = 343 K, number-average molecular weight $\overline{M}_n$ = 2112 (10,11), and polydispersity $\overline{M}_w/\overline{M}_n$ = 1.99 .

2) Procedure (9)
All sorption-desorption experiments were carried out in a Cahn RG microbalance (sensitivity = 10^{-6} g) connected to a conventional vacuum line.
Samples of 6x6x0.6 mm were placed on the balance pan, and to attain volume relaxation, they were conditioned by degassing for 65 h at high vacuum (10^{-6} torr) at the temperature of the experiment.
The level of water concentration required was provided by allowing different water vapor pressures in the system by means of a mercury differential manometer.

3) Results
Sorption isotherms as a function of temperature and water vapor concentration are shown in Figure 2 for temperatures below Tg. Water content w_1 (grams of water per 100 g of dry polymer) decreases with increasing temperature and decreasing water vapor concentration.
At 303 K it has the form of a BET type II isotherm (12) with which the ratio of the polymer specific (internal) area accesible to the water molecules, to the geometrical (external) area is S_{BET}/S_G = 1500, being a clear sign of the presence of some porosity or microvoids frozen in the polymer slabs.

A) Heat and Entropy of Mixing

Were computed with the Flory-Huggins equation (13) and results are shown in Figure 3.
The heat of mixing (ΔH_M) comprises (8) the energy of attachment of the sorbed molecu-

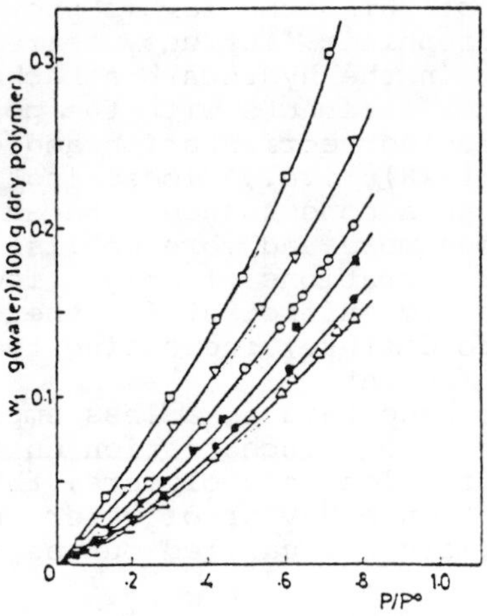

Figure 2. Sorption curves of water in epoxide prepolymers below Tg: □, 303 K; ▽, 310 K; ○, 315 K; ■, 320 K; ◓, 326 K; Δ̇, 333 K; .... computed for dual mode of sorption. Courtesy of Macromolecules 15, 1145 (1982)

les to the polymer, the energy of swelling or reorientation caused by penetration of the water molecules into the polymer, and the mutual interaction between sorbate molecules, plus the sorption of water upon water, which occurs as polymer sites are occupied. For epoxide prepolymers the dipole and swelling forces will decrease (ΔH_M will increase since it is negative), while the other forces will increase (decreasing ΔH_M) with increasing sorption.
At low water concentrations (less than 0.04 g/100 g), $-\Delta H_M$ first increases and then levels off, the behavior being explained by

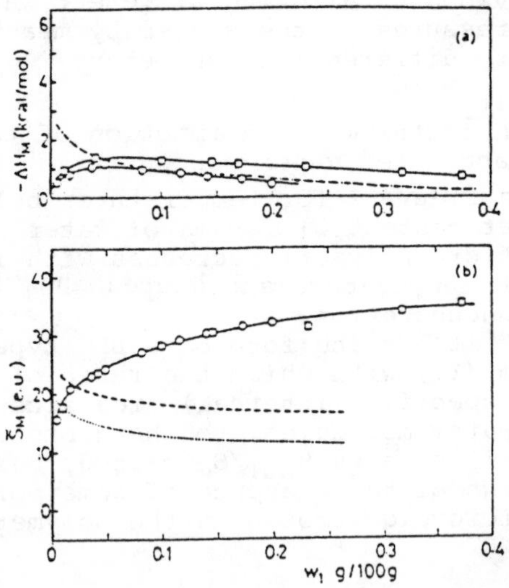

Figure 3. (a) Heat of mixing: □, 303 K; ○, 315 K; (-·-) theoretical heat of mixing according to Freundlich. (b) Entropy of mixing: □, 303 K; ○, 315 K; (...) theoretical assuming for the water molecule one rotation; (---) Idem two rotations. Courtesy of Macromolecules 15, 1145 (1982).

a dual mode of sorption (see part B later on) composed of an exothermic (increasing ΔH_M) process of trapping water in preexisting holes (which therefore do not require the expenditure of energy to create them), opposed by a dissolution of the water in the polymer and a corresponding reorientation of the segments and creation of new holes, which is an endothermic process and therefore makes ΔH_M decrease.
At higher water concentrations (greater than 0.04g/100g), $-\Delta H_M$ decreases steadily toward 0 kcal/mol, in good agreement (Figure 3) with the theoretical heats of sorption computed with Freundlich model.
With the entropy of mixing (ΔS_M) the total entropy ($\bar{S}_M$) was computed and results are shown in Figure 3 as well as the theoretical values computed (14) from statistical thermodynamics. At low concentrations the theoretical and experimental entropies coincide, indicating that the water molecules are adsorbed at the active sites, with a heat of mixing of ca. 1.0 kcal/mol. As the water content increases, other layers are adsorbed. At higher concentrations the entropy values are close to these of the vapor phase, indicating clustering of the water.

B) Dual Mode of Sorption and Cluster Formation

Sorption at low water concentrations is consequently a dual mode of sorption provided by sorbed molecules of the Langmuir type and by free molecules of the Henry type, being the equilibrium concentration (C in (g of water)/(g of dry polymer)) of water in the polymer at pressure P (15):

$$C = k_D \cdot P + C_H' \cdot bP/(1 + bP) \qquad (5)$$

in which k_D is the Henry's law solubility coefficient ((g of water)(g of dry polymer)$^{-1}$(torr)$^{-1}$, b is the Langmuir affinity constant (torr)$^{-1}$, and C_H' is the Langmuir sorption capacity (g of water)(g of dry polymer)$^{-1}$). Since Henry's coefficient (k_D) is much larger than Langmuir's ($C_H'b$) from the very beginning of the adsorption most af the water is in the liquid state, which is an indication that cluster formation begins to take place at very low water concentration.
The effect of temperature on $C_H'b$ shows that the glass transition temperature of the polymer in the presence of water decreases to 323 K (the Tg of the pure polymer is 343 K).
The Zimm-Lumberg clustering parameters (16) G_{11}/v_1 are shown in Figure 4. Since $G_{11}/v_1 > -1$ denotes the tendency for the water to cluster, it confirms that cluster formation takes place at very low water activities. At water volume fractions of 0.40 or higher, the behavior is ideal ($G_{11}/v_1 = -1$).

The cluster size, i.e., the average number

of water molecules in a cluster, is (9):

$$c_1G_{11} + 1 = \emptyset_2 \left[\frac{\ln\emptyset_1}{\ln a_1}\right] \qquad (6)$$

where G_{11} is the cluster integral, a_1 is the water activity, v_1 is the partial molar volume of water, and $\emptyset_1$ and $\emptyset_2$ are the volume fractions of water and polymer.

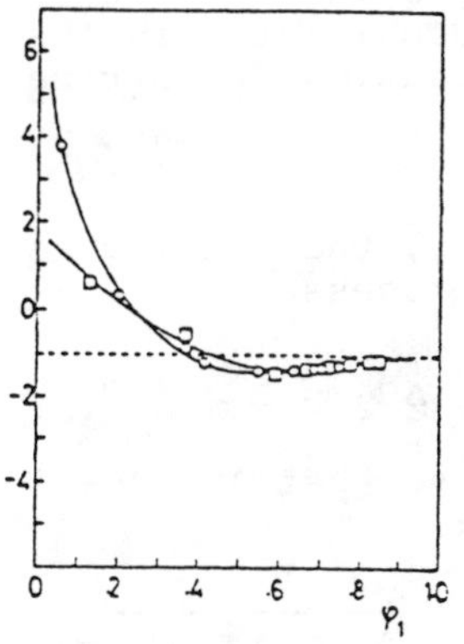

Figure 4. Clustering function of the water-epoxide system: □, 303 K; ○, 315 K. Courtesy of Macromolecules 15, 1145 (1982)

The number of clusters in a given amount of polymer is proportional to

$$w_1/(c_1G_{11} + 1) \qquad (7)$$

A plot of this quantity vs. w_1 is shown in Figure 5. Below 0.02g/100g, the slope is +1 (dashed line) since $c_1G_{11} = 0$, and, therefore, there is a balance between the water-polymer (active sites) and water-water contacts (the water molecules are accommodated in preexisting holes, the cluster size being greater than two).

As the water interacts with the polymer chains, it may produce some of the following effects in this order: (i) reorientation and chain displacement, i.e., reversible loosening or effective plasticization of the structure; (ii) solvation or reversible rupture of weak interchain bonds; (iii) irreversible disruption of the polymer matrix (microvoids). In the glassy state segmental rotations are restricted, tneding to immobilize the holes, and only interchain separation can take place. Above Tg there is free segmental rotation, and the average chain separation of closely packed regions and the number and size of the holes present adjust themselves to yield a true equilibrium state. As a consequence, the polymer slabs have a large increase in free volume.

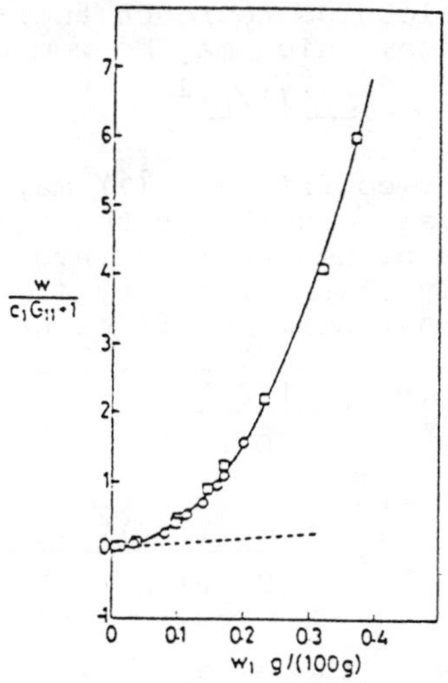

Figure 5. Number of clusters: □, 303 K; ○, 315 K. Courtesy of Macromolecules 15 (1982)

III) DIFFUSION COEFFICIENTS

To know them (19), sorption isotherms were measured as a function of temperature and water vapor concentration not only at temperatures below Tg (See Fig. 2) but also at temperatures above Tg.

The ratio of the amount of water ΔW_t (See below) adsorbed at time t to the amount ΔW_∞ adsorbed at equilibrium is shown in Figure 6: at temperatures below Tg (Figure 6-a) the system behaves linearly at all concentrations, while at the temperatures slightly above Tg used (Figure 6-b) an anomalous sigmoidal behavior is observed, because at these temperatures slightly over Tg, and more so when large clustering takes place simultaneously, the microbrownian motion of polymer chain segments may not have a mobility great enough to achieve the rapid establishment of equilibrium chain conformations.

With these data the diffusion coefficients of water were computed as follows:

A) Effect of Water Concentration

1) Initial Diffusion Coefficient

As the slab of thickness l, area A, and volume V, hold in vacuum, it is suddenly exposed to a water vapor pressure (P) equivalent to a water concentration C_o, a quasi-equilibrium concentration C_{S1} is instantaneously created at the surface of the polymer, followed by the establishment of the same equilibrium concentration C_∞ throughout the polymer slab by simple diffusion ($C_{S1} = C_\infty$). This equilibrium ($t = \infty$) concentration in the bulk of the polymer (C_∞ in kg/m^3) may be computed from the sorption isotherms considering the density ρ of the epoxide prepolymer equal to 1.187, though its exact value is not easy to ascertain (17):

$$C_\infty = w_1 \times \frac{10^3}{(100/\rho)\times 10^{-6}} = k \times \frac{W_\infty}{W_o} \qquad (8)$$

in which W_∞ is the weight of polymer

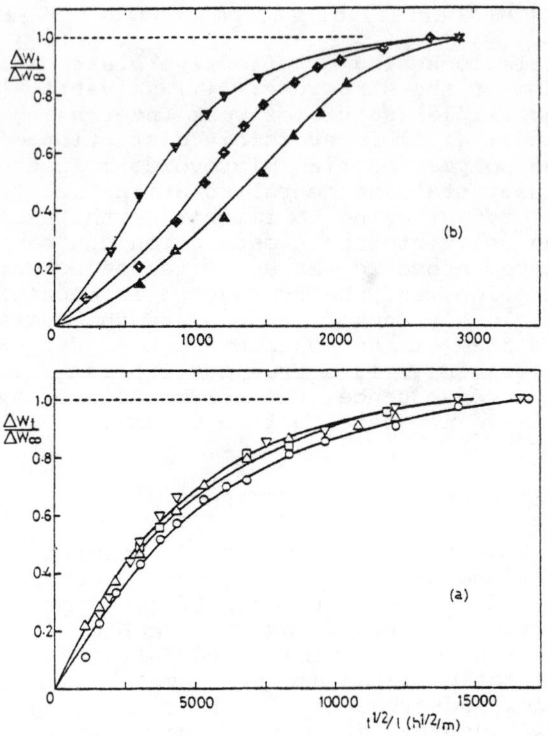

Figure 6. Sorption curves; (a) below Tg and water vapor pressure of 8.9 torr: ○, 298 K; △, 303 K; ▽, 309 K; □, 315 K. (b) above Tg at experimental temperature of 348 K: ◆, 18.8 torr; ▼, 8.9 torr; ▲, 4.1 torr. Courtesy J. Polym. Sc. (19)

plus water at time $t = \infty$, W_o is the weight of dry polymer ($t = 0$), $\Delta W_\infty = W_\infty - W_o$ is the weight of water sorbed at time $t = \infty$, and $k = 1{,}187$.

In this first stage the range of concentration is usually very small, and so it can be assumed that the diffusion coefficient is constant over the small concentration range.

The driving force for this initial diffusion toward the inside of the slab is the concentration difference ΔC_1:

$$\Delta C_1 = C_{S1} - 0 = C_{S1} \tag{9}$$

in which 0 is the concentration at absolute vacuum. With it, the flux density j_1 is according to Fick´s law:

$$j_1 = \frac{1}{A} \cdot \frac{d(\Delta W_1)}{dt} = \sigma_1 \times \frac{\Delta C_1}{\Delta(1/2)} \tag{10}$$

in which σ_1 is the diffusivity, equal to the initial diffusion coefficient (D_1) in an ideal solution.

Combining equations 8, 9, 10, and calling $\Delta Q = \Delta W_{t(n)} - \Delta W_{t(n-1)}$, the following equation is obtained:

$$D_1 = \frac{\Delta Q}{\Delta t} \times \frac{(1/2)}{A} \times \frac{1}{C_\infty} \tag{11}$$

2) Increasing diffusion coefficient

Once the equilibrium water content (ΔW_1) characteristic of the initial stage of the sorption curves is achieved, the water content of the slab increases with time more slowly, making the sorption kinetic to be no longer Fickean. Since now the polymer does not responds now rapidly enough to changes in conditions it is better to consider the system as being in a non-equilibrium state in which the amount of water ΔW_t accumulates in the slab making its concentration C_t to rise to the value:

$$C_t = k \times \frac{\Delta W_t}{W_o} \tag{12}$$

Consequently, the driving force for the diffusion process:

$$\Delta C_2 = C_\infty - C_t \tag{13}$$

decreases, so that $\Delta C_2 < \Delta C_1$, and equation 11 becomes:

$$D_2 = \frac{\Delta Q}{\Delta t} \cdot \frac{1/2}{A} \cdot \frac{1}{C_\infty - C_t} \tag{14}$$

3) Decreasing diffusion coefficient

Simultaneously with the increase in water content inside the slab, the relaxation of the chain segments takes place, swelling the polymer mass, increasing its free volume, and modifying the driving force for the diffusion process to the new value:

$$\Delta C_3 = (C_\infty - C_t) \cdot e^{-t/\tau} \tag{15}$$

such as $\Delta C_2 < \Delta C_3 < \Delta C_1$, and therefore:

$$D_3 = \frac{\Delta Q}{t} \cdot \frac{1/2}{A} \cdot \frac{1}{(C_\infty - C_t) \cdot e^{-t/\tau}} \tag{16}$$

i.e., the diffusion coefficient decays to 1/e of its initial value in the relaxation time τ, i.e., the time required for the macromolecules to reach their random configuration, which may be computed as:

$$\tau = \frac{\gamma \cdot (1/2)^2}{\sigma} \tag{17}$$

The activity coefficient (γ) may be considered equal to unity at the relatively small water concentrations involved, and consequently the diffusivity ($\sigma = D \cdot \gamma$) equals the diffusion coefficient, resulting:

$$\tau_{(1)} = \frac{(1/2)^2}{D_1} \tag{18}$$

and:

$$\tau_{(2,3)} = \frac{(1/2)^2}{D_{eff}(t_n - 1)} \tag{19}$$

The inverse of the exponent in equation 16 is the Deborah number:

$$De = \frac{\text{relaxation time}}{\text{experimental time}} =$$

$$= \frac{\tau}{t} = \frac{(1/2)^2}{D \cdot t} \qquad (20)$$

and its value (Figure 7) is always much smaller than one, except at the beginning of the experiments.

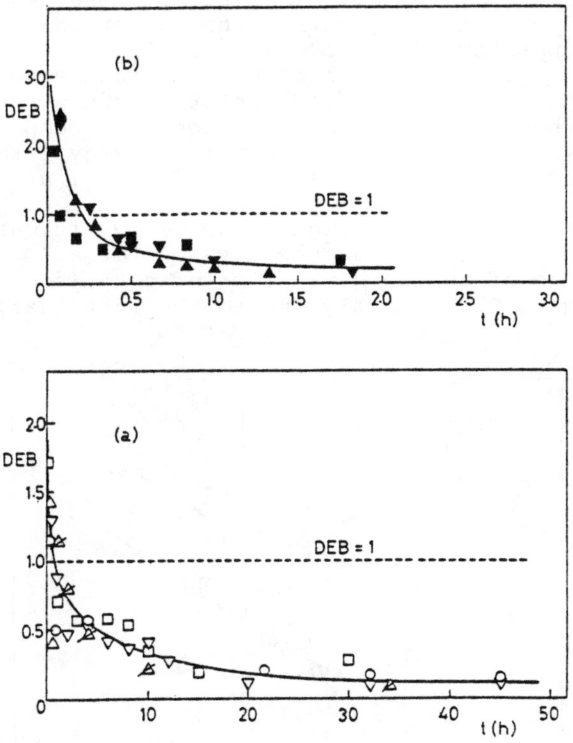

Figure 7. Deborah number versus experimental time at water vapor pressure of 18.8 torr and 12.9 torr (tag triangles), a) below Tg: ○, 298 K; △, 303 K; ▽, 309 K; □, 315 K. b) above Tg: ●, 338 K; ▼, 343 K; ■, 347 K. Courtesy J. Polym. Sc. (19)

It confirms the postulate previously stated that relaxation effects (the molecules diffusing down the concentration gradient are regarded as pushing their way past the polymer segments, i.e., pushing aside neighbouring chains of the polymer) are to be taken into account: on increasing the solvent concentration the increased ease of polymer segmental motion leads to an immediate decrease in the diffusion coefficient which opposes to the increasing diffusion coefficient D_2, due to the concentration changes inside the polymer, leading to the results shown in Figure 8 (data have been plotted versus time $\underline{t}$ because $\underline{t}$ is a directly measured variable, while C_t is a magnitude computed from w_1 -see Figure 2-; they are compared in Table II at P = 18.8 torr).

4) Overall or efficient diffusion coefficient

The overall diffusion coefficient (D_{eff}) is, in agreement with irreversible thermodyna-

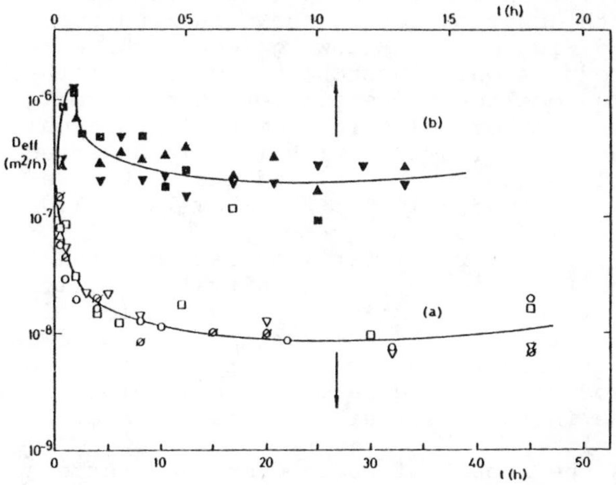

Figure 8. Effective diffusion coefficient (D_{eff}) versus experimental time (t). a) below T_g at water vapor pressure of 18.8 torr: ○, (ø, 12.9 torr) 298 K; △, 303 K; ▽, 309 K; □, 315 K. b) above Tg at water vapor pressure of 18.8 torr: ▲, 338 K; ▼, 343 K; ■, 347 K. Courtesy of J. Polym. Sc. (19)

TABLE II

Relation between time and water concentration at water vapor pressure of 18.8 torr

time (hours)	C_t (kg/m^3)	
	338 K	298 K
0.03	0.33	0.00
0.08	0.57	----
0.25	0.77	1.57
0.42	1.08	----
0.50	1.21	1.93
0.67	1.32	----
1.00	1.45	2.30
1.33	1.50	----
1.50		2.53
3.00		3.20
6.00		3.96
10.00		4.55
22.00		5.22
32.00		5.52
45.00		5.87

mic predictions (18):

$$D_{eff} = \frac{\Delta Q}{\Delta t} \cdot \frac{1/2}{A} \cdot \frac{1}{(C_\infty - C_t)(1 - e^{-t/\tau})} \qquad (21)$$

which reduces to equation 14 when $\tau \ll t$, and to equation 11 when is also $C_t = 0$.
Diffusion coefficients (D_{eff}) computed with equation 21 are shown in Figure 8-a and Figure 8-b. Below Tg (where the water produces only interchain separation) they are some ten times smaller than above Tg (where there also is segmental rotation).
The initial diffusivity (D_1) is observed only at $t \leqslant 0.1$ hours, and its value varies from 3×10^{-6} m^2/h below Tg to 1.0×10^{-6} m^2/h above Tg.
D_1, D_2 and D_3 differ from each other in the following way: the initial diffusivity D_1 is produced mainly by dissolution of the water in the polymer matrix, followed by reorientation of the polymer segments producing at the same time new holes (i.e., the initial diffusion may go on indefinetly).
As the amount of water dissolved increases, its rate of diffusion inside the polymer increases (decreasing diffusivity D_2) making the diffusion come to a halt when the resultant resistence due to the accrued accumulation of water (ΔW_t) builds up an internal concentration $C_t = C_\infty$ equal and opposed to the initial concentration C_S.
However, the complete halting (i.e., the final true equilibrium sorption) of the diffusion process is delayed by the improved diffusivity (decreasing diffusion coefficient D_3) provided by the relaxation of the chain segments.

B) Combined effect of temperature and sorption rate

The rate at which the water concentration in the slab increases with time is (19):

$$\dot{N} = \frac{1}{W_o} \frac{\Delta Q}{\Delta t} \tag{22}$$

A log-log plot of D_{eff} versus $\dot{N}$ is shown in Figure 9 at different temperatures

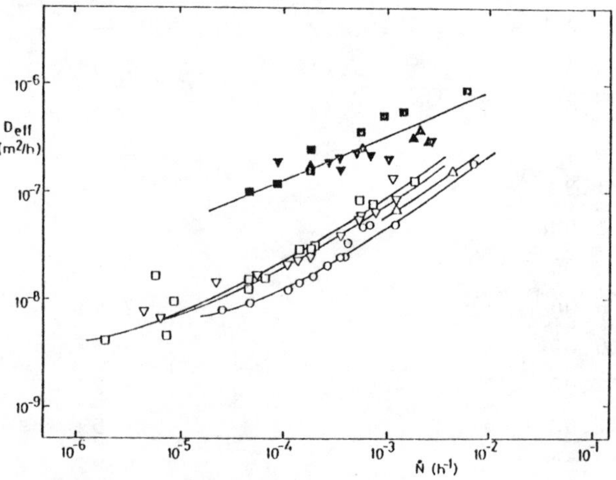

Figure 9. Effective diffusion coefficient (D_{eff}) versus sorption rate ($\dot{N}$) at water vapor pressure of 18.8 torr: a) below Tg: ○, 298 K; △, 303 K; ▽, 309 K; □, 315 K. b) above Tg: ▲, 338 K; ▼, 343 K; ■, 347 K.
Courtesy J. Polym. Sc. (19)

and constant pressure of 18.8 torr.
Below Tg (there is only interchain separation) it can be shifted parallel to the abscissa axis to provide a single curve for all temperatures (Figure 10): it is equivalent to multiply the sorption rate $\dot{N}$ by a temperature dependent shift factor a_T. A semi-logarithmic plot of these shift factors a_T versus the reciprocal temperature (1/T) provides a straight line, indicating that the temperature dependence of the shift factor a_T obeys the Arrhenius equation with a constant temperature independent activation energy for diffusion (ΔE) equals to 52.3 kJ/mol at P = 18.8 torr.
At temperatures slightly above Tg, it was not possible to deduce a value for a_T because of the large scatter of data (see Figure 9), possibly due to the fact that,

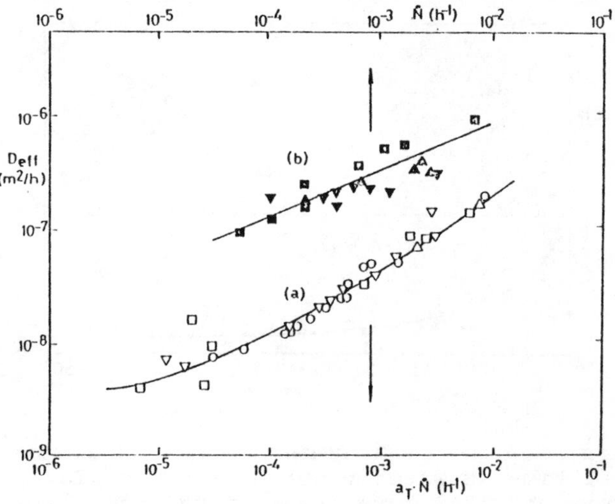

Figure 10. Effective diffusion coefficient (D_{eff}) at water vapor pressure of 18.8 torr: a) below Tg and reference temperature T_o = 298 K: ○, 298 K; △, 303 K; ▽, 309 K; □, 315 K. b) above Tg: ▲, 338 K; ▼, 343 K; ■, 347 K. Courtesy J. Polym. Sc. (19)

in spite of being present rotation of the chain segments plus interchain separation, as already said, the microbrownian motion of polymer segments may not have a mobility great enough to achieve the rapid establishment of the equilibrium chain conformations, i.e., the average chain separation of closely packed regions, and the number and size of holes present cannot adjust themselves to yield a true equilibrium state, with the water being associated in completly random fashion throughout the polymer slab.
The simultaneous effect of temperature and sorption rate on the diffusion coefficient of epoxide prepolymers can be described by the plot of log D_{eff} versus log $a_T \cdot \dot{N}$ (Figure 10). A statistical ana-

lysis of these results, using the method of least squares, provided the following third order polynomial (19):

$$\log D_{eff}(T,\dot{N}) = A_o + A_1(\log a_T.\dot{N}) + + A_2(\log a_T.\dot{N})^2 + A_3(\log a_T.\dot{N})^3 \quad (23)$$

in which the coefficients A_o - A_3 have the following values: A_o = - 4.700048, A_1 = = 1.277945, A_2 = 0.1533245, A_3 = 0.0085299.

REFERENCES

1. J.V. Aleman, RAPRA Translations, No. 1787, London (1972).
2. J.V. Aleman, Rev. Plast. Mod., 232, 509 (1975).
3. J. Crank, C.S. Park, eds.,"Diffusion in Polymers", Academic Press (1968).
4. J.C. Laria and J.V. Aleman, Polym. Eng. Sc. 21(12), 804 (1981).
5. O. Krischer, W. Kast and K. Kroll, "Trocknunstechnik", Springer-Verlag, (1978).
6. S. Whitacker, IEC Fundam., 16, 408 (1977).
7. D. Stockburger and F.R. Faulhaber, Chem. Eng. Technol., 41, 456 (1969).
8. A.G. Day, Trans. Faraday Soc. 59, 1218 (1963).
9. J.L.G. Fierro and J.V. Aleman, Macromolecules 15(4), 1145 (1982).
10. J.V. Aleman, J. Polym. Sc.,Polym.Chem. Ed. 18, 2567 (1980).
11. J.V. Aleman, Polym. Eng. Sc. 18(15), 1160 (1978).
12. S. Brunauer, L.S. Deming, W.E. Deming and E. Teller, J. Am. Chem. Soc. 62, 1723 (1940).
13. P.J. Flory,"Principles of Polymer Chemistry", Cornell University Press, Ithaca, N.Y. (1953).
14. A. Clark,"The Theory of Adsorption and Catalysis", Academic Press, New York, (1970).
15. A.S. Michaels, W.R. Vieth and J.A. Barrie, J. Appl. Phys. 34, 1 (1963).
16. B.H. Zimm and J.L. Lumberg, J. Phys. Chem. 60, 425 (1956).
17. N. Hatta and N. Kumanotani, J. Appl. Polym. Sc. 17, 3545 (1973).
18. H.L. Frisch, J. Chem. Phys. 41, 3679 (1964).
19. J.L.G. Fierro and J.V. Aleman, J. Polym. Sc. (submitted for publication).

MODELLING OF MASS TRANSFER WITHIN THE SOLID AS A BASIS FOR THE MATHEMATICAL DESCRIPTION OF CONVECTION DRYING

B. Dressel
K.E. Militzer

Technische Universität Dresden,
Dresden, German Democratic Republic

ABSTRACT

The modelling of mass transfer processes within the solid in drying makes it possible to consider the parameters of influence on process kinetics more exactly as this has been possible by means of customary calculation methods. The model is based on a diffusion statement with applying normalized effective transfer coefficients, which may be determined from the classic kinetic experiment. They are dependent on moisture, but independent of temperature and of linear dimension of solid. It becomes possible to design and to model technical dryers by making use of these transfer coefficients, which characterize the material to be dried.

NOMENCLATURE

A_i	surface of the material, m
B_{Gl}	slope of the line of equilibrium, kg/kg
$Bi = \frac{\sigma t^* B_{Gl}}{D^* S \varrho_x}$	BIOT - modulus
$D^* = D_{eff} t^*/S^2$	normalized effective transfer coefficient
D_{eff}	effective transfer coefficient m^2/s
D_{XS}	diffusion coefficient of water in the solid, m^2/s
D_{YS}	diffusion coefficient of vapour in the solid, m^2/s
h	specific enthalpy of air, kJ/kg
$\dot{m}$	mass flow rate, kg/s
s	space-coordinate in the solid, m
S	linear dimension of the solid, m
t	process time, s
$t^* = \frac{X'-X^*}{(dX/dt)_k}$	time constant of the process, s
$u = \frac{X - X^*}{X_k - X^*}$	normalized moisture of the material
X	moisture content of the material, kg/kg
X^*	equilibrium moisture of the material, kg/kg
Y	moisture content of the air, kg/kg
Y_S	saturation moisture of the air, kg/kg
σ	evaporating coefficient, $kg/(m^2 s)$
ψ	true drying rate of the material
$\tau = \frac{t - t_k}{t^* - t_k}$	normalized process time
$\xi = s/S$	normalized linear dimension of the solid
ϱ	density, kg/m^3

Subscripts

H_2O - water
k - critical point
x - solid
y - air

Superscripts

' - inlet of dryer

1. The mass transfer within the solid as the main parameter of influence on process kinetics

Convection drying as a typical process of solid-fluid mass transfer is determined by two main partial processes:

- mass transfer within the solid and
- exterior mass transfer from the surface of the solid into the drying medium.

As is already well known in the field of fluid-fluid mass transfer processes the influence of exterior mass transfer on process kinetic is only small due to the great mobility of fluid phase. What determined the process kinetics are the transfer processes within the solid. Since the drying mechanisms are quite complex and the recording of measured values is very expensive, the most simple way of modelling process kinetics of convection drying is that of using an integral exterior drying rate, which has been experimentally determines under constant exterior conditions /5/, /6/. The phenomena determining the drying rate are only considered implicitly by the experimental determination of the surface moisture on the side of the air Y_i.
Since mistakes may occur when assuming experimental results which have been obtained under constant exterior conditions to be valid for varying exterior conditions (small amount of air, oscillating drying) the explicit detection of the processes of interior mass transfer appears to be favourable.

2. Mechanisms of transport and models derived from them

Depending on the special behaviour of matter to be handled and transport mechanisms caused by it various models have been proposed for the description of mass transfer within the solid.
The transport mechanisms of vapour diffusion in air-filled pores of the solid is taken for granted in the extreme case of handling capillary-porous solids. This process is described by means of the shell model, already known e.g. from adsorption:

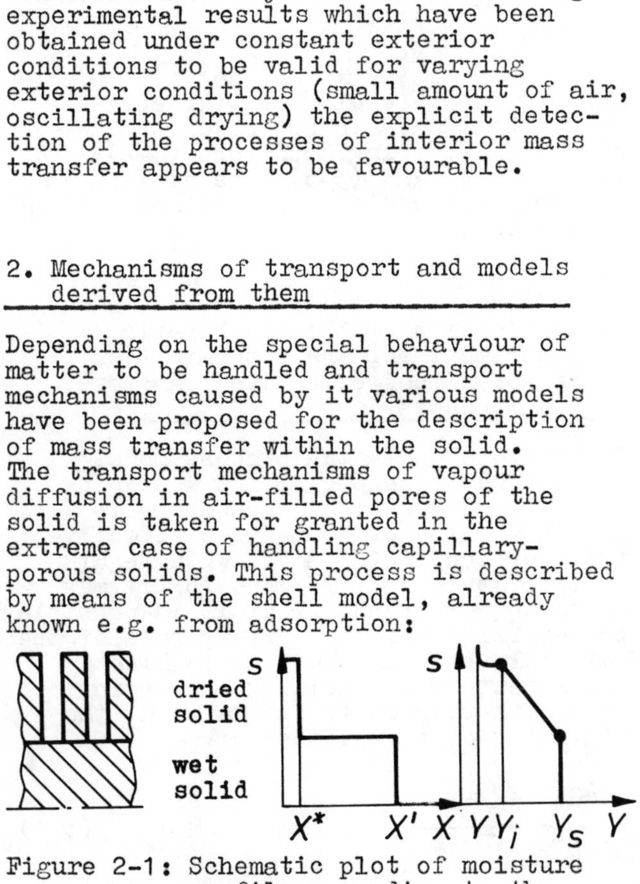

Figure 2-1: Schematic plot of moisture profils according to the shell model

The evaporating front, on which constant saturation of air is assumed as a condition of equilibrium, sinks into the inside of the solid with increasing drying during the second drying phase. An interconnection between the first law of FICK and the equation of exterior convectional mass transfer provides a model statement:

$$\dot{m}_{H_2O} = A_i \frac{\varrho_y D_{YL}}{S - S_o} (Y_S - Y_i) =$$

$$= \sigma A_i (Y_i - Y) \qquad (2\text{-}1)$$

If the coefficients of transport D_{YL} and σ are known, an information about the location of evaporating front S_o is needed for the application of the above-mentioned model. Liquid diffusion inside the material may be assumed for gel-like colloidal solids. In this case, evaporation of moisture occurs on the surface of the body, permanent equilibrium between the moisture of the material and that of air being presumed.

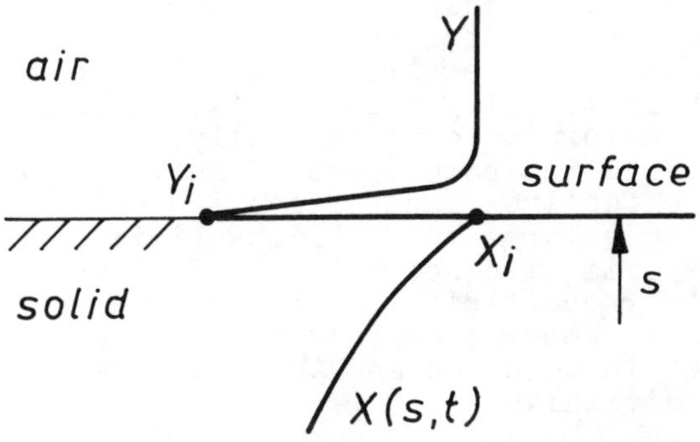

Figure 2-2: Schematic plot of moisture profils according to the diffusion model

The moisture transfer within the material may be described mathematically by means of the second law of FICK:

$$\frac{\partial X}{\partial t} = \nabla^2 (D_{XS}\, X) \qquad (2\text{-}2)$$

Here the determination of the transfer coefficient D_{XS} appears as a problem, since it also depends on the solid. Since in practice the properties of the matter to be dried will but rarely correspond to one of the two extreme cases, it is to be expected that for the mathematical description of the process one of the model concepts will hold exactly only in the ideal case.

Thus, the different portions of liquid and vapour diffusion are taken into consideration by means of a mean effective value of the transfer coefficient in the model statement for moisture diffusion. This mean effective value D_{eff} is identical with the transfer coefficient of liquid within the solid in the case of pure liquid diffusion. Though the phenomenological justification of the model decreases with the portion of vapour diffusion increasing this simplification is considered to be reasonable in the interest of a uniform description of the process and because of the satisfactory results already obtained by using this model

3. Model statement for technical drying

The following model components are needed for the designing of technical dryers:

1. Kinetics of the process
2. Equilibrium
3. Moisture balance.

The kinetics of the process are described by means of the model introduced in chapter 2. Since the effective transfer coefficient must include the interoir transport processes in their totality and variety, it has to be determind experimentally. In order to reduce the parameters of influence on this effective transfer coefficient and thus to enlarge the validity limit of an experiment a normalized effective transfer coefficient D^* is defined, which in general depends only on the matter to be dried and the moisture of the material.

$$D^* = D_{eff}\frac{t^*}{S^2} = f(X,\ \text{type of matter}) \quad (3\text{-}1)$$

The conditional equation for the process kinetics may now be written in the following form

$$\frac{\partial X}{\partial t} = \frac{S^2}{t^*}\nabla^2 (D^* X) \quad (3\text{-}2)$$

boundary condition

$$\left(\frac{\partial X}{\partial s}\right)_{/s=S} = \frac{\sigma t^*}{\varrho_x D^* S^2}(Y_i - Y). \quad (3\text{-}3)$$

In the case of constant enthalpy of air (h = const) for the description of the equilibrium state on the surface of the material the sorption isenthalps are made use of. It is possible to calculate them by using the experimentally determined sorption isothermes for the materials to be dried or by taking **them** from suitable reference books. Sorption isenthalps describe the equilibrium state for the whole drying process, if constant enthalpy of air is assumed. The sorption isenthalp displays the following relationship.

$$Y_i = f(X_i) \quad (3\text{-}4)$$

The balance of the process can take various forms according to the present phase current and the process control. In general, it serves the purpose of determining the local and/or temporal variation of the air moisture in drying medium corresponding to the mass conservation law. As a simple example, the differential balance for a continous countercurrent channel dryer in which the material is overflown will be given here:

$$dY = \frac{\dot{m}_x}{\dot{m}_y}\, d\bar{X} \quad (3\text{-}5)$$

The now complete differential equation system will have to be solved numerically, using equations (3-2), (3-3), (3-4) and (3-5). The application of very simple difference methods often proves to be successful.

4. Application of classic kinetic experiment for the determination of normalized effective transfer coefficients

In order to apply the process model introduced in chapter 3 for designing technical dryers it is necessary to know the value of normalized effective transfer coefficients D^*. With regard to literature /3/, /5/ and /6/ it may be taken for granted that it is possible to determine the true drying rate of the material $\varphi(\bar{u})$ in the classic kinetic experiment by normalizing the drying rate with respect to its value at the beginning of the process:

$$-\frac{d\bar{u}}{d\tau} = \frac{d\bar{X}/dt}{(d\bar{X}/dt)_k} = \varphi(\bar{u}) \quad (4\text{-}1)$$

This parameter as a property of the material is independent of the choice of the exterior conditions of the process. This independence arises as a main result of normalization. Thus the same normalization should be used for the determination of the normalized effektive transfer coefficient D^* from the classic kinetic experiment. The following further assumptions are necessary:

- classic kinetic experiment
  $Y = const = Y'$
  $\sigma = const$
- D^* is constant in sections
- one-dimensional transfer of moisture (plate, cylinder, sphere)
- linear approximation of the sorption isenthalp

$$Y_i = A_{Gl} + B_{Gl} X_i \qquad (4-3)$$

- homogeneous initial moisture distribution inside the material.

Thus it follows:

$$\frac{\partial u}{\partial \tau} = D^* \frac{\partial^2 u}{\partial \xi^2} \qquad (4-4)$$

$$\left(\frac{\partial u}{\partial \xi}\right)_{\xi=1} = Bi\, u(\xi = 1) \qquad (4-5)$$

$$u(\tau = 0, \xi) = 1 \qquad (4-6)$$

The differential equation (4-4) may be solved by applying the boundary condition (4-5). A conditional equation for the normalized effective transfer coefficient is deducible from the results. These equations are listed in table 1. In general, D^* may be calculated from the following initial quantities:

- true drying rate $\nu(\bar{u})$
- kinetics of the process $\tau(\bar{u})$
- slope of equilibrium line B_{Gl}
- evaporation coefficient σ
- process time constant t^*
- linear dimension S and shape of body.

Though the iterative method necessary for the calculation generally converges after two or three iterations a computer-aided evaluation is recommended, the calculation being very complicated. Finally, D^* is approximated as a function of the moisture of the material:

$$D^* = f(\bar{u}) \qquad (4-7)$$

5. Characteristics of the normalized effective transfer coeffecient

The normalized effective transfer coefficient D^* may be regarded as the ratio between maximum diffusion flow inside the solid and maximum mass flow from the surface of the material into the fluid:

$$D^* = \frac{D_{eff}\, t^*}{S^2} = \frac{\frac{D_{eff}}{S^2}(X' - X^*)}{\sigma A_i (Y_S - Y')} \qquad (5-1)$$

As can be deduced from the conditional equation as well as from phenomenological considerations D^* represents a function of the moisture of the material. This conclusion could be confirmed by the interpretation of kinetic experiments. Moreover, the effective transfer coefficient has been shown to be independent of the air state and the linear dimension of body in general.

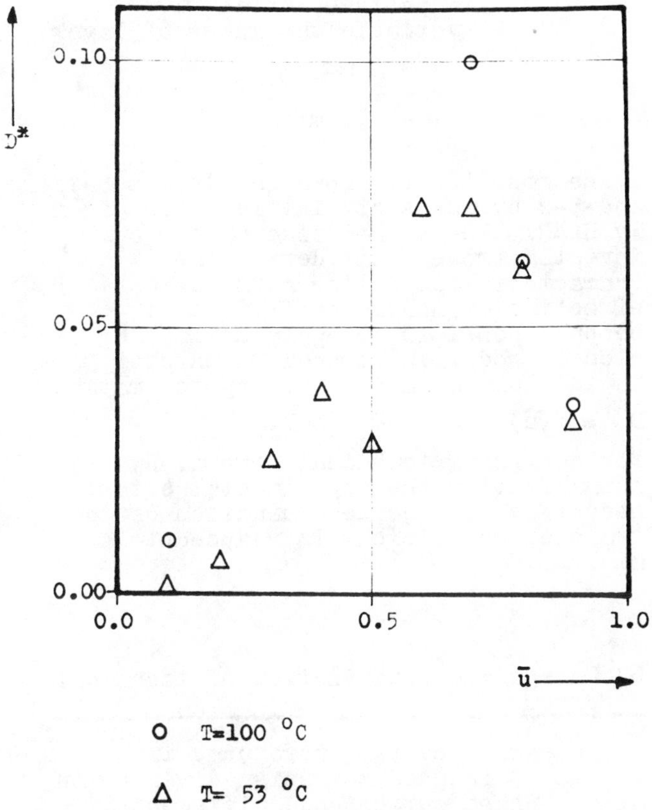

Figure 5-1: Normalized effective transfer coefficients for spheres of gaseous concrete at different drying temperatures. Measuring data according to /4/.

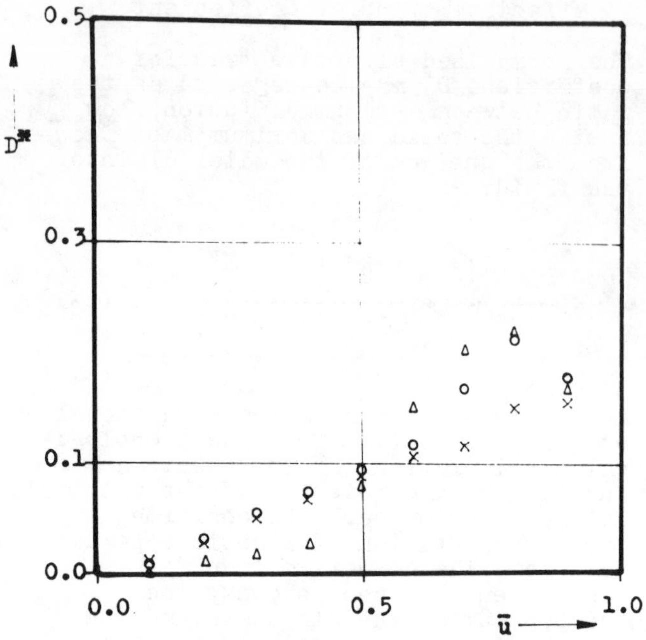

Figure 5-2: Normalized effective transfer coefficients for a pasty material with variable thickness of layer:

o - 3 mm
× - 2 mm
△ - 1,5 mm

These results are shown in figures 5-1 and 5-2 by means of kinetic dates given by HAERTLING /4/ and also by our own investigations. Considering the characteristics of the normalized effective transfer coefficient, it becomes possible to apply it in the process model introduced in chapter 3 in the form of a suitable approximation

$D^* = f(\overline{u})$

for designing technical dryers. By normalization the experimental effort necessary for the determination of the transfer coefficient is reduced to a minimum.

6. Design and optimization of technical dryers

The equation system introduced in chapter 3 is used to design dryers with various phase current and different process progress controls, taking into account the normalized effective transfer coefficient determined according to chapter 4.

The evaluation model thus created permits the design of dryers on the basis of transfer phenomena which explicitly influence the process kinetics. Like in hitherto existing methods of design the necessary kinetic data can be determined from the classic kinetic experiment.

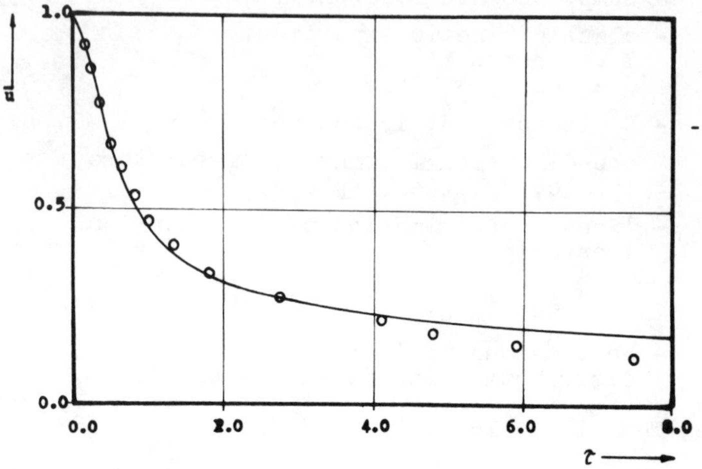

Figure 6-1: Checking up of a parallel flow dryer for spheres of gaseous concrete.
o - measuring data according to /4/.

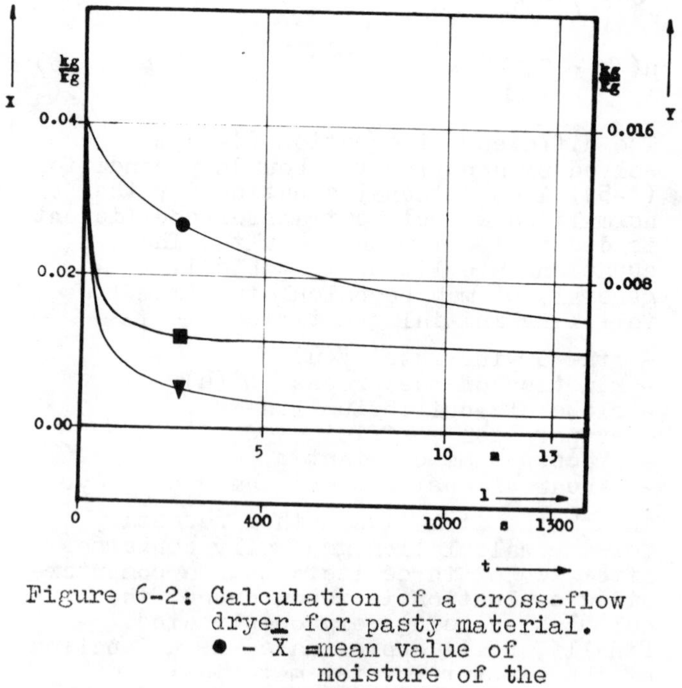

Figure 6-2: Calculation of a cross-flow dryer for pasty material.
● - $\overline{X}$ =mean value of moisture of the material
■ - Y"=moisture of exit air
▼ - X_i =moisture at the sur-face of the material

Table 1: Conditional equations for D^*

form of solid	D^*	μ_n
smooth plate of infinite extent (thickness 2 S)	$D^* = \dfrac{\psi(\bar{u})}{(1+\frac{A^*}{B_{Gl}})\sum\limits_{n=1}^{\infty} \dfrac{\mu_n \sin\mu_n \exp(-\mu_n^2 D^* \tau)}{(\frac{3\mu_n}{2Bi}+\frac{\mu_n}{2})\sin\mu_n-(1-\frac{\mu_n^2}{2Bi})\cos\mu_n}}$	$\dfrac{Bi}{\mu_n} = \tan\mu_n$
cylinder of infinite length (radius S)	$D^* = \dfrac{\psi(\bar{u})}{(1+\frac{A^*}{B_{Gl}})\sum\limits_{n=1}^{\infty} \dfrac{4Z(\mu_n)\exp(-\mu_n^2 D^* \tau)}{\frac{1}{\mu_n} J_1(\mu_n) + \frac{1}{Bi} J_o(\mu_n)}}$ $Z(\mu) = \sum\limits_{\nu=0}^{\infty} \dfrac{(-1)^{\nu}}{\nu!\,\nu!\,(2\nu+2)} \left(\dfrac{\mu}{2}\right)^2$	$\dfrac{J_1(\mu_n)}{J_o(\mu_n)} = \dfrac{Bi}{\mu_n}$
sphere (radius S)	$D^* = \dfrac{\psi(\bar{u})}{4(1+\frac{A^*}{B_{Gl}})\sum\limits_{n=1}^{\infty} \dfrac{(\cos\mu_n-\sin\mu_n/\mu_n)\exp(-\mu_n^2 D^* \tau)}{\cos\mu_n(\frac{1}{Bi}+\frac{1}{2}) - \frac{\sin\mu_n}{\mu_n}(\frac{1}{Bi}+\frac{\mu_n^2}{2Bi}-1)}}$ $A^* = \dfrac{A_{Gl} + B_{Gl}X^* - Y'}{X' - X^*}$ (correcting term for the equilibrium curve)	$\dfrac{\mu_n}{\tan\mu_n} = 1-Bi$

In contrast to customary dryer designing methods by means of the true drying rate of the material /5/, /6/, the advantage results, that changed dimensions of the material as well as repercussions of the exterior process conditions on the interior transport processes may be considered in the calculation.
The aim of the design of dryers is to choose the operating point of the dryer in such a way that the desired quality of the product may be reached by a minimum expense and to dimension the dryer in such a way, that it offers optimal prerequisites to solve the drying problem economically. For this purpose it is necessary to connect the process model which has been found with a suitable cost model. Since the process model cannot be explicitly reduced to parameters which influence costs, it appears recommendable to carry out a step-by-step computer simulated search method. Along with the optimization of available simple types of dryers this method makes it possible to optimize more complicated types of dryers such as dryers with an oscillating regime as to operating point and construction size. Thus, a wide field for the practical application of the calculation method is opened up, which is mean to save energy and reduce costs in the drying process.

REFERENCES

1. DRESSEL, B.: Beitrag zur verfahrenstechnischen Modellierung der Konvektionstrocknung. Diss. Dresden 1982.

2. DRESSEL, B.: Nutzung effektiver Transportkoeffizienten zur Trocknermodellierung. Wiss. Z. TU Dresden, to be published 1983.

3. GINSBURG, A.S., KRASNIKOW, W.W., and MILITZER, K.E.: Durchführung und Auswertung von Trocknungsversuchen. Wiss. Z. TU Dresden, Vol. 31, no. 3, p. 19-24, 1982.

4. HAERTLING, M.: Messung und Analyse von Trocknungsverlaufskurven als Grundlage zur Vorausberechnung von Trocknungsprozessen. Diss. Karlsruhe 1978.

5. KRISCHER, O., KAST, W.: Trocknungstechnik Bd. 1: Die wissenschaftlichen Grundlagen der Trocknungstechnik. Springer, Berlin-Heidelberg-New-York, 1978

6. WEISS, S., ADOLPHI, H.V. and MILITZER, K.E.: Lehrwerk Verfahrenstechnik: Thermische Verfahrenstechnik I. DVG Leipzig 1978.

NATURAL DRYING IN CONCRETE PAVEMENTS

C. L. D. Huang

Department of Mechanical Engineering
Kansas State University
Manhattan, Kansas 66506

ABSTRACT

The need to understand and predict the process of heat and moisture transfer in concrete is essential in studying ways to prevent frost damage in concrete highways and airfield runways subjected to freezing and thawing cycles. In this paper, the theory based on the non-equilibrium thermodynamics and laws of physics is applied to the investigation of the moisture distribution during drying in a light-weight concrete pavement with a moisture barrier between the concrete slab and the base structure under pre-icing conditions. The slab is 10^{-1} m thick and subjected to a temperature gradient to simulate normal weather conditions in winter. An implicit finite difference scheme is employed. The numerical solution yields the moisture, temperature, and pressure distributions in the concrete pavement. The histories of the averaged moisture content and of the moisture content for various locations within the slab are also obtained.

INTRODUCTION

The need to understand and predict the process of heat and moisture transfer in concrete is essential in studying ways to prevent frost damage in concrete highways and airfield runways subjected to freezing and thawing cycles. Although the introduction of air-entrained concrete several years ago has significantly aided in the reduction of frost damage, the problem is neither fully solved nor completely understood. In concept, frost damage can be prevented by the installation of an embedded heat source in the concrete slab [1]. The small temperature gradient generated by such a heat source will cause the moisture in the saturated slab to migrate outward. This will reduce the internal moisture content in the concrete pavement to a level sufficient so that when freezing does occur, the resulting expansion will not fracture the concrete structure. The power required by the heat source is a function of both the environmental conditions and the imposed time period during which the moisture content is to be decreased. To prevent damage from freezing the internal moisture content of the concrete must be reduced to at least 75-80% of the saturated value (depending on the concrete, aggregate and other parameters [2, 3, 4]) prior to freezing.

To predict the moisture distribution within a concrete pavement as a function of time, both the relevant material characteristics and the transport properties must be considered. A diffusion theory with a linear or a nonlinear coefficient of diffusivity, is believed inadequate for the description of the behavior of the mass transfer within a pavement. The transfer of mass in all phases and the transfer of heat are coupled to each other and must be considered simultaneously. During the funicular stage (liquid saturated stage), diffusion is the major mechanism of moisture transport. However, in the pendular stage (unsaturated liquid flow stage), experiments have shown that diffusion, capillary and evaporation-condensation actions are the governing mechanisms in the mass transfer process. Experimental evidence indicates that the pore size distribution within a porous medium is an important parameter affecting mass transfer [5]. Theoretical analysis previously published confirms this fact [6]. Therefore, a general mathematical model to predict the simultaneous multiphase moisture and heat transfer in porous media was needed. This generalized model was developed based on a system of differential equations for simultaneous heat and mass transfer in a porous media using the principles of non-equilibrium irreversible flow of heat and mass with linear phenomenological equations and the laws of conservation applied in a macroscopic sense [7].

In this paper, the theory previously developed is applied to the investigation of the moisture distribution during drying in a light-weight concrete pavement with a moisture barrier between the slab and the base structure under pre-icing conditions. The slab is 10^{-1}m (100mm) thick and subjected to a temperature gradient generated by imbedded heat sources at the bottom surface of the concrete to simulate normal weather conditions in winter, (see Fig. 1).

The basic description of heat and mass transport in a concrete slab can be given by three nonlinear partial differential equations. An analytical solution for these differential equations is difficult and as yet has not been obtained. In this paper, an implicit finite difference numerical scheme is employed in obtaining an approximate numerical solution. The numerical solution of the simultaneous algebraic equations yields the values of moisture concentration, temperature, and pressure at the pre-assigned grid points throughout the domain investigated. The average moisture content stored in the slab is calculated as a function of time. The moisture content is also calculated for various locations within the slab.

STRUCTURE OF HYDRATED CEMENT PASTE

The mechanisms for moisture movement within a concrete slab are dominated by the structure of the cement paste of concrete, which forms a porous capillary system [8, 9]. Use of a single-phase diffusion model to describe the mass transfer mechanism has been shown to be an oversimplification. For coarse grained soils, the capillary flow theory which is more intricate than the diffusion theory yields adequate results for predicting the migration of moisture. However, for a porous medium with a fine texture, such as concrete, the surface energy of the pore space, which is ignored in both the diffusion and capillary flow theories, significantly affects the movement of moisture [10]. Therefore the capillary flow theory is insufficient to predict the moisture migration in a fine textured porous medium such as concrete.

The results of a series of studies concerning the structure and the physical properties of hardened portland cement was published by Powers and Brownyard [11]. The pore structure of the cement paste can be divided into gel pores and capillary pores. The average diameter of a gel pore is 18 Å. The structure of gel pores results from the growth of very small irregular cryptocrystalline particles, which may have the shapes of fibers, rolled foils, tubes and plane sheets. The capillary pores are scattered throughout the mass of the cement gel with a diameter usually greater than 200 Å. The magnitude of the capillary porosity of cement pastes depends on the water-cement ratio with an increase in water-cement ratio causing an increase in capillary porosity. Structural differences among pastes are primarily the result of the differences in capillary porosity, which in turn depends upon the water-cement ratio. An increase in water-cement ratio causes an increase in the capillary porosity. Structural differences among pastes can also result from changes in the chemical composition of the cement. Various sizes of gel and capillary pores exist in cement pastes thus causing the porous system to have a large specific surface of the order of $2 \times 10^5 m^2/kg$ by dry weight. The distribution of these pore sizes is a major characteristic of the porous system [12].

EQUILIBRIUM SORPTION

In a wet porous body both liquid and gaseous phases of water exist with the surface of the solid matrix as their boundaries. In this paper, the solid matrix is considered, in a macroscopic sense, as a rigid body with homogenous and isotropic pore structures. Within the porous system the physical properties of the water in the liquid phase are different from that of free water. Layers of water molecules next to the surface of the solid matrix are tightly held by Van der Walls attraction. The adsorbed water film has limited mobility and moves primarily by a process of surface diffusion. The liquid beyond the adsorbed layer in capillary pores is subjected to capillary forces which results in a higher degree of mobility than the adsorbed water. Therefore, the primary transport mechanism for the non-adsorbed water is convection.

In the gaseous phase, on a macroscopic scale, both the air and water vapor can be transported by molecular diffusion and molar convection. The concentration of water vapor in the mixture is determined from the equilibrium-sorption relationship which depends upon the topological characteristics of the system, such as the porosity, specific surface and pore size distribution. Vassilious and White [13] have proposed that the equilibrium sorption relation can be expressed as a functional relation between the moisture content θ and the "equivalent pore diameter" r.

$$r = r(\theta) \tag{1}$$

The moisture content is approximated by

$$\theta = \frac{m}{\varepsilon} \simeq 1 - \frac{\varepsilon_g}{\varepsilon} \tag{2}$$

assuming that the water content in the gaseous phase is negligible in comparison with that in the liquid phase. The porosity of the system is denoted by ε. ε_g denotes the volume fraction of the gaseous phase, and m the volumetric moisture content. From the Kelvin equation, r can be expressed as a function of temperature and relative vapor pressure.

$$r = - \frac{2\sigma M_w}{\rho_w RT} \left| \ln \left(\frac{P_v}{P^\circ_v}\right) \right|^{-1} \tag{3}$$

where σ denotes the surface tension, M_w the molecular weight of water, ρ_w density of water, and R the gas constant. Hence, a functional relationship of moisture content θ, temperature T, and the relative vapor pressure (P_v/P°_v) can be established. The variable r in Equation (3) should be interpreted as a characteristic length of pore sizes. All portions of the pore space with pore sizes larger than those given by Equation (3) are accessible to the gaseous phase. Experimental evidence supporting this assumption has been given by Corey and Brooks [14]. Since only the drying processes will be considered in this paper, the curves shown in Figure 2 are for the desorption equilibrium of a light weight concrete. From Equations (1, 2, 3) and the desorption vapor equilibrium curve given in Figure 2, a relationship between the effective porosity and the water content has been established as

$$\varepsilon_g = \varepsilon\{1 - \theta[r(\phi, P, T)]\}. \tag{4}$$

BASIC EQUATIONS FOR HEAT AND MASS TRANSFER IN CONCRETE

By assuming the existence of a local thermal equilibrium in the porous system, and that the gaseous vapor phase is the major factor in the mass transfer process, the basic equations for heat and mass transfer were derived using the laws of conservation of mass, momentum, energy, and the kinetic theory of ideal gases [7]. For one-dimensional mass and heat transfer in a slab of thickness L, the equations are given in the following form:

$$A_i \frac{\partial\phi}{\partial t} + B_i \frac{\partial P}{\partial t} + C_i \frac{\partial T}{\partial t} = D_i \frac{\partial^2\phi}{\partial x^2} + E_i \frac{\partial^2 P}{\partial x^2} + F_i \frac{\partial^2 T}{\partial x^2}$$

$$+ G_i \left(\frac{\partial\phi}{\partial x}\right)^2 + H_i \left(\frac{\partial P}{\partial x}\right)^2 + I_i \left(\frac{\partial T}{\partial x}\right)^2 + J_i \frac{\partial\phi}{\partial x} \cdot \frac{\partial P}{\partial x}$$

$$+ K_i \frac{\partial\phi}{\partial x} \cdot \frac{\partial T}{\partial x} + L_i \frac{\partial P}{\partial x} \cdot \frac{\partial T}{\partial x} \qquad (5)$$

where the coefficients $A_i, \ldots, L_i$ are functions of the dependent variables ϕ, P, T, and ε_g. The coefficients are defined as follows:

$A_1 = \overline{Y}\,(\partial\varepsilon_g/\partial\phi) + \varepsilon_g$

$A_2 = (1 - \phi)\,(\partial\varepsilon_g/\partial\phi) - \varepsilon_g$

$B_1 = \overline{Y}\,(\partial\varepsilon_g/\partial P) + \varepsilon_g\,(\phi/P)$

$B_2 = (1 - \phi)\overline{O}_p$

$C_1 = \overline{Y}\,(\partial\varepsilon_g/\partial T) - \varepsilon_g\,(\phi/T)$

$C_2 = -(1 - \phi)\overline{O}_T$

$D_1 = D\varepsilon_g\,(M_a/M)$

$D_2 = -D\varepsilon_g\,(M_w/M)$

$E_1 = \zeta\phi\,(\mathring{k}_g/\eta_g)$

$E_2 = \zeta\,(1 - \phi)\,(\mathring{k}_g/\eta_g)$

$F_1 = 0$

$F_2 = 0$

$G_1 = -D\overline{O}_\phi\,(M_a/M)$

$G_2 = D\overline{O}_\phi\,(M_w/M)$

$H_1 = \phi\overline{W}_p\,(\mathring{k}_g/\eta_g)$

$H_2 = (1 - \phi)\,\overline{W}_p\,(\mathring{k}_g/\eta_g)$

$I_1 = 0$

$I_2 = 0$

$J_1 = D\overline{O}_P\,(M_a/M) + (\mathring{k}_g/\eta_g)\,[\phi(\partial\zeta/\partial\varepsilon_g) \cdot (\partial\varepsilon_g/\partial\phi) + \zeta]$

$J_2 = -D\overline{O}_P\,(M_w/M) + (\mathring{k}_g/\eta_g)\,[(1 - \phi)\,(\partial\zeta/\partial\varepsilon_g) \cdot (\partial\varepsilon_g/\partial\phi) - \zeta]$

$K_1 = -D\overline{O}_P\,(M_a/M)$

$K_2 = D\overline{O}_T\,(M_w/M)$

$L_1 = -\phi\overline{W}_T\,(\mathring{k}_g/\eta_g)$

$L_2 = -(1 - \phi)\,\overline{W}_T\,(\mathring{k}_g/\eta_g)$

$A_3 = \rho_w Q\,(\partial\varepsilon_g/\partial\phi)$

$B_3 = \rho_w Q\,(\partial\varepsilon_g/\partial P) - \varepsilon_g$

$C_3 = \rho_w Q\,(\partial\varepsilon_g/\partial T) + [(\varepsilon - \varepsilon_g)\rho_w(C_p)_w + \rho_s(1 - \varepsilon)(C_p)_s] + \varepsilon_g P[\phi M_w(C_p)_v + (1 - \phi)M_a\,(C_p)_a]/RT$

$F_3 = K$

$I_3 = (\partial K/\partial\varepsilon_g)\,(\partial\varepsilon_g/\partial T)$

$K_3 = (\partial K/\partial\varepsilon_g)\,(\partial\varepsilon_g/\partial\phi) - DM_aM_w\varepsilon_g P[(C_p)_a - (C_p)_v]/RTM$

$L_3 = (\partial K/\partial\varepsilon_g)\,(\partial\varepsilon_g/\partial P) + (\mathring{k}_g\zeta P)[M_w(C_p)_v\phi\;M_a\,(C_p)_a\,(1 - \phi)]/RT$

in which

$\overline{O}_\phi = \varepsilon_g[(M_w - M_a)/M] - (\partial\varepsilon_g/\partial\phi)$

$\overline{O}_p = (\varepsilon_g/P) + (\partial\varepsilon_g/\partial P)$

$\overline{O}_T = (\varepsilon_g/T) + (\partial\varepsilon_g/\partial T)$

$\overline{Y} = \phi - (\rho_w RT/M_w P)$

$\overline{W}_p = (\zeta/P) + (\partial\zeta/\partial\varepsilon_g)\,(\partial\varepsilon_g/\partial P)$

$\overline{W}_T = (\zeta/T)\;(\partial\zeta/\partial\varepsilon_g)\;(\partial\varepsilon_g/\partial T)$

The four variables in Equation (5) ϕ (x, t), P (x, t), T (x, t) and ε_g require that this equation be incorporated with Equation (4) to form the complete set of governing differential equations.

The boundary conditions on the surfaces of the slab as shown in Figure 1 are;

(i) moisture barrier at x = 0

$$\frac{\partial\phi}{\partial x} = 0\,, \quad \frac{\partial P}{\partial x} = 0\,, \text{ and } q = q_x = 0 \qquad (6)$$

(ii) Pavement surface at x = L

$$\frac{\partial \phi}{\partial x} = \alpha \left(\frac{RT}{PD}\right) (\phi_\infty - \phi_{x=L})$$

$$P = P_{atm}, \text{ and } \frac{\partial T}{\partial x} = \frac{h}{k} (T_\infty - T_{x=L}) \qquad (7)$$

where α denotes the mass transfer coefficient, K the diffusion coefficient, h the heat transfer coefficient and k the thermal conductivity ϕ_∞ and T_∞ are the concentration fractions of water vapor and temperature immediately surrounding the porous system. In this paper the value of ϕ_∞ is 0.0048, since this value corresponds to a relative humidity of 60% with T_∞ at 4°C and q_o of 5 x 10^3 g/s^3. For a slab surrounded by air at atmospheric pressure and temperature between 0° and 60°C the transfer coefficients can be calculated from the following equations;

$$\alpha = \alpha_F + 1.63 \times 10^{-5}[(T_\infty - T)/B]^{1/4} \qquad (8)$$

$$h = 2.75 \times 10^8 \alpha + \sigma_s e (T_e^4 - T^4)/(T_e - T) \qquad (9)$$

$$K = [K_g^n \varepsilon_g + K_\ell^n \varepsilon_\ell + K_s^n (1 - \varepsilon)]^{1/n} \qquad (10)$$

where α_F denotes the mass transfer coefficient resulting from forced convection. B is the characteristic length of the slab, σ_s the Stefan-Boltzman constant, T_e the temperature of the enclosure of the system (Te = 4°C) and n the topological constant of the porous system (n = 0.25). K_g, K_ℓ and K_s are the conductivities for gas, liquid, and solid matrix respectively; ε, ε_g and ε_ℓ are the porosity of the porous system, the volume fraction for gas, and the volume fraction for liquid respectively.

The initial conditions for the present study are given as follows;

$$\phi (x,o) = 0.007 \ (m = 0.106), \ T(x,o) = 2°C, \text{ and}$$

$$P = P_{atm} \qquad (11)$$

Equations (4, 5, 6, 7 and 11) can be used to form a nonlinear boundary value problem for the simultaneous transfer of heat and mass within a concrete slab.

NUMERICAL ANALYSIS AND RESULTS

Obtaining an analytical solution for the nonlinear partial differential equation given by Equations (4, 5, 6, 7 and 11) was not possible. However, in this study, an implicit finite difference scheme was employed to obtain approximate numerical results. The first step was to replace the equations with a set of algebraic finite backward-in-time equations, which gave the relationships among the dependent variables ϕ, P, and T at neighboring points in an (x, t) space. The numerical solution of these simultaneous algebraic equations yielded the values of the dependent variables at the pre-assigned grid points throughout the domain investigated. For the set of algebraic equations developed, there was only one space increment Δx and one temporal increment Δt. The use of a finer mesh for Δx and Δt may have resulted in a smaller error, however, the magnitude of Δx and Δt cannot be chosen arbitrarily since to achieve a stable computation, a considerably smaller value of Δt must be used at the beginning of the computational simulation than is used in the more advanced stages. This fact results from the errors associated with the initial assumption used for the sorption equilibrium.

It should be noted that the coefficients $A_i, \ldots, K_i$ (i = 1, 2, 3) are functions of the three dependent variables and the following material characteristics; porosity and permeability of the solid matrix; diffusion coefficient of water vapor in air; viscosity of the gaseous mixture; sorption equilibrium relation and heat of sorption. The physical properties and empirical parameters corresponding to a light-weight concrete are listed in Table 1.

The numerical solutions for the average drying history of a light weight concrete slab as well as for the drying history of various points within the slab were obtained by the use of a digital computer. The results have been reproduced graphically in Figure 3. As can be seen in Figure 3, the drying rates at the various locations within the slab are significantly different during the funicular stage while all the drying rates approach asymptotically the same value at the pendular stage. While the moisture content within the slab approaches a hygro-equilibrium situation with time, the moisture content near the surface of the slab quickly approaches an equilibrium state. This fact is to be expected since the drying rate on the surface of the slab is much faster than that at the bottom of the slab.

The distributions of moisture content, temperature and gage pressure with time are represented graphically in Figure 4 for a light-weight concrete slab. During the funicular stage, the moisture content distribution curve is smooth and convex, which signifies that the mechanism of evaporation-condensation plays an insignificant role. However, during pendular stage, the moisture distribution curve is rough and wavy, which agrees qualitatively with experimental data for sand layers [15, 16]. As can be seen in Figure 4, the average temperature in the slab rises rapidly and approaches a steady state condition within a short period of drying time, while the gage pressure in the slab builds up quickly at the bottom of the slab and is steadily released.

It should be noted that at the beginning of the drying stage an extremely large pressure gradient may exist beneath the surface of the slab resulting from the application of the heat impulse. Such a high pressure gradient may cause fissures or cracks to form on the surface of a pavement. Therefore, the design and installation of the embedded heat sources must be properly conducted.

CONCLUSIONS AND REMARKS

The moisture distribution curves revealed that at high pore saturation (funicular saturation), the moisture movement is relatively independent of the properties of the porous system and dependent on the parameters characterizing the surroundings, such as temperature, velocity and relative humidity of the ambient air. As the internal moisture content is lowered, the liquid threads in the porous medium break down and capillary action begins which results in the internal characteristics of the porous system playing the primary role in controlling the migration of moisture. When capillary action begins the moisture distribution curves are no longer smooth and convex but rather rough and wavy. During the pendular stage, the liquid moisture has to be vaporized to move from one location to another. This results in a reduction and a variation in the drying rate with both time and location within the slab. At low moisture contents, vapor flux is the only mechanism controlling the transfer of moisture within the system thus the drying rate becomes very slow.

The sorption equilibrium curve is one of the most important factors in determining the mechanisms of moisture transfer since it determines the drying rate in the desorption processes. Moisture migration in a cement paste with low porosity is much slower that in a paste with high porosity. A good cement paste has a high specific surface, and can hold a significant amount of moisture in the hygroequilibrium state.

In the theory presented in this paper the solid matrix was assumed to be a rigid structure which corresponds to a constant porosity within the solid matrix. Therefore the problems associated with creep, hydration and carbonation in the concrete were not considered in this study

REFERENCES

1. George, D. and Wiffen, C.S.: "Snow and Ice Removal from Road Surfaces by Electrical Heating." Highway Research Record No. 94, 1965

2. MacInnis, C. and Beaudoin, J.J.: "Effects of Degree of Saturation on the Frost Resistance of Mortar Mixes." ACI Journal, March 1968.

3. Klieger, P. and Landgren, R.: "Performance of Concrete Slabs in Outdoor Exposure." Highway Research Record, No. 268, 1969.

4. Buth, E. and Ledbetter, W.B.: "Influence of the Degree of Saturation of Coarse Aggregate on the Resistance of Structural Lightweight Concrete to Freezing and Thawing." Highway Research Record, No. 328, 1970.

5. Bazant, Z.P. and Thonguthai, W., "Pore Pressure and Drying of Concrete at High Temperature." Jl of the Engineering Mechanics Division, ASCE, Vol. 104, No. EM5, pp. 1059-1079, Oct. (1978).

6. Huang, C.L.D., Siang, H.H., and Best, C.H.: "Heat and Moisture Transfer in Concrete Slabs." Int. J. Heat Mass Transfer, Vol. 22, pp. 257-266, 1979.

7. Huang, C.L.D.: "Multi-Phase Moisture Transfer in Porous Media Subjected to Temperature Gradient." Int. J. Heat Mass Transfer, Vol. 22, pp. 1295-1307, 1979.

8. Powers, T.C.: "Physical Properties of Cement Paste." Proc. Fourth Int. Symposium of Chemistry of Cement, Washington, 1960; Nat. Bur. Standards, Monograph 43 - Vol. II, Sept. (1962).

9. Powers, T.C.: "Mechanisms of Shrinkage and Reversible Creep of Hardened Cement Pastes, TThe Structure of Concrete and Its Behavior Under Load," Proc. Intl. Conf., London, Sept. (1965)

10. Bazant, Z.P.: "Constitutive Equation for Concrete and Shrinkage Based on Thermodynamics of Multiphase Systems," Materioux Et Constructions, Vol. 3, 13, (1970).

11. Powers, T.C. and Brownyard, T.L.: "Studies of the Physical Properties of Hardened Parland Cement Paste," Research Laboratories of Portland Cement Association, Bulletin 22, Chicago, (1948).

12. Wittmann, G.H.: "Hydraulic Cement Pastes: Their Structure and Properties." Proceedings of a Conference, University of Sheffield. Cement and Concrete Association, pp. 96-117, (1976).

13. Vassilious, B. and White, J.: "Vapor Pressure Cappilarity Relationships in Clays and Their Application to Certain Aspects of Drying." Transactions, the British Ceramic Society, 47, pp. 351-378, (1948).

14. Corey, A.T. and Brooks, R.H.: "Drainage Characteristics of Soils", Soil Science Society of American Proceedings, Vol. 39, 2, (1975).

15. Ccaglske, N.H. and Hougen, O.A.: "Drying Granular Solids," Ind. Engg. Chem. 29, pp. 805-813, (1937).

16. Hougen, O.A., McCauley, H.J., and Marshall, Jr. W.R.: "Limitations of Diffusion Equations in Drying," Trans, Am. Inst. Chem. Eng. 36, pp. 183-210, (1940).

NOMENCLATURE

a,b,n	empirical parameters
e	emissivity
h	heat transfer coefficient
k	permeability, $[m^2]$

K	effective thermal conductivity tensor, $[kg\ m/s^3\ K]$
K_i	thermal conductivity of i-component, $[kg\ m/s^3\ K]$
m	volumetric moisture content per unit volume of porous body, $[m^3/m^3]$
m_o	initial volumetric moisture content per unit volume of porous body, $[m^3/m^3]$
M	averaged mass for mixture
M_i	molecular weight of i-component, $[Kg/m\ s^2]$
P	total macroscopic pressure $[kg/m\ s^2]$
P_1	microscopic local pressure of i-component, $[Kg/m\ s^2]$
P_v	vapor pressure, $[Kg/m\ s^2]$
P_v^o	equilibrium vapor pressure of bulk water, $[Kg/m\ s^2]$
q	heat flux, $[Kg/s^3]$
r	hydraulic porous radius or characteristic length of a porous medium, $[m]$
R	gas constant, $[m\ s^2/Kg]$
T	absolute temperature, $[K]$
t	time $[s]$
x	x coordinate $[m]$
α	mass transfer coefficient, $[mol/m^2 s]$
β	empirical constant, $[K_g/s^2\ K]$
ε	porosity of the porous system
$\varepsilon_i(t)$	volume fraction of the i-component, $[m^3/m^3]$
ζ	relative permeability $= K_g/K_g^{\circ}$
η_i	shear viscosity of i-component, $[Kg/m\ s]$
θ	moisture (water) content, see Equation (2)
λ	latent heat of vaporization from the bulk liquid $[m^2/s^2]$
ρ_i	density of the i-component, $[kg/m^3]$
σ	surface tension of gas-liquid interface, $[kg/s^2]$
σ_s	Stefan-Boltzman constant, $[Kg/s^3\ K^4]$
ϕ	mole fraction of water vapor of the gaseous mixture, $[mol/mol]$

SUBSCRIPTS

g	of gaseous mixture
ga	of air in a gaseous mixture
gv	of vapor in a gaseous mixture
i	i-component of the mixture
ℓ	of liquid
s	of solid
w	of water

TABLE 1 VALUES OF THE CONSTANTS USED IN THE EQUATIONS (Light Weight Concrete)

Physical Constants

R	$= 8.3149 \times 10^{-14}$ $[Kg\ m^2/s^2\ Kmol]$
σ_s	$= 5670 \times 10^{-8}$ $[Kg/s^3\ K^4]$
$(C_p)_a$	$= 1.0063 \times 10^3$ $[m^2/s^2\ K^2]$
$(C_p)_s$	$= 0.879 \times 10^3$ $[m^2/s^2\ K^2]$
$(C_p)_v$	$= 1.8646 \times 10^3$ $[m^2/s^2\ K^2]$
$(C_p)_w$	$= 4.1793 \times 10^3$ $[m^2/s^2\ K^2]$
D	$= 0.256 \times 10^{-4}$ $[m^2/s]$
e	$= 0.8$
k_g	$= 0.02613$ $[Kg\ m/s^3\ K]$
k_s	$= 1.4422$ $[Kg\ m/s^3\ K]$
k_w	$= 0.616$ $[Kg\ m/s^3\ K]$
M_a	$= 28.952 \times 10^{-3}$ $[Kg/mol]$
M_w	$= 18.016 \times 10^{-3}$ $[Kg/mol]$
ε	$= 0.3$ $[m^3/m^3]$
η_g	$= 1.83 \times 10^{-5}$ $[Kg/m\ s]$
K_g^o	$= 2.50 \times 10^{-4}$ $[m^2]$
λ	$= 2.4418 \times 10^6$ $[m^2/s^2]$
ρ_s	$= 2.6 \times 10^3$ $[Kg/m^3]$
ρ_w	$= 0.99707 \times 10^3$ $[Kg/m^3]$

Geometric Constants

L	$= 0.10$ $[m]$
B	$= 2.00$ $[m]$

Empirical Constants

$a = 1.209658 \times 10^{-11}$ [ms^2/Kg]

$b = 5080$ [K]

$n = 0.25$

$\beta = 0.167 \times 10^{-3}$ [Kg/s^2 K]

$v = 1.0$

$\sigma_s = 121.2 \times 10^{-3}$ [Kg/s^2]

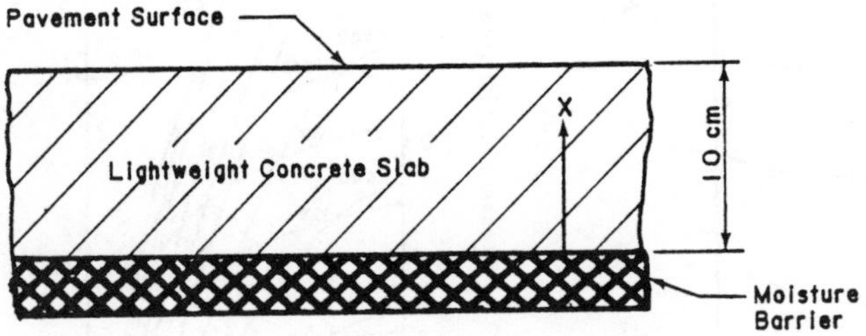

Figure 1: Concrete Slab in Natural Drying

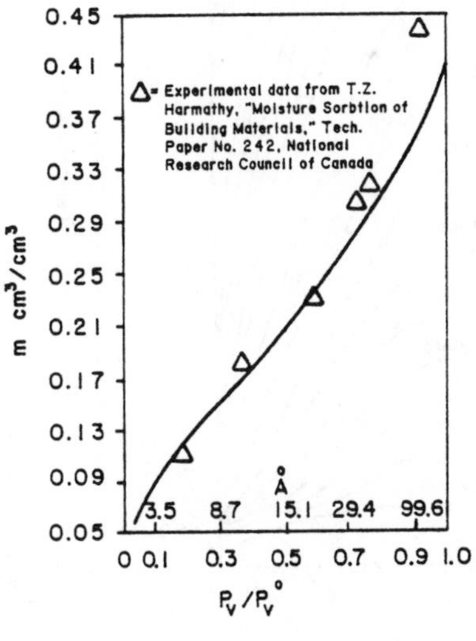

Figure 2: Liquid-Vapor Equilibrium Curve for a Light-Weight Concrete.

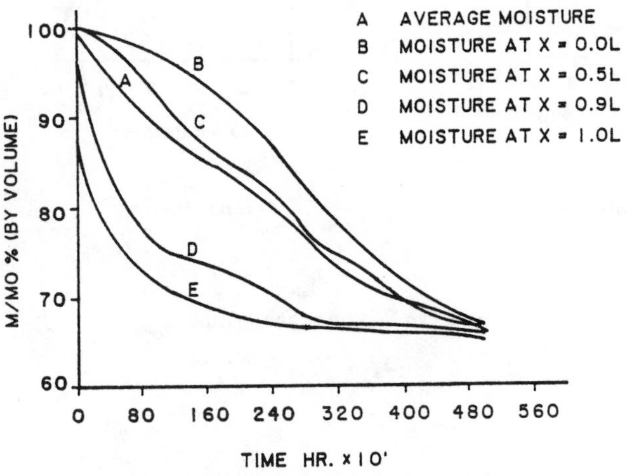

Figure 3: Histories of Natural Drying at Various Locations of Concrete Slab, and of the Average Moisture in the Slab.

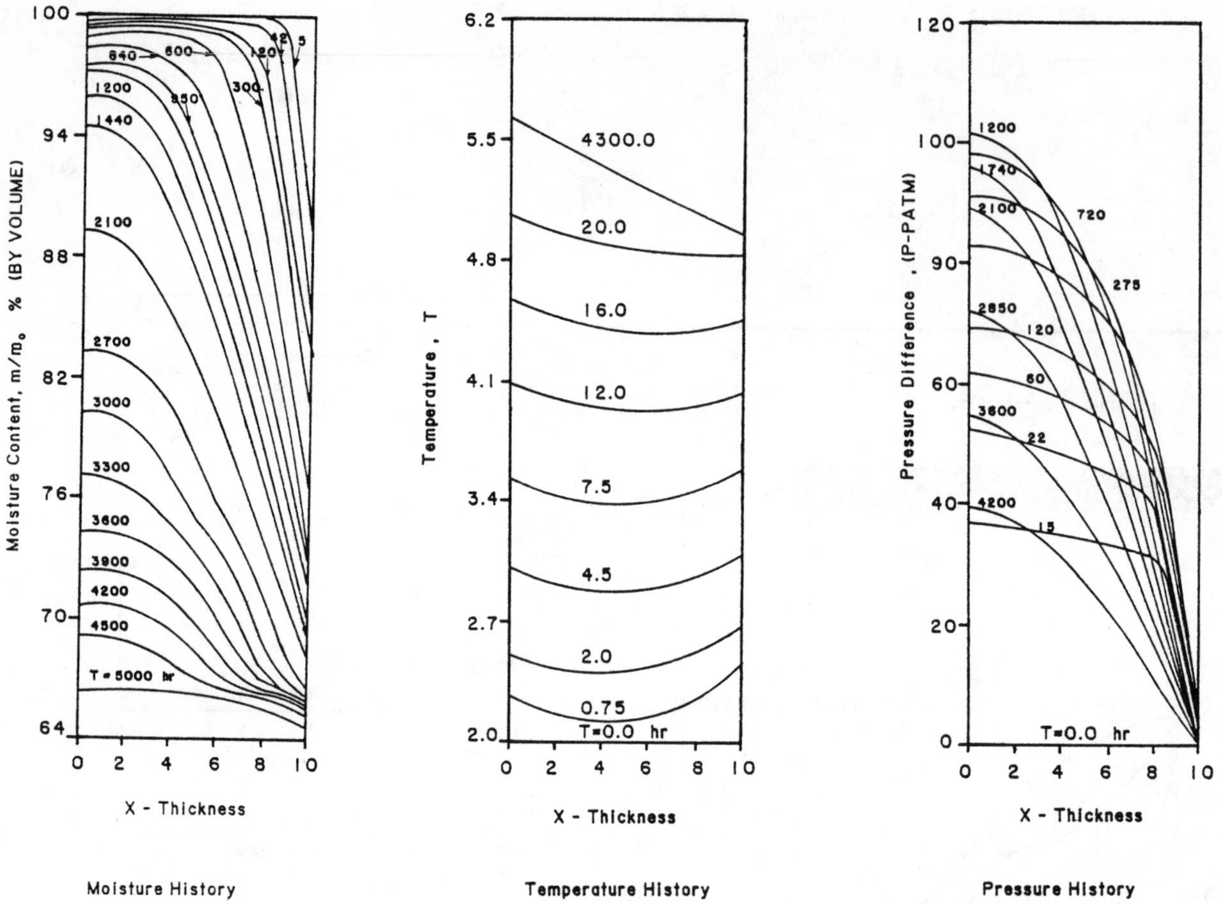

Figure 4: Distributions of Moisture, Temperature, and Pressure in a Concrete Slab.

RELAXATION EFFECT OF HEAT AND MASS FLUX AS THE CRITERION OF CLASSIFICATION OF DRYING MATERIALS

E. Mitura
S. Michałowski

Łódź Technical University, Łódź, Poland

ABSTRACT

A new conception of the classification of drying material based on the magnitude of relaxation of heat and mass flux has been proposed. It has been concluded from the considerations upon drying kinetics during falling drying rate period.

NOMENCLATURE

- a = thermal diffusivity, m^2/s
- c = propagation velocity of wave, m/s
- c_p = specific heat, J/kgK
- C = constant
- D = diffusion coefficient, m^2/s
- F = function in equations (16) - (19)
- G = localization function
- g = time function, dimensionless temperature of material
- h = time function, dimensionless moisture content of material
- I = functional
- K = constant
- L = effective linear dimension, m
- q = heat flux, W/m^2
- t = time, s
- T = temperature, ^oC, K
- V = volume, m^3
- X = moisture content, kg/kg
- y_i = experimental value
- x, y, z = coordinates
- Δx = parameter deviation
- Z = constant (Z_2/Z_1), $1/m^2$
- Δ = discriminant of differential equation
- ε = porosity
- λ = conduction coefficient, W/mK
- ϱ = density, kg/m^3
- τ = relaxation time

Subscripts

- d = diffusion
- e = equilibrium
- f = drying agent
- h = heat
- i = local value
- m = wet bulb
- o = critical value
- I, II = first, second drying period

INTRODUCTION

Drying considered as a physical process is very complicated because it depends on such factors as material structure, kind of the bond of the moisture with the solid, geometry of drying system, and parameters of the drying agent. Their influence on the process is nearly impossible to predict, especially with regard to the falling drying rate period. That is the reason probably that until now the generalization of the drying process, theoretical or analytical, phenomenological or empirical, has been still an open problem.
From the point of view of every designer the first reconnaisance of the problem is to know the kinetics of drying. This provides him with basic information for the choice of the proper drying apparatus and technique, both yielding good quality of the final product and low energy consumption.
Equally important is the classification of drying materials. Its aim is more ambitious and it tends to the generalization of the knowledge of the behaviour of materials during drying and sometimes tries to connect the material properties with drying kinetics. The simplest way is to take into account the general properties of the solid. According to Łykow (1) the behaviour of colloidal or capillary-porous material is essential, nevertheless Krisher (2) and Keey (3) represent the opinion that its hygroscopicity is more important. More sound approach to this problem is given by Kaminskij (4) and

Sażin and Szadrina (5). Kaminskij using Rebinder approach (6) divides drying material on the basis of the energy of moisture bond with a material and Sażin and Szadrina on the basis of pore sizes (this classification was limited to the capillary-porous material only). Closer to the drying practice are the classifications of Kisielew (5) and Łykow (1) based upon the shape of sorption isotherms and of drying curves, respectively. From the analysis of unsteady state heat and mass transfer during drying two further classifications by Sażin (5) and by Kwasza (7) emerged. Quite new approach was presented by Łykow (8). On the basis of the rule of Coleman and Noll (9,10) the drying material is treated as the object with "memory".

Summarizing, most of the existing classifications have only qualitative character or are limited to the narrow material group, so the problem of dried materials classification is still open.

A NEW MATHEMATICAL DESCRIPTION OF THE FALLING DRYING RATE PERIOD

The kinetics of drying, after definition, represents changes in time of the mean temperature and mean moisture content of the drying material under constant conditions. From the two existing drying periods the constant drying rate one is nothing more than an "ordinary", steady state, heat/mass transfer from the interfacial surface. In spite of this, the falling drying rate period is an unsteady state process and, in fact, it is "responsible" for the behaviour of the material during drying, drying problems in practice, and theoretical difficulties in modelling. Taking this into account, our considerations have been concentrated on this period.

The paper presents a completely new approach based on the variational calculus and the action functionals for dissipative processes (11,12), just introduced into science. The heat and mass transfer has been examined separately. For heat transfer the conduction, dissipation, and relaxation are assumed to be decisive. The heat flux in a three dimensional temperature field is given by the relationship

$$q = -\lambda \nabla T - \tau_h \frac{\partial q}{\partial t} \tag{1}$$

and unsteady state heat conduction by differential hyperbolic equation

$$\frac{\partial T}{\partial t} = a\left(\nabla^2 T - \frac{1}{c_h^2}\frac{\partial^2 T}{\partial t^2}\right) \tag{2}$$

The action functional is used as reported by Sieniutycz (12)

$$I(T) = \frac{1}{2}\int_0^{t_e}\iiint_V \left\{\varrho c_p \tau_h \left(\frac{\partial T}{\partial t}\right)^2 - \lambda (\text{grad } T)^2\right\} \exp\left(\frac{t}{\tau_h}\right) dV dt \tag{3}$$

Assuming constant ϱ, c_p, τ_h, λ and introducing $c_h^2 = a/\tau_h$ one has its another form

$$I(T) = \frac{\lambda}{2}\int_0^{t_e}\iiint_V \left\{\frac{1}{c_h^2}\left(\frac{\partial T}{\partial t}\right)^2 - \left[\left(\frac{\partial T}{\partial x}\right)^2 + \left(\frac{\partial T}{\partial y}\right)^2 + \left(\frac{\partial T}{\partial z}\right)^2\right]\right\} \exp\left(\frac{t}{\tau_h}\right) dV dt \tag{4}$$

For drying by the flow of drying agent over the material surface and evaporation only from one side, the following temperature field is assumed

$$T = T_f + (T_m - T_f) \cdot G(x,y,z)\, g(t) \tag{5}$$

where time function $g(t)$ fulfils the following boundary conditions

$$g(0) = 1, \quad g(\infty) = 0, \quad \frac{dg}{dt}(0) = 0 \tag{6}$$

From the point of view of drying kinetics it represents dimensionless temperature changes in time and in space and when $G(x,y,z) = 1$, means dimensionless temperatures in time, which after definition, is equivalent to the temperature curve. Introducing (5) into (4) after some rearrangements, one has

$$I(T) = \frac{\lambda}{2}\int_0^{t_e}\left[\frac{Z_1}{c_h^2} g'^2(t) - Z_2 g^2(t)\right]\exp\left(\frac{t}{\tau_h}\right) dt \tag{7}$$

From the kinetics point of view Z_1 and Z_2 are constant and the functional (7) can be converted into

$$I(g) = \int_0^{t_e} U\left[t, g(t), g'(t)\right] dt \tag{8}$$

The functional (8), which was defined in a set of functions $g(t)$, $0 \leq t \leq \infty$ with continuous derivates and conditions (6), should have an extreme. For this it is necessary the function $g(t)$ to fulfil Euler equation

$$U_g - \frac{d}{dt} U_{g'} = 0 \tag{9}$$

which, for the process being considered, assumes the form

$$\tau_h\, g''(t) + g'(t) + \frac{Z_2}{Z_1}\, a\, g(t) = 0 \tag{10}$$

Now the solution is easy and can be obtained trough so called characteristic equation of relation (10)

$$\tau_h r^2 + r + \frac{Z_2}{Z_1}\, a = 0 \tag{11}$$

with its discriminant

$$\Delta = 1 - 4\, Z\, a\, \tau_h \tag{12}$$

as the set of four relationships for different values of τ_h and Δ . For $\tau_h = 0$ it was found

$$g_4(t) = e^{-Zat} \tag{13}$$

The next solutions are given in Table 1. The procedure for the mass transfer is analogical and only adjusted to this process. Instead of heat flux q the mass flux is used, temperature is exchanged into moisture content X and equations rearranged taking into account that mass diffusivity stands now for thermal one. The boundary conditions are

$h(0) = 1$

$h(\infty) = 0$

$$\frac{dh}{dt}(0) = \frac{\left(\frac{dX}{dt}\right)_I}{X_o - X_e} = K \tag{14}$$

to stress that during first drying period the drying rate is constant. All next changes are formal: h, C, index d are used instead of g, Z, index h.

Table 1. Principles of Drying Material Classification

Class	Heat Transfer
I	$g_1 = \left[\cos\left(\frac{\sqrt{-\Delta}}{2\tau_h}t\right) + \frac{1}{\sqrt{-\Delta}}\sin\left(\frac{\sqrt{-\Delta}}{2\tau_h}t\right)\right] e^{-\frac{t}{2\tau_h}}$ $\tau_h > 0$ $\Delta < 0$ $0.25 < Fo_{rh}$
II	$g_2 = \left(1 + \frac{t}{2\tau_h}\right) e^{-\frac{t}{2\tau_h}}$ $\tau_h > 0$ $\Delta = 0$ $Fo_{rh} = 0.25$
III	$g_3 = \frac{\sqrt{\Delta} - 1}{2\sqrt{\Delta}}\, e^{\frac{-\sqrt{\Delta}-1}{2\tau_h}t} + \frac{\sqrt{\Delta}+1}{2\sqrt{\Delta}}\, e^{\frac{\sqrt{\Delta}-1}{2\tau_h}t}$ $\tau_h > 0$ $\Delta > 0$ $0 < Fo_{rh} < 0.25$

Class	Mass Transfer
I	$h_1 = \left[\cos\left(\frac{\sqrt{-\Delta}}{2\tau_d}t\right) + \frac{2K\tau_d + 1}{\sqrt{-\Delta}}\sin\left(\frac{\sqrt{-\Delta}}{2\tau_d}t\right)\right] e^{-\frac{t}{2\tau_d}}$ $\tau_d > 0$ $\Delta < 0$ $0.25 < Fo_{rd}$
II	$h_2 = \left(1 + Kt + \frac{t}{2\tau_d}\right) e^{-\frac{t}{2\tau_d}}$ $\tau_d > 0$ $\Delta = 0$ $Fo_{rd} = 0.25$
III	$h_3 = \frac{\sqrt{\Delta} - 2K\tau_d - 1}{2\sqrt{\Delta}}\, e^{\frac{-\sqrt{\Delta}-1}{2\tau_d}t} + \frac{\sqrt{\Delta} + 2K\tau_d + 1}{2\sqrt{\Delta}}\, e^{\frac{\sqrt{\Delta}-1}{2\tau_d}t}$ $\tau_d > 0$ $\Delta > 0$ $0 < Fo_{rd} < 0.25$

The obtained final equations are similar: for $\tau_d = 0$

$$h_4 = e^{-CDt} \tag{15}$$

and next ones are shown in Table 1.

EXPERIMENTAL VERIFICATION

The measurements were performed in a typical tunnel dryer. The materials were selected to represent a variety of properties and structures. The conditions of drying and sizes of material samples were the same for all runs: air temperature 53.5°C, air linear velocity 2.2 m/s, air relative humidity 17%, and drying material in the form of 40 x 40 x 10 mm rectangulars. The results are given as dimensionless temperature and drying curves in Figures 1 and 2.

These experimental data were used for determination of the constants Δ , τ and Z or C of the previously obtained equations by means of the method of minimization of expressions

$$F_1(\Delta,\tau) = \sum_{i=1}^{n} t_i\left[y_i - g_1(\Delta,\tau,t_i)\right]^2 \tag{16}$$

$$F_2(\Delta,\tau) = \sum_{i=1}^{n} t_i\left[y_i - g_2(\tau,t_i)\right]^2 \tag{17}$$

$$F_3(\Delta,\tau) = \sum_{i=1}^{n} t_i\left[y_i - g_3(\Delta,\tau,t_i)\right]^2 \tag{18}$$

$$F_4(Z) = \sum_{i=1}^{n} t_i\left[y_i - g_4(Z,t_i)\right]^2 \tag{19}$$

for heat transfer, and similar relationships (g changed into h) for mass transfer. The calculations were performed by means of computer ODRA 1305 in FORTRAN 1900.

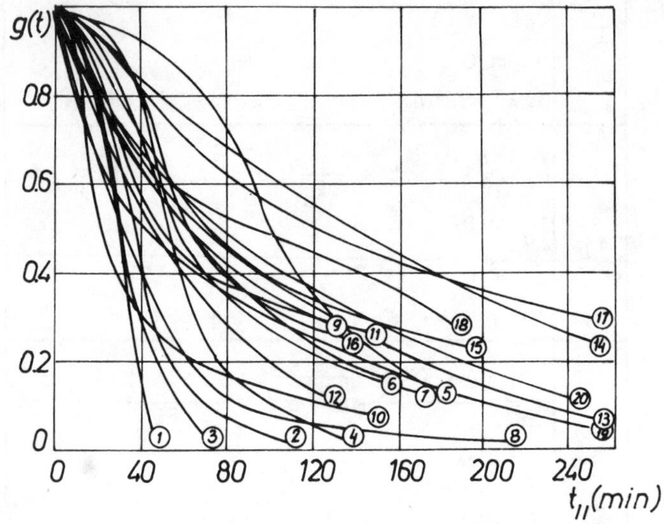

Figure 1. Experimental dimensionless temperature curves
Materials: 1.glass spheres, 2.sand, 3.calcium carbonate, 4.faience, 5.cotton wool, 6.suporex, 7.molecular sieves, 8.cherry tree wood, 9.lignin, 10.ceramics I 11.ceramics II, 12.poppy seeds, 13.natrium perborate, 14.rice, 15.peat, 16.manna grits, 17.macaroni, 18.bread, 19.wheat seeds, 20.barley grits

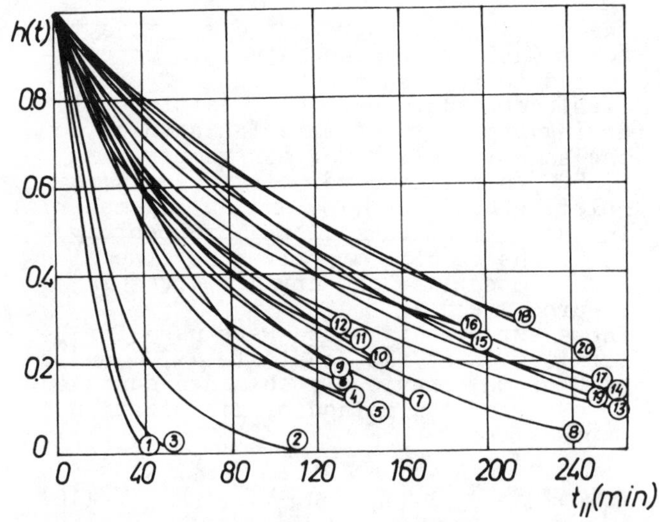

Figure 2. Experimental dimensionless drying curves
Materials as in Figure 1

The sensitiveness of the proposed equations was tested by calculating the deviation (from the optimum value) of the obtained equation parameters Δx induced by the deviation of function $|\Delta F|$ from the relationships

$$\Delta x_1 = \sqrt{\frac{2\,|\Delta F_1|}{\left|\frac{\partial^2 F_1}{\partial \tau'^2}\right| + 2\left|\frac{\partial^2 F_1}{\partial \tau' \partial \Delta'}\right| + \left|\frac{\partial^2 F_1}{\partial \Delta'^2}\right|}} \tag{20}$$

$$\Delta x_2 = \sqrt{\frac{2|\Delta F_2|}{\left|\frac{d^2 F_2}{\partial \tau'^2}\right|}} \tag{21}$$

$$\Delta x_3 = \sqrt{\frac{2\,|\Delta F_3|}{\left|\frac{\partial^2 F_3}{\partial \tau'^2}\right| + 2\left|\frac{\partial^2 F_3}{\partial \tau' \partial \Delta'}\right| + \left|\frac{\partial^2 F_3}{\partial \Delta'^2}\right|}} \tag{22}$$

$$\Delta x_4 = \sqrt{\frac{2|\Delta F_4|}{\left|\frac{d^2 F_4}{dZ'^2}\right|}} \quad \text{or} \quad \Delta x_4 = \sqrt{\frac{2|\Delta F_4|}{\left|\frac{d^2 F_4}{dC'^2}\right|}} \tag{23}$$

For the assumed deviation of function ΔF of 10%, the deviations of equation constants were within 0.03 to 7.7%, which can be treated as a good result.
The comparison of calculated curves (eqs. (13) and (15), Table 1) and experimental data led to the conclusion that only one from four curves fits to the experimental point, or more precisely, in the set of four equations there is always only one with the lowest standard deviation. One example of this is shown in Figures 3 and 4, where in fact only the equation for $\tau > 0$ and $\Delta < 0$ is valid.

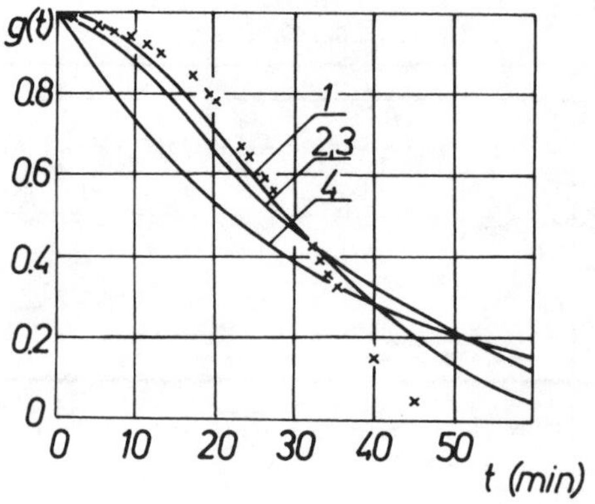

Figure 3. Dimensionless temperature curves
Glass spheres $\varepsilon = 0.363$
x - experimental data; 1,2,3,4-theoretical curves

These figures are also a good illustration for the accuracy of equations selected by the experiment, which for heat transfer is nearly good and for mass transfer fairly good (standard deviation for all runs ranged from 0.04 to 7%). It is interesting that eqs. (13) and (15) were never confirmed by experiments. For these relationships the time of relaxation was omitted ($\tau = 0$) and as concluded from $c^2 = a/\tau$, $c = \infty$ which meant they are charged with the paradox of infinite propagation velocity of heat and mass wave.

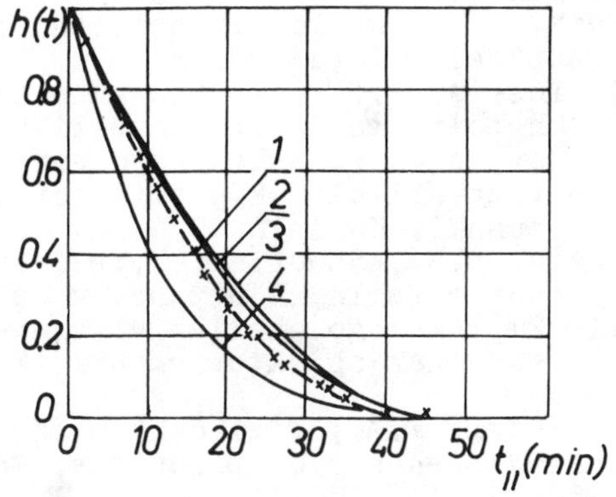

Figure 4. Dimensionless drying curves Glass spheres $\varepsilon = 0.363$ x - experimental data; 1,2,3,4 - theoretical curves

PROPOSED CLASSIFICATION

Thus, the usefulness of the variational technique has been proved. The elaborated method can be adapted not only for the mathematical description of the temperature of drying curves but also as a basis for the new classification of drying materials. The criterion of the latter was proposed in the form of a modified Fourier number connected with relaxation effects: heat relaxation Fourier number

$$Fo_{rh} = \frac{a\tau_h}{L^2} \qquad (24)$$

and mass relaxation Fourier number

$$Fo_{rd} = \frac{a\tau_d}{L^2} \qquad (25)$$

which are similar to the Łykow (1) conception of Fo_{rq}. Because the constants Z and C have the dimension of $1/m^2$ it was proposed to use as a linear dimension $L = 1/\sqrt{Z}$ or $L = 1/\sqrt{C}$ for heat and mass transfer, respectively. Taking into account the relationship (12) one has

$$Fo_{rh} = Z\, a\, \tau_h = \frac{1 - \Delta_h}{4} \qquad (26)$$

$$Fo_{rd} = C\, D\, \tau_d = \frac{1 - \Delta_d}{4} \qquad (27)$$

Fundamentals of the proposed classification are given in Table 1, its application for the examined materials in Table 2, and its graphic illustration in Figures 5 and 6.

Table 2. Classification of Examined Materials

Class	Heat Transfer
I	τ_h = 821.67 to 1026.26 s Δ = -0.891967 to -0.172129 Fo_{rh} = 0.293 to 0.473 $t_{1\ to\ 0.3}$ = 25 to 102 min Materials: glass spheres, sand, calcium carbonate, faience
II	τ_h = 919.49 to 1173.64 s Δ = 0 Fo_{rh} = 0.250 $t_{1\ to\ 0.3}$ = 76 to 130 min Materials: cotton wool, suporex, poppy seeds
III	τ_h = 709.42 to 996.53 s Δ = 0.034001 to 0.650368 Fo_{rh} = 0.087 to 0.241 $t_{1\ to\ 0.3}$ = 58 to 240 min Materials: molecular sieves, cherry tree wood, lignin, ceramics I, ceramics II, natrium perborate, rice, peat, manna grits, macaroni, bread, wheat seeds, barley grits
Class	**Mass Transfer**
I	τ_d = 726.40 to 1792.54 s Δ = -0.939990 to -0.440000 Fo_{rd} = 0.360 to 0.485 $t_{1\ to\ 0.3}$ = 18 to 31 min Materials: glass spheres, sand, calcium carbonate
II	τ_d = 1135.27 to 2562.25 s Δ = 0 Fo_{rd} = 0.250 $t_{1\ to\ 0.3}$ = 68 to 110 min Materials: faience, cotton wool, suporex, molecular sieves, cherry tree wood
III	τ_d = 985.02 to 9900.10 s Δ = 0.412160 to 0.706399 Fo_{rd} = 0.073 to 0.147 $t_{1\ to\ 0.3}$ = 100 to 210 min Materials: lignin, ceramics I, ceramics II, poppy seeds, natrium perborate, rice, peat, manna grits, macaroni, bread, wheat seeds, barley grits

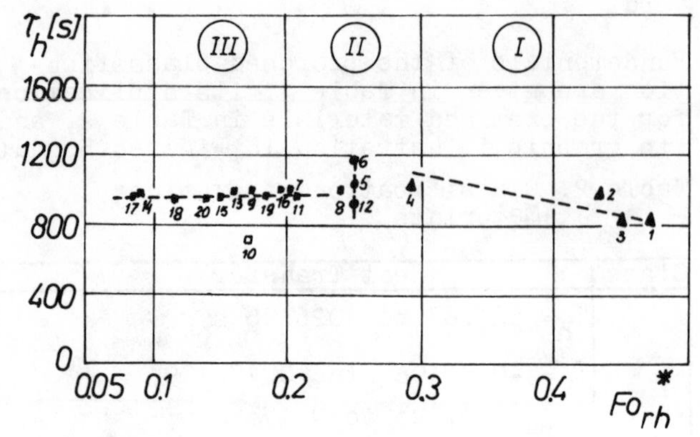

Figure 5. Heat flux relaxation effect as a criterion of drying materials classification

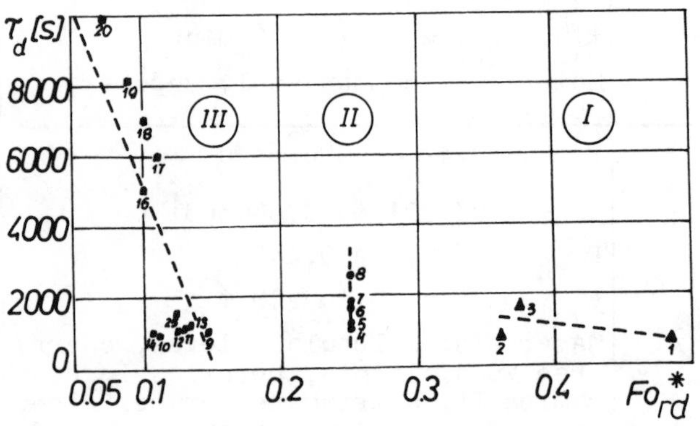

Figure 6. Mass flux relaxation effect as a criterion of drying material classification

To the first class of the greatest Fourier numbers (>0.25) belong the materials which are heated easily and become dry quickly. The opposite end occupy the materials of the lowest Fourier numbers (0 to 0.25). They make warm and become dry with difficulty. It is class III. Of course, class II is intermediate.

It is worth to stress that the above approach can be applied to the drying process with regard to the recent developments of physical theory of dissipative processes.

CLOSURE

The application of variational calculus creates new possibilities for a mathematical description of unsteady state heat and mass transfer processes. The heat and mass processes in dissipative systems of unavoidable relaxation effects run with finite wave propagation velocity and should be described by hyperbolic differential equations. The consideration of relaxation effects gives the basis for the classification of drying materials.

REFERENCES

1. Łykow, A.W., Teoria sushki, Moscow 1968
2. Krischer, O., Die wissenschaftlichen Grundlagen der Trocknungstechnik, Berlin 1956
3. Keey, R.B., Drying Principles and Practice, Oxford 1972
4. Kaminskij, Ł.P., Issledovanie dispersnykh polimernykh khimicheskikh produktov, sodierzhashchikh organicheskoe raztvoriteli, kak obiektov glubokoj sushki, Kalinin 1970
5. Sażin, B.S., Szadrina, N,Ye., Vybor i razchot sushilnykh ustanovok na osnovie kmpleksnogo analiza vlazhnykh materialov kak objektov sushki, Moscow 1979
6. Rebinder, P.A., Fiziko-khemicheskije osnovy pishchevykh produktov, Moscow 1962
7. Kwasza, W.B., Gelperin, N.J., Ajnsztiejn, W.G., Chim.Prom., 6, 1971
8. Łykow, A.W., Inż.Fiz. Ż., 26, 5, 781, 1974
9. Astarita, G., Marrucci, G., Principles of Non-Newtonian Fluid Mechanics, London 1974
10. Coleman, B.D., Noll, W., Rev.Mod.Phys. 33, 239, 1961
11. Vujanovič, B., Djukič, D.J., Int.J. Heat Mass Transfer, 15, 1111, 1972
12. Sieniutycz, S., Int.J.Heat Mass Transfer, 20, 1222, 1977

OPTIMAL MULTISTAGE CROSSCURRENT FLUIDIZED DRYING WITH PRODUCT RECYCLE

S. Sieniutycz
Z. Szwast

Warsaw Technical University, Warsaw, Poland

ABSTRACT

The general theory developed in the associated work [1] is used to formulate and solve an optimization problem of minimum economic costs for steady-state multistage crosscurrent fluidized drying with product recycle. The optimized controls related to the inlet gas are the temperature and the flow of gas for each stage of the cascade. The discrete algorithm with a constant Hamiltonian [1] is applied to compute the optimal control.
The optimal results are obtained numerically for the family of drying processes corresponding with various values of the recycle ratio. The role of parameter λ characterising the investment costs is also investigated. The properties of the optimal solid and gas parameters are analyzed.

NOMENCLATURE

A_a	apparatus cross-sectional area, m^2
A	coefficient appearing in eqn. (17)
b_g	exergy of drying gas, kJ/kg
c_g	unit price of drying gas, \$/kg
c_p, c_w	heat capacity of dry air and moisture vapour, respectively, kJ/kgK
e	economic value of unit exergy, \$/kJ
e_{el}	unit price of electrical energy, \$/J
F	exergy performance index, kJ/kg
F', F"	economic performance index, \$/kg
G^n, G^N	gas flow rate up to stage n and total gas flow rate, respectively, kg/h
ΔG^n	dry gas flow rate through stage n, kg/h
$g^n = \Delta G^n/S$	dimensionless gas flow at stage n, kg/kg
H	Hamiltonian
I_s	solid enthalpy per unit mass of dry material, kJ/kg
i_g	gas enthalpy per unit mass of absolute dry inlet gas, kJ/kg
i_s	enthalpy of gas in equilibrium with solid, kJ/kg
J, j_o	total cost of cascade and fixed limiting value of J, respectively, for a one-stage process when $A_a \rightarrow 0$, \$
M	mixing function
N, n	total number of stages and current stage number, respectively
ΔP	pressure drop in fluidized layer, Pa
p	fluidized drying apparatus price per unit area of sieve bottom, \$/m^2
R	recycle solid flow rate, kg/h
r	recycle ratio
S	dry solid flow rate, kg/h
T_g, T_s	inlet gas temperature and solid temperature, respectively, K
T_o	ambient temperature, K
T_{gr}	maximum acceptable payout time, year
T_u	utilization time of dryer during year, h/year
v	superficial gas velocity, m/h
W_s, W_k	absolute moisture content of solid/mass of moisture per unit mass of dry solid/and final value of W_s, respectively, kg/kg

X_s humidity of gas in equilibrium with solid, kg/kg

X_o ambient humidity, kg/kg

$\tilde{z}$ factor describing freezing of capital costs

z adjoint variable at stage n

Greek symbols

β coefficient describing renovations, $year^{-1}$

η pumping efficiency

θ^n $= \frac{\Delta G^n}{S + R}$ dimensionless gas flow at stage n, kg/kg

λ $= \lambda_e / e$,exergy coefficient of investment and gas pumping cost,kJ/kg

λ_e economic coefficient of investment and gas pumping costs, \$/kg

ρ gas density, kg/m^3

Auxiliary variables appearing in computer program

DELTA,DELTAP,KR1,KR2,KR3,i,s,L,P

Superscripts

n,N stages n and N,respectively

* highest allowable value

Subscripts

g gas

p dry air

s solid,gas in equilibrium with solid

w moisture

1. INTRODUCTION

In view of uneconomical nature of drying processes, in an attempt to decrease total costs of drying (i.e. the sum of investment and operational costs) the use of unconventional, optimally controlled, drying processes is recommended where the parameters of the drying agent may vary with the residence time of solid [1,2]. As the rigorous principles of the rational control are relatively unknown in the drying theory,there is a need for formulating and solving the mathematical optimization problems in drying. Therefore,a general discrete optimization theory is developed in [1],applicable to the various multistage and continuous drying processes of cocurrent,countercurrent and crosscurrent type where the state changes occur in a single direction of space or time. A general optimization problem for the multistage fluidized drying process involves the variations of temperatures,humidities and flows of inlet gas for each stage of cascade [1,2]. To evaluate the role of variations of the controls along the process path,the optimization results were found in [2] for the two conventional variants of drying(in which the optimal controls were identical for every stage) and these results were compared with those obtained for unconventional (control) drying processes without product recycle.For variant I the decision variables were the temperature, humidity and flow of the inlet gas.For variant II the decision variables were only the temperature and flow of the inlet gas,the inlet gas humidity being equal to the ambient humidity. It was found that the optimal costs for variant I of the conventional process were several per cent higher than those for the unconventional process,whereas the optimal costs for variant II were considerably higher of the order of 20 percent. Since, in industry, even a small percentage increase in profit is usually important the use of unconventioned(control) processes is recommended.

This paper deals with the multistage fluidized drying processes with product recycle, these processes being characterized by variable temperature and flow of inlet gas for each stage. The inlet gas humidity is assumed to be equal to the ambient humidity. The optimization computations are performed for silica gel-water-air system.

Thermodynamic equilibrium between the gaseous and solid phases leaving the fluidized stage is assumed as it is commonly observed in bed of heights such as those used in industry(of the order of 0.5 m). Consequently the simplest(equilibrium-stage) model is accepted,replacing the more complicated bubble kinetic model with an error of the order of 5 %. Attention is directed towards explaining how the recycle ratio affects the properties of optimal solid and gas parameters as well as values of drying performance criterion.

2. OPTIMIZATION PROBLEM

Consider the N th stage crosscurent cascade of fluidized dryers in which a moist solid (silica gel) is to be dried by a gas (air),Fig.1. The air from every stage of the cascade is released into the atmosphere. The mass flow rate of the dry solid in feed stream is equal to S; in recycle stream it is equal to R ,hence the mass flow rate of the dry solid in stream passing through the cascade is equal to S + R . The initial state of the solid, defined by moisture content W_s^f and enthalpy I_s^f (or temperature T_s^f), is prescribed. For the final solid state the only requirement pertains to the final moisture content W_s^N which has to be preassigned. The final solid enthalpy I_s^N (or temperature T_s^N) is free . It should assume the value which optimizes the performance index. The state of the solid before the first stage of the cascade is described by following mixing functions:

$$I_s^o = \frac{SI_s^f + RI_s^N}{S + R} \qquad (1)$$

$$W_s^o = \frac{SW_s^f + RW_s^N}{S + R} \qquad (2)$$

The dry gas mass flow rate at every stage of the cascade, ΔG^n, and the total gas flow rate, G^N, are undetermined. The maximum allowable inlet gas temperature is equal to T^* and the inlet gas humidity is equal to the ambient humidity. The thermodynamic equilibrium between the gaseous and solid phases leaving the fluidized stage is described by the semiempirical equations below:

$$i_s^n(I_s{}^n, W_s{}^n) = 4,2\left[a_1 W_s{}^n - b_1 + (c_1 W_s{}^n + d_1)/Y^n\right]^{-2} \qquad (3)$$

$$X_s^n(I_s{}^n, W_s{}^n) = \left[a_2 W_s{}^n - b_2 + (c_2 W_s{}^n + d_2)/Y^n\right]^{-2} \qquad (4)$$

where

$$Y^n = Y^n(I_s{}^n, W_s{}^n) = 0.24\, I_s{}^n + 32,6 W_s{}^n / (0.094 + W_s{}^n) \qquad (5)$$

which hold true for the silica gel-water -air system considered here.
The coefficients in.eqns.(3)and (4) are given in Table 1. These equations were found to describe the tabular data [3] to an accuracy of 10 % . They hold in the range $0,1 \leq W_s \leq 0.2$ kg/kg and $0 \leq T_s \leq 100^oC$. The effects of heat of

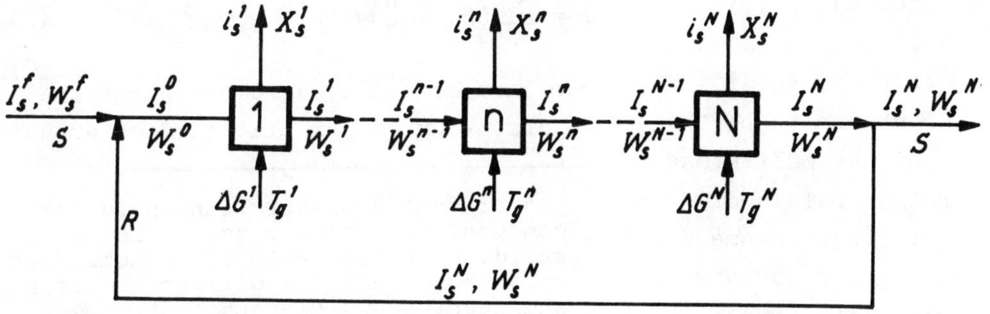

Fig.1.The multistage fluidized drying with product recycle.

TABLE 1

Coefficients in expressions for drying equilibrium

i	a_i	b_i	c_i	d_i
1	7.1114×10^{-2}	1.3913×10^{-1}	1.6448×10^{1}	5.0111×10^{0}
2	5.2355×10^{0}	5.6037×10^{0}	4.5728×10^{2}	1.8365×10^{2}

moisture sorption and of drying equilibrium are also taken into account in these equations.

Under the above-mentioned equilibrium-stage assumption the process state equations obtained as result of elementary balances of energy and mass of arbitrary stage $n(n=1,\dots N)$ are

$$I_s^{\,n} - I_s^{\,n-1} = \left[i_g^{\,n}(T_g^{\,n}) - i_s^{\,n}(I_s^{\,n}, W_s^{\,n})\right]\theta^n \tag{6}$$

$$W_s^{\,n} - W_s^{\,n-1} = \left[X_0 - X_s^{\,n}(I_s^{\,n}, W_s^{\,n})\right]\theta^n \tag{7}$$

where $i_g^{\,n}(T_g^{\,n})$ is the inlet air enthalpy (at stage n) computed as

$$i_g^{\,n}(T_g^{\,n}) = (c_p + X_0 c_w)(T_g^{\,n} - 273.15) + 2502.7X_0 \tag{8}$$

and

$$\theta^n = \frac{\Delta G^n}{S+R} \tag{9}$$

is the gas flow through n-th stage per unit flow of solid passing through the cascade.

The optimization problem is characterized by the following variables (for $n = 1,\dots N$), state variables $I_s^{\,n}, W_s^{\,n}$, controls (decision variables) T_g^n, ΔG^n. The total economic cost, i.e. the sum of total investment costs and total exploitation costs, was chosen to be the performance criterion of the multistage process under consideration. This criterion involves, amongst others, some special costs (related e.g., to administration, lighting, nonenergetic imputs or the feed and recycle transmission) which are independent of the process itself and[*], as such, can be neglected in the final form of the performance index. Thus, we formulate the performance criterion F' as the sum of the variable part of the investment and exploitations costs, i.e.

$$F' = \left(\frac{\tilde{z}}{T_{gr}} + \beta\right)\frac{J}{T_u S} + \sum_{n=1}^{N} \frac{e_{el}\,\Delta P\,\Delta G^n}{\eta\,\varrho\,S} + \sum_{n=1}^{N} c_g(T_g^{\,n})\,\frac{\Delta G^n}{S} \tag{10}$$

In the above expression the first term describes the investment costs, the second the gas pumping costs, and the third the costs of isobaric preparation of drying gas of the temperature $T_g^{\,n}$. All costs refer to unit mass of flowing feed stream. For a particular drying problem the quantities $\tilde{z}$, T_{gr}, β, T_u, S, e_{el}, ΔP, η and ϱ are known constants. The explicit form of the function $c_g(T_g^{\,n})$ is also known. Whereas the total value J of the N-stage cascade can be described by formula (11)

$$J = \sum_{n=1}^{N}(j_0 + pA_a^n) = Nj_0 + \frac{p}{\varrho v}\sum_{n=1}^{N}\Delta G^n \tag{11}$$

For industrial fluidized equipment the apparatus price per unit surface area

[*] The present considerations hold for constant feed and recycle streams of solid. Hence the costs of transmission of these streams as well as additional equipment costs are constant.

of sieve bottom, p, is a characteristic measure of equipment value [4]. The constant j_0 is the value of the fixed costs of every stage N which appears even if the process is not conducted. For the concrete problem the unit apparatus price p and the superficial gas velocity v are known constants. We assume that the total number of stages N, is fixed. To find the suitable expression for performance index we substitute eqn. (11) into eqn. (10) and simplify the result, rejecting additive constants. Consequently for constant N the performance index can be written as

$$F'' = \sum_{n=1}^{N} (1 + r) \left[c_g(T_g) + \lambda_e \right] \theta^n \qquad (12)$$

where

$$r = \frac{R}{S} \qquad (13)$$

is recycle ratio of solid and λ_e is the so-called economic coefficient of the investment costs, including the cost of gas pumping.* On the basis of the eqns. (11) and (12) it is easy to see that λ_e is expressed by the economic and technical quantities as

$$\lambda_e = \left(\frac{\tilde{z}}{T_{gr}} + \beta \right) \frac{p}{T_u \varrho v} + \frac{e_{el} \Delta P}{\eta \varrho} \qquad (14)$$

i.e., it is related to both investment and gas pumping costs. All quantities appearing on the right-hand side of eqn. (14) are known constants. Therefore λ_e is also known.

The drying air price c_g can be computed with the help of a so-called exergy tariff, i.e. as the product of the unit exergy price e, which is estimated as e = 8 \$ GJ^{-1} [4] and the unit exergy $b_g(T_g^n)$

$$c_g^n = e\, b_g^n (T_g^n) \qquad (15)$$

* Pumping cost was formally included into λ_e due to the constancy of ΔP, see however [1].

For the constant air humidity equal to the ambient humidity X_0 the exergy of air as a function of its temperature T_g^n is described [4] by the formula

$$b_g^n(T_g^n) = (c_p + X_0 c_w) \left(T_g^n - T_0 - T_0 \ln \frac{T_g^n}{T_0} \right) \qquad (16)$$

Here we shall use a convenient simplified form of exergy that is obtained after Taylor expansion of the exact exergy function around the ambient temperature T_0, neglecting third order and higher terms. Then the gas unit price is

$$c_g^n = \frac{eA}{2} \left(T_g^n - T_0 \right)^2 \qquad (17)$$

where

$$A = \left(c_p + X_0 c_w \right) / T_0 \qquad (18)$$

The use of eq. (17) in formula (12) yields the following final performance index

$$F = \sum_{n=1}^{N} (1 + r) \left[\frac{A}{2} (T_g^n - T_0)^2 + \lambda \right] \theta^n \qquad (19)$$

where

$$F = F'' / e \qquad (20)$$

is expressed in exergy units, kJ/kg and

$$\lambda = \lambda_e / e = \text{constant} \qquad (21)$$

Our problem comes down to finding the optimal decision variables θ^n, eqn. (9), and T_g^n, for n=1,...N, which minimize the performance index F (eqn. (19)) under the constraints resulting from the state equations (6) and (7), the mixing functions (1) and (2), the boundary conditions for I_s^f, W_s^f and W_s^N, and the algebraic constraints imposed on the decisions T_g^n. The maximum allowable air temperature is equal to T^*. The decisions θ^n and the total air flow are free. As it was mentioned earlier the humidity of the inlet gas at every stage is equal to the ambient humidity X_0.

3. APPLICATION OF OPTIMIZATION ALGORITHM WITH A CONSTANT DISCRETE HAMILTONIAN

The drying process equations considered here (the performance index (19) and the state equations (6) (7)) are linear with respect to the unconstrained decision variables θ^n. This fact enables the use of the special discrete optimization algorithm of the maximum principle with a constant discrete Hamiltonian [1,5]. This algorithm exists for a class of multistage processes which exhibits a linearity property in relation to same important decision variable θ^n describing the increment of the idependent variable t at stage n for n=1,...N

$$\theta^n = t^n - t^{n-1} \tag{22}$$

When the processes considered are autonomous with respect to t the general performance index and discrete state equations for n=1,...N and i = 1,..s are, respectively, of the form

$$F = \sum_{n=1}^{N} \theta^n f_o^n(\underline{x}^n, \underline{u}^n) \tag{23}$$

$$x_i^n - x_i^{n-1} = \theta^n f_i^n(\underline{x}^n, \underline{u}^n) \tag{24}$$

Here $\underline{x}^n$ is the s-dimensional state vector at stage n, x_i^n is the i-th coordinate of $\underline{x}^n$, $\underline{u}^n$ is the r-dimensional decision vector at stage n, and t^n is the idependent variable at stage n. The mathematical model, eqs.(22)- (24), is linear with regard to the decision variable θ^n and can be arbitrary (e.g., nonlinear) in relation to the decision vector $\underline{u}^n$ - hence, the special character of the decision variable θ^n.

For the processes with product recycle [5,6] the mixing functions describing the state vector before the first stage are of the form

$$x_i^o = M_i(\underline{x}^N) \tag{25}$$

for i = 1,... s (here we present the algorithm for the prescribed state of the feed stream).

By introducing the adjoint vector $\underline{z}^n$, it was ahown [1,5,6] that a suitable definition of the Hamiltonian leading to a striking analogy between the discrete optimization algorithm and the weak continuous one, is as follows

$$H^{n-1}(\underline{x}^n, \underline{z}^{n-1}, \underline{u}^n) = f_o^n(\underline{x}^n, \underline{u}^n) + \sum_{i=1}^{s} z_i^{n-1} f_i^n(\underline{x}^n, \underline{u}^n) \quad n=1,..N \tag{26}$$

(the sign of z_i^{n-1} is changed in contrast to [1]). With the use of the Hamiltonian the necessary optimality conditions take the form (n=1,...N, i=1,...s), as well as l=1,...r

$$\frac{x_i^n - x_i^{n-1}}{\theta^n} = \frac{\partial H^{n-1}}{\partial z_i^{n-1}} \tag{27}$$

$$\frac{z_i^n - z_i^{n-1}}{\theta^n} = -\frac{\partial H^{n-1}}{\partial x_i^n} \tag{28}$$

$$\frac{\partial H^{n-1}}{\partial u_l^n} = 0 \tag{29}$$

$$H^{n-1} = H^n = \text{constant} \tag{30}$$

For our drying problem $x_1^n = I_s^n$, $x_2^n = W_s^n$ and $u^n = T_g^n$.

Eqn. (27) is the state equation (24) in canonical representation, Eqn.(28) is the adjoint equation. The prescribed coordinates of the vector $\underline{x}^N$ correspond to the undeterminal coordinates of the vector $\underline{z}^N$, whereas the undetermined coordinates of the vector $\underline{x}^N$ correspond to the coordinates of the vector $\underline{z}^N$ satisfying the following equation

$$z_i^N = \sum_{j=1}^{s} z_j^o \frac{\partial M_j(\underline{x}^N)}{\partial x_i^N} \tag{31}$$

Eqn.(29) allows the determination of the stationary optimal decision vectors $\underline{u}^n$

i.e. the decision vectors in the interrior of the allowable region. When some coordinate of the vector $\underline{u}^n$, at an arbitrary stage, obtained from eqn.(29) lies outside the allowable region, then eqn.(29) should be rejected and the optimal boundary value of $\underline{u}^n$, extremizing Hamiltonian (26), should be accepted.

From eqn.(30) it results that, for the considered processes, the discrete Hamiltonian (26) is constant along the optimal trajectory. This constancy is the result of the optimal choice of the stationary unconstrained decision variable θ^n. Thus it pertains to the case when the eventual constraints on θ^n are not operative. When the cumulative independent variable increment $t^N - t^o = \sum_{n=1}^{N} \theta^n$ is unconstrained the constant optimal value of the Hamiltonian is equal to zero, whereas the prescribed $t^N - t^o$ corresponds to the undetermined constant value of the Hamiltonian.

In [5] and [6] other multistage processes with product recycle and linear versus θ^n (of more complicated topology as well as the nonautonomous type) are searched for.

For the drying process investigated in this paper, the enthalpy and moisture content of the solid play the role of the components of the two-dimensional state vector, and the temperature of the air becomes one-dimensional decision vector. The dimensionless air flow at the stage n is the decision variable θ^n. The Hamiltonian function is (see eqs. (26), (19) and (6)-(8))

$$H^{n-1}(I_s^n, W_s^n, z_1^{n-1}, z_2^{n-1}, T_g^n) = (1+r)\left[\frac{A}{2}(T_g^n - T_o)^2 + \lambda\right] + z_1^{n-1}\left[(c_p + X_o c_w)(T_g^n - 273{,}15) + 2502{,}7\, X_o - i_s^n(I_s^n, W_s^n)\right] + z_2^{n-1}\left[X_o - X_s^n(I_s^n, W_s^n)\right] = 0 \qquad (32)$$

for n=1,..N.

Since the total gas flow is undetermined (free $\Delta G^N/(S+R) = \sum_{n=1}^{N} \theta^n$) the constant value of the Hamiltonian (32) is equal to zero.

The adjoint equations related to enthalpy I_s^n and moisture content W_s^n are respectively

$$\frac{z_1^n - z_1^{n-1}}{\theta^n} = -\frac{\partial H^{n-1}}{\partial I_s^n} = z_1^{n-1}\frac{\partial i_s^n(I_s^n, W_s^n)}{\partial I_s^n} + z_2^{n-1}\frac{\partial X_s^n(I_s^n, W_s^n)}{\partial I_s^n} \qquad (33)$$

$$\frac{z_2^n - z_2^{n-1}}{\theta^n} = -\frac{\partial H^{n-1}}{\partial W_s^n} = z_1^{n-1}\frac{\partial i_s^n(I_s^n, W_s^n)}{\partial W_s^n} + z_2^{n-1}\frac{\partial X_s^n(I_s^n, W_s^n)}{\partial W_s^n} \qquad (34)$$

As I_s^N is undetermined and W_s^N is prescribed we have the following final boundary conditions associated with eqns. (33) and (34), respectively (see eqns. (31), (25), (1) and (13)):

$$z_1^N = \frac{R}{S+R} \qquad z_1^o = \frac{r}{1+r} \qquad z_1^o \qquad (35)$$

$$z_2^N \text{ - undetermined} \qquad (36)$$

Computation of the partial derivatives in eqs.(33) and (34) with the use of eqs. (3) and (4) and the solving of these equations with respect to z_1^{n-1} and z_2^{n-1} yields

$$z_1^{n-1} = \frac{z_1^n - \theta^n(z_1^n\beta_2^n + z_2^n\beta_1^n)}{1 + \theta^n(\alpha_1^n - \beta_2^n) + (\theta^n)^2(\alpha_2^n\beta_1^n - \alpha_1^n\beta_2^n)} \qquad (37)$$

$$z_2^{n-1} = \frac{z_1^n\theta^n\alpha_2^n + z_2^n(\alpha_1^n\theta^n + 1)}{1 + \theta^n(\alpha_1^n - \beta_2^n) + (\theta^n)^2(\alpha_2^n\beta_1^n - \alpha_1^n\beta_2^n)} \qquad (38)$$

where

$$\alpha_1^n = 2Y^n(c_1 W_s^n + d_1)E_1^n \qquad (39)$$

$$\beta_1^n = 2Y^n(c_2 W_s^n + d_2)E_2^n \qquad (40)$$

$$\alpha_2^n = 2Y^n\left[a_1(Y^n)^2 + c_1Y^n - \frac{3.064(c_1W_s^n + d_1)}{(0,094+W_s^n)^2}\right]E_1^n \quad (41)$$

$$\beta_2^n = 2Y^n\left[a_2(Y^n)^2 + c_2Y^n - \frac{3.064(c_2W_s^n + d_2)}{(0.094+W_s^n)^2}\right]E_2^n \quad (42)$$

$$Y^n = Y^n(I_s^n, W_s^n) = 0.24I_s^n + 32.6W_s^n/(0.094 + W_s^n) \quad (43)$$

$$E_i^n = \left[(a_iW_s^n - b_i)Y^n + c_iW_s^n + d_i\right]^{-3} \text{ for } i=1,2 \quad (44)$$

The stationary optimal gas temperatures T_g^n are determined from the relation

$$\frac{\partial H^{n-1}}{\partial T_g^n} = A(1 + r)(T_g^n - T_o) + z_1^{n-1}(c_p + X_oc_w) = 0 \quad (45)$$

Solving for T_g^n we obtain

$$T_g^n = T_o - \frac{z_1^{n-1}(c_p + X_oc_w)}{A(1 + r)} \quad (46)$$

If the stationary value of T_g^n, computed from eqn.(46) is higher than the maximum allowable temperature T^*, then the optimal temperature is equal to T^*, i.e.

$$T_g^n = T^* \quad (47)$$

The equations derived above enable the preparation of the computational block scheme Fig.2 . In the computational procedure the unknown quantities z_1^N, z_2^N and I_s^N are assumed, The subsequent assumed values z_1^N differ one from another by KR1 (Fig.2) and they belong to the computational range determined by z_{1*}^N and z_1^{N*}. Similarly the values KR2, z_{2*}^N and z_2^{N*} pertain to z_2^N as well as KR3 I_{s*}^N and I_s^{N*} pertain to I_s^N in the computer programme. After use of eqs.(37), (38) and (46) (or, respectively, eqs. (37) (38) and (47) if the optimal air temperature is T^*) in eq.(32), the resulting formula contains exclusively the single unknown variable θ^n. However, when solving this rearranged eqn. (32) the use of a trial-and-error procedure is required. The optimal flow θ^n is found in the preassigned computational range determined by θ_* and θ^*. The subsequent values of θ^n differ by DELTA and the initial value of this quantity is designated DELTAP . The accuracy of computations of the optimal θ^n is DELTAP x 10^{-s} The block "NO SOLUTION" means that the optimal decision value θ^n does not belong to the range θ_* to θ^* for the definite set of values z_1^N, z_2^N and I_s^W. It should be stressed that the optimal trajectory and the related optimal decisions cannot be obtained for arbitrary values of variables z_1^N, z_2^N and I_s^N. The optimal data of trajectories and decisions are described only by those solutions which meet the condition (35). Such solutions should be selected and, then, the corresponding states of the solid feed I_s^f, W_s^f, should be computed. For this purpose eqs.(1) and (2) transformed into the form

$$I_s^f = (1 + r)\, I_s^o - r\, I_s^N \quad (48)$$

$$W_s^f = (1 + r)\, W_s^o - r\, W_s^N \quad (49)$$

should be exploited. The solutions which meet the condition (35) describe the family of optimal trajectories heaving the common final moisture content $W_s^N = W_k$ and various initial states I_s^f, W_s^f. The initial state I_s^f, W_s^f of every computed trajectory is optimal in relation to the assumed values I_s^N and z_2^N (i.e. for each pair of values I_s^N , z_2^N such a value of z_1^N is assigned which obeys the condition (35)). Among the various trajectories obtained as a result of

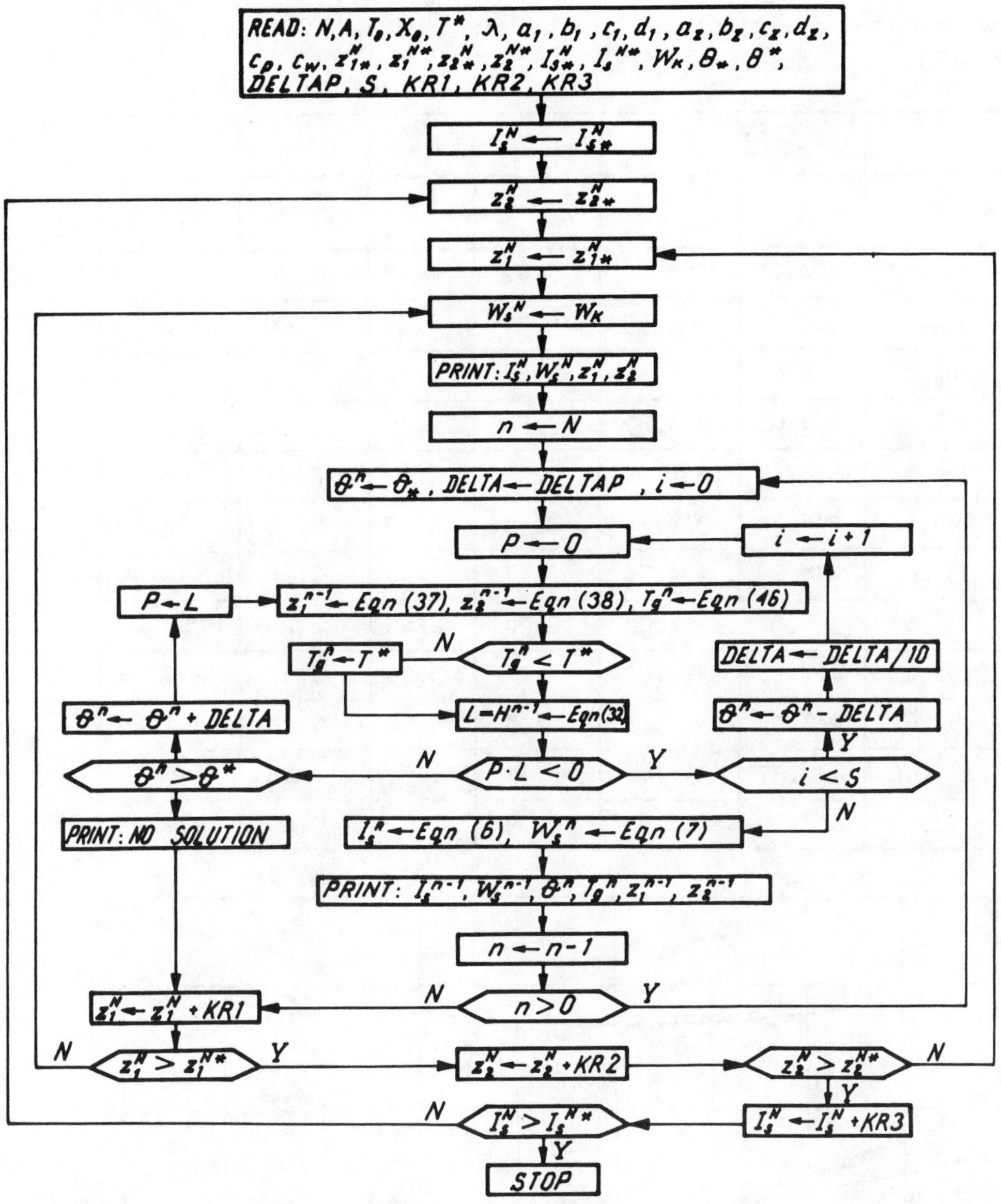

Fig. 2 The computational scheme of the problem

the computations, the specific trajectory should be found that has the initial state equal to the preassigned state I_s^f, W_s^f

4. RESULTS

In computations a relatively high upper limit of the allowable temperature $T^* = 500$ K, was assummed to enable an investigation of unrestricted stationary control. The optimal controls and optimal trajectories were obtained for the following numerical data: $N = 3$, $W_s^f = 0{,}2$ kg/kg, $I_s^f = -11{,}2$ kJ/kg (and hence $T_s^f = 293.15$ K), $W_s^N = W_k = 0.1$kg/kg, $c_p = 1.00$ kJ/kgK, $c_w = 1.88$ kJ/kg K, $T_0 = 293.15$K and $X_0 = 0.008$ kg/kg. Various values of recycle ratio, r, and various values of λ (various unit apparatus prices) were tested in the optimization computations. The results are shown in Table 2, and in Fig. 3.

Table 2 Data of the optimal decisions and of the performance index for various λ and **r** .

λ	r	T_g^1	T_g^2	T_g^3	g^1	g^2	g^3	F
kJ/kg	kg/kg	K	K	K	kg/kg	kg/kg	kg/kg	kJ/kg
0.42	0	304.31	304.75	304.65	9,82	8,32	7.48	16.53
	0.1	304.37	304.79	304.65	9.77	8.41	7.62	16.69
	0.2	304.43	304.82	304.66	9.66	8.44	7.71	16.73
	0.3	304.48	304.85	304.67	9.56	8.46	7.78	16.75
4.20	0	338.45	338.88	334.52	3.51	2.77	2.21	64.73
	0.1	338,54	338.90	334.64	3.45	2.79	2.25	64.82
	0.2	338.62	338.90	334.74	3.41	2.81	2.26	64.90
	0.3	338.69	338.91	334.83	3.38	2.83	2.29	65.03
42.0	0	452.63	447.45	378.02	1.48	0.96	0.18	217.04
	0.1	452.66	447.95	387.29	1.44	0.98	0.20	217.29
	0.2	452.66	448.37	394.86	1.41	0.99	0.23	217.57
	0.3	452.71	448.90	401.92	1.39	1.00	0.26	217.92

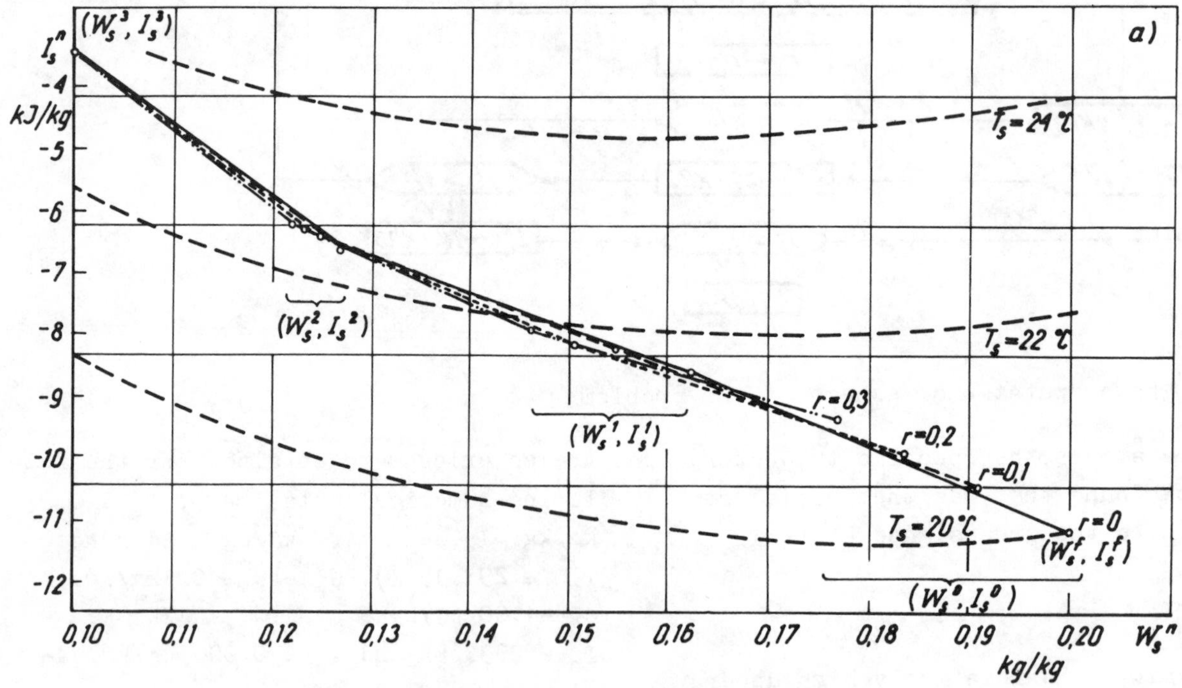

Fig. 3. Optimal trajectories of the process

a) λ = 0.42 kJ/kg

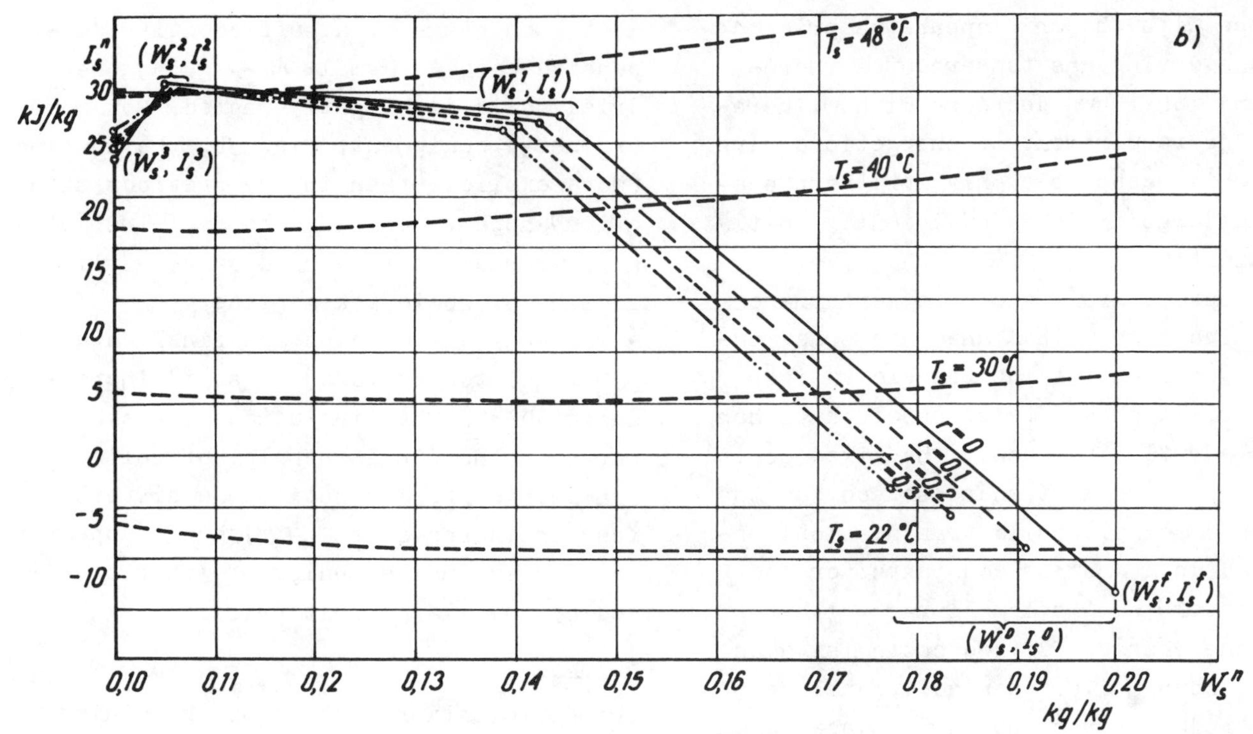

Fig.3 ctd. Optimal trajectories of the process b/ λ = 42.0 kJ/kg

In order to simplify the discussion of gas flows these flows were referred to the unit flow of feed solid i.e. the decision g^n, eq.(50), was additionally introduced, this decision resulting synonymously from the original decision θ^n as

$$g^n = \frac{\Delta G^n}{S} = (1 + r)\,\theta^n \qquad (50)$$

In analysis of the computational data the following conclusions are found:

- When unit apparatus price is sufficiently low (small λ) then the optimal temperatures of gas supplied to subsequent stages of cascade are close one to another and the highest T_g^n is at the first stage and the lowest at the second stage. However, the temperatures T_g^n increase with λ, this effect being most significant for T_g^1 and T_g^2 as well as less significant for T_g^3. As a result for intermediate and large λ, the temperatures T_g^1 and T_g^2 attain approximately the same values, whereas the temperature T_g^3 comparable with T_g^1 and T_g^2 for small λ becomes the lowest. When recycle ratio becomes larger an increase of the temperatures T_g^n is observed. This concerns mainly the third stage of cascade, especially for equipments characterized by large unit prices.
- Optimal drying gas flows decrease when λ increases. For the apparata characterized by small unit prices (small λ) a weak tendency for decreasing of gas flows through the subsequent stages appears; this tendency becoming more explicit when λ increases. This observation concerns mainly the last (third) stage where the lowest gas flows are obtained for sufficiently costly equipment. It is of interest that, when recycle ratio increases, the total dimensionless gas flow, $\sum_{n=1}^{N} g^n$, remains practically unchanged and its allocation between the various stages of the cascade becomes slightly more uniform.
- It results from the above conclusions that the optimal process intensity in-

creases with the unit apparatus price λ (as the drying gas temperatures increase and its flows decrease with λ). Therefore the requirement of energetic drying agents is acceptable only for the processes undergoing in a sufficiently costly equipment.

- The optimal values of the performance index increase with λ and are practically independent of the recycle ratio r. However, it should be stressed that when recycle ratio increases, the fixed costs of the process must increase too as the costs of recycle solid transmission (considered up to now as constants for the constant r) will become larger with r. When the change of fixed costs is taken into account (i.e. when additional term is introduced into the performance index) it results that, for our process, the economical calculations not justify the use of the variant with the product recycle. Such process should be, however, realized under some special technical circumstances e.g. when too large moisture content of feed solid makes fluidization difficult or even impossible. Furthermore it should be remembered that due to the equilibrium nature of the single stage model an important virtue of recirculation, namely its positive influence on the process kinetics, was lost in our case. Therefore the role of recycle will become more significant when the kinetical terms will predominate in drying.

In Fig. 3 the optimal states of dried solid (i.e. discrete optimal trajcetories) are presented. It is observed that:

- A tendency appears for the largest changes of the solid moisture content at the first stage and for the lowest ones at the third stage of cascade. This tendency becomes stronger when the unit price of apparatus, λ, increases and when the recycle ratio, r, is larger.
- At each stage of a sufficiently inexpensive cascade (small λ -Fig. 3a) the dried solid is slightly heated. For more expensive equipment, Fig. 3b, the heating (more explicit than in the previous situation) appears only at the two first stages, whereas at the third stage a slight cooling of solid takes place.
- The optimal value of the final enthalpy (final temperature) of solid increases with λ. The influence of recycle ratio on the final enthalpy of solid is insignificant. The noticeable differences (an increase of I_s^N with r) appear for an expensive equipment, characterized by the large unit price.

SUMMARY

The optimization problem of the minimum economic costs i. e. the sum of operational and investment costs was formulated and solved for the control crosscurrent fluidized drying with product recycle. The decision variables were the temperatures and flows of inlet gas.
The original discrete optimization theory, developed in part 1 of this work [1] was applied to the multistage process with product recycle and on its basis a computer program was developed. This program made it also possible to obtain an optimization solution for the continuous process of drying in a horizontal exhanger treated as a limit of the multistage process investigated when the total number of stages becomes sufficiently large.
The optimization results i.e. the optimal decisions and the optimal trajectories were obtained and presented in terms of the recycle ratio r as well as the parameter λ characterising the unit price of the apparatus (c.f. eq. (14)) in this paper.
These results are of considerable gene-

rality as amongst the whole family of optimal solutions a particular solution can be found which pertains to the special values of λ and r calculated for the specified drying process.

Datailed computations were made for the air-water-silica gel system and a three stage cascade was comprehensively investigated. On the basis of an analysis of the optimal decisions and the optimal trajectories the following conclusions were obtained:

1. For the cascade of inexpensive apparata (small λ) the optimal gas temperatures are practically identical for each stage of cascade, whereas for expensive apparata the optimal gas temperatures decrease when the stage number increases (for λ = constant). The optimal gas temperatures increase with λ. This means that the use of highly energetic drying agent is acceptable only for sufficiently expensive equipment. When recycle ratio becomes larger an increase of temperatures T_g^n is observed, this increase being significant only for the third stage of cascade composed of expensive apparata

2. The optimal gas flows decrease with the number , n, (for λ=constant) as well as with the unit apparatus cost λ (for n = constant). When the recycle ratio increases (for λ= constant) the allocation of the optimal gas flows between the various stages tends to the uniform one and the total gas flow remains practically unchanged.

3. The largest decrease of solid moisture content is in the first stage of cascade and the lower is in the third one. This property becomes more explicit when λ increases and r decreases. For small λ, the solid temperatures increase at every stage whereas for large λ the solid temperatures (higher than those for small λ) increase in the first and the second stages as well as decrease in the third stage.

4. Optimal values of performance index increase with λ and they are practically unaffected by recycle ratio. However the total costs (which also include the costs of the recycle transmission) increase in our case of the equilibrium stage model. The drying process with a recycle can be economical only then when the role of kinetic terms becomes significant or when the special technical difficulties (as e.g. agglomeration of the fresh moist feed) can be omitted due to the dry product recycle.

REFERENCES

1. Sieniutycz S., A general theory of optimal discrete drying processes with a constant hamiltonian, Drying 1983.
2. Sieniutycz, S., Szwast, Z., The Chem. Eng. Journal, 25, 63 (1982)
3. Sieniutycz, S., Rep. Inst. Chem. Eng., Warsaw Tech. Univ., 2(3) 17 (1973)
4. Sieniutycz, S., Optimization in Process Engineering, WNT, Warsaw, 1978.
5. Szwast, Z., Ph.D. Thesis, Warsaw Tech. Univ., 1979
6. Sieniutycz, S., Szwast, Z., Inż. Chem. Proc., 2 , 393 (1981).

A GENERAL THEORY OF OPTIMAL DISCRETE DRYING PROCESSES WITH A CONSTANT HAMILTONIAN

S. Sieniutycz

Warsaw Technical University, Warsaw, Poland

ABSTRACT

For multistage decision processes with variable time increment a general optimization theory is studied, the theory comprising the dryers in which the state changes are in a single direction. As an example, a multistage fluidized drying is introduced for the purpose of the numerical optimization studies to be presented in the supplementary paper [1]. Two performance indices K^p and K^c are investigated, corresponding to the production costs and total costs, respectively. Properties of the Lagrangian multiplier λ [1,2] which relate both types of costs are studied and its numerical value as a Hamiltonian is discussed. The analogy between the description of multistage and continuous processes is pointed out and a stationary algorithm closely analogous to the weak form of Pontryagin's maximum principle algorithm [3] is obtained for the discrete process case. It is shown that a Hamiltonian function has constant value along the autonomous process trajectory of the discrete processes with optimal continuous time increment. The significance of theory for optimization of various drying processes is pointed out.

NOMENCLATURE

$b_g[i_{gn}, X_{gn}]$ gas exergy per unit mass of absolute dry gas as a function of inlet gas state i_{gn}, X_{gn} at stage n, kJ/kg

c_g inlet gas price per mass unit of absolute dry gas, \$/kg

$\tilde{F}^c_n[I_{sn}, W_{sn}, \lambda]$ optimal overall exergy costs expressed as a function of variables I_{sn}, W_{sn}, λ, n, kJ/kg

$\tilde{F}^p_n[I_{sn}, W_{sn}, \lambda]$ optimal exergy cost of production expressed as above, kJ/kg

f^1, f^2 functions describing energy exchange rate and mass exchange rate, Eqs. (1) and (2), respectively, kJ/kg

G overall flow rate connected with dimensionless time τ, kg/h

ΔG_n gas flow rate through stage, n, kg/h

$g_n = \Delta\tau_n = \frac{\Delta G_n}{S}$ dimensionless gas flow at stage n, termed as dimensionless time increment

I_s solid enthalpy per unit mass of absolute dry substance, kJ/kg

I_{sn} outlet solid enthalpy from stage n, kJ/kg

I_{so}, I_{sN} initial and final solid enthalpy, respectively, kJ/kg

i_g gas enthalpy per mass unit of absolute dry gas, kJ/kg

i_{gn} gas inlet enthalpy to stage n, kJ/kg

i_s enthalpy of gas in equilibrium with material, kJ/kg

K_n^p cost of production expressed in exergy terms corresponding to n stage battery, kJ/kg

K_{en}^p cost of production expressed in economic terms corresponding to n stage battery, \$/kg

$\tilde{K}_n^p[I_{sn}, W_{sn}, \tau_n]$ optimal cost of production in exergy units expressed as a function of variables I_{sn}, W_{sn}, τ_n, kJ/kg

$\tilde{K}_n^c[I_{sn}, W_{sn}, \tau_n]$ overall optimal exergy costs expressed as above, kJ/kg

n current stage number

N overall number of stages

P function in the expression describing cost of production at stage, Eq.(8), kJ/kg

S, (G) dry solid (gas) flow rate, kg/h

W_s absolute moisture content of solid, kg/kg

W_{sn} solid moisture content after stage n, kg/kg

W_{so}, W_{sN} initial and final solid moisture content, respectively, kg/kg

X_g gas humidity per mass unit of absolute dry gas, kg/kg

X_{gn} absolute inlet gas humidity to stage n, kg/kg

X_s humidity of gas at equilibrium with solid, kg/kg

μ_s chemical potential of moisture in solid, kJ/kg

$\tau_n = \sum_{k=1}^{k=n} g_k = \frac{G}{S}$ dimensionless time (gas flow) in multistage fluidization process

τ_N dimensionless final time

λ the Lagrangian multiplier, kJ/kg

Brackets in equations: ( ) and { } refer to multiplication and minimum operation, respectively

Brackets in equations: [ ] refer to functional dependence

Subscripts

g gas, gas inlet

n stage n, n-stage battery

N final stage, final state

o initial state (solid phase)

s solid, equilibrium with solid

Superscripts

c overall cost

p production cost

~ functional minimum

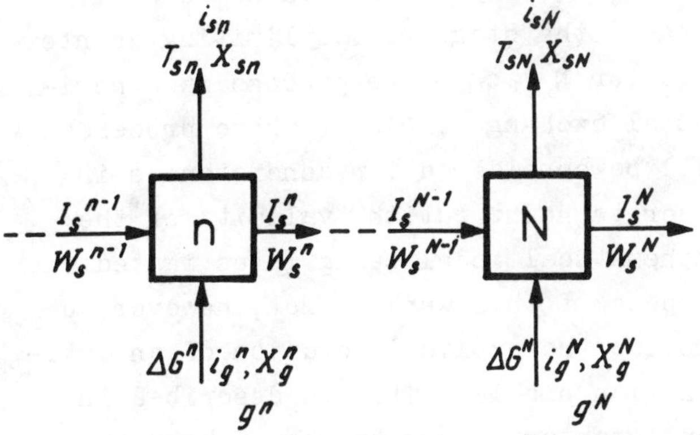

Fig. 1 The block scheme of a general multistage decision drying process.

PROBLEM FORMULATION AND SCOPE

When considering an optimization of various dryers continuous or multistage (i.e. discrete) models can be used. In this work a general discrete theory is developed which has turned out to be very suitable for optimization of the class of dryers and other equipment in which the state changes occur in a single distinguished direction of space or time. To this class belong the cocurrent and countercurrent dryers as well as those crosscurrent dryers in which an ideal mixing occurs in a perpendicular direction as e.g. fluidization dryer of the horizontal exchanger type. The processes investigated can be either continuous or discrete physical nature. In the former case the discretization is only a mathematical procedure applied to the continuous process equations for the purpose of computation of the optimal control.

We will present below original theory which has turned out to be very suitable for optimization of various dryers of the above mentioned type as well as many other apparata as e.g. heat and mass exchangers and chemical reactors [2] .

The objective of our study will be the isobaric-adiabatic multistage fluidization decision process with an ideal mixing at the stage or its limiting counterpart for $N \rightarrow \infty$ - the process in a horizontal exchanger, Fig. 1. These processes will be considered for concreteness in order to point out the validity of the mathematical model being investigated. The purpose of this work is not, however, to provide the explicit solution of an optimization problem. This is described in a supplementary paper by the author and Szwast [1] . The present work is devoted basically to the examination of the general properties of a mathematical model of optimization which can be suitably developed on the basis of consideration of a multistage fluidized drying process. Such examination is meaningful since the model being investigated can also be applied to many other processes with an ideal mixing of the stage e.g. those with heat transfer and chemical reaction.

In the process shown on Fig. 1, solid flows horizontally and gas flows vertically across the battery. The total flow of drying gas per unit flow of solid is specified as equal to τ_N. In further analysis, the variable τ will be termed as the dimensionless process time since its numerical value is proportional to the residence time of the solid if gas superficial velocity is the same at every stage of the process.

Due to simultaneous energy and mass exchange between gaseous and solid phases, the solid can change from one thermodynamic state to another. The optimization problem pertains to an optimal transition of solid from the constant inlet solid state I_{so}, W_{so} to the specified outlet solid state I_{sN}, W_{sN} when total dimensionless time τ_N is preassigned. The solid thermodynamic state I_s, W_s and the dimensionless time τ are state variables, the gas thermodynamic parameters i_g, X_g and the dimensionless time increment g are decisions in our optimization model. The discrete process state equations can be obtained as the result of energy and mass balance of the stage. For the purpose of this work, it will be sufficient to assume equilibrium between outlet gas and outlet solid. This case occurs commonly in practice. Furthermore, it will not be necessary to examine the explicit form of the state equations for the specific gas and solid considered.

Therefore, we will consider the state equations in the form:*

$$I_{sn}-I_{sn-1}=g_n f^1 [I_{sn}, W_{sn}, i_{gn}, X_{gn}]$$
$$\text{with } f^1 = i_{gn} - i_s \; I_{sn}, W_{sn} \qquad (1)$$

$$W_{sn}-W_{sn-1}=g_n f^2 [I_{sn}, W_{sn}, i_{gn}, X_{gn}]$$
$$\text{with } f^2 = X_{gn} - X_s [I_{sn}, W_{sn}] \qquad (2)$$

$$\tau_n - \tau_{n-1} = g_n \qquad (3)$$

Equations (1) and (2) result directly from the energy and mass balance. They conserve their implicit from also in the nonequilibrium case as it was shown in the author´s previous work [2] . The functions f^1 and f^2 in Eqs. (1) and (2) describe the discrete rates of energy and mass exchange, respectively

*Brackets [ ] in equations refer to functional dependence; brackets ( ) and { } -to multiplication.

The process performance index was accepted originally as the cost of the controlling gas. This optimization criterion involves the economic value of the inlet gas stream as related to the unit mass flow of dry solid:

$$K_{eN}^{p} = \sum^{N} c_g g_n . \qquad (4)$$

The above criterion should be interpreted as the production costs of the drying gas preparation.

In view of the uncertainity of gas unit economic values as functions of its thermodynamic state occurring due to local fluctuation and the fluctuation in time of these values, attention is directed toward that formulation of our optimization problem which is independent of these fluctuations. It is known from thermodynamic and economic considerations [4] that unit values of chemical substances are often accepted as proportional to the unit available energies. Such an assumption is a consequence of available energy properties [2,4].

Available energy, termed also exergy, is the semipositive function of the thermodynamic state of a substance and the thermodynamic parameters of the environment. It is equal to zero in the state of equilibrium of a substance with the environment, and it is always positive outside of this equilibrium. Thus, exergy has properties of price. The establishment of the economic values scale on the exergy basis leads to the relationships:

$$c_g = \iota \, b_g \qquad (5)$$

$$c_s = \iota \, b_s \qquad (6)$$

The proportionality coefficient ι describes the money paid when unit exergy is consumed. Consequently, it represents the unit exergy costs.

Substituting Eqs. (5) and (6) into Eq. (4) gives

$$K_{eN}^{p}/\iota = K_N^p = \sum_1^N b_g[i_{gn}, x_{gn}] g^n \qquad (7)$$

of exergy performance index, Eq. (7), is the task of an optimization.

In the present studies, it will be useful to examine a generalized from of Eq. (7) which is:

$$K_N^p = \sum_1^N P[i_{gn}, x_{gn}, I_{sn}, W_{sn}]\, g_n \qquad (8)$$

(The variables I_{sn} and W_{sn} appear explicitaly in P when the outlet gas is exploited [2]).

The problem dealing with minimization of the performance index, Eq. (8), with the difference restrictions, Eqs. (1) through (3), is termed as an original optimization problem.

Since the dynamic programming optimization method is very effective in the investigation of general properties of the optimal processes, this method is being used in our considerations. The Bellman's recurrence equation is three-dimensional in the original problem case. In order to reduce the problem dimensionality, the Lagrangian multiplier is introduced into Eq. (8). This leads to transformation of the performance index which is:

$$K_N^c = \sum_1^N \left(P[i_{gn}, x_{gn}, I_{sn}, W_{sn}] + \lambda\right) g_n \qquad (9)$$

The problem dealing with minimization of the performance index, Eq. (9), is termed as a transformed optimization problem. The Bellman's recurrence equation is two-dimensional in the case of the transformed problem.

Although the optimization computations are usually performed for the above described two-dimensional problem, our three-dimensional problem will be given equal consideration. The equivalence of both formulations of optimization will be shown in the stationary optimum case, and properties of functions describing the optimal costs $\tilde{K}_N^p$ and $\tilde{K}_N^c$ will be analyzed. It turns out that although this kind of analysis is occasionally done, several important relations have not been noticed yet. This will concern among others, close analogies between the descriptions of continuous and discrete optimal processes, which hold when the continuous time increment g_n is chosen optimally in the discrete process case. It will be shown that costs functions can be recovered one from another with the help of the Legendre transform. The optimal process time will be obtained as a partial derivative of the optimal cost function $\tilde{K}^c$ versus the Lagrangian multiplier playing the role of a modified Hamiltonian. Also, the Hamilton-Jacobi equation will be obtained for discrete process case. Finally, it will be proved that a discrete optimization algorithm exists which turns out to be fully analogous to the stationary form of the continuous version of the maximum principle. This algorithm is based on a different definition of the Hamiltonian than those given by Rozonoer [5], Katz [6], Halkin [7], and Fan [8]. As a consequence of this difference, the Hamiltonian function has a constant value property along the discrete process trajectory when the process is autonomous with respect to the continuous time.

The relationships discussed above are known so far only in the case of the continuous processes. In this case, they can be found in the optimization theory [3] and in classical mechanics, [9]. The optimization problem considered in this article gives an adequate model to present a discrete algorithm with the constant Hamiltonian function. In this work, only the stationary form of the algorithm will be considered. The more general considerations are given in book [2].

INTERPRETATION OF COSTS K_N^p AND K_N^c AND RECURRENCE EQUATIONS

The two performance indices discussed in this work are given by Eqs. (8) and (9). A comparison of these equations leads to the conclusion that the transformed costs K_N^c are related to the original costs K_N^p by the expression:

$$K_N^c = K_N^p + \lambda \sum_{1}^{N} g_n \qquad (10)$$

As it is shown in the paper [1] the second component of the sum in the formula, Eq. (10), is related to the process investment cost. That component describes the investment cost expressed in the exergy units relative to the unit flow of the dry solid. Therefore, the performance index K^c describes the overall exergy cost of the process also relative to the unit flow of the dry solid. The quantity λ in the formula, Eq. (10), is the apparatus cost expressed in exergy units relative to the unit dimensionless time τ, and to the unit amount of solid.

So far, only the economic interpretation of the Lagrangian multiplier λ has been considered. In order to get a picture of the mathematical interpretation of λ, we must compare the conditions necessary to obtain the absolute extremum of the right sides of the recurrence equations for the original three-dimensional problem, as well as for the transformed two-dimensional problem.

Since the systems considered are autonomous with respect to the variables n and τ, the recurrence equation which describes the three-dimensional problem, can be written in the general form

$$\tilde{K}_n^p[I_{sn},W_{sn},\tau_n]=\min_{g_n,i_{gn},X_{gn}}\Big\{P[I_{sn},W_{sn},i_{gn}X_{gn}]g_n + \tilde{K}_{n-1}^p[I_{sn}-g_nf_n^1, W_{sn}-g_nf_n^2, \tau_n-g_n]\Big\} \quad (11)$$

which applies to the function $\tilde{K}^p$ defined by the relationship:

$$\tilde{K}_n^p[I_{sn},W_{sn},\tau_n]=\min_{\frac{i_{gl},X_{gl},g_l}{I_{gn},X_{gn},g_n}} K_n^p[i_{gl},X_{gl},g_l\ldots] \quad (12)$$

For the transformed version of the optimization problem, the corresponding recurrence equation has the form

$$\tilde{F}_n^c[I_{sn},W_{sn},\lambda]=\min_{g_n,i_{gn},X_{gn}}\Big\{\big(P[i_{gn}X_{gn}I_{sn},W_{sn}]g_n + \lambda g_n\big) + \tilde{F}_{n-1}^c[I_{sn}-g_nf_n^1, W_{sn}-g_nf_n^2]\Big\} \quad (13)$$

which applies to the function $\tilde{F}_n^c$ defined as:

$$\tilde{F}_n^c[I_{sn},W_{sn},\lambda]=\min_{\frac{i_{gl},X_{gl},g_l}{i_{gn},X_{gn},g_n}} K_n^c[i_{gl},X_{gl},g_l\ldots] \quad (14)$$

STATIONARY TIME INCREMENT CONDITIONS

The necessary condition to obtain the stationary extremum of the right sides of Eqs. (11) and (13) with respect to the decision g_n is obtained by setting to zero the respective partial derivatives. As a result, we obtain the relationships:

$$\frac{\partial\tilde{K}_{n-1}^p}{\partial\tau_{n-1}}=\frac{\partial\tilde{K}_{n-1}^p}{\partial I_{sn-1}}f_n^1+\frac{\partial\tilde{K}_{n-1}^p}{\partial W_{sn-1}}f_n^2-P_N \quad (15)$$

$$\lambda=\frac{\partial\tilde{F}_{n-1}^c}{\partial I_{sn-1}}f_n^1+\frac{\partial\tilde{F}_{n-1}^c}{\partial W_{sn-1}}f_n^2-P_n \quad (16)$$

Now, our task is to investigate relations between optimal costs $\tilde{K}_n^p$ and $\tilde{F}_n^c$ on the basis of Eqs. (15) and (16).

TRANSFORMATIONS OF FUNCTIONS $\tilde{K}^p$ AND $\tilde{F}^c$

From the definition of the functionals K^p and F^c, it follows that the functions of the optimal costs $\tilde{K}_n^p$ and $\tilde{F}_n^c$ fulfill the relation:

$$\tilde{F}_n^c[I_{sn},W_{sn},\lambda]=\tilde{K}_n^p[I_{sn},W_{sn},\tau_n]+\lambda\tau_n \quad (17)$$

From the formula describing the total differential of the function $\tilde{K}_n^p[I_{sn},W_{sn},\tau_n]$ it follows that:

$$d\left(\tilde{K}_n^p-\frac{\partial\tilde{K}_n^p}{\partial\tau_n}\tau_n\right)=\frac{\partial\tilde{K}_n^p}{\partial I_{sn}}dI_{sn}+\frac{\partial\tilde{K}_n^p}{\partial W_{sn}}dW_{sn}+\tau_n d\left(-\frac{\partial\tilde{K}_n^p}{\partial\tau_n}\right) \quad (18)$$

The total differential of the function $\tilde{F}_n^c[I_{sn},W_{sn},\lambda]$ is

$$d\tilde{F}_n^c=\frac{\partial\tilde{F}_n^c}{\partial I_{sn}}dI_{sn}+\frac{\partial\tilde{F}_n^c}{\partial W_{sn}}dW_{sn}+\frac{\partial\tilde{F}_n^c}{\partial\lambda}d\lambda \quad (19)$$

Comparing Eqs. (15) through (19) yields:

$$\left(\frac{\partial\tilde{K}_n^p}{\partial I_{sn}}\right)_{W_{sn},\tau_n}=\left(\frac{\partial\tilde{F}_n^c}{\partial I_{sn}}\right)_{W_{sn},\lambda} \quad (20)$$

$$\left(\frac{\partial\tilde{K}_n^p}{\partial W_{sn}}\right)_{I_{sn},\tau_n}=\left(\frac{\partial\tilde{F}_n^c}{\partial W_{sn}}\right)_{I_{sn},\lambda} \quad (21)$$

$$\tau_n=\left(\frac{\partial\tilde{F}_n^c}{\partial\lambda}\right)_{I_{sn},W_{sn}} \quad (22)$$

$$\lambda=-\left(\frac{\partial\tilde{K}_{n-1}^p}{\partial\tau_{n-1}}\right)_{I_{sn-1},W_{sn-1}} \quad (23)$$

For the optimal process we can omit the minimization sign in Eq. (11). Differentiating both sides of the obtained relationship with respect to the time τ_n, leads to a conclusion that in the optimal process the following equality holds:

$$\frac{\partial \tilde{K}_n^p}{\partial \tau_n} = \frac{\partial \tilde{K}_{n-1}^p}{\partial \tau_{n-1}} \tag{24}$$

Equation (24) is valid in the systems which are autonomous with respect to continuous time τ.

Each of the equations, Eqs. (20) through (24), is important. Equations (20) and (21) reveal that the functions $\tilde{K}_n^p$ and $\tilde{F}_n^c$ have the same optima with respect to the enthalpy I_{sn} and moisture content W_{sn} after stage n. Equation (22) permits us to compute the optimal process time τ_n if known is the function $\tilde{F}_n^c[I_{sn}, W_{sn}, \lambda]$, obtained as a solution of the transformed problem. The relation, Eq. (23), gives the mathematical interpretation of λ, whereas the equality, Eq. (24) justifies the assumption of constant λ along the process trajectory.

Multiplying the corresponding side of Eq. (20) by the partial derivative $\partial f_n^1/\partial i_{gn}$, and Eq. (21) by the partial derivative $\partial f_n^2/\partial i_{gn}$, then adding the appropriate sides of the obtained equations and subtracting the partial derivative $\partial P_n/\partial i_{gn}$ from both sides of the resulting expression, we obtain the relation:

$$\frac{\partial \tilde{K}_n^p}{\partial I_{sn}} \frac{\partial f_n^1}{\partial i_{gn}} + \frac{\partial \tilde{K}_n^p}{\partial W_{sn}} \frac{\partial f_n^2}{\partial i_{gn}} - \frac{\partial P_n}{\partial i_{gn}} = \frac{\partial \tilde{F}_n^c}{\partial I_{sn}} \frac{\partial f_n^1}{\partial i_{gn}} + \frac{\partial \tilde{F}_n^c}{\partial W_{sn}} \frac{\partial f_n^2}{\partial i_{gn}} - \frac{\partial P_n}{\partial i_{gn}} \tag{25}$$

Conducting a similar operation for the partial derivatives $\frac{\partial f_n^1}{\partial X_{gn}}$, $\frac{\partial f_n^2}{\partial X_{gn}}$, and $\frac{\partial P_n}{\partial X_{gn}}$ we get the analogous result

$$\frac{\partial \tilde{K}_n^p}{\partial I_{sn}} \frac{\partial f_n^1}{\partial X_{gn}} + \frac{\partial \tilde{K}_n^p}{\partial W_{sn}} \frac{\partial f_n^2}{\partial X_{gn}} - \frac{\partial P_n}{\partial X_{gn}} = \frac{\partial \tilde{F}_n^c}{\partial I_{sn}} \frac{\partial f_n^1}{\partial X_{gn}} + \frac{\partial \tilde{F}_n^c}{\partial W_{sn}} \frac{\partial f_n^2}{\partial X_{gn}} - \frac{\partial P_n}{\partial X_{gn}} \tag{26}$$

It is easy to notice by setting both sides of Eqs. (25) and (26) to zero that we obtain the necessary conditions of the stationary extremum for the expressions in the large brackets { } of Eqs. (11) and (13) with respect to the decisions i_{gn} and X_{gn}. This conclusion as well as the relations, Eqs. (20) through (23), prove the equivalency of the algorithms described by Eqs. (11) and (13). This equivalency also applies when the decision variables are constrained. The should be shown by considering the relations, Eqs. (25) and (26), as well as the Kuhn-Tucker's theorem [11] applied to the problem in which the minimum of the left-hand sides of Eqs. (11) and (13) is sought.

Let us assume that the function $\tilde{F}_n^c[\lambda, I_{sn}, W_{sn}]$ describing the overall exergy costs is known as a result of solving the two-dimensional problem. If it becomes a necessity to know the optimal costs of production then, Eq. (22) helps us to compute these costs expressed as a function of the variables I_{sn}, W_{sn}, λ; that is, to obtain the function $\tilde{F}_n^p[I_{sn}, W_{sn}, \lambda]$. The discussed function is described by the following expression resulting from the definition of the overall costs, Eq. (10), and from Eq. (22):

$$\tilde{F}_n^p[I_{sn}, W_{sn}, \lambda] = \tilde{F}_n^c[I_{sn}, W_{sn}, \lambda] - \lambda\left(\frac{\partial \tilde{F}_n^c[I_{sn}, W_{sn}, \lambda]}{\partial \lambda}\right)_{I_{sn}, W_{sn}} \tag{27}$$

In the case of the three-dimensional problem we obtain the data concerning the optimal production costs as the function $\tilde{K}_n^p[I_{sn}, W_{sn}, \tau_n]$. Then, there exists a relation analogous to Eq. (27). This relation is used to find the function $\tilde{K}_n^c[I_{sn}, W_{sn}, \tau_n]$ describing the overall process costs:

$$\tilde{K}_n^c[I_{sn}, W_{sn}, \tau_n] = \tilde{K}_n^p[I_{sn}, W_{sn}, \tau_n] - \tau_n \left(\frac{\partial \tilde{K}_n^p[I_{sn}, W_{sn}, \tau_n]}{\partial \tau_n} \right)_{I_{sn}, W_{sn}} \tag{28}$$

Equations (27) and (28) are important because they indicate that the transformation from the optimal overall costs to the optimal production costs as well as the reverse transformation are performed by means of the respective Legendre's transformations. This fact allows for a flexibility of computing either of the two costs from the other. The reader can check that similar relations exist for the continuous systems considered in theoretical mechanics [9], They concern then the optimal action function as well as the optimal abbreviated action function. Other analogies used in the description of the optimal discrete process and the optimal continuous process should also be pointed out. The form of the stationary time increment conditions, Eqs. (15) and (16), shows that the right-hand sides of these equations have the structure of Hamiltonian functions. The form of such Hamiltonians is fully analogous to the Hamiltonian known in the continuous system case. Equation (16) indicates that the numerical value of its Hamiltonian is equal to the value of the Lagrangian multiplier. Since the left-hand sides of Eqs. (15) and (16) are equal and relations described by Eqs. (20) and (21) exist, Eq. (15) and Eq. (16) describe the same Hamiltonian. Due to the constancy of λ, the discussed Hamiltonian has a constant value along the autonomous process trajectory, similar to the continuous version. This fact suggests that we investigate our Hamiltonian more comprehensively.

THE DISCRETE HAMILTON-JACOBI EQATION

Properties of the Lagrangian Multiplier.

A dual interpretation of the Lagrangian multiplier (as the apparatus price λ and as the negative partial derivative of the optimal costs $\tilde{K}_n^p$ with respect to time $-\frac{\partial \tilde{K}_n^p}{\partial \tau_n}$) can be used for the continuous process as well as for the discrete process. In the considered problem, the derivative $-\frac{\partial \tilde{K}_n^p}{\partial \tau_n}$ tends to zero for processes with rates approaching zero and increases with the increasing driving forces of the process, $i_{gn} - i_{sn}$, $x_{gn} - x_{sn}$. Therefore, the derivative $-\frac{\partial \tilde{K}_n^p}{\partial \tau_n}$ is a measure of the process intensity. The relationship $\lambda = -\frac{\partial \tilde{K}_n^p}{\partial \tau_n}$ implies that in the optimal process the apparatus price is a function of the process intensity if the solid state after the process (I_{sn}, W_{sn}) is specified.

Quite interesting is the investigation of the Lagrangian multiplier properties when the same process is described by different state variables. It is only reasonable to consider the thermodynamic transformations of the variables which do not include time explicitely.

For the sake of an example, the relationship considered below is a transformation from the variables $t_s - \mu_s$ (material temperature - chemical potential of moisture in a solid phase) to the variables $I_s - W_s$ (solid enthalpy - solid moisture content). The function describing the optimal costs of production was designated for the first type of variables as $\tilde{K}_n^{pI}[t_{sn}, \mu_{sn}, \tau_n]$, and for the other one as $\tilde{K}^{pII}[I_{sn}, W_{sn}, \tau_n]$. The total derivative of the function K_n^{pI} with respect to time τ_n is described by the relation:

$$\frac{d\tilde{K}_n^{pI}}{d\tau_n} = \left(\frac{\partial\tilde{K}_n^{pI}}{\partial\tau_n}\right)_{t_{sn},\mu_{sn}} + \left(\frac{\partial\tilde{K}_n^{pI}}{\partial t_{sn}}\right)_{\tau_n,\mu_{sn}} \frac{dt_{sn}}{d\tau_n} + \left(\frac{\partial\tilde{K}_n^{pI}}{\partial\mu_{sn}}\right)_{\tau_n,t_{sn}} \frac{d\mu_{sn}}{d\tau_n} \qquad (29)$$

The transformation to the variables (I_{sn}, W_{sn}) permits us to express the considered derivative in the form of a relationship:

$$\frac{d\tilde{K}_n^{pI}}{d\tau_n} = \frac{\partial\tilde{K}_n^{pI}}{\partial\tau_n}_{t_{sn},\mu_{sn}} + \left(\frac{\partial\tilde{K}_n^{pI}}{\partial t_{sn}}\right)_{t_{sn},\mu_{sn}} \left\{\left(\frac{\partial t_{sn}}{\partial I_{sn}}\right)_{W_{sn}} \frac{dI_{sn}}{d\tau_n} + \left(\frac{\partial t_{sn}}{\partial W_{sn}}\right)_{I_{sn}} \frac{dW_{sn}}{d\tau_n}\right\}$$
$$+ \left(\frac{\partial\tilde{K}_n^{pI}}{\partial\mu_{sn}}\right)_{\tau_n,t_{sn}} \left\{\left(\frac{\partial\mu_{sn}}{\partial I_{sn}}\right)_{W_{sn}} \frac{dI_{sn}}{d\tau_n} + \left(\frac{\partial\mu_{sn}}{\partial W_{sn}}\right)_{I_{sn}} \frac{dW_{sn}}{d\tau_n}\right\} \qquad (30)$$

The formula pertaining to the derivative of the function $\tilde{K}^{pII}$ is a follows:

$$\frac{d\tilde{K}_n^{pII}}{d\tau_n} = \left(\frac{\partial\tilde{K}_n^{pII}}{\partial\tau_n}\right)_{I_{sn},W_{sn}} + \left(\frac{\partial\tilde{K}_n^{pII}}{\partial I_{sn}}\right)_{\tau_n,W_{sn}} \frac{dI_{sn}}{d\tau_n} + \left(\frac{\partial\tilde{K}_n^{pII}}{\partial W_{sn}}\right)_{\tau_n,I_{sn}} \frac{dW_{sn}}{d\tau_n} \qquad (31)$$

The functions $\tilde{K}_n^{pI}$ and $\tilde{K}_n^{pII}$ describe the same physical quantity but are expressed only by means of different variables.

Therefore, $\tilde{K}_n^{pI} = \tilde{K}_n^{pII}$ and $\frac{d\tilde{K}_n^{pI}}{d\tau_n} = \frac{d\tilde{K}_n^{pII}}{d\tau_n}$.

Equating the formulas, Eqs. (30) and (31), we conclude the following:

$$\left(\frac{\partial\tilde{K}_n^{pII}}{\partial\tau_n}\right)_{I_{sn},W_{sn}} = \left(\frac{\partial\tilde{K}_n^{pI}}{\partial\tau_n}\right)_{t_{sn},\mu_{sn}} \qquad (32)$$

$$\left(\frac{\partial\tilde{K}_n^{pII}}{\partial I_{sn}}\right)_{\tau_n,W_{sn}} = \left(\frac{\partial\tilde{K}_n^{pI}}{\partial t_{sn}}\right)_{\tau_n,\mu_{sn}} \left(\frac{\partial t_{sn}}{\partial I_{sn}}\right)_{W_{sn}} + \left(\frac{\partial\tilde{K}_n^{pI}}{\partial\mu_{sn}}\right)_{\tau_n,t_{sn}} \left(\frac{\partial\mu_{sn}}{\partial I_{sn}}\right)_{W_{sn}} \qquad (33)$$

$$\left(\frac{\partial\tilde{K}_n^{pII}}{\partial W_{sn}}\right)_{\tau_n,I_{sn}} = \left(\frac{\partial\tilde{K}_n^{pI}}{\partial t_{sn}}\right)_{\tau_n,\mu_{sn}} \left(\frac{\partial t_{sn}}{\partial W_{sn}}\right)_{I_{sn}} + \left(\frac{\partial\tilde{K}_n^{pI}}{\partial\mu_{sn}}\right)_{\tau_n,t_{sn}} \left(\frac{\partial\mu_{sn}}{\partial W_{sn}}\right)_{I_{sn}} \qquad (34)$$

Since in the optimal process $\lambda = \frac{-\partial\tilde{K}_n^p}{\partial\tau_n}$, Eq. (32) can be written in the form:

$$\lambda^I = \lambda^{II} \qquad (35)$$

The last relation proves the important property of the Lagrangian multiplier as a quantity which is invariant with respect to thermodynamic state variables transformations. It follows from Eqs. (29) through (35) that Eq. (35) applies to both autonomous system and to non-autonomous ones.

From the physical standpoint, the invariance of the multiplier λ signifies a natural phenomenon that the process optimal intensity does not depend upon the choice of state variables by means of which a definite process is being described. From the economical viewpoint, the invariance of λ means that the optimal apparatus price cannot depend upon the type of variables describing the process. Thus, the Lagrangian multiplier has properties which indicate its usefulness in applications. It should be noticed that partial derivatives of the criterion functions $\tilde{K}_n^p$ and $\check{F}_n^c$, with respect to any of the thermodynamic state variables, generally do not have similar properties* and therefore they are not as useful in applications as the Lagrangian multiplier λ.

Relation Between Price λ and the Modified Hamiltonian of the Optimization Problem.

When no restrictions are imposed on gas inlet parameters and gas inlet flow, many interesting results can be obtained analytically. In particular (as it is shown below), it is possible to derive a discrete stationary algorithm closely related to the discrete maximum principle algorithm but having a form different from the form of the discrete maximum principle algorithms known so far [5,6,7,8]. Such an algorithm is based on the Hamiltonian definition which is different from the one used so far. It turns out that it demonstrates a very close analogy to the Pontriagin maximum principle [3]. For autonomous systems the discussed algorithm is characterized by the constancy of the modified Hamiltonian along the discrete process trajectory, similar to the case of the continuous maximum principle. The nature of the algorithm discussed below suggests that it is possible to estabilish a formalism for discrete prcess similar to the one known in theoretical mechanics [9]. This aspect of the problem is discussed in another publication [12].

The derivation of the considered algorithm termed a stationary Hamiltonian principle algorithm will be discussed in the example of minimization of functionals K_n^p and F_n^c, which have a generalized form for nonautonomous systems*.

The method of derivation can be quickly adopted by the reader for any arbitrary discrete process with optimal time increment.

The recurrence dynamic programming equation leading to evaluation of the function $\tilde{K}_n^p[I_{sn}, W_{sn}, \tau_n]$ is described by Eq. (11). The generalized from of this equation for a nonautonomous system is:

$$\tilde{K}_n^p[I_{sn}, W_{sn}, \tau_n] = \min_{g_n, i_{gn}, X_{gn}} \Big\{ P_n[I_{sn} W_{sn} \tau_n \ldots] g_n + \tilde{K}_{n-1}^p \Big[I_{sn} - g_n f^1 [I_{sn}, W_{sn}, \tau_n, i_{gn}, X_{gn}], W_{sn} - g_n f^2 [I_{sn}, W_{sn}, \tau_n, i_{gn}, X_{gn}], \tau_n - g_n \Big] \Big\} \quad (36)$$

The above recurrence equation is associated with minimization of the sum:

$$\sum_1^N P_n [I_{sn}, W_{sn}, \tau_n, i_{gn}, X_{gn}] g_n \quad (37)$$

when the state equations have the form:

$$\frac{I_{sn} - I_{sn-1}}{g_n} = f^1 [I_{sn} W_{sn}, \tau_n, i_{gn}, X_{gn}] \quad (38)$$

$$\frac{W_{sn} - W_{sn-1}}{g_n} = f^2 [I_{sn}, W_{sn}, \tau_n, i_{gn}, X_{gn}] \quad (39)$$

$$\frac{\tau_n - \tau_{n-1}}{g_n} = 1 \quad (40)$$

It is essential that the problem described by Eqs. (28) through (40) can be treated as a discrete approximation of related continuous problem. In such a problem one should minimize the integral

$$\int_0^{\tau_N} P[I_s, W_s, \tau, i_g X_g] d\tau \quad (41)$$

---

* See for example, Eqs. (33) and (34).

* The assumption that the system is nonautonomous was introduced purposefully to generalize our considerations. The fact that the investigated process is autonomous will be applied later.

when the continuous state equations have a form:

$$\frac{dI_s}{d\tau} = f^1\left[I_s, W_s, \tau, i_g, X_g\right] \quad (42)$$

$$\frac{dW_s}{d\tau} = f^2\left[I_s, W_s, \tau, i_g, X_g\right] \quad (43)$$

Let us assume differentiable functions P_n and $\tilde{K}_n^p$ with respect to each state variable and each decisional variable. The necessary conditions for stationary minimum of K_n^p are obtained by differentiating the expression in the brackets on the right-hand side of Eq. (36) sequencially with respect to variables g_n, i_{gn}, X_{gn}. The resulting equations have the following form:

$$P_n\left[I_{sn}, W_{sn}, \tau_n, i_{gn}, X_{gn}\right] - \frac{\partial \tilde{K}_{n-1}^p}{\partial I_{sn-1}} f_n^1\left[I_{sn}, W_{sn}, ..\right]$$

$$- \frac{\partial \tilde{K}_{n-1}^p}{\partial W_{sn-1}} f_n^2\left[I_{sn}, W_{sn}, \tau_n, i_{gn}, X_{gn}\right] - \frac{\partial \tilde{K}_{n-1}^p}{\partial \tau_{n-1}} = 0 \quad (44)$$

$$\frac{\partial P_n}{\partial i_{gn}} - \frac{\partial \tilde{K}_{n-1}^p}{\partial I_{sn-1}} \frac{\partial f_n^1}{\partial i_{gn}} - \frac{\partial \tilde{K}_{n-1}^p}{\partial W_{sn-1}} \frac{\partial f_n^2}{\partial i_{gn}} = 0 \quad (45)$$

$$\frac{\partial P_n}{\partial X_{gn}} - \frac{\partial \tilde{K}_{n-1}^p}{\partial I_{sn-1}} \frac{\partial f_n^1}{\partial X_{gn}} - \frac{\partial \tilde{K}_{n-1}^p}{\partial W_{sn-1}} \frac{\partial f_n^2}{\partial X_{gn}} = 0 \quad (46)$$

In order to obtain a complete system of equations descrbing the stationary solution of the optimization problem, it is necessary to add to the system, Eqs. (44) through (46), Eq. (36) in which the minimization sign was omitted:

$$\tilde{K}_n^p\left[I_{sn}, W_{sn}, \tau_n\right] = P_n\left[I_{sn}, W_{sn}, i_{gn}, X_{gn}\right] g_n + \tilde{K}_{n-1}^p\left[I_{sn} - g_n f_n^1, W_{sn} - g_n f_n^2, \tau_n - g_n\right] \quad (47)$$

It is worthwhile to notice that Eqs. (45) and (46) describe the necessary conditions for stationary extremum with respect to the variables i_{gn} and X_{gn} of the following function:

$$H_{n-1}\left[I_{sn}, W_{sn}, n, i_{gn}, z_{n-1}^1, z_{n-1}^2\right] \equiv z_{n-1}^1 f_n^1 + z_{n-1}^2 f_n^2 - P_n \quad (48)$$

in which the adjoint variables z_{n-1}^1, z_{n-1}^2 are defined as follows:

$$z_{n-1}^1 \equiv \frac{\partial \tilde{K}_{n-1}^p}{\partial I_{sn-1}} \quad (49)$$

$$z_{n-1}^2 \equiv \frac{\partial \tilde{K}_{n-1}^p}{\partial W_{sn-1}} \quad (50)$$

Equation (44) describes properties of the optimal function H_{n-1}. Using the definition of the Hamiltonian, Eq. (48), the definition of the adjoint variables, Eqs. (49) and (50), as well as expressing the optimal decisions i_{gn}, X_{gn} as functions of time, state and adjoint variables, Eqs. (45) and (46), leads to the relation:

$$H_{n-1}\left[I_{sn}, W_{sn}, \tau_n \frac{\partial \tilde{K}_{n-1}^p}{\partial I_{sn-1}}, \frac{\partial \tilde{K}_{n-1}^p}{\partial W_{sn-1}}\right] = -\frac{\partial \tilde{K}_{n-1}^p}{\partial \tau_{n-1}} \quad (51)$$

which is a generalization of the Hamilton-Jacobi equation for discrete systems. The basic property of the Hamiltonian H_{n-1} for nonautonomous system is obtained by differentiating both sides of Eq. (47) with respect to the variable τ_n. Then:

$$\frac{1}{g_n}\left(\frac{\partial \tilde{K}_n^p}{\partial \tau_n} - \frac{\partial \tilde{K}_{n-1}^p}{\partial \tau_{n-1}}\right) = \frac{\partial P_n}{\partial \tau_n} - \frac{\partial \tilde{K}_{n-1}^p}{\partial I_{sn-1}} \frac{\partial f_n^1}{\partial \tau_n} - \frac{\partial \tilde{K}_{n-1}^p}{\partial W_{sn-1}} \frac{\partial f_n^2}{\partial \tau_n} \quad (52)$$

or, on the basis of the Hamiltonian definition and time increment definition:

$$\frac{H_n - H_{n-1}}{\tau_n \quad \tau_{n-1}} = \frac{\partial H_{n-1}}{\partial \tau_n} \qquad (53)$$

similarily, as in the continuous version of the Pontriagin principle. For the systems autonomous with respect to time, the following result is obtained:

$$H_n = H_{n-1} = \lambda \qquad (54)$$

on the basis of Eqs. (23), (51) and (53). The last relation indicates that in the case of autonomous systems the modified Hamiltonian is constant along a discrete process trajectory contrary to the discrete Hamiltonians used so far in current literature. A numerical value of the discussed Hamiltonian is equal to the Lagrangian multiplier λ. When the Lagrangian multiplier is used in optimization of any arbitrary autonomous system, then a family of optimal trajectories for the same numerical value of the process Hamiltonian is computed. In theoretical mechanics [9] an analogous family of trajectories for the continuous process is called a family of natural trajectories.

The Eq. (54) also indicates that in the optimization problem considered in this work, the apparatus price expressed in approprate units is equal numerically to the optimal Hamiltonian value. This conclusion serves as an economic interpretation of the Hamiltonian for continuous and multistage systems.

THE STATIONARY HAMILTONIAN PRINCIPLE

In order to determine the relationship between variables Z_n and Z_{n-1}, it suffices to compute the partial derivatives of both sides of the relation, Eq. (47) with respect to variables I_{sn} and W_{sn}. As a result we obtain the equations:

$$\frac{\partial \tilde{K}^p_n}{\partial I_{sn}} = \frac{\partial P_n}{\partial I_{sn}} + \frac{\partial \tilde{K}^p_{n-1}}{\partial I_{sn-1}}\left(1 - \frac{\partial f^1}{\partial I_{sn}} g_n\right) - \frac{\partial \tilde{K}^p_{n-1}}{\partial W_{sn-1}} \frac{\partial f^2_n}{\partial I_{sn}} g_n \qquad (55)$$

$$\frac{\partial \tilde{K}^p_n}{\partial W_{sn}} = \frac{\partial P_n}{\partial W_{sn}} + \frac{\partial \tilde{K}^p_{n-1}}{\partial W_{sn-1}}\left(1 - \frac{\partial f^2_n}{\partial W_{sn}} g_n\right) - \frac{\partial \tilde{K}^p_{n-1}}{\partial I_{sn-1}} \frac{\partial f^1_n}{\partial W_{sn}} g_n \qquad (56)$$

which can be expressed in the terms of adjoint variables as:

$$\frac{Z^1_n - Z^1_{n-1}}{g_n} = \frac{\partial P_n}{\partial I_{sn}} - Z^1_{n-1} \frac{\partial f^1_n}{\partial I_{sn}} - Z^2 \frac{\partial f^2_n}{\partial I_{sn}} \qquad (57)$$

$$\frac{Z^2_n - Z^1_{n-1}}{g_n} = \frac{\partial P_n}{\partial W_{sn}} - Z^1_{n-1} \frac{\partial f^1_n}{\partial W_{sn}} - Z^2 \frac{\partial f^2_n}{\partial W_{sn}} \qquad (58)$$

Using in Eqs. (57) and (58) the Hamiltonian definition, Eq. (48), as well as Eq. (40), give:

$$\frac{Z^1_n - Z^1_{n-1}}{\tau_n - \tau_{n-1}} = - \frac{\partial H_{n-1}}{\partial I_{sn}} \qquad (59)$$

$$\frac{Z^2_n - Z^2_{n-1}}{\tau_n \quad \tau_{n-1}} = - \frac{\partial H_{n-1}}{\partial W_{sn}} \qquad (60)$$

From the Hamiltonian definition as well as from the discrete state equations, Eqs. (38) to (40), result the following equations:

$$\frac{I_{sn} - I_{sn-1}}{\tau_n \quad \tau_{n-1}} = \frac{\partial H_{n-1}}{\partial Z^1_{n-1}} \qquad (61)$$

$$\frac{W_{sn} - W_{sn-1}}{\tau_n \quad \tau_{n-1}} = \frac{\partial H_{n-1}}{\partial Z^2_{n-1}} \qquad (62)$$

The taking into account of the Hamiltonian definition in Eqs. (45) and (46) leads to the equations:

$$\frac{\partial H_{n-1}}{\partial i_{gn}} = 0 \qquad (63)$$

$$\frac{\partial H_{n-1}}{\partial X_{gn}} = 0 \qquad (64)$$

which holds for an unconstrained extremum. When contraints are present H must be locally maximal at the boundary [2].
In order to obtain a complete system of equations characterizing the optimal process to the system, we must also add equation (53) to the system of Eqs. (59) through (64). Recirculation of the product can also be taken into account [1,2].
The system of Equations, (59) through (64) describes a stationary version of the Pontryagin principle [3] for discrete systems. The essential difference between the continuous and the discrete version re lies on the fact that the Eq. (53) does not result (as it does in the continuous version) from canonical equations, but constitutes an independent condition of optimality. The equation is necessary since in the discrete version we have one more decision-the time interval $\Delta\tau_n = g_n$.
As a result, the system of discrete equations for the optimal process consists of seven equations because it has seven variables: $I_s, W_s, Z^1, Z^2, i_g, x_g,$.
For discrete systems which are autonomous with respect to time τ_n, an additional condition of optimality is constant Hamiltonian condition. Then, on the basis of Eqs. (23) and (54), the Hamiltonian has the same constant value in both continuous and discrete processes when these processes are joined. This fact has some usefulness in computations. The chemical engineering applications of the theory presented are widely disscussed in the book [2]. The numerical solution of our fluidized drying problem is obtained and analyzed in [1].

CONCLUSIONS

Using the discrete verion of dynamic programming it was shown that the considered drying problem is a good example revealing the existence of a original discrete algorithm which is analogous to the stationary version of the continuous maximum principle. This algorithm applies if the continuous time increments on the stage are chosen to be optimal.
In such a case, the Hamiltonian function has a constant value property along the discrete process trajectory for the systems which are autonomous with respect to the continuous time. The discussed algorithm turns out to be more analogous to the continuous version of the maximum principle than any other similar algorithms so far known. For the discrete processes analogous to the ones which are considered in mechanics, this algorithm leads to a generalization of the energy conservation principle as applied to the discrete systems.
For the fluidized drying problem formuleted in this article, and solved in the supplementary paper [1] the Hamiltonian is connected with the optimal apparatus price, meaning that its numerical value is equal to the mentioned price expressed in the appropriate units. In terms of λ the properties of the optimal drying trajectories and the optimal decisions are numerically investigated and rigorous principles of optimal control are estabilished in [1].

The continuous version is obtained as a limiting case when $n \to \infty$.

REFERENCES

1. Sieniutycz,S,;Szwast Z.,Optimal Multistage Fluidized Drying with Product Recycle, in Mujumdar A.S.(ed.), "Drying 83" ,Hemisphere Publ.Corp.,New York (1983).
2. Sieniutycz,S.,Szwast Z., Practics of Optimization Calculations,WNT,Warsaw, Poland, 1982 .
3. Pontryagin,L.S.,Boltyanskii,V.G., Gamkrelidze,R.V.,Mischenko,E.F., Mathematical Theory of Optimal Processes . English translation by K.N. Trirogoff, Interscience,New York,N.Y., 1962 .
4. Szargut,J.,Petela,R.,Exergia,WNT,Warsaw, Poland, 1965 .
5. Rozonoer,L.I.,Automation and Remote Control.,20,1288 (1959) ,also 20, 405 (1959), 20, 1517 (1959).
6. Katz,S.,Ind.Eng.Chem.Fundamentals, 1 , 226 (1962).
7. Halkin,H.,S.I.A.M. J.Control,4, 90, (1966).
8. Fan,L.T.,Wang,C.S.,The Discrete Maximum Principle: A Study of Multistage System Optimization. Wiley, New York, 1964 .
9. Whittaker,E.T.,A Treatise on the Analytical Dynamics of Particles and Rigid Bodies with an Introduction to the Problem of Three Bodies, Dover New York 1944 .
10. Sieniutycz,S.,ChemEng.Sci.,37,1557 (1982).
11. Kuhn,H.W.,Tucker,A.W.,"Nonlinear Programming" in J.Neyman (ed.),Proceedings of the Second Berkeley Symposium on Mathematical Statics and Probability, University of California Press, Bekeley, California, (1951) .
12. Sieniutycz,S.Rep.Inst.Chem.Eng., Warsaw Techn.Univ., 3, 27 (1974).

CALCULATION OF DRYING PARAMETERS FOR THE PENETRATING EVAPORATION FRONT

S. Szentgyorgyi*
K. Molnar**

*Kelti K. utca 27, H-1024 Budapest, Hungary
**Gyorgy A. ut 13, H-1125 Budapest, Hungary

ABSTRACT

An accurate and an approximate method is described for the determination of moisture content and temperature as a function of time for the specific drying process, when the evaporation intensity suddenly changes within the wet region and where the front of this sudden change penetrates with varying velocity into the deeper lying regions of the material. The essence of the process is demonstrated by a simple model: the relative intensity of evaporation is the phase-changing criterion. Its sudden change can be dealt with as a moving boundary condition. In the course of approximate calculations, the lower and upper approach of the exact solution has been developed for the demonstrated model. The numerical example shows that the upper approach is the most accurate one, but its calculation takes the most effort. The lower approach being much simpler and the still simpler so-called linear approach seems to satisfy practical requirements in many cases, too.

NOMENCLATURE

A	surface area, m^2
c	specific heat, J/kgK
Δh_v	evaporation heat, J/kg
L	thickness, m
$\underset{\approx}{M}$	mass, kg
$\tilde{M}_W$	mole weight of moisture, kg/kmol
$\dot{m}$	mass flux density (drying velocity), kg/m^2s
$\Delta\bar{\bar{m}}_L$	specific moisture reduction, kg/m^2
P	pressure, N/m^2
$\dot{q}$	heat flux density, W/m^2
$\tilde{R}$	universal gas constant, 8,3443 J/mol K
s	slope of the tension curve, $N/K\ m^2$
T	temperature, K
t	time, s
V	volume, m^3
W	mass of moisture, kg
X	moisture content referring to the dry material, kg/kg
Z	thickness of a non-moving dry bed, m
z	co-ordinate, m

Greek Letters

α	heat transfer coefficient, W/m^2K
β	mass transfer coefficient, m/s
δ	eigenvalue
δ_e	effective diffusion coefficient, m^2/s
ε	phase-changing criterion
χ	thermal diffusivity, m^2/s
Γ	temperature factor, according to Ea. (25)
λ	thermal conductivity, W/mK
ξ	location co-ordinate,m
Ω	determined by Eqs (34) and (35)
Ψ	porosity, m^3/m^3
η	saturation coefficient
ν	dimensionless temperature, according to Eq. (26)

Dimensionless numbers

Bi	Biot number
Fo	Fourier number

Subscripts(considering)

I	dried region
II	wet region
c	critical
d	dew-point
e	steady-state
F	surface area
G	gas-phase
hm	simultaneous heat and mass transfer
L	liquid-phase
m	mass transfer
O	initial state, and for surface: value at the location z=0, resp.
P	pores, at constant pressure
V	vapor-phase
W	moisture
Z	thickness, referring to non-moving dry bed
ξ	evaporation front of layer with varying thickness

INTRODUCTION

Following the energy-balance of porous wet material (1, 2):

$$\frac{\partial T}{\partial t} = \varkappa \frac{\partial^2 T}{\partial z^2} + \frac{\Delta h_v \varepsilon}{c\varrho}\frac{\partial X}{\partial t} \qquad (1)$$

is true, where

$$\varepsilon = \frac{\operatorname{div} \dot{m}_V}{\operatorname{div}(\dot{m}_V + \dot{m}_L)} = \frac{\operatorname{div} \dot{m}_V}{\operatorname{div} \dot{m}_W} \qquad (2)$$

in the so-called phase-changing criterion

Experience (3) shows that ε suddenly changes in many cases, when a certain value X_c of the moisture content has been reached. The front of the sudden change of the phase-changing criterion $\Delta\varepsilon$ penetrates into the wet material at a varying velocity depending on time. In such cases the analytical solution of the differential equations describing the drying process becomes rather difficult. The problem is caused by the moving front, arising due to the sudden change of ε. The description of a method for the computer aided solution of the problems owing to the moving (penetrating) front, as well as for their analytical approximate solution is given hereunder. For the sake of easy consideration of a plane of finite thickness and a simple model, too.

Mathematical model

Let us examine the drying of liquid-phase moisture in a macroporous system for the specific case, when at the beginning of the drying process a cetain portion η of the pores of the bed is saturated. The model assumed to be a plane of infinite frontal surface and thickness 2L, is to be dried from both sides by means of flowing gas. The conductive and convective transport coefficients, resp., are considered to be constant.

With the increase of drying time the plane can be divided into the dried region I and the wet region II, as demonstrated in Fig. 1. In region I: X=0 and in region II: $\dot{m}_v=0$; thus the energy-balance for both portions can be written as:

$$\frac{\partial T^{\rm I}}{\partial t} = \varkappa_{\rm I}\frac{\partial^2 T^{\rm I}}{\partial z^2} \qquad (3)$$

and

$$\frac{\partial T^{\rm II}}{\partial t} = \varkappa_{\rm II}\frac{\partial^2 T^{\rm II}}{\partial z^2}, \qquad (4)$$

respectively.

Uniqueness equations

The initial condition is:

$$T^{\rm I}(z, 0) = T^{\rm II}(z, 0) = T_0 = \text{const.} \qquad (5)$$

The boundary conditions are:
1. At the median (symmetry) plane, at the location z=L

$$\left.\frac{\partial T^{\rm II}}{\partial z}\right|_{z=L^-} = 0; \qquad (6)$$

2. At the boundary (interphase) surface of the two regions, at the location $z=\xi$

$$\dot{q}|_{z=\xi^-} = -\lambda_{\rm I}\left.\frac{\partial T^{\rm I}}{{\rm I}\partial z}\right|_{z=\xi^-} = \lambda_{\rm II}\left.\frac{\partial T^{\rm II}}{\partial z}\right|_{z=\xi^+} + \dot{m}_{W,z=\xi}\,\Delta h_v \qquad (7)$$

at the location z=0

$$\dot{q}|_{z=0^-} = \alpha_G(T_G - T_F) = -\lambda_{\rm I}\left.\frac{\partial T^{\rm I}}{\partial z}\right|_{z=0^+}. \qquad (8)$$

The relationship between the total moisture content of the material and the coordinate ξ, is:

$$M_L = M_{L0} - \eta V_P \frac{\xi}{L}\varrho_L \qquad (9)$$

and hence, applying the porosity term:

$$\Psi = \frac{V_P}{LA} \qquad (10)$$

yields:

$$\frac{M_{L0} - M_L}{A} = \Delta\overline{m}_L = \eta\Psi\varrho_L. \qquad (11)$$

For drying velocity, expressed by the front velocity, we obtain:

$$\dot{m}_{W\xi} = -\frac{1}{A}\frac{dM_L}{dt} = \eta\Psi\varrho_L\frac{d\xi}{dt} \qquad (12)$$

Thus, the boundary condition (7) can also be expressed by the penetration velocity of the front ξ as:

$$\dot{q}|_{z=\xi^-} = -\lambda_{\rm I}\left.\frac{\partial T^{\rm I}}{\partial z}\right|_{z=\xi^-} = -\lambda_{\rm II}\left.\frac{\partial T^{\rm II}}{\partial z}\right|_{z=\xi^+} + \eta\Psi\varrho_L\frac{d\xi}{dt}. \qquad (13)$$

Numerical solution

The differential equations Eqs (3) and (4) may be solved numerically with the aid of a computer, considering boundary conditions (5), (6), (8) and (13) and prescribed initial conditions. To obtain the numerical solution, the method of finite differences and the so-called explicit method cannot be applied directly, therefore, their modified form had to be developed (4). The grid, chosen for the numerical solution, is shown in Figure 2.

The location distributions are plotted along the horizontal axis, while the time distributions appear along the vertical axis of the demonstrated co-ordinate system. Numerical computation requires a correlation system in order to satisfy drying boundary conditions (13) at point ξ of the penetrating drying front, so that an ever occurring front-location should coincide with a location-grid point. The penetration velocity of the front is, however, varying, therefore either the distances of the front-grid points, or the time steps must be chosen as variable. For practical calculation purposes the time steps Δt

were chosen as variable, while the front steps ΔZ were kept constant. The number of front steps, of course, must correspond to the number of time steps.

Accordingly, the grid composed of front steps (auxiliary grid) is not suitable in general to serve as location-grid, at the same time, since in this case no temperature change would be obtained for the wet zone behind the front. The distances between the points of the location-grid (main grid) must be large enough to display the temperature changes within the total wet region, even in case of a relatively poor penetration-rate of the front. The instantaneous location of the above-mentioned auxiliary grid, forming one unique new grid-location, will contribute to this main grid. The front (always situated at the auxiliary grid) divides distribution n between the subsequent and its preceding main grid point in two -generally not equal - portions. Therefore, the derivatives involved in the solution procedure had to be expressed in an uneven division, too.

The plate of half thickness L was divided in k portions as illustrated in Fig. 2, this grid being the so-called main grid with respect to the locations. The width of the applied auxiliary grid is ΔZ, its magnitude being one n-th portion of the main grid distance. Thus, the distribution number of the auxiliary grid can be expressed by the product: k n. For the case of a chosen number of k the number of time steps after which temperature change will occur at the location: z=L, depends only on the number of the auxiliary grid distribution n. The distribution according to locations is indicated by the lines i-const. at the main grid. The time distribution of the chosen grid changes from time level to time level in such a way that with the chosen time step Δt and with the penetration velocity of the front calculated at the former time level, the dry front will just remove to a distance ΔZ at the end of the time step and thus coincide with a location-grid point (lying on an auxiliary grid, and from time to time on a main grid). With the correct choice of the distribution number n of the auxiliary grid, the stability of the explicit solution method might also be guaranteed for the changing values of Δt.

At a given j-th instant of time the penetration velocity of the front can be written as:

$$\left(\frac{d\xi}{dt}\right)_j$$

and this velocity changes as a function of time. In the course of the numerical calculations this velocity is regarded as constant for each elementary time step. Let the penetration velocity of the front be:

$$\left(\frac{d\xi}{dt}\right)_{j-1}$$

at the time level j - 1 and assuming the penetration velocity of the front to be constant until time level j, Δt_j can be determined, i.e.:

$$\Delta t_j = \Delta Z \frac{1}{\left(\frac{d\xi}{dt}\right)_{j-1}}$$

Using the difference formulas of equal and unequal location distributions required for the numerical calculations the differential equation system and the boundary conditions can be rewritten to difference equations with the aid of which the new temperature values can be expressed at the new time levels for every location-grid distribution (at all the main grid points and at the sole auxiliary grid point concerned). Detailed correlations for calculation are referred to in (4).

Approximate solution

The described numerical solution method is practically accurate and can also be applied in the case when the mass and transport coefficients are considered to depend on temperature and on moisture content, too and thus the differential equations will turn to non-linear. Engineering practice, however, also needs correlations, which can be easier dealt with and are relatively simple and on the basis of which one of the main parameters of the drying process, i.e. the moisture content as a function of time can be determined with satisfactory accuracy. The sequence of the procedure is, as follows:

Let us assume a material of two layers, comprising the dry region I of thickness Z at both sides and the wet region II extending from both sides until the median (symmetry) plane having the thickness L - Z, Drying by convection induces this kind of material (of two layers) to approach to steady-state, when:

$$\dot{q}_e = \dot{m}_W \Delta h_v. \tag{14}$$

In this case:

$$\dot{q}_e = -\lambda_I \left(\frac{\partial T^I}{\partial z}\right)_e = \text{const.} = \lambda_I \frac{T_{Ze} - T_F}{Z} = \alpha_G(T_G - T_F), \tag{15}$$

thus, heat flux density can be expressed as:

$$\dot{q}_e = \frac{T_G - T_{Ze}}{\frac{1}{\alpha_G} + \frac{Z}{\lambda_I}} = K_H(T_G - T_{Ze}). \tag{16}$$

If vapor is saturated in the dry region and the tension curve of vapor is considered to be linear, then the drying velocity can be written as:

$$\dot{m}_{We} = \frac{1}{\frac{\tilde{R}T_G}{\beta \tilde{M}_W} G + \frac{\tilde{R}T_{Ze}}{\delta_e \tilde{M}_W}} (p_{VZ} - p_{VG}) \tag{17}$$

and the slope of the tension curve considered as linear will be:

$$s = \frac{p_{VZ} - p_{VG}}{T_Z - T_d} \tag{18}$$

Hence, the drying velocity expressed by means of the temperature difference as a driving force, yields:

$$\dot{m}_{We} = \frac{s(T_{Ze} - T_d)}{\frac{\tilde{R}T_G}{\beta \tilde{M}_W} + Z\frac{\tilde{R}T_{Ze}}{\delta_e \tilde{M}_W}} = K_m s(T_{Ze} - T_d). \tag{19}$$

Thus, the temperature at the location of the boundary surface of the dry-wet (interface) region, which has reached steady-state, will be:

$$T_{Ze} = \frac{T_d + T_G \frac{K_H}{\Delta h_v s K_m}}{1 + \frac{K_H}{\Delta h_v s K_m}} \tag{20}$$

The substitution of Eqs (16) and (19) into Eq. (4) gives:

$$\dot{m}_{We}\left[\frac{1}{\alpha_G} + \frac{\tilde{R}T_G}{\beta \tilde{M}_W s \Delta h_v} + \left(\frac{1}{\lambda_1} + \frac{\tilde{R}T_Z}{\delta_e \tilde{M}_W s \Delta h_v}\right) Z\right] = \frac{T_G - T_d}{\Delta h_v} \tag{21}$$

and with the application of Eqs (12) and (21), in the case of Z=ξ

$$\frac{\dot{m}_{W\xi}}{\dot{m}_{We}} = \eta \Psi \varrho_L \frac{d\xi}{dt} \frac{\Delta h_v}{T_G - T_d} \left[\frac{1}{\alpha_G} + \frac{\tilde{R}T_G}{\beta \tilde{M}_W s \Delta h_v} + \left(\frac{1}{\lambda_1} + \frac{\tilde{R}T_\xi}{\delta_e \tilde{M}_W s \Delta h_v}\right)\xi\right] \tag{22}$$

is obtained.

Let us assume that in the case of the penetrating front, the condition due to Eq. (15) is valid, too, i.e. the temperature profile of the dried layer is linear, then the drying velocity can be written as:

$$\dot{m}_{W\xi} = \frac{1}{\frac{\tilde{R}T_G}{\beta \tilde{M}_W} + \xi \frac{\tilde{R}T_\xi}{\delta_e \tilde{M}_W}} s(T_\xi - T_d). \tag{23}$$

According to Eqs (23) and (19), for the case of Z=ξ

$$\frac{\dot{m}_{W\xi}}{\dot{m}_{We}} = \frac{T_\xi - T_d}{T_{Ze} - T_d} = 1 + \frac{T_\xi - T_{Ze}}{T_{Ze} - T_d} = 1 - \frac{T_0 - T_{Ze}}{T_d - T_{Ze}} \frac{T_\xi - T_{Ze}}{T_0 - T_{Ze}} = 1 - \Gamma_{Ze} \nu_\xi \tag{24}$$

can be obtained, where

$$\Gamma_{Ze} = \frac{T_0 - T_{Ze}}{T_d - T_{Ze}} \tag{25}$$

$$\nu_\xi = \frac{T_\xi - T_{Ze}}{T_0 - T_{Ze}}. \tag{26}$$

The comparison of Eqs (22) and (24) gives:

$$\eta \Psi \varrho_L \frac{d\xi}{dt} \frac{\Delta h_v}{T_G - T_d} \left[\frac{1}{\alpha_G} + \frac{\tilde{R}T_G}{\beta \tilde{M}_W s \Delta h_v} + \left(\frac{1}{\lambda_1} + \frac{\tilde{R}T_\xi}{\delta_e \tilde{M}_W s \Delta h_v)}\right)\xi\right] = 1 - \Gamma_{Ze}\nu_\xi. \tag{27}$$

With the integration of Eq. (27) between the limits 0÷t and 0÷ξ, respectively, we obtain:

$$\frac{\eta \Psi \varrho_L \Delta h_v}{T_G - T_d}\left[\left(\frac{1}{\alpha_G} + \frac{\tilde{R}T_G}{\beta \tilde{M}_W s \Delta h_v}\right)\xi + \left(\frac{1}{\lambda_1} + \frac{\tilde{R}T_\xi}{\delta_e \tilde{M}_W s \Delta h_v}\right)\frac{\xi^2}{2}\right] = t - \int_0^t \Gamma_{Ze}\nu_\xi \, dt. \tag{28}$$

Hence, the drying time as a function of the location ξ of the front will be:

$$t = \frac{\eta \Psi \varrho_L \Delta h_v}{T_G - T_d}\left(\frac{1}{\alpha_G} + \frac{\tilde{R}T_G}{\beta \tilde{M}_W s \Delta h_v}\right)\zeta + \frac{\eta \Psi \varrho_L \Delta h_v}{T_G - T_d}\left(\frac{1}{\lambda_1}\frac{\tilde{R}T_\xi}{\delta_e \tilde{M}_W s \Delta h_v}\right)\frac{\xi^2}{2} + \int_0^t \Gamma_{Ze}(t)\nu_\xi \, dt = t_1 + t_2 + t_3. \tag{29}$$

The term $\Gamma_{Ze}(t)$ in Eq. (29) represents value Γ_{Ze} related to ξ=Z which varies in the course of integration, therefore, $\Gamma_{Ze}(t)$ is variable, too, and cannot be factored out and set before the integral.

According to the relationship between thickness of the dried front and the moisture decrease of the material (see: Eq. (11), Eq. (29) can also be expressed by $\Delta\bar{m}_L$ which yields:

$$t = \frac{\Delta h_v}{T_G - T_d}\left(\frac{1}{\alpha_G} + \frac{\tilde{R}T_G}{\beta \tilde{M}_W s \Delta h_v}\right)\Delta\bar{m}_L + \frac{\Delta h_v}{T_G - T_d}\frac{1}{2\eta\Psi\varrho_L}\left(\frac{1}{\lambda_1} + \frac{\tilde{R}T_\xi}{\delta_e \tilde{M}_W s \Delta h_v}\right)\Delta\bar{m}_L^2 + \int_0^t \Gamma_{Ze}(t)\nu_\xi \, dt = t_1 + t_2 + t_3. \tag{30}$$

Total time requirement can be imagined as consisting of three parts, i.e.:

- t_1 the time required to evaporate $\Delta\bar{m}_L$ if only the resistance of the drying gas would exist;
- t_2 the time necessary for evaporation if only the resistance of the dried layer would exist; (t_1+t_2: total time requirement if the

region-interphase (ξ) would always correspond to steady temperature, according to Eq. (20));

- t_3 represents the time requirement due to deviation from steady temperature.

For the last term of Eq. (29) and Eq. (30) resp., the following unequality seems to be evident, i.e.:

$$\Gamma_{0e}\int_0^t v_0\,dt \leq \int_0^t \Gamma_{Ze}(t)\,v_\xi\,dt \leq \Gamma_{Ze}\int_0^t v_Z\,dt\,, \tag{31}$$

where

subscript 0 refers to the drying of the non-moving front; and subscript Z refers to the model of the material of two layers ξ=Z, with a constant thickness.

The unequality Eq. (31) states that in the case of the penetrating front, the time requirement for moisture-decrease $\Delta\bar{m}_L$ surpasses the one considered for moisture evaporation at the surface, but it is lower than the one considered for moisture evaporation from the beginning, through the dry layer ξ=Z, corresponding to $\Delta\bar{m}_L$.

Therefore, in the unequality Eq. (21) the left hand side represents the lower approach, while the right hand side represents the upper approach of the exact value. The solution for the lower approach (2) is:

$$\Gamma_{0e}\int_0^t v_0\,dt = \Gamma_{0e}\frac{L_2}{\varkappa_{II}}\int_0^{Fo_0} v_0\,dFo_0 = \Gamma_{0e}\frac{L_2}{\varkappa_{II}}\frac{1}{Bi_{hm0}}\Omega_0 \tag{32}$$

where

$$\Gamma_{0e} = \frac{T_0 - T_{0e}}{T_d - T_{0e}} \tag{33}$$

and

$$\Omega_0 = f(Fo_0, Bi_{hm0}); \tag{34}$$

the following function is:

$$\Omega = \sum_{k=1}^{k=\infty} 2\frac{\sin^2\delta_k}{\delta_k^2 + \delta_k\sin\delta_k\cos\delta_k}(1 - e^{-\delta^2 k\,Fo_h})\,. \tag{35}$$

The eignvalues δ_k are the roots of the transcendent equation

$$\cot\delta = \frac{\delta}{Bi_{hm}}\,. \tag{36}$$

The set of curves demonstrating the above function (35) is shown in Figure 3.

$$Fo_0 = \frac{\varkappa_{II}\,t}{L^2}\,, \tag{37}$$

$$Bi_{hm0} = \frac{\alpha_G L}{\lambda_{II}}\left(1 + \frac{\beta M_W s\Delta h_v}{\tilde{R}T_G\alpha_G}\right) \tag{38}$$

The solution for the upper approach can be written as:

$$\Gamma_{Ze}\int_0^t v_\xi\,dt = \Gamma_{Ze}\frac{(L-Z)^2}{\varkappa_{II}}\int_0^{Fo} v_Z\,dFo = \Gamma_{Ze}\frac{(L-Z)^2}{\varkappa_{II}}\frac{1}{Bi_{hmZ}}\Omega_Z \tag{39}$$

where

$$\Gamma_{Ze} = \frac{T_0 - T_{Ze}}{T_d - T_{Ze}} \tag{40}$$

and

$$Fo_Z = \frac{\varkappa_{II}\,t}{(L-Z)^2} \tag{41}$$

and furthermore

$$Bi_{hmZ} = \frac{1}{\dfrac{\lambda_{II}}{L\alpha_G} + \dfrac{Z}{L}\dfrac{\lambda_{II}}{\lambda_I}}\left[1 + \frac{s\Delta hv}{\dfrac{\tilde{R}T_G}{\beta\tilde{M}_W} + Z\dfrac{\tilde{R}T_\xi}{\delta_e\tilde{M}_W}}\left(\frac{1}{\alpha_G} + \frac{Z}{\lambda_I}\right)\right] \tag{42}$$

The term $\Omega=f(Fo_Z, Bi_{hmZ})$enclosed in Eq. (39) can also be taken from Figure 3 referred to in (5) as function (35).

Comparison of the recommended solutions

Applicability of the approximation method described in the foregoing is demonstrated by a numerical example. A tray dryer is used to dry wet, granular material. The depth of the bed is 0,08 m. The drying air passes over the frontal surface of the bed which can be regarded as having infinite length, the bed is completely insulated everywhere else. The initial condition is:

The data system of the calculation is:

T_G = 318 K
T_d = 287,1 K
α_G = 10,467 W/m^{2}K
Δh_v = 2388,3 kJ/kg
$\tilde{M}_W$ = 18 kg/knol
β = 0,00811 m/s
c_{pV} = 1,926 kJ/kg K
ϱL = 1000 kg/m^3
L = 0,08 m
x_I = 2,61 x 10^{-7} m^2/s
x_{II} = 4,22 x 10^{-7} m^2/s
λ_I = 0,9304 W/mK
λ_{II} = 2,035 W/mK
Ψ = 0,2 m^3/m^3
T_0 = 291 K
s = 0,015499 bar/K
$\tilde{R}$ = 8,3143 J/molK
δ_e = 2,97 x 10^{-5}m^2/s
η = 1

The drying characteristics of the bed were determined numerically by a computer, as described in the foregoing. The temperature profile of the bed as a function of the location at different instants of time is represented in Figure 4, showingthat for the completely dried portion I the temperature profile can be considered as linear.

Penetration velocity.of the front (proportional to drying velocity) as a function of time is shown in Figure 5.

On the basis of Figure 5 it can be stated that after the initial section, drying velocity decreases. This decrease begins where diffusion resistance of the dried region taking control over the drying phenomenon.

Further on the value of the function $t/\xi=f(\xi)$ was determined by computer aided calculus (exact solution) as well as by lower and upper approximation as has already been described in the foregoing. The so-called linear approach assuming the case $t_3=0$ and denominated regular region in our referred paper (5) - is also plotted in the following Figure 6, showing the results.

On the basis of Figure 6 it can be stated that the obtained approximate solutions yield approximations of satisfactory accuracy for practical calculations. It always depends on the kind of task and the required accuracy of the drying process, which of the approximate solutions seems the most reasonable to be applied. The upper approach shows rather good coincidence with the **exact** solution which can be seen in Figure 6; this fact proves that after the initial instance of time the main resistance of the drying process is the resistance of the already dried region.

Thus, the previously described-numerical computer aided, and approximate analytical, resp.- methods are also suitable to produce the source-function of the network model for the analysis of a real, non-moving bed type drying equipment.

REFERENCES

1. Luikov, A.V.: Heat and Mass Transfer in Capillary-porous Bodies, Pergamon Press, Oxford, 1966.
2. Szentgyorgyi, S.:Szaritasi kezikonyv (Drying Handbook), (Editor: Imre, L.) Muszaki Konyvkiado, Budapest, 1974.
3. Lukov, A.V.: Int. J. Heat Mass Transfer, Vol.18 No 1-A, p. 1-14, (1975)
4. Szentgyorgyi, S.-Molnar K.: Periodica Polytechnica, Mechanical Engineering, 20.1. (1976) 20. 2 (1976)
5. Molnar, K.: Muszaki Tudomany, 52, (1976), 93-111

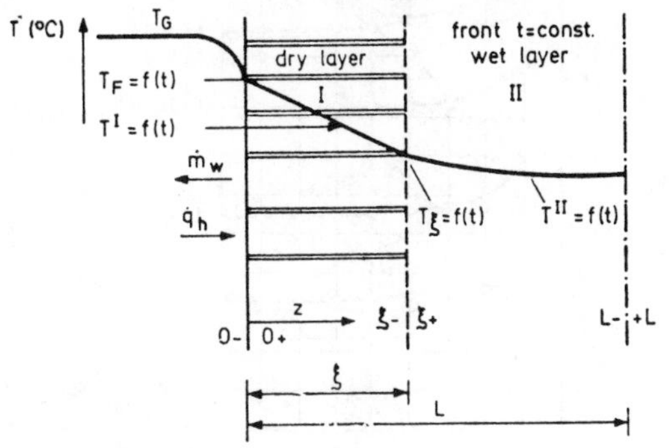

Fig. 1. Examined model

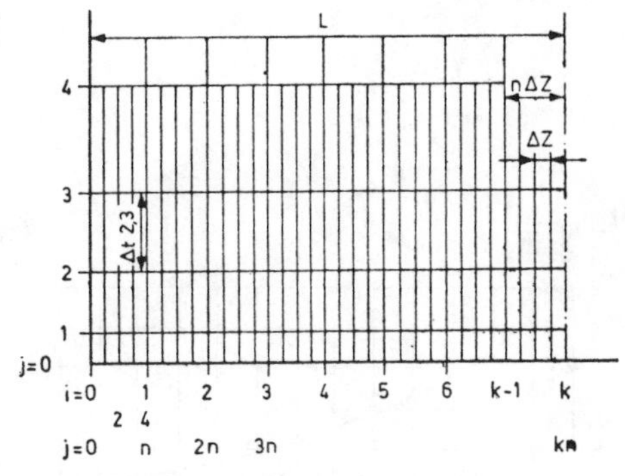

***Fig. 2.* Grid applied for the numerical solution**

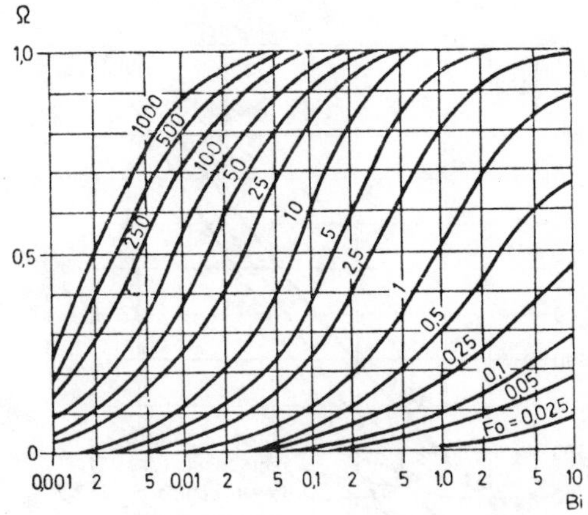

Fig. 3. Ω vs. Bi; Fo

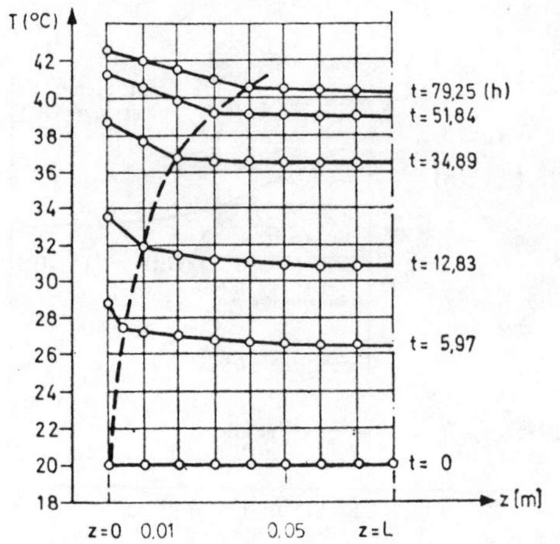

Fig. 4. **Temperature profile of the bed at various instants of time**

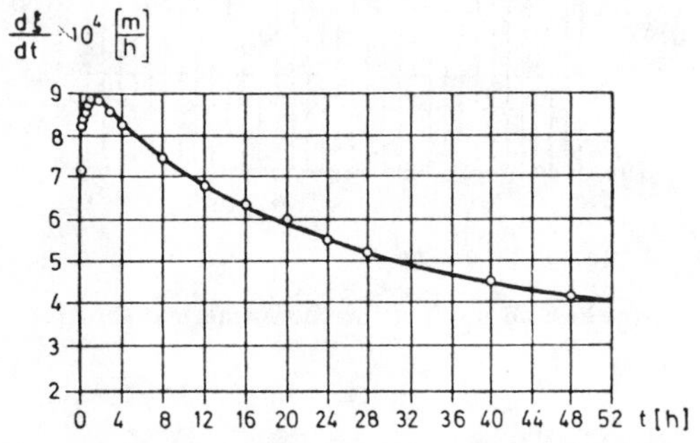

Fig. 5. **Penetration velocity vs. time**

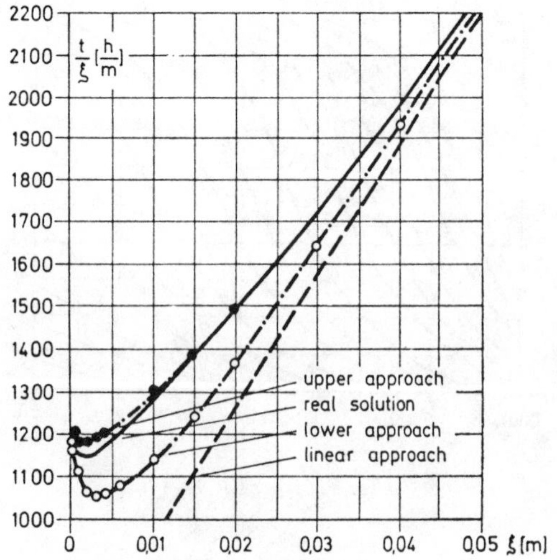

Fig. 6. **Comparison of solutions**

ACKNOWLEDGEMENT

The editor is grateful to Purnima Mujumdar for preparing this typescript.

COMPUTER CALCULATION METHOD OF THE FALLING RATE PERIOD OF DRYING

S. Szentgyorgyi, K. Molnar and M. Orvos
Technical University
Budapest, Hungary

ABSTRACT

Large number of experimental work and computer calculations indicate that during the falling rate period the governing mechanism of the drying follows the evaporation-condensation model.

That means that evaporation zone penetrates from the surface into the deeper region.

An exact computer calculation method is developed for the determination of moisture content, temperatures and the front velocities.

Comparison of the experimental works and computer calculations showed good agreement.

NOMENCLATURE

A	surface area, m^2
Δh_V	evaporation heat, J/lg
L	thickness, m
$\underset{\sim}{M}$	mass, kg
$\tilde{M}_W$	mole weight of moisture, kg/knol
m	mass flux density (drying velocity), kg/m^2s
p	pressure, N/m^2
$\underset{\sim}{q}$	heat flux density, W/m^2
$\tilde{R}$	universal gas constant, 8.3443 J/molK
s	slope of the tension curve, N/Km^2
T	temperature, K
t	time, s
Δt	time step, s
v	velocity, m/s
X	moisture content of dry material, kg/kg
ΔZ	front step
z	co-ordinate, m

Greek letters

α	heat transfer coefficient, W/m^2K
β	mass transfer coefficient, m/s
δ_{eff}	effective diffusion coefficient, m^2/s
κ	thermal diffusivity, m^2/s
λ	thermal conductivity, W/mL
ξ	location co-ordinate, m
ρos	density of the dry material, referred to the volume of the wet material, kg/m^3

SUBSCRIPTS

I	dried region
II	wet region
b	dew point (modified)
cr	critical
F	surface area
G	gas-phase
j	leap of moisture content
0	initial state and value at location z=0, resp
V	vapour-phase
W	moisture
$\underset{\sim}{\xi}$	evaporation front

SUPERSCRIPTS

I	dried region;	$\tilde{\Delta}$	average
II	wet region;	*	equilibrium

INTRODUCTION

Experimental measurement carried out on capillary porous materials have proved for the major part of cases that the density of source of internal evaporation is $\sigma_V=0$ during the drying period of free moisture, i.e. the moisture arrives to the surface of the wet material in the state of liquid phase and evaporates there. This enables to determine drying velocity and drying time in this region by aid of a relatively simple procedure (1). It could be stated that in the region of the so-called "falling rate period of drying" an unstable, intermediate zone will develop first, where local recondensations occur, too; $\sigma_V<0$. Following to this intermediate zone of relatively short duration, the wet material separates into an outer zone of low and an inner zone of higher moisture content which we consider as second falling rate period. Evaporation proceeds at the interphase boundary surface, lying between both zones. Moisture arrives from the inner parts of the material to the interphase surfaceby liquid diffusion and passes after evaporation by vapor diffusion into the main mass of the drier zone. The boundary surface of the wetter zone moves towards the inside of the material as a function of time and the drying process slows down.

There exists a certain group of materials which when submitted to drying, evaporation front penetrates into the deeper lying layers of the drying material, during the falling rate period of drying.

Figure 1 demonstrates a plate of gypsum-pearlite of 80 to 20% by volume where the location of the front at different time values can be well observed by temperature distribution.

From the starting time t_0 of the front on, the relationship between the average moisture content of the material and the location of the front is linear for a while, see (2. 3).

Mathematical model of the linear section

The energy-balance of both zones can be written as (1):

$$\frac{\partial T^{I}}{\partial t} = \kappa_{I} \frac{\partial^2 T^{I}}{\partial z^2} \quad (1)$$

and

$$\frac{\partial T^{II}}{\partial t} = \kappa_{II} \frac{\partial^2 T^{II}}{\partial z^2} \quad (2)$$

according to the denotations of Fig. 2.

Uniqueness conditions

According to Fig. 1, the initial condition is:

$$T^{I}(z,0) = T^{II}(z,0) = T_0 \cong \text{const} \quad (3)$$

The boundary conditions are:

1. In the median (symmetry) plane, at the location z=L

$$\left.\frac{\partial T^{II}}{\partial z}\right|_{z=L} = 0 \quad (4)$$

2. At the evaporation front, at the location z=ξ

$$q|_{z=\xi} = -\lambda_{I} \left.\frac{\partial T^{I}}{\partial z}\right|_{z=\xi^-} = -\lambda_{II} \left.\frac{\partial T^{II}}{\partial z}\right|_{z=\xi^+} + m_W \Delta h_V \quad (5)$$

3. At the surface of the material, at the location z=0

$$q|_{z=0} = \alpha_G (T_G - T_F) = -\lambda_{I} \left.\frac{\partial T^{I}}{\partial z}\right|_{z=0^+} \quad (6)$$

On the basis of (3) the drying velocity can be written as

$$m_W = \frac{Ms}{A}\frac{d\bar{X}}{dt} = \rho_{OS} (\Delta X)_j \frac{d\xi}{dt} \quad (7)$$

where

$$(\Delta X)_j = \bar{X}_{cr} - \bar{X}^*$$

Thus, boundary condition (5) can also be expressed by the penetration velocity of the front, i.e.

$$q|_{z=\xi^-} = \lambda^{I} \left.\frac{\partial T^{I}}{\partial z}\right|_{z=\xi^-} = \lambda^{II} \left.\frac{\partial T^{II}}{\partial z}\right|_{z=\xi^+} + \rho_{OS} (\Delta X)_j \frac{d\xi}{dt} \Delta h_V$$

Considering the uniqueness conditions (3), (4), (6) and (8), the Eqs (1) and (2) cannot be solved analytically. For this reason, we have developed a quite accurate numerical method for the solution of the differential equations. (We developed an approximate calculation, too(2, 3)).

Numerical solution

To obtain the numerical solution, the method of finite differences, a modified variety of the so-called explicit method will be applied (4). The grid, chosen for the numerical solution, is shown in Fig. 3.

Numerical computation, requires an equation-system to satisfy the boundary condition (8) at point ξ_t of the penetrating drying front, so that the location of the front should lie in a location -grid point. The penetration velocity of the front is, however, varying (decelerating), therefore either the distances between the front grid points, or the time steps must be variable. For practical computational reasons we have chosen the time steps Δt to be variable and the front steps ΔZ to be constant.

The number of front steps must correspond to the number of time steps. Accordingly, the grid composed of front steps (auxiliary grid) is not suitable to serve as a location grid at the same time, since in this case no temperature change would be obtained for the wet zone behind the front.

The distances between the points of the location grid (main grid) must be large enough to display the temperature changes within the total wet front, even in case of a relatively poor penetration-rate of the front. The instantaneous location of the above mentioned auxiliary grid, forming one single new grid location will contribute to this main grid. The front grid point divides distribution n between the subsequent and the preceding main grid point into two (generally not equal) portions. Therefore, the derivatives involved in the solution procedure must be expressed in an unequally divided form, too.

The plate of half thickness L was divided into k portions, illustrated in Fig. 3, this grid formed with respect to the locations being the so-called main grid. The width of the applied auxiliary grid was ΔZ and one main grid distance was divided into n portions.

According to the preceding considerations the distribution number of the auxiliary grid is:

$$\frac{L}{\Delta Z} = kn$$

The distribution according to locations, is indicated by the lines i=const., at the maind grid. The time distribution of the chosen grid is not constant (Δt≠const.), changing from time level to time level(the time level is indicated by the lines j=const.) in such a way that with the chosen time step Δt and with the penetration velocity of the front calculated at the former time level, the drying front will just remove to a distance Δξ=ΔZ at the end of the time step, and thus coincide with a location grid point (lying on an auxiliary grid, and from time to time on a main grid).- With the right choice of the distribution number n of the auxiliary grid, the stability of the explicit solution method might also be guaranteed for changing values of Δt.

On the basisof the chosen grid, the number of time steps required for complete drying out (ξ=L) will be kn.

According to the foregoing, the time level j=n also means this auxiliary grid location of order n where the drying front is just located at the given instant of time, therefore further on j will denote the number of the auxiliary grids.

Hence, the chosen grid represents:
- main grid locations: i=0,n,2n,...,kn;
  grid distribution: nΔZ
- auxiliary grid locations j=0,1,2,....,kn;
  grid distribution: ΔZ
- time grid locations: j=0,1,2,...,kn;
  grid distribution: Δt≠const.

and if j=n,2n,...,kn. then j=i.

he difference-formulas of equal location distributions and unequal location distributions required for the numerical calculations, will be recapitulated in (5). With the knowledge of difference-formulas, the differential equations describing the boundary conditions may be transcribed to difference equations.

Difference formulas

Accordint to (1), (2), (4) and (8)

$$\frac{T'_i - T_i}{\Delta t_j} = \kappa_I \frac{T_{i+1} - 2T_i + T_{i-1}}{(n\Delta Z)^2} \tag{9}$$

$$\frac{T'_i - T_i}{\Delta t_j} = \kappa_{II} \frac{T_{i+1} - 2T_i + T_{i-1}}{(n\Delta Z)^2} \tag{10}$$

$$\left.\frac{\partial T^{II}}{\partial z}\right|_{z=L^-} = \frac{T'_k - T'_j}{(kn-j)\Delta Z} = 0 \tag{11}$$

$$\alpha_G(T_G - T_F) = -\lambda_I \frac{T'_j - T'_0}{j\Delta Z} \tag{12}$$

and

$$-\lambda_I \frac{T'_j - T'_0}{j\Delta Z} =$$

$$= -\lambda_{II} \frac{-n[2(i-j)+3n]T'_j + (i+2n-j)^2 T'_{i+1} - (i+n-j)^2 T'_{i+2}}{n(i+n-j)(i+2n-j)\Delta Z} + \frac{s}{d+j\Delta Z e}(T'_j - T_b)\Delta h_V \tag{13}$$

where

$$d = \frac{\tilde{R} T_G}{\beta \tilde{M}_W} \tag{14}$$

$$e = \frac{\tilde{R} T_\xi}{\delta_{eff} \tilde{M}_W} \tag{15}$$

Moreover, the detailed description of the difference formulas required for calculation of the internal points of the dried region I and the wet region II, are included in (5).

At a given instant of time the penetration velocity of the front can be written as:

$$\left(\frac{dZ}{dt}\right)_j$$

This velocity is changing as a function of time. In the course of the numerical calculations this velocity is regarded as constant for each elementary time step.

Let the penetration velocity of the front be:

$$\left(\frac{dZ}{dt}\right)_{j-1}$$

at the time level j-1 and assume the penetration velocity of the front not to change until the time level j, i.e. until reaching the next auxiliary grid point. Hence, time Δt_j required to cover the auxiliary grid distribution ΔZ, can be determined, according to Fig. 4 as:

$$\Delta t_j = \Delta Z \frac{1}{\left(\frac{dZ}{dt}\right)_{j-1}} \tag{16}$$

The flow-chart of computation can be found in (5).

Comparison of measured and calculated results

The studied model has been a plate of gypsum-pearlite of 80 to 20 percents per volume, as already mentioned above its size being 0.1 x 0.1 x 0.03 m.

The plate was dried by means of warm air in a drying channel, while its weight and temperature distribution according to thickness has been measured as a function of time.

The state characteristics of drying air are:

$$T_G = 354 \text{ K}$$

$$P_{VG} = 3\,166.57 \text{ N/m}^2$$

$$v_G = 1.96 \text{ m/s}$$

The state characteristics of the wet material are:

$$T_0 = 316.5 \text{ K}$$

$$\tilde{X}_0 = 0.288 \text{ kg/kg}$$

$$\rho_{os} = 780 \text{ kg/m}^3$$

Temperature distribution obtained from the measured results is demonstrated in Fig. 5. Calculations were carried out by using the following data

α_G =26.74 W/m^{2}K

β =0.0319 m/s

λ_I =0.2758 W/mK

λ_{II}=0.78 W/mK

δ_{eff}=2.63x10^{-5}m^2/s

κ_I = 2.44x10^{-7}m^2/s

κ_{II}=7x10^{-7}m^2/s

R =8.3143 J/molK

s =394.3 N/m^{2}K

L =0.015 m

T_b =326.16 K

Δh_V=2375.172 kJ/kg

M_w=18 kg/kmol

The transport and material characteristics have been determined by measurements. The results of the calculations are contained in Fig. 5, too.

It is evident from Figure 5 that the calculations methods agree well with the measurement results.

References

1. Lukov, A.V.:Heat and Mass Transfer in Capillary -Porous Bodies, Pergamon Press, Oxford, 1966.

2. Szentgyorgyi, S. Molnar, K.: Calculation of Drying Parameters for the Penetrating Evaporation Front, Proceedings of the First International Symposium on Drying, Science Press, Montreal, 1978.

3. Szentgyorgyi, S Molnar, K Orvos, M.: Modelling and Calculation of the Drying Process of Capillary Porous Materials, Second Internatio-

nal Symposium on Drying, 1980

4. Carnahan, B Luther, H.A. Wilkes, J.D.: Applied Numerical Methods, John Wiley Incs., New York, 1969.

5. Szentgyorgyi, S.-Molnar, K.: Per. Polytechn 20. 1-2, (1976).

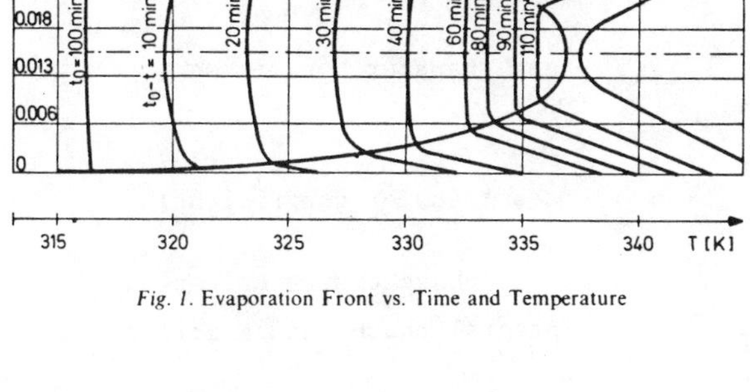

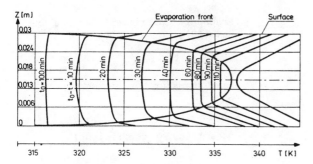

Fig. 1. Evaporation Front vs. Time and Temperature

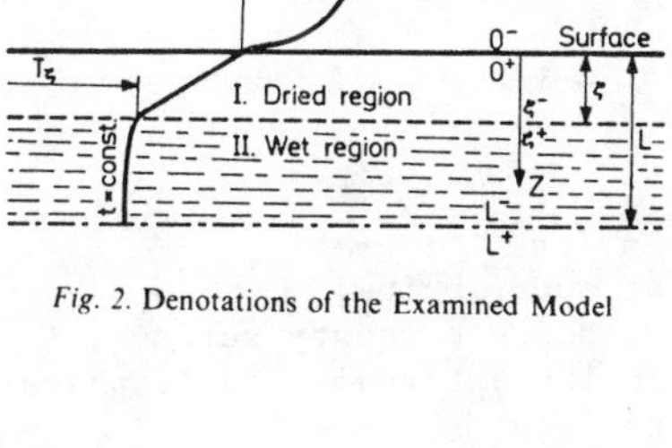

Fig. 2. Denotations of the Examined Model

ACKNOWLEDGEMENT

The Editor is grateful to Purnima Mujumdar for preparing this typescript.

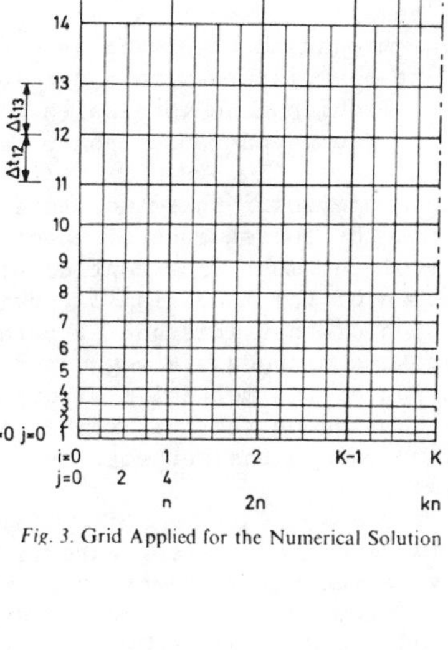

Fig. 3. Grid Applied for the Numerical Solution

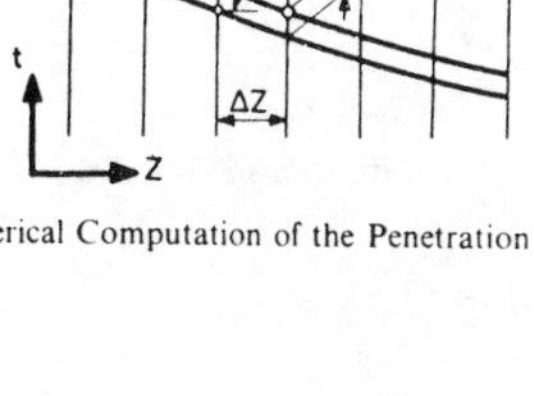

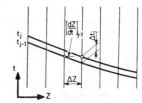

Fig. 4. Numerical Computation of the Penetration of the Front

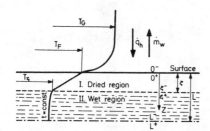

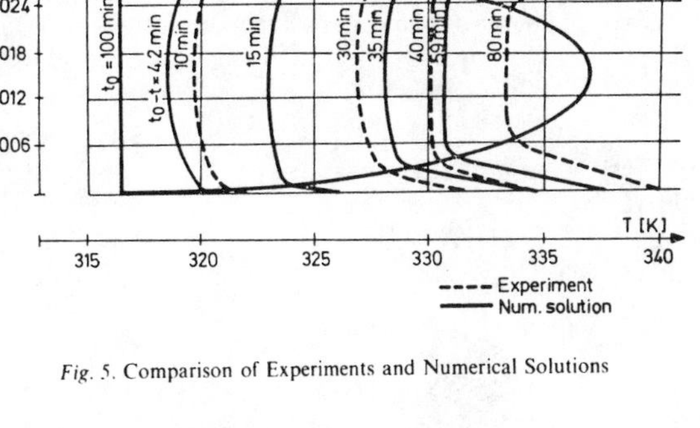

Fig. 5. Comparison of Experiments and Numerical Solutions

METHOD FOR INVESTIGATING THE MECHANISM OF DRYING WET SUBSTANCES

S. Szentgyorgyi
Technical University
Budapest, Hungary

ABSTRACT

For solving the problems of drying, among other factors, the mechanism of drying within the wet substance and its mathematical model have to be known. The mechanism of drying is considered as the movement of moisture within the substance. In the case of liquid phase wetness this means the movement of liquid and vapour phase moisture.

The presence of vapour phase moisture within the substance is the consequence of internal evaporation. One of the most important details in the investigation of the mechanism of drying is the clarification of this internal evaporation. In this paper some methods are suggested for the investigation of the mechanism of drying. These methods are based on the measurement of temperature distribution and on the mathematical model of batchwise drying.

A method is described for the experimental study of the drying mechanism of wet substances. This approach permits to determine the phase change criterion during drying. Knowledge of the phase change criterion provides information on the rate of evaporation within the substance but it is also required for solving some problems of drying by means of calculations.

NOMENCLATURE

b	$\frac{m}{lt^2T}$	-factor defined by Eq. (20);
c	$\frac{l^2}{t^2T}$	-specific heat of the wet substance;
h	l	-length of the pole;
j_m	$\frac{m}{l^2t}$	-mass flow density;
j_q	$\frac{m}{t^3}$	-heat flow density;
m	m	-mass;
p	$\frac{m}{lt^2}$	-partial pressure;
r	$\frac{l^2}{t^2}$	-latent heat;
r	l	-position vector;
t	T	-temperature;
t^0	T	-pole temperature coordinate;
V	l^3	-volume;
z	l	-coordinate along layer thickness;
Z	l	-half-thickness of the plane plate;
X	-	-moisture per unit dry mass
α_m	$\frac{m}{t^3T}$	÷transport coefficient defined by Eq. (22);
α_q	$\frac{m}{t^3T}$	-heat transfer coefficient;
β	$\frac{t}{l}$	-mass transfer coefficient;
ε	-	-local phase change criterion;
ε_f	-	-integral phase change criterion;
λ_q	$\frac{lm}{t^3T}$	-thermal conductivity;
ρ_o	$\frac{m}{l^3}$	-mass per unit volume of wet substance;
σ	$\frac{m}{l^3t}$	-source density of mass;
τ	t	-time

SUBSCRIPTS

e	-at equilibrium;
f	-on phase boundary surface;
$\|f^+$	-on the external side of phase boundary surface;
$\|f^-$	-on the internal side of phase boundary surface;
G	-refers to main mass of drying gas or to decelerating period of drying;
kr	-critical value;
L	-liquid phase moisture;

n -at wet bulb temperature, or in decelerating period of drying;
s -dry substance;
V -vapour phase moisture;
0 -initial value;

SUPERSCRIPT

- -average value

Mathematical model of batch drying

According to Likov (1), the mathematical model of batch drying can be established as follows:

The material balance equation for liquid moisture in the wet substance is

$$\frac{\partial \rho_{0L}}{\partial \tau} = -\operatorname{div} \mathbf{j}_{mL} + \sigma_L .$$

In the balance equation of the vapour phase moisture the left side is in general negligible, thus

$$0 = -\operatorname{div} \mathbf{j}_{mV} + \sigma_V . \tag{1}$$

According to the mass conservation law:

$$\sigma_L + \sigma_V = 0$$

thus the total material balance of the moisture is:

$$\frac{\partial \rho_{0L}}{\partial \tau} = -\operatorname{div} \mathbf{j}_m$$

where

$$\mathbf{j}_m = \mathbf{j}_{mL} + \mathbf{j}_{mV} . \tag{2}$$

Introducing the wetness related to the dry material, i.e.

$$X = \frac{m_L}{m_s} = \frac{\rho_{0L}}{\rho_{0s}}$$

the total material balance becomes

$$\rho_{0s} \frac{\partial X}{\partial \tau} = \operatorname{div} \mathbf{j}_m \tag{3}$$

The conditions of unambiguity are, with respect to time:

$$X|\mathbf{r}, \tau_0| = X_0|\mathbf{r}|$$

and with respect to boundary conditions in Fig. 1:

$$\mathbf{j}_{mV}|_{f^-} + \mathbf{j}_{mL}|_{f^-} = \mathbf{j}_{mV}|_{f^+} = \mathbf{j}_{mf} . \tag{4}$$

The energy balance is

$$\rho_{0s} c \frac{\partial t}{\partial \tau} = -\operatorname{div} \mathbf{j}_q - r\sigma_V \tag{5}$$

where the time condition is

$$t|\mathbf{r}, \tau_0| = t_0|\mathbf{r}|$$

and the boundary condition is

$$\mathbf{j}_q|_{f^+} = \mathbf{j}_q|_{f^-} - r\mathbf{j}_{mL}|_{f^-} . \tag{6}$$

Let us introduce the following expression using Eqs. (2) and (3):

$$\varepsilon = \frac{\operatorname{div} \mathbf{j}_{mV}}{\operatorname{div} \mathbf{j}_m} = \frac{\sigma_V}{-\rho_{0s} \dfrac{\partial X}{\partial \tau}} \tag{7}$$

where ε is the so-called local phase change criterion.

The energy balance and the phase change criterion yield:

$$\rho_{0s} c \frac{\partial t}{\partial \tau} = -\operatorname{div} \mathbf{j}_q + r\varepsilon\rho_{0s} \frac{\partial X}{\partial \tau} .$$

The boundary condition, using Eq. (2) is:

$$\mathbf{j}_q|_{f^+} = \mathbf{j}_q|_{f^-} - r/\mathbf{j}_{mf} - \mathbf{j}_{mV}|_{f^-}/ .$$

For the sake of simplicity, let us investigate the boundary condition in the case of an infinite plane plate subject to drying on both sides (Fig. 2):

The mass flow density can point only in direction Z, therefore there is no need to supply the vector sign, but the mass flow is regarded as positive if it proceeds in the positive direction of the Z-axis, i.e, outwards from the material.

For reasons of symmetry, at the half-thickness of the plane, i.e. at z=0

$$j_m|_0 = j_{mV}|_0 = 0$$

thus

$$j_{mV}|_{f^-} = \int_0^Z \operatorname{div} \mathbf{j}_{mV} dz = \int_0^Z \sigma_V dz = z\sigma_V$$

further on

$$j_{mf} = \int_0^Z \operatorname{div} \mathbf{j}_m dz = -\rho_{0s} \int_0^Z \frac{\partial X}{\partial \tau} dz = -\rho_{0s} Z \frac{d\bar{X}}{d\tau}.$$

In the above equations $\bar{\delta}_V$ and $\bar{x}$ are the integral mean values of the variables in question. Upon introducing the integral phase change criterion,

$$\varepsilon_f = \frac{j_{mV}|_{f^-}}{j_{mf}} = \frac{j_{mV}|_{f^-}}{j_{mV}|_{f^-} + j_{mL}|_{f^-}} = \frac{\bar{\sigma}_V}{-\rho_{0s}\frac{d\bar{X}}{d\tau}} \quad (9)$$

the boundary condition equation can be written in the following form:

$$j_q|_{f^+} = j_q|_{f^-} - r(1-\varepsilon_f) j_{mf}. \quad (10)$$

Source densities and phase change criteria

The condition for solving differential equation systems (3) and (5) or (8) is the knowledge of vapour source density σ_V, local phase change criterion ε or of integral phase criterion ε_f.

The physical meanings of these parameters are following: σ_V: amount of vapour formed in unit volume and time at one point of the wet substance. As $\sigma_V = -\sigma_L$, the vapour is formed by the evaporation of the liquid substance. Physically ε means (according to Eq. (7)) the ratio of the rates of internal evaporation at one point to the decrease of relative wetness.

The integral phase change criterion ε_f is according to Eq. (9) the ratio of mass flow densities of vapour phase on the interface of substance and drying gas to that of total moisture.

Obviously, the functions mentioned are not simple material characteristics, but variables strictly dependent on the rules of moisture movement within the wet substance during drying. Today comprehensive knowledge of these rules is still missing, therefore differential equation systems (3) and (5) or (8) have solutions only with assumed-and in general trivial-functions or with irreal boundary conditions (2, 3, 4).

Without the knowledge of the moisture transport process of and the mechanism of internal evaporation, we cannot even decide about the methods to be applied for the solution of these differential equation systems. If namely, $\sigma_{\bar{V}}$ or ε is constant for a longer period, or a continuous and differentiable function of time and place, then analytical methods can be used for calculating the drying free moisture (1, 2, 5), while in cases involving also the decelerating period, numerical methods (3, 6, 7, 8) developed from numerical solution methods of differential equations of thermal conduction under so-called boundary conditions of third order are available. If, however, in a period of drying an internal evaporation front proceeds inwards into the substance, this means a discontinuity to exist in the function σ_V or ε, and the place of this discontinuity proceeds into the substance. In this case the above methods cannot be used; only numerical methods developed for the moving boundary condition place can be applied (9, 10, 11, 12).

Thus it can be stated that without the knowledge of the mathematical model relating to concrete cases of moisture transfer processes, the differential equation system expressing the heat and mass balance of drying cannot be solved for technically important cases. Therefore, the possibilities for the measurement of σ_V, ε and ε_f revealing the main features of moisture transfer will be studied here.

Measurement of σ_V and ε

Restricting us to the plane plate mentioned above, it can be written:

$$j_q = -\lambda_q \frac{\partial t}{\partial z}$$

thus from Eq. (5)

$$\sigma_V \delta_V = \left(\frac{\lambda_q}{\rho_{0s}c}\frac{\partial^2 t}{\partial z^2} - \frac{\partial t}{\partial z}\right)\frac{\rho_{0s}c}{r}, \quad (11)$$

consequently, σ_V can be measured by determining the temperature distribution in time and place. To the measurement of the local phase change criterion ε, as it is seen from Eq. (7), besides σ_V, also the knowledge of the local moisture distribution in time and place is needed.

Measurement of $_f$

The heat and mass flow densities in Eq. (10) written in the usual way:

$$j_q|_{f^+} = \alpha_q(t_f - t_G) \tag{12}$$

$$j_q|_{f^-} = -\lambda_q \left.\frac{\partial t}{\partial z}\right|_{f^-} \tag{13}$$

and

$$j_{mf} = \beta(p_{Vf} - p_{VG}). \tag{14}$$

transform Eq. (10) to:

$$\alpha_q(t_f - t_G) = -\lambda_q \left.\frac{\partial t}{\partial z}\right|_{f^-} - r(1-\varepsilon_f)\beta(p_{Vf} - p_{VG}) \tag{15}$$

The energy balance, Eq. (5) substituting Eq. 1 and the Gaussian theorem, becomes:

$$\rho_{0s} c \frac{d}{d\tau}\int_V t\,dV = -\int_A (\mathbf{j}_q|_{f^-} + r\mathbf{j}_{mV}|_{f^-})d\mathbf{A}$$

i.e.

$$\rho_{0s} c A \int_0^Z \frac{\partial t}{\partial \tau} dz = (-j_q|_{f^-} - r j_{mV}|_{f^-})A. \tag{16}$$

where flow dinsities j are the surface integral mean values. As from boundary conditions (4) and (6)

$$-j_q|_{f^-} - r j_{mV}|_{f^-} = -j_q|_{f^+} - r j_{mf}$$

from this and Eq. (16) using also Eqs (12) and (14)

$$\alpha_q(t_f - t_G) = -\rho_{0s} c \int_0^z \frac{\partial t}{\partial \tau} dz - r\beta(p_{Vf} - p_{VG}).$$

During drying the temperature distribution of the substance tends to the steady state, where δt/δτ=0. Let us denote the characteristics of the drying gas in this state by subscript e.

The above equation for this state becomes

$$\alpha_q(t_{fe} - t_G) = -r\beta(p_{Ve} - p_{VG}) \tag{17}$$

In this steady state Eq. (15) can be given as:

$$\alpha_q(t_{fe} - t_G) = -\lambda_q \left.\frac{\partial t_e}{\partial z}\right|_{f^-} - r\beta(1-\varepsilon_f)(p_{Vfe} - p_{VG}) \tag{18}$$

The difference of Eqs (15) and (18) is:

$$\alpha_q(t_f - t_{fe}) = -\lambda_q \left| \left.\frac{\partial t}{\partial z}\right|_{f^-} - \left.\frac{\partial t_e}{\partial z}\right|_{f^-} \right| - r\beta(1-\varepsilon_f)(p_{Vf} - p_{Vfe}) \tag{19}$$

Introducing the notation:

$$b = \frac{p_{Vf} - p_{Vfe}}{t_f - t_{fe}} \tag{20}$$

transforms Eq. (19) into

$$\alpha_q(t_f - t_{fe}) = -\lambda_q\left(\left.\frac{\partial t}{\partial z}\right|_{f^-} - \left.\frac{\partial t_e}{\partial z}\right|_{f^-}\right) - r\beta b(1-\varepsilon_f)(t_f - t_{fe}) \tag{21}$$

If α_m=rβb, then from Eq. (21) (22)

$$h = \frac{t_f - t_{fe}}{-\left(\left.\frac{\partial t}{\partial z}\right|_{f^-} - \left.\frac{\partial t_e}{\partial z}\right|_{f^-}\right)} = \frac{\lambda_q}{\alpha_q + (1-\varepsilon_f)\alpha_m} \tag{23}$$

"h", a quantity having length dimension is called pole. From Eq. (23):

$$t^0 = t_{fe} + h\left.\frac{\partial t_e}{\partial z}\right|_{f^-} = t_f + h\left.\frac{\partial t}{\partial z}\right|_{f^-} \tag{24}$$

which means that the surface tangents of the temperature distribution for the drying substance intersect at a distance h from the surface, and at a temperature ^{0}t.

From Eq. (16), substituting Eqs (13), (9) and (14)

$$-\rho_{0s} c \int_0^Z \frac{\partial t}{\partial \tau} dz = -\lambda_q \left.\frac{\partial t}{\partial z}\right|_{f^-} + r\beta\varepsilon_f(p_{Vf} - p_{VG}).$$

For the steady state, considering Eq. (17):

$$\left.\frac{\partial t_e}{\partial z}\right|_{f^-} = \varepsilon_f \frac{\alpha_q}{\lambda_q}(t_G - t_{fe}) \tag{25}$$

Substituting this into Eq. (24):

$$t^{\circ}=t_{fe}+h\varepsilon_f\frac{\alpha_q}{\lambda_q}(t_G-t_{fe}). \tag{26}$$

Drying period of free moisture

In this period the equilibrium pair t_{fe}-Pv_{fe} lies on the saturation curve of the vapour ($Pv^n(t)$). In this case $t_{fe}=t_n$, $Pv^{fe}=Pv^n$, where t_n is the wet bulb temperature. According to Eq. (17) the pairs t_n, Pv^n are cut out of the saturation curve by the straight line with the slope:

$$\frac{p_{V^n}-p_{V^G}}{t_n-t_G}=-\frac{\alpha_q}{\beta}. \tag{27}$$

and crossing point G with coordinates t_G, Pv^G (Fig. 3).

The expression according to Eq. (20)

$$b_n=\frac{p_{Vf}-p_{V^n}}{t_f-t_n} \tag{28}$$

is here the secant of the saturation curve between points f and n.

In the period tending to the wet bulb temperature:

$$\alpha_{mn}=r\beta b_n.$$

or

$$h_n=\frac{t_f-t_n}{-\left(\left.\frac{\partial t}{\partial z}\right|_{f^-}-\left.\frac{\partial t_e}{\partial z}\right|_{f^-}\right)}=\frac{\lambda_q}{\alpha_q+(1-\varepsilon_f)\alpha_{mn}} \tag{29}$$

and

$$t^{\circ}=t_n+h_n\varepsilon_f\frac{\alpha_q}{\lambda_q}(t_G-t_n) \tag{30}$$

From Eq. (29):

$$\varepsilon_f=\frac{\alpha_q+\alpha_{mn}}{\alpha_{mn}}-\frac{\lambda_q}{h_n\alpha_{mn}}$$

hence

$$t^{\circ}=t_n+\frac{\alpha_q+\alpha_{mn}}{\lambda_q}\frac{\alpha_q}{\alpha_{mn}}(t_G-t_n)h_n-\frac{\alpha_q}{\alpha_{mn}}(t_G-t_n) \tag{31}$$

Thus there is a linear correlation between the temperature ^0t at the intersection and pole distance h while between h_n and $_f$ there is not, namely from (30), substituting (29):

$$t^{\circ}=t_n+\frac{\varepsilon_f\alpha_q}{\alpha_q+(1-\varepsilon_f)\alpha_{mn}}(t_G-t_n). \tag{32}$$

The correlations are shown in Fig. 4. From Eqs (32) and (28) it is apparent that

$$\text{if } \varepsilon_f=0, \text{ then } t^{\circ}=t_n \text{ and } h_n=\frac{\alpha_q}{\alpha_q+\alpha_{mn}};$$

$$\text{and if } \varepsilon_f=1, \text{ then } t^{\circ}=t_G \text{ and } h_n=\frac{\lambda_q}{\alpha_q}.$$

Eq. (24) relating to the pole in the drying period of the free moisture is:

$$t^{\circ}=t_f+h_n\left.\frac{\partial t}{\partial z}\right|_{f^-}. \tag{33}$$

In Fig. 4, the pole vs. temperature distribution has been plotted for the case $\varepsilon_f=0.4$, as an example. As it is known (1, 2), in the case $\varepsilon_f\neq 0$ the surface temperature t_f tends to t_n, but the temperature will not be constant in various layers of the wet substance.

The drying of free moisture comes to an end at a critical value of the substance wetness. After that, the so-called decelerating period follows.

Decelerating period of drying

After the free moisture is consumed, the temperatures tend to the drying gas temperature t_G, thus $t_{fe}=t_G$, and $Pv^{fe}=Pv^G$; in this case, according to Eq. (25)

$$\left.\frac{\partial t_e}{\partial z}\right|_{f^-}=0$$

and (20) can be written as

$$b_G = \frac{p_{Vf} - p_{VG}}{t_f - t_G} \tag{34}$$

thus

$$\alpha_{mG} = r\beta b_G$$

and

$$h_G = \frac{t_f - t_G}{-\left.\dfrac{\partial t}{\partial z}\right|_{f^-}} = \frac{\lambda_q}{\alpha_q + (1-\varepsilon_f)\alpha_{mG}}. \tag{35}$$

According to Eq. (26), in every case

$$t^0 = t_G$$

while if

$$\varepsilon_f = 0, \text{ then } t^0 = t_G \text{ and } h_G = \frac{\lambda_q}{\alpha_q + \alpha_{mG}}$$

$$\varepsilon_f = 1, \qquad t^0 = t_G \text{ and } h_G = \frac{\lambda_q}{\alpha_q}$$

The value of b_G defined in Eq. (34) varies vary much during drying, as the pair t_f-Pv^f depends very much on the substance wetness in the decelerating period.

Anyway, its value is known at the beginning of the decelerating period:

$$b_{Gkr} = \frac{p_{Vn} - p_{VG}}{t_n - t_G} = \frac{\alpha_q}{r\beta}$$

i.e.

$$\alpha_{mGkr} = -\alpha_q.$$

This means that from Eq. (35):

$$h_{Gkr} = \frac{\lambda_q}{(\varepsilon_{fkr})_G \alpha_q} = \frac{t_G - t_n}{\left.\dfrac{\partial t}{\partial z}\right|_{f^-}}$$

i.e. at the beginning of the decelerating period:

$$\left.\frac{\partial t}{\partial z}\right|_{f^-} = \frac{t_G - t_n}{h_{Gkr}} = (\varepsilon_{fkr})_G \frac{\alpha_q}{\lambda_q}(t_G - t_n).$$

At the same time, at the end of the drying of free moisture it can be written from Eqs (33) and (30):

$$\left.\frac{\partial t_e}{\partial z}\right|_{f^-} = \frac{t^0 - t_n}{h_{nkr}} = (\varepsilon_{fkr})_n \frac{\alpha_q}{\lambda_q}(t_G - t_n).$$

From the above equations the following conclusions can be drawn:

If at the end of the drying of free moisture, upon transition to the decelerating period, the surface tangent of the temperature profile changes suddenly, then this is due to the sudden change of ε_{fkr}.

If ε_{fkr} were to remain unchanged, the surface tangent would not change either; then extension of the straight line $t_n t^0$ would cut off the initial pole of the decelerating period at t_G.

As b_G and, consequently α_{mG} too, much change in the decelerating period, it is apparent from Eq. (35) that the pole distance h_G only remains constant during drying if $\varepsilon_f = 1$.

REFERENCES

1. Luikov, A.V.: Heat and Mass Transfer in Capillary-Porous Bodies. Pergamon Press, Oxford 1966.
2. Szentgyorgyi, S: Handbook of Drying*(Ed. Laszlo Imre), Muszaki Konyvkiado, Budapest 1974.
3. Szentgyorgyi, S.-Molnar, K.: Gep, 26, (3), 95 1974.
4. Michailov, M.D.:Int. J. Heat Mass Transfer, Vol. 18, 797, 1975.
5. Szentgyorgyi, S.: Acta Technica Acad. Sci. Hung., 71, 407, 1971.
6. Molnar, K.: Handbook of Drying*(Ed. Laszlo Imre) Muszaki Konyvkiado, Budapest, 1974.
7. Bimbenet, K.J.-Depevre, D.-Le Maguer, M.: Chimie et Industrie-Génie Chimique, 104, No. 15 (Sept). 1971.
8. Bonacina, C.-Comini, G.: III. Conference on Drying, Budapest, 1971, Section A, No. 21.
9. Hill, I.E.-Sunderland, I.E.: Int. J. Heat and Mass Transfer, 14, 625, 1971.
10. Szentgyorgyi, S.-Molnar, K.: Periodica Polytechnica, Mech. Engng. 20, NO. 1, 48, no. 2, 98, 1976.

11. Szentgyorgyi, S.-Molnar, K.: Proceedings of the First Int. Symposium on Drying, Science Press, Montreal, 1978.
12. Szentgyorgyi, S.-Molnar, K.: Acta Technica Acad. Sci. Hung., 79, 3 1979. (in press).

* In Hungarian.

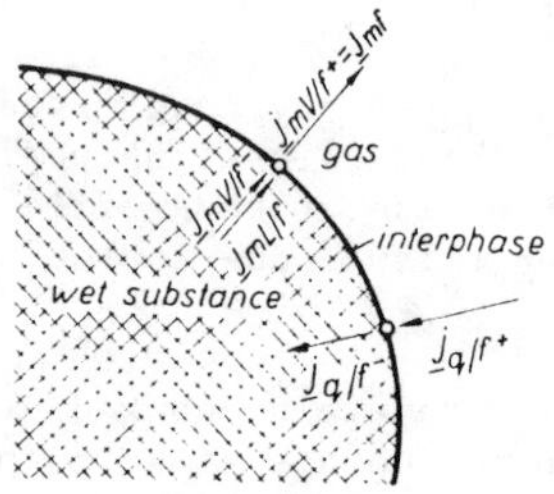

Fig. 1 Illustration to boundary condition equations (4) and (6)

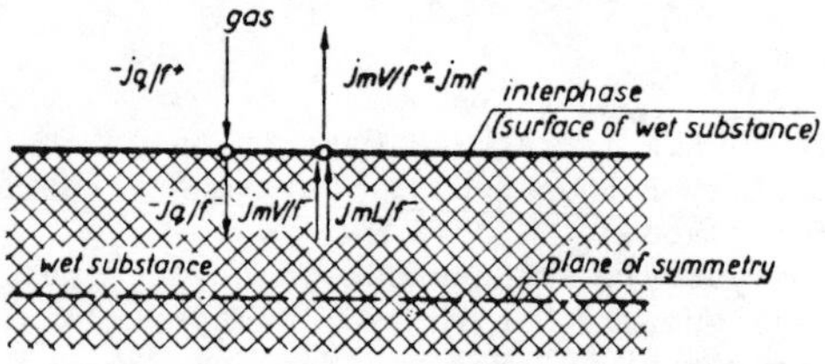

Fig. 2 Illustration to eqs (9) and (10)

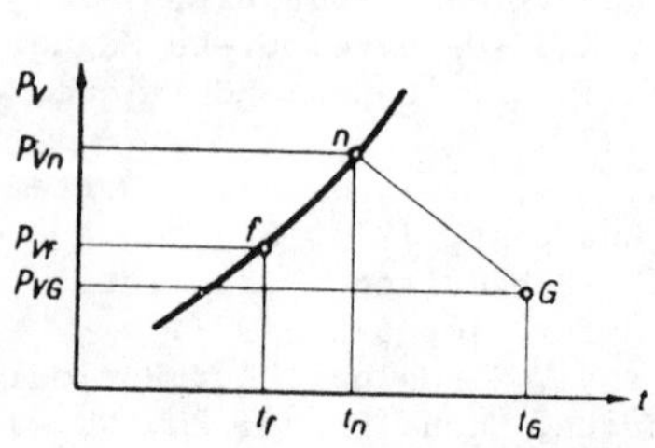

Fig. 3 Illustration to Eqs (27) and (28)

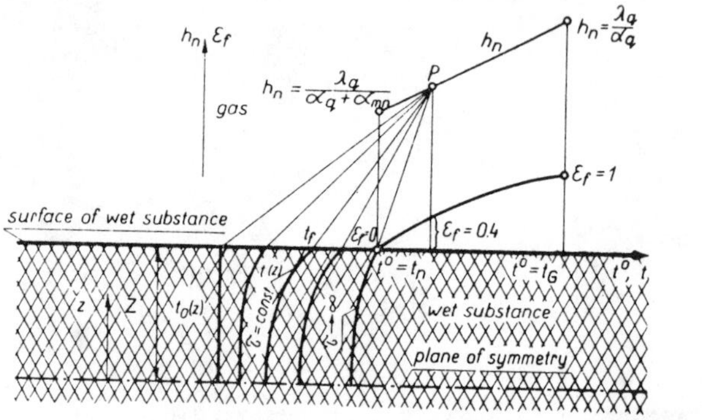

Fig. 4 Relationship between the pole coordinate and the integral phase change criterion for drying of free moisture

ACKNOWLEDGEMENT

The Editor is grateful to Purnima Mujumdar for preparing this typescript.

PROGRESS IN FREEZE DRYING

H. Çeliker
B. Kısakürek

Department of Chemical Engineering
Middle East Technical University
Ankara, Turkey

ABSTRACT

This paper presents a review of the works done in the field of mathematical modeling and experimentation of the freeze drying process. The aim is to establish the fundamental works and researchers in this field together with their proposed experimentation and mathematical modeling techniques.

The development of the experimentation and theory of the freeze drying process is given by subdividing the fundamental works in this field into three main sections. In the first section, the early development of the process is summarized. The second part covers the fundamental studies on effective experimental parameters and the third section gives the various mathematical models developed in this field.

In each fundamental work that is mentioned;

i. The product that is investigated
ii. The mode of heat transfer that is applied
iii. The solution technique for the prevailing expressions,

are emphasized. In addition, general transport equations that are widely used by several researchers are mentioned.

INTRODUCTION

Freeze drying is an ideal method for drying of several food products, medical and pharmaceutical products. In this way, the desirable functional and structural characteristics are largely maintained. The shelf life of drugs has been increased considerably by freeze-drying. Bacterial growth and enzyme action in aqueous pharmaceutical preparations can be greatly reduced or eliminated by this method. Freeze drying can retard the oxidation of biological substances. Proteins are not denatured during freeze-drying and heat sensitive materials retain their viability after being freeze dried. Due to these superior aspects of the method, the process attracted the attention of several researchers and many experimental and theoritical studies are done trying to investigate and simulate the mechanism of freeze drying.

The widely accepted mechanism for freeze drying is the uniformly retreating ice front model. According to this model, when a slab of frozen material is dried an outer dried region surrounds a central frozen core. The heat necessary for sublimation could be transferred to the ice-front in several different ways and the water vapor generated is transferred out across the already dried layer.

The modes of heat transfer from the heating medium to the drying sample could be analyzed as following:

i. Radiant heat transfer-the case where the radiant energy could penetrate the sample. In this way the drying times are considerably reduced and a more uniform heat supply is achieved.
ii. Radiant heat transfer-the case where the radiant energy cannot penetrate the sample.
iii. Conduction heat transfer-when the sample is kept directly in touch with the heating plates.
iv. Heat transfer by convection-in this way, the drying times are reduced.
v. Suitable combinations of the above listed modes of heat transfer.

The mode of heat transfer within the sample could be either through the dried layer or the frozen layer.

The type of flow within the dried layer could be in different forms. The dried part can be treated as a bundle of capillaries due to its porous structure. The drying rates are determined by the interaction of combined mechanisms of heat flow, hydrodynamic vapor flow and diffusion. The flow regimes through these bundle of capillary tubes could be characterized by the ratio of the mean free path to the diameter of capillary which is known as the Knudsen number. At high pressures continious flow regime exists for Kn less than 0.001. If pressure is sufficiently lowered free molecule flow regime will exist at Kn greater than two. Between these two extremes the so called slip flow regime exists.

EARLY DEVELOPMENT

Up to World War II. freeze drying had been regarded as a scientific engagement and large scale application had not practiced. The importance of the process was realized in 1942 when Greaves produced freeze dried blood plasma for the army. It was after this time that several patents are followed for freeze drying of food products. Later on Flosdorf (5,6) realized the fact that freeze drying could be used for food products and other bulky materials. He has also given a good review of the early development of the process.

W.H. Zamzow and W.R. Marshall, Jr. (22) made a preliminary experimental study of freeze drying with infrared radiation which can penetrate the sample. The variables that are investigated in drying time measurements were the cake thickness, initial moisture content and energy concentration. They concluded that drying times are considerably reduced because of the penetration of radiation into the sample. In addition to these Zamzow and Marshall had given an excellent review of the work done in the field of freeze drying by infrared radiation.

Harper and Tappel (8) and Ginette (7) developed theories representing the heat and vapor transport in simple terms by making use of quasi-steady state models. Harper and Tappel were the first to synthesize the process of freeze drying thoroughly. In their paper they have also discussed specific processing conditions for specific foods. The analytically obtained equation for the total drying time by Harper and Tappel is as such:

$$\theta = \frac{\rho_m (X_o - X_e)}{a\ P(P_m + b)} \frac{L_o^2}{2} \qquad (1)$$

where; θ=drying time; X_o=initial water content; X_e=equilibrium water content; L_o=total thickness of sample; P_m=arithmetic avg. pressure and ΔP=pressure drop.

FUNDAMENTAL STUDIES ON EFFECTIVE EXPERIMENTAL PARAMETERS

The possibility that freeze drying rate could be increased by varying the vacuum pressure was proven when Ehlers et.al. and Harper (9) measured the heat transfer coefficients in freeze dried foods at different vacuum pressures. Samples of freeze-dried beef, apples and peaches were investigated. He has outlined a theory to show the effect of pressure on permeabilities and thermal conductivities of gases in porous solids in the slip flow and free molecule flow regimes. The mean pore diameters that are calculated from the permeability data were in agreement with the microscopic observations. Thermal conductivities followed the predicted behavior of a constant value at high pressures, with a gradual decrease to another constant value at very low pressures.

Later on, Harper and El-Sahrigi (10) made similar measurements as Harper did but this time by making use of gases such as Freon 12, CO_2, N_2,Ne, He and H_2 which are of high conductivity. They made use of freeze dried samples of apple, pear and beef. They have developed a semi-theoritical relationship to express the effective thermal conductivity in terms of series-parallel combinations of individual conductivities of gas and solid. The gas conductivity is found to be inversely proportional with the pressure which is expressed as 1+C/P where C is a constant for a particular gas and solid.

Lusk, Karel and Goldblith (14) presented a paper in which a method for the determination of the thermal conductivities of salmon, haddock and perch were described. The determination is based on the experimentally observed drying rates and corresponding temperature gradients by making use of quasi-steady state assumption. During experiments radiation heat transfer is utilized.

Triebes and King (21) worked on the factors influencing the rate of heat conduction in freeze drying of turkey meat. The effect of pressure, temperature, relative humidity and grain orientation on thermal conductivity is investigated. The heating was by conduction where the sample was directly in touch with the heating plates.

Clark and King (2) devised an improved freeze drying process in which food pieces and water absorbent material are placed in a bed in mixed or layered form. The light gas that is passed through the bed transports water from food to desiccant and at the same time transports heat in the opposite direction, from desiccant to food. The test material was turkey's outer breast which was cooked and cut into cubes. They have presented both experimental and theoritical results of their study.

An experimental study of the equilibrium vapor pressure of frozen bovine muscle is presented by Dyer, Carpenter and Sunderland (3). The temperature range investigated is -23°C to -1°C. Experimental results showed that the equilibrium vapor pressure of beef is approximately 20 % lover than the vapor pressure of pure ice at the same temperature. They have attributed this fact to the vapor pressure depression because of the dissolved solute species in the frozen liquid phase of beef muscle. Also experiments are done only by the frozen liquid phase which is mechanically removed from the meat to determine the effect of solid meat matrix on vapor pressure.

Unidirectional freeze-drying studies on turkey meat are done by Sandall, King and Wilke (19) in which they have measured the drying rates in the presence of water vapor alone and also in the presence of 0 to 760 mmHg of nitrogen or helium. Both stagnant gas phase and circulating gas phases are studied. They have investigated the effect of pressure and humidity of drying chamber and also the effect of meat surface temperature and grain orientation. It is proposed that for the transitional flow regime the internal mass transfer coefficient D' can be related to the Kn diffusivity as such:

$$D' = \frac{D_{AB}}{1 + \frac{D_{AB}}{D_{KA}}} \qquad (2)$$

where D_{AB} is the bulk diffusivity and D_{KA} is the Knudsen diffusivity.

Burke and Decerau (1) presented a survey of the literature. In one way it could be regarded as an updating of the Harper and Tappel's work in 1957. The freezing step of the freeze drying process and the freeze drying equipment is discussed in detail.

G.D. Saravacos (20) evaluated the freeze drying characteristics of simple gels. He has kept the heat transfer conditions constant and has given the main attention to water transfer in freeze-drying.

MATHEMATICAL MODELING TECHNIQUES

Dyer and Sunderland (4) presented an analytical study, the primary aim of which is to investigate the important mechanisms involved in the process. The equations of continuity, momentum and energy are simultaneously solved. The boundary conditions are all controllable external conditions. Trial and error procedure is required to calculate the drying times. Results for freeze drying of bovine muscle are also presented.

Hill and Sunderland (11) presented a theoritical analysis for heat and mass transfer during sublimation dehydration. The analytically solved differential energy equations in the dried and frozen regions are as such.

$$\frac{d^2T_d}{dx^2} + \frac{C_p\rho_i}{k_d}\frac{dX}{dt}\frac{dT_d}{dx} = 0 \qquad (3)$$

for the dried layer where d=dry layer and f=frozen layer.

$$\frac{d^2T_f}{dx^2} = 0 \qquad (4)$$

The energy equation which is written over the dried layer is as such:

$$k_f \frac{dT_f}{dx}\bigg|_{x=X} - k_d \frac{dT_d}{dx}\bigg|_{x=0} - C_pN_w(T_o-T_X) = -N_w\Delta H \qquad (5)$$

where N=flux of water vapor, T=temperature, C_p=specific heat, ΔH=heat of sublimation.

The subscripts w, 0 and X denotes the water vapor, surface position and interface position respectively.

Equations for the free molecule, transition and continuum flow regimes are coupled with the energy equations to get closed form solutions.

Infact these equations of energy are used by many researchers to describe energy transport in freeze drying process. Some researchers utilized these equations in the differential form by solving them simultaneously with the other transport equations by numerical means while some researchers solved them analytically and then utilized.

Massey and Sunderland (16) made an investigation of the heat and momentum transfer of a binary mixture of gases flowing in a parallel plate channel where mass injection occurs at one wall. The walls are at different temperatures and applications to freeze drying are presented.

Dominic Meo III and J.C. Friedly (17) formulated the problem of operating a freeze drier to obtain a fixed final moisture content in minimum time as an optimal control problem.

Ma and Peltre (15) by making use of unsteady state analysis they have derived a mathematical model using microwave heating. The model takes into account the variations of the transport and dielectric properties in the sample with both time and location as a function of temperature and pressure. Besides this, experimental drying curves are obtained for the freeze dehydration of beef meat with microwave energy.

A non-steady state mathematical model is proposed by Liapis and Litchfield (13) and by making use of this model they have formulated the problem of operating a freeze drier to obtain a fixed final moisture content in minimum time as an optimal control problem.

Mujumdar, Li and Jog (18) developed a mathematical model to simulate the freeze drying process for the materials in the form of slabs. The drying times for radiant supply of heat are computed by a computer code. The governing equations are solved numerically using the Regula Falsi method with the ice front temperature as the iteration variable. The mathematical model is seen to be in good agreement with the experimental results.

A mathematical model which describes the non-steady state heat and mass transfer operations during the freeze drying process is presented by Liapis (12). Both sublimation and adsorption are permitted to occur simultaneously so that a sorbed moisture concentration gradient is predicted to occur across the porous layer as drying proceeds. This adsorption-sublimation model equations are solved numerically. He claimed that this model predicts the terminal drying rates better than the uniformly retreating ice front model.

DISCUSSION

During freeze drying different modes of heat transfer could be applied and the vacuum pressure could be varied from 0-760 mmHg which causes different types of mass transfer phenomena within the dried portion. Due to the fact that the combination of the modes of heat and mass transfer applied could be many and varied the mathematical models that result are also many and varied. As a result of this the accumulated freeze drying literature contains different studies of many researchers under different practical conditions.

Although in this study it has been tried to classify the fundamental freeze drying studies this was only possible in a broader sense as it is presented. The classification in a narrower sense is avoided because of the dissimilarities among the various studies. But we are already working an a more detailed classification of the accumulated freeze drying literature and this will be published in the near future.

Some researchers tried to generalize the transport equations for all types of heat and mass transfer phenomena that are applicable but of course this lead to complex transport equations which are difficult to solve. Many of these models contain variables which are difficult to measure experimentally and difficult to estimate theoritically.

Meanwhile, some other researchers made studies in which they made use of simple transport equations containing experimentally obtainable parameters.

There is a well accumulated literature on the different process variables and on physical properties that influence the freeze drying rate of meat products. But the same thing cannot be said for several other food products such as fruits, fruit juices and vegetables.

Several freeze drying models have been appeared in the literature up to now. The earliest and the most widely used of these is the uniformly retreating ice front model (URIF) which was postulated by King. URIF model is intended to describe the drying process drying the removal of 90 % of the moisture.

Lately, Liapis proposed an adsorption-sublimation model which he claims that it predicts the experimental terminal drying rates more accurately than the prior models.

The earlier studies on modeling of the freeze drying process such as those of Dyer and Sunderland (4) and Hill and Sunderland (11) were at quasi-steady state. The more lately proposed mathematical models e.g. Ma and Peltre (15), Liapis and Litchfield (13) and Liapis (12) assumes non-steady state heat and mass transport to occur during the freeze drying process.

REFERENCES

1. Burke, R.F. and R.V.Decerau, "Advances in Food Research, Academic Press, N.Y.", V13, pp 1-88, 1964.
2. Clark, J.P. and C.J.King, Chemical Engineering Progress Symposium Series, No.108, V67, pp 102-111, 1969.
3. Dyer, D.F., D.K.Carpenter and J.E.Sunderland, J.Food Sci., V31, pp 196-201, 1966.
4. Dyer, D.F. and J.E.Sunderland. Journal of Heat Transfer, Trans ASME, November, pp 379-384, 1968.
5. Flosdorf, E.W., Chem. Eng. Progr., No.7, V43, pp 343-348, 1947.
6. Flosdorf, E.W., "Freeze-Drying, Reinhold, New York", 1949.
7. Ginette, L.F., R.P.Graham and A.I.Morgan, Jr, Natl. Symp. Vacuum Technol. TRans., V5, pp 268, 1958.
8. Harper, J.C. and A.L.Tappel, "Advances in Food Research, Academic Press, N.Y.", V7, pp 171-234, 1957.
9. Harper, J.C., AIChE, No.3, V8, pp 298-302, 1962.
10. Harper, J.C. and A.F.El Sahrigi, IEC Fund., No.4, V3, pp 318-323, 1964.
11. Hill, J.E. and J.E.Sunderland, Int. J. Heat Mass TRansfer, V14, pp 625-637, 1970.
12. Liapis, A.I. Drying '80, Proceedings of the Second Int.Symp.,Edit.A.S. Mujumdar, V2, pp 224-228, 1980.
13. Liapis, A.I. and R.J.Litchfield, Chem. Eng. Science, No.7, V34, pp 975-981, 1978.
14. Lusk, G., M.Karel and S.A.Goldblith, Food Technol., October, pp 121-124, 1964.
15. Ma, Y.H. and P.R.Peltre, AIChE, No.2, V21, pp 335-350, 1975.
16. Massey, Jr., W.M. and J.E.Sunderland, Int. J. Heat Mass Transfer, V15, pp 493-502, 1972.
17. Meo III, D. and J.C.Friedly, AIChE Symp. Series, No.132, V69, pp 55-62, 1973.
18. Mujumdar, A.S., Y.K.Li and V.Jog, Drying '80 Developments in Drying, Edit. A.S. Mujumdar, V1, pp 233-241, 1980.
19. Sandall, O.C., C.J.King and C.R.Wilke, AIChE, No.3, V13, pp 428-437, 1967.
20. Saravacos, G.D., Food Technol., V19, pp 625-629, 1965.
21. Triebes, A.T. and C.J.King, IEC Process Design and Dev., No.4, V5, pp 430-436, 1966.
22. Zamzow, W.H. and W.R.Jr.Marshall, Chem. Eng. Progr., No.1, V48, pp 21-32, 1952.

A PARAMETRIC STUDY OF THE FREEZE-DRY PROCESS FOR PRESERVATION OF ACTIVITIES OF BIOLOGICAL SUBSTANCES

Sui Lin[1] and Ta-hsu Chou[2]

1. Concordia University, Montreal, PQ, Canada H3G 1M8

2. Wayne State University, Detroit, Michigan 48201 U.S.A.

ABSTRACT

An analytical model of the freeze-dry process for preservation of activities of biological substances is defined. It is a coupled heat and mass transfer problem with a variable temperature and a variable vapor concentration at the moving sublimation front. In view of the fact that the temperature at the sublimation front is very close to the temperature at the triple point state of the moisture, the Clausius-Clapeyron equation describing thermodynamic property relations during the sublimation process can be linearized. Analytical solutions of the problem are obtained by making use of the quasi-steady-state approximation.

NOMENCLATURE

a	thermal diffusivity, m^2/s
D	diffusion coefficient, m^2/s
D^*	dimensionless parameter defined by Eq. (3.11)
Δh_{sv}	specific latent heat of sublimation, J/kg
H	dimensionless parameter defined by Eq. (5.8)
ℓ	thickness of porous plate, m
P	pressure, N/m^2
R_i	individual gas constant, J/(kg.K)
s	position of sublimation front, m
S	dimensionless position of sublimation front defined by Eq. (3.9)
t	time, s
T	temperature, K
x	space coordinate, m
y	dimensionless space coordinate defined by Eq. (3.8)
α	heat transfer coefficient, $W/(K.m^2)$
β	mass transfer coefficient, m/s
$\theta, \theta_s, \theta_i$	dimensionless temperatures defined by Eqs. (3.5), (3.6) and (3.7), respectively
λ	thermal conductivity, W/(K·m)
ν	dimensionless latent heat of sublimation defined by Eq. (3.12)
ρ_m	mass concentration of moisture, kg/m^3
τ	dimensionless time defined by Eq. (3.10)
$\phi, \phi_s, \phi_\infty, \phi_i$	dimensionless mass concentration of moisture defined by Eqs. (3.1), (3.2) (3.3) and (3.4) respectively.

NOTATION OF SUBSCRIPTS

i	initial condition in frozen region
s	sublimation front
∞	surrounding condition
3	triple point state

INTRODUCTION

The preservation of biological activity of a variety of natural substances by the lyophilization (freeze-dry) process is well documented and the method has been extensively employed [1,2]. Lyophilization is a very useful method in dealing with short or long term storage of biological substances without substantial alteration of their activities, although precautions have to be exercised [3,4,5]. It is known that the constituents of the lyophilization solution play important roles in determining how successful the preservation will be resulted. However, the physical conditions for this process are less than precise and in this sense, the lyophilization process is still limited at an empirical stage.

Lyophilization can be considered as a sublimation process taking place in a porous body. During sublimation, the processes of heat and moisture transfer cannot generally be matched, thus to maintain a constant temperature or a constant vapor concentration at the sublimation front. Therefore, the temperature and vapor concentration at the sublimation front has to be established as a part of the solution in the analysis. It is evident, however, the number of equations describing the heat and mass transfer processes for sublimation is insufficient to solve the problem. In order to resolve this difficulty and give a complete description of this problem an additional thermodynamic equation describing the phase change relation is required.

The coupled heat and mass transfer problem with unknown boundary conditions at the moving phase front is a nonlinear problem and its solution involves considerable mathematical complexities and difficulties. In order to obtain an analytical solution, a quasi-steady state approximation was used in [6] for solving the problem. Due to the complexity of the problem, the temperature and vapor concentration at the sublimation front have to be calculated numerically. For a large number of parameters, there is the disadvantage of numerical calculation which makes it difficult to obtain an overall view of the interaction among the different parameters. Numerical results obtained from [6] however indicate that the variation of the temperature at the moving sublimation front is relatively small. The average value of the temperature at the sublimation front can be taken as 99% of the temperature at the triple point state of moisture with an error of 1%. In view of this result, simplifications have been made in the present paper to obtain a closed form solution of the problem. An important result obtained in the present paper is that the number of parameters determining the location of the moving sublimation front is reduced to two in comparison with six which appeared in [6].

STATEMENT OF THE PROBLEM

We consider a biological substance dissolved in aqueous solution, with or without buffer. The solution is placed into a commercially available lyophilizing vessel which is placed into a freezing mixture made of acetone-ethanol (1:1) and crushed dry ice. A swirl motion of the vessel is conducted in the freezng mixture so that a uniform and thin layer of the frozen sample can be formed on the inside wall of the vessel. The vessel is then taken out of the freezing mixture and connected to a lyophilizing machine with a high performance vacuum pump to sublimate the moisture from the frozen sample.

For the purpose of formulation of the sublimation process, the following assumptions are made:

1. The thin layer of the frozen sample formed on the inside wall of the vessel can be considered as a flat porous plate containing frozen moisture. During the sublimation process, the sublimation front divides the porous plate into two regions; the vapor region $(0<x<s(t))$ and the frozen region $(s(t)<x<\ell)$, where $s(t)$ locates the sublimation front and ℓ is the thickness of the porous plate.
2. The frozen moisture is uniformly distributed in the porous plate which is initially at a uniform temperature T_i and at a uniform mass concentration of moisture, ρ_{mi}.
3. There is no moisture movement in the frozen region.
4. The thickness of the thin porous plate, ℓ, is small. The heat capacity of the porous plate can be neglected in comparison with the latent heat of sublimation. Hence, the condition of no temperature gradient in the vapor region may be used and the temperature distribution in the frozen region may be considered as linear. Similarly, a linear distribution of the moisture concentration in the vapor region may also be used.
5. The vapor formed at the sublimation front is the pure vapor of moisture. Because sublimation takes place at a rather low pressure, the vapor may be regarded as an ideal gas.
6. The convective terms in the vapor region and the heat transferred by radiation are small and may be neglected.
7. The latent heat of sublimation and the thermophysical properties of each region remain constant, but may differ from region to region.
8. The Soret effect, or thermal diffusion, gives rise to a mass flux which is normally very small relative to the normal Fickian flux, and may be neglected.

During the lyophilization process, the latent heat of sublimation is supplied from the surroundings of the lyophilizing vessel, which has to pass through the frozen region of the frozen sample to reach the sublimation front as shown in Fig. 1. The maximum temperature at the inside wall of the lyophilizing vessel has to be limited to the temperature at the triple point state of the moisture, T_3, in order to avoid melting. Numerical results obtained from [6] indicate that

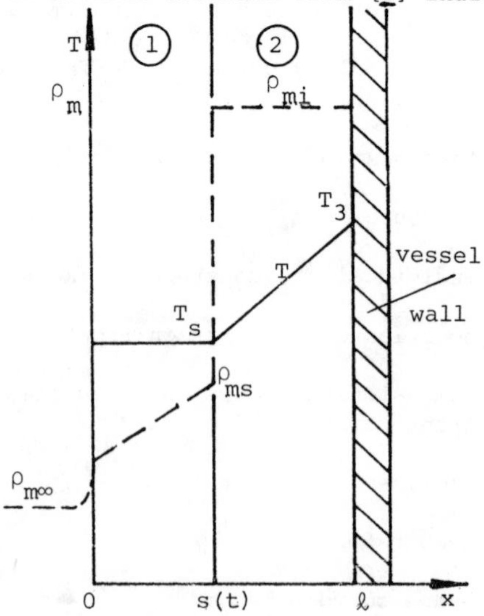

Figure 1: Schematic Diagram of the Analytical Model of the Sublimation Process for Preservation of Activities of Biological Substances.
1) Vapor Region 2) Frozen Region

the average value of the temperature at the sublimation front is about 99% of the temperature at the triple point state of the moisture. For analysis, it is reasonable to assume that the temperature at the inside wall of the lyophilizing vessel is equal to the temperature at the triple point state of the moisture.

The sublimation process can then be presented by the following system of equations:

$$\frac{\partial \rho_m}{\partial t} = D \frac{\partial^2 \rho_m}{\partial x^2}, \quad 0<x<s(t), \quad t>0 \tag{2.1}$$

$$\frac{\partial T}{\partial t} = a \frac{\partial^2 T}{\partial x^2}, \quad s(t)<x<\ell, \quad t>0 \tag{2.2}$$

where D is the volume averaged diffusion coefficient of the moisture in the vapor region and a is the volume averaged thermal diffusivity in the frozen region. For evaluation of values of D and a, the method described by Slattery [7] may be used. The initial and boundary conditions can be described as

$$T(x,0) = T_i, \quad 0<x<\ell \tag{2.3}$$

$$s(t=0) = 0 \tag{2.4}$$

$$D \frac{\partial \rho_m(0,t)}{\partial x} = \beta[\rho_m(0,t) - \rho_{m\infty}], \quad t>0 \tag{2.5}$$

$$\rho_m(s,t) = \rho_{ms}(t), \quad t>0 \tag{2.6}$$

$$T(s,t) = T_s(t), \quad t>0 \tag{2.7}$$

$$T(\ell,t) = T_3, \quad t>0 \tag{2.8}$$

where β in Eq. (2.5) is the vapor mass transfer coefficient, $\rho_{m\infty}$ is the ambient vapor mass concentration.

The moisture mass and heat balance at the sublimation front is expressed as follows:

$$D \frac{\partial \rho_m(s,t)}{\partial x} = [\rho_{mi} - \rho_{ms}(t)] \frac{ds(t)}{dt}, \quad t>0 \tag{2.9}$$

$$\lambda \frac{\partial T(s,t)}{\partial x} = \rho_{mi} \Delta h_{sv} \frac{ds(t)}{dt}, \quad t>0 \tag{2.10}$$

Where λ is the volume averaged thermal conductivity in the frozen region and Δh_{sv} is the specific latent heat of sublimation of moisture. It should be noted that in the system of equations the values of the vapor mass concentration $\rho_{ms}(t)$ and the temperature $T_s(t)$ at the sublimation front and the location of the sublimation front s(t) are unknown and should be determined as a part of the solution.

DIMENSIONLESS REPRESENTATION OF THE SYSTEM OF EQUATIONS

For the purpose of simplifying the system of equations, the following dimensionless variables and parameters are introduced:

$$\phi = \rho_m/\rho_{m3} \tag{3.1}$$

$$\phi_s = \rho_{ms}(t)/\rho_{m3} \tag{3.2}$$

$$\phi_\infty = \rho_{m\infty}/\rho_{m3} \tag{3.3}$$

$$\phi_i = \rho_{mi}/\rho_{m3} \tag{3.4}$$

$$\theta = T/T_3 \tag{3.5}$$

$$\theta_s = T_s(t)/T_3 \tag{3.6}$$

$$\theta_i = T_i/T_3 \tag{3.7}$$

$$y = x/\ell \tag{3.8}$$

$$S = s(t)/\ell \tag{3.9}$$

$$\tau = Dt/\ell^2 \tag{3.10}$$

$$D^* = D/(\beta\ell) \tag{3.11}$$

$$\nu = D\rho_{m3}\Delta h_{sv}/(\lambda T_3) \tag{3.12}$$

where T_3 and ρ_{m3} are the temperature and vapor density of moisture at triple-point state respectively.

The system of equations (2.1) to (2.10) can then be expressed in the following dimensionless form:

$$\frac{\partial \phi}{\partial \tau} = \frac{\partial^2 \phi}{\partial y^2} \quad 0<y<S(\tau), \quad \tau>0 \tag{3.13}$$

$$\frac{\partial \theta}{\partial \tau} = \frac{a}{D}\frac{\partial^2 \theta}{\partial y^2}, \quad S(\tau)<y<1, \quad \tau>0 \tag{3.14}$$

$$\theta(y,0) = \theta_i \quad 0<y<1 \tag{3.15}$$

$$S(\tau=0) = 0 \tag{3.16}$$

$$D^* \frac{\partial \phi(0,\tau)}{\partial y} = \phi(0,\tau) - \phi_\infty \tag{3.17}$$

$$\phi(S,\tau) = \phi_s(\tau) \tag{3.18}$$

$$\theta(S,\tau) = \theta_s(\tau) \tag{3.19}$$

$$\theta(1,\tau) = 1 \tag{3.20}$$

$$\frac{\partial \phi(S,\tau)}{\partial y} = [\phi_i - \phi_s(\tau)] \frac{dS}{d\tau} \tag{3.21}$$

$$\frac{\partial \theta(S,\tau)}{\partial y} = \nu\phi_i \frac{dS}{d\tau} \tag{3.22}$$

THE QUASI-STEADY-STATE APPROXIMATION

Under assumption (4), the sublimation process can be solved by the quasi-steady-state approximation, in which the temperature distribution in the frozen region and the distribution of the vapor mass concentration in the vapor region are assumed to be instantaneously in steady states. Under this assumption, the differential equations (3.13) and (3.14) become

$$\frac{d^2\phi}{dy^2} = 0 \tag{4.1}$$

and $$\frac{d^2\theta}{dy^2} = 0 \tag{4.2}$$

and the initial condition, equation (3.15), may be ignored.

The solutions of equations (4.1) and (4.2), which satisfy the boundary conditions, equations (3.17) (3.18), (3.19) and (3.20), can be expressed respectively as:

$$\frac{\phi-\phi_\infty}{\phi_s(\tau)-\phi_\infty} = \frac{D^*+y}{D^*+S(\tau)} \qquad 0<y<S(\tau), \quad \tau>0 \tag{4.3}$$

and

$$\frac{\theta - 1}{\theta_s(\tau)-1} = \frac{1-y}{1-S(\tau)}, \qquad s(\tau)<y<1, \quad \tau>0 \tag{4.4}$$

where $\phi_s(\tau)$, $\theta_s(\tau)$ and $S(\tau)$ are three unknowns which have to be determined. There are, however, only two equations (3.21) and (3.22) available. An additional equation is therefore required for solving the problem. The following is the Clausius-Clapeyron equation describing thermodynamic property relationships during phase changes of a pure substance.

THE CLAUSIUS-CLAPEYRON EQUATION

Under the quasi-steady-state approximation, the Clausius-Clapeyron equation may be applied to the sublimation front [8], thus

$$\frac{dP}{dT} = \frac{\Delta h_{sv}}{T(v''' - v')} \tag{5.1}$$

where P is the pressure, Δh_{sv} is the specific heat of sublimation, v' and v''' are the specific volumes of the frozen moisture and vapor respectively. Since the vapor pressure is low, v''' is much larger than v' which may be neglected. Then equation (5.1) can be expressed as

$$\frac{dP}{dT} = \frac{\Delta h_{sv}}{Tv'''} \tag{5.2}$$

We consider the vapor at the sublimation front as an ideal gas so that

$$v''' = \frac{R_i T}{P} \tag{5.3}$$

where R_i is the individual gas constant. Substitution of equation (5.3) into equation (5.2) yields

$$\frac{dP}{dT} = \frac{\Delta h_{sv} P}{R_i T^2} \tag{5.4}$$

Integration of equation (5.4) gives

$$\ell n \frac{P}{P_3} = \frac{\Delta h_{sv}}{R_i}\left(\frac{1}{T_3} - \frac{1}{T}\right) \tag{5.5}$$

where P_3 and T_3 are the pressure and temperature respectively at the triple point state of moisture. By making use of the equation of state for ideal gases, equation (5.5) can be applied to the sublimation front at $x = s(t)$ as follows:

$$\frac{\rho_{ms}(t)}{\rho_{m3}} = \frac{T_3}{T_s(t)} \text{Exp}\left[\frac{\Delta h_{sv}}{R_i T_3}\left(1 - \frac{T_3}{T_s(t)}\right)\right] \tag{5.6}$$

Equation (5.6) can then be written in diemsnionless form as follows:

$$\phi_s(\tau) = \frac{1}{\theta_s(\tau)} \text{Exp}\left[H\left(1 - \frac{1}{\theta_s(\tau)}\right)\right] \tag{5.7}$$

where $$H = \frac{\Delta h_{sv}}{R_i T_3} \tag{5.8}$$

The exponential function in equation (5.7) can be expanded by using the Maclaurin series as:

$$e^z = 1 + z + \frac{z^2}{2} + \ldots \tag{5.9}$$

where $z = H\left[1 - \frac{1}{\theta_s(\tau)}\right]$

Since the numerical result obtained from [6] indicates that the average value of θ_s can be taken as 0.99 with an error of 1%, it can be concluded that

$$\left|1 - \frac{1}{\theta_s(\tau)}\right| << 1$$

The exponential function in equation (5.7) may then be approximated by neglecting the terms of order higher than one,

$$\text{Exp}\left[H\left(1 - \frac{1}{\theta_s(\tau)}\right)\right] \simeq 1 + H\left[1 - \frac{1}{\theta_s(\tau)}\right] \tag{5.10}$$

Substituting equation (5.10) into equation (5.7) yields

$$\phi_s(\tau) = \frac{1}{\theta_s^2(\tau)}\left[(1+H)\theta_s(\tau) - H\right] \tag{5.11}$$

Furthermore, because $\theta_s(\tau) \simeq 0.99$, equation (5.11) may be simply approximated by

$$\phi_s(\tau) = (1+H)\theta_s(\tau) - H \tag{5.12}$$

SOLUTIONS OF $\phi_s(\tau)$, $\theta_s(\tau)$ and $S(\tau)$

Equations (3.21), (3.22) and (5.12) are the three equations which are required for solving the three unknowns: $\phi_s(\tau)$, $\theta_s(\tau)$ and $S(\tau)$.

Substituting equation (4.4) into equation (3.22) yields:

$$\frac{dS}{d\tau} = \frac{1}{\nu\phi_i}\left[\frac{1-\theta_s(\tau)}{1-S(\tau)}\right] \tag{6.1}$$

Since the vapor mass concentration $\phi_s(\tau)$ at the sublimation front is much smaller than the initial frozen moisture concentration ϕ_i, it may be neglected from equation (3.21) which can be simplified as follows:

$$\frac{\partial\phi(S,\tau)}{\partial y} = \phi_i \frac{dS}{d\tau} \qquad (6.2)$$

Substituing equation (4.3) into equation (6.2) and then eliminating the term $dS/d\tau$ from equations (6.1) and (6.2) gives

$$\phi_s(\tau) = \phi_\infty + \frac{D^*+S(\tau)}{\nu}\left[\frac{1-\theta_s(\tau)}{1-S(\tau)}\right] \qquad (6.3)$$

Substituting equation (5.12) into equation (6.3) yields:

$$\theta_s(\tau) = 1 - \frac{\nu(1-\phi_\infty)[1-S(\tau)]}{D^*+\nu(H+1)+[1-\nu(H+1)]S(\tau)} \qquad (6.4)$$

Substituting equation (6.4) into equation (5.12) gives:

$$\phi_s(\tau) = 1 - \frac{\nu(H+1)(1-\phi_\infty)[1-S(\tau)]}{D^*+\nu(H+1)+[1-\nu(H+1)]S(\tau)} \qquad (6.5)$$

Equations (6.4) and (6.5) show that $\theta_s(\tau)$ and $\phi_s(\tau)$ are functions of $S(\tau)$ which can be obtained by substituting equation (6.4) into equation (6.1) as follows:

$$\frac{dS(\tau)}{d\tau} = \left(\frac{1-\phi_\infty}{\phi_i}\right)\frac{1}{[D^*+\nu(H+1)]+[1-\nu(H+1)]S(\tau)} \qquad (6.6)$$

Integrating equation (6.6) and using the initial condition, equation (3.16) gives the dimensionless location of the moving sublimation front,

$$S(\tau) = \frac{D^*+\nu(H+1)}{1-\nu(H+1)}\left[\sqrt{1+\frac{2[1-\nu(H+1)]}{[D^*+\nu(H+1)]^2}\left(\frac{1-\phi_\infty}{\phi_i}\right)\tau}-1\right] \qquad (6.7)$$

Let $H^* = \nu(H+1)$ (6.8)

and

$$\tau^* = \left(\frac{1-\phi_\infty}{\phi_i}\right)\tau \qquad (6.9)$$

Eqauation (6.7) becomes

$$S(\tau^*) = \frac{D^*+H^*}{1-H^*}\left[\sqrt{1+\frac{2(1-H^*)\tau^*}{(D^*+H^*)^2}} - 1\right] \qquad (6.10)$$

It can be seen that the location of the moving sublimation front S is a fucntion of τ^* with only 2 parameters, D^* and H^*. The time required for the complete sublimation process can be predicted from equation (6.10) by letting $S(\tau^*_{S=1}) = 1$,

$$\tau^*_{S=1} = \frac{1}{2}(2D^* + H^* + 1) \qquad (6.11)$$

DISCUSSION AND CONCLUSIONS

As indicated in [6], the average value of the temperature at the sublimation fron is about 99% of the temperature at the triple point state of the moisture. By making use of this result, the Clausius-Clapeyron equation has been linearized from which the values of the dimensionless temperature θ_s and the dimensionless vapor mass concentration ϕ_s at the sublimation front are determined. As an illustration, Figure 2 shows the variation of θ_s and ϕ_s as functions of the dimensionless location of the sublimation front S for $D^*=5.0$, $\nu=0.02$, $H=21.47$ and $\phi_\infty=0.01$. It is shown that the variation of θ_s is very small. However, the small variation of θ_s results in a relatively large variation of the vapor-mass concentration at the sublimation front.

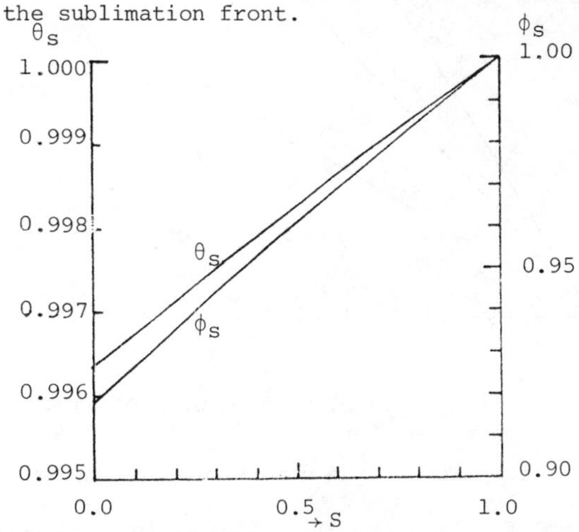

Figure 2: θ_s and ϕ_s as functions of S for $\phi_\infty = 0.01$, $\nu = 0.02$, $D^* = 5.0$ and $H = 21.47$.

For the sublimation process, the most important variable is the location of the sublimation front which is shown in Figs. 3 and 4 as functions of τ^* with D^* and H^* respectively as parameters. By the definition of D^* in equation (3.11), D^* represents the ratio of the convective mass transfer resistance at the surface $x = 0$, to the diffusion mass transfer resistance in the porous medium. Consider the case that the diffusion mass transfer resistance in the porous medium is kept as constant. An increase of the value of the convective mass transfer resistance at $x = 0$ results in an increase of the value of D^*. Therefore the higher the value of D^*, the higher is the convective mass transfer resistance, and in turn less mass is transferred out of the surface $x = 0$. Hence, it requires more time for sublimation as shown in Fig. 3.

The dimensionless parameter $H^* = \nu(H+1)$ is nearly proportional to the square of the latent heat of sublimation (see equations (3.12) and (5.8)). Hence H^* is essentially governed by the latent heat of sublimation. It is clear that any process having a larger latent heat of sublimation requires more heat to be absorbed at the sublimation front and hence a longer time is required for sublimation as shown in Fig. 4. However, the effect of the variation of H^* on the location of the sublimation front is very limited. Figures 3 and 4 show that the dimensionless location of the sublimation front S is practically

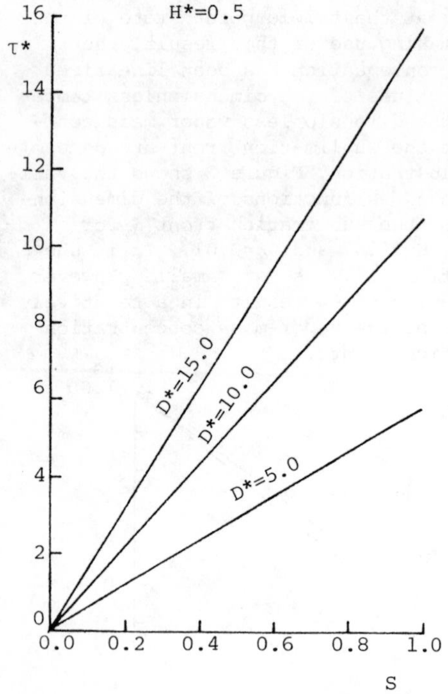

Figure 3: τ^* as a function of S with D* as a parameter for $H^* = 0.5$

a linear function of the dimensionless time τ^*.

The simplification and linearization process introduced in the present paper reduces significantly the number of parameters required for determining the location of the sublimation front. From equation (6.10), it is seen that S is a function of τ^* with only two parameters D* and H*. In comparison with the result obtained in [6], there are six parameters required for determining the location of the sublimation front.

ACKNOWLEDGEMENT

The present work is supported in part by the Natural Sciences and Engineering Research Council of Canada, under Grant No. A7929.

REFERENCES

1. Anderson, J.O., and Nei, T., The Effects of Freezing and Freeze-Drying on Adenosine Triphosphatase and Acetylchlinesterase of Human Red Cell, in Freeze-Drying of Biological Materials, International Institute of Refrigeration Bulletin, Annex 5, pp. 109-111, 1973.

2. Hanafusa, N., Freezing and Drying of Enzyme Proteins, in Freeze-Drying of Biological Materials, International Institute of Refrigeration Bulletin, Annex 5, pp. 9-18, 1973.

3. Deisseroth, A., and Dounse, A.L., Nature of the Change Produced in Catalase by Lyophilization, Arch. Biochem., Biophys. 120, pp. 671-692, 1967.

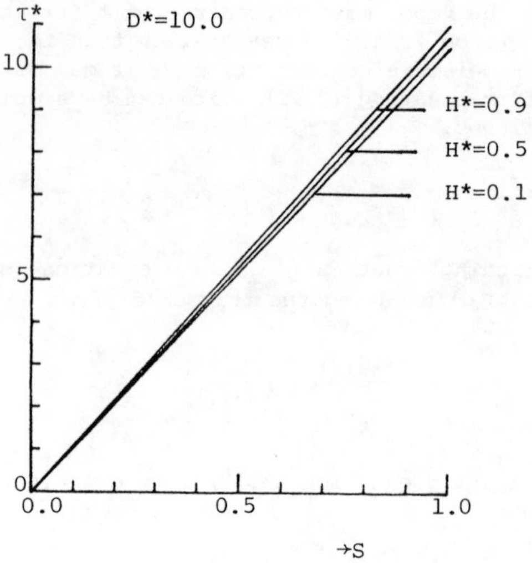

Figure 4: τ^* as a function of S with H* as a parameter for $D^* = 10.0$.

4. Litt, M. and Boyd, W., Preservation of Haemocyanin, Nature (London) 181, p. 1078, 1958.

5. Lea, C.H. and Hawke, J.C., Lipovitellin II, The Influence of Water on the Stability of Lipovitellin and Effects of Freezing and Drying, Biochem. J. 52, pp. 105-114, 1952.

6. Lin. S. and Chou, T.H., The Heat and Mass Transfer Characteristics of the Sublimation Process for Preservation of Activities of Biological Substances, Proceedings of the 7th International Heat Transfer Conference, Sept. 1982, Munchen, West Germany, Vol. 6 pp. 129-134

7. Slattery, J.C., Momentum, Energy and Mass Transfer in Continua, McGraw-Hill, New York, 1972, Section 7.3 and 10.3.

8. Holman, J.P., Thermodynamics, 3rd Edition, McGraw-Hill, New York, Section 7-10, 1981.

DRYING OF GRANULAR MATERIALS AT FORCED FLOW OF DRYING AGENT THROUGH THE BED

J. Pikoń

Silesian Technical University, Gliwice, Poland

ABSTRACT

The paper presents construction and principles of convection drier with a slidable bed. Heat exchange and resistance of flow are also presented.

Convection drier with a slidable bed is a new type of invention for drying granular materials of mono or polydisperse structure. Direction of realizability is based on forced flow of a drying agent through the grainy bed poured off between perforated or louver boards.

Its shift during the process of drying can be carried on gravitationally, dependent on specific quality of bed. Applied dissimetrically gapping system of walls as well as the travel of the bed which can be forced by means of a special shelf conveyor greatly facilitates the process.

The idea of drying a compact grainy bed with the flow of drying agent through the bed comes down from a series of data. Bernstein had studied the criterion of similarity for the heat flow process with the air flow through globule deposit and worked out the following formula:

$$Nu = A \cdot Re^m$$

the exponent m amounting to 0,6. The Reynolds number has been defined basing itself on the flow velocity calculated for unristricted section whereas the characteristic lineal dimension is identified with the spherical diameter of the element. The quantity A is dependent on the bed porosity and can be read off by the Fig.1

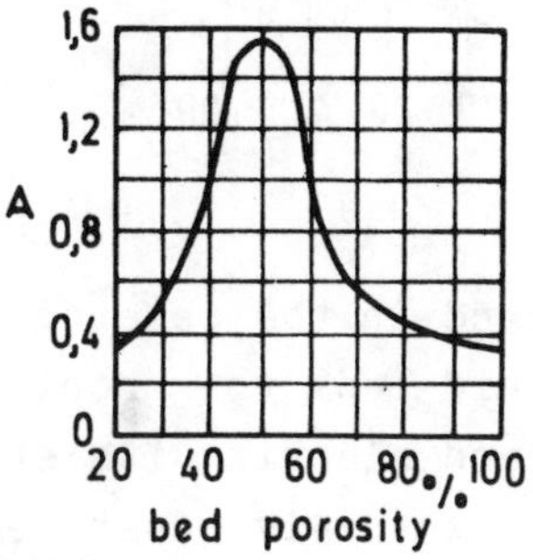

Fig.1

The investigetions carried out have prowed that the compact stationary beds of 0,3 + 0,6 porosity are characteristic of intensive thermic process.

Fretting of the bed brought about by the increase in porosity leads to reduction of flow velocity in the bed and the fall of intensity of heat flow process as well as bulk material flow one.

After bed fluidization has been carried out the reduction of gas relative velocity in the ratio to grains taken away is followed. The results of American studies relating to heat transfer into a compact grainy bed and into fluid one are presented in the Fig.2.The studies corroborate unfavourable influence of a fretting bed on the intensity of heat flow process.

Compact grainy beds make a specific filtering structure to afford possibilities for high dust removal of aerosol. That is why in the case of a fine-grained bed the drying process can go together with that of dust extraction. In the case of polydisperse beds and applied flow velocities of $V_0 = 0,3 \div 1,5$ m/s the rise minute fractions is practically eliminated.

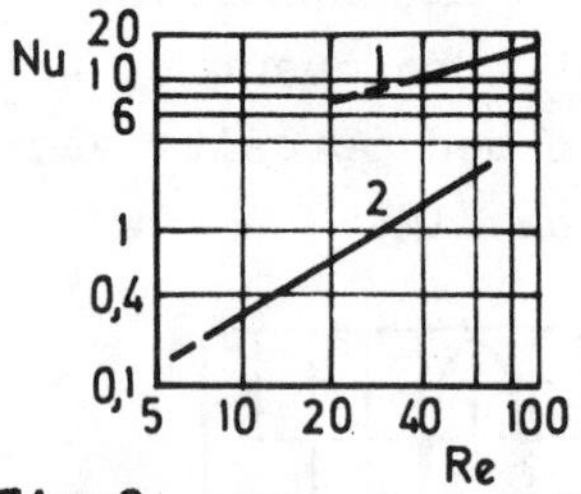

Fig.2

On the other hand in the case of drying in loose or fluid beds the problem is very laborius and clearing its effects requires using expens-ive dust removal arrangements.

The bed of loose or fluid states involves overcoming the resistance of flow Fig.3 and so of higher running costs.

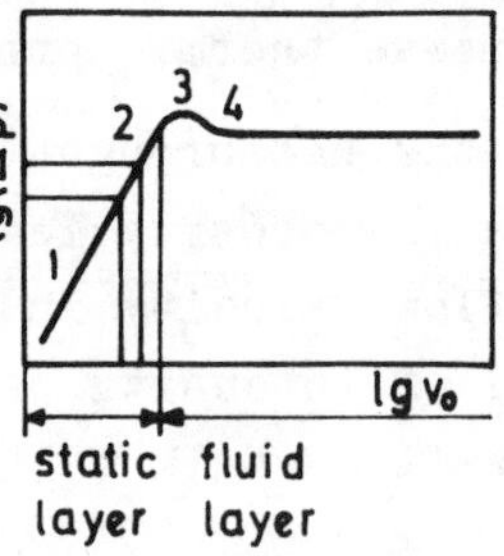

Fig.3

At the fluid phase we get full development of interphasic surface, however, what positively improves the intensity of process represented by the specific evaporation factor.

In the case of a compact bed subjected to a forced movement during the process of drying, the displacement with the rotation of elements of bed occurs and leads to full utulization of interphasie surface.

The specific evaporation factors have been calculated on the grounds of experimental data. They exceed the values which are characteristic for drying process of loose layers /fluid or pneumatic/. It corroborates the argument of economical advantages of drying coarse and fine-grained bulk materials in the compact slidable bed with the flow of drying agent through the layer.

The referred evidence has directed the work at the new construction of the drier, putting into practice the drying grainy materials in a compact bed with the flow of drying agent througt the layer. The elaborated construction named: "Convection drier with slidable bed" obtained a home patent and several foreign ones.

The drawing in the Fig 4 is a reproduction of the original on a reduced scale.

The drier of this type is resolved upon a binary system /Two-component system/ and owing to it one can yield a number of simplifications in a handling range of bulk materials and drying gases and further economical reasons of this type of realization comes down from reducing heat loss as well as limitation of overall dimensions of the drier.

On the grounds of the device some studies on kinetic process of grainy drying beds with a drying agent flow through the layer have been carried out in order to elaborate a reasonable design based on that type of driers.

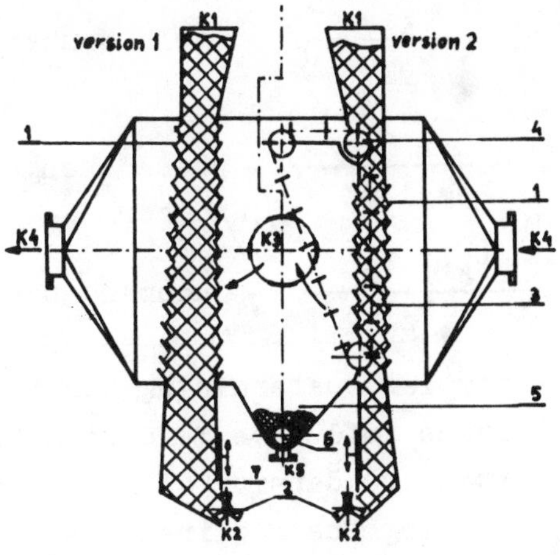

Fig.4

STUDIES ON KINETIC DRYING PROCESS

While analysing the drying process of a grainy material with the drying agent flow through the bed it was found that the said passes off in accordance with the direction of the drying agent flow causing a continuous drying of elementary layers dx from 0 to the thicknees of H bed Fig 5. That process is characterized by such a great intensity that a complete drying of an inlet zone is observed at a simultaneus insignificant initiated drying of the outlet zone. For bulk materials characterizing themselves by significant drying velocities during the first period, some moisture outdropping from a drying gas at a continuation is observed. That unfavourable effect can be eliminated by the specific selection of drying agent parameters.

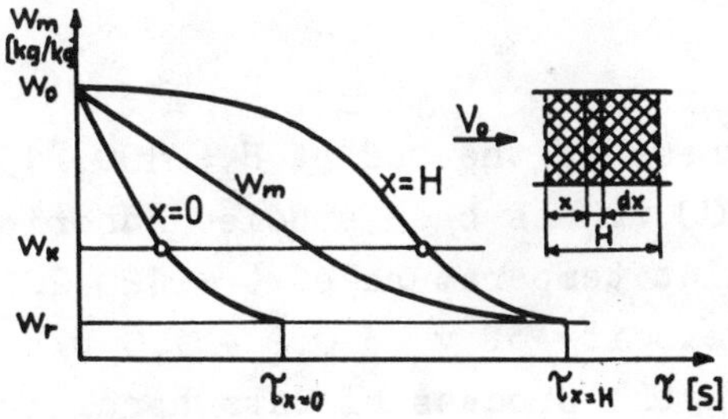

Fig.5

Trying to describe the process based on the heat and mass exchange has not brought yet the anthor to a usable form in an analitical point. Our own studies have been carried out on average moisture measurement of the bed w_m, on temperature measurements in the bed as well as on the inlet and outlet of the drying agent. A typical exchange course of the average moisture of the bed and temperature obtained at the quick coke drying is presented in Fig 6.

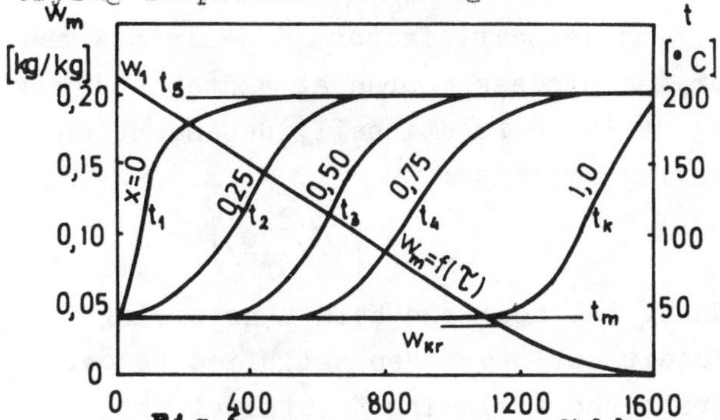

Fig.6

On the grounds of measurement results it was found that average moisture change lines show a specific quality of drying process so one can divide them into the first period of drying at a rectilinear course in the range from w_1 to w_{kr}, and the other one in the range from w_{kr} to w_r at a curvilinear one. The temperature courses show a progressing drying process in the direction of drying agent flow. The temperature rise under the layer /t_k of $x = 1,0$/ will have started only when critical moisture appears, so during the first period of drying there is a peak temperature potential at our disposal. It determines on high intensity of a satisfactory drying process. The carried out studies have proved author apt idea forecasting possibility of analitic version of the process through the average moisture change of bulk metarial during the time $w_m = f/\tau/$.

The time of drying at stable drying agent parameters is calculated according the formula:

$$\tau = \frac{1}{\varkappa}\left[/w_1-w_{kr}/+/w_{kr}-w_r/ \ \ln\frac{w_{kr}-w_r}{w_2-w_r}\right] [s] \quad /1/$$

taking: $\varkappa \left[\frac{1}{s}\right]$ - as drying velocity factor

w_1, w_{kr}, w_r, w_2 [kg/kg] - appropriate material moisture initial, critical, equilibrium and final with reference to dry substance.

Drying velocity factor $\varkappa$ describes the drying process course at a stable drying agent it is functionally dependent on drying velocity:

$$\varkappa = f^{*}\left[\Psi^{*}/\frac{dw_m}{d\tau}/\right] \quad /2/$$

Whent the heat and bulk mass drying conditions have been satisfied we can write the following functional influence of basic parameters on drying velocity factor: $\varkappa = \Psi\left[v_o, \varrho_g, d_z, \eta_g, M, \delta_A, H, a, \varrho_u, c_{pg}, \lambda_g, r, /t_s-t_m/, t_s\right]$ /3/

Following the dimensional analysis method basing on Fig 3 circumscribing the grainy bed drying process at a drying agent flow through the bed a dimensioless equation has been obtained:

$$/Pi/ = C\ /Re/^A /Le/^B /K/^D /Gu/^E /\frac{H}{d_z}/^F / \frac{\varrho_u}{\varrho_g}/^G \quad /4/$$

taking:

$Pi = /\frac{\varkappa \cdot d_z}{\gamma_g \cdot a}/$ - nondimensional number characterizing the intensity of grainy bed convection drying process at a drying agent through the bed

$Re = /\frac{v_o \cdot d_z \cdot \varrho_g}{\eta_g}/$ - Reynolds number

$Le = /\frac{\lambda_g}{\varrho_g \cdot c_{pg} \cdot D}/$ - Lewis number

$K = /\frac{r}{c_{pg} \cdot \Delta t}/$ - phase change number

$Gu = /\frac{t_s \cdot t_m}{t_s}/$ - Guchman number

$/\frac{H}{d_z}/$ - simplex, determining bed thickness influence referring to standard density of the layer meeting the requirements of the bed d_z

$/\frac{\varrho_u}{\varrho_g}/$ - simplex, defining heaped up bed density referring to that of the drying gas

Calculation the constant and dimensioless equation regression factor reguired taking up the studies which have been carried out on a laboratory system /Fig 7/, puting into practice grainy bed drying process with drying agent flow through the bed. The installation has been designed like this way, that it made possible for permanent measurement of the average moisture of the bed.

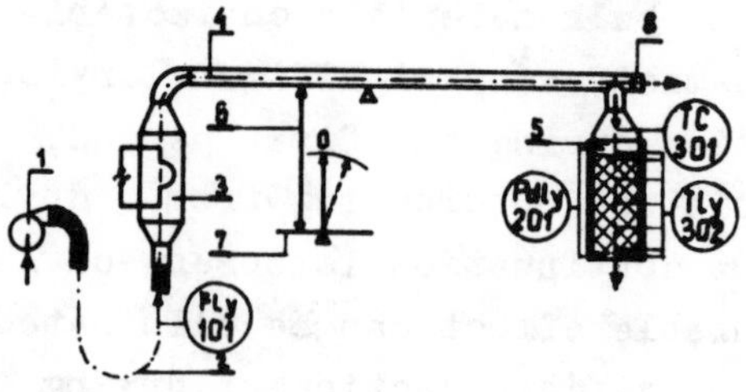

Fig.7

Thickness of the bed of H = 100, 150 and 200 mm has been studied adhering to inlet temperature of t = 120÷200°C and velocity of v_o = 0,2 ÷ 1,0 m/s. During the process of data handling it was found out that Lewis /Le/ and

Guchmans /Gu/ numbers are characterized by over-low variation, that is why laying down the influence of those numbers on studied effect would require an extensive programe of studying in the line of higher temperature.

Experimental findings have been elaborated by the least square method in cracovian confinement assuming probability of p = 0,95.

Dimensionless equation circumscribing the quick coke drying process is of the form:

$$Pi = 0{,}469/Re/^{0{,}896}/K/^{-0{,}633}/\frac{H}{d_z}/^{-1}/\frac{\varrho_u}{\varrho_g}/^{-1} \qquad /5/$$

The equation has been worked out at the following parameter changes: /Re/ = = 50 ÷ 125, /K/ = 14 ÷ 31, $/H/d_z/$ = = 20 ÷ 41, $/\varrho_u/\varrho_g/$ = 670 ÷ 830. Referring to the number of /Pi/, relative root-mean square error of $\delta_t = \pm 10{,}28\%$.

All the physical parameters of drying gas composing the dimensionless equation of /5/ are referred to the over layer temperature as to the characteristic quantity since that of under the layer is a stable one and near the temperature of the wet thermometer. The vaporization heat r is referred to the temperature of the wet thermometer of t_m corresponding the inlet one of t_s, while $\Delta t = t_s - t_m$ determines characteristic, psychrometric temperature potential of the gas.

ON DRYING OF MATERIALS IN THROUGH CIRCULATION SYSTEM

T.S. RAJAN AND S.H. IBRAHIM

DEPARTMENT OF CHEMICAL ENGINEERING
REGIONAL ENGINEERING COLLEGE
TIRUCHIRAPALLI-620015

ABSTRACT

Drying characteristics in a through circulation system have been studied and reported. Five different systems: vanadium pentoxide, silica gel, molecular sieves, activated charcoal and bituminous coal have been investigated and individual correlations of gas phase transfer units N_{OG} with the moisture content (X) have been obtained for the falling rate period. Parameters chosen are: flow rate, temperature, particle size and bed height. Statistical significance test has been applied not only to design the experiments but also help in analysis and formulation of results. A generalised equation to cover all the systems has been arrived at as follows:

$$N_{OG} = (11.25 - 28.12(x_0\varepsilon_s))(d_p/z_p)^{-0.544}(X)^D$$

The above equation, which can predict drying rates within an accuracy of 25 percent, is valid for temperature range of 50 to 100^0C and a flow regime of particle Reynolds number 30-300, the variation in particle size being from 0.89 to 5.55 mm.

NOMENCLATURE

C - constant in the equation $N_{OG} = C(X)^D$

C_1, C_2 - constnats in the equation $C = C_1(d_p/z_p)^{C_2}$

D - exponent in the equation $N_{OG} = C(X)^D$

d_p - particle diameter, m

G_s - mass flow rate, kg/m^3s

T - temeprature ^0K

N - drying rate , kg/m^3s

N_{max} - maximum drying rate under adiabatic saturated conditions, kg/m^3s

N_{OG} - number of gasphase transfer units

X - moisture content, kg moisture/kg dry solid

X_0 - initial moisture content, kg moisture/kg dry solid

Y_1 - humidity of ambient air, kg moisture/kg dry gas

Y_2 - outlet humidity, kg moisture/kg dry gas

Y_s - saturation humidity, kg moisture/kg dry gas

z_p - bed height, m

ε_s - bulk density of bone dry material, kg/m^3

INTRODUCTION

Granular materials are best dried using through circulation system. The process of drying using through circulation system resembles that of batch or tray drying in that, part of drying takes place under constant rate drying condition and the rest under falling rate condition. The mechanism of removal of moisture under constant rate drying condition is fairly well understood and design procedures are made available by Gilliland (1938) and Sherwood (1930). But similar conclusions cannot be made for the falling rate period, for which data available is very little. The situation is still worse with reference to through circulation system. With a view to fill in this void and make available suitable correlation to help design of drying equipment, the present work has been taken up.

The earliest investigation in through circulation drying dates back to 1942, when Marshall and Hougen (1942) have shown exponential type of equations to fit the drying curve during the falling rate period. This is applicable for certain materials only. In 1949, Joseph Allerton et al. studying the mechanism of through circulation drying of filter cakes, have referred to possible use of number of gas phase transfer units to define drying characteristics. Recently, Japanese workers, Atsushi et al. (1977) in trying to predict the critical point and falling rate curves, have shown that diffusivity assumed constant by earlier workers depends much on the moisture content and the diffusion equation employed to explain moisture movement becomes non-linear and requires application of perturbation technique or a numerical finite difference method. The authors have therefore adopted approximate expressions developed by Suzuki et al. (1975) in their work. Thus it is seen that the work with reference to through circulation drying and in particular about the problems of falling rate period has been scarce and there is scope for further investigation.

The problem of drying is complicated by the fact that different mechanisms viz., capillary flow and diffusion mechanisms as pointed out by Hougen et al. (1939), gravitational and frictional effect as observed by Pearce et al. and thermal gradient as identified by De Vries (1958) operate and these factors in turn depend upon (i) nature of particle-hygroscopic or non-hygroscopic, porous or nonporous (ii) the size of the particle and its pore structure (iii) the height and compactness of bed (iv) the thermophysical properties of the solid material and finally (v) the instantaneous moisture content. Naturally suitable approach has to be evolved for solution to this problem.

In this paper, investigation carried out on five different systems - vanadium pentoxide, silica gel, molecular sieves, activated charcoal and bituminous coal has been reported. Concept of number of gas phase transfer units has been applied and correlations are made. Effect of parameters - temperature and particle size - is taken into account and the choice of these parameters has been made using statistical significance test by Rajan (1981). It must be emphasized here that the statistical test of significance, in addition to pinpointing on the parameters of significance has enabled finding effect of interactions of parameters and also helped in identifying cause for errors in experimentation and analysis.

EXPERIMENTAL SET UP:

Fig. 1 shows the experimental facility used in the work. It consists of a through circulation dryer, a blower, heating and flow measuring devices. A cantilever system with an inductive pick up is used for continuous monitoring of weight loss during drying. The drying curve is obtained on the pen-chart recorder. Inlet and outlet temperatures are measured using mercury-in-glass thermometers. A wet and dry bulb thermometer gives the humidity of ambient air.

The air from the blower passes through a surge tank and is measured by a calibrated rotometer. It then passes through a section of pipe surrounded with an electric heating tape. The heated air passes through the perforated bottom of the dryer tube made of aluminium and thermally insulated. As the moisture is removed by the heated air, the dryer tube suffers a loss in weight, which is sensed by the inductive pick up. The signal from the inductive pick up is processed in the displacement meter and finally recorded on the penchart recorder.

EXPERIMENTAL PROCEDURE:

Particles in the size range of 0.89 mm to 5.55 mm were used employing flow regime corresponding to particle Reynolds number 30-300 and temperatures 50^0 to 100^0C.

The particles, soaked over night, were filled in the dryer tube after shaking off loose moisture. Heater having been switched on earlier and air flow adjusted, heated air at the required temperature and flow velocity was made available. The dryer tube was then transferred to the pan of the automatic weighing set up and the displacement meter was made to read zero with the zero-adjustment nut. Immediately the penchart recorder was put on. The drying was continued until the drying curve became parallel to the time axis indicating the end of drying. The material after drying was weighed and its weight checked with the original dry weight taken.

ANALYSIS OF RESULTS:

Hot air, when passed through thoroughly wet material, becomes almost saturated and removal of moisture takes place under steady state condition. When the particles start developing dry patches on their surface, the drying rate falls and the falling rate period sets in. At any instant, drying rate is given by

$$N = G_s(Y_2-Y_1) \qquad (1)$$

where Y_2 and Y_1 are the outlet and inlet humidity values and G_s is the mass flow rate of heated air. The theoretical maximum rate of drying under saturated condition is given by

$$N'_{max} = G_s(Y_s- Y_1) \qquad (2)$$

where Y_s = saturation humidity.

Combining (1) and (2)

$$\frac{N}{N'_{max}} = \frac{Y_2 - Y_1}{Y_s - Y_1}$$

$$= \frac{(Y_s - Y_1) - (Y_s - Y_2)}{(Y_s - Y_1)}$$

$$= 1 - \frac{Y_s - Y_2}{Y_s - Y_1}$$

Since N_{OG} is defined as: $\ln \frac{Y_s - Y_1}{Y_s - Y_2}$ (3)

$$\frac{N}{N'_{max}} = 1 - e^{-N_{OG}} \quad \text{or}$$

$$N_{OG} = \ln(1 - N/N'_{max})^{-1} \quad (4)$$

Equation (4) is used to predict the drying rate and the number of gasphase transfer units. (N_{OG}) itself is correlated with moisture content (X) based on experimental data. The relationship obtained is expressed as follows:

$$N_{OG} = C(X)^D \quad (5)$$

Drying curves obtained with the pen-chart recorder were analysed. The critical point was established by drawing a tangent to the initial portion of the curve representing the constant rate drying process. Time-moisture data ware collected from the drying curve commencing from the critical point and tabulated. Curves obtained for bituminous coal could be fitted with an exponential type of equation indicating that the loss of moisture occurred very slowly following an exponential decay pattern. Fig. 2 shows a set of typical drying curves obtained in drying bituminous coal. Curves of other four systems - vanadium pentoxide, silica gel, molecular sieve and activated charcoal - could be represented by power equations indicating comparatively a shorter falling rate period. Fig. 3 shows a set of drying curves for silica gel, which is typical of the other four systems. The rate values (N) were obtained by differentiating the equation to the drying curve. The theoretical maximum rate of drying (N'_{max}) was calculated using equation (2) and the N_{OG} values found using equation (4). For each particle size, the calculated values of N_{OG} were plotted against moisture content (X) on logarithmic coordinates for different temperature-flow combinations. It was found that a linear relationship existed between (N_{OG}) and (X) and an equation of the type $N_{OG} = C(X)^D$ could be formed. The exponent D assuming a characteristic value for each substance irrespective of particle size, the constant C was found to be varying with particle size. Again a logarithmic plot of C versus ratio of particle size (d_p) to bed height (z_p) showed a linear relationship yielding an equation of type $C = C_1(d_p/z_p)^{-C_2}$. The general equation depicting drying of each material could therefore be expressed as

$$N_{OG} = C_1(d_p/z_p)^{-C_2}(X)^D$$

The results obtained in the above form are given in Table 1 for each of the systems studied.

The percentage error in estimating N_{OG} values using the above correlations was found to be about 25 to 30% in some cases. With a view to reduce the error, statistical significance tests were made and the parameters of significance were decided upon. Correlations, which now include temperature as additional parameter of significance give considerably lower errors and are given in Table 2.

TABLE 1

N_{OG} Correlation for different systems

S.No.	System	Equation
1.	Vanadium pentoxide	$1.3(d_p/z_p)^{-0.557}(X)^{0.78}$
2.	Silica gel.	$1.2(d_p/z_p)^{-0.213}(X)^{0.51}$
3.	Molecular sieve	$4.8(d_p/z_p)^{-0.404}(X)^{0.89}$
4.	Activated charcoal	$1.9(d_p/z_p)^{-0.22}(X)^{0.7}$
5.	Bituminous coal	$7.5(d_p/z_p)^{-0.544}(X)^{1.30}$

However, to simplify the situation, a generalised correlation applicable for all systems has been attempted. Fig. 4 shows a logarithmic plot of C (in equation $N_{OG} = C(X)^D$) versus (d_p/z_p) for all the systems and it appears that the resulting straight line could have a common slope $C_2 = 0.544$. Further plotting of the new values of C_1 as obtained from Fig. 4, against the product of initial moisture content (X) and the bulk density indicates a straight line relationship resulting in

$C = 11.25 - 28.12(X_0\varepsilon_s)$ (Fig. 5)

TABLE 2

Generalised N_{OG} correlation for different systems

System	Equation	Error %
Vanadium pentoxide	$808(X)^{0.78}e^{-(0.013T+154d_p)}$	20
Silica gel	$3.7(X)^{0.51}e^{-227(d_p)}$	15
Molecular sieve	$52(X)^{0.89}e^{-(0.003T+73d_p)}$	15
Activated charcoal	$27(X)^{0.89}e^{-(0.005T+50d_p)}$	20
Bituminous coal	$11.44(X)^{1.33}e^{-(0.01T+95d_p)}$	20

Thus, we may write a generalised correlation applicable for all systems in the following form

$$N_{OG} = (11.25-28.12(X_0\varepsilon_s))(d_p/z_p)^{-0.544}(X)^D$$

The above equation can predict drying rates of falling rate period of through circulation systems within an accuracy limit of ± 20% and is valid for a temperature range of 50 to 100°C, flow defined by particle Reynolds number 30-300 and particle size from 0.89 mm to 5.55 mm.

CONCLUSION

In view of the automatic weighing set up made use of, results obtained are more reliable and the N_{OG} approach to simplify correlations and provide useful information for the design of through circulation systems.

REFERENCES

1. Atsushi Endo et al., AIChE Symposium series No. 183, Vol. 73, 57-82 (1977).

2. De Vries, D.A. Trans. Am. Geophys. Union 39, 909-915 (1958).

3. Gilliland, E.R., Ind. Engng. Chem. 30, 506 (1938).

4. Hougen, D.A., McCauley, H.J. and Marshall W.R., Am. Inst. Chem. Engrs. 35, 183 (1939)

5. Joseph Allerton, Llyod B. Brownell and Donald L. Katz, Chem. Engng. Prog. 1, 45, 10, 619-635 (1949).

6. Marshall, W.R., Jr., and Hougen O.A., Trans Am. Inst. Chem. Engrs., 38, 91 (1942).

7. Pearce, J.F. and T.R. Olive and Newitt D.M., Trans. Inst. Chem. Engrs (London) 27, 1 (1949).

8. Sherwood, T.K. Ind. Engng. Chem., 21, 12, 976 (1929), 22 138 (1930).

9. Rajan, T.S., Simultaneous Heat & Mass Transfer in through circulation drying - Ph.D. Thesis submitted to Madras University (India) Aug. 1981.

10. Suzuki, M., R.B. Keey, and S. Meeda Paper XXC AIChE meeting Boston (1975).

ACKNOWLEDGEMENT

The Editor is grateful to Purnima Mujumdar for retyping the paper submitted by the authors and to Victor Jariwalla for redrafting all figures.

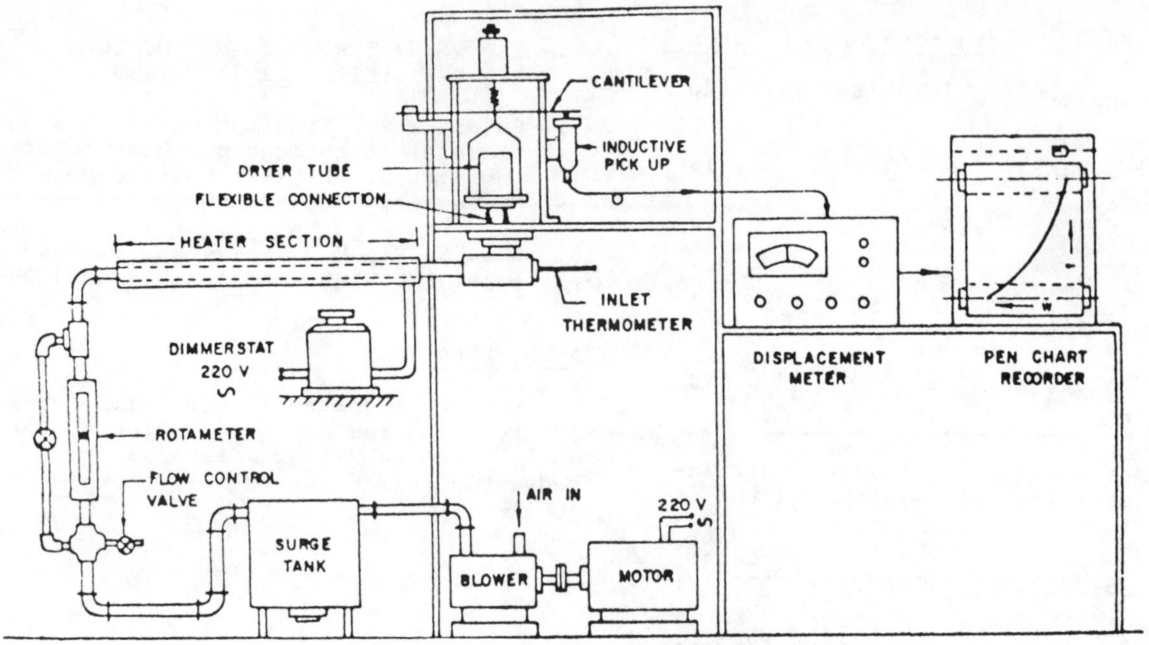

FIG 1 EXPERIMENTAL SET UP

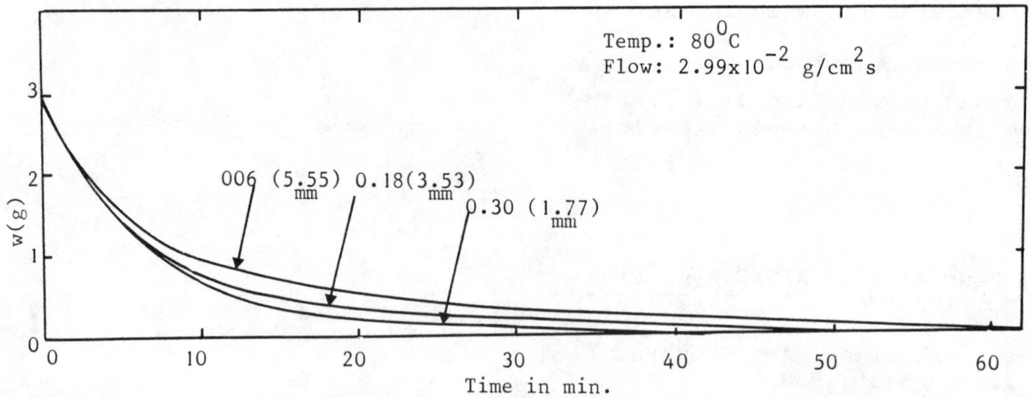

Fig. 2 Drying Curves, Bituminous Coal.

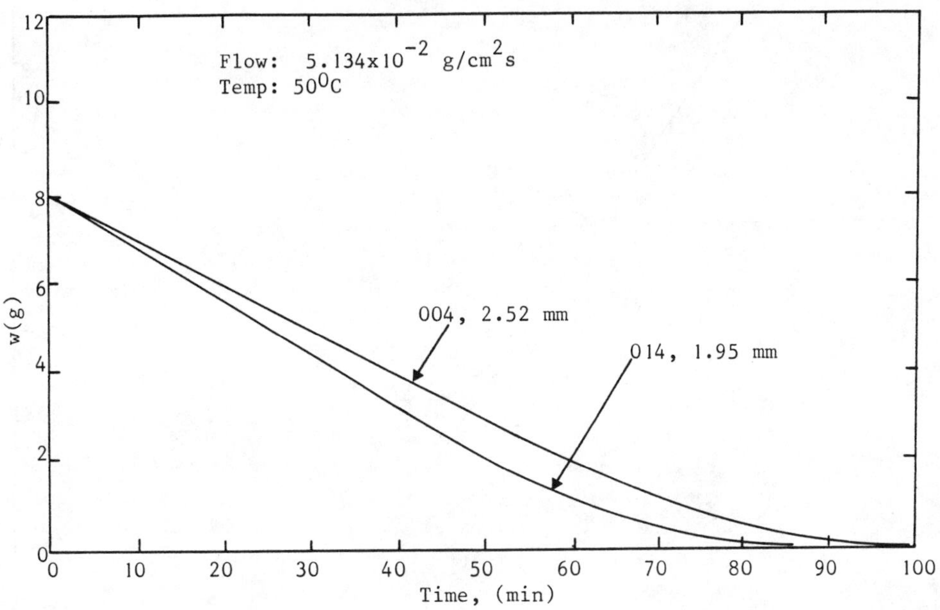

Fig. 3 Drying Curves, Silica Gel.

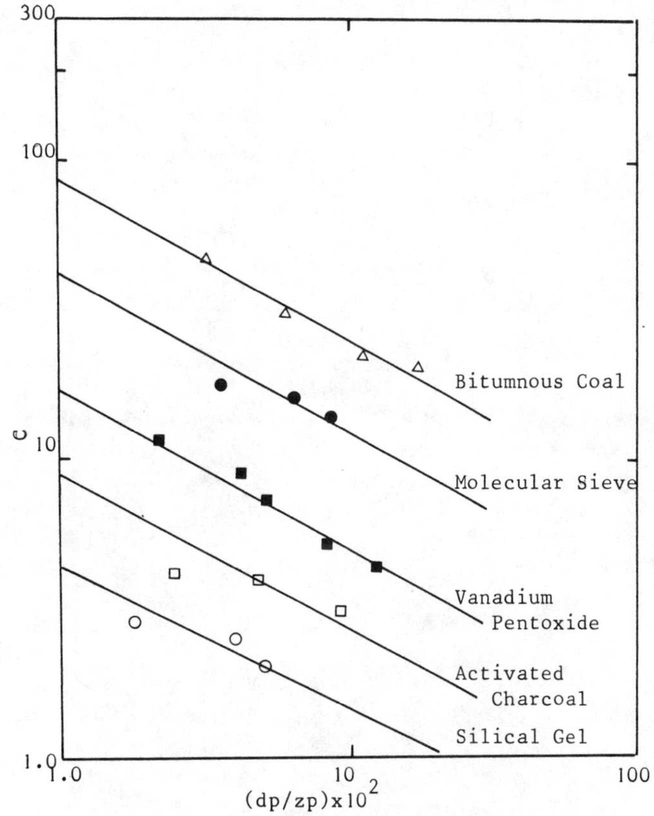

Fig. 4 C vs dp/Zp

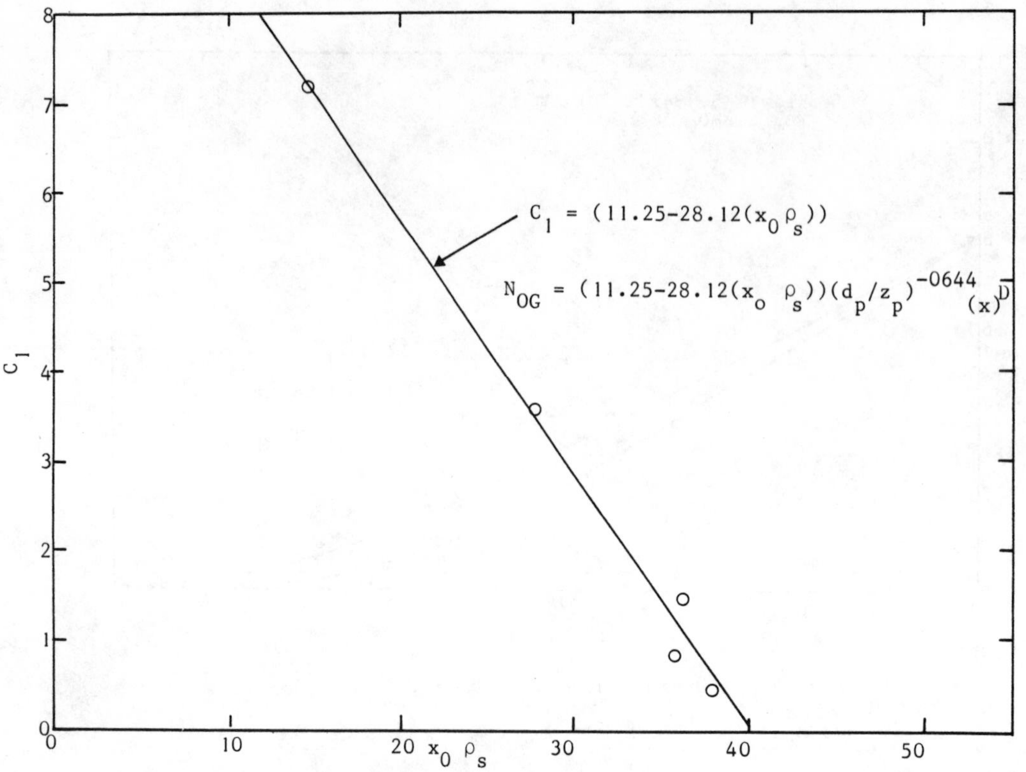

Fig. 5 C_1 vs (X_0P_0)

ANALYTICAL EVALUATION OF THE INFLUENCE OF PARTICLE MASS CHANGE DURING DRYING IN PNEUMATIC TRANSPORT OF PARTICLES.

Mushtayev V.I., Timonin A.S., Tyrin N.V.,
Chernyakov A.V., Ryauzov A.T.,

Moscow Technical University
Moscow, USSR.

Abstract

Analytical solution of the equations of particle motion in pneumatic transport in straight ducts of pneumatic dryers with mass change of particle in time considered, is presented. Analytical relationships reflecting the interdependence of hydrodynamics and heat and mass transfer in the drying process are given.

Nomenclature

d_p - particle diameter
Δh_v - latent heat of vaporization
i - slip ratio
ℓ - distance from the entrance of duct
m - mass
p - pressure
Q - mass flowrate of solid
S_p - particle area
T - temperature
X - moisture content

Greek Symbols

α - heat transfer coefficient
λ - heat conductivity
ζ - drag coefficient
τ - time, ρ - density

Subscripts

e - equilibrium
g - gas
k - final
m - average
max - maximum
0 - initial
p - particle
v - vapor

High drying rates for fine granular materials are achieved in dryers which offer high relative gas-particle velocities. The material being dried is carried along the dryer either in the fluidization or the pneumatic transportation regimes. In either one of these regimes the change of particle mass has a substantial significance on the drying process as it influences both the hydrodynamic and and immediate drying conditions.

Two stages of the hydrodynamics-drying relationship can be distinguished:
1. the influence of mass change on the reduction of the relative velocity of phases.
2. resultant reduction of the heat and mass transfer rates.

The second stage, unlike the first one, is observed only at high slip ratios, $i = \frac{u_g - u_p}{u_g}$, as Nu at small slip ratios is virtually independent of the Reynolds number, Re [1], and the mass transfer is due mainly to free convection. The first influencing factor is significant even at relatively low values of i thus affecting the drying time.

In consequence, in the pneumatic transport regime in straight ducts two stages can also be distinguished. During the first stage both aforementioned factors are significant; hence simultaneous integration of the hydrodynamic as well as the heat and mass transfer equations is necessary. During the second stage, only the first factor is important so that the drying equations can be integrated separately and the result substituted into the hydrodynamic equations.

Consequently, neglect of the interdependence of the hydrodynamics and heat and mass transfer processes and, separate solution of these problems is physically invalid and does not yield an adequate mathematical model.

In this paper an approximate analytical solution of the equations of hydrodynamics and heat and mass transfer in the acceleration zone of the dryer and for straight ducts is presented.

The following assumptions are made:
1. Particles are spherical
2. Particle motion is described by the following equation [2, 4]

$$m_p \frac{du_p}{d\tau} = \zeta \frac{\pi d_p^2}{8} \rho_g (u_g - u_p)^2 \qquad (1)$$

Using the following relationships

$$d\ell = u_p d\tau \quad ; \quad i = (u_g - u_p)/u_g$$

we obtain

$$-\frac{di}{d\ell} = \frac{K_i}{m_p} \frac{i^2}{1-i} \qquad (1a)$$

where

$$K_i = \zeta \frac{\pi d_p^2}{8} \rho_g$$

Other terms of the full hydrodynamic equation of particle motion [3] are small and as such are neglected [2].

3. Constant drying rate period is assumed.
4. Concentration of dry particles is small enough that the rate of change of temperature difference between particle surface and gas $\Delta(T_g - T_e)/\Delta\ell$ is negligible at every stage and one can describe $\Delta T(\ell)$ in terms of a discrete function.

Despite these simplifications the solution of the problem provides information on the dependence of i and X on the basic technological parameters, u_g, X_0, d_p, T_0, Q and ℓ_k.

The form of solution searched for is the following:

$$i = i_0 - \Psi(\Delta m_p / m_{p0}) \qquad (2)$$

where i_0 - solution for constant particle mass
Ψ - correction factor for variable mass of particle.

Using the method of perturbations [5] the following form of solution is assumed for Ψ.

$$\Psi = a_1 \tilde{m} + a_2 \tilde{m}^2 + a_3 \tilde{m}^3 + \ldots \qquad (3)$$

where $\tilde{m} = \Delta m_p / m_{p0}$

Introducing Eq. (3) and (2) into (1a) and neglecting terms of order of magnitude $(\tilde{m}^3)$ and higher one obtains:*

$$a_1 = 0 \qquad (4)$$

$$a_2 = \frac{K_i i_0^2}{2 K_M m_{p0}(A + i_0^n)} \qquad (5)$$

$$a_3 = \frac{K_i i_0^2}{3 K_M m_{p0}(A + i_0^n)} \left[1 - \frac{K_i i_0 (3 - 4 i_0)}{K_M m_{p0}(A + i_0^n)(1 - i_0)} \right] \qquad (6)$$

To obtain f itwas necessary to integrate the drying equation which for the constant -drying-rate period, has the following form:

$$\Delta h_v \frac{dm_p}{d\tau} = S_p \alpha_p (T_g - T_e) \qquad (7)$$

The dependence of α_p on i is subsequently taken as

$$\alpha_p = \frac{\lambda_g B (A + i^n)}{d_p} \qquad (8)$$

where A, B and n are dimensionless constants. Using Eq. (1a) we now have

$$f = \frac{d\tilde{m}}{d\ell} = \frac{K_M (A + i^n)}{1 - i}$$

where

$$K_M = \frac{S_p \lambda_g (T_g - T_e) B}{m_{p0} d_p \Delta h_v u_g}$$

As a rough approximation one can take $i \simeq i_0$. Solving Eq. (1a) we now get

$$\ln i_0 + \frac{1}{i_0} - 1 \approx \frac{2 K_i \ell}{m_{p0}} \qquad (9)$$

The Equation (9) cannot be solved analytically to yield i_0. Hence the left hand side of it must be approximated by a discrete function of i_0.

Transforming the left hand side using $(1-i_0)$ for larger i_0's and i_0 for smaller i_0's and obtaining the coefficients by the least squares method yields:

$$\frac{2K_i \ell}{m_{p0}} = \begin{cases} (1-i_0)^2/2 ; & 0.8 \leqslant i_0 < 1 \\ 0.1(1-i_0) + (1-i_0^2)/2 ; & 0.5 \leqslant i_0 < 0.8 \\ 0.77/i_0 - 1.2 ; & 0.1 \leqslant i_0 < 0.5 \\ 1/i_0 - 4.5 ; & 1 \leqslant i_0 \end{cases} \qquad (10)$$

* Translator's note: Equations (4) through (6) contain constants (K_M,A,n) which are defined later on this page.

To obtain the relationship, $\tilde{m}(\ell)$, the expression (10) is substituted into Eq. (8). Introducing $\gamma=m/\ell$ and subsequently

$$\frac{d\tilde{m}}{d\ell}=\frac{1}{m_{po}}\frac{d\gamma}{d\ell}\ell+\gamma$$

allows one to separate variables of Eq. (8).

As a result of this integration the following expressions are obtained:

$$\tilde{m}=\frac{K_M\left[2(A+1)\ell^{1/2}+n(2K_i/m_{po})^{1/2}\ell\right]m_{po}^{1/2}}{2(2K_i)^{1/2}}$$

$$\tilde{m}=1-\frac{1}{m_{po}}\left\{\ell_{Imax}\left[B_{II}+\left(\frac{\tilde{m}_{Imax}}{K_i\ell_{Imax}}\right)^2\right]\ell-B_{II}\ell^2\right\}^{1/2} \qquad (11)$$

$$\tilde{m}=K_M A_m\left[\ell-\ell_{IImax}-\ln\frac{2K_i m_{po}+\ell}{2K_i m_{po}+\ell_{IImax}}\right]$$

$$\tilde{m}=K_M A_m\left[\ell-\ell_{II\,max}-\ln\frac{7K_i m_{po}+\ell}{7K_i m_{po}+\ell_{IIImax}}\right]$$

where

$$B_{II}=0.25\,K_M m_{po} A_m$$

To integrate the Eq. (8) the averaged value of $A_m=A+i_m^n$ was used.

The value of T_g-T_e was considered constant for each stage of approximation. Here T_g may be expressed by the heat balance equation.

Substitution of Equations (11) and (4) through (6) in Equations (2) and (3) gives the solution of Equation (1) for the conditions of mass change during the drying process.

Analysis of the solution leads to the conclusion that the dependence of the final solution of the characteristic parameters of the process changes substantially as compared to the solution assuming constant mass. e.g. in the relationship $i(d_p)$ besides the term in $\sqrt{1/d_p}$ another term of $1/d_p^2$ appears, the weight of this term being proportional to ℓ.

The general form of this relationship is given by the following equation:

$$i=1-\frac{N_1\ell^{1/2}}{d_p^{1/2}}-\frac{N_2\ell}{d_p^2} \qquad (12)$$

This relationship for $i(d_p, \ell)$ allows one to simplify significantly the simultaneous solution of Equation (1a) and the mass transfer equation when polydispersity of the material is considered. This gives reduction of the computer time by 50 to 100 times. This is due to the fact that the numerical computation of the function $i(d_p, \ell)$ requires separate computation for each diameter for any given location in the duct. To obtain the function $i(d_p)$ it is necessary to perform the calculations for at least 50 diameters (range of particle size: 10-1000 μm)

A comparison of the analytical solution with the numerical one obtained by integration of Equations (1a) and (8) with the relationship $\Delta T_m(\ell)$ shows satisfactory agreement of both. For this ΔT_m in the analytical solution was taken after one iteration.

It has been proven that the accuracy of the present method allows one to use it for engineering dryer calculations. The range of application of this method is as follows:

$$a_3\tilde{m}^3 \ll i_0 \qquad \text{i.e.}\ \frac{K_i}{2K_M A_m}\left(\frac{X_m}{1+X_m}\right)\ll 1 \qquad (13)$$

Analysis of the accuracy of the expression Equation(12), proves that after correction of the coefficients for two points along the curve, 0.5% accuracy can be obtained.

Acknowledgements

Translated by Dr. Z. Pakowski, Łódź Technical University, Łódź, Poland. Typescript prepared by Purnima Mujumdar, Montreal, Canada
Edited by Dr. A.S. Mujumdar, Montreal, Canada

References

1. Babukha G.L., Rabinovich N.I.- "Mechanics and heat transfer in polydisperse gas supension flows", Naukova Dumka, (1969) (in Russian).
2. Timonin A.S., Mushtayev V.I., Planovskii A.N., Prygunov V.F., Tekh. Osn. Khim. Tekhn., XIV, 3 (1980)
3. Soo S.L. "Hydrodynamics of multiphase systems", Mir (1971)
4. Busroyd R. "Gas flow with suspended particles", Mir, (1975)
5. Kozdoba L.A. "Methods of solving nonlinear heat conduction problems",Nauka (1975)

DRYING OF GAS AND SOLIDS IN FLUIDIZED BEDS

K. Viswanathan and D. Subba Rao

Department of Chemical Engineering
Indian Institute of Technology,
Hauz Khas, New Delhi-110016, INDIA.

ABSTRACT

A comprehensive mathematical model is developed to describe drying of gas and solids in a fluidized bed taking into account the varying bubble characteristics along the length of the fluidized bed and the gas to particle mass transfer resistance in the dense phase. The predictions of the general model are compared with those of a simplified model developed earlier which leads to a criterion for the validity of the simplified model. A simple graphical procedure is developed to obtain the equilibrium relationship from fluidized bed experiments itself. Experiments are performed on the drying of air and silica gel in a batch fluidized bed. Experiments cover two bed weights, three particle sizes and six fold variation in the fluidization number (u_o/u_{mf}). The predictions of the model on the variation of outlet moisture concentration and outlet temperature with time compare excellently well with the experimental results. The Murphree efficiency in the present study varies from 1 to 0.5.

NOMENCLATURE

a = equals $\rho u_o A_t/M_p$, 1/s

a_o = equals $u_b/(K_{bc}L_f)$, dimensionless

a_1 = equals $(\alpha+\beta)\varepsilon_{mf} u_b/L_f$, 1/s

a_2 = equals $(1-\delta-\alpha\delta-\beta\delta)u_e\varepsilon_{mf}/(\delta L_f)$, 1/s

A = a parameter in the equilibrium relationship, $(g/kg)^{1-n}$

A_t = area of cross section of the bed, cm^2

b = equals aE, 1/s

b_o = equals $(1+\gamma_c K_r/K_{ce})$, dimensionless

b_1 = equals$(K_{bc}+K_{ce}+\gamma_c K_r+a_1/a_o)a_o/K_{ce}$, dimensionless

b_2 = equals $a_o\, a_1/K_{ce}$, dimensionless

b_g = equals $b_o(1+g)$, dimensionless

B = temperature independent parameter in the equilibrium relationship, $(g/kg)^{1-n}\,(^oC)^{-m}$

C = concentration of moisture, g/kg

C_{in} = inlet moisture concentration, g/kg

C_{out} = outlet moisture concentration, g/kg

C_{pg} = mean specific heat of air, Kcal/(kg deg. C)

C_{ps} = mean specific heat of solids, Kcal/(kg.deg.C)

C_s = equilibrium surface moisture concentration on solids, g/kg

d_1, d_2 = constants in homogenious solution, dimensionless

d_b = bubble diameter, cm

d_{bm} = maximum attainable bubble diameter, cm

d_{bo} = bubble diameter at the distributor, cm

$\bar{d}_b$ = representative average bubble diameter in a compartment, cm

d_p = particle diameter, cm

d_t = diameter of the fluidized bed, cm

D = diffussivity of moisture in air, cm^2/s

E = Murphree efficiency of the dryer, dimensionless

E_m = maximum attainable Murphree efficiency when gas to particle mass transfer resistance is absent, dimensionless

f = throughflow factor, dimensionless

f_1 = equals $K_{ce}(1+a_o m_1)[\exp(m_1 x_i)-\exp(m_1 x_{i-1})]/(a_2 m_1)$, dimensionless

f_2 = equals $K_{ce}(1+a_o m_2)[\exp(m_2 x_i)-\exp(m_2 x_{i-1})]/(a_2 m_2)$, dimensionless

g = equals $\gamma_c K_r (x_i - x_{i-1})(\gamma_e/\gamma_c + 1/b_o)/a_2$, dimensionless

g' = acceleration due to gravity, cm/s^2

h = height along the fluidized bed, cm

Δh = compartment height, cm

K_{bc} = gas exchange co-efficient from bubble to cloud-wake, 1/s

K_{ce} = gas exchange co-efficient from cloud-wake to emulsion, 1/s

K_{pg} = gas to particle mass transfer co-efficient, cm/s

K_r = effective rate constant based on volume of solids $6\ K_{pg}/(\varphi_s d_p)$, 1/s

L_f = length of expanded bed, cm

L_{mf} = length of bed at minimum fluidization, cm

m = exponent on temperature in the equilibrium relationship, dimensionless

m_1, m_2 = equal $(-b_1 \pm \sqrt{b_1^2 - 4b_o b_2})/(2b_2)$, dimensionless

M_p = mass of solids in the fluidized bed, gm

n = exponent on Q in the equilibrium relationship, dimensionless

N_E = exchange number in Eqn. 17, dimensionless

N_R = reaction number in Eqn. 18, dimensionless

p = equals $0.3/d_t$, 1/cm

Q = quantity of moisture/adsorbate on the solids, g/kg

r_i = equals $K_{bc} L_f/u_b$ evaluated at $d_b = \bar{d}_{bi}$, dimensionless

Sc = Schmidt number, dimensonless

t = time, s

T = temperature of the fluidized bed equals that of outlet air ^oC

u_b = absolute bubble rise velocity, cm/s

u_e = interstitial gas velocity in emulsion, cm/s

u_f = equals u_{mf}/ε_{mf}, cm/s

u_{mf} = minimum fluidization velocity, cm/s

u_o = inlet superficial gas velocity, cm/s

x = equals h/L_f, dimensionless

Δx_i = equals $\Delta h_i/L_f$, dimensionless

X = equals C_s/C_{so}, dimensionless

y = mole fraction of non-diffusing component, dimensionless

Y = equals C_s/C_{in}, dimensionless

Z = equals Q/Q_o, dimensionless

Greek Symbols

α = volume of wake per volume of bubble, dimensionless

β = volume of cloud per volume of bubble, dimensionless

γ = volume of solids per volume of bubble, dimensionless

δ = volume of bubbles per volume of bed, dimensionless

ε_{mf} = porosity at minimum fluidization, dimensionless

ρ = density of air, gm/cm^3

ρ_p = density of particles, gm/cm^3

φ_s = sphericity, dimensionless

Subscripts

b = bubble

c = cloud-wake

e = emulsion

i = compartment

N = topmost compartment

INTRODUCTION

Drying is an important and often critical operation in many chemical, food and pharmaceutical industries. In many instances, fluidized beds are used as dryers of solids and gases (1-10) (Drying of gas is also referred to as adsorption). Due to their remarkable temperature uniformity, fluidized beds allow easy temperature control and operation of the bed upto the highest temperature acceptable in relation to thermal degradation of solids is possible. Multiple stages are often used (1-7) to achieve uniform residence time and product quality.

In spite of the many industrial applications, theory and modelling of fluidized bed dryers have lagged far behind. Design procedures presented (6,7) for fluidized dryers are based on total heat and mass balance of the whole apparatus with hardly any attention being paid to the role of bubbles. Reinterpretations of published data on heat and mass transfer according to bubbling bed models (8,11,12) have been made by many investigators. But none of these models tried to analyse the unsteady (time-varying) characteristics that is necessary to describe drying of gas and solids in a fluidized bed.

However, some recent models available in literature (5,13-15) do consider the unsteady behaviour of fluidized bed adsorbers and dryers. The model of Vanderschuren (5) is not based on any of the bubbling bed models but is based on Cholette and Cloutier model (16) and contains two parameters which have to be estimated from experimental results. The model of Cranfield and Gliddon (15) is valid only for slow rising bubbles without clouds and is highly numerical. The model of Hoebink and Rietema (13,14) assumes a constant bubble size which is a parameter to be estimated and bubble growth is not considered. Thus all these models (5,13-15) either have some parameters to be estimated or are valid only under constrained situations. Furthermore, all these models require prior knowledge of equilibrium relationship between equilibrium surface concentration of the adsorbate (C_s) and the moisture/adsorbate loading (Q) on the particles. Often relationship between C_s and Q is assumed to be linear (8,14,15) which may not be quite valid (3,5,17).

Recently, a model was developed (18) to describe adsorption in a batch fluidized bed. It was demonstrated that the model can also be used for the estimation of the equilibrium relationship from simple batch experiments. The model was based on an assumption that the gas to particle mass transfer resistance in the dense phase is negligible.

In the present paper, a rigorous model is developed for fluidized bed dryers whereby the previously used assumption that gas to particle mass transfer resistance in the dense phase is negligible is removed. Further, unlike the previous model (18) which was developed only for fluidized bed adsorbers, the present model would be applicable for both fluidized bed adsorbers and dryers. A much simpler method of obtaining the equilibrium relationship than the one previously described (18) is also given. Finally, comparison of predictions of the model with experimental data would be presented.

MODEL FOR A FLUIDIZED BED DRYER

The model is based on semicompartmental approach (19) where the bed is divided into four phases as shown in Figure 1. The equations describing the hydrodynamics of the fluidized bed would then be (19-21) as given in Table 1. The model is based on the following assumptions.

(1) The solids are assumed to be well mixed. That is, the temperature (T), moisture loading on the particles (Q) and hence the equilibrium surface concentration (C_s) are independant of position in the fluidized bed and vary only with time.

(2) The bed temperature equals exit gas temperature (8).

(3) The wake fraction is given by (8)

$$\alpha = \frac{0.785 \exp(-66.3\, d_p)}{\rho_p(1-\varepsilon_{mf})} \qquad (1)$$

(4) The mass transfer co-efficient from gas to particle in the dense phase is given by (8)

$$K_{pg} = \frac{D}{\varphi_s d_p\, y}\left[\,2+1.8\, Re_{mf}^{0.5}\, Sc^{0.33}\right] \qquad (2)$$

(5) The equilibrium relationship is given by (3,18)

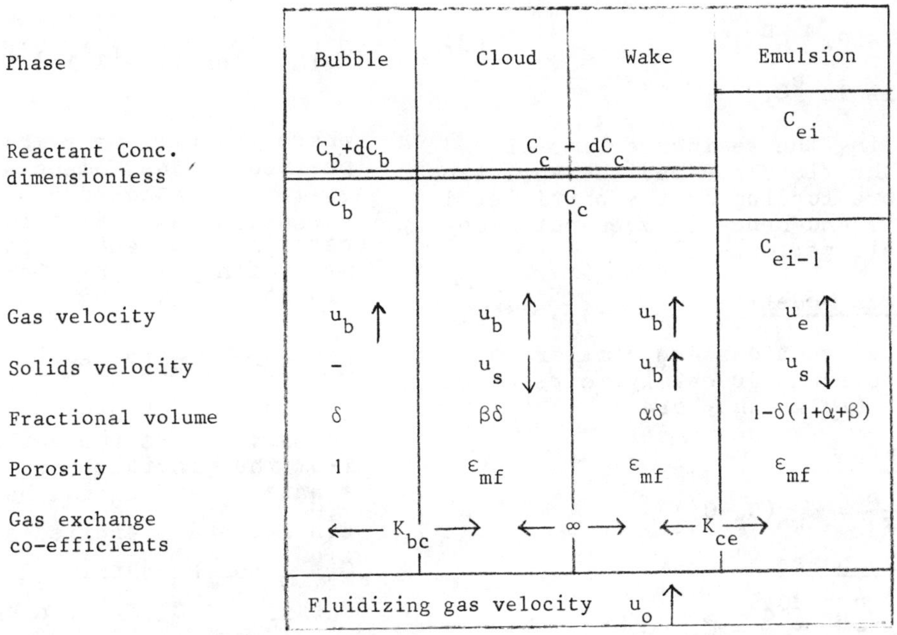

Figure 1 - Representation of the Fluidized Bed According to Semi-compartmental Model.

Table 1 - Hydrodynamic Equations Used in the Semicompartmental Model

Variable	Equation	Ref.
Bubble size	$d_b = d_{bm} - (d_{bm} - d_{bo})\exp(-ph)$	
	$d_{bm} = 0.65\ [A_t(u_o-u_{mf})]^{0.4}$	23
	$d_{bo} = 0.00376(u_o-u_{mf})^2$	
Compartment size	$\Delta h_i = \bar{d}_{bi} = \frac{1}{p}\ln[F + (1-F)\exp(pd_{bm})]$ for $i = 1$ to $N-1$, where $F = (1-d_{bo}/d_{bm})\exp(-ph_{i-1})$	19
	$\bar{d}_{bN} = d_{bm}/[1-\ln(d_{bN-1}/d_{bN})/(p\Delta h_N)]$	
Bubble rise velocity	$u_b = u_o - u_e\ \varepsilon_{mf} + 0.71\sqrt{g\ \bar{d}_{bi}}$	19,21
Emulsion gas velocity	$u_e\varepsilon_{mf} = u_{mf} - \alpha\ \delta\ u_b\ \varepsilon_{mf}/(1-\delta-\alpha\delta)$	19,21
Cloud size	$\beta = 3u_f/(u_b - u_f)$	19,20,21
Bubble fraction	$\delta = \dfrac{u_o - u_e\ \varepsilon_{mf}}{[1 + (\alpha + \beta)\varepsilon_{mf}]\ u_b - [1 + \alpha + \beta]\ u_e\ \varepsilon_{mf}}$	
Gas exchange co-efficients	$K_{bc} = 4.5\ \dfrac{u_{mf}}{d_b} + 5.85\ \dfrac{D^{0.5}\ g^{0.25}}{d_b^{1.25}}$	22
	$K_{ce} = 6.78\ \left(\dfrac{D\varepsilon_{mf}\ u_b}{d_b^3}\right)^{0.5} + 3\ \dfrac{(1-\delta)\ u_{mf}}{d_b(1-\delta-\alpha\delta)(1-u_f/u_b)}$	8,19
Gas to particle mass transfer coefficient	$K_{pg} = \dfrac{D}{\varphi_s d_p y}[2 + 1.8\ Re_{mf}^{0.5}\ Sc^{0.33}]$	8

$$C_s = AQ^n = BT^m\,Q^n \quad (3)$$

where T is in ^oC.

(6) During the residence time of gas in the fluidized bed the variation of moisture loading on the particles is negligible and hence C_s does not change during this period.

The General Model

The equations are similar to those described for catalytic reactors elsewhere (19). They are

Bubble phase

$$-\frac{u_b}{L_f}\frac{dC_b}{dx} = K_{bc}(C_b - C_c) \quad (4)$$

Cloud-wake phase

$$-(\alpha+\beta)\,\varepsilon_{mf}\frac{u_b}{L_f}\frac{dC_c}{dx} = K_{ce}(C_c - C_{ei}) - K_{bc}(C_b - C_c) + \gamma_c K_r (C_c - C_s) \quad (5)$$

Emulsion phase

$$-\frac{[1-\delta(1+\alpha+\beta)]}{\delta L_f} u_e\,\varepsilon_{mf}\frac{(C_{ei}-C_{ei-1})}{(x_i - x_{i-1})} = \gamma_e K_r (C_{ei} - C_s) - K_{ce}(\overline{C}_c - C_{ei}) \quad (6)$$

where K_r in Eqns. 5 and 6 is the effective rate constant for mass transfer from gas to particles based on unit volume of particles and is given by

$$K_r = 6K_{pg}/(\varphi_s\, d_p) \quad (7)$$

The equations have been written for upflow of gas in emulsion which only can be expected for low wake fractions for the present particles predicted by Eqn.1. However, for downflow of gas in emulsion, the procedure can be developed as described for catalytic reactors elsewhere (19).

The solution to the above equations can be obtained following exactly the same procedure as described elsewhere (19) as

$$C_b = C_{ei-1}/b_g + (1-1/b_g)C_s + d_1(\exp(m_1 x) + f_1/b_g) + d_2(\exp(m_2 x) + f_2/b_g) \quad (8)$$

$$C_c = C_b + a_o[m_1 d_1 \exp(m_1 x) + m_2 d_2 \exp(m_2 x)] \quad (9)$$

$$C_{ei} = (C_{ei-1} + f_1 d_1 + f_2 d_2 + gC_s)/(1+g) \quad (10)$$

where the various symbols used are explained in the Nomenclature. It may however be mentioned here that all the variables would vary for all the compartments and subscript 'i' has not been used with them just for the purpose of clarity.

Boundary Conditions

The inlet concentration to each compartment is the outlet concentration from the previous compartment. The constants d_1 and d_2 can be estimated from Eqns. 8 and 9 applied at $x = x_{i-1}$. Then C_{ei-0} can be estimated from Eqn. 10 and $C_b(x_i)$ and $C_c(x_i)$ can be estimated from Eqns. 8 and 9.

Since α is assumed to be constant, β decreases with compartment number, and δu_b remains constant, emulsion phase gets concentrated by gas shed out by clouds at compartments' interfaces according to

$$C_{ei+0} = \{\delta u_b(\beta_i - \beta_{i+1})C_c(x_i) + u_e\,\varepsilon_{mf}[1-\delta(1+\alpha+\beta_i)]\,C_{ei-0}\} / \{\delta u_b(\beta_i - \beta_{i+1}) + u_e\,\varepsilon_{mf}[(1-\delta(1+\alpha+\beta_i)]\} \quad (11)$$

$C_b(x_i)$, $C_c(x_i)$ and C_{ei+0} are the boundary conditions to (i+1)th compartment. At the grid, the initial boundary condition is

$$C_{in} = C_b(0) = C_c(0) = C_{e0} \quad (12)$$

Then proceeding from the first compartment concentration profiles can be obtained. The Murphree efficiency is defined by

$$E = (C_{in} - C_{out})/(C_{in} - C_s) \quad (13)$$

where the outlet concentration is given by

$$u_o\,C_{out} = \delta_N u_{bN}[C_b(1) + (\alpha+\beta_N)\,\varepsilon_{mf}\,C_c(1)] + [1-\delta(1+\alpha+\beta_N)]u_e\,\varepsilon_{mf}\,C_{eN} \quad (14)$$

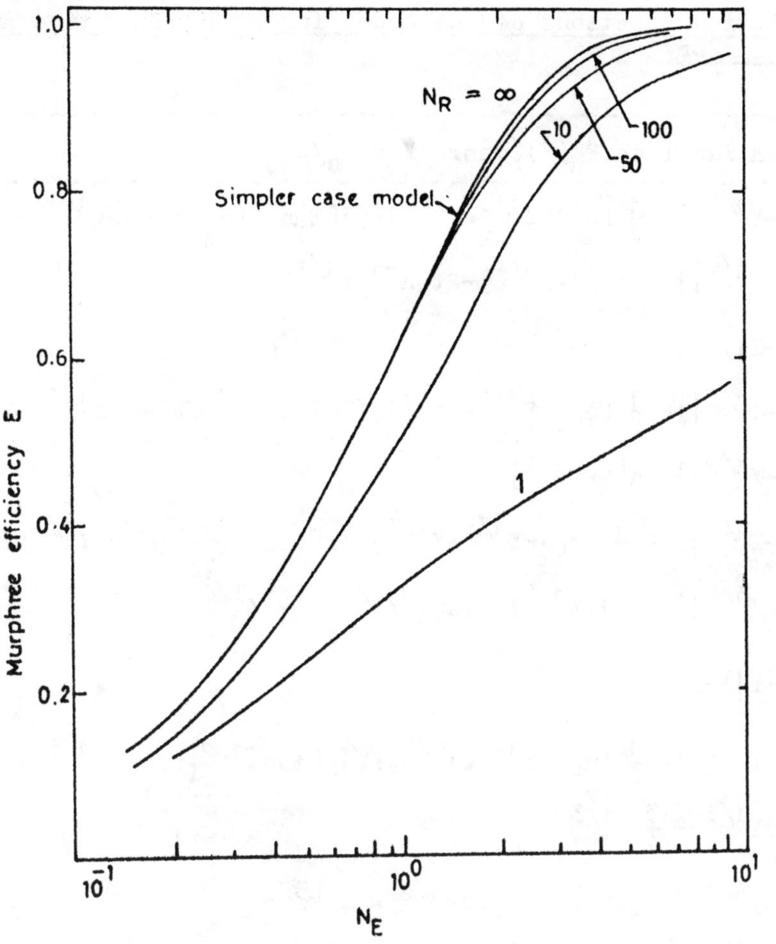

Figure 2 - Predictions of the General Model as a Function of Dimensionless Reaction and Exchange Numbers.

A Simplified Model

This model has been described elsewhere (18). Here gas to particle mass transfer resistance in the dense phase is assumed to be negligible. That is

$$C_c = C_e = C_s \quad (15)$$

The Murphree efficiency can then be proved (18) to be

$$E_m = 1 - \exp(-N_E) \quad (16)$$

$$N_E = \sum_{i=1}^{N} r_i \Delta x_i - \ln(\delta_N u_{bN}/u_o) \quad (17)$$

The Murpbree efficiency predicted by this simplified model is actually upper bound (maximum possible value at this bubbling condition). The general model reduces to the simplified model when N_R approaches infinity.

The predictions of the general model are given in Figure 2. Here N_R is the reaction number given by

$$N_R = K_r(1-\varepsilon_{mf})L_{mf}/u_o \quad (18)$$

It can be seen from Figure 2 that the predictions of the simplified model are never more than 5 per cent away from the predictions of the general model when the condition

$$N_R \geqslant 50 \quad (19)$$

is satisfied.

Unsteady State Equation

The material balance equation is

$$M_p \frac{dQ}{dt} = \rho u_o A_t (C_{in} - C_{out}) \quad (20)$$

From Eqns. 3,13 and 20 one obtains

$$\frac{dQ}{dt} = aE(C_{in} - BT^m Q^n) \quad (21)$$

Table 2 - The Solution Functions of the Model for both Adsorption and Desorption/Drying

n	Solution Function $F_n(Y)$ where $Y = C_s/C_{in}$
3/2	$-\ln[\lvert 1-Y^{1/3}\rvert]+\frac{1}{2}\ln[1+Y^{1/3}+Y^{2/3}]-\sqrt{3}[\tan^{-1}(\frac{1}{\sqrt{3}}+\frac{2}{\sqrt{3}}Y^{1/3})-\tan^{-1}\frac{1}{\sqrt{3}}]$
4/3	$-\ln[\lvert 1-Y^{1/4}\rvert]+\ln[1+Y^{1/4}]-2\tan^{-1}Y^{1/4}$
1	$-\ln[\lvert 1-Y\rvert]$
3/4	$-\ln[\lvert 1-Y^{1/3}\rvert]+\frac{1}{2}\ln[1+Y^{1/3}+Y^{2/3}]+\sqrt{3}[\tan^{-1}(\frac{1}{\sqrt{3}}+\frac{2}{\sqrt{3}}Y^{1/3})-\tan^{-1}\frac{1}{\sqrt{3}}]-3Y^{1/3}$
2/3	$-\ln[\lvert 1-Y^{1/2}\rvert]+\ln[1+Y^{1/2}]-2Y^{1/2}$
3/5	$-\ln[\lvert 1-Y^{1/3}\rvert]+\frac{1}{2}\ln[1+Y^{1/3}+Y^{2/3}]-\sqrt{3}[\tan^{-1}(\frac{1}{\sqrt{3}}+\frac{2}{\sqrt{3}}Y^{1/3})-\tan^{-1}\frac{1}{\sqrt{3}}]-\frac{3}{2}Y^{2/3}$
4/7	$-\ln[\lvert 1-Y^{1/4}\rvert]+\ln[1+Y^{1/4}]-2\tan^{-1}Y^{1/4}-\frac{4}{3}Y^{3/4}$
1/2	$-\ln[\lvert 1-Y\rvert]-Y$
3/7	$-\ln[\lvert 1-Y^{1/3}\rvert]+\frac{1}{2}\ln[1+Y^{1/3}+Y^{2/3}]+\sqrt{3}[\tan^{-1}(\frac{1}{\sqrt{3}}+\frac{2}{\sqrt{3}}Y^{1/3})-\tan^{-1}\frac{1}{\sqrt{3}}]-3Y^{1/3}-\frac{3}{4}Y^{4/3}$
2/5	$-\ln[\lvert 1-Y^{1/2}\rvert]+\ln[1+Y^{1/2}]-2Y^{1/2}-\frac{2}{3}Y^{3/2}$
1/3	$-\ln[\lvert 1-Y\rvert]-Y-\frac{1}{2}Y^2$

Similarly the energy balance equation is given by

$$\frac{dT}{dt} = \frac{a}{C_{ps}}\left[\frac{(C_{in}-C_s)E\,H_a}{1000} - (T-T_{in})C_{pg}\right] \tag{22}$$

Since Eqns. 21 and 22 are coupled differential equations they can be solved only numercially. However, assuming the variation in T^m to be negligible compared to variation in Q^n, Eqn.21 can be solved first analytically, the solution of which can be used later in Eqn. 22 to obtain the temperature variation. The solution to Eqn.21 can then be proved to be (18)

$$F_n(Y)-F_n(Y_o) = \frac{naEB^{1/n}}{C_{in}^{\frac{1}{n}-1}}\; t\, T^{m/n} \tag{23}$$

The solution functions were given elsewhere (18) for adsorption. It can be generalized to include the case of drying of solids (or desorption) as well whereby the solution functions become slightly modified and these are given in Table 2. However, when the inlet gas is perfectly dry then the following equation should be used.

Drying When Inlet Gas is Perfectly Dry

Here C_{in} equals zero. Then Eqn.21 can be solved to give

$$F_n(X) = \begin{cases} 1-X^{1-1/n} = (n-1)aEtA^{1/n}C_{s0}^{1-1/n} & \text{if } n \neq 1 \\ \ln(1/X) = aEAt & \text{in } n = 1 \end{cases} \tag{24}$$

where X equals C_s/C_{so}. In terms of moisture loading Eqn. 24 becomes

$$F_n(Z) = \begin{cases} 1-Z^{n-1} = (n-1)aEAtQ_o^{n-1} & \text{if } n \neq 1 \\ \ln(1/Z) = aEAt & \text{if } n=1 \end{cases} \tag{25}$$

where Z equals Q/Q_o.

Table 3 - Properties of Solid Particles and the Range of Experimental Conditions Studied

Fluidized bed diameter	=	7.0 cm
Range of u_o/u_{mf} studied	=	1 to 6
Bed weights	=	25gm;50 gm
Bed material	=	Silica gel

Size Range mm	d_p mm	ε_{mf}	u_{mf} Expt.	cm/s Calc.[27]
0.355-0.425	0.390	0.45	5.5	4.75
0.425-0.500	0.462	0.50	6.9	6.59
0.500-0.600	0.550	0.56	9.1	9.16

Since the solution functions given in Eqns. 23 to 25 are very general in the sense that they can be used for the design and scale-up of fluidized dryers, numerical values of these functions have been tabulated elsewhere (10) for ready usage.

For constant rate drying period, C_s is a constant (C_{so}) equal to the equilibrium concentration at the prevailing wet bulb temperature. Then Eqns. 13 and 20 give

$$Q = Q_o - aE(C_{so} - C_{in})t \qquad (26)$$

EXPERIMENTAL AND PARAMETERS ESTIMATION

The details of the experimental set up and the procedure of experimentation have been described elsewhere (18). However, the experiments that are to be reported here cover a much wider range of variables than before (18). The properties of solid particles and the operating conditions studied are summarised in Table 3.

The experiments on drying of air (adsorption) were carried out as explained before (18). Then the wet silica gel was dried in a stream of partially dry air. After the desorption run was over the silica gel was regenerated in an electrically heated fluidized bed for extended periods.

Determination of Exchange Co-efficient, K_{bc}

In order to test the validity of the model, the first step is to choose an expression for bubble to cloud exchange co-efficient, K_{bc}, which critically affects the efficiency through the parameter r in Eqn. 17. It is given by

$$K_{bc} = 1.5 \frac{f\, u_{mf}}{d_b} + 5.85 \frac{D^{0.5}\, g^{0.25}}{d_b^{1.25}} \qquad (27)$$

where f is the throughflow factor which affects the cloud size β as

$$\beta = f\, u_f/(u_b - u_f) \qquad (28)$$

and the parameter f is predicted to be widely different by different models as

$$f = \begin{cases} 3 & \text{Davidson-Harrison } (22) \\ 1 & \text{Murray } (23) \end{cases} \qquad (29)$$

and empirical correlations (24,25) predict f to be 1.3. Since so much uncertainty exists, it is proposed to obtain f in the following manner.

Since bed expansion is critically affected by the throughflow term, bed expansion measurements were made for the present silica gel. The probe used to measure expanded bed length L_f was a 6 mm diameter hollow copper tube with one end sealed and having four equispaced holes of 1 mm diameter on its circumference. The other end was connected to an inclined manometer with water as the liquid. The vertical position in the fluidized bed at which the pressure indicated by the manometer equals zero gives the expanded bed length L_f. The per cent expansion was calculated as

$$\text{Per cent expansion} = 100(L_f - L_{mf})/L_{mf} \qquad (30)$$

The experimental per cent expansion for the present silica gel is shown in Figure 3. The per cent expansion was calculated for the two models in Eqn.29 and the predictions are also shown in Figure 3. It can be clearly seen that Davidson-Harrison model (22) is much better than Murray model (23). To confirm this further, bed expansion measurements were made with sand as the bed material as shown in Figure 4. It can be concluded from Figures 3 and 4 that f equals 3 and hence Eqn. 27 becomes

$$K_{bc} = 4.5 \frac{u_{mf}}{d_b} + 5.85 \frac{D^{0.5}\, g^{0.25}}{d_b^{1.25}} \qquad (31)$$

Estimation of Equilibrium Relationship

A mathematical procedure was evolved earlier (18) to determine the equilibrium relationship from fluidized bed experiments itself. Here a much simpler method is given.

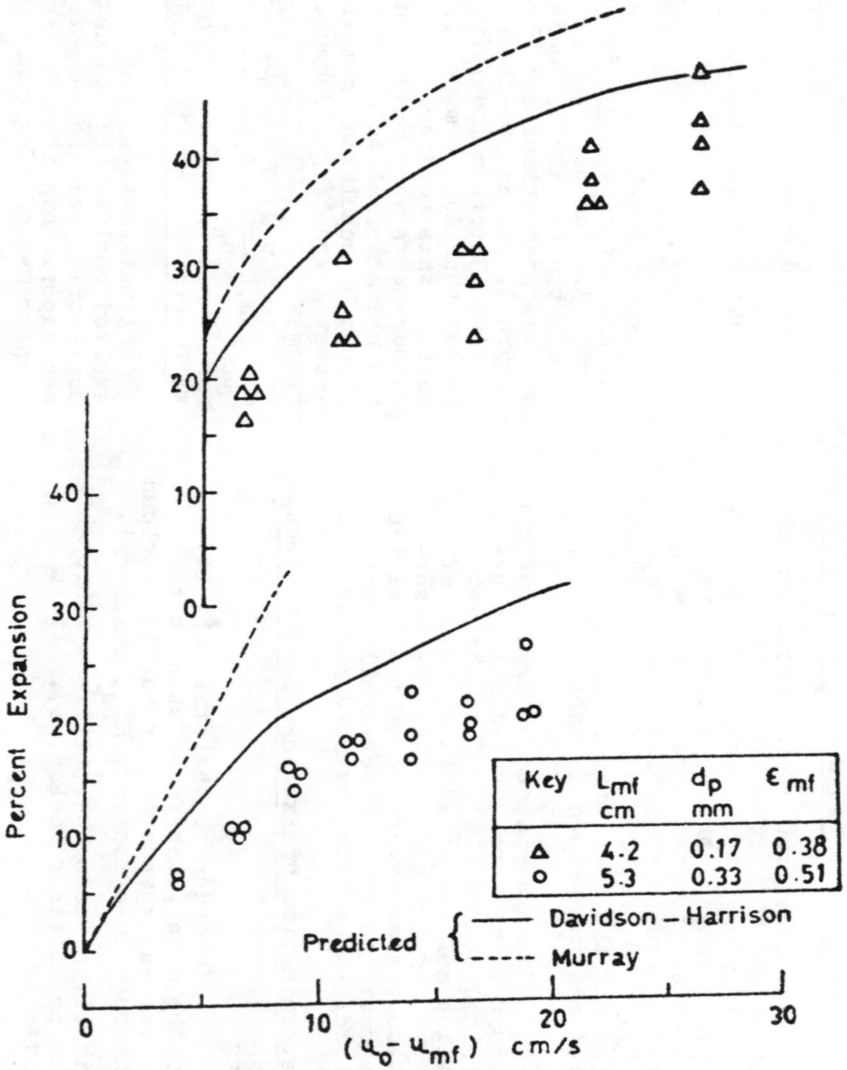

Figure 4 - Per cent Expansion for Sand Bed

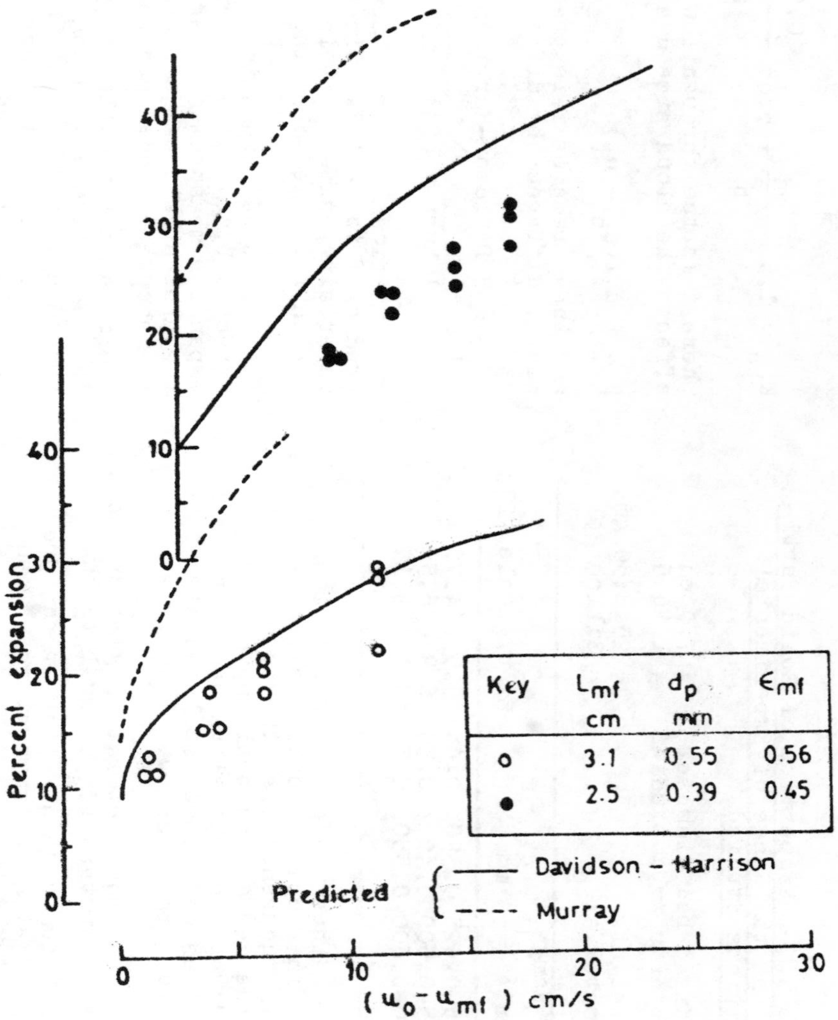

Figure 3- Per cent Expansion for Silica Gel Bed

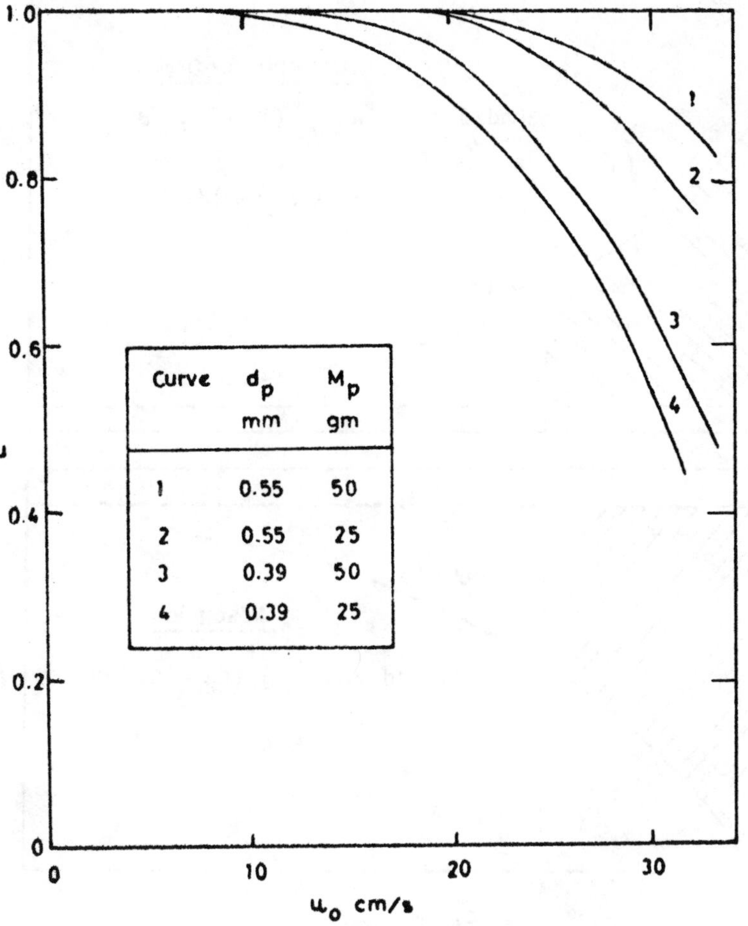

Figure 5 - Predicted Variation of Murphree Efficiency With Fluidizing Velocity With Particle Size and Bed Mass as Parameters

Initially, using the relationship obtained for K_{bc}, the Murphree efficiency is calculated using the model and plotted as a function of the operating conditions as shown in Figure 5. From here, the conditions corresponding to E equal to unity are chosen as the operating conditions for determining the equilibrium relationship. This is because under these conditions the effect of gas to particle mass transfer characteristics and bubbling characteristics are eliminated, and Eqn. 13 indicates C_{out} to be equal to C_s. Then the moisture loading on the particles Q can be obtained as a function of time by graphical integration as explained in Figure 6. Eqn. 20 gives

$$Q = Q_o + a \int_0^t (C_{in} - C_{out})dt \qquad (32)$$

where Q_o equals zero for the adsorption runs in which the initial adsorbent is perfectly dry. Since E equals unity, Eqn. 13 shows that C_{out} equals C_s. Thus a set of values (C_s, Q, T) can be obtained which can be used to obtain the parameters B,m and n in Eqn. 3 by logrithmic linear regression. Since analytical expressions for the solution function in Eqn. 23 as given in Table 2 exist only for certain values of n, the n-value nearest to the value obtained by linear regression and for which analytical solution exists (in Table 2) can be chosen. The m and B values may be recalculated using this n-value. The equilibrium relationships thus obtained are

$$C_s = 0.0165\ T^{1.54}\ Q^{3/7} \quad \text{Adsorption}$$
$$C_s = 0.0219\ T^{1.54}\ Q^{4/7} \quad \text{Desorption} \qquad (33)$$

Typical experimental runs from which these relationships were obtained are shown in Figures 7 and 8. The calculated variation of outlet humidity and

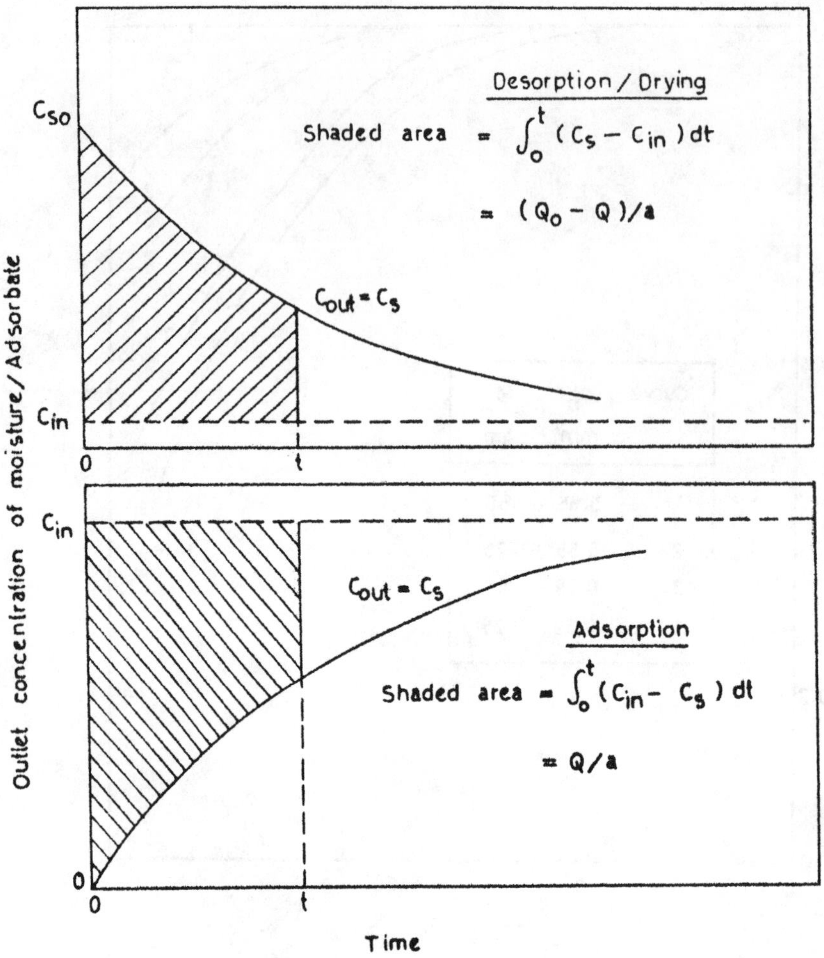

Figure 6 - Illustration of the Method of Determination of Moisture Loading on Particles for Obtaining the Equilibrium Relationship.

temperature with time is also shown in these figures. The good comparison of experimental and predicted profiles indicates that the model representation of equilibrium relationship is adequate and the method of estimating the equilibrium relationship from fluidized bed experiments is quite elegant and hence can be applied to any system. In the calculated profiles, the specific heat of solids C_{ps} was assumed (26) to be 0.316 Kcal/(Kg deg C) and the differential heat of adsorption/drying was estimated by a method to be described below.

In the previous publication (18), the equilibrium relationship for adsorption was found to be

$$C_s = 0.035\ T^{1.1}\ Q^{4/7} \tag{33a}$$

The temperature varied between roughly 25°C and 40°C earlier (18) whereas in the present study it varied in a lower range. Hence there exists a small difference in the forms of equilibrium relationship obtained then and now. The predictions of Eqns. 33 and 33a are shown graphically in Figure 9 in which for 10°C and 20°C Eqn. 33 is used and for 30°C and 40°C Eqn. 33a is used. Thus it can be concluded that the fluidized bed can be an effective tool in obtaining the equilibrium relationships.

Estimation of Heat of Adsorption H_a

The heat transfer balance can be expressed as (from Eqns. 13 and 22)

$$\frac{dT}{dt} = \frac{a}{C_{ps}}\left[\frac{(C_{in} - C_{out})\,H_a}{1000} - (T - T_{in})\,C_{pg}\right] \tag{34}$$

The thermocouples do have a response time which means a correction has to be applied (6) to obtain the actual tem-

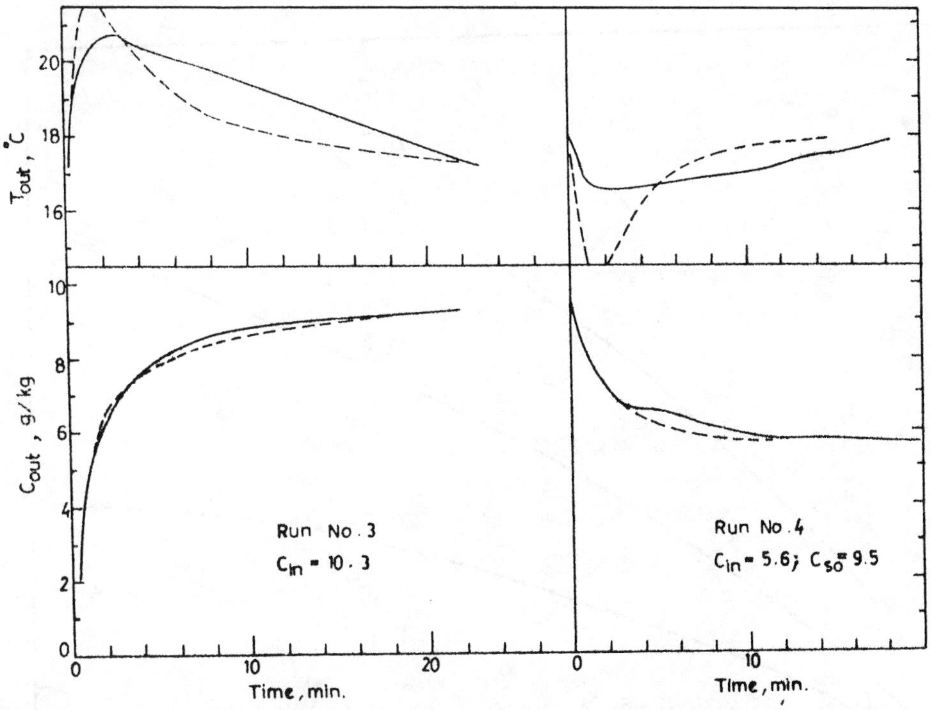

Figure 7 - Comparison of Predicted and Experimental Variation of Outlet Moisture Concentration and Temperature with Time. d_p = 0.55 mm; u_o = 15.8 cm/s; M_p = 25 gms; E = 1.

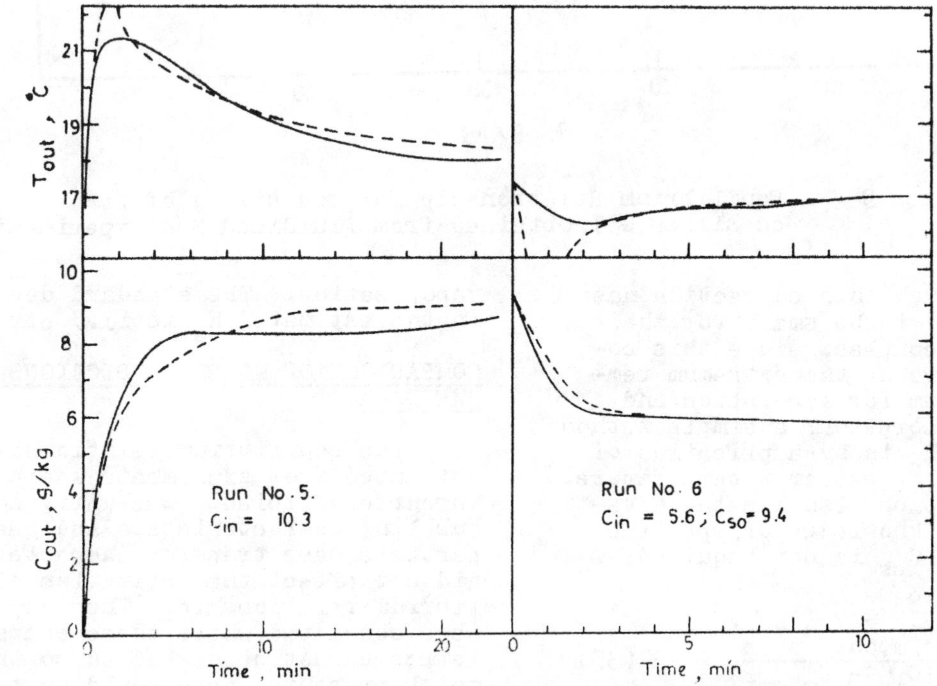

Figure 8 - Comparison of Predicted and Experimental Variation of Outlet Moisture Concentration and Temperature With Time. d_p = 0.55 mm; u_o = 18.4 cm/s; M_p = 25 gms; E = 1.

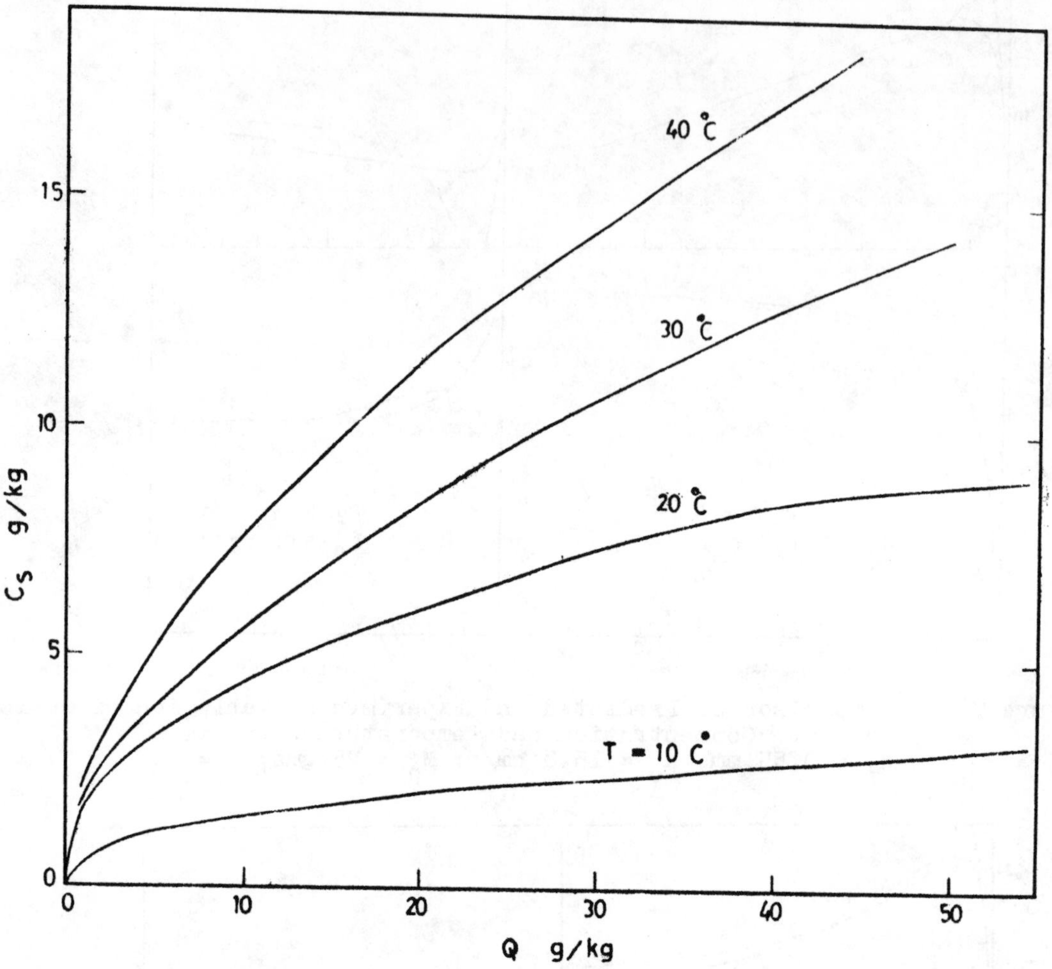

Figure 9 - Equilibrium Relationship for the Drying of Air on Silica Gel Obtained from Fluidized Bed Experiments.

perature. Though this correction has been found (18) to be small for the present thermocouples, since this correction is zero at the extremum temperature-maximum for adsorption and minimum for desorption, a simple method of estimating H_a is by application of Eqn. 34 at the extremum temperature. Such an estimation also has the advantage that the knowledge of specific heat of solids C_{ps} is not required.

Eqn. 34 leads to

$$H_a = 1000\ C_{pg} \frac{(T_{m,out} - T_{in})}{(C_{in} - C_{m,out})} \qquad (35)$$

Substituting C_{pg} equal to 0.25 Kcal/(Kg. deg C) (26) the heat of adsorption was estimated and estimations for 18 typical runs for the two extreme particle sizes are summarised in Table 4. It can be seen that the average value of 490 Kcal/Kg. is close to heat of vaporisation. The standard deviation in the estimated H_a is 13.5 per cent.

COMPARISON OF MODEL PREDICTIONS WITH THE DATA

The equilibrium relationship was estimated from expriments where the Murphree efficiency was unity so that bubbling characteristics and gas to particle mass transfer characteristics did not affect the estimation of equilibrium relationship. Then experiments were conducted where these characteristics cannot be neglected so that the model presented here could be tested. Typical experimental runs alongwith the predictions are shown in Figures 10 to 13. Since these runs were not used for estimating the equilibrium relationships, the excellent comparison of the predictions with the experimental results attests to the essential correctness of the assumptions of the model.

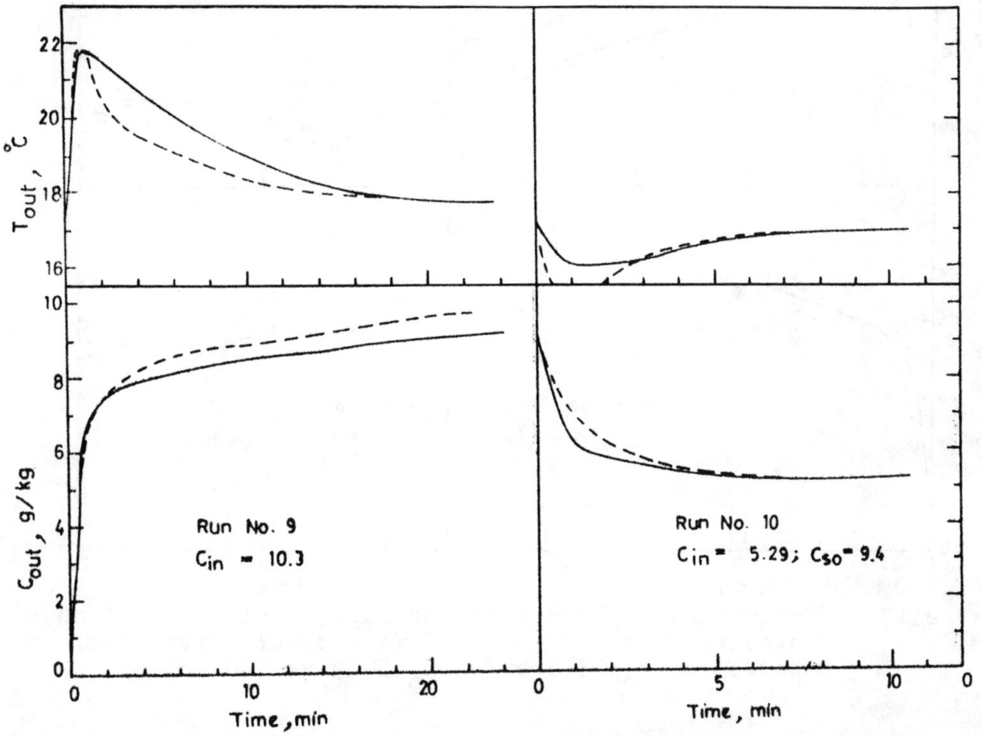

Figure 10 - Comparison of Predicted and Experimental Variation of Outlet Moisture Concentration and Temperature With Time. d_p = 0.55 mm; u_o = 23,6 cm/s; M_p = 25 gms; E = 0.96.

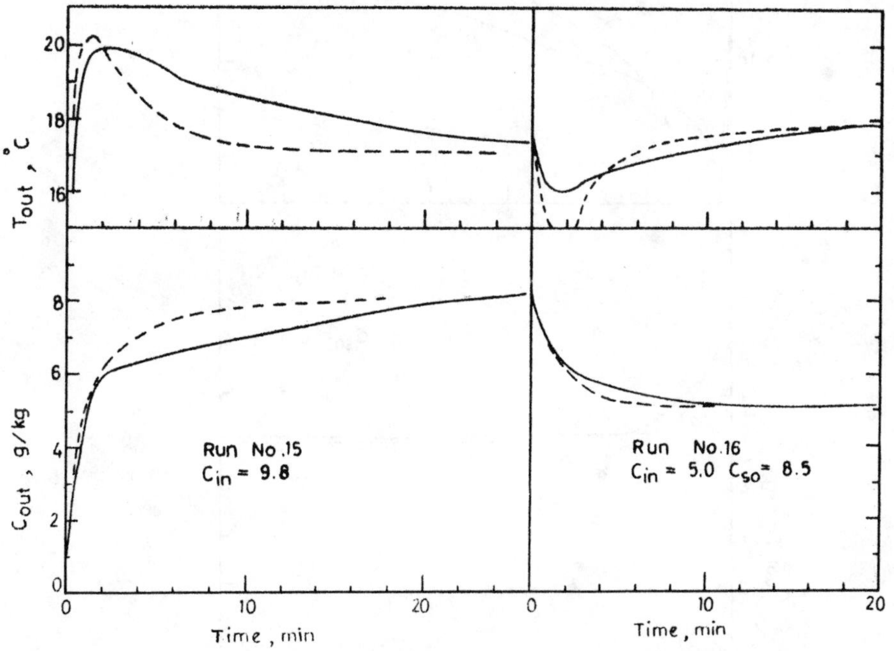

Figure 11 - Comparison of Predicted and Experimental Variation of Outlet Moisture Concentration and Temperature With Time. d_p = 0.39 mm; u_o = 21.0 cm/s; M_p = 25 gms; E = 0.87.

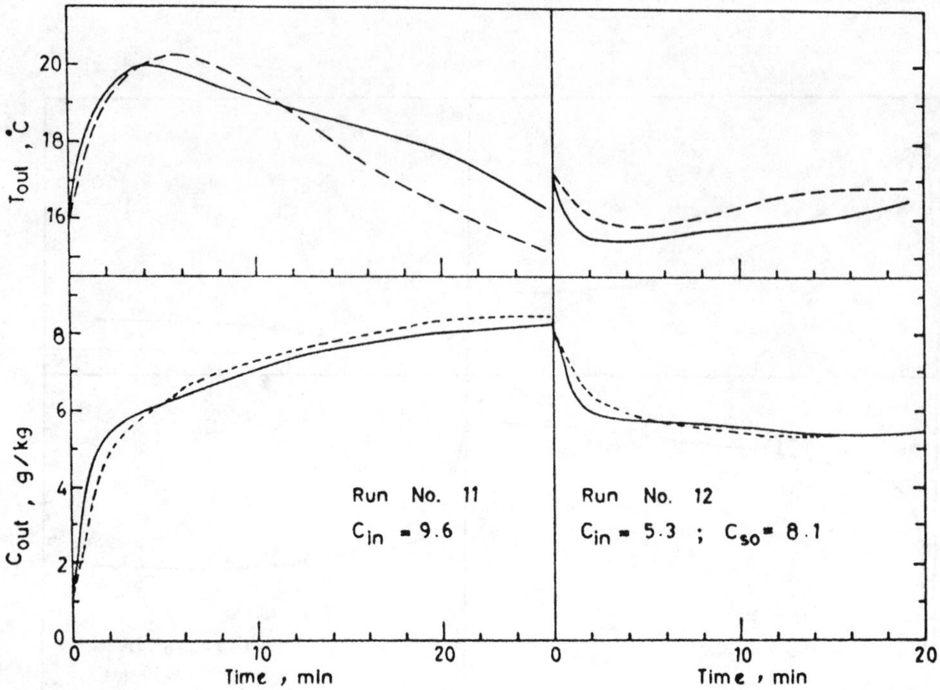

Figure 12 - Comparison of Predicted and Experimental Variation of Outlet Moisture Concentration and Temperature With Time. d_p=0.39mm; u_o=23.6 cm/s; M_p=50 gms; E=0.86.

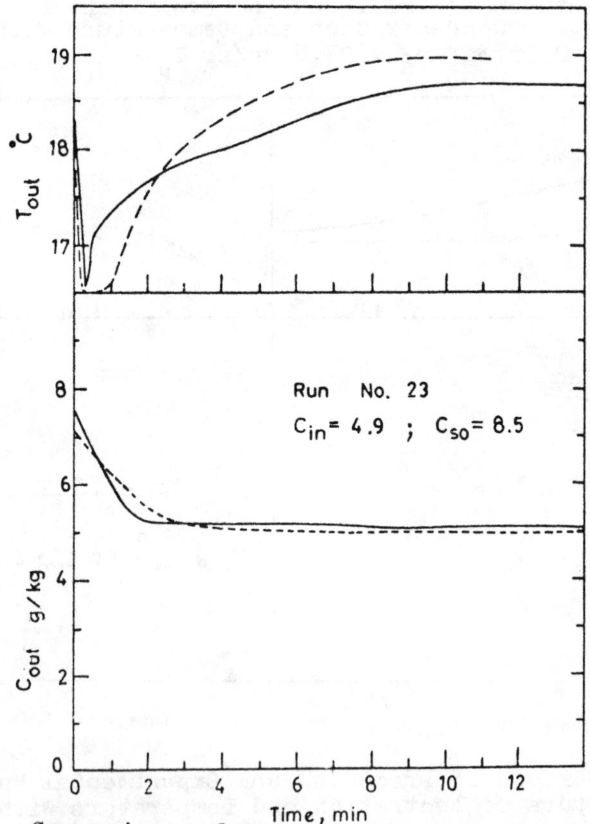

Figure 13 - Comparison of Predicted and Experimental Variation of Outlet Moisture Concentration and Temperature With Time. d_p=0.39 mm; u_o= 31.1 cm/s; M_p = 25 gms; E = 0.5.

Table 4 - Summary of Heat of Adsorption for 18 Typical Runs with the Maximum and Minimum Particle Sizes

Run No.	$T_{m,out}$ °C	T_{in} °C	$\bar{C}_{m,out}$ g/kg	C_{in} g/kg	H_a Kcal/kg
			ADSORPTION RUNS		
1.	20.6	14.5	7.5	10.3	545
3.	20.8	14.5	6.8	10.3	450
5.	21.3	14.5	6.4	10.3	436
7.	20.7	14.5	6.8	10.3	443
9.	21.8	14.5	7.0	10.3	553
11.	20.0	13.5	6.0	9.6	452
13.	19.1	13.1	6.8	9.5	565
15.	20.0	13.8	6.0	9.8	408
17.	19.1	13.7	7.1	9.7	519
			DESORPTION RUNS		
2.	16.5	18.0	6.4	5.7	536
4.	16.5	18.0	6.6	5.6	375
6.	16.2	17.0	6.0	5.6	500
8.	16.6	18.0	6.0	5.3	500
10.	16.0	17.5	6.2	5.2	375
12.	15.5	17.0	6.0	5.3	536
14.	15.3	16.5	5.1	4.6	600
16.	16.0	16.0	6.1	5.0	455
18.	16.2	18.0	6.8	6.0	563

Further, the good comparison is further proof of the validity of the exchange co-efficient (given by Davidson-Harrison model (22), f equal to 3 in Eqn. 27) obtained here. This means that the method of obtaining the throughflow factor f from bed expansion measurements can be effectively used for many systems.

CONCLUSIONS

A comprehensive mathematical model was evolved to describe drying of gas and solids in a batch fluidized bed. The predictions of the general model were compared with that of a simplified model developed earlier (18) which led to the development of a criterion for the validity region of the simpler model. All the present experiments satisfy the criterion given in Eqn.19. The mathematical procedure developed earlier (18) to solve the non-linear material balance differential equation for adsorption was extended and generalized to include the case of drying (or desorption) of solids as well.

The experiments on bed expansion were used to ascertain the value of the throughflow factor predicted differently by different models (Figures 3 and 4). Then using the throughflow factor so estimated in the exchange equation (Eqn.27) the Murphree efficiency was calculated as a function of the operating parameters (Figure 5) and this was used to determine the conditions corresponding to E equal to unity under which experiments needed to be performed to obtain the equilibrium relationship.

A new method, much simpler than the one developed earlier (18), was developed to determine the equilibrium relationship from simple fluidized bed experiments itself. The special property that enables fluidized bed to be an effective tool in establishing equilibrium relationships is the well mixed character of solids (because of which Q is a constant for all the particles at any time). A new method was also given to estimate the heat of adsorption/drying.

Experiments were conducted under conditions where bubbling characteristics could not be ignored. The predictions of the model showed excellent agreement with the data on the variation of both the outlet moisture concentration and temperature with time. This means that the bubbling characteristics assumed in the model are quite adequate and the model can be reliably used for the design and scale-up of fluidized bed dryers.

ACKNOWLEDGEMENT

The financial assistance for a part of this project provided by the Department of Science and Technology is gratefully acknowledged.

REFERENCES

1. Cox, M., Trans. Inst. Chem. Engrs., 36, 29 (1958).
2. Ermenc, E.D., Chem. Eng. Progr., 52, 488 (1956).
3. Ermenc, E.D., Chem. Eng., May 29, 68, 87 (1961).
4. Rowson, H.M., Brit. Chem. Eng., 8, 180 (1963).
5. Vanderschuren, J., Chem. Eng. Jnl., 21, 1 (1981).
6. Vanecek, V., M. Markvart and R. Drbochlov, 'Fluidized Bed Drying', English Translation by H. Landau, Leonard Hill, London, 1966.

7. Romankov, P., in 'Fluidization' edited by J.F. Davidson and D. Harrison, Academic Press, London, 1971.

8. Kunii, D. and O. Levenspiel, 'Fluidization Engineering', John Wiley and Sons Inc., New York, 1969.

9. Nonhebel, G. and A.A.H. Moss, 'Drying of Solids in the Chemical Industry', Butterworth, London, 1971.

10. Viswanathan, K., D.S. Rao and B.C. Raychaudhury, Ind. Chemical Engr., Accepted for publication (1982).

11. Kato, K and C.Y. Wen, Chem. Eng. Progr. Symp. Ser. No. 105, 66, 100 (1968).

12. Beek, W., in 'Fluidization' edited by J.F. Davidson and D. Harrison, Academic Press, London, 1971.

13. Hoebink, J.H.B.J. and K. Rietema, Chem. Eng. Sci., 35, 2135 (1980).

14. Hoebink, J.H.B.J. and K.Rietema, Chem. Eng. Sci., 35, 2257 (1980).

15. Cranfield, R.R. and B.J. Gliddon, Inst. Chem. Engrs. Symp. Ser. No. 38, Paper H4 (1974).

16. Cholette, A. and L. Cloutier, Can. J. Chem. Eng., 37 , 105 (1959).

17. Carter, J.W. and D.J. Barett, Trans. Inst. Chem. Engrs., 51, 75 (1973).

18. Viswanathan, K., D. Khakhar and D.S. Rao, Accepted for publication in Chem. Eng. Commun., (1982).

19. Viswanathan, K., Accepted for publication in Ind. Eng. Chem. Fundam., (1982).

20. Viswanathan, K. and D.S. Rao., Accepted for publication in Int. J. Multiphase Flow, (1982).

21. Viswanathan, K., and D.S. Rao., Proc. 33rd IIChE Conference at New Delhi, 1, 61 (1980).

22. Davidson, J.F., and D. Harrison, 'Fluidized Particles', Cambridge University Press, Cambridge, 1963.

23. Mori, S. and C.Y. Wen, AIChE J, 21, 109 (1975).

24. Sit, S.P. and J.R. Grace, Chem. Eng. Sci., 36, 327 (1981).

25. Grace, J.R., ACS Symp. Ser. No.168, Chemical Reactors, p-1 (1981).

26. Perry, J.H.,'Chemical Engineers Handbook', McGraw Hill Kogakusha Ltd., 5th ed., 1973.

27. Wen, C.Y. and Y.H. Yu, AIChE J, 12, 610 (1966).

FLUID MECHANICS, HEAT TRANSFER AND DRYING IN SPOUTED BEDS WITH DRAFT TUBES

J. K. Claflin and A. G. Fane

School of Chemical Engineering and Industrial Chemistry
University of New South Wales, Kensington, Australia 2033

ABSTRACT

This paper compares fluid mechanical and thermal performance of spouted beds with and without draft-tubes. The draft-tube beds are shown to provide well regulated particle residence times, lower pressure drop, and annular air flows similar to conventional beds. Contact efficiency, based on batch heating and drying of wheat, was found to be in the range 70 to 95% depending on conditions. The conventional beds were marginally more efficient.

NOMENCLATURE

c_p, c_{ps}	heat capacity of gas, of solids (J kg^{-1}°C^{-1})
K	heat loss coefficient (eqn. A.1) (W°C^{-1})
L_E	separation distance (m)
$\dot{m}$	mass flow rate of air (kg s^{-1})
M	mass of bed solids (kg)
P	pressure (Pa)
Q	volumetric flow rate of air (m³ s^{-1})
t	time (s)
T	temperature (°C)
Δw	moisture loss (kg)
Z	bed depth (m)
λ	latent heat of water (J kg^{-1})
η, η_D	contacting efficiency (defined by eqns. A.2 or (5))

Subscripts

a	ambient
A	annulus
b	bed
e	exit
i	inlet
o	bed at t = o
S	spout
T	total

INTRODUCTION

Spouted beds offer an alternative to fluidisation for gas/solids contacting of particles >1mm diameter. Conventional spouting is achieved by introducing a high velocity jet of fluid into the bottom of a bed of particles. Potential advantages of the spouted bed over fluidisation are the lower pressure drop (typically 0.7 x ΔP_f), the more regular particle movement, and the slightly lower superficial velocities required for spouting (particularly for vessels >0.5m diameter) (1).

When draft-tubes are introduced into the spouted bed, these advantages may be accentuated (2-6). In this paper we compare the fluid mechanics of draft-tube spouted beds with conventional spouted beds, and describe their gas/solid heat transfer and drying capabilities.

EXPERIMENTAL

Spouted beds of 0.3m diameter fitted with a 60° cone and air inlet diameter of 0.05m were used. Most studies were in a cylindrical bed but a few experiments used a half cylinder bed fitted with a perspex front face. Draft tubes were aligned axially and varied in length from 0.8 to 1.3m with diameters from 0.05 to 0.08m. Three types of draft-tube were used, solid wall, porous wall and composites with a porous lower section and solid wall upper section. System pressures were measured by manometers and temperatures by copper-constantan thermocouples. Bed materials were usually wheat (5.5mm x 3.2mm, equal volume sphere diameter of 3.7mm) but some experiments used Celcon (acetyl copolymer with mean diameter 2.5mm).

Several fluid mechanical parameters were investigated. Particle motion was observed through the perspex front face of the half cylinder bed, and the downward velocity of particles in the annulus was measured by noting the time of passage of individual grains. Air flow distribution was examined by comparing annular air flow (obtained from differential pressure measurements) with the known input flow. Minimum operating velocities were measured by reducing gas flow until the bed reached minimum spouting conditions for the conventional bed, or until choking occurred in the draft-tube beds. System pressure drops and axial pressure profiles were measured for a range of operating conditions.

Thermal studies involved the measurement of batch heating kinetics and batch drying kinetics. Batch heating kinetics were measured with a bed depth of 0.4m, being equivalent to a 15-18 kg load. The draft-tube used was copper, 1.3m long x 0.05m diameter, with a separation distance (air inlet to bottom of draft-tube) of 0.15m. Inlet air temperatures in the range 80°C to 120°C were used and air flowrates were 1.2 or 1.5 x minimum operating conditions. Bed temperatures were measured using a bare thermocouple located in the annulus, 0.26m above the inlet, and exit gas temperatures were measured above the bed.

Batch drying kinetics were measured for wheat with an initial moisture content of about 13 wt% (dry basis). Samples taken from the bed during each run were analysed for moisture by evaporation to dryness (according to AACC Method 44-15A).

RESULTS AND DISCUSSION

A. Fluid Mechanics

Particle movement in a conventional spouted bed follows a regular cycle from spout, fountain to annulus. However, because particles may enter the spout at any point along its length the residence times may vary widely; this does not happen if a draft-tube is used. Figure 1 shows this by comparing the conventional and draft-tube beds in terms of the radial variation of the annular residence time ratio (= residence time for particle at outer wall/residence time of particle). The factors controlling the magnitude of the particle residence times in the draft-tube system are gas flow, bed depth and draft-tube geometry, as detailed elsewhere (2).

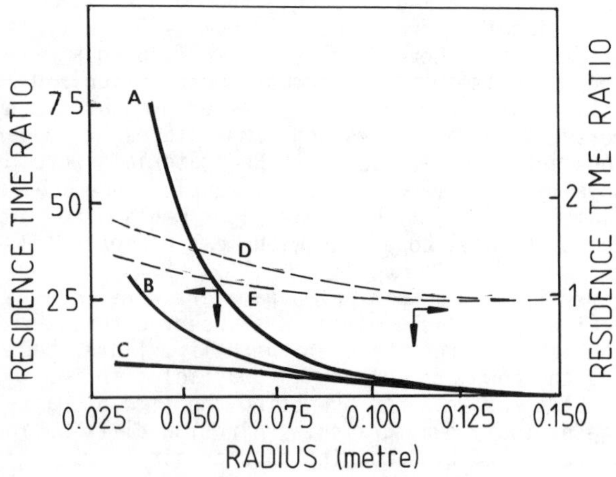

Figure 1. Residence Time Ratio vs Radius (- Conventional Bed; --- Draft Tube; Airflow A=0.035, B=0.032, C=0.028, D=0.035, E=0.021m³ s⁻¹)

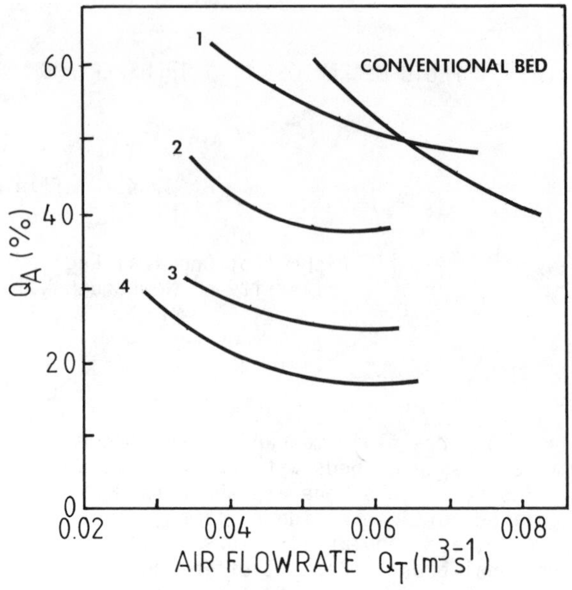

Figure 2. Percent Annular Airflow vs Total Airflow ((1) 1.3m draft-tube, L_E=0.15; (2)-(4), 0.8m tube, L_E=0.20, 0.15, 0.10m)

From a gas/solid contacting viewpoint the distribution of gas between spout and annulus is important. Figure 2 shows the percent annular airflow ($Q_A\% = (Q_A/Q_T) \times 100$) for a conventional bed and various beds with solid-wall draft-tubes. Annular flow is more readily achieved in conventional beds unless the draft-tube bed has a long draft-tube or a large separation distance (L_E = distance from air inlet to bottom of draft-tube). Enhanced annular air flow can also be achieved by use of a porous wall draft-tube, which permits gas exchange between spout and annulus and at the same time maintains control of particle movement. Figure 3 compares annular air profiles for solid-wall, porous wall and a composite draft-tube. The maxima in the profile for the porous tube was characteristic of this type of draft-tube, and can be explained by variations in axial pressure gradients (2). Flow reversal from annulus to spout can be avoided by using a composite-tube with a solid wall upper section.

The axial pressure gradient dP/dZ for the conventional bed increased towards the top of the bed, and this is because the higher up the spout the greater the particle load. With a porous draft-tube dP/dZ decreased with height as it would in a pneumatic conveyor. At the bottom of the bed the spout pressure was greater than the annulus pressure, but further up the bed the position was reversed. These data agree with the observed maxima in the profiles of gas distribution for porous draft-tubes (Figure 3). Included in Figure 3 is the annular gas profile for the porous draft-tube system predicted by a fluid-mechanical model (2). This model assumes

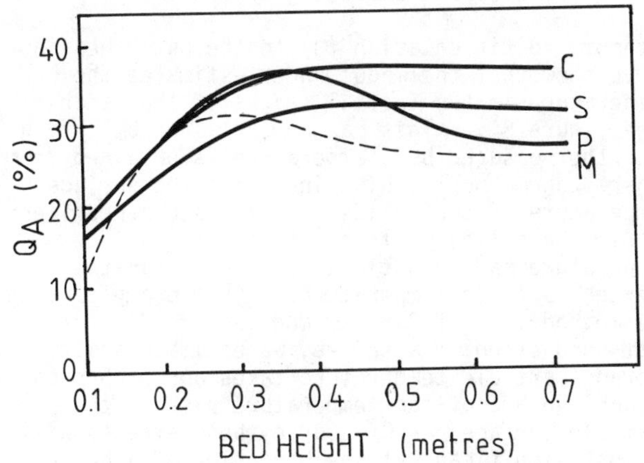

Figure 3. Percent Annular Airflow for Draft-Tube Spouted Beds
(C = composite, $Q_T = 0.032 m^3 s^{-1}$;
S = solid, Q_T = 0.034; P = porous,
Q_T = 0.035, M = model porous)

that at any level the axial pressure gradients in the spout and annulus are equal, i.e.

$$(dP/dZ)_A = (dP/dZ)_s \quad (1)$$

and the gas distributes itself to maintain this equality. The spout pressure gradient is given by pneumatic transport theory as

$$(dP/dZ)_s = (dP/dZ)_a + (dP/dZ)_g + (dP/dZ)_f \quad (2)$$

where the three terms in equation (2) are due to particle acceleration, gravity and friction. The model predicts the observed maxima in Q_A; details are given elsewhere (2).

An important operational parameter is the overall system pressure drop, ΔP_T. Figure 4 shows that pressure drops with draft-tubes were 40 to 80% of the values for a conventional spouted bed. A comparison between solid-wall and porous draft-tubes showed that for the same overall pressure drop the porous tube system tended to have 10 to 20% more annular airflow (taken as the maximum in Q_A).

Comparing conventional and draft-tube spouted beds we have seen that the latter provides far more regular particle movement, lower pressure drops and a tendency to less annular airflow (although the use of longer, porous or composite draft-tubes can obviate this effect).

B. Thermal Performance

The thermal performance of the spouted beds was assessed initially in terms of batch heating kinetics, using beds of predried wheat and beds of polymer granules (Celcon). Figure 5 compares the heat-up kinetics for a conventional bed with that of a draft-tube bed, for similar bed weights and airflow rates. Heat-up in the conventional bed was slightly more rapid than in the draft-tube

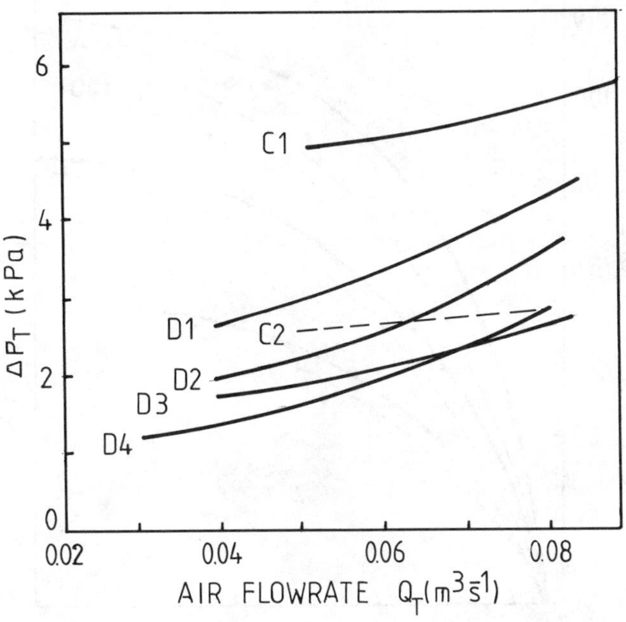

Figure 4. Total Pressdrop vs Airflow rate
(Conventional Beds, C1 with Z = 0.8m;
C2 with Z = 0.4m; Draft-Tubes (0.8 m)
with Z = 0.8m, D1 has L_E = 0.2;
D2, L_E = 0.15; D4, L_E = 0.10;
D3 is Draft-Tube (1.3m), Z = 0.4,
L_E = 0.15)

bed. This implies more efficient contacting in the conventional bed.

An estimate on the contact efficiency, η can be obtained by matching the following equation to the data (see Appendix for details),

$$t = \frac{M\, c_{ps}}{(\dot{m}c_p\, \eta + K)} \ln \left\{ \frac{\dot{m}c_p \eta (T_i - T_o) + K(T_a - T_o)}{\dot{m}c_p \eta (T_i - T_b) + K(T_a - T_b)} \right\} \quad (3)$$

where K allows for heat losses from the equipment to the atmosphere. K was measured in a series of independent runs and found to have a value of 3.0 W °C^{-1}. It should be noted that at steady state

$$\eta = K\{(T_b - T_a)/(T_i - T_b)\}/\dot{m}\, c_p \quad (4)$$

For the data shown in Figure 5 the efficiencies for the conventional and draft-tube beds averaged 0.8 and 0.7 respectively. The higher efficiency in the conventional bed probably arises because at a given airflow rate this bed was operated at only 1.2 x minimum spouting whereas the draft-tube bed operated at 1.5 x minimum spouting. Under these conditions the draft-tube bed had a lower percentage airflow through the annulus. Nevertheless both systems achieved high efficiency which could probably be improved by optimisation

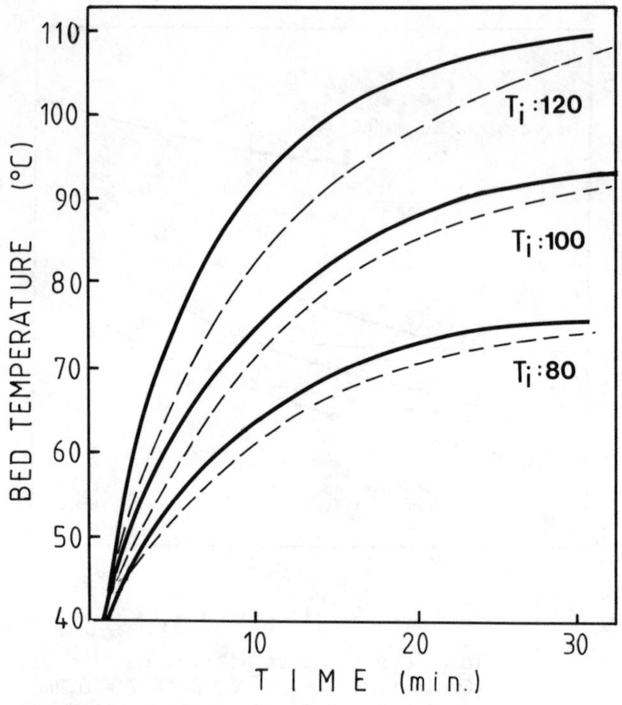

Figure 5. Batch Heat-up Kinetics for Dry Wheat (- Conventional Bed; --- Draft-Tube)

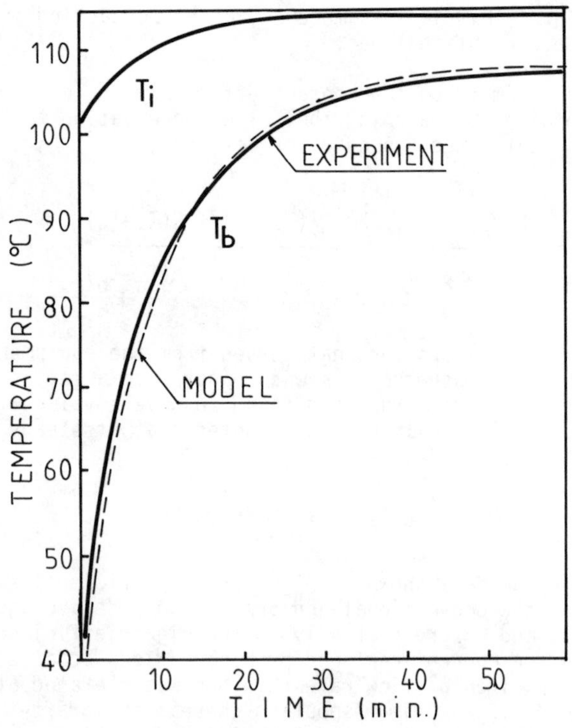

Figure 6. Batch Heat-up Kinetics for Celcon : Model vs Experimental Data

of bed geometry and operating conditions.

Attempts to fit equation (3) to the batch heating curve show that the model under estimates the bed temperature in the initial stages of the batch (see Figure 6). This can be explained by noting that the measured bed temperature is obtained from a bare thermocouple which indicates the surface temperature of the solids. Temperature profiles within the solids will mean that this surface temperature may be noticeably higher than the average particle temperature. (For example using the methodology of Carslaw and Jaeger (7), for heat conduction in a sphere, we estimate that average particle temperature takes about 25 seconds to achieve 90% of the temperature rise at the particle surface). Of course these effects will be most significant at the beginning of a batch. run.

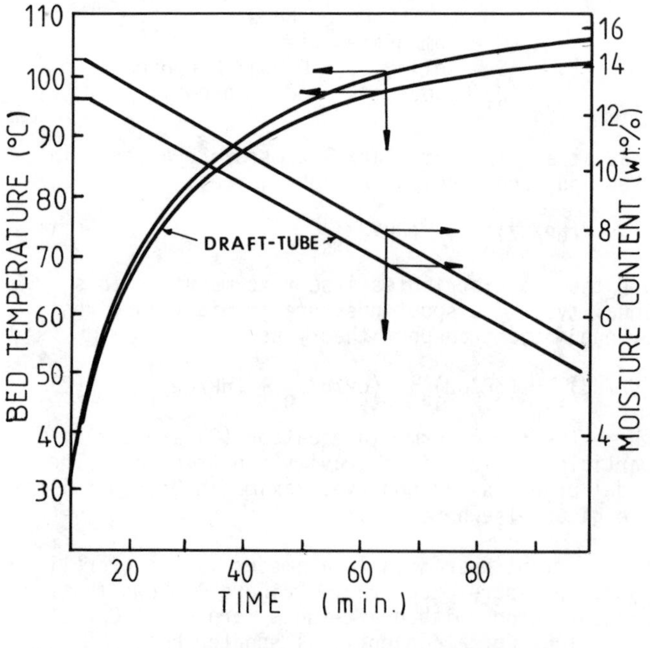

Figure 7. Batch Drying Curves for Wheat in Conventional and Draft-Tube Beds (T_i = 112.5°C, Q_T = 0.05m³ s^{-1})

Figure 7 compares batch drying data for wheat in the conventional and draft-tube beds. The curves were characterised by a brief initial period without moisture loss, as the bed temperature rose from ambient. Subsequent drying is typical of the falling rate period, with the semilog of the moisture content being proportional to time (Figure 7). Falling rate drying would be expected for wheat with initial moistures around 13 wt%, as found by Simmons et al (8). In contrast, Khoe and van Brakel (6) reported constant rate drying for rice in a draft-tube spouted bed, although they had moisture contents in the range 30 to 15%. Differences between the conventional and draft-tube beds for drying purposes were small and can be explained by the slightly better contacting efficiency of the conventional bed. An estimate of contacting efficiency based on the batch drying kinetics is obtained by defining, η_D

(the drying efficiency) as,

$$\eta_D = \frac{\text{(heat to moisture and wheat + ambient losses)}}{\text{(heat by air if exit temp = bed temp)}}$$

$$\text{i.e. } \eta_D = \frac{\Delta W\lambda + M\bar{c}_{ps}(T_{bt}-T_o) + K\int_0^t (T_b-T_a)dt}{\int_0^t \dot{m}\, c_p (T_i-T_b)dt} \qquad (5)$$

Applying equation (5) to the data in Figure 7 gives $\eta_D > 0.8$ for the draft-tube system and $\eta_D > 0.9$ for the conventional bed.

In summary the conventional and draft-tube spouted beds operate close to the ideal well-mixed system, with exit air in thermal equilibrium with the bed. The conventional bed has marginally better thermal contacting efficiency, but this is achieved at higher pressure drop and with less regulation of particle movement. The use of the draft-tube(porous or solid)provides additional design variables which can be manipulated to control both gas and particle movement. This may be particularly advantageous in continuous flow applications, and work is continuing in this direction.

ACKNOWLEDGEMENTS

The authors gratefully acknowledge financial support from the Wheat Industry Research Council of the Department of Primary Industry, Australia.

APPENDIX : Batch Heating Kinetics

A heat balance on the system gives,

[Heat in] = [Heat out] + [Heat accumulation] + [Heat losses]

$$\dot{m}\, c_p T_i = \dot{m}\, c_p T_e + Mc_{ps}\, dT/dt + K(T_b-T_a) \qquad (A.1)$$

where the coefficient K combines overall heat transfer coefficient and vessel surface area for external heat losses. We define the contacting efficiency as,

$$\eta = (T_i-T_e)/(T_i-T_b) \qquad (A.2)$$

which gives $\eta = 1.0$ for exit air temperature (T_e) equal to the average bed temperature (T_b). Eliminating T_e from equation (A.1) and rearranging gives

$$dT_b/dt + c_1 T_b + c_2 = 0 \qquad (A.3)$$

where $c_1 = (K + \dot{m}\, c_p\, \eta)/M c_{ps}$

$$c_2 = -(KT_a + \dot{m}\, c_p\, \eta\, T_i)/M c_{ps}$$

Integration of (A.3) from t = 0 to t, $T_b = T_o$ to T_b gives

$$t = \frac{M c_{ps}}{(\dot{m}\, c_p \eta + K)} \ln\left\{\frac{\dot{m}\, c_p\, \eta(T_i-T_o) + K(T_a-T_o)}{\dot{m}\, c_p\, \eta(T_i-T_b) + K(T_a-T_b)}\right\} \qquad (A.4)$$

Equation (A.4) gives the batch heat-up kinetic curve in the absence of mass transfer.

REFERENCES

1. Mathur, K. and Epstein, N. (1974) "Spouted Beds", Academic Press, New York.
2. Claflin, J.K. and Fane, A.G. (1983) "Spouting with a porous draft-tube" Can. Jnl. Chem. Eng. (in press).
3. Buchanan, R.H. and Wilson, B. (1965) "The Fluid-Lift Solids Recirculator" Mech. and Chem. Eng. Trans. (Inst. Eng. Australia), 117-124.
4. Pallai, E.V. and Nemeth, J. (1972) "Residence Time Distribution in Spouting Beds", CHISA (1972), paper C3.11, 1-18.
5. Kambrock, W. (1976) "Mixing and Homogenising of Granular Bulk Material in a Pneumatic Mixer Unit", Powder Technol. 15, 199-206.
6. Khoe, G.K. and von Brakel, J. (1980) "A Draft-tube Spouted Bed as Small Scale Grain Drier", Solids Sep. Processes, I.Chem.E. Symp. Ser. No. 59, 6/1-6/13.
7. Carslaw, H.S. and Jaeger, J.C. (1959) "Conduction of Heat in Solids", 2nd Edn. Oxford University Press.
8. Simmons, W.H.C., Ward, G.T. and McEwan, E. (1953) "The Drying of Wheatgrain. Part I: The Mechanism of Drying". Trans. Inst. Chem. Eng., Vol. 31, p.265-278.

SPOUT-FLUID BED AND SPOUTED BED HEAT TRANSFER MODEL

A.Chatterjee and U.Diwekar

Department of Chemical Engineering
Indian Institute of Technology, Powai, Bombay 400076, India

ABSTRACT

Spout-Fluid Bed is a relatively new type of fluid bed which is aimed to overcome some limitations of spout and fluidized beds. Very little work is done on modeling for heat transfer in spouted bed and no work is done on heat transfer modeling in spout-fluid beds. A heat transfer model helps in understanding the fundamentals of transport mechanism invloved and aids in scaling up of several operations including drying and heating from laboratory data. A model to estimate gas-to-particle heat transfer co-efficent utilising bubbling bed concept has been developed for both spout-fluid bed and spouted bed. Gas-to-particle heat transfer is of special interest for drying applications. Recently published experintal spouted bed heat transfer data were used to check the proposed model and was found in good agreement with the experimental values. Subsequently the model was extended to apply for spout-fluid bed operations. Effect of various parameters as particle diameter, bed height, fluid flow rate on heat transfer coefficient utilizing the proposed model were compared for spouted bed and spout-fluid bed.

NOMENCLATURE

a' = specific surface of solids, m^2/m^3
a = annulus

C_p = specific heat, J/Kg.K

d = diameter, m

D = diffusivity, m^2/sec.

De = Equivalent diameter

g = acceleration due to gravity, m/s^2

H = bed height, m

(H_{Jc}) = volumetric heat transfer coefficient, $W/m^3.K$

h = heat transfer coefficient between gas and particles, $W/m^2.K$

k = thermal conductivity, W/m.K

Nc = rate of mass transfer

q = volumetric flow rate, m^3/s

R = radius, m

s = surface for transfer, m^2

t = time, sec

T = temperature, K

u = superficial gas velocity, m/sec

v = volume, m^3

Greek symbols

γ_J = fraction of soild in the cylindrical bubble

ϵ = bed voidage

μ = viscosity, N s/m^2

ρ = density, Kg/m^3

δ = volume fraction of bubble

θ = angle of conical base, degree

ϕ = shape factor of parcicle

η_h = heat transfer effectiveness factor

Subscripts

b = bulk solid

c = drag

e = emulsion i.e. annulus

G, g = gas

J = spout jet

O = orifice

p = particle

s = solid

INTRODUCTION

Solid-fluid contacting system has very significant role to play in chemical, metallurgical and allied industries. Importance of such solid-fluid contacting techniques are tremendously increasing due to shortage of energy sources which necessitates improvement in various solid-fluid contact systems as in coal conversion, drying, combustion, gasification and several other operations. Fluidization and spouting operations are well established methods for fluid-particle contact system because of their several useful characteristics such as ease of transfer of solids, uniformity of temperature and concentration conditions resulting in high and uniform heat and mass transfer characteristics. However, these two techniques have also certain limitations which have affected their wider applications.

Spout-Fluid bed (S-F bed) technique was developed by Chatterjee (1) to overcome some limitations of the spouted and the fluidized bed by simultaneously maintaining both spouting and fluidization in the same bed. S-F bed has the added advantage of the oppurtunity to regulate with the spouting fluid flow rate and the fluidizing fluid flow rate, best suited during the course of its operation.

Extensive information on both heat transfer experimental data and heat transfer models for fluidized beds are available in published sources. Only limited information on spouted bed heat transfer experiments and very little information on spouted bed heat transfer models are available. Malek and Lu (2) studied wall-to-bed heat transfer in spouted bed and Uemkai and Kugo(3) have computed fluid-to-bed heat transfer coefficient by passing cool air in a spouted bed with continuous through-feed of particles. Fluid-particle heat and mass transfer studies in air spouted beds of silica gel and activated coal particles were investigated recently by Kmiec (4,5) and the results obtained were extended to develop models for the particles circulation and heat transfer between spouting gas and particles.

However, the available literature date on heat transfer characteristics for S-F bed is very scanty. Several papers on spout-fluid bed were presented in the recent 2nd International Symposium on spouted bed at 32nd Canadian Chemical Engineering Conference, Vancouver, British Columbia, and the proceeding is a good single source to obtain great deal of information about various aspects of spout-fluid beds. Chatterjee et al (6) has investigated wall-to-bed heat transfer characteristics in a rectangular S-F bed segment column and found that under indentical flow conditions 'h' value in S-F bed was about 30% more than the corresponding fluidized bed. They further correlated an empirical equation for the heat transfer coefficient in S-F bed. Donadono and Masimilla (7) reported some heat transfer data for a S-F system in a split column utilizing alumina, bronze, limestone and plastic beads as particles and air as the fluid medium. A theoretical model that described the flow pattern through spout-fluid beds was developed by Heil and Tels(8) from fundamental relationships of gas flow through porous media. Practically no information is available about fluid-to-particle heat transfer in S-F bed and no work on formulation of heat transfer model in S-F bed is reported so far.

Gas-to-particle heat transfer is an important aspect of heat transfer study which affects process and equipment design and are of particular relevance to solve growing energy crisis, through possible option of S-F bed utilization for drying, combustion, gasification, roasting, calcination etc.,

The hyrodynamics of pressure and flow characteristics, solid and fluid movement behaviour, mass and heat transfer data, bubble and jet mechanism, their relationships with different bed parameter are helpful requirements for theoretical treatment of heat transfer studies in the S-F bed. There is very little information on these fields available for S-F bed

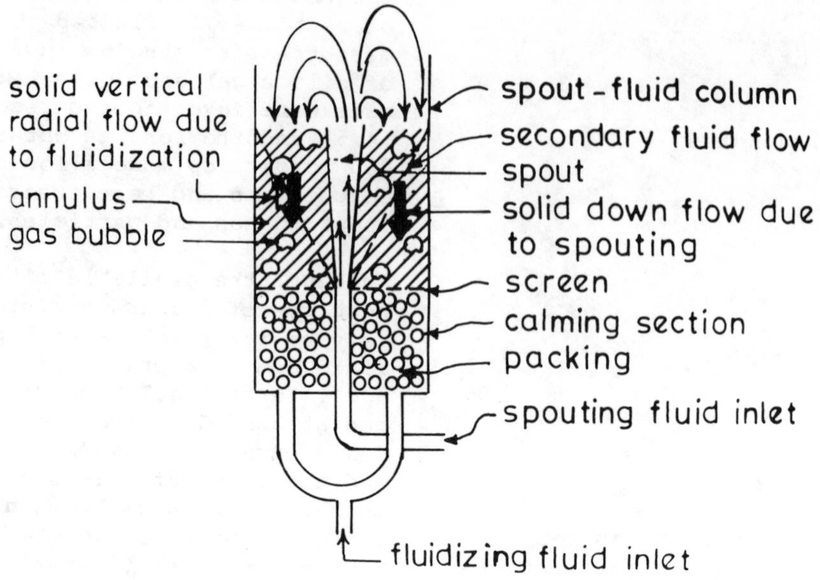

Fig.1 Schematic diagram of one type of spout-fluid bed

operation and only limited information available for spouted beds. Davidson's bubbling bed model (9) has been established as weatherproof and versatile for many bubbling bed operations, even in absence of several of the above relationships. In the present investigation Davidson's bubbling bed model was modified and extended to use in gas-to-bed heat transfer for spouted bed to verify its applicability. When it was found that such modification has given good agreements with experimental data, that modified model was further extended to use in the case of spout-fluid bed for which no experimental data exists.

DESCRIPTION OF SPOUT-FLUID BED

Figure 1 shows a schematic diagram of one type of a typical S-F bed. Spout-fluidization is obtained by imparting a spout in a fluidizing bed or fluidizing the annulus of an already spouted bed. The excess fluid is advantageously utilized to achieve the merits of both these techniques combined giving a high mixing rate of particle and fluid. In the annulus of the S-F bed the fluidizing fluid moves upwards and the solids move downward because of gravity and thus a counter-current solid-fluid contact is established. The bubble and the emulsion phase exchange mass sufficiently often for almost all vertical and radial concentration gradients to approach zero that is perfect mixing and isothermal condition. While solid feeding is a difficult operation in the fluidized bed it can conveniently be done through the spout channel in the S-F bed. However, a number of varieties in the geometry and in operation are possible in spout-fluidization operation as we have seen in the case of spouted bed and the fluidized bed operations and as described by Chatterjee et al (6). In a spouted bed, the central spout jet behaves more or less like an elongated cylind-rical bubble. This situation is more correct in the case of minimum spouting condition where it is like chains of coalescing trail of bubbles. Their functioning as regarding volumetric exchange conditions and solid and fluid movements have similarity with a bubbling bed operation. This is more true in case of spout-fluid beds as major portion of the bed is in bubbling condition. Hence concept of Davidson's bubbling bed model was extended for the case of cylindrical bubble with an aim to use the model for spouted beds and spout-fluid beds.

DERIVATION OF MASS AND HEAT TRANSFER COEFFICIENT FROM SURFACE OF A CYLINDRICAL BUBBLE

With reference to Fig.2 there are 3 coordinate axes. Z - vertical axis, Y - Radial axis, Ø - angle with X-Y line. It is required to calculate the rate of transfer of a diffusing substance from

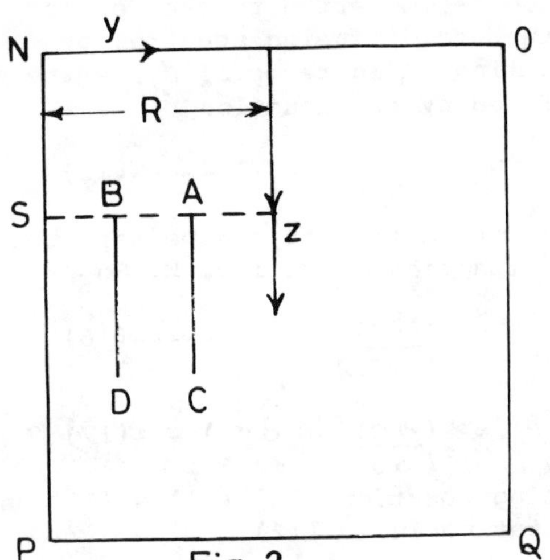

Fig. 2

the surface of cylindrical bubble NOPQ, where the concentration is C^* to the interior of the cylindrical bubble where the concentration is C_o

As a first approximation we can assume that the layer of the gas within the cylindrical bubble near the surface NOPQ is having same velocity as that of fluid at the surface. So for a given height Z, the velocity at a point which is at a distance Y from S will be same, denoted by W_s (in fact boundary layer effect will give variation). This W_s, velocity of fluid at a surface can be calculated from fluid mechanics.

Applying Bernoulli's equation

$$W_s = (2\ gz)^{\frac{1}{2}} \qquad ----(1)$$

To find the equation governing diffusion within the bubble, we consider the balance on the element formed by rotating the infinitesimal area ABCD (Fig.2) about the axix Z and AC & BD are adjacent streamlines. Because these lines are streamlines, there can be no convection across them. Since the film within which diffusion occurs is very thin, diffusion along streamlines can be neglected in comparison with diffusion across them. In writing down a material balance for the elements ABCD there are therefore only two terms, namely

(a) The difference between diffusion across BD & across AC, this difference must balance.

(b) The difference between convection across AB & across CD

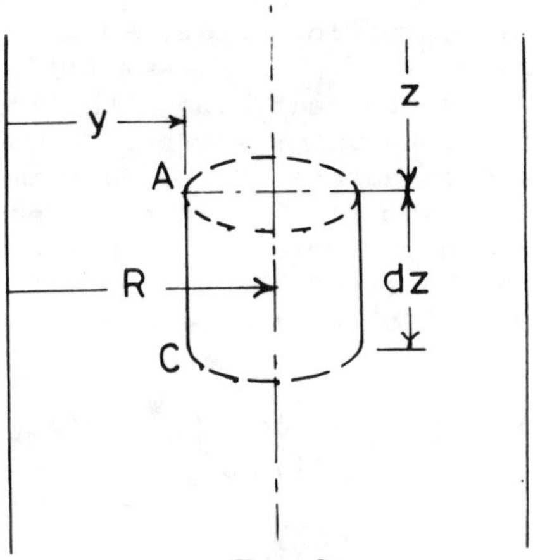

Fig. 3

Fig.3 When AC rotated across Z axix

To derive term (a), we have to find diffusion across AC. Referring to Fig.3 diffusion across AC can be written as

$$2\pi(R-Y)\ dz \times D_G\ \partial C/\partial Y \qquad -----(2)$$

where

$2\pi(R-Y)\ dz$ = Area

D_G = diffusion coefficient

from (2) the difference between the diffusion across AC & across BD is then

$$\frac{\partial}{\partial Y}\left(2\pi(R-Y)\ dz.\ D_G \partial C/\partial Y\right) \qquad --(3)$$

Simplifying

$$2\pi D_G\left((R-Y)\frac{\partial C}{\partial Y^2} - \frac{\partial C}{\partial Y}\right) dy.dz \qquad --\ (4)$$

The convection across AB is $C x W_s$ x area of transfer.

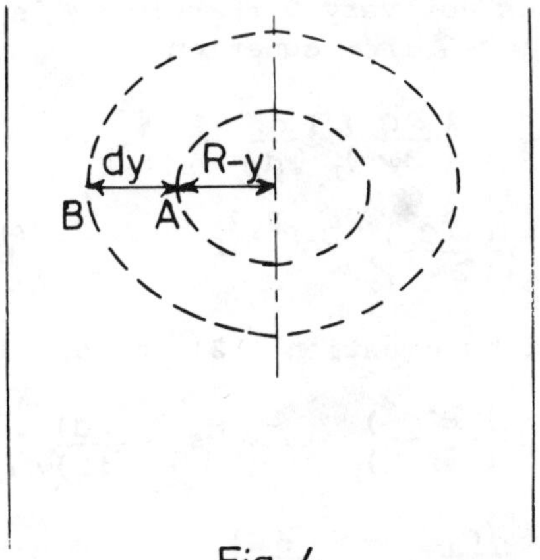

Fig. 4

From Fig.4 convection across AB is

$C\, W_S\, 2\pi(R-Y)\, dy$ -----(5)

and if C were constant along the streamlines, the convection across CD would be the same as across AB, because the flow across each line is equal. Hence when there is a gradient of concentration along the streamlines, the net convection out of element, term (b) above must be

$$2\pi(R-Y)\, W_S\, dY\left(C + \left(\frac{\partial C}{\partial z}\right)_\psi dz\right) - 2\pi(R-Y) \times W_S C\, dy \qquad \text{-----(6)}$$

Neglecting Y as R>>Y

$$2\pi R\, W_S \left(\frac{\partial C}{\partial z}\right)_\psi dz dy \qquad \text{-----(7)}$$

ψ being streamfunction, which must of course be constant along a streamline. Similarly neglecting Y terms from equation (4)

as $R \dfrac{\partial^2 C}{\partial Y^2} >> \dfrac{\partial c}{\partial y}$ so neglecting $\partial c/\partial Y$

also & equating with equation (7)

$$dz\, 2\pi D_G\, R\left(\frac{\partial^2 C}{\partial Y^2}\right)_z dy = 2\pi R W_S \left(\frac{\partial C}{\partial z}\right)_\psi \times dz\, dy \qquad \text{------(8) and}$$

$$D_G\left(\frac{\partial^2 C}{\partial Y^2}\right)_z = W_S \left(\frac{\partial C}{\partial z}\right)_\psi \qquad \text{------(9)}$$

This equation can be put into an integrable form by elimination of Y in terms of ψ, the two variables being related by the equation

$$2\pi\psi = 2\pi R W_S Y \qquad \text{-------(10)}$$

each side of this equation representing the total flow across SA. Now at constant Z, W_S does not vary & therefore ψ is proportional to Y & from equation (10) giving

$$\text{as } \frac{\partial^2 C}{\partial Y^2} = \left(\frac{\partial^2 C}{\partial \psi^2}\right)_z \left(\frac{d\psi}{dy}\right)^2 \qquad \text{------(11)}$$

$$\left(\frac{\partial^2 C}{\partial Y^2}\right)_z = \left(\frac{\partial^2 C}{\partial \psi^2}\right)_z D_G R^2 W^2{}_S \qquad \text{------(12)}$$

Substituting equation (12) in equation (9)

$$D_G\, R^2 W_S{}^2 \left(\frac{\partial^2 C}{\partial \psi^2}\right)_Z = W_S \left(\frac{\partial C}{\partial z}\right)_\psi \qquad \text{--(13)}$$

$$D_G R^2 W_S \left(\frac{\partial^2 C}{\partial \psi^2}\right)_z = \left(\frac{\partial C}{\partial z}\right)_\psi \qquad \text{---(14)}$$

This can be converted to the form of the standard diffusion equation by eliminating W_s in terms of $\emptyset$, where $\emptyset$ is defined by the equation

$$\frac{d\emptyset}{dz} = R^2 W_S \qquad \text{-----(15)}$$

From equation (15) it can be seen that $\emptyset$ is a function of W_S i.e. Z. So

$$\left(\frac{\partial^2 C}{\partial \psi^2}\right)_Z = \left(\frac{\partial^2 C}{\partial \psi^2}\right)_\emptyset \qquad \text{-----(16)}$$

and

$$\left(\frac{\partial C}{\partial z}\right)_\psi = \left(\frac{\partial C}{\partial \emptyset}\right)_\psi \left(\frac{d\emptyset}{dz}\right)_\psi \qquad \text{---(17)}$$

Combining equation (15) (16) & (17) and substituting in eq (12)

$$D_G\left(\frac{\partial^2 C}{\partial \psi^2}\right)_\emptyset = \left(\frac{\partial C}{\partial \emptyset}\right)_\psi \qquad \text{----(18)}$$

Boundary conditions

(a) Adjacent to the surface of the bubble NOPQ the concentration is assumed to be C, and therefore $Y = 0$ i.e. $\psi = 0 \quad C = C^*$ from equation (10)

(b) it is assumed that $c = Co$ at large values of ψ & for all values of $\partial\psi$ at $Z = 0$
So $\psi \to \infty,\ C = C_o$

So equation (18) can be solved using the above conditions

Suppose

$$\eta = \psi/2(D_G\emptyset)^{\frac{1}{2}} \qquad \text{-----(19)}$$

$$\text{Then } \left(\frac{\partial\eta}{\partial\psi}\right)_\emptyset = (D_G\ \emptyset)^{-\frac{1}{2}} \qquad \text{-----(20)}$$

$$\text{and } \left(\frac{\partial\eta}{\partial\emptyset}\right)_\psi = \psi/2 \times -\tfrac{1}{2}\,(D_G)^{-\frac{1}{2}}\emptyset^{-3/2} \qquad \text{--(21)}$$

So equation 18 can be written as

$$D_G\left(\frac{d^2 c}{d\eta^2}\right)\left(\frac{\partial\eta}{\partial\psi}\right)^2_\emptyset = \left(\frac{dc}{d\eta}\right)_\psi \left(\frac{\partial\eta}{\partial\emptyset}\right)_\psi \qquad \text{---(22)}$$

From (20) & (21)

$$\frac{d^2 c}{d\eta^2} = -2\eta \frac{dc}{d\eta} \qquad \text{-----(23)}$$

Intergrating factor $= e^{\eta^2}$

so equation (23) becomes

$$\frac{d}{d\eta}\left(\frac{dc}{d\eta} e^{\eta^2}\right) = 0 \qquad \text{-----(24)}$$

solution of equation (24)

$$C = \int_0^{\eta} A e^{-\eta^2} d\eta + B \quad \text{-----(25)}$$

putting boundary conditions

$\psi = 0 \quad \eta = 0 \quad C = C^*$

$\psi = \infty \quad \eta = \infty \quad C = C_o$

So equation (25) becomes

$$\frac{C - C_o}{C^* - C_o} = 1 - \frac{2}{\sqrt{\pi}} \int_0^{\eta} e^{-\eta^2} d\eta \quad \text{-----(26)}$$

Rate of transfer across the curved surface refering to Fig.2

$$Nc = -D_G \int_0^H 2\pi R \left(\frac{\partial C}{\partial Y}\right)_{Y=0} dZ \quad \text{---(27)}$$

H being max.value of Z. In this equation (27) Y & Z can be replaced by ψ & ϕ. Using equation (10) & (15)

$$\left(\frac{\partial C}{\partial Y}\right)_{Y=0} = \left(\frac{\partial C}{\partial \psi}\right)_{\psi=0} \left(\frac{d\psi}{dY}\right) = \left(\frac{\partial C}{\partial \psi}\right)_{\psi=0} R W_s$$

From (27)

$$Nc = -D_G \int_0^H 2\pi R^2 W_s \left(\frac{\partial C}{\partial \psi}\right)_{\psi=0} dZ \quad \text{---(28)}$$

$$-D_G \int_0^{\phi_1} 2\pi \left(\frac{\partial C}{\partial \psi}\right)_{\psi=0} d\phi \quad \text{---(29)}$$

where ϕ_1 can be obtained from

$$\phi_1 = \int_0^H R^2 W_s \, dZ \quad \text{-----(30)}$$

from equation (1) $W_s = \sqrt{2gZ}$

so $\phi_1 = (2g)^{\frac{1}{2}} R^2 \times \frac{2}{3} \times H^{3/2}$ ----(31)

$$= \frac{2\sqrt{2}}{3} g^{\frac{1}{2}} R^2 H^{3/2} \quad \text{-------(32)}$$

from equation (26) & (29)

$$\left(\frac{\partial C}{\partial \psi}\right)_{\psi=0} = (C^* - C_o) \times \frac{2}{\sqrt{\pi}} \frac{1}{(D_G \phi_1)^{\frac{1}{2}}} \quad \text{-(33)}$$

$$Nc = 4(\pi D_G \phi_1)^{\frac{1}{2}} (C^* - C_o) \quad \text{---- (34)}$$

Now an overall mass transfer coefficient may be defined in terms of the surface area of a sphere of diameter of bubble.

Volume of the bubble = $\pi R^2 H$

$Nc = K_G \pi D_e^{\,2} \, (C^* - C_o)$ -----(35)

if H/R = r then equation (32) can be written as

$$\phi_1 = \frac{2\sqrt{2}}{3} \times g^{\frac{1}{2}} R^2 (rR)^{3/2}$$

$$= \frac{2\sqrt{2}}{3} \times r^{3/2} g^{\frac{1}{2}} R^{7/2} \quad \text{-----(36)}$$

substituting (36) in equation (34)

$$Nc = 4 \left(\pi \times \frac{2\sqrt{2}}{3} \times r^{3/2} g^{\frac{1}{2}}\right)^{\frac{1}{2}} R^{7/4} D_G^{\frac{1}{2}} (C^* - C_o) \quad \text{---(37)}$$

Volume of bubble = $\pi R^3 r$

by defination of De, $4/3\,\pi (De/2)^3 = \pi R^3 r$

or $R = \left(\frac{1}{6r}\right)^{1/3} De$ ------(38)

Substituting in equation (37)

$$Nc = \frac{4\left(\pi \frac{2\sqrt{2}}{3} r^{3/2}\right)^{\frac{1}{2}} g^{1/4} De^{7/4} D_G^{1/2} (C^* - C_o)}{(6r)^{7/12}} \quad \text{--(39)}$$

equating (39) with equation (35)

$$K_G = 0.7707 \, r^{2/12} D_G^{\frac{1}{2}} De^{-1/4} g^{1/4} \quad \text{--(40)}$$

Conversion of this equation (40), in terms of jet height H and diameter of spout ds, simplifies as

De is related to H & ds by

$$\frac{\pi}{4} d_s^{\,2} H = \frac{\pi De^3}{6} \quad \text{----(41)}$$

$$De = 1.145 \, H^{1/3} \times ds^{2/3} \quad \text{---(42)}$$

substituting in equation (40), the mass transfer coefficient is

$$K_G = 0.8363 \, D_G^{\frac{1}{2}} (H)^{1/12} d_s^{-1/3} g^{\frac{1}{4}} \quad \text{---(43)}$$

By analogy, the heat transfer coefficient for cylindrical bubble is

$$h_{Jc} = 0.8363 \, \rho_g C_{pg} \left(\frac{K_g}{\rho_g C_{pg}}\right)^{\frac{1}{2}} \times H^{1/12} d_s^{-1/3} g^{1/4} \quad \text{----(44)}$$

DERIVATION OF MODIFIED BUBBLING BED MODEL FOR SPOUTED BED GAS TO PARTICLE HEAT TRANSFER

The central jet in the spouted bed has been considered as a long cylindrical bubble with diameter ds and height H. Thus we can estimate the heat transfer from spout (jet) to particles as follows:

(heat lost by gas in jet) = (heat taken by solid in jet) +

(heat transfered to)
(cloud in annulus)
(where it is absorbed) , OR

$$-\rho_g C_{pg} \frac{dT_{gJ}}{dt} = -\rho_g C_{pg} U_J \frac{dT_{gJ}}{dH} = \eta_h \nu_J h_p^*$$

$$a'(T_{gJ} - T_s) + (H_{Jc})_J (T_{gJ} - T_s) \quad \text{--(45)}$$

h_p^* is local single sphere heat transfer coefficient. $(H_{Jc})_J$ is volumetric heat transfer coefficient and can be calculated as

$$(H_{Jc})_J = \frac{q_J C_{pg} \rho_g + h_{Jc} S_{Jc}}{V_J} \quad \text{--(46)}$$

$$\text{where } q_J = \frac{U_{sup}}{\epsilon_a} \times \frac{\pi}{4} ds^2 \quad \text{--(47)}$$

$$\text{and } V_J = \pi/4 \, ds^2 H \quad \text{--(48)}$$

substituting equation (46) with eq.(44)

$$(H_{Jc})_J = 0.25\left(\frac{U_{SUP} \rho_g C_{pg}}{\epsilon_a H}\right) + 0.8363\left[\frac{(K_g \rho_g C_{pg})^{1/2} H^{1/2} g^{1/4}}{ds^{4/3}}\right] \quad \text{--(49)}$$

Heat lost by jet can also be written in terms of overall heat transfer coefficient based on temperature of gas in cylindrical bubble.

$$(Nu)_{overall} = \frac{h_{p\,oven}\, d_p}{K_g}$$

$$\cong \frac{\delta}{1-\epsilon_a}\left[\nu_J \eta_h Nu^*_{pt} + \frac{\phi_s d_p^2}{6 K_g}(H_{Jc})\right] \quad \text{--(50)}$$

The time average temperature of cyclindrical bubble and annulus gas when measured by thermocouple, it gives 'apparent' gas temperature and corresponding coefficient is the apparent coefficient. Then we have

$$\Delta T_{app} = T_{g\,app} - T_s = [\delta T_{gb} + (1-\delta) T_{ge}] - T_s \quad \text{--(51)}$$

under normal conditions $T_{ge} \cong T_s$. Thus

$$\Delta T_{app} = \delta (T_{gb} - T_s) \quad \text{--(52)}$$

$$\text{Hence } (Nup)_{app} = \left[\frac{1}{1-\epsilon_a} \nu_J \eta_h Nu^*_{pt} + \frac{\phi_s d_p^2}{6 K_g}(H_{Jc})_J\right] \quad \text{--(53)}$$

(Nu^*_{pt}) can be calculated from Ranz & Marshall (10) equation for single sphere

$$(Nu^*_{pt}) = 2 + 0.6\, P_r^{1/3} R_e^{1/2} \quad \text{---(54)}$$

Factor ϵ_a was calculated as loosely packed particles for the experimental condition and diameter of the spouted bed by Abdelrazek's equation (11, page - 99)

$$d_s = 0.315\, d_c (u_s/(gH)^{1/2})^{0.33} \quad \text{---(55)}$$

$\nu_J \eta_h$ was estimated to be 0.006 forthe system of experiments by Kmiec (4). Soild lines in Fig.5, 6 and 7 show experimental values of fluid-to-particle heat transfer coefficient for different experimental conditions as reported by Kmiec (4). When Eq. (53) along with Eq.(54), and (55) were utilized for the same conditions and parameters as utilized by Kmiec, the calculated 'h' values were found to be in good agreement with the corresponding experimental 'h' values reported by Kmiec. The dashed line in Fig.5, 6 and 7 represents the 'h' values calculated by the present proposed model and compares with the experimental values. The close proximity of the two lines shows that modified bubbling bed model is applicable for spouted bed system.

EXTENSION OF THE MODIFIED MODEL FOR APPLICATION IN SPOUT-FLUID BED

In the previous sections we have seen that the extended bubbling bed model fits well for spouted bed operations. Hence attempt was taken to apply the modified model by incorporating appropriate corrections for the case of spout-fluid beds. Looking at the physical view of the spout-fluid bed, it can be observed that the annulus is in more expanded condition than the spouted bed and hence secondary fluid flow to the annulus would be more. This is due to lesser resistance offered by the increased voidage of the expanded fluidized annulus. Also because of fluidization in the annulus, more solid movement in the vertical and radial direction and gross downward movement due to spouting action, would result in the annulus solids to move vigorously. Fluidization of the annulus would result more solid bleeding into the spout jet for turbulent contact with hot jetting gas for better heat transfer to occur. The present model, at this stage, would be used only to find the heat transfer coefficient from the central spout jet to the fluidizing annulus of the S-F bed. Understandably such 'h' values are not for the whole S-F bed conditions.

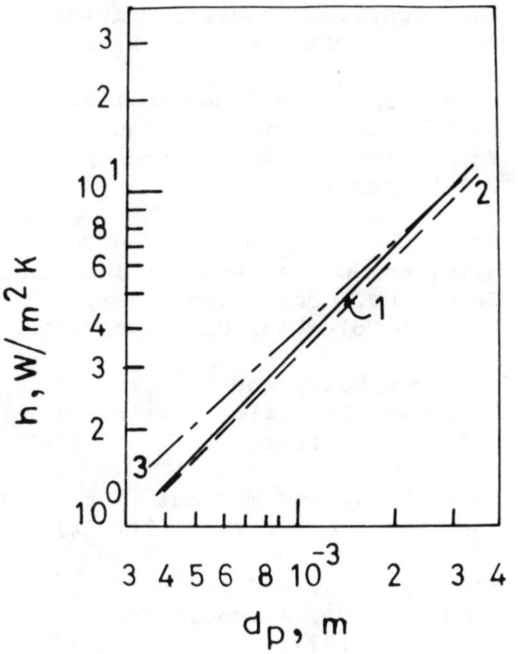

Fig.5 Gas-to-particle heat transfer coefficient vs particle diameter. Silica gel particle, H = 0.0707 m, u = 0.450 m/s, cone angle = 30°, d_c = 0.09 m, d_o = 0.013, 1 = experimental values for spouted bed from Ref.4, 2 = values from proposed model for spouted bed, 3 = values from proposed model for spout-fluid bed

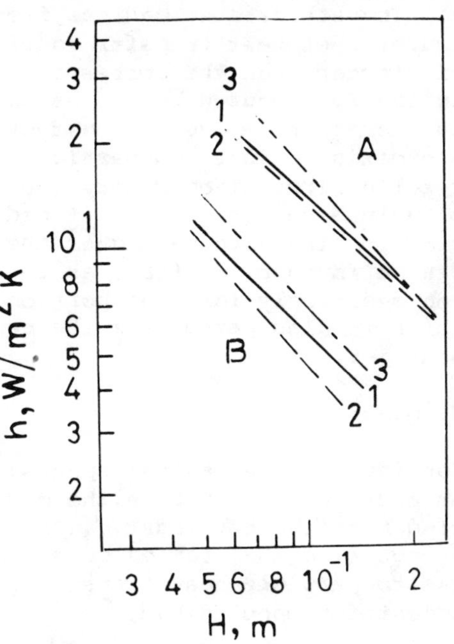

Fig.6 Gas-to-particle heat transfer coefficient vs bed depth. Conditions same as Fig.5 except for B:- d_p = 0.000874, u = 0.455 m/s, and for A:- d_p = 0.002233 m, u = 0.89 m/s.

In absence of information required for hydrodynamic relationship of this complicated system, a rather simple and logical approach would be to consider eq.(53) with porosity factor for fluidized bed and spout diameter as reported by Hadžismajlović et al (12) for S-F bed. Final form of the equation from the extended model for spout-fluid gas-to-particle heat transfer is

$$(Nu)_{app} = \frac{1}{1-\epsilon_{SF}}\left[\nu_J \eta_h Nu^*_{pt} + \frac{\phi_s d_p^2}{6 K_g}(H_{JC})_{SF}\right] \quad --- (56)$$

Eqn.(56) was utilized for all the data points for which the spouted bed correlation was utilized. The dash-dot lines are graphical representation of Eq (56)

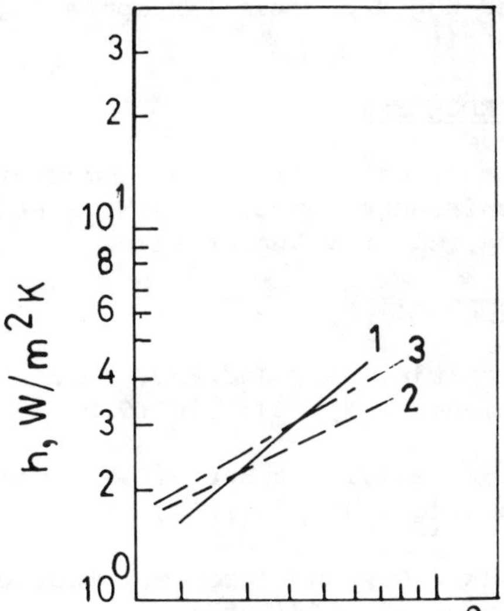

Fig.7 Gas-to-particle heat transfer coefficient vs superficial gas velocity. Conditions same as Fig.5 except cone angle = 90°, d_p = 0.000874

on each of the three plots on Fig.5, 6 and 7. In each case we can see from the figures that heat transfer coefficient is larger than the corresponding coefficient for spouted bed. As envisaged earlier this is due to the fact of intimate mixing of gas and particle and higher solid circulation rates. However, 'h' values for the total S-F bed operation is likely to be higher than that can be found by Eq.(56), as this equation deals only for that portion of heat which is dissipated from the central spout jet.

CONCLUSIONS

1. Considering the central spout-jet as a cylindrical bubble, the modified bubbling bed model can be used by equation (53) for calculation of gas-to-particle heat transfer coefficient for spouted bed.

2. Spout jet-to-bed particle heat transfer coefficient for spout-fluid bed can be correlated by equation (56) derived from the modified bubbling bed equation (53).

3. Spout jet-to-bed particle heat transfer coefficient for spout-fluid bed is higher than spout-to-bed particle heat transfer coefficient for spouted bed under similar operating condition.

ACKNOWLEDGEMENTS

The authors acknowledge the computational assistance rendered by Mr.P.Rajadhyaksha and Mr.V.Venkatraman.

REFERENCES

1. Chatterjee, A., Ind.Eng.Chem., Proc. Des. Dev., 9, (2), 370(1970)

2. Malek, M.A., and Lu, B.C.Y., Can.J. Chem.Engg. 42, 14(1964)

3. Uemaki, O., and Kugo, M., Kagaku Kogaku, 31, 348(1967)

4. Kmiec, A., Can. J. Chem.Engg. 53, 18 (1975)

5. Kmiec, A., Chem. Eng. J. 19, 189(1980)

6. Chatterjee, A., Adusumilli, R.S.S., and Deshmukh, A.V., 2nd Int.Symp. on Spouted Beds at 32nd Can. Chem. Eng. Conf., Vancouver, British Columbia, Oct. 3-6, (1982)

7. Donadono, S. and Massimilla, L., Fluidization, Proc. of 2nd Eng. Foundation Conf., Cambridge, P.375, April 1978

8. Heil, C. and Tels, M., 2nd Int. Symp. on Spouted Beds at 32nd Can. Chem. Eng. Conf., Vancouver, British Columbia, Oct. 3-6, (1982)

9. Davidson, J.F, and Harrison, D., Fluidized Particles, Cambridge University Press, New York, 1963.

10. Ranz, W.E., and Marshall, Jr., W.R., Chem.Eng.Progr., 48, 141(1952)

11. Mathur, K.B., and Epstein, N., Spouted Beds, Academic Press, New York, 1974.

12. Hadžismajlović, Dž. E., Grbavčić, Z.B., Vuković, D.V., and Littman, H., 2nd Int. Symp. on Spouted Beds at 32nd Can. Chem.Eng. Conf., Vancouver, British Columbia, Oct. 3-6, (1982)

SPOUTED BED TECHNOLOGY- A BRIEF REVIEW

A.S. Mujumdar

McGill University
Montreal, Canada

Abstract

A brief survey is made of the technology of spouted beds, newer variations of the spouted bed technique along with the limitations and advantages of the different types currently being investigated. A new classification scheme is presented for spouted beds based on various criteria. Current and potential applications of the spouted bed technique are identified and some recommendations made for further research in selected areas with special reference to drying of grains. Advantages of the modified spouted beds are discussed where appropriate and areas requiring further work are identified.

Introduction

Spouted bed (SB) technology has come a long way since its discovery by Mathur and Gishler at the National Research Council in Ottawa, Canada in the early 1950's (1). Like most discoveries it was accidental; its origin was in unsuccessful fluidization of wheat. They noted that by blocking most of the distributor area and leaving only a small orifice in the centre of a 6-inch diameter column a bed of wheat could be maintained in a stable recirculatory (not fluidized) motion. In the words of Gishler (1), "ignorance (about fluidization) proved to be a boon for spouting". The first application contemplated for SB was drying of wheat. Indeed a pilot scale dryer for wheat was designed and operated successfully. Possibly owing to the low cost of energy in the early days, this application never became commercial although now its potential is again quite high especially if modified spouting techniques are employed to overcome some of the inherent limitations of the conventional SB. A variety of possible modifications is discussed in this article.

The conventional SB consists essentially of a cylindrical vessel with a conical bottom (angle about 60^0). An air jet is injected through a round orifice (or nozzle) at the bottom with a velocity which is sufficient to penetrate the entire bed of particles and generate a spout (or fountain or geyser) at the top surface of the bed. Clearly, the minimum spouting velocity depends on the particle as well as geometric characteristics of the bed. Numerous correlations abound in the literature which permit reasonable calculation of the flow parameters and selection of the SB geometry. It is not the intent of this paper to list all the numerous correlations on the maximum spoutable bed height, peak pressure drop, minimum spouting velocity etc. It is noteworthy that for particles of highly nonspherical shape and for wet, sticky or rough particles the existing correlations may not yield reliable results.

Essentially, the SB technique is an efficient solid-gas or solid-liquid contactor; both research and applications in the former area exceed those in the latter. We will therefore concern ourselves exclusively with the former. The conventional SB results in a dilute phase pneumatic transport of particles entrained by the spouting jet in the central core region and a dense phase (voidage nearly equal to that of a packed bed) downward notion of the particles along the annular (sometimes termed downcomer) region bounded by the cylindrical wall. Thus the particle-gas contact is cocurrent in the core (or spout) and countercurrent in the downcomer or annulus. This recirculatory motion is a characteristic of the SB. The particle residence time can be controlled within wide limits by letting the particles go through a desired number of cycles prior to their withdrawal. Both batch and continuous modes of operation are possible.

Applications

Numerous applications have been envisaged for SB's. Among these are: drying, grannulation, coating, heterogeneous reactions, coal carbonization and gasification, heating or cooling of particles etc. More recently a commercial application consists of a SB of inert particles (glass) in which a slurry or suspension is sprayed; the fine coating on the inert particles dries quickly and is ground off during transport within the bed before it is entrained by the carrier gas and collected. Modified versions of this grinder-dryer are in commercial operation in Hungary. In this application SB replaces conventional spray dryer. SB dryer can also dry difficult-to-handle sticky pastes.

Among specific applications actually investigated in the laboratory or pilot scale one may cite: drying and coating of tablets, coating of nuclear fuel microspheres with pyrolytic carbon or silicon carbide with high temperature thermochemical reactions, coating of prilled urea with sulphur to reduce its water solubility, coating of biocompatible materials, combustion of coal or peat, eletrowinning of metals, mixing, blending or agglomeration of solids, drying of grains, pastes and slurries etc. Mathur and Epstein (2) and the Proceedings of the 2nd International Symposium on Spouted Beds contain references to various applications. Although current industrial applications appear to be rather limited the potential is high (especially with modified spouted beds) and we may see more extensive applications in the next decade as the SB technology matures.

Limitations of the Conventional Spouted Bed Technique

Perhaps one of the reasons for the limited commercial exploitation of the SB technology is the limitations among which one may list: (1) high pressure drop prior to onset of spouting, (2) limits on geometry of the SB for efficient operation, (3) air flow rate governed by requirements of spouting rather than the heat/mass transfer requirements, (4) limited operating range, (5) limited capacity per unit space (due to limits on size of the chamber and maximum spoutable height), and (6) difficulty of scale-up. In view of these limitations the conventional SB technique is especially suited to handling particles that are too large to be readily fluidized and for capacities that are not very high e.g. pharmaceutical applications. Also, for handling high value, low tonnage materials the conventional SB technique appears suitable.

Classification of SB's

Following Mathur and Epstein (1) we treat recirculating beds as variants of the spouted bed technique although their characteristics can be quite different from those of the classical SB. SB's can be classified in various ways depending upon the criterion chosen. Following are various criteria and classification schemes developed by the author after surveying the recent literature in the field.

(1) Mechanism of Spouting

a.	Pneumatic	e.g. conventional SB; draft-tube S.B.
b.	Mechanical	e.g. screw-SB; particles in the central core conveyed upwards by means of a mechanical screw conveyor.
c.	Vibratory	e.g. vibro-spouted bed; a part of the conical bottom is flexible and vibrated to impart a vertical motion to the core region.

(2) Vessel geometry

- (1) Conical (popular in USSR, Hungary, Poland etc)
- (2) Cylindrical with conical bottom (conventional)
- (3) Flat bottom (use of optimally located draft tube can reduce or eliminate dead zones near bottom)

(3) Cross-section of bed

- (1) axisymmetric
- (2) two dimensional (large aspect ratio)
- (3) three dimensional (small aspect ratio rectangular, half or sector columns, triangular etc)

(4) Air entry

- (1) single nozzle at bottom
- (2) central nozzle with auxilliary flow along annulus
- (3) tangential or swirling air entry along periphery without central nozzle (e.g. mechanically-assisted spouting with conveyor screw)
- (4) no air (e.g. vibro-spouted bed for vacuum drying)
- (5) multiple nozzles.

(5) Internals

- (1) With draft tube or divider plates (passive internals) (draft tube may be partially or fully permeable, nonuniform cross section).
- (2) No internals (classical SB)
- (3) With active internals (e.g. mechanical screw).

(6) Operation Mode

- (1) batch
- (2) continuous

(7) Bed particles

- (1) active (i.e. undergoing heat/mass transfer)
- (2) passive or inert (e.g. drying of pastes or slurries)

(8) Combined Modes

- (1) Spouted bed
- (2) Spout-fluid bed (several variations possible)
- (3) Fluid-spout bed (several variations possible)
- (4) Jet in fluid bed.

It is impossible to cover in a brief survey all the various types of SB's and operating modes. Only the more important types with emphasis on potential applications in drying of particulate solids will be discussed in the following sections.

Some selected topics in spouted beds

A. Draft tube spouted beds.

Introduction of a cylindrical tube some distance above the air nozzle of a circular SB influences the characteristics of the SB remarkably. Buchanan and Wilson (3) and Taskaev and Kozhina (4) were perhaps the first to develop this concept to overcome limitations of SB in respect of allowable range and size of particles and system geometry. They called it the Fluid-Lift Solids Recirculator (FLSR). Like SB it exhibits stable dense-phase solids recirculation in the wall region and lean-phase conveying up the enntre with disengagement at the top. Unlike the SB it prevents cross-flow of solids between the two counterflowing streams, but under certain conditions it makes the air flow in the annulus go down rather than upwards.

Since the draft tube avoids diffusion of the jet, proper location of the draft tube (of optimal diameter) allows in principle an unlimited bed height since there is no limit on the height to which one can convey the particles. Once disengaged at the top the particles must descend along the annulus to be re-entrained by the air jet at the bottom. The draft tube may be higher, equal to or lower than the bed height in the dense-phase annulus. The only effect of bed height on aerodynamics of FLSR is to control the amount of gas sucked in by the eductor action at the bottom of the tube. Buchanan and Wilson have given design information on the basis of extensive data, which allows selection of appropriate geometric and operating parameters.

Among the advantages of the draft-tube SB are: (1) can be applied to any solids that flow from a hopper while spouting can be applied only to fairly large and uniform size particles, (2) solids recirculation initiated at lower pressure drops, (3) bed depth and diameter can be altered substantially, (4) lower air requirement for given solids recirculation, (5) amount of recirculation can be varied independently of column diameter, bed height and particle size etc.
Among its disadvantages are (1) reduced mixing, (2) more complex design, (3) some tendency to clog or choke during start-up or shut-down and (4) lower heat and mass transfer rates due to a more regulated particle motion. The FLSR is especially suitable for multi-spout operation; one possible way to increase the capacity of a spouting unit.

Recently Khoe and Van Brakel (5) and Claflin and Fane (6) have studied draft-tube SB for drying and for thermal disinfestation of wheat respectively. Better thermal economy and better control of residence time is reported in the presence of draft tubes. For disinfestation it is important to control particle time-temperature history; for wheat a surface temperature of 65^{0}C is needed for not less than 4 minutes. Longer exposure damages the grain. Presence of draft tubes necessarily limits the annular air flow. Pallai and Nemeth (7) used a porous draft tube made of metal screen to permit some gas flow from the spout to the annulus without any crossflow of particles. Clearly such a device would limit the maximum spoutable bed height simultaneously.

In view of the advantages of the draft-tube it seems that all further work on drying of grains in circular spouted beds should use draft tubes. For larger capacities a novel device to be discussed later (a bank of two dimensional spouted beds with internal heating for combined direct-indirect drying) is recommended.

According to Claflin and Fane (6) use of porous screen draft tube does not provide dramatic increase in the annular (dense-phase) gas flow as a fraction of the total flow as compared with solid draft tubes. The effect of porosity of the insert is more significant on the flow distribution which shows a maximum in the case of porous tubes. Annulus to spout flow can be constrained by tapering the draft tube with its cross-section decreasing with height. However, there is danger of choking in this case.

One comment about particle residence time is in order. Since the particle velocities are much higher in the spout than in the annulus, the mean cycle time of particles in the bed is a close approximation to the annular residence time which increases with spacing between the lower edge of the draft tube and the inlet air nozzle and decreases with air flow rate. Draft tube porosity variation has little effect on the particle cycle time (6). Its effects may be more appreciable on the transfer rates.

B. Multi-Spouted Beds

Since the diameter of a round SB cannot be made indefinitely large (1.5 m appears to be maximum practical size) one obvious way to increase capacity per unit floor area is to use a multiplicity of spouts with or without draft tubes. This leads to complex design of the bottom of the vessel and independently controlled air flow lines to each nozzle. If all nozzles are supplied through one common line even the slightest maldistribution or minor variations in pressure drop results in unstable operation. Some of the nozzles will not spout at all while some may coalesce as the jets seek paths of least resistance in the bed. Besides, it is impossible to avoid dead zones between neighbouring spouts. It appears therefore that multiple round spouts in a vessel of any geometry is not a desirable configuration in general. Alternate means must be sought to increase the capacity of a SB. Use of draft tubes helps minimize interspout interaction and reduce tendency to instability.

Slotted or Two Dimensional Spouted Beds

Spouted beds of rectangular section with aspect ratio (ratio of the dimensions of the two sides) of the order of unity have apparently been studied in the USSR. These do not seem to offer much advantage over the axisymmetric bed. Such beds are strongly three dimensional and scale-up of such devices is probably more difficult than that for the conventional SB because of wall effects.

Much greater flexibility, simplicity of construction and ease of design is afforded by a slotted or two-dimensional SB first proposed and studied by the author (8, 9). Here the aspect ratio is large (greater than 10). The bed walls are plane with a converging section at the inlet through which spouting air is introduced in the form of a uniform slot jet. Thus, there is no scale-up problem if the laboratory scale model utilizes the same smaller dimension as the field unit. The field unit will then simply have a larger aspect ratio. Variation of the slot opening near the walls is feasible if wall effects must be accounted for. Another advantage is the fact that the bed walls can be heating panels to supply supplemental heat for drying. Thus, a low grade heat source may be employed as the spouting air (e.g. solar heated air, waste heat etc). This way one can combine direct and indirect drying resulting in better thermal economy especially for heat sensitive products.

For high capacities it is also possible to use a bank of two-dimensional beds with common walls between adjacent beds which may be heating or cooling panels. As a further variant, one may use half-beds with a common slot nozzle supplying spouting air to both halves. In this case (currently under study in author's laboratory) the pressure drop is expected to be lower than for the full column. Since the frictional resistance for flow of particles against a smooth wall is lower than that against counterflowing particles, the overall bed pressure drop is reduced. This is true especially for rough surface particles as noted by Swaminathan and Mujumdar (10). This is likely to be the case also for wet and sticky particles during the initial period of drying.

D. Vibro-Spouted Beds

Here vertical vibration of a flexible diaphragm which forms the central bottom piece of a conical vessel, which is stationary, results in generation of net upward motion of the central core forming the so-called spout. Since no aeration exists such a device can be used only with contact or radiative heat transfer. It has some potential as a vacuum dryer. It has been investigated at the Hungarian Academy of Sciences, Veszprem, Hungary.

E. Ring-spouted Bed

One modified spouted bed variant uses a ring-shaped air inlet for introduction of air to increase the bed volume without increasing the bed height (i.e. pressure drop). Such a device has been tried successfully on a pilot scale for drying pasty solids. USDA has also used a similar concept for a hot air grain popper. Here the air-solids contact time is only 15-30 seconds.

F. Mechanical Conveyor Spouted Bed

As noted earlier, the fact that the air flow in a SB is ditermined not by process requirements but by the fluid mechanical requirement to sustain spouting, is one of the limitations of the SB. This in turn limits the residence time in the bed. Pallai et al at the Hungarian Academy of Sciences, Veszprem, Hungary, overcame this difficulty by mounting a screw conveyor device in the core which replaces the conventional lean-phase core and carries solids to the top. The circulation time can thus be controlled indefinitely from minutes to hours. The air flow can be independently set depending on process (e.g. drying) constrainst. Air is introduced at the bottom in a tangential, swirling motion. It has been shown that this device (already in commercial use) also results in uniform (not highly skewed as in SB) air distribution in the annular dense phase. Using inert glass spheres this device can act as a dryer-grinder for pastes and slurries. By using appropriate combination of operating variables the product characteristics can be altered over a wide range.

G. Noncircular Spouted Beds

Half-section beds are commonly studied as they permit visual observation of particle motion despite wall effects. It was shown recently in Ref. (10) that half beds represent full beds adequately for smooth particles. For rough particles half beds display a lower pressure drop and higher spoutable height i.e. these are better than full columns. This may be taken advantage of in the design of direct/indirect SB dryers by introducing vertical partitions which could be heating panels. More than one partitions may be used for extra heat transfer surface leading to the so-called sector beds of the type studied by Green and Bridgewater (11). Triangular beds with spouting air introduced in one corner may also be commercially viable if arranged in a honeycomb pattern. The pressure drop is expected to be lower for such beds as the spouting is confined between two adjacent smooth walls.

H. Conical Spouted Beds

Spouted beds of fully conical cross-section are popular in the USSR and East Europe. One of their principal advantages is the fact that they can be operated at higher air velocities if higher transfer rates are needed. The particle flow pattern, lean-phase and dense-phase zones etc are quite different from those of the conical-cylindrical SB. See Strumillo et al (12, 13) for details. Markowski and Kaminski (13) have given some aerodynamic data on the jet-spouted bed e.g. minimum spouting velocity, bed expansion, maximum pressure drop etc. Conical vessels with bed heights approximately equal to diameter at the top are especially useful for drying pasty materials at high feed rates. Higher air velocities mean high intensity of particle motion and higher expansion of bed. The bed voidage can be 0.85-0.90. Further, unlike the classical SB the boundary between core and annulus is not distinct and the bed voidage changes insignificantly from zone to zone.

I. Spout-Fluid Beds

Spout-fluid beds were originally developed to overcome some of the limitations of SB and fluid beds. Here the total air flow is split between the spout (nozzle) inlet tube and the annular region surrounding the spout. Very recently Vukovic et al (14) have presented flow regime maps for two-phase fluid-solid mobile beds. By using appropriate combinations of spout and annular fluid flow, and depending upon the particle properties and bed height for a vessel of uniform cross-section, their flow regime maps display areas for existence of different types of two-phase mobile beds viz. fluidized, spouted, spout-fluid, fluidized with local spout and packed bed. Furthermore, they defined the boundaries between different types by theoretically based and experimentally determined equations.

The following table summarizes the essential criteria. (Here u=superficial velocity, H=bed height, V=superficial air velocity. Subscripts: mf=minimum fluidization, ms=minimum spouting, T=terminal, A=annulus, N=nozzle (spout)).

Type	Basic Criteria
1. Spouted bed	$H<H_m$, $u<u_{mf}$, $V_A=0$, $V_T>V_N>V_{ms}$
2. Fluidized bed	$u>u_{mf}$, $V_A>V_{mf}$, $V_A<V_T$, $V_N=0$.
3. Spout-fluid bed	$H<H_{msf}$, $u_{AH}<u_{mf}$; $V_A<V_{AmSF_{max}}$ $(V_T-V_A)>V_N>(V_N)_{mSf}$; $V_A>V_{AmSF}$
4. Fluid-spout bed	$H>H_{mSf}$, $u>u_{mf}$, $u=u_{mf}$ (in annulus) $(V_T-V_A)>V_N>V_{mf}-(V_{ASf})_{max}$ $V_A>(V_{ASf})_{max}$ but< V_{mf}
5. Fluid bed with local spout	$u>u_{mf}$, $V_A>V_{mf}$ but$<(V_T-V_N)$ $A_i u_t<V_N<(V_T-V_A)$ (A_i=spout inlet area)

In spout-fluid beds spouting air is introduced through central nozzles while fluidizing air is supplied through a flat or conical distributor which may be a porous plate or a perforated plate. Spout-fluid beds display improved solids mixing in the annulus and hence may be more suited than fluid beds for solids that tend to agglomerate. In general, in the appropriate operating range spout-fluid beds can overcome the limitations of spouted beds (e.g. maximum spoutable height, spout stability, poor mixing) and fluid beds (e.g. slugging, stratification tendencies).

The application of a spout fluid bed with

a draft tube was perhaps first described by Taskaev and Kozhina in 1956 (4). This device has been used for low temperature coal pyrolysis to handle the plastic stage of coal. Recently, Westinghouse utilized this concept for first stage coal devolatalizer with caking coals (15, 16). They called it a recirculating fluidized bed. Such devices have been used for large-scale hydrogenator gasifying heavy hydrocarbon oils. A reactor for oil gasification using a multiplicity of draft tubes has also been described in the literature (17). Commercial blenders utilizing the recirculating bed with tubular inserts are available for blending powders. Other industrial applications include: coating tablets, granulation and agglomeration of fine powders, mixing and blending etc. The draft tube may be operated as a dilute-phase pneumatic transport tube or one can fluidize the solids in the draft tube.

Yang and Keairns (16) have examined the important design parameters for spouted fluid beds with draft tube. The important design parameters are: distributor plate gas bypassing characteristics, area ratio between downcomer and draft tube, diameter ratio between distributor plate and draft tube inlet, area ratio between draft tube gas supply and concentric solids feeder. They have also given design equations supported by experimental data.

Use of spout-fluid beds for drying has not been studied to author's knowledge. It appears that there may be some merit to studying this type of device for solids drying since it combines advantages of spouted and fluid beds. There is also evidence in the literature which suggests much higher transfer rates in spout-fluid beds than in spouted beds. This is especially beneficial in indirect-type dryers for heat sensitive materials.

Heat and Mass Transfer

Mathur and Epstein (2) have reviewed most of the work in this area. It is beyond the scope of this article to review these aspects. Some general comments are in order, however.

Fluid-to-particle heat transfer rates in the annulus, spout region and in the fountain region are different. Heat transfer coefficients are in the order of 50 W/m^2K in the annulus and about 8 times this value in the spout. In the spout region, for larger particles the Biot numbers can be large and hence intra-particle temperature gradients may be significant while in the annulus, which is like a loose packed bed, the gas reaches the particle temperature within a short distance of entry in the spout. Heat transfer in the fountain turns out to be mor efficient than in the spout because of longer residence times of the particles.

For indirect heat transfer bed-to-wall heat transfer data are needed. Malek and Lu (21) have presented an empirical correlation to estimate bed-to-wall heat transfer rates. For immersed heating elements very limited data appear to be available. For vertically mounted cylindrical heaters maximum heat transfer coefficient is displayed at the spout-annulus interface and it increases with particle diameter.

Generally the heat transfer data are similar to those for submerged objects in moving packed beds. It is noteworthy that location of heating elements is very crucial since transfer rates can easily vary by 10 or more. More work is needed in this area - particularly for heated or cooled draft tubes.

As in the case of heat transfer a two-region model is used for particle mass transfer. Use of a mass transfer coefficient for the loosely packed annulus gives a conservative estimate of the overall transfer rate. For drying in the falling rate the internal diffusion rate controls overall drying rate. If the moisture diffusivity variation with temperature and moisture is known one can use the well known Becker and Sallans (22) equations to estimate drying times in a spouted bed apparatus. Much remains to be done in this area, however.

Concluding Remarks

A brief overview is presented of recent developments in spouted bed technology. Various modified versions of SB's and some novel design configurations are outlined and discussed. Draft-tube SB's and two-dimensional beds appear to be especially viable commercially for drying and other applications requiring controlled circulation time of particles. The screw-conveyor type SB is an attractive device for drying of paste-like materials and slurries to replace the conventional spray dryer economically for low tonnage applications.

Numerous problems worthy of further study remain to be tackled. Aerodynamics, immersed surface heat transfer and drying in various configurations of SB's remain to be investigated further with a view to generating generalized empirical correlations based on phenomenological models. Scale-up criteria and design procedures need to be developed for the newer variants of SB's (20). Segregation of polydisperse particles in spouted beds has not been investigated in much detail (18, 19). Drying in spout-fluid beds needs to be investigated as yet. Aerodynamics, immersed surface heat transfer and drying of grains in two dimensional spouted beds are being investigated in author's laboratory at McGill University. Drying and heat treatment of grains in SB's fitted with draft tubes which are impermeable/permeable to air and also with uniform and/or variable cross-section offer definite potential for commercialization especially for small-scale farm drying. These areas need to be studied in depth in the near future. On a fundamental scale, theoretical studies and computational models for gas-particle flow in various types of SB's are needed to reduce the current dependence on empiricism. Mathematical models, even if approximate and semi-empirical, will reduce the cost of experimentation especially in the development of modified spouted beds for drying as well as other applications. For applications which require a narrow-band residence time distribution multi-staging of spouted beds should be considered. Several beds may be spouted with same air if the beds are stacked one over the other.

Finally, it may be worthwhile examining the energy efficiency advantages of a combined direct-indirect spouted bed dryer using hot air as well as superheated steam.

Acknowledgements

The assistance of Purnima Mujumdar in the preparation of this typescript is acknowledged. This paper is based partially on Plenary Lecture given by the author at the Brasilian Symposium on Transport Processes in Particulate Systems, São Carlos, Brasil, October 1982.

References

1. Gishler, P.E., Proc. 2nd Int. Symp. Spouted Beds, Vancouver, Canada, Oct. 3-6, 1982. pp. 1-3.

2. Mathur, K.B. and Epstein, N., Spouted Beds, Academic Press, N.Y. (1974)

3. Buchanan, R.H. and Wilson, B., Mech. & Chem. Eng. Transactions, The Instn. Eng., Australia, 1965, pp. 117-124.

4. Taskaev, N.D. and Kozhina, M.I., Sec. Ind. Eng. Chem., Vol. 50, No. 9, pp. 1402 (1958)

5. Khoe, G.K. and Van Brakel, J., Solids Sep. Precesses, I. Chem. Eng., 59, 6/1 - 6/13 (1980).

6. Claflin, J.K. and Fane, A.G., CHEMECA 81 - 9th Australasian Conference on Ehem. Eng., Christchurch, New Zealand, Aug. 30-Sept. 4, 1981.

7. Pallai, E. and Nemeth, J., CHISA paper C3.11, 1-18, (1972).

8. Mujumdar, A.S., DRYING'83, Hemisphere-McGraw-Hill, N.Y.

9. Raghavan, V. and Mujumdar, A.S., DRYING'83, Hemisphere-McGraw-Hill, N.Y. (1983)

10. Swaminathan, R. and Mujumdar, A.S., ibid.

11. Green, M.S. and Bridgewater, J., Proc. 2nd Int. Symp. Spouted Beds, Vancouver, Canada, Oct 3-6, 1982.

12. Strumillo, C. and Kaminski, W., DRYING'80, Vol 2, Ed. A.S. Mujumdar, Hemisphere-McGraw-Hill, N.Y. (1980)

13. Markowski, A. and Kaminski, W., Proc. 2nd Int. Symp. Spouted Beds, Vancouver, B.C., Oct. 3-6, 1982.

14. Vukovic, D.V. et al, ibid

15. Yang, W.C. and Keairns, D.L., ibid.

16. Yang, W.C. and Keairns, D.L., AIChE Wymp. Series, Vol. 74, No. 176, 1978.

17. McMahon, J.F. U.S. Patent 3,825,477 (1972)

18. Rovero, G., Proc. 2nd Int. Symp. Spouted Beds, Vancouver, Oct. 1982.

19. Ishikura, J., Tanaka, I. and Shinohara, H., ibid. (1982).

20. Nemeth, J., Pallai, E. and Aradi, E., ibid (1982)

21. Malek, M.A. and Lu, B.C.Y., Can.J. Chem. Eng., 42, 14-20 (1964)

22. Becker, H.A. and Sallans, H.R., Chem. Eng. Sci., 13, 97-112 (1961).

DEVELOPMENT OF SPOUTED BED DRYER

E. Pallai[1]
J. Németh[1]
E. Aradi[2]

1. Research Institute for Technical Chemistry of the Hungarian Academy of Sciences, Budapest, Hungary
2. Univ.of Agricultural Sciences, Keszthely, Hungary

ABSTRACT

A brief account is given of the considerations involved in the scale-up of spouted bed dryers on the basis of laboratory scale tests. Criteria for specification of spouted bed dryer geometry are discussed in the light of published literature and authors own work with conventional as well as newer types of spouted bed dryers . Advantages of the novel modifications of the spouted bed are identified.

NOMENCLATURE

D_c - Column diameter, m
D_i - Inlet nozzle diameter, m
H_c - Length of the cylindrical device, m
Q_a - volumetric flow-rate in the annulus, m^3/s
Q_i - volumetric flow-rate through the inlet nozzle, m^3/s
ΔP_i - initial pressure drop for spouting a fixed bed, N/m^2
d_p - particle diameter, m
g - gravitational acceleration, m/s^2
h - bed depth, m
h_{max} - maximum spoutable bed depth, m
u_i - superficial fluid velocity in the inlet, m/s
u_s - spouting gas velocity, m/s
u_{ms} - gas velocity at incipient spouting, m/s
v_T - terminal velocity of solid particles, m/s
α - bevel angle
ε_m - void fraction at minimum fluidization, m^3/m^3
ρ_g - density of fluid, kg/m^3
ρ_s - density of solid particles, kg/m^3

INTRODUCTION

For a long time investigation of the flow pattern of spouted beds and extension of the areas of their application have been the main activities in the area of spouted beds. The problem of scale-up of spouted bed equipment has been treated only by a relatively small number of authors /Mathur and Gishler,1955; Mathur,1971; Pallai, 1970/.

The scale-up of spouted bed dryers entails two basic problems:

/1/ to find the limits of applicability of known correlations of flow,

/2/ to establish the ways of eliminating or at least minimizing the various drawbacks of the conventional spouted bed dryer by appropriate modifications of construction along with pertinent rules.

Limitations of the conventional spouted bed apparatus include the high pressure drop across the central air nozzle, the high value of the peak pressure drop prior to spouting, nonuniform distribution of residence time of particles, and lastly the fact that the air requirement is determined, not by drying constraints but rather by requirements of good spouting

SCALE-UP OF CONVENTIONAL SPOUTED BED DRYERS

From the viewpoint of their effects on the formation of the spout and associated circulating flow pattern the following dimensions of equipment are the most significant D_c, D_i, α, H_c /see Nomenclature/. The length of the cylindrical part of the equipment depends on the bed height as a further dependent variable. Consequently, only the first three dimensions can be chosen independently. Of these, the choice of the bevel angle is the simplest because an angle of 60° is suitable for most materials to be dried. Both diameters can be analyzed separately or as the ratio, D_c/D_i. At a given value of D_c the lower limit of D_i is not limited in principle. By decreasing D_i the bed height can be increased which is equivalent to an increase of the residence time in the dryer. However, it is not desirable to decrease D_i below a certain value for the following reasons:

- the pressure drop of air flowing through a narrow nozzle at a high velocity is very high. The increased demand of ventilation power means higher capital and operating costs /Pallai and Halász,1978/
- the air-particle contact is limited in case of extremely high velocities of the entering gas to the inner spout. From the aspect of heat and mass transfer the case when a significant portion of the drying medium flows counter-currently to the downward sliding annulus is more favourable.

On the basis of our experimental results, for $u_i/v_T < 150$, the ratio of the volumetric flow rate of air flowing in the annulus to the total entering air can be described as a fair approximation by the following empirical relationship:

$$Q_a/Q_i = A - \left[1-(h/h_{max})^{1/2}\right]^3 \qquad 1.$$

where $A = 0.4$ for $D_c/d_p = 20 - 50$, and
$A = 0.5$ for $D_c/d_p = 50 - 200$.

The validity of Eq.1 has been checked experimentally up to $D_c/d_p = 200$ in the range $D_c = 50 - 600$ mm.
For $u_i/v_T > 150$ ranges i.e. for small values of D_i and at a significant gas feed rate the value of Q_a/Q_i is very low, attaining the maximum value of 0.2 only at a level of $h/h_{max} = 0.7$, which for $u_i/v_T < 150$ ranges between 0.4 and 0.6 /see Fig.1/.

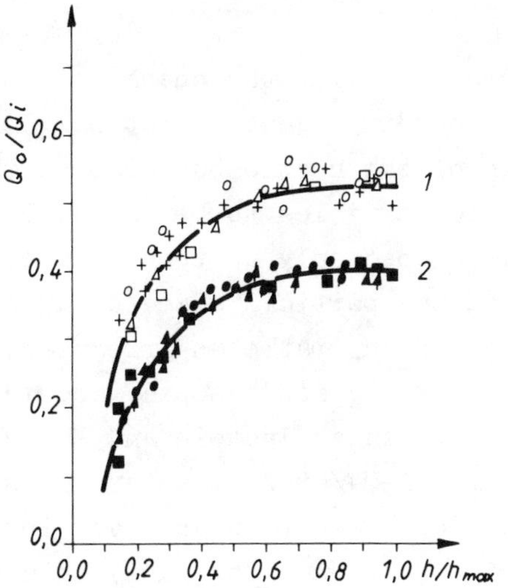

Fig.1 Distribution of gas velocity along the height of the annulus.
Curve 1: D_c= 60 mm, d_p= 0.7 mm /activated carbon/, +D_i= 4 mm, oD_i= 4 mm, △D_i= 6 mm, □D_i= 8 mm.
Curve 2: D_c= 60 mm, d_p= 1.7 mm /activated carbon/, ●D_i= 4 mm, ▲D_i= 6 mm, ■D_i= 8 mm,
D_c= 60 mm, d_p= 2.55 mm /glas spheres/, ●D_i= 6 mm, ▲D_i= 10 mm.

Returning now to the choice of the dimension D_i, it should be noted that the increase of D_i, for given D_c, is limited by practical reasons. The greater is the diameter D_i the lower will be the applicable bed height. However, in order to ensure optimum conditions of drying the bed height must be maintained above a definite value. According to literature data /Mathur,1971;

Rowe and Claxton,1975/, the air leaving the spouted bed dryer may be considered as saturated when the bed height exceeds one meter. An appreciably high peak pressure drop is required prior to spouting /Rjabinowitsch,1977/, due to the formation of a loosely packed bed and of an inner spout. On loosening a fixed bed the peak of pressure drop may attain a multiple /2-3 times/ of the bed weight. On view of the opposing effects mentioned about the variation of the dimension D_i, the optimum value of the ratio D_c/D_i is in the range: $D_c/D_i = 6 - 10$.

As mentioned earlier the change of the diameter of the dryer depends not only on the dimension of the nozzle but also on the size of the particles to be dried. The bed can be developed only up to a D_i value depending on the particle size /D_i/d_p=3-30/. At a higher D_i/d_p ratio an irregular, aggregative slugging is developed. According to literature data /Epstein and Mathur, 1971/, /Imre,1974/ the diameters of spouted bed dryers operated in plants are in the range 600 to 1000 mm. Obviously this dimension also limits the unit efficiency. Also the ratio D_c/d_p plays a decisive role in the formation of the flow pattern of spouted bed. On the basis of our experiments the optimal ratio of D_c/d_p is: $D_c/d_p = 25 - 200$.

Possibilities of scale-up of a spouted bed dryer with nozzle.

On testing several constructional variations for increasing of the efficiency of the spouted bed dryer equipped with nozzle, the following three types are recommanded.

Combination of units with an oblong-shaped cross section /Fig.2/

Here the body of the equipment with an extended oblong cross section is fitted with a conical extension piece. A partially or fully impermeable draft tube may be located above the nozzles. Use of an impermeable draft tube can increase the bed height by about 35 %.

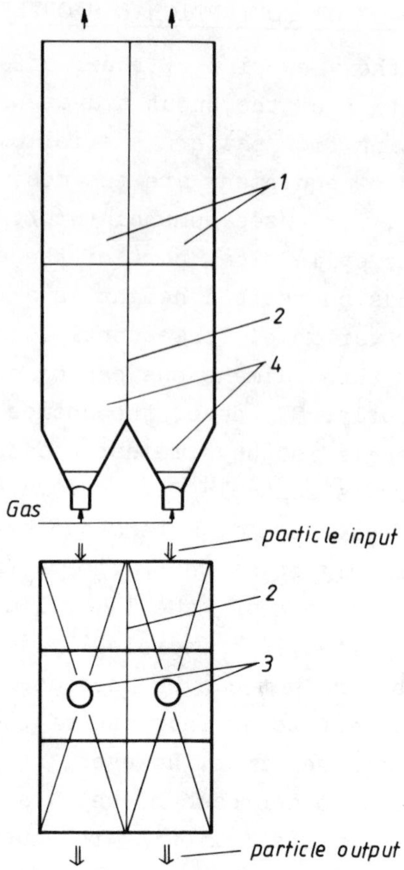

Fig.2 Twin-body spouted bed dryer with common material feed and discharge arrangements and with individual air inlets /Type 1/.1- spouted beds, 2-divider sheet, 3-nozzles,4-draft tube

The longer side of the oblong apparatus is 3-30-fold as long as the shorter one. On shaping the device in this way it was possible to obtain a distribution of residence times of particles which was more uniform and better controlled than for devices of circular cross section /Németh et al.,1973/.

Devices of circular cross section,equipped with four nozzles /Fig.3, Type 2/.

Air enters the spouted bed device as shown in Fig.3, from a separate or from a common air duct in a way that the individual spout annuli should contact each other

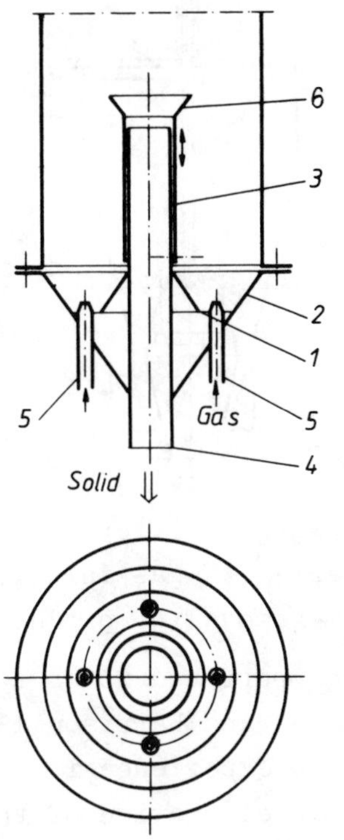

Fig.3 Spouted bed dryer with four nozzles. 1-pad, 2-nozzles, 3-movable draft tube, 4-tube for removal of material, located within the movable draft tube, 5-nozzles for air inlet, 6-cone-shaped draft piece.

within the space of the annulus cross section. The number of spout channels decreases within the bed, the residence-time distribution curve of particles takes up a more and more assymetric shape in the course of continuous operation with a maximum in the domain of longer times. We will here refrain from further detailed description of particle movement and refer to published literature /Khoe,1980/. The removal of material could be ensured by means of a collecting tube adjustable to a desired height and located concentrically in the bed. The diameter of the device with four nozzles can be increased to nearly double the diameter of the conventional spouted bed with one nozzle. This resulted for one case in a fourfold increase of the drying capacity.

Spouted bed dryer with accessory air feeding /Fig.4, Type 3/.

As shown in Fig.4 the spouted bed can be equipped with a central nozzle through which compressed air is introduced. Since the air leaving the nozzle plays a role only in the formation of spout, it requires only a small volumetric flow and hence the pressure drop can thus be maintained at an economical level. The bulk of the drying air enters through the gas distributor located below the annulus. By means of this device it was possible to readily increase the maximum spoutable bed depth by 20 - 25 %.

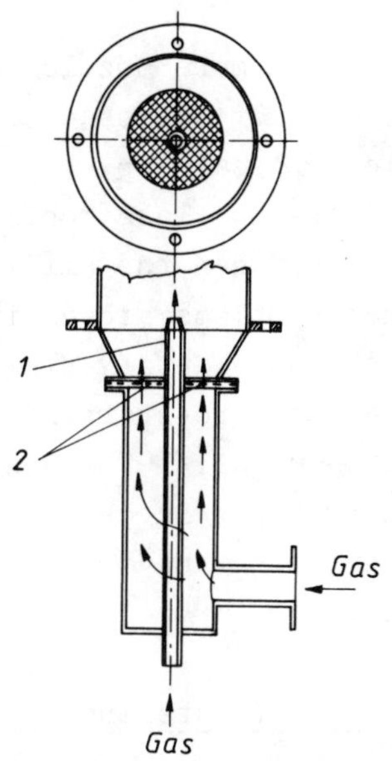

Fig.4 Spouted bed dryer with double gas inlet arrangement. 1-nozzle, 2-gas distributor.

On the basis of plant experience, following is a summary of key results about these three modified spouted beds:

a/ The maximum spoutable bed depth can be increased with types 1 and 3. In this way the amount of removable moisture could be raised in drying.

b/ The use of the principle of building blocks made possible, in case of type 1 an increase of the unit efficiency. The volumetric flow rate on the gas side of a double unit having a cross section of 2 x x 0.2/1.6 m could be raised to 3000-3600 m^3/h i.e. up to the capacity range of conventional dryers of industrial scale. The same statement is valid in case of the unit with accessory air feeding, from the standpoint of unit efficiency.

c/ Continuous removal of the granular material in case of type 2 presented a possibility for a significant scale-up which could be utilized not only in the drying operation. On this principle also cooling-ventilating warehouses may be equipped.

Scale-up possibilities of the modified spouted bed dryers

The gas inlet with more nozzles did not in itself prove a fully satisfactory solution to the problem of scale-up. It was difficult to eliminate the less mobile or stagnant zones between the sliding layers and additional problems arosed in the control of the gas quantity flowing through the nozzles and the re-starting after each stand. Romankov et al./1979/ went around this problem by employing slits instead of nozzles for entering the drying air into the apparatus.

Dryer with slit for entering the gas

According to the first variation /see Fig. 5/, the air enters through the annulus shaped channel and exits at the top of the dryer. The material is removed, as shown in Fig.3 at the middle of the duct. The removal of particles is adjustable, whereas adjusment of the gas inlet into scaled-up equipment cannot yet be considered as satisfactorily solved. It was no doubt this problem that led soviet researchers to developed the second variation /Fig.6/, where the air enters through horizontal slits with such a high velocity as to make the entire bed recirculate intensively. This type of equipment

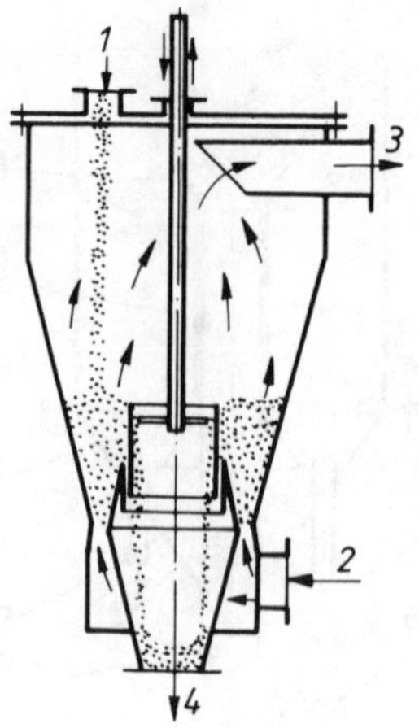

Fig.5 Dryer with slit for entering the gas 1-material, 2-gas inlet, 3-gas outlet, 4-removal of product.

is some-times reffered to as a vortex bed dryer /Davidson and Harrison,1979/. On the basis of plant experience it has been ascertained that elongation of the slits, with vertical partition cards possible dividing the bed into sections, as well as the multiplication of the number of vortex beds are both safe procedures which yield greater results.
According to Fig.6 gas feeding may be refered to as tangential.

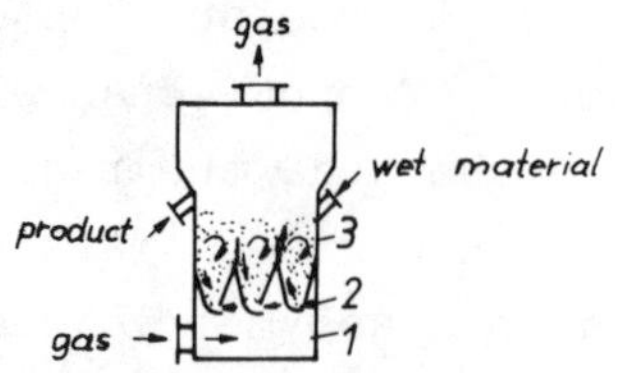

Fig.6 Dryer with slit for gas inlet 1-mixing chamber, 2-slit, 3-drying conpartment.

Devices with tangential air feeding

In the last few years efforts were made to find novel types of construction by means of which the introduction of air through nozzles can be avoided i.e. the entire pressure drop on the side of air can be better approached to the pressure drop of the spout itself which is known to be smaller than the pressure drop of a fluidized layer of the same height. The problem was solved by tangential air feeding with a so-called "swirling ring"according to Fig.7 /Aradi et al.,1976/.

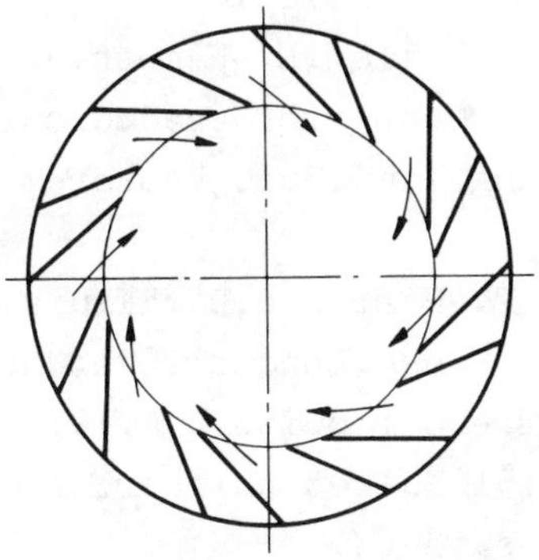

Fig.7 Schematic of the swirl inlet ring.

According to our experiments this device helps maintain uniform spouted bed particle circulation. The flow pattern of particles promoting deseggregation in combined axial and radial directions is preferable in the case of drying of adhesive materials containing greather amounts of surface moisture. Soviet researcher /Elperin and Dolidiwitsch,1973/ also ascertained conducting laboratory tests /D_c=45 to 60, D_i=15 to 20 and d_p=1.5 to 5 mm/ tangential gas feeding to be more advatageous to gas inlet nozzles. They found that the gas velocity of initial spouted bed movement decreased to an extent. Their explanation for this was that the nearly concentric particle movement at the bottom of the sliding layers /annulus/ favoured the development of the spout channel. A further thermological phenomenon also makes it advisable to avoid the use of a centrally located air-inlet nozzle. For in the traditional, spouted bed dryer with nozzle there is a significant difference in the temperatures of the gas and particles leaving the bed. Note that with fluidized bed dryers there is practically no temperature difference, in fact it may even happen at times that the temperature of the exiting gas is lower than the temperature of the dried product, leaving the dryer. This of course does not mean that there is no thermal-equilibrium between the gas and particles in the spout channel. The reason behind this particular phenomenon is that the volumetric concentration of particles flowing at a high velocity /few m/s/ in the inner channel is quite small, usually 3-5%, consequently the temperature of the bed is determined by the temperature of the particles in the annulus. The afore-described inhomogenity in traditional, spouted bed dryers was reduced by soviet researchers /Zabrodskij,1971/ by feeding air pulsated with 1.5 - 15 Hz frequency into the apparatus. In the course of drying wheat, for example in the vicinity of the velocity of initial spouted bed movement /u_s= 1.1 to 1.4u_{ms}/ therely increasing the heat transfer coefficient between the gas and particles by 12-15 %.

Of the basis of our own and published data refered to previously, in the course of dryer-development we understook to develop construction - types with tangential gas feeding through more ducts and abandoned the gas inlet nozzle-construction, including the fairly complex pulsating mechanism of pulsating-air flowing through the inlet. Fig.8 shows a novel type spouted bed dryer equipped with an inner cylinder, a draft tube and an attachment permeable to air. The start of the dryer equipment became simpler with the use of the draft tube whose shape was similar to that of the spout. In case of materials with high moisture content /such as washed and centrifuged sowing seeds/ the demand of the water evaporation efficiency is significant, and this requires large volumes of drying air. Using a device of the type shown in Fig.8 it was possible to operate with 3-5 times the flow rates used in conventional spouted bed dryers provided the upper portion of the device was constructed of an air-permeable material e.g. linen. A seed dryer of industrial scale /D_c= 540 mm/based upon this principle has been put into successful operation. The evaporation capacity of this dryer was 270 kg water/m^2h whereas its specific power consumption was 3000 kJ/kg water.

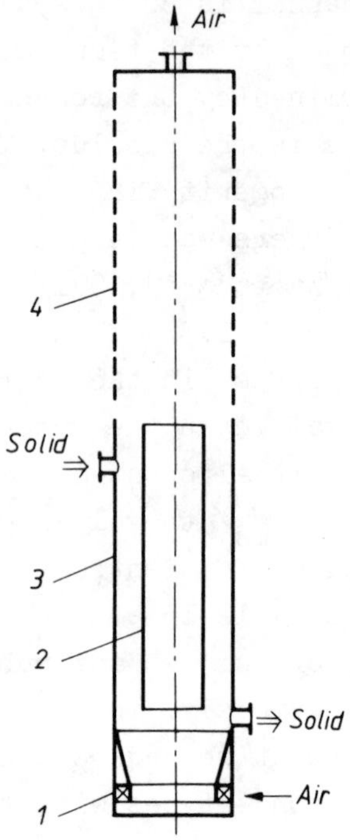

Fig.8 Spouted bed dryer with a draft tube and an air-permeable attachment. 1-tangential air inlet, 2-draft tube 3-dryer body, 4-air permeable extension.

A spouted bed dryer with tangential air feeding equipped with an inner conveyor screw is also developed. Along the vertical axis of the device a conveyor screw without any casing can be operated which is capable of ensuring, quite independently of the air flow rate, the recirculating motion of the spout even in the presence of small grain materials /where $D_c/d_p \geq 500$/ for which the spout does not develop in the conventional way /Németh and Pallai, 1970/,/Blickle et al., 1978/.

REFERENCES

Aradi,E.-Pallai,E.-Blickle,T.-Monostori,E.-Németh,J.-Varga,J.:Contacting and drying of material in a spouted system, Hung.Patent,No.176.030 /1976/.

Blickle,T. - Pallai,E.-Németh,J.-Aradi,E. Varga,J.: Monogr.of Acad.Group at Veszprém, Hung.Acad.Sci.,Veszprém,Hungary, /1978/.

Davidson,J.F.-Harrison,D.:"Fluidization", Academic Press,London,1971.Chapter XII.

Elperin,I.T.-Dolidowitsch,A.F.: Avt.svid. No.385.151 Bulletin izobr. 25, 1973 /Soviet patent/.

Epstein,N. -Mathur,K.B.: Heat and Mass Transfer in Spouted Beds. A review. Can.Journal of Chem.Eng.,49, 467 /1971/

Epstein,N.-Lim,C.J.-Mathur,K.B.: Data and Models for Flow Distribution and Pressure Drop in Spouted Beds. Can.Journal of Chem.Eng., 56, 436 /1978/

Imre,L.:'Száritási kézikönyv', Műszaki KK. Budapest, Hungary, 1974.

Khoe,G.K.:"Mechanics of Spouted Beds", Delft University Press, 1980.

Mathur,K.B.: Spouted Bed Drying; The Problem of Scale-up, Can.Journal of Chem. Eng., 49, 476 /1971/

Mathur,K.B. - Gishler,P.E.: Technique contacting gases with coarse particles, Am. Inst.Chem.Eng., 1, 157 /1955/

Németh,J.-Pallai,E.: Spouted bed technique and its application, Magyar Kémikusok Lapja, 25, 74 /197o/

Németh,J. - Pallai,E. - Blickle,T.-Győry, J.: Hung.Patent Off. No.160.333 /1973/

Pallai,E.: Doctoral thesis, Hung.Acad. Sciences, Budapest, 1970.

Pallai,E. - Aradi,E.-Halász,G.:Examination of the cost-function of spouted bed dryers, CHISA'78,Prague, Czeho - slowakie, 1978.

Rjabinowitsch,M.J.:"Teplovije procecci v fontanirujuschcem cloje", Naukova Dumka, Kiev, USSR. 1977.

Romankov,P.G. - Rashkovckaja,N.B.:"Cuska

vo vzveshennom coctojanii", Himija, Leningrád, USSR, 1979.

Rowe,P.N.-Claxton,K.T.: Trans.Instn.Chem. Engrs./U.K./ 43, T 321 /1965/

Zabrodskiy,C.C.:"High-temperature experiments in fluidized bed", $E_n$ergia, Mockau, USSR, 1971.

USEFULNESS OF VIBRO-FLUIDIZED BED FOR DRYING OF WETTED AND AGGLOMERATED MATERIAL

Kanichi Suzuki, Asao Fujigami and Kiyoshi Kubota

Department of Applied Biological Science,
Hiroshima University, Hiroshima, Japan

ABSTRACT

Wetted and agglomerated material was dried in a vibro-fluidized bed. Diameter of drying chamber was 110 mm. Influences of intensity of vibration, air velocity and uniformity of moisture content in the bed were investigated. From these results, usefulness of vibro-fluidized bed for drying of wetted and agglomerated materials which cannot be treated in a conventional fluidized bed dryer was discussed.

NOMENCLATURE

a = amplitude of vibration [cm]

D_p = diameter of material [mm]

d.m. = bone dry material

f = frequency of vibration [1/s]

g = acceleration due to gravity [cm/s^2]

H = humidity [kg-H_2O/kg-dry air]

L = bed height [cm]

t_i = air temperature at inlet of bed [°C]

t_w = wet-bulb temperature [°C]

u = superficial air velocity [cm/s]

W = moisture content [kg-H_2O/kg-d.m.]

W_o = initial moisture content [kg-H_2O/kg-d.m.]

θ = drying time [min], [h]

ω = angular frequency of vibration [1/s]

INTRODUCTION

Most of powders and granular solids, especially of foodstuffs, have a tendency to agglomerate when they are wetted or moistened. In such cases, the fluidization does not occur, and a conventional fluidized bed dryer cannot be used. Therefore, it was tried to investigate the drying processes of wetted and agglomerated material in a vibro-fluidized bed dryer.

The sample used was okara which is a isolated cake in the expression process of the production of soybean milk. It is considered that the usefulness of the vibro-fluidized bed for drying of wetted and agglomerated materials can be discussed passably by the drying results of this substance.

EXPERIMENTAL

The schematic diagram of the experimental apparatus used in this study is shown in Fig.1. The drying chamber was made of an acrylic resin tube. The inner diameter was 110 mm and the height was 500 mm. The exterior wall of the drying chamber was covered with a 20 mm-thick foaming polystyrol resin plate for heat insulation. The distributor was of the perforated-plate type which was covered with a 300-mesh stainless steel wire net. The system of the vibration generating device is as shown in this figure. That was very easy to adjust both of amplitude and frequency of vibration.

Okara was procured from the factory at the time of the experiment. The initial moisture contents were in the range from 3.5 to 4.0 kg-H_2O/kg-d.m. Two kinds of okara samples were prepared for the experiments. One was the sieved okara by the sieve of 12 mesh, and the other was the original one. The size of the sieved okara was thus equal to or smaller than 1.4 mm. Hereafter, the former and the latter sample are designated as sieved okara and no sieved okara respectively.

The charge weight of sample was 140 grams, and the initial bed height L was 4 cm. The bed height was decreased as the drying process proceeded because of a shrinking property in the drying process of the material. To measure the change and the uniformity of moisture content in the bed, about 0.5 grams of sample was collected from the upper portions of the bed at every fixed time.

EXPERIMENTAL RESULTS AND DISCUSSION

The examples of the drying curves of sieved okara are shown in Fig.2. One was the measured value when the bed was vibrated and the other was the

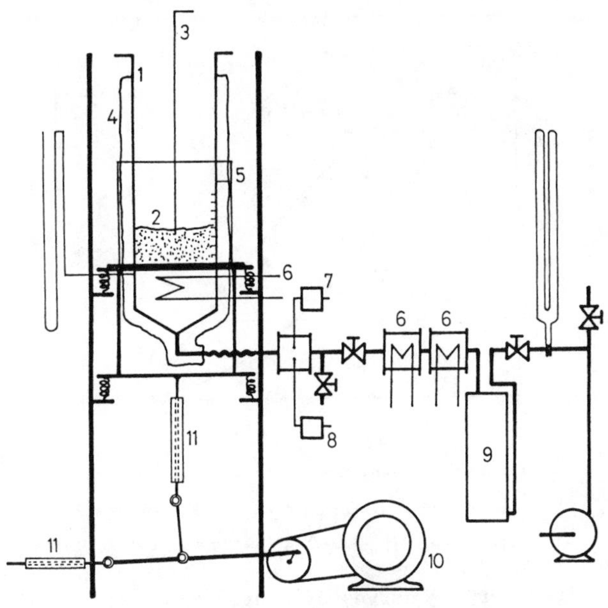

Fig.1 Schematic diagram of experimental apparatus

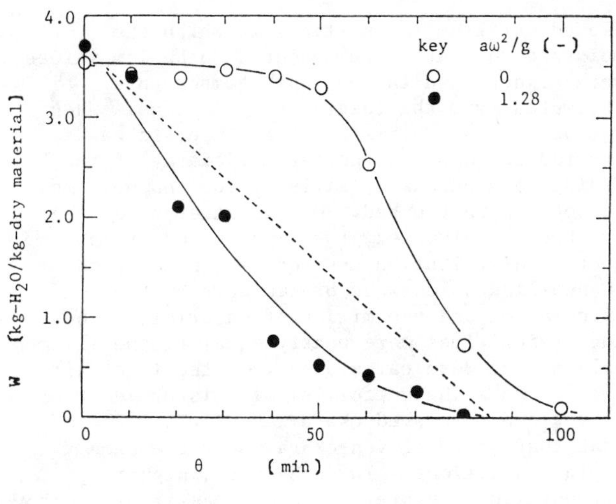

Fig.2 Examples of drying curves of sieved okara (dotted values are estimated drying curve from inlet air conditions)

result when the bed was not vibrated, where the superficial air velocity u was 10.0 cm/s, the air temperature was 72 °C, and the air humidity was 0.011 kg-H_2O/kg-dry air. Although the sample was sieved okara, it had still bulky and irregular shapes and tended to agglomerate easily. Therefore when the bed was not vibrated, the sample was not fluidized in the air velocity conditions used, and the drying process was the same as in a through-flow drying. In such case, a drying zone rised gradually from the bottom toward the top of the bed. Thus the moisture content in the upper portion of the bed did not change untill the drying zone reached to that portion.

The dotted curve in this figure indicates the theoretical drying curvr calculated by assuming that the whole bed is dried with uniform moisture content, and all drying stages are in a constant rate period, and further there are no heat losses.

Even though the air velocity was very low as shown, the drying time was still longer than the calculated value when the bed was not vibrated. On the other hand, when the bed was vibrated, the material in the bed was circulated and mixed according to the intensities of vibration, and thus the uniformity of moisture content in the bed increased consequently. The drying curve for the vibrated bed showed that the circulation or the mixing of the bed was sufficient for the drying with uniform moisture content in the bed under the condition of vibration shown in the figure. Furthermore, the results showed that there were no heat losses in this case.

As already mentioned in the previous papers[1,2], the mixing rate or the uniformity of moisture content in the bed depended upon the properties of materials, the air velocity, the bed height, and the intensity of vibration. And when the air velocities were constant, the circulation rate of sample increased as the intensity of vibration increased. Thus the drying curves as in a through-flow drying or in a complete mixing drying were obtained according to the intensities of vibration as shown in Fig.3. Similar results were also observed in no sieved okara. However, higher intensities of vibration than for sieved okara had to be used in these cases. Thus, as descrived above, the vibro-fluidized bed has a special characteristics for fluidization or mixing of materials under the conditions in which the bed cannot be fluidized in the conventional fluidized beds.

If the humidity in the air flowed through the bed saturates in the bed whether the bed is vibrated or not, the drying times in both conditions must be identical. However, when the bed was not vibrated, the drying time was longer than the value for vibrated bed even at the low air velocity as showm in Fig.3. As a principal reason of these results, it was considered that the hot air might flow out selectively through rough void portions in the bed when the bed was not vibrated. Once the channelling holes are formed, the substance around them may be dried quickly and selectively. But after this, the heat of hot air flowed through the channelling holes will not used so much for the drying.

Then the ranges of the change in moisture content distribution in the upper portion of the bed were measured to verify the assumption. The results are shown in Fig.4 and Fig.5. It is evident from these figures that the moisture content distributed widely when the bed was not vibrated. Furthermore, the temperatures of air flowed out from the portions in which the moisture contents were measured to be low values were higher than the wet-bulb temperature estimated from the inlet air conditions. When the bed was not vibrated, therefore, it was considered that the thermal efficiency of the drying was decreased fairly by the channelling of hot air in the bed. On the other hand, when the bed was vibrated, the range of moisture content distribution was reduced, because the channelling holes were broken down by the circulation and the mixing of materials. Thus the drying times were nearly equal to the minimum values that were calculated from the inlet air conditions. The uniformity of moisture content in the bed of sieved okara was slightly better than that of no sieved okara under the same drying conditions. In such cases, higher intensities of vibration were necessary to improve the uniformity of moisture content in the bed of no sieved okara.

From these experimental results, two advantages of the vibro-fluidized bed were recognized. One is that the drying process can be carried out at uniform moisture content in the bed. The other one is that high thermal efficiency of the drying can be obtained.

SUMMARY

Okara was dried in a vibro-fluidized bed dryer. Even though okara had bulky and irregular shapes and tended to agglomerate, it was able to be circulated and mixed in the bed under the appropriate conditions of vibration at the air velosities used. Further, the bed was able to be dried at uniform bed moisture content under suitable air velocities and the intensities of vibration. The uniformity of moisture content in the bed increased as the intensity of vibration increased. Since the hot air flowed out selectively through rough void portions in the bed when the bed was not vibrated, the drying time took longer than the values estimated from the inlet air conditions of the bed. When the bed was vibrated, the channelling holes were broken down, and thus the drying times were nearly equal to the theoretical values, and the thermal efficiency of the drying was improved.

REFERENCES

1. Suzuki, K., Hosaka, H., Yamazaki, R. and Jimbo, G. : J. Chem. Eng. Japan, 13, 117 (1980).
2. Suzuki, k., Fujigami, A., Yamazaki, R. and Jimbo, G. : J. Chem. Eng. Japan, 13, 493 (1980).
3. Suzuki, K., Fujigami, A., Kubota, K. and Hosaka, H. : J. Jap. Soc. Food Sci. Tech., 27(8), 393 (1980).

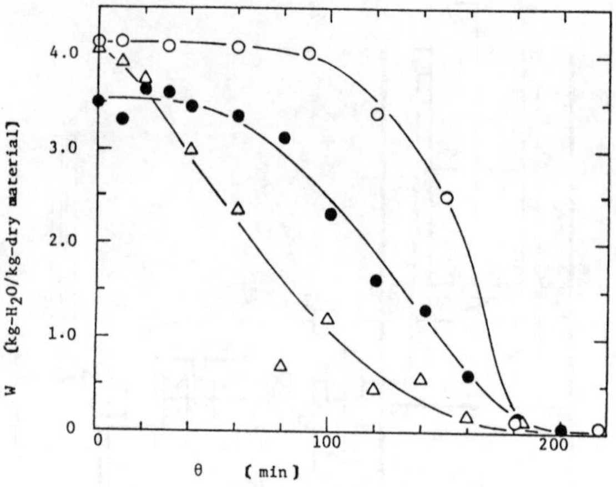

(under 12 mesh, u = 5.0 cm/sec, t_a = 71 - 78 °C, L = 4 cm, -○-○- : $a\omega^2/g$ = 0.49, -●-●-: $a\omega^2/g$ = 0.87, -△-△- : $a\omega^2/g$ = 1.28)

Fig.3 Drying curves of sieved okara at various intensities of vibration

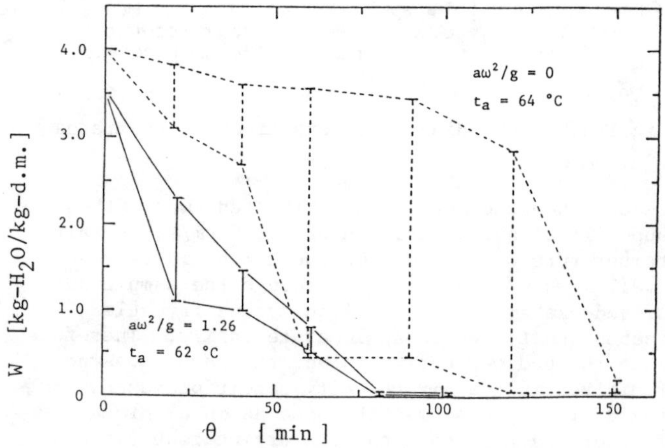

Fig.4 Effect of vibration on distribution range of moisture content in the top portion of bed (Sieved okara, u = 12.4 cm/s, L = 4 cm)

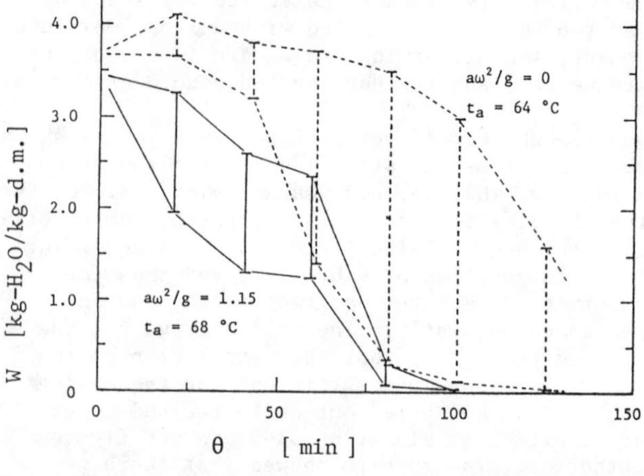

Fig.5 Effect of vibration on distribution range of moisture content in the top portion of bed (No sieved okara, u = 12.4 cm/s, L = 4 cm)

BED EXPANSION AND PRESSURE DROP IN VIBROFLUIDIZED LAYERS

K.ERDÉSZ and Z.ORMÓS

Research Institute for Technical Chemistry of the Hungarian Academy of Sciences, Veszprém, Hungary

ABSTRACT

This paper summarizes our results of tests carried out to lay the groundwork of vibrofluidization operations. Experiments were carried out in a specially designed, batch-operated vibrofluidization system equipped with a swinging plate. The energy requirement of vibrofluidization was also determined. It was concluded that compared with simple gas fluidization, vibrofluidization improved the flow pattern of the particulate material, and that the optimum range of vibrational acceleration - with respect to energy consumption - was between 1 g an 5 g.

INTRODUCTION

Use of vibrofluidization increases in technologies dealing with particulate materials. Essentially, in suitably designed fluidization systems extra energy is introduced into the fluidized layer via mechanical vibration. Vibrofluidization is often advantageous because the intensive motion of the layer creates uniform, controlled and directed mixing [1], the pressure drop across the layer and the minimum fluidization velocity are decreased [2,3], heat and mass transfer are considerably improved [4,5], vibration promotes the fluidization of so-called hard-tofluidize materials, etc.

These advantages can be utilized in a number of drying, granulation, separation and other operations common in food processing and chemical technologies. At present several major companies (e.g. AEROMATIC, ANHYDRO, WEDAG, etc.) produce in series continuous vibrofluidization systems for the food processing industry. The number of special equipment used in the chemical industries, which take advantage of vibration, also increases (rectification columns, absorbers, catalytic reactors, etc.).

Theoretical study of vibrofluidization began some 20 years ago, and a number of publications followed which were recently reviewed in [6] and [7]. A monograph [8] and several texts [8,9,12] were also published.

Primiraly these investigation dealt with the flow patterns and heat transfer characteristics of vibrofluidized layers. Practical application of vibrofluidizers proved that considerable energy savings were possible. Therefore, based on hydrodynamic measurements, the system was energetically analysed using the model of swinging mass.

EXPERIMENTS

Most hydrodynamic measurements were carried out in systems which were filled with particulate material and the entire apparatus was vibrated [1,2,3]. On the other hand, from a technological point of view it is more advantageous if the system is stationary and only the layer supporting plate is vibrated. This reduction in the moving masses decreases the energy consumption. Such a system is shown schematically in Fig.1.

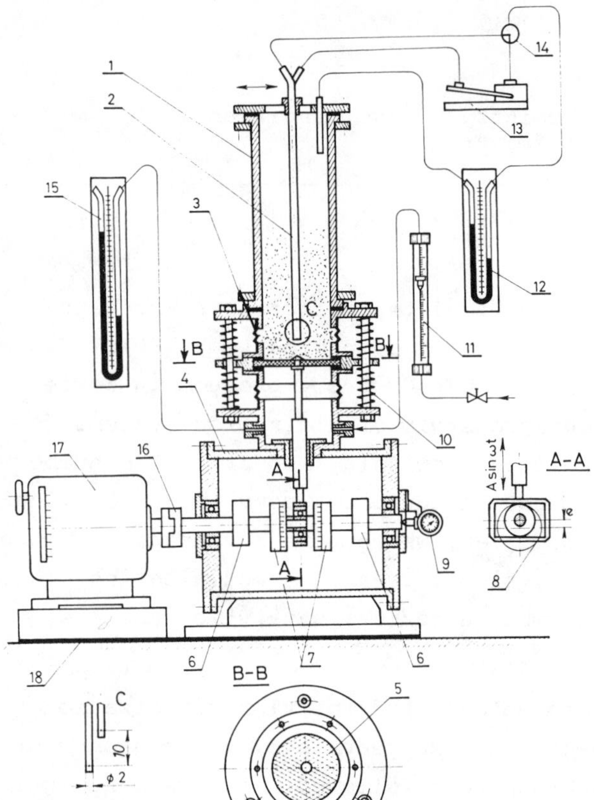

Fig.1 Experimental apparatus
1 - cylindrical glass column, D = 0.108 m; 2 - probe, 3 - flexible gasket; 4 - drive mechanism; 5 - gas distributor plate; 6 - balancing mass; 7 - excenter; 8 - crank mechanism; 9 - tachometer; 10 - spring; 11 - rotameter; 12,15 - U-tube manometer; 13 - micromanometer; 14 - valve; 16 - flexible cluth; 17 - variable speed electromotor; 18 - rubber plate

The supporting plate is vibrated by a sinusoidal drive mechanism which insures the harmonic vibration of the plate. The cylindrical springs located between the supporting plate and the drive mechanism establish the eigen-frequency of the system selected in such a manner that the system proper will not resonate. To decrease the amount of vibration transferred to the base the system is mounted to a heavy welded supporting frame.

Primarily, the hydrodynamic measurements were designed to reveal the relationship of vibration energy input and fluidization characteristics such as minimum fluidization pressure and bed expansion.

Experiments were carried out with sand of varying particle size, bed height and supporting plates of varying flow resistance (the sand was previously dried).

Vibration parameters were changed as follows:

- Amplitude: A = from 0.6, to 0.9 and 1.85 mm
- Frequency: ω = from 0 to 267 s^{-1}.

Accordingly, the vibrational characteristic, $\Gamma = A\omega^2/g$ was changed in the

Γ = from 0 to 13.4 range.

Pressure drop across the layer, Δp as a function of fluidization gas velocity u" was measured along with bed expansion Y/Y_o. The characteritic fluidization curves are shown in Figs.2 and 3. It can be seen that in the presence of vibration the characteristic pressure peak of gas fluidization (Fig.2) is missing and relative bed expansion increases (Fig.3). When the intensity of vibration is lower then 1g the layer is at first compacted, followed by bed expansion at higher gas velocities. Vibration has no significant effect on the magnitude of minium fluidization gas velocity.

The effect of increasing vibration amplitude is shown in Fig.4. Increasing vibration amplitude results in decreased pressure drop across the layer.

The minimum fluidization pressure drop (Δp_{mf}) was determined by visually

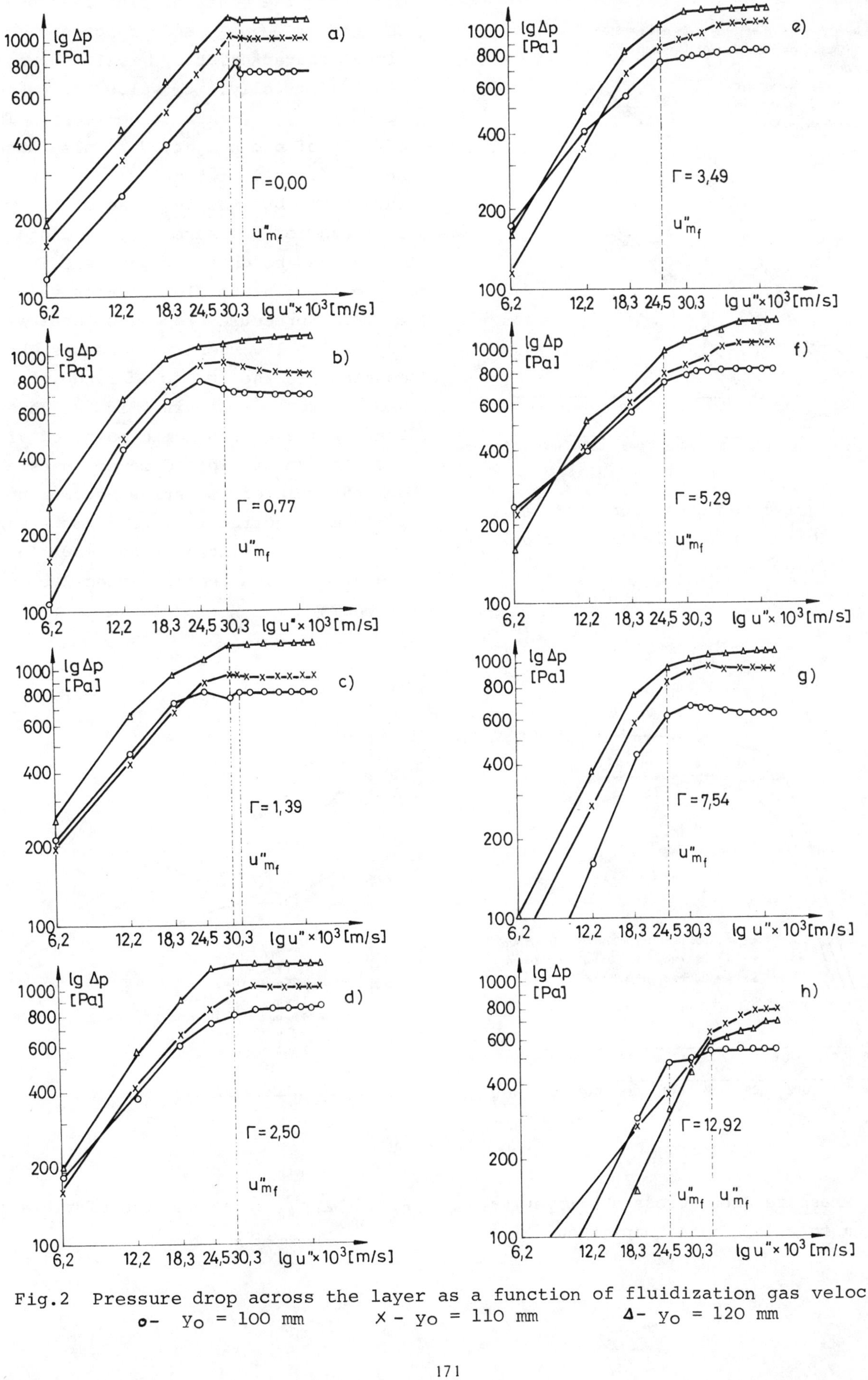

Fig.2 Pressure drop across the layer as a function of fluidization gas velocity
o- y_o = 100 mm x - y_o = 110 mm Δ- y_o = 120 mm

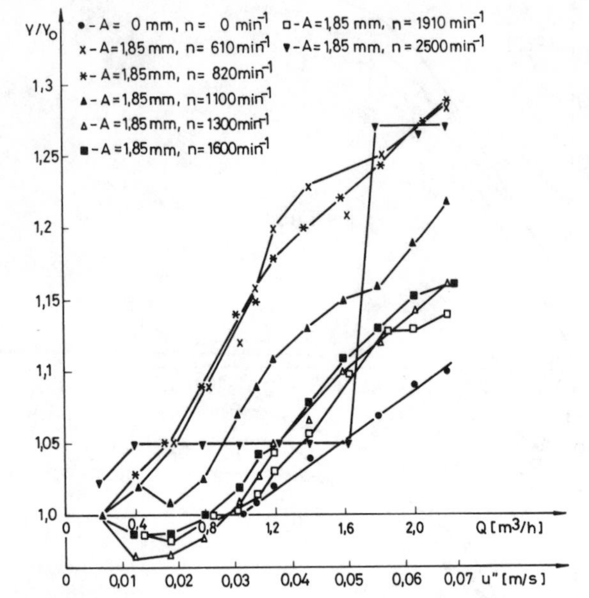

Fig.3 Relative bed expansion, Y/Y_o as a function of fluidization gas velocity, u"

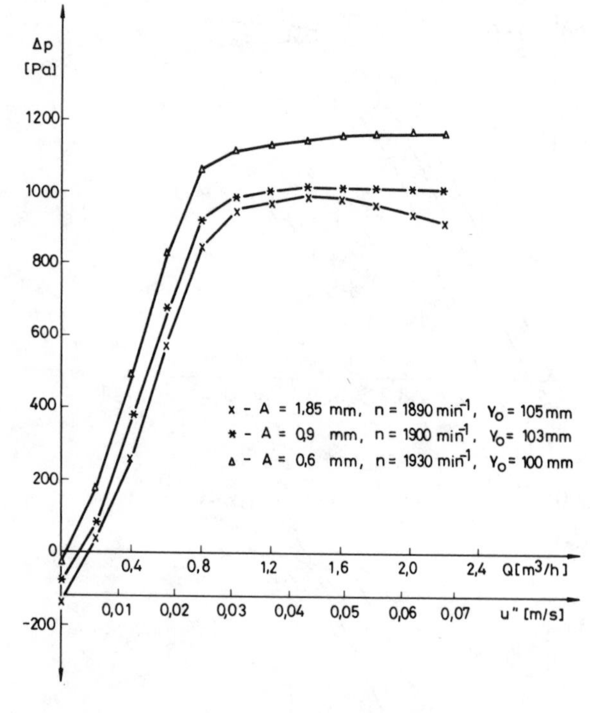

Fig.4 Pressure drop across the layer as a function of the amplitude

observing the onset of fluidization. Changes of Δp_{mf} as a function of vibrational characteristic, Γ, is shown in Fig.5. Initially, its value increases (up to Γ= 1 to 1.5), then it decreases. The effects of bed height, H is also apparent in the figure. It too increases the pressure drop, but the decrease of minimum fluidization pressure drop is faster at higher bed heights. At gas velocities above the minimum fluidization velocity, u''_{mf}, the pressure drop, Δp, decreases (up to Γ = 1), then it increases, becomes constant in the Γ= 2 to 5.5 range, then slowly decreases again (Fig.6). This means that from a hydrodynamic point of view once the gas-fluidized layer is established, the role of vibration becomes negligible more correctly, its effects manifest only at higher intensities resulting in an operation which demands excessive amounts of energy.

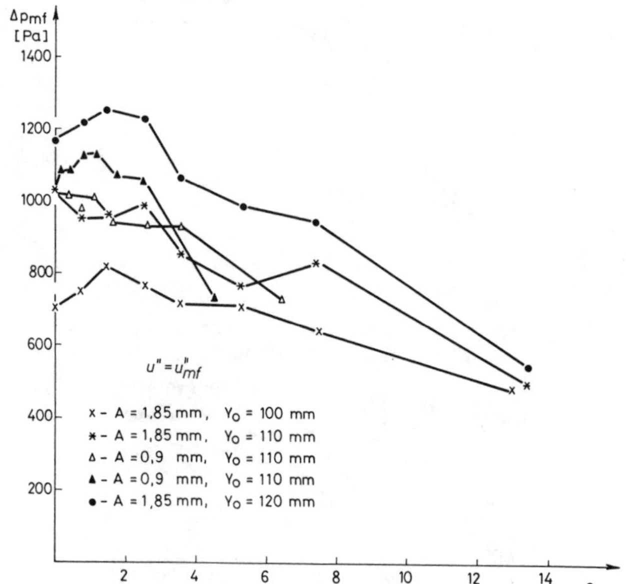

Fig.5 Δp_{mf} as a function of vibration parameter

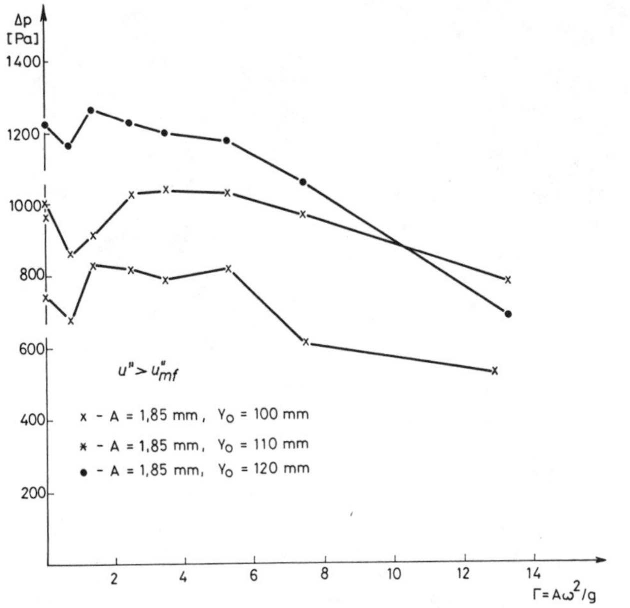

Fig.6 Pressure drop across the layer at a $u'' > u''_{mf}$ gas velocity

The following polynomial regression equation was obtained from the statistical analysis of measured values for the vibration efficiency, K_v (standard error: 0.036; T value: 29.24):

$$K_v = 1.06 - 0.03452\,\Gamma + 0.00056\,\Gamma^2 - 0.00007\Gamma^3 \quad (1)$$

where:

K_v is the vibration efficiency, the ratio of minimum fluidization pressure in the presence and absence of vibration:

$$K_v = \frac{(\Delta p_{mf})_v}{(\Delta p_{mf})_s}$$

K_v values derived from measured data are shown in Fig.7 against Γ. The full curve is the regression function corresponding to Eq.1. It can be seen that due to layer compaction vibration initially hampers fluidization (in the Γ = 0-1 range), but above this limit significant improvement occurs, i.e. vibration almost halves the minimum fluidization pressure. A deficiency of the regression equation is that it does not account for the initial layer compaction, correspondingly it is valid only for $\Gamma > 1$.

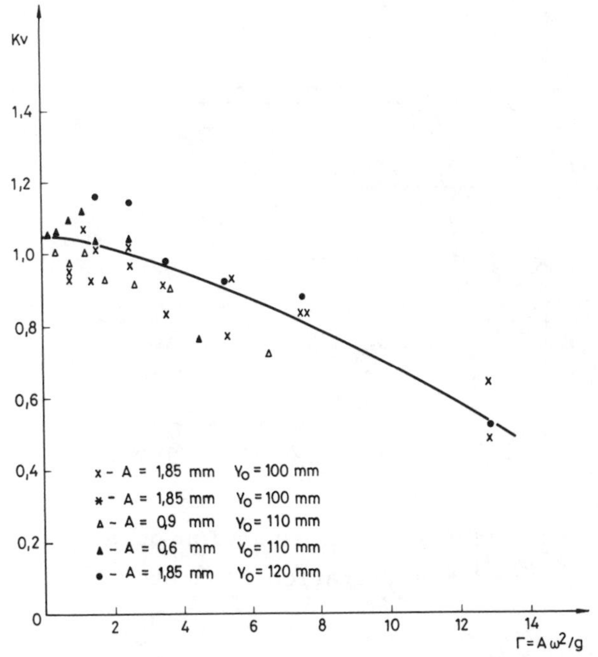

Fig.7 Vibration efficiency, K_v as a function of the vibration parameter

Changes of relative bed expansion are shown in Fig.8 in the case of vibration only, and in Fig.9 for simultaneous vibration and fluidization. It is apparent in Fig.8 that vibration initially causes a compaction of the bed (up to $\Gamma \sim 1$ to 2), followed by bed expansion (up to Γ = 5), then bed expansion remains practically constant. At a gas flow rate which corresponds to the minimum fluidization gas velocity the layer contracts somewhat up to Γ = 5, then at more intense vibration it expands but it will not exceed the value in the absence of vibration (i.e. gas fluidization only).

The effect of support plate resistance is shown in Fig.10. It can be seen that with less permeable support plates (higher support resistance) the minimum fluidiza-

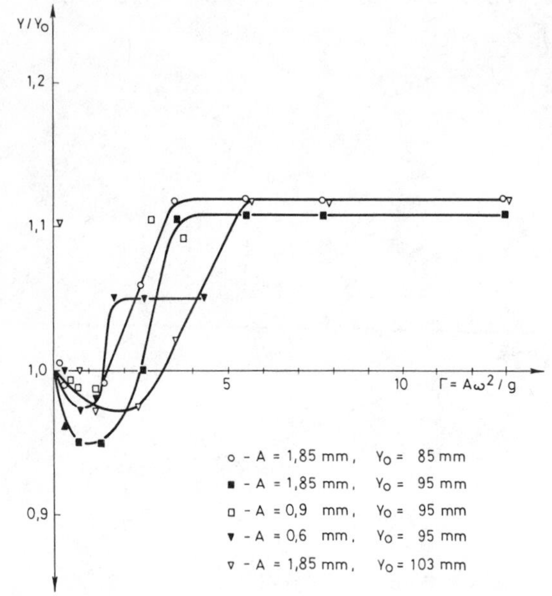

Fig.8 Relative bed expansion as a function of the vibration parameter

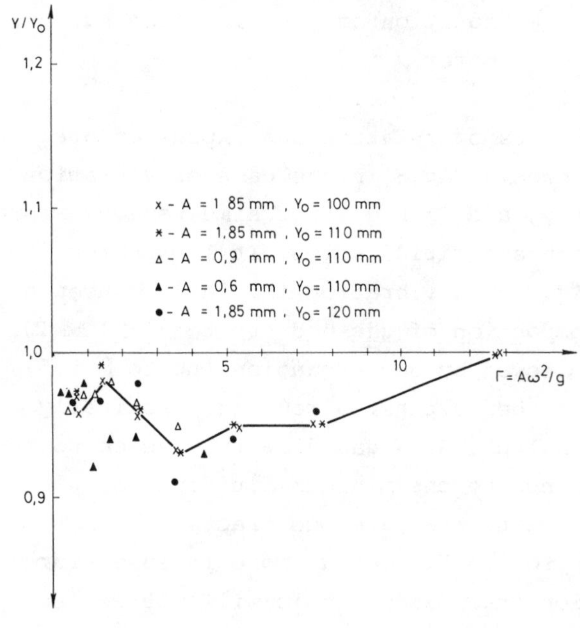

Fig.9 Relative bed expansion at $u'' = U''_{mf}$ gas velocity

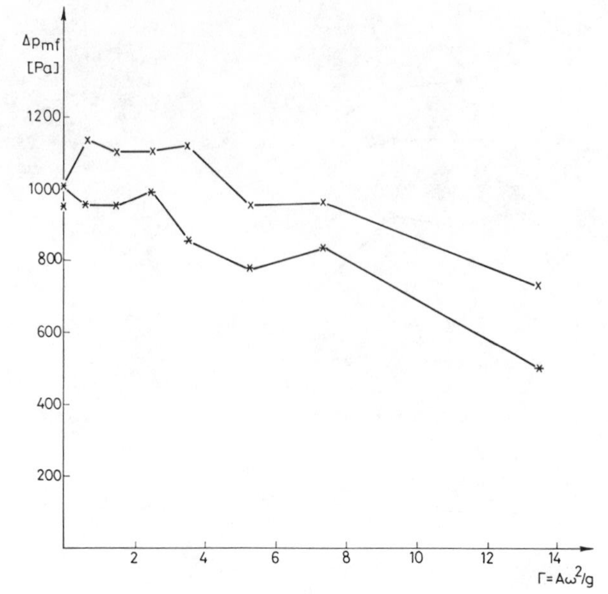

Fig.10 Effect of support plate resistance upon pressure drop across the layer (y_o = 100 mm, $u'' = u''_{mf}$)
x - sieve mesh (d< 1μ)
✱ - porous bronze (3 mm wide)

tion pressure is lower over the entire range tested. Data obtained with various particle size fractions are shown in Fig. 11. It can be seen that with increasing particle size the pressure decreasing effects of vibration are hindered.

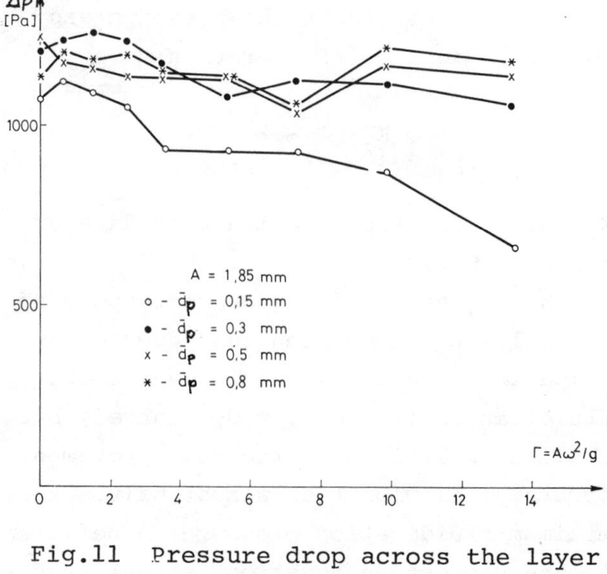

Fig.11 Pressure drop across the layer as a function of particle size $\bar{d}_p = (d_{pmax} + d_{pmin})/2$

The Energy Requirement of Vibrofluidization

The energy requirement of vibrofluidization consists of two parts: the kinetic energy of fluidization gas flow and the vibration energy transmitted to the layer. Pressure drop decrease brought about by vibration decreases the first component of energy consumption, but vibration itself consumes energy. The economy depends on the balance of these processes.

For a swinging period the kinetic energy of gas flow can be calculated by Eq.2:

$$E_p = \Delta p_{mf} \times U''_{mf} \times F/n \qquad (2)$$

In a swinging system with a single mass the energy required to maintain vibration can be calculated by Eq.3 [11]:

$$E_v = m\pi A^2\omega^2 \times \left[\frac{\omega^2}{\sqrt{(\omega_o^2 - \omega^2)^2 + 4h^2\omega^2}}\right]\sin\varphi \qquad (3)$$

where:

$$\omega_o = \sqrt{\frac{1}{mc}} \qquad (4)$$

$$h = \frac{K}{2m} \qquad (5)$$

$$\varphi = \mathrm{arctg}\left(\frac{2h\omega}{\omega_o^2 - \omega^2}\right) \qquad (6)$$

$$K = D_1 \times 2\sqrt{\frac{m}{c}} \qquad (7)$$

$$c = \frac{8.D^3.i}{d^4.G} \qquad (8)$$

Calculations were carried out with the following constants:

- swinging mass: $m = 3.68$ kg
- slip modulus of spring steel: $G = 8.3 \times 10^{10}$ N/m^2
- outer diameter of screw-spring: $D = 2.2 \times 10^{-2}$ m
- diameter of spring-wive: $d = 0.2 \times 10^{-2}$ m
- number of turnus: $i = 6$

From these data the spring constant becomes:

$c = 3.92 \times 10^{-2}$ m/N, and

the eigen-frequency of the swinging system is:

$\omega_o = 26.38\ s^{-1}$.

The damping coefficient of the swinging system, D_1 which is influenced by the characteristics of the particulate material was not determined, rather two extreme values were tested as $D_{1,max} = 0.1$ and $D_{1,min} = 0.01$. These data are based on values measured with various test materials.

The results of the calculations are shown in Fig.12.

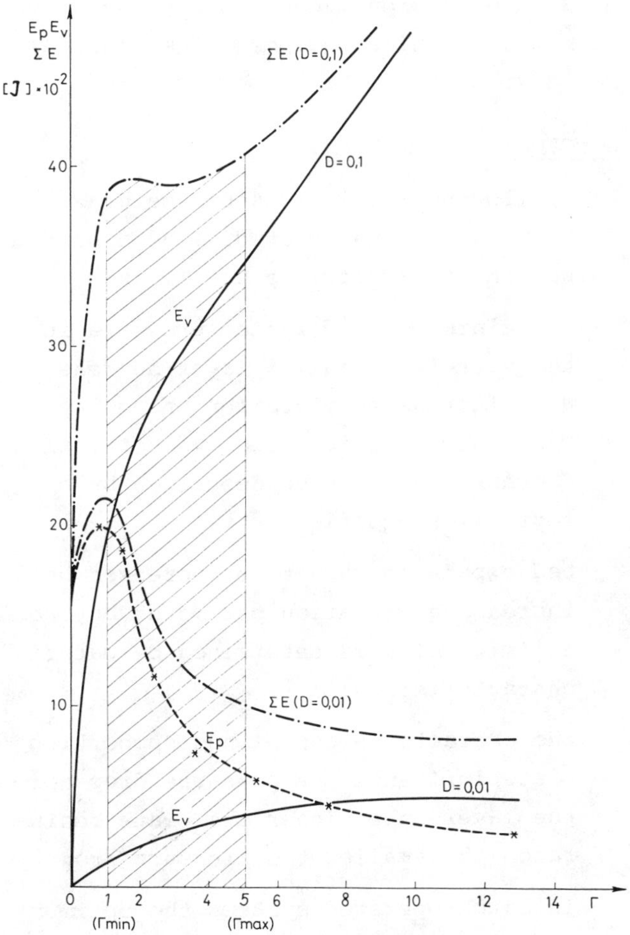

Fig.12 Energy consumption as a function of the vibration parameter, Γ.
--- kinetic energy of fluidizing gas
—— vibration energy
-.- overall energy requirement

The full line depicts the vibration energy, the dotted line the kinetic energy of gas flow determined from measured data. The dot/dash line represents the resultant energy consumption as a function of Γ.

It can be seen that when the damping of the swinging system is small (free flowing powders) increasing vibration intensity enhances the overall energy consumption up to $\Gamma = 1$, then it decreases. Above $\Gamma = 5$ the decrease is not significant anymore. If the damping is large (e.g. with cohesive powders) then the overall energy consumption increases rapidly up to $\Gamma = 1$, then it is practically constant between $\Gamma = 1$ and 5, then it increases again. Therefore, it is not advantageous to increase beyond 5 because no significant energy saving can be realized while equipment wear increases rapidly.

CONCLUSIONS

The following conclusions can be drawn from the hydrodynamic studies of vibrofluidization of particulate materials:

- the minimum fluidization pressure in the vibrofluidization layer decreases with increasing vibration intensity. The minimum fluidization velocity also decreases and can be described by a regression equation;
- bed expansion cannot be increased by increasing vibration parameter beyond a limit which is determined by material characteristics;
- the characteristics of bed supporting plate influence the pressure drop across the layer, the higher the plate resistance the smaller the pressure drop.
- in batch-operated systems the optimum vibration parameters are in the Γ = from 1 to 5 range. At first the maximum vibration amplitude has to be selected with respect to the strength of the mechanical units, then the vibration frequency has to be optimalized to avoid resonance of the apparatus (or its components). With known swinging masses resonance can be avoided by adjusting the spring coefficient.

SYMBOLS

- A - amplitude, m
- c - spring constant, m/N
- d - diameter of spring wire, m
- d_p - particle diameter, m
- D - outer diameter of spiral spring
- D_1 - damping coefficient
- E_p - kinetic energy of gas flow, J
- E_v - vibration energy of the swinging system, J
- F - bed cross section area, m^2
- g - gravity constant, $g = 9.81$ m s^{-2}
- G - slip modulus, N/m^2
- h - damping frequency, s^{-1}
- i - number of turns of the spring
- K - damping coefficient, kg/s
- K_v - vibrational efficiency
- m - mass, kg
- n - frequency, s^{-1}
- Ω - volumetric flow rate of fluidizing gas, m^3/h
- u'' - linear velocity of fluidizing gas, $u'' = \Omega/F$, m/s
- u''_{mf} - minimum fluidization gas velocity, m/s
- Y - bed height, mm
- Y_o - initial bed height, mm
- Δp_{mf} - pressure drop at minimum fluidization velocity, Pa
- Γ - vibration parameter
- φ - phase angle, rad
- ω - frequency, s^{-1}
- ω_o - eigen-frequency, s^{-1}

REFERENCES

[1] Kroll,W.: Chem.Ing.Techn. 1, 27,(1955)

[2] Bratu,R., Jinescu,J.: B.Chem.Eng. 8, 16 (1971)

[3] Mustajev,V.I., Korotkov,B.M.: Khimicheskoe i neftanoe mashinostroenie 12 (1973)

[4] Yamazaki,R., Kanagawa,Y., Jimbo,G.: Journ.Chem.Eng. Japan, 5, 7, (1974)

[5] Gutman,R.G.: Trans.Inst.Chem.Engr. 54 (1976)

[6] Rahul Gupta, Mujumdar,A.S.: The Canad. Journ. of Chem.Engg. 6, 58, (1980)

[7] Strumiłło,C., Pakowski,Z.: DRYING'80 ed. Mujumdar,A.S.: Hemisphere Publishing Co. New York, 1980. pp. 211-226.

[8] Chlenov,V.A., Mihailov,N.V.: Vibrokipyashchii Sloi, Nauka Moscow, 1972.

[9] Karmazin,V.D.: Tekhnika i primenenie vibrohipyashchevo sloia. Naukova Dumka Kiev, 1977.

[10] J.Dörnyei: Production of Instant Food Products, Agricultural Publ.House, Budapest, 1981 (in the Hungarian)

[11] Bikovski,I.: Fundamental of Vibration Engineering Mir Publishers, Moscow, 1972.

[12] Németh,J.: Száritási kézikönyv. Technical Publ. House Budapest. Part 7 pp.281-291. (Handboock of Drying ed. by Imre,L., in the Hungarian)

AERODYNAMIC CHARACTERISTICS OF A
VIBRATED BED OF PARTICLES

Arun S. Mujumdar

Department of Chemical Engineering
McGill University, Montreal, Canada

ABSTRACT

Results of an experimental study of the effects of vertical mechanical vibration on the flow behaviour of beds of nearly spherical particles will be summarized. Influence of the following parameters was studied: vibration frequency (0-160 Hz), amplitude (0-4.25 mm) and bed height (25-125 mm), using polyethylene and molecular sieve particles. The effect of vibrational parameters on the bed structure, pressure drop, a newly defined minimum mixing velocity and overall aerodynamic behaviour of the bed is discussed on the basis of quantitative measurements as well as visual observations.

NOMENCLATURE

a - half-amplitude of vibration
d_p - particle diameter (average)
g - acceleration due to gravity
H - bed height (static)
U - superficial air velocity in bed

Greek Letters

ρ - density of solid
ω - angular frequency
Γ - nondimentional vibrational acceleration ($= a\omega^2/g$)
ΔP - bed pressure drop

Subscripts

mf - at minimum fluidization
mm - at minimum (solids) mixing

INTRODUCTION

Vibrated fluidized beds have found numerous industrial applications for the drying or thermal processing of difficult-to-fluidize particulate solids. If the particulates when wet tend to agglomerate mechanical vibration may be employed to break up the lumps as they are formed; this assists both the fluidization and drying operations. Gentle handling, better control of the residence time distribution, ability to fluidize sticky material of a wide particle size distribution, improved drying and heat transfer rates and reduced power consumption are among the advantages of the vibrated fluid bed (VFB) over the conventional fluid bed. As a result of their unique characteristics VFB's have also found industrial use as heterogeneous catalytic reactors, granulators, coolers, for coating of particles etc.

No attempt is made in this paper to review the literature on VFB's or to compare the results with those published in the open literature. The interested reader is referred to the reviews of Strumillo and Pakowski (1) and Mujumdar and Gupta (2). Prior work on the flow and immersed-surface heat transfer in VFB's is presented by Gupta and Mujumdar (3), Leung, Gupta and Mujumdar (4) and Ringer and Mujumdar (5). The objective of this paper is to present primarily our recent studies on the characteristics of relatively deep beds (i.e. greater than 50 mm depth) which appear to behave differently from shallow beds studied in References 3 and 4. Attempts were also made to fluidize with aid of mechanical vibration such difficult-to-fluidize materials as corn, wood bark and hog fuel (waste product of lumber and pulp and paper industries). Even at very high vibrational accelerations and high air velocities beds of corn could not be fluidized. The wood products had an extremely wide particle size and shape variation which made it impossible to achieve stable fluidization of the material.

EXPERIMENTAL APPARATUS AND PROCEDURE

The experimental apparatus is shown schematically in Ref. 3. For a detailed description the reader may refer to Ref. 6. The VFB is a chamber of cross-section 0.2 x 0.2 m fitted with a perforated plate distributor of 23 percent open area. The chamber is fitted with two glass windows to permit visual observation (needed especially to observe the expanded bed height for calculation of the bed voidage and the minimum mixing velocity first defined by Gupta and Mujumdar (3)). An SCR-controlled motor is used to vary the frequency of vibration. The vibration amplitude could be varied stepwise by means of a specially designed variable-eccentricity arrangement. The entire vibratory part is supported on resonant springs. Note that the entire bed along with the distributor plate is vibrated vertically in a sinusoidal manner. The parameter ranges studied are listed in Ref. 3 along with the expected uncertainties. The bed height was varied over a wider range than that investigated earlier in the same apparatus. The various particulate materials used along with their physical properties are listed in Ref. 3. Although measurements were also made of heat transfer between the vibrated bed and a cylinder immersed in the bed these results will form the subject of a separate communication.

The bed pressure drop was obtained by subtracting the pre-calibrated pressure drop across the distributor plate which varied with the air velocity, frequency and amplitude of vibration. As noted earlier the velocity-pressure drop curves did not display any hysterisis within the experimental error. The pressure drop at minimum fluidization under vibratory conditions was estimated from the velocity at which the pressure drop was essentially independent of the velocity. In the presence of intense vibration this "definition" failed particularly over the velocity range achieved with the present set-up.

Indeed, the concept of a minimum fluidization velocity does not seem appropriate for the case of a vibrated fluid bed because of the extended range of the transition section between the packed bed regime and the fluidized state. As noted by Gupta and Mujumdar (3) if the concept of U_{mf} is applied rigorously for a VFB the results for shallow beds indicate an increase in U_{mf} with vibrational acceleration, which is contrary to physical reasoning and practical experience with shallow beds.

a. Structure of the Bed

In general bed homogeneity has been reported to improve with application of vibration. On the basis of extensive experiments and visual observations with numerous materials it is noted that the improvement of homogeneity occurs only for relatively shallow beds ($H \lesssim$ 25-50 mm) of moderate size particles (1000 μm < d_p < 3000 μm) vibrated at modest vibratory accelerations ($\Gamma \lesssim 3$). These figures are only approximate since the bed structure is also influenced by the air velocity.

Beds of large particles, such as dry corn, could not be fluidized at all even at Γ values of up to 4 and superficial air velocities in excess of 1.5 m/s. At high vibration intensities the bulk of the bed moves as a slug with hardly any interparticle mixing; only particlesin the top layer undergo vigorous up and down motion.

With the nearly spherical glass beads ($\bar{d}_p$ = 700 μm) application of moderate vibration (Γ - 1) did not visually influence the bubbling phenomenon. At higher vibrational levels, however, the bubbling became more erratic and more intense. These observations were made with H = 100 mm. In this case vibration appeared to promote bubble coalescence rather than to impede it as observed by some of the earlier workers. This observation is also at variance with that of Morse (8) who reported improvement in bed structure for beds of finer particles. Mushtaev et al. (9) have noted deterioration of bed structure for Γ > 4. On the basis of present work such deterioration could occur at much lower Γ values.

b. Bed Pressure Drop

Gupta and Mujumdar (3) have reported on the three distinct shapes of the pressure drop-velocity curves that can be obtained in a VFB. For deep beds only Type A curves were obtained i.e. the curves were similar to those for nonvibrated beds. It was noted earlier that Type A curves are obtained generally under conditions which tend to damp out the propagation of vibration through the bed.

Figure 1 shows the flow curve for a bed of corn for Γ > 1; the effect of vibration on ΔP is mixed i.e. no systematic influence is discernible. For Γ > 1 the solids mixing within the bed was observed to be negligible with the entire bed moving as a piston with hardly any particle movement. Clearly vibration cannot always help fluidize granular solids of large size. It was also confirmed that beds of fibrous, loosely packed solids cannot be fluidized even with the aid of vibration.

Figures 2 through 7 show the ΔP vs U curves for 2.5 mm molecular sieve particles (ρ = 887 kg/m^3). The bed height varied from 25 mm to 100 mm with vibration amplitudes of 2 and 4.25 mm. The vibrational acceleration varied from 0 to 48. At lower superficial air velocities the bed pressure drop displays an increase over the nonvibrated bed (i.e. Γ = 0) for Γ < 1.0. The increase in ΔP is rather dramatic for lower Γ's. For H = 25 mm, from Figure 2 it is readily seen that ΔP for Γ = 0.40 is nearly 40 percent higher at U = 0.55 m/s but above this value of U the ΔP is lower. Except for H = 100 mm at Γ = 2.25 the bed pressure drop is lower over the entire velocity range. For deep beds (H = 100 mm) the effect of vibration is understandably damped out (Figure 4) and hence ΔP reductions are attained only at very high vibrational accelerations. Figure 7 displays the reduction in ΔP for Γ = 4.8 for H = 100 mm. For shallower beds (H = 25 mm and 50 mm, Figures 5 and 6 respectively) Γ = 1.7 also yields ΔP's which are always lower than those for Γ = 0.

From Figures 4 and 7 it may be noted that ΔP_{mf} is essentially unaffected by Γ. This is consistent with the observations of Ref. 3 on the effect of bed height on ΔP_{mf}. In general it is noteworthy that the Γ effect is more pronounced at lower U values when the bed moves as a whole and there is little mixing of particles taking place in the bed. At very high Γ's (Γ > 4) it was noted that the bed mixing is once again impaired as the bed moves as a piston with little particle movement even at high superficial velocities.

Figures 8 through 15 are for polyethylene particles (d_p = 3.5 mm, ρ = 630 kg/m^3) which were also nearly spherical in shape. From Figures 8 and 12 it is readily seen that for shallow beds (H = 2.5 mm) vibration reduces ΔP_{mf}. However, the effect of Γ on P_{mf} is minimal for Γ > 0.4 (Figure 9 through 11, 13 through 15). The effect of Γ is generally less than that found for the heavier molecular sieve particles. Another remarkable feature of these data is that ΔP for the vibrated bed is always lower than that for the nonvibrated bed over most of the velocities for H = 25 mm. As the bed height increases the value of Γ at which the crossover from higher ΔP to lower ΔP occurs, also increases. For H = 125 mm (Figure 11) there is only a marginal effect of vibration on ΔP for Γ = 2.25.

Although $\Gamma (= a\omega^2/g)$ is generally used to lump the effect of vibration amplitude and frequency comparison of Figures 2-4 and 8-11 with Figures 5-7 and 12-15 respectively shows a systematic effect of vibration amplitude, a. The former sets of figures are for a = 2.5 mm while the latter set are for a = 4.25 mm. In general, the effect of amplitude appears to be to intensify the effect of Γ i.e. if the same Γ value is attained using two different amplitudes (and frequencies) then the vibrational acceleration effect is more pronounced at the higher amplitude. For example, at higher amplitudes reduction in ΔP due to vibration seems to occur at lower Γ's.

Finally, Figures 16a and 16b represent plots of U_{mm}/U_{mf} versus Γ for molecular sieve and polyethylene particles. U_{mf} was computed using the Wen and Yu (10) correlation. In the design of a VFB the designer is interested in determining the minimum air velocity at which the particles in the bed begin to move relative to each other. Unfortunately this air velocity could be determined only visually and in a subjective manner. No discernible effect was noted in the ΔP vs U curves at $U = U_{mm}$ as defined here. Hence, the results must be considered with caution. For shallow beds of spherical particles Ref. 3 has presented an empirical correlation for U_{mm}/U_{mf}. No such correlation is attempted here in view of the limited data. For shallow beds the U_{mm} is only about 20 percent of U_{mf} for Γ = 1.40. For deeper beds, as can be seen from Figure 16, a higher aeration rate is needed even at higher Γ values. This reflects the progressive damping of vibrational effects with bed height.

It is also noteworthy that beyond $\Gamma \approx 0.40$ U_{mm} is essentially unaffected by the magnitude of the vibrational acceleration. Indeed, at very high Γ's the bed may move as a piston and the U_{mm} required to mix solids would be much greater than at lower Γ values. Because of the blower rating restrictions U_{mm} at very high Γ's could not be determined in this study. In contrast with the results for shallow beds which display a nearly linear dependence of U_{mm}/U_{mf} on Γ, the results for deep beds show a rapid decrease in U_{mm}/U_{mf} over $0 < \Gamma < 0.4$ and then a very gradual fall over $0.4 < \Gamma < 1.7$. For higher Γ's it is speculated that U_{mm} would increase with Γ.

CLOSURE

Results of an experimental study to determine the influence of vertical sinusoidal vibration on the flow characteristics of deep vibrated beds are presented. Both reductions and increases in bed ΔP are noted as a result of vibration. The ΔP at minimum fluidization is, however, nearly always reduced by vibration (Γ); the extent of reduction depends on the bed height, particle density and, to some extent, on the vibration amplitude. The minimum mixing velocity decreases with Γ, at first rapidly ($0 < \Gamma < 0.4$) and then gradually. Very high vibrational accelerations ($\Gamma > 4$) were found to be detrimental to the performance of the bed particularly for large size particles. The effect of low open area distributor grid needs to be investigated to determine if the present results (with a rather large open area distributor) are influenced by the distributor design.

ACKNOWLEDGEMENTS

The author acknowledges the assistance of the following in carrying out the experimental program and data reduction: Janet Amos, Victor Lee and Vibhakar Jariwala.

REFERENCES

1. Strumillo, C. and Pakowski, Z., DRYING '80, Vol. 1, Ed. A.S. Mujumdar, Hemisphere, N.Y. (1980).
2. Gupta, R. and Mujumdar, A.S., ibid.
3. Gupta, R. and Mujumdar, A.S., Can. J. Chem. Eng., 58, 1980, pp. 332-338.
4. Leung, P., Gupta, R. and Mujumdar, A.S., DRYING '80, Vol. 2, Ed. A.S. Mujumdar, Hemisphere, N.Y. (1980).
5. Ringer, D. and Mujumdar, A.S., DRYING '82, Ed. A.S. Mujumdar, Hemisphere, N.Y. (1982).
6. Gupta, Rahul, M.Eng. Thesis, McGill University, Montreal, Canada, 1979.
7. Bratu, E.A. and Jinescu, G.I., Brit. Chem. Eng., 16 (8), 691-95 (1971).
8. Morse, R.D., Ind. Eng. Chem., 47 (6), 1170-80 (1955).
9. Mushtaev, V.I. et al. Chem. & Petrol. Eng., 12, 1083-85 (1973).
10. Wen, C.J. and Yu, Y.H., AIChE J., 12, 610 (1966).

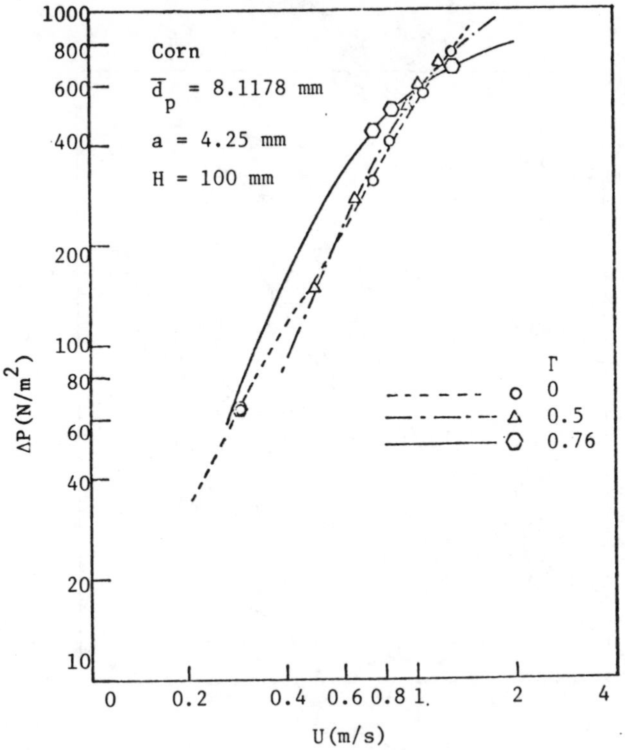

Figure 1 Effect of Vibration on ΔP-U Curves for Beds of Corn

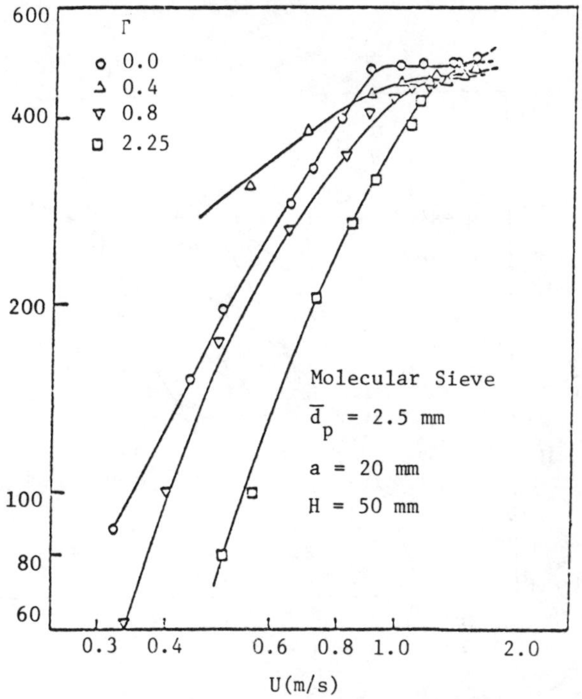

Figure 3 ΔP-U Curves for Molecular Sieve Particles, H = 50 mm, a = 2.0 mm

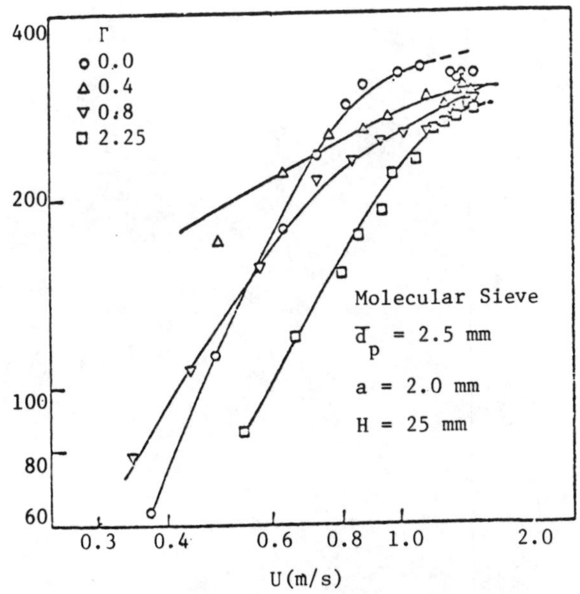

Figure 2 ΔP-U Curves for Molecular Sieve Particles, H = 25 mm, a = 2.0 mm

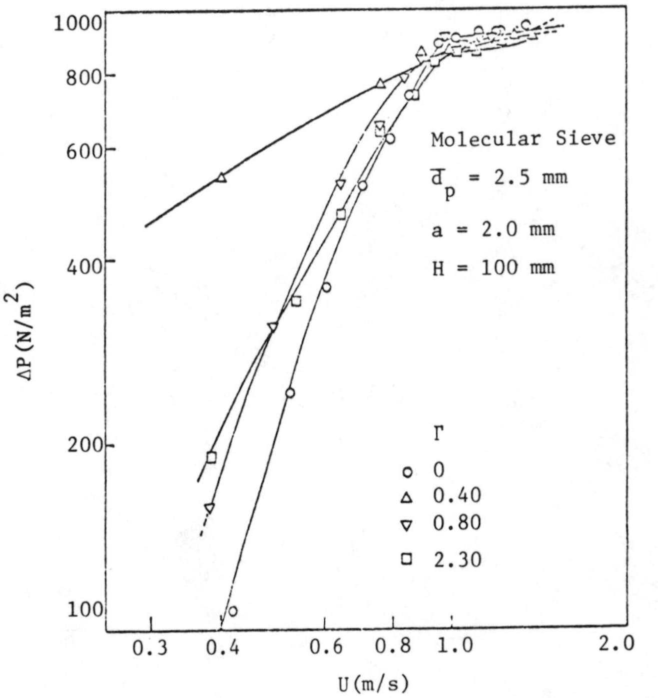

Figure 4 ΔP-U Curves for Molecular Sieve, H = 100 mm, a = 2 mm

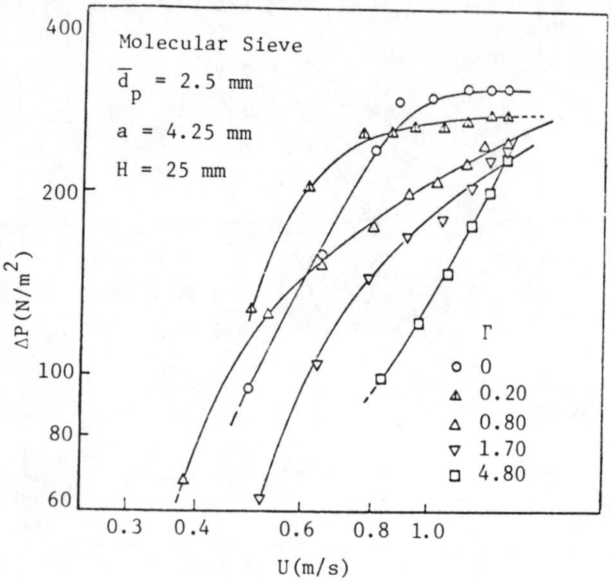

Figure 5 Pressure Drop Curves: Molecular Sieve, H = 25 mm, a = 4.25 mm

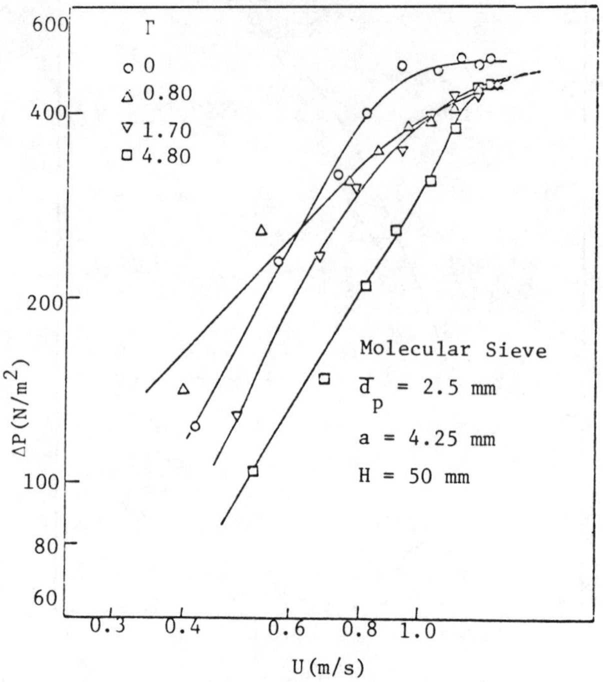

Figure 6 Pressure Drop Curves: Molecular Sieve, H = 50 mm, a = 4.25 mm

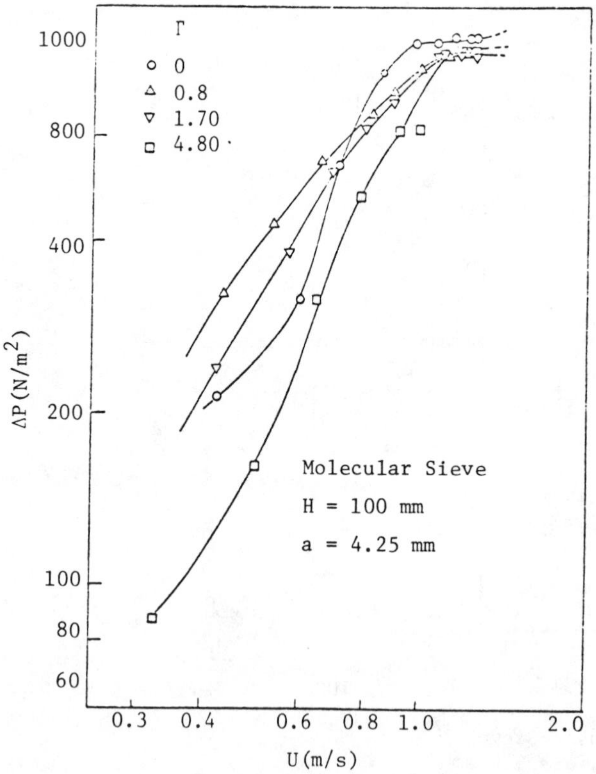

Figure 7 Pressure Drop Curves for Molecular Sieve, a = 4.25 mm.

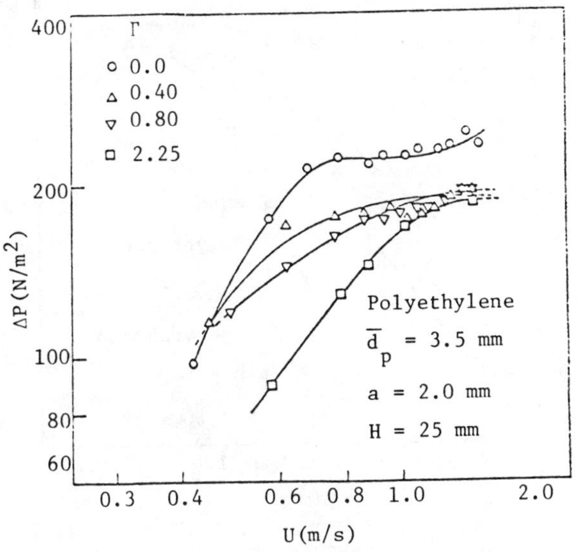

Figure 8 Pressure Drop Curves for Polyethylene Particles, H = 25 mm, a = 2 mm

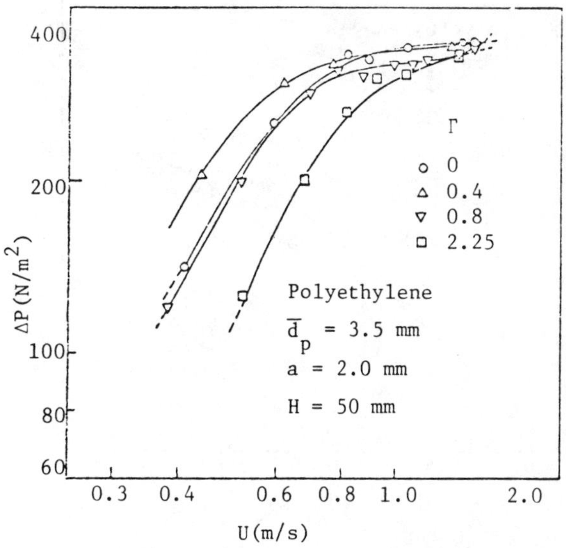

Figure 9 Pressure Drop Curves for Polyethylene Particles, H = 50 mm, a = 2 mm

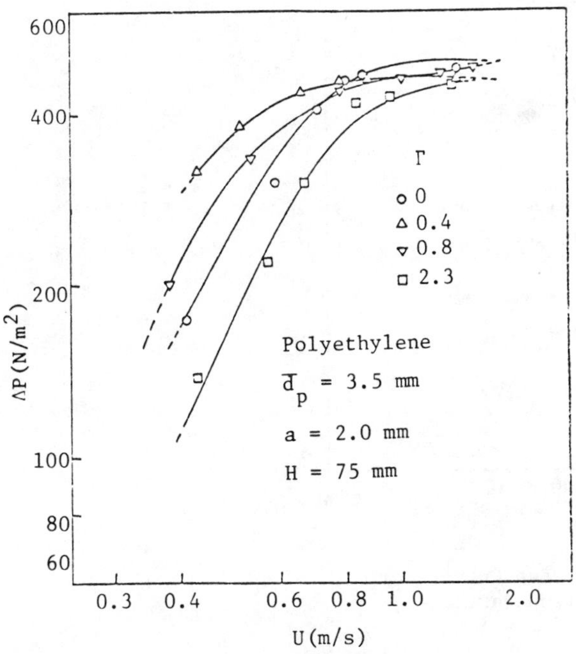

Figure 10 Pressure Drop Curves for Polyethylene, H = 75 mm, a = 2 mm

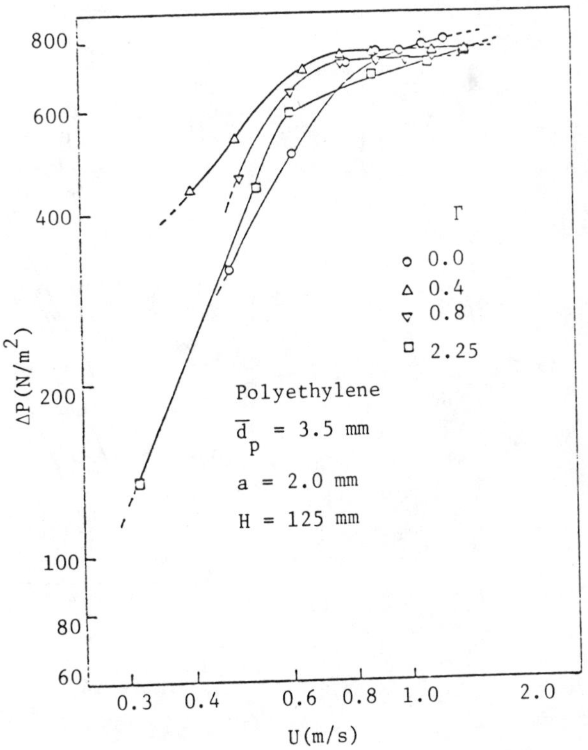

Figure 11 Pressure Drop Curves for Polyethylene, H = 125 mm, a = 2 mm

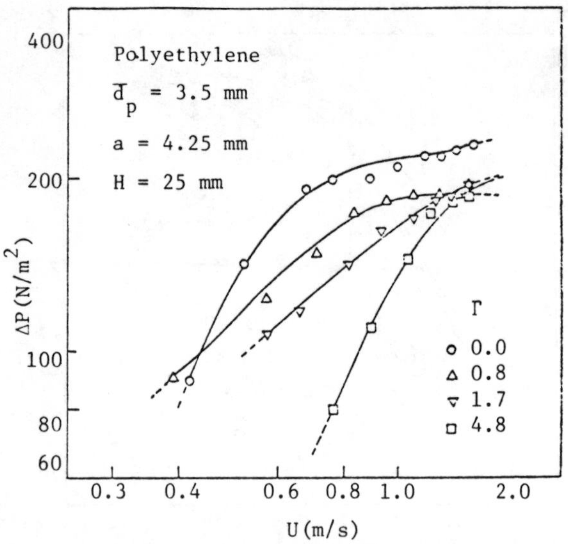

Figure 12 Pressure Drop for Polyethylene, H = 25 mm, a = 4.25 mm

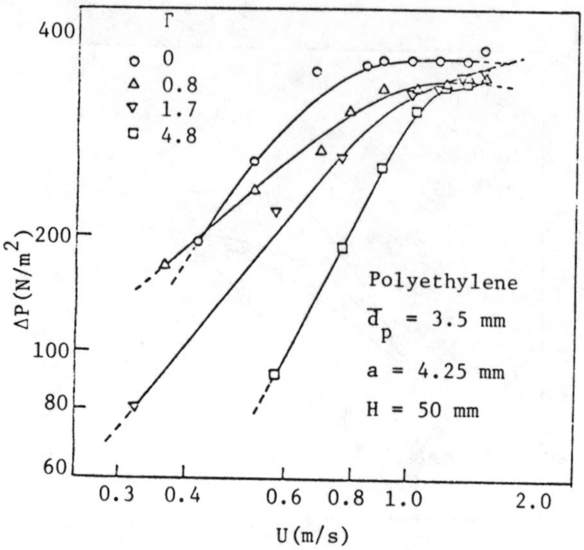

Figure 13 Pressure Drop Curves for Polyethylene, H = 50 mm, a = 4.25 mm

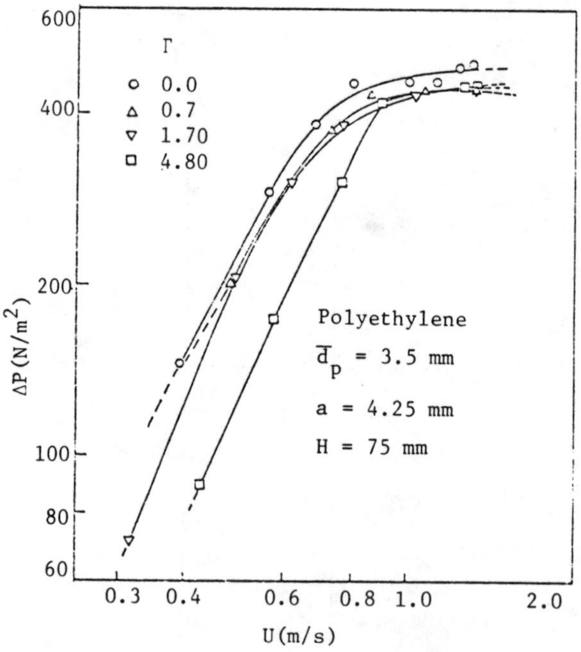

Figure 14 Pressure Drop Curves for Polyethylene Particles, H = 75 mm, a = 4.25 mm

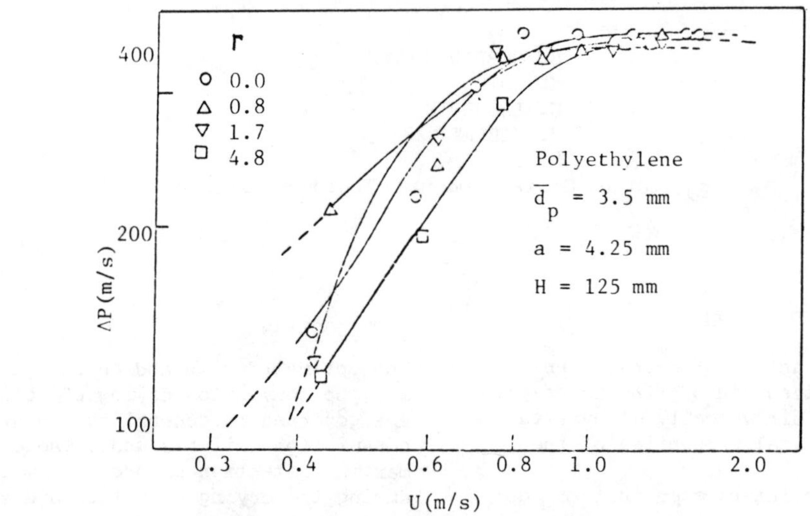

Figure 15 Pressure Drop Curves for Polyethylene Particles, H = 125 mm, a = 4.25 mm

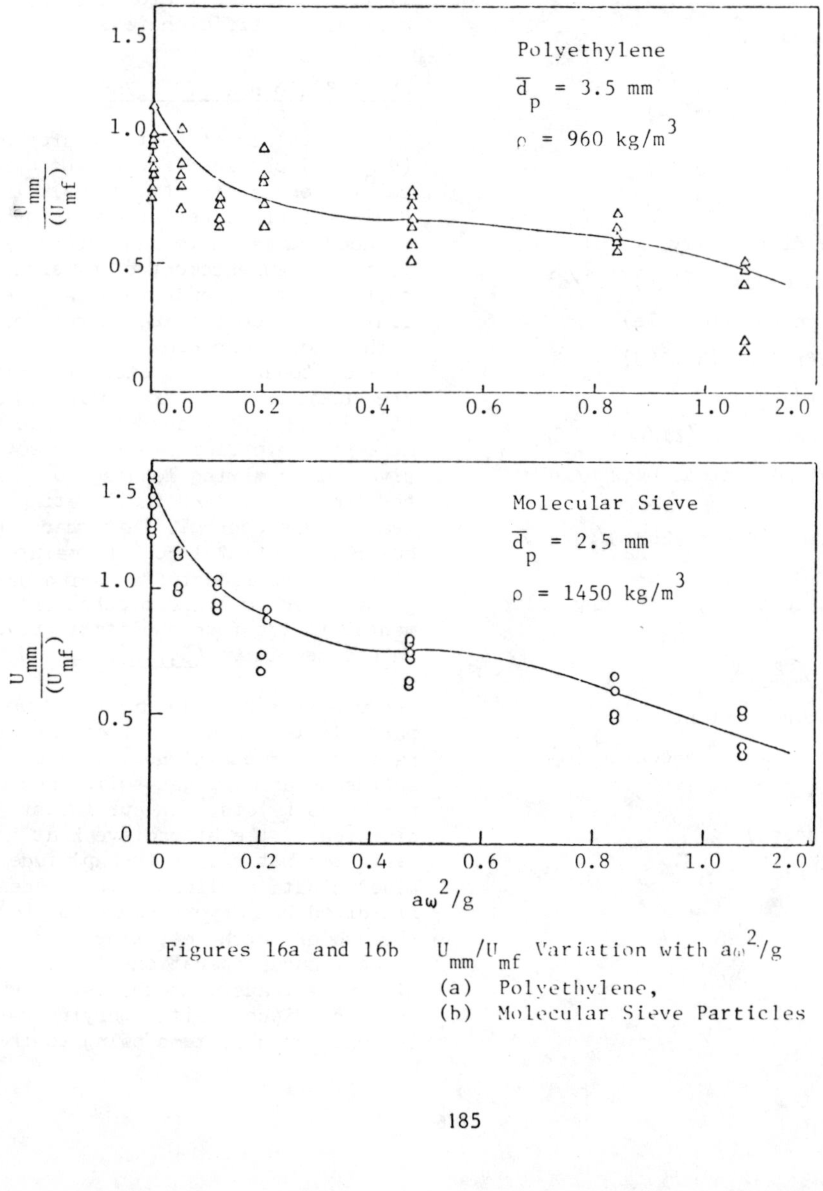

Figures 16a and 16b U_{mm}/U_{mf} Variation with $a\omega^2/g$
(a) Polyethylene,
(b) Molecular Sieve Particles

KINETICS OF WHOLE CORN GRAINS DRYING IN A FLUIDIZED BED OF FINE INERT PARTICLES

S. COBBINAH
M. ABID
C. LAGUERIE
H. GIBERT

Institut du Génie Chimique, Toulouse, France

ABSTRACT

This paper concerns the kinetics of coarse corn grains drying in a fluidized bed of fine inert particles. The influence of air velocity, temperature and humidity and the physical properties of the fine particles were investigated.
Experimental results were interpreted in terms of a Fickian diffusional mechanism.

NOMENCLATURE

A - particle surface area (m^2)

$Bi_h (=\frac{hR}{K_s})$ heat transfer Biot number (-)

C_p - specific heat capacity (J/kg °C)

d_p - particle diameter

D_{eff} - effective moisture diffusivity (m^2/s)

D_o - factor in Arrhenius equation (9) (m^2/s)

E_a - drying activation energy (kJ/mole)

h - heat transfer coefficient ($W/m^2 °C$)

H - air relative humidity (-)

ΔH_v - latent heat of vaporization (KJ/kg)

$\bar{m}$ - average solid moisture content (kgH_2O/kg dry solid)

m_e - equilibrium moisture content (kgH_2O/kg dry solid)

m_o - initial moisture content (kgH_2O/kg dry solid)

m_s - mass of grain (Kg)

m - drying rate ($kg\ H_2/m^2 s$)

M - dimensionless moisture content

r - radial distance from particle center (m)

R - particle radius (m)

t - time (sec, min or hours)

T - temperature (°C or K)

Subscripts

d - dry solid

g - air

w - wet solid

INTRODUCTION

The optimal design and operation of equipments for the convection drying of solid foodstuffs are governed in general by a minimum energy consumption criteria under the constraint of maximum retention of product qualities both during the drying operation and storage.
Energy consumption reduction methods are invariably aimed at (a) higher fluid-solid heat transfer rates when the external transfer is the limiting kinetic factor and (b) maximum energy utilisation efficiencies.

FLUID-SOLID HEAT TRANSFER

The low values of heat transfer Biot number ($Bi_h < 1.0$) obtained for hygroscopic porous particulate food stuffs (1,2) imply that the solid-fluid interface resistance is of a greater significance compared to heat conduction within the particle. Enhancement of external heat transfer coefficients depends to a large extent on the fluid-solid contacting technique. In comparison with other techniques (fixed bed dryers, rotary dryers, tower dryers etc.) fluidized bed dryers are known to offer the following advantages (3,4,5) (a) higher heat transfer coefficients due to a more intimate fluid-solid contact (b) more intensive solids mixing leading to an almost homogenous bed temperature and facilitating a reliable bed temperature control. This characteristic is particularly beneficial for the maintenance of the quality of the foodstuffs during drying since the physico-chemical and biochemical reactions detrimental to the food qualities are extremely temperature sensitive (6,7,8).

The use of a hot fluidized medium of fine inert particle technique, a promising one with a wide range of industrial applications (eg dry medium solids separator, gas-solids and catalytic reactions (9)etc.) in our investigation was not dictated by the attractive heat transfer characteristics but rather the aptitude to furnish kinetic data applicable to the design of continuous fluidized bed dryers. Romankov (4) pointed out that the use of batch test kinetic data for the design of continuous operations is hazardous owing to the different transfer mechanisms prevailing in the two systems. Higher initial drying rates are obtained in continuous systems owing to the interparticle heat transfer phenomena provoked by the temperature difference between hot dry particles and cold wet particles.

A few studies, using a simultaneous heat and mass transfer method (11,12,13) estimated the inter-particle heat transfer mode contribution to the overall transfer in the range 80-95 % while the transient transfer method (14) yielded an estimation of 30 %. In brief, interparticle transfer developed in continuous operations is beneficial but can lead to overdrying and energy wastage if neglected.

MAXIMUM ENERGY UTILISATION EFFICIENCY

Available literature on the energy consumption reduction strategies based on the thermodynamic conditions (temperature and humidity) of the fluid medium reveals that although superheated steam offers higher drying rates and greater thermal efficiencies compared to air as a drying medium, it is unsuitable for temperature sensitive materials like pharmaceuticals and biological products. High superheat temperatures are necessary for an appreciably economic operation. Further more, air-steam mixtures are considered to be poor drying media in the falling rate period due to the resulting higher equilibrium moisture content (24,25,26).

In spite of the unfavourable criticisms against air-steam mixture media, it appeared expedient to reexamine its influence on the drying of hygroscopic porous and the possibility of striking a compromise between the lowered drying rates and the advantages envisaged in humid drying conditions :

(a) reduction of stress cracking due to high initial drying rates in more severe conditions of high temperatures and low humidities,

(b) possibility of operating in a closed circuit with a heat pump to improve the thermal efficiency, (27),

(c) possibility of recovering volatile aromatic compounds in the dryer exhaust gas and safer drying of materials liable to oxidation in a dry medium. (24).

INTERNAL MOISTURE MIGRATION KINETICS

An accurate prediction of the grain moisture content and temperature-time behaviour is not only necessary for design purposes but particularly for the retention of product quality. It has been established that, of the various factors affecting the qualities (microbial, nutritional and organoleptic) of foods during drying and storage periods, temperature and moisture content are the most critical (6,7,8). In this connection, several mathematical models based on either of the two main mechanisms, namely capillary flow (15,16) and the diffusional flow mechanism (liquid-diffusion (17,18,19) and evaporation condensation (20,17)theories have been extensively applied to food materials.

As pointed out by Rossen and Hayakawa (22)in their literature review, food systems are extremely complex and moisture may be transfered simultaneous by several different mechanisms depending on the nature of the material, the type of moisture linkage, the moisture content and temperature, etc. This complexity renders the generalization of predictions from the mathematical models based on a particular mechanism unreliable.

The search for more general methods led to the simultaneous heat and mass transfer theories of Luikov (23)derived from the thermodynamics of irreversible processes and that of Krischer (16) based on a combination of the capillary and diffusional mechanisms.

Simulation of cereal grains drying assuming a Fickian diffusional mechanism with either constant or temperature and moisture content dependent diffusivities have been found to yield results quite compatible with the reality.

EQUIPMENT AND EXPERIMENTAL PROCEDURES

The pilot plant employed (Fig. 1) in our investigations consists essentially of a drying unit and an air conditioning unit.

The dryer consists of a rectangular cross sectioned (12 x 36 cm) stainless steel column ① containing fine inert particles of lucern or sand (10 cm high). A vibrating basket ②, with holes large enough to allow the free movement of the fine particles, is incorporated to prevent the agglomeration of humid particles at the initial drying stages and to facilitate the discharge of the finished dry product.

The air conditioning unit consists of an air-blower (not shown in the figure) equipped with an air filter, a 12 kw electric air-preheater ③ and a boiler (not shown) with a maximum hourly capacity of 40 kg under 6 bars. Water droplets entrained by the steam are eliminated by means of a baffled steam trap ④.

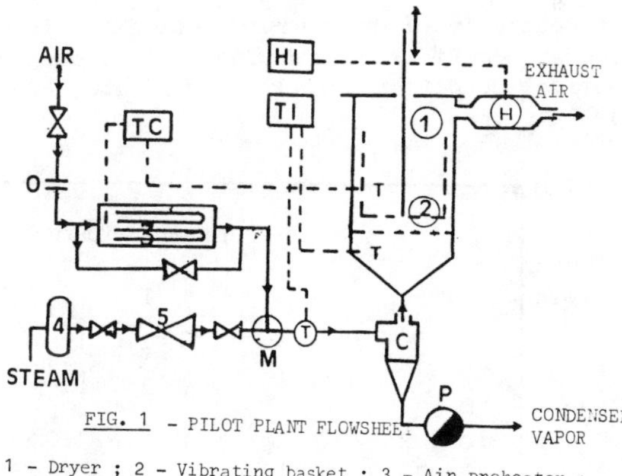

FIG. 1 - PILOT PLANT FLOWSHEET

1 - Dryer ; 2 - Vibrating basket ; 3 - Air preheater ; 4 - Steam-trap ; 5 - Pressure reducer C-cyclone , HI - Humidity indicator ; H-LiCl probe ; T - Thermocouple TI - Temperature indicator ; TC - Temperature controller ; P - Purge.

OPERATION AND MEASURENTS

Prior to the introduction of steam in the circuit, the dryer and its contents are sufficiently warmed up to prevent any condensation, by passing only preheated air through the drying chamber. Dry steam is then injected into the preheated air at the point M some few centimeters from the cyclone separator. The humidified air is cleaned of any condensed vapour during its passage through the cyclone before entering the drying chamber. The system was then allowed to attain steady conditions before the introduction of corn grains.

Measurement and control of air, steam and humid air temperatures were effected by a Philips multichannel indicator and a 2mm diameter iron-constantan thermocouples.

Solid temperature was directly measured by means of a 0.5 mm diameter chromel alumel thermocouple.

Air humidity was measured by a Philips lithium chloride probe with a sensitivity of 0.5°C in the directly measured dew point.

Steam flow rate was determined indirectly from the measured air humidity and flow rates by a material balance assuming that the additional moisture released into the humid air by the drying phenomena is negligible.

Drying curves were obtained by determining the moisture content of corn samples withdrawn at regular time intervals. Sample moisture content was determined by a vacuum oven drying gravimetric method at 70°C with vacuum pressure not exceeding 100 mm Hg for 16 hrs.

RESULTS AND DISCUSSION

Reproductivity of drying runs were established by the close agreement between drying runs conducted under identical conditions of air temperature, humidity and flow rate (Fig. 2)

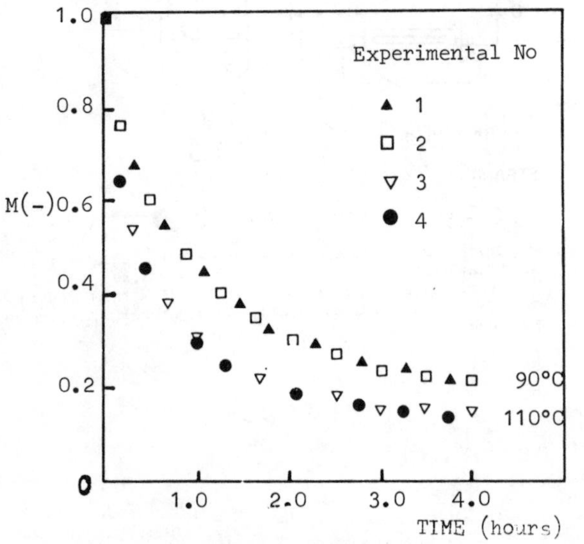

FIG. 2 REPRODUCTIBILITY TESTS

INFLUENCE OF BED HYDRODYNAMICS AND PHYSICAL PROPERTIES OF FINES

As expected for hygroscopic materials exhibiting solely the falling rate drying characteristics, the drying kinetics was neither influenced by the fluidization number (Fig. 3) nor the physical properties (size and nature) of the fine particles constituting the fluidized medium (Fig. 4). The limiting kinetic factor was the internal moisture movement

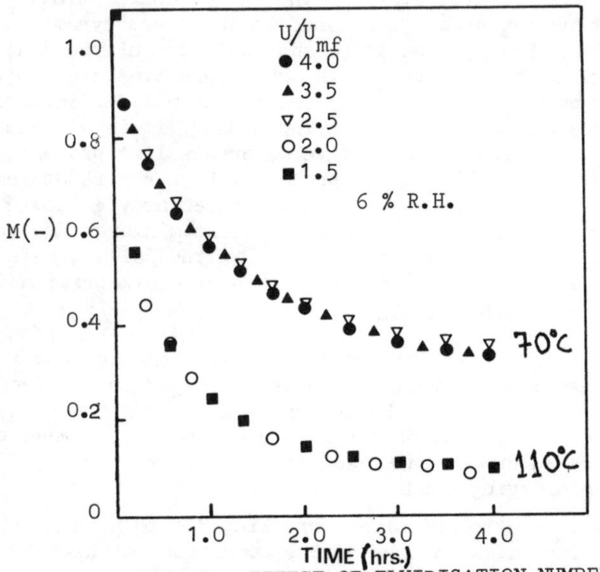

FIG. 3 - EFFECT OF FLUIDISATION NUMBER (U/U_{mf}) ON DRYING KINETICS

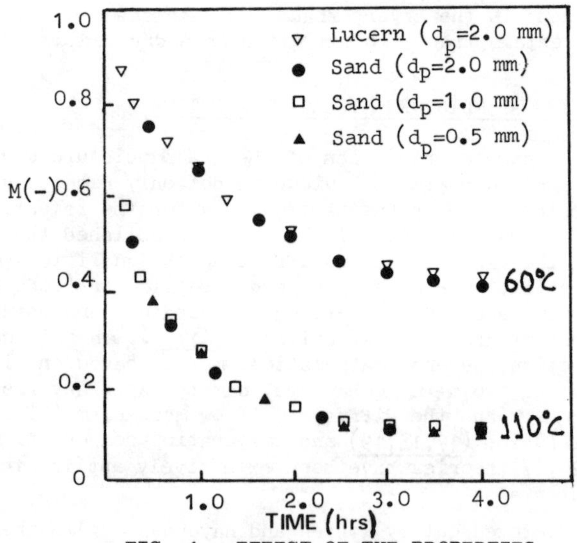

FIG. 4 - EFFECT OF THE PROPERTIES (NATURE AND SIZE) OF THE FLUIDIZED SOLIDS ON DRYING KINETICS (U/U_{mf} = 2.0, R.H=6 %)

INFLUENCE OF AIR TEMPERATURE

Figures 5 and 6 depict the effect of the air temperature on the drying kinetics in air and air-steam mixture fluid media respectively. The drying rate-moisture content curves were obtained by fitting of the experimental moisture content-fine curves to a polynomial function using a Gauss-Newton method followed by an analytical differentiation of the best fitting curve to obtain the drying rate at various moisture contents or times. This method circumvents the rather laborious numerical differentiation.

The observed increase in drying rate with temperature can be explained in terms of an Arrhenius type of dependency of the diffusivity on the material temperature.

INFLUENCE OF AIR HUMIDITY

A comparison of the shape of the drying curves (Fig. 7a) shows that the use of an air-steam mixture rather than air as a drying medium does not alter the drying charactesristics.

A comparison of the drying rates (Fig. 7b) reveals a decrease in the air-steam mixtures in accordance with previous researches (24,25,26) but not alarmingly significant as claimed. To give this observation a more practical dimension, estimated drying times corresponding to a final moisture content of 15 % required in practice for a safe post drying storage were plotted as a function of air temperature (Fig. 8). An interesting feature of practical importance is that the increase in drying times in humid atmospheres decreases with temperature. It is thus possible through a rational choice of air temperature to limit the drying rate drops to levels compatible with the overall process economy.

MODELISATION AND SIMULATION

Basic assumption of mathematical model: The internal moisture movement mechanism is diffusional with flux given by Fick's law :

$$N = \rho_s D_{eff} (m, T) \nabla m \qquad (1)$$

where D_{eff}, the effective diffusivity is actually a function of material temperature and moisture content.

MATHEMATICAL FORMULATION

(a) Transient mass transfer equation

A microscopic moisture balance based on the Fiction diffusional flux (eq.1) yields a general equation :

$$\frac{\delta m}{\delta t} = \nabla \left(D (m,T) \nabla \right) \qquad (2)$$

(b) Transient heat transfer equations

Given the low values of heat transfer Biot number for granular foodstuffs (1,2) it appeared expedient to neglect the internal temperature gradients in order to effect a simplified macroscopic heat balance :

$$m_s C_p \frac{dT}{dt} = A.h (T_g - T) - A \dot{m} H_v \qquad \ldots (3)$$

A rigorous resolution of the coupled heat and mass transfer equations (2) and (3) by analytical methods encounters two obstacles :

(a) The non-linearity introduced by the functional dependence of the diffusivity on material temperature and moisture content.

(b) Scarcity of data on the variation the diffusivity as a function of temperature and moisture content.

In order to surmount the above problems in addition to that of irregular geometric solid shapes, a less rigorous procedure consisting of a preliminary identification of isothermal diffusivities followed by a more rigorous solution involving a temperature effect term was employed. The following simplifying assumptions were necessary :

(a) In the low moisture content range (29-10 % dry basis) the diffusivity is independent of moisture content at a given temperature.

(b) Moisture movement is an isothermal process. This is justified in practice by the rapid rise of grain temperature to a fairly constant value close to air dry bulb temperature during the drying :

(c) The grains are spherical with equivalent diameter defined by

$$d_{grain} = (6 / \pi \cdot V_{grain})^{1/3} \qquad (4)$$

(d) The grains are rigid and isotropic

Under these assumptions eq (2) reduces to a one-dimensional diffusion process in spherical coordinates

$$\frac{\delta m}{\delta t} = \frac{D_{eff}(T)}{r} \left(\frac{2}{r} \cdot \frac{\delta m}{\delta r} + \frac{\delta^2 m}{\delta r^2} \right) \qquad (5)$$

with initial and boundary conditions :

$$m(r,o) = m_o \qquad (6a)$$

$$m(R,t) = m_e \qquad (6b)$$

$$\frac{dm}{dr}(0,t) = 0 \qquad (6c)$$

The boundary condition (eq 6b) specifying a constant moisture content value at the particle-fluid interface is justified by the high mass transfer Biot. This constant value is assumed to be equal to the equilibrium moisture content given by the sorption isotherm :

$$H = 1 - \exp (1{,}39.10^{-5} \cdot T^{4,3} \cdot M^{3,2}) \qquad (7)$$

Equation 7 was obtained by fitting our experimental results with Henderson's model (28) by means of a multi-linear regression method with coefficient of

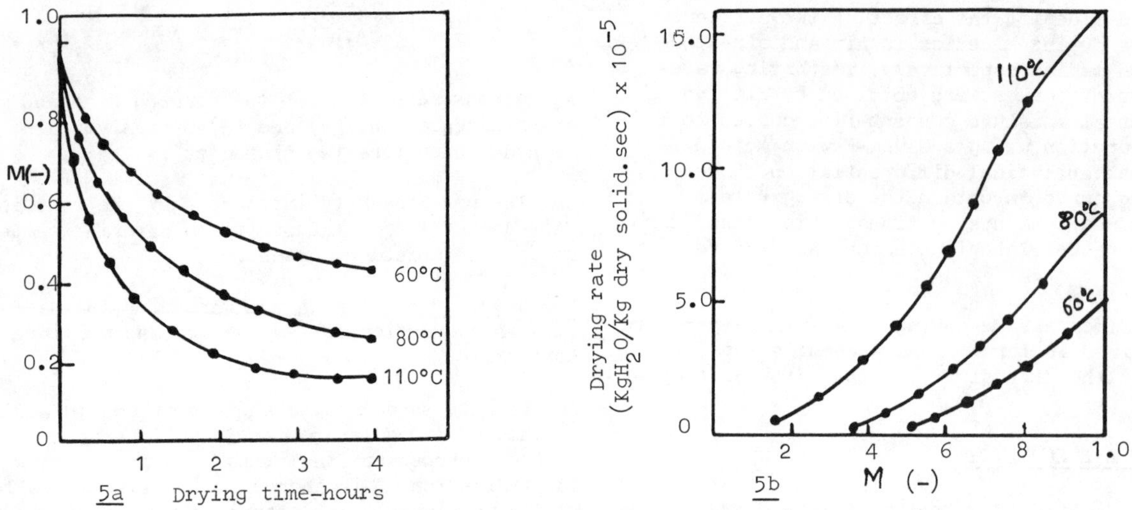

FIG. 5 - EFFECT OF TEMPERATURE IN AIR DRYING MEDIUM (R.H. = 6 %)

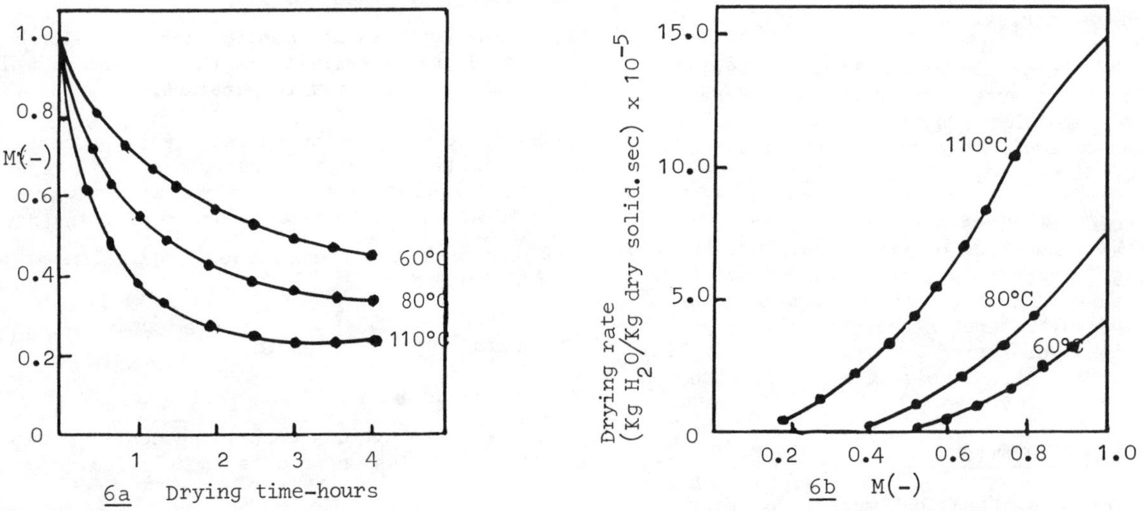

FIG. 6 - EFFECT OF TEMPERATURE IN AIR-STREAM MIXTURE DRYING MEDIUM (H.R. = 40 %)

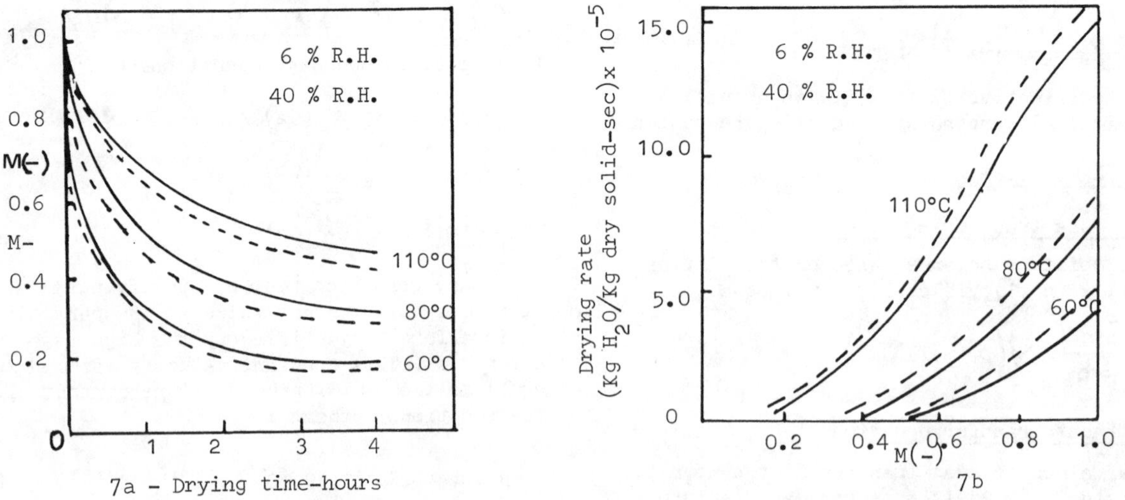

FIG. 7 - COMPARISON OF DRYING KINETICS IN AIR AND AIR-STEAM MIXTURE FLUID MEDIA

multiregression of 0,99 and standard deviation of 0.11 %.

Equation 5 in conjunction with the initial and bound-a y conditions (Eqs 6) permits an analytical solution of the average dimensionless moisture content as a function of time (10)

$$M = \frac{\bar{m} - m_e}{m_o - m_e} = \frac{6}{\pi^2}\sum_{n=1}^{\infty} 1/n^2 \cdot \exp(-\pi n)^2 \qquad (8)$$

Moisture diffusivity, the only unknown model parameter, at a given temperature was calculated by fitting the isothermal model (eq. 8) to the experimental drying curves by means of a Gauss-Newton method minimizing the square of the deviation between the experimental curve and eq. 8.

The calculated diffusivities at several temperatures were correlated assuming an Arrhenius type of temperature dependency :

$$D_{eff} = D_o \exp\left(-E_a/RT\right) \qquad (9)$$

with corresponding estimated activation energy and the factor D_o respectively of 2.650 KJ/mole and 1.946 x 10^{-6} m^2/s.

NON-ISOTHERMAL SIMULATION

Since the grain temperature is no more considered constant, the diffusivity by virtue of its dependence on temperature becomes a function of drying time. Integration of eq. 5 involves a time integral term

$$\frac{\bar{m} - m_e}{m_o - m_e} = \frac{6}{\pi^2}\sum_{n=1}^{\infty} 1/n^2 \cdot e^{-(\pi n)^2 \cdot \int_0^t \frac{D(t')dt}{R^2}} \qquad (10)$$

with the drying rate given by dm/dt

$$\frac{d\bar{m}}{dt} = -\frac{6}{R^2} D(t) \cdot (M_o - M_e) \sum_{n=1}^{\infty} e^{-\pi n^2 \cdot \int_0^t \frac{D(t')\,dt}{R^2}} \qquad (11)$$

The simulation was effected by solving the coupled heat and mass transfer equations (3) and (10) together with auxillary equations (7) and (8) by means of a computer program using a forth order Runge-Kutta-Merson integration algorithm with an automatic time-step adjustement and relative error checks.

Figure 9 shows that a slightly more accurate moisture content-time prediction is obtained on considering the varying temperature effects during drying.

Figure 10 also illustrates the fairly accurate grain temperature rise predicted by the rather simplified macroscopic heat transfer equation.

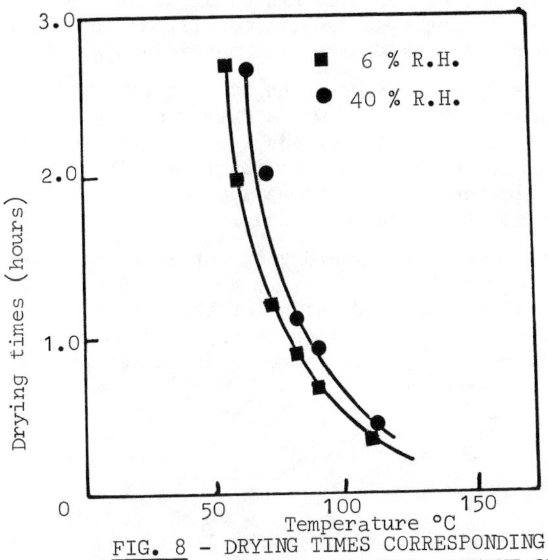

FIG. 8 - DRYING TIMES CORRESPONDING TO A FINAL MOISTURE CONTENT OF 15 % d.b. AS A FUNCTION OF TEMPERATURE

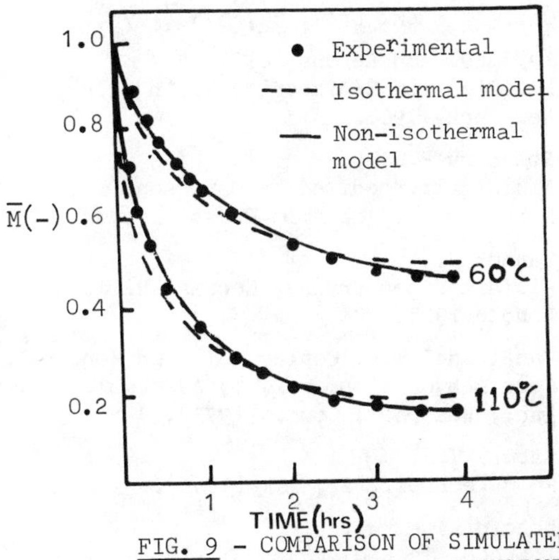

FIG. 9 - COMPARISON OF SIMULATED AND EXPERIMENTAL DIMENSIONLESS : MOISTURE CONTENT-TIME CURVES

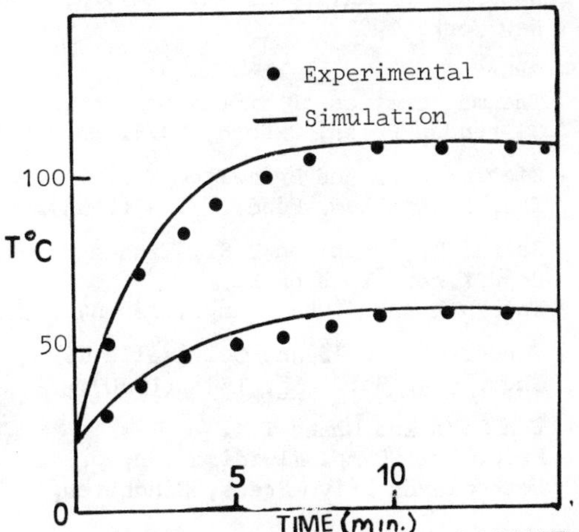

FIG. 10 - COMPARISON OF SIMULATED AND EXPERIMENTAL TEMPERATURE-TIME CURVES

CONCLUSION

(a) The drying kinetics exhibited no dependence neither on the hydrodynamic conditions nor the physical properties of the fine particles.

(b) Drying rates were found to be strongly influenced by air temperature and to a fairly negligible extent, in the early drying periods, by the air humidity. The influence of the higher equilibrium moisture content in humid conditions becomes appreciable only in the lower moisture content ranges.

(c) Although simulated drying curves were quite compatible with the reality, it does not in our opinion constitute a validification of the diffusional mechanism. It should rather be regarded as an interpretation of experimental results in terms of a selected mechanism.

REFERENCES

(1) - Alzamora, S.M., Chirife, J. and Violaz P.E., J. Fd Technol., 14, 369 (1979).

(2) - Vaccarezza L.M., Lombardi J.L., Chirife J. Can. J. Chem. Eng., 52, 576 (1975).

(3) - Kunii D. and Levenspiel O. Fluidization Engineering, John Wiley, New York, 1969.

(4) - Romankov P.G. Fluidization edited by Davidson J.F. and Harrison D., Academic Press, London 1971.

(5) - Vaneck V. Fluidized bed drying, Leonard Hill, London 1965.

(6) - Van Arsdel W.B. Copley M.J. and Morgan A.I. Food Dehydration, vol. 1, AVI Publ. Co., Inc., Westpart, Conn. (1973).

(7) - Labuza T.P. Food Technol., 22, 263 (1968).

(8) - Loncin M. et al. J. Food. Technol., 3, 131 (1968).

(9) - Hetsnoni Gad Handbook of multiphase systems, Mc Graw-Hill New York, 1982.

(10) - Crank J. The mathematics of diffusion Clarendon Press, Oxford, 1975.

(11) - Ziegler E.N. and Brazelton W.J. Ind. Engng Chem. Fund., 3, 94(1964).

(12) - Shirai T., Yoshiton Y.S., Tanaka S. Hoju K. and Yoshida S. Kagaku Kogaku (Chem. Eng., Japan) 29, 880 (1965).

(13) - Vanderschuren J. and Delvosalle C. Chem. Eng. Sci., 35, 1741 (1980).

(14) - Wen C.Y. and Chang T.M. Proc. Int. Symp. Fluidization, p. 491 Netherlands Univ. Press, Eindhoven, (1967).

(15) - Gordling P. Physical Phenomena during the drying of Foodstuffs in Fundamental Aspects of the dehydration of foodstuffs, Soc. Chem. Ind., London, 1958.

(16) - Krischer O. Die Wissenschaft Grund Lagender Trocknungstechnik, Springer, Berlin, 1963.

(17) - Pabis J. and Hendersen J.M. J. Agric. Eng. Res., 6, 272 (1961).

(18) - Bakker-Arkema F.W., Brook F.W. and Hall C.W. Drying cereal grains, AVI Publ. Co Westport CN (1974).

(19) - Young J.H. and Whitaker T.B. Trans ASAE, 14, 1051 (1971).

(20) - Henry, P.S.H. Discussions Faraday Soc., 3, 243 (1948).

(21) - Harmathy T.Z. Ind. Org. Chem. Fundam., 8, 92(1969), pp. 92-103, 1969.

(22) - Rossen J.L. and Hayakawa K. AIChE Symposium Series, 163, 71 (1973).

(23) - Luikov A.V. Int. J. Heat Mass Transfer, 9, 139 (1966).

(24) - Reay D. The Chemical Engineer, July/August, 507 (1976)

(25) - Wenzel L. and White R.R. Ind. Eng. Chem., 43, 1829(1951).

(26) - Chu J.C., Finelt S., Hoerrner W. and Lin M. Ind. Eng. Chem., 51, 275 (1958).

(27) - Hodgett D.L. The Chem. Engineer, July/August, 510(1976)

(28) - Henderson S.M. Agric. Eng., 33, 29 (1952).

DESIGN OF A CONTINUOUS FLOW TUNNEL GRAIN DRYER UTILIZING A RECIRCULATING GRANULAR MEDIUM

G.S.V. Raghavan[1] and R. Langlois[2]

[1]Department of Agricultural Engineering, Macdonald Campus of McGill University, Ste. Anne de Bellevue, Quebec

[2]Energy Technology, John Abbott College, P.O. Box 2000 Ste. Anne de Bellevue, Quebec Canada H9X 1C0

ABSTRACT

A grain dryer using a novel heat transfer concept was designed and constructed. The grain is submerged in a heated recycled medium which prevents high thermal gradients across the kernal surface. The recycled medium is augered through the interior of the cylindrical dryer and then returned through the exterior shell while being heated. Initial mixing of the grain with the medium, transport-forward and drying of the grain, and separation are all accomplished within the drum. Preliminary tests showed good mechanical and thermal performance.

INTRODUCTION

A solid, granular, inert medium can be used in combination with air to greatly enhance the heat transfer rates, resulting in faster drying. In such a technique, grain is mixed with the granular material and vigorously agitated until it is dried and then separated through a simple sieving process. The agitation is done by fluidizing the mixture (Savoie and Désilets (1)) or by mechanical action (Raghavan and Harper (2)). Raghavan and Harper (2), Khan et al. (3), and Lapp et al. (4) have dried grains using particle-particle heat transfer technique in a rotary drum. They investigated the effects of various parameters such as residence time, drum rotating speed and angle, type and particle size of granular medium, and mass ratio of

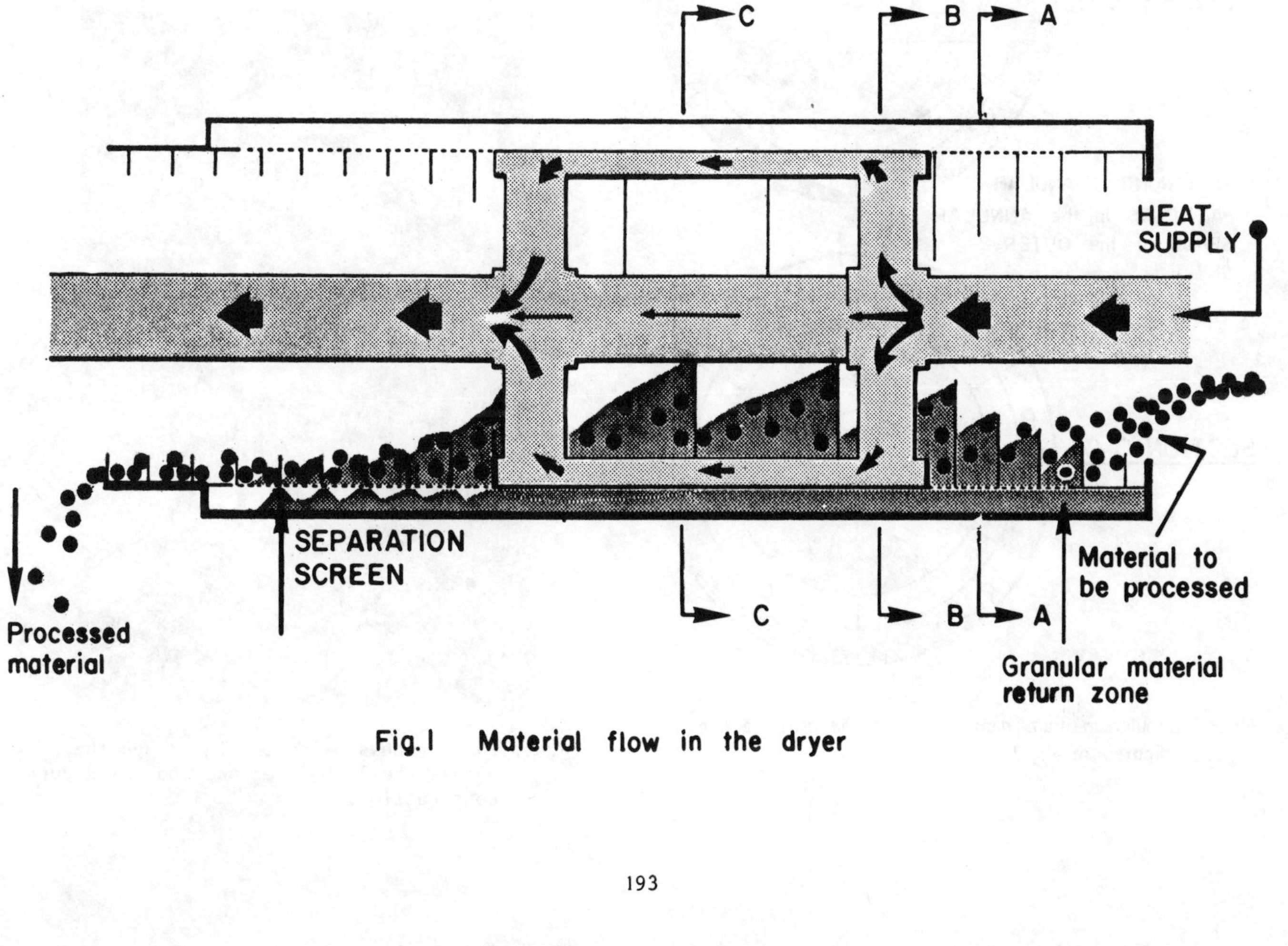

Fig.1 Material flow in the dryer

cereal to granular medium on the drying performance.

In addition to the experimental studies mentioned above, theoretical aspects pertaining to particle-particle heat transfer have been studied by Raghavan et al. (5), Richard and Raghavan (6-8), Sullivan and Sabersky (9) and Wunschmann and Schlünder (10).

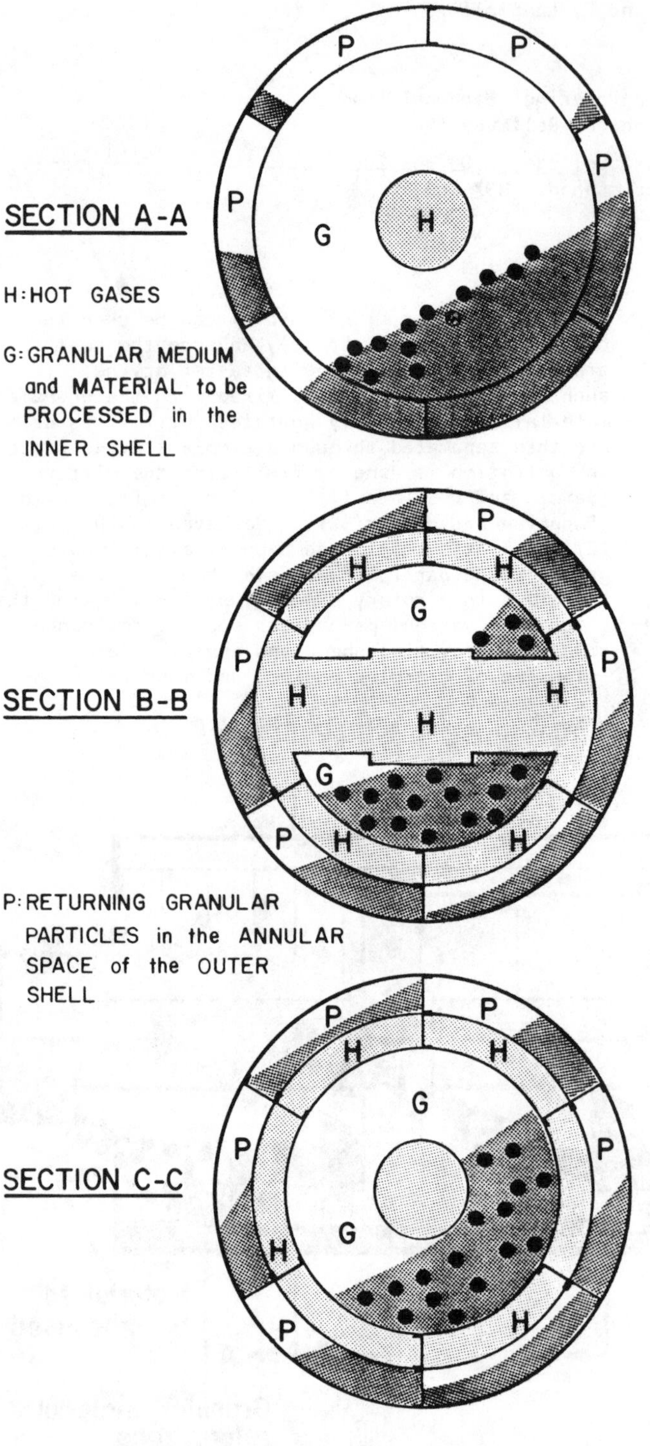

Fig. 2 Granular medium distribution at the cross-sections indicated in Fig. I

From the literature it is clear that the granular medium can be used for drying, processing and heat treating. However, the number of designs explored are too few. It is proposed in this paper to design build and test a machine which uses particle-particle heat transfer technique for grain drying.

The Tunnel Grain Dryer

The machine was designed to contain the following features in order to satisfy the particle-particle heat transfer concept.

(a) Maintenance and circulation of the proper amount of granular medium within the dryer.
(b) Suitable inlet for the grain entry.
(c) Mixing of the grain with the granular medium.
(d) Heating of the granular medium.
(e) Transport of the grain-granular medium mixture
(f) Separation of the grain from the granular medium.
(g) Variable grain-granular medium mass ratio.
(h) Variable residence time and temperature.

The schematic of the dryer is illustrated in Figure 1. The material to be processed (grain) enters at the right hand side. It mixes with the heated medium and is transferred through the innermost shell by a helix. At the left hand side it enters a separation zone where a fine screen separates the grain from the medium. The grain is removed by the helix, but the medium returns through a coarse screen. The heat is supplied axially with the use of a gas burner. All the tubes carrying the flue gases were designed to be contained within the shell to maximize heat transfer capabilities. The

Fig. 3. Photograph showing the inner and the middle shell, baffles and the helix during construction.

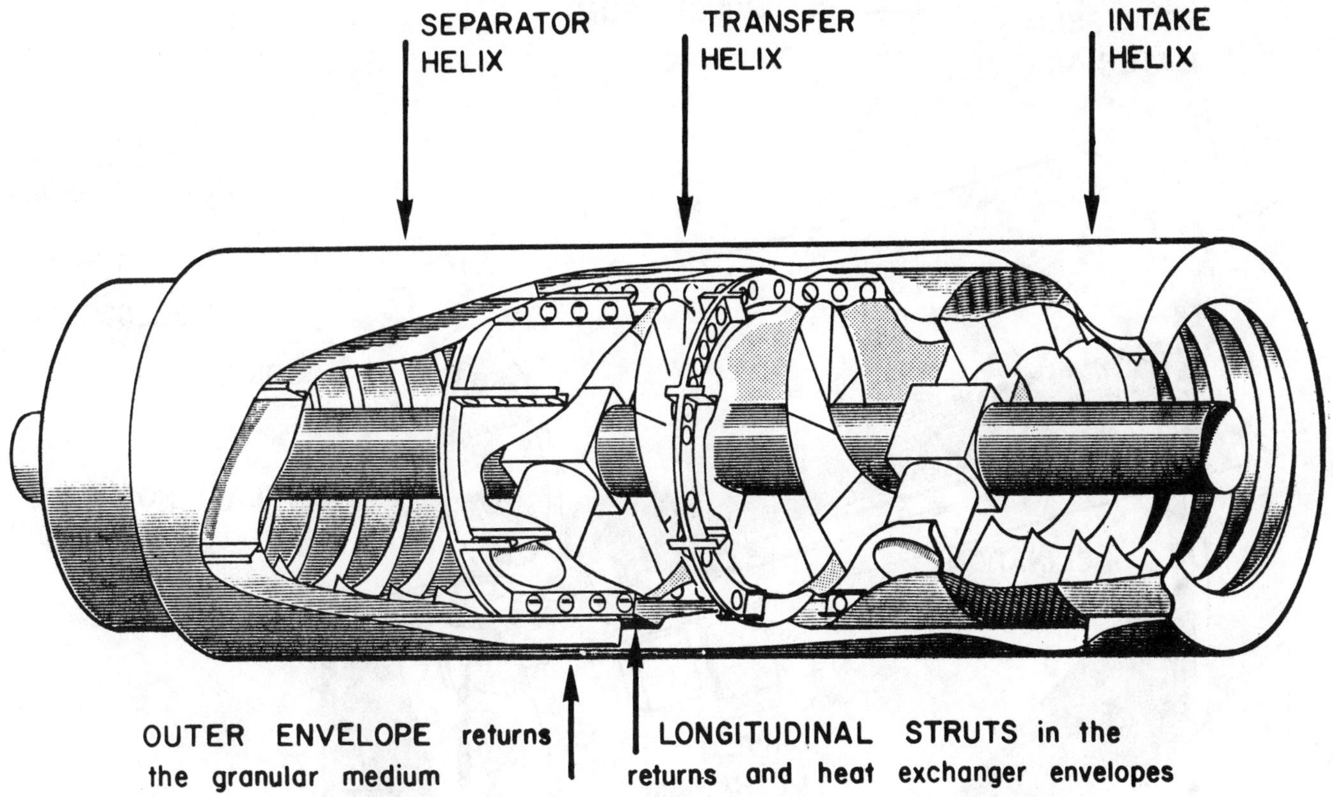

Fig. 4 Cross-sectional view of the dryer

gas path and the various components are labelled in Fig. 1. The distribution of the granular material is different at various longitudinal positions of the drum. This is illustrated in the diagram (Fig. 2).

The hot gases enter the plenum where a choke diverts most of the heat to two side tubes (Figs. 1 and 3). The tube at the center of the dryer was used to support the helix during the initial construction. The inner shell wrapped around the helix supports the mixture and acts as a surface for the heat exchange. To extract the maximum amount of heat from the hot gases, a network of baffles with perforations was secured to the shell (Fig. 3). A second sheet metal cover was then rivetted, which not only encloses the flue gases, but also acts as a second heat transfer surface. A number of longitudinal struts was then secured to the heat exchanger providing a support for the wire mesh separator/return screens, and to channel the returning medium. Screens of 6 mm mesh and 3 mm mesh were installed for the return or entrance zone and separator or exit zone and secured to the entrance and exit helixes, respectively.

A third sheet metal cover was wrapped around to complete the dryer. A pictorial-sectional view of the dryer is presented in Fig. 4. The intake, transfer and separator helix, heat supply zone, longitudinal struts, outer envelope which carries the return granular medium are all shown in the figure.

To support and rotate the dryer, a wooden stand, idlers and a 0.75 kW motor were assembled. The cradle on which the dryer rests is adjustable for any tilt angle which along with the variable speed motor can control the residence time and grain/medium mass ratio. The set-up is shown in Fig. 5. The surface heat loss was reduced by providing a suitable insulation around the drum.

The machine was tested for suitability of its operation and sizing. The grain feed rate obtained with the machine was about one tonne per hour. It was possible to vary the drum speed from 15 rpm to 40 rpm. The overall dimension of the dryer was 70 cm in diameter by 1.8 m in length. The entrance, mixing and exit sections were respectively 48 cm, 81 cm and 38 cm in longitudinal length.

SUMMARY AND CONCLUSIONS

A self contained continuous flow tunnel grain dryer was designed and built to comply with the particle-particle heat transfer technique. The machine is capable of drying/processing grain using such a technique. The design of the machine was such that the granular medium recycled well within the unit. The heat exchange capabilities were

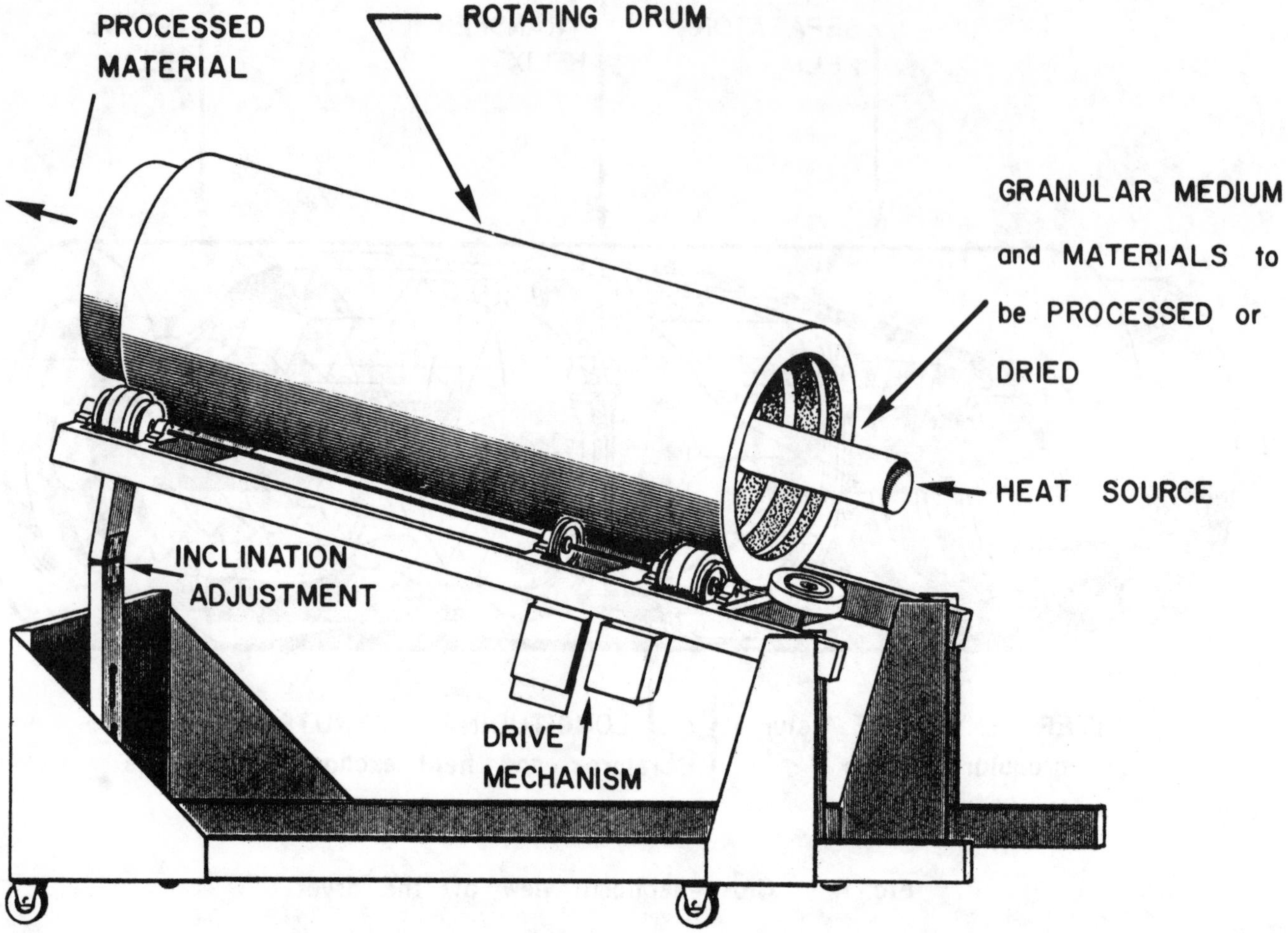

Fig. 5 A pictorial view of the experimental dryer/processor

maximized in the dryer. The material (grain) to be processed mixed well with the granular medium and separated easily at the exit. It was possible to vary the residence time and grain/medium mass ratio. The machine capacity was found to be one tonne per hour.

ACKNOWLEDGEMENT

The financial support of the NSERC, without which this study would not have been feasible, is gratefully acknowledged.

REFERENCES

1. Savoie, P. and D. Désilets. Corn drying in a fluidized corn-sand mixture. ASAE paper No. NA78-304 (1978).
2. Raghavan, G.S.V. and J.M. Harper. High temperature drying using a heated bed of granular salt. Trans. ASAE 17(1): 108-111 (1974).
3. Khan, A.E., A. Amilhussin, J.R. Arboleda, A.S. Manola and W.J. Chancellor. Accelerated drying of rice using heat conduction media. ASAE Paper No. 73-321 (1973).
4. Lapp, H.M., G.S. Mittal and J.S. Townsend. Cereal grain drying and processing with solid heat transfer media. ASAE Paper No. 76-3524 (1976).
5. Raghavan, G.S.V., J.M. Harper and R.D. Haberstroh Heat transfer study using granular media. Trans. ASAE 17(6): 589-592. (1974).
6. Richard, P. and G.S.V. Raghavan. Heat transfer between flowing granular materials and immersed objects. Trans. ASAE 23(6): 1564-1568, 1572 (1980).
7. Richard, P. and G.S.V. Raghavan. Drying and processing by immersion in a heated particulate medium. Advances in Drying, Hemisphere Publ., N.Y. (1983).
8. Richard, P. and G.S.V. Raghavan. A study of the heat transfer parameters for drying by immersion in a heated granular medium. Drying '80. Vol. II. 2nd Int. Symp. on Drying. pp.272-281 (1980).
9. Sullivan, W.N. and R.H. Sabersky. Heat transfer to flowing granular media. Int. J. Heat Mass Transfer 18: 97-107 (1975).
10. Wunschmann, J. and E.V. Schlünder. Heat transfer from heated plates to stagnant and agitated beds of spherical shaped granules. Proc. 5th Int. Heat Transfer Conf., Vol. 5, pp. 49-53 (1974).

SOME AERODYNAMIC ASPECTS OF SPOUTED BEDS OF GRAINS

R. Swaminathan and A.S. Mujumdar

Department of Chemical Engineering
McGill University, Montreal
Quebec, Canada.

ABSTRACT

Results of an experimental study of the aerodynamic characteristics of a flat bottomed spouted bed with internals are presented. The two types of internals that were used are: (i) a vertical partition and (ii) a draft tube results obtained for the half sectional column are compared with those for a full column. For barley, the peak pressure drop across the full column bed was found to be appreciably higher than that across a half column bed. Spouted beds of corn and oat showed less significant increases in the pressure drop. Whereas the pressure drop curves for beds without draft-tube (for both half column and full column beds) displayed significant hysterisis, the hysterisis displayed by beds with a draft-tube was much weaker. Also, the bed pressure drop required to spout and the extent of the "dead" zone around the bottom were found to be significantly reduced in the presence of a draft-tube.

INTRODUCTION

A Spouted Bed is a solid-fluid contacting device in which granular solids are carried rapidly upwards by the incoming fluid, forming a central core surrounded by a dense annulus of downward moving solids. The particles return in the form of a fountain at the top of the bed. The downward moving annulus and the air flaring from the central spout provide a dense phase counter current contact between the fluid and the solids (Fig. 1).

Coarse particles are handled effectively in a spouted bed which, when subjected to fluidization, show a marked tendancy toward slugging. Khoe and Van Brakel (3) have summarized potential advantages of spouted beds for drying of free flowing granular materials. Among these are: (i) possibility of operation with air at higher temperatures and hence greater thermal efficiency because of short residence time in the spout; (ii) possibility of simultaneous operations such as dehusking, roasting, coating and parboiling (of rice); (iii) lower capital and operating costs compared with fluid bed drying; (iv) lower pressure drop operation (as compared to fluid beds) etc.

However, the maximum spoutable bed height, difficulty of scale-up, and spout instability for deep beds are limitations of a conventional Spouted Bed (without internals) that can be overcome by the so-called draft-tube spouted beds. Furthermore, use of a draft-tube offers potential for some lowering of the bed pressure drop required to spout by placing a smooth surfaced draft-tube to separate the spout-annulus interface of a conventional spouted bed.

Buchanan and Wilson (4) were the first to study the draft-tube spouted bed of conical base design. They obtained a two-fold reduction in the minimum airflow requirement for spouting by reducing the Tube separation from 18.3 cm. to 3.3 cm. This was accompanied by greatly reduced solids circulation rate. Mathur and Epstein (1) have discussed the advantages of the draft-tube system and the absence of restrictions due to particle properties and bed depth which apply to conventional spouted beds. Also, the use of a draft-tube for drying of heat sensitive particles (such as paddy, in which the drying rate should be carefully controlled) prevents product degradation and yet maintains high efficiency (5). Khoe and Van Brakel (6) studied the draft-tube spouted bed for a small scale grain dryer. They noted that high dryer efficiency is obtainable with large separation distances and low air flux at inlet. For low pressure drop, the tube separation distance can be reduced to minimize the air flux. Particle mixing in draft-tube beds has been studied by Schwedes and Otterbach (7) and by Krambrock (8).

EXPERIMENTAL FACILITY AND PROCEDURE

Fig. 1 shows schematically the spouted bed geometries used in this study. A plexiglass pipe was used as the main column of the apparatus. The vertical partition was so fabricated that when installed it divided the column and the inlet orifice into two equal halves i.e. the bed can be operated as a set of two half-columns. The draft-tube (which was used only with full columns) could be moved up or down so that the separation distance between the flat bottom and the lower end of the draft-tube could be varied. In this facility the ratio D_c/D_i was varied by using different inlet plates with different orifice diameters. Note that the bed geometry is different from that used conventionally; the bottom is flat rather than conical. The flat bottom geometry is much cheaper to fabricate commercially. The ranges of operation are listen in Table 1. Corn, oat and barley were the three grains selected. The physical properties of these test materials are tabulated in Table 2.

The ΔP-U curves for beds of oat displayed more than one peak in the upper part of the curve which

Table 1

Parameter ranges

Type	D_c (m)	D_i (m)	Material	Bed Height	Separation distance (m)
Full column	0.23	0.078	corn	0.15, 0.20, 0.25	-
	0.23	0.038	barley	0.15, 0.25	-
	0.23	0.038	oat	0.23, 0.30, 0.46	-
	0.23	0.051	corn	0.15, 0.235, 0.30	-
	0.23	0.051	barley	0.15, 0.23, 0.30	-
	0.23	0.051	oat	0.15, 0.23, 0.30	-
Half column	0.23	0.038	corn	0.25	-
	0.23	0.038	barley	0.25	-
	0.23	0.038	oat	0.30	-
Draft-tube	0.23	0.038	barley	0.91	2.3×10^{-2}
Spouted bed	0.23	0.038	barley, corn, oat	0.95	6.0×10^{-2}
	0.23	0.038	barley, oat	0.99	10.0×10^{-2}

Table 2

Material characteristics

Material	d_p (m)	particle shape* factor	Density ρ_s (kg/m^3)
Corn	8.12×10^{-3}	1.31	1119.09
Barley	3.90×10^{-3}	1.18	1275.30
Oat	4.06×10^{-3}	1.26	948.21

* The particle shape factor ϕ is defined as

$$\phi = \frac{\text{Surface area of particle}}{\text{Surface area of an equal volume sphere}}$$

RESULTS AND ANALYSIS

PRESSURE DROP CHARACTERISTICS

Spouted bed without internals

The three typical shapes of pressure drop curves obtained in this study for spouted beds without internals are sketched in Fig. 2. Only shapes similar to those shown in Fig. 2(A) have been reported in the published literature.

For barley the pressure drop curves were always found to display the shape shown in Fig. 2(B). It is characterized by a loop at high air flow rates. Along this loop the pressure drop for increasing flow is lower than that for decreasing flow. It may be noted that barley grains used had a rough surface texture as compared to corn and oat.

is for increasing air flow (Fig. 2(C)). This behaviour was peculiar to the bed with an inlet orifice diameter of 0.038 m. Only one peak was observed with D_i = 0.051 m. No physical explanation could be offered for this phenomenon at this time. It is probably a result of successive compaction and expansion of the bed with increase in the air flow rate.

Haft column spouted bed

The shapes of the ΔP-U curves obtained for half column beds of barley, oat and corn (Fig. 3 through Fig. 5) were not different from the general shapes obtained for full column beds shown Fig. 3.

In a half column bed, whereas the superficial velocity when the spouting commences is higher than the value when the spouting stops, the pressure drop at the onset of spouting is less than the value when the spouting stops.

Draft-tube spouted bed

From data obtained for beds of barley, the difference between the superficial velocity when spouting starts and the velocity when spouting stops was found to be large when the separation distance (between the flat bottom and the lower end of the draft-tube) was 0.1 m. For smaller values of the separation distance the differences were less significant. The difference was more significant for beds of oat and corn than for barley i.e. for the smoother surface grains.

FULL COLUMN AND HALF COLUMN - PRESSURE DROP BEHAVIOR

Table 3 compares the peak pressure drops required for spouting in full and half columns for all three grains used. Full columns of barley, oat and corn display 27%, 7% and negligibly higher peak pressure drops, respectively, than those for half columns of the same materials and bed heights. Thus spouted beds of barley behave quite differently depending upon the presence or absence of the internal partition.

This large difference in the peak pressure drop for full and half columns of barley may be attributed to the rough texture of the barley grains used. One expects a higher coefficient of friction between two kernels of barley than between the smooth plexiglas partition and barley kernels. Thus, the force required to overcome the friction in the former case is less than that in the latter case; i.e. barley kernels would flow move readily along a smooth wall than along a layer of barley kernels. On the basis of the limited data obtained in this study, it is suggested that half-columns, commonly used to simulate full-column spouted beds, can give representative results only for relatively smooth solid particles. The favorable comparison reported in the prior literature in mainly due to the fact that most studies with spouted beds have been confined to beds of smooth particles e.g. wheat, peas, corn etc. It is postulated that moist particles would also display differences in full-column and half-column behaviour.

MINIMUM SPOUTING VELOCITY

Table 4 presents a comparison of the predicted values of the minimum spouting velocity using the well known Mathur and Gishler correlation and the experimental results of this study for a spouting column without internals.

The predicted and the measured values show close agreement except for oat for one inlet diameter (D_i = 0.038 m) which, as discussed earlier, showed erratic behaviour. Measured values of the minimum spouting velocity for barley was also found to be always higher than the predicted values. This may once again be attributed to the rough surface of the barley kernels. In view of this observation, it is suggested that correlations may be improved by incorporation of suitable roughness parameters.

DRAFT-TUBE SPOUTED BED

Experiments were performed using a plexiglas draft tube, 0.89 m long and 0.076 m internal diameter. Only one spout inlet diameter was used in the study viz: D_i = 0.038 m. Fig. 6(A) and Fig. 6(B) show the schematic of the spouted bed without and with a draft-tube respectively. For a flat-bottomed spouted bed, the dead region along the bottom plate was found to be considerably less with draft-tube than without.

From the results of this experimental study it is concluded that it is possible to effectively eleminate the dead zone in a flat-bottom spouted bed by adjusting the geometric parameters such as D_c/D_i, D_c/D_t and D_t/D_i (where D_c= column diameter, D_i= inlet orifice diameter and D_t= diameter of the draft tube). The flat bottom spouted bed could then replace the conventional conical bottom spouted bed.

It was observed visually, that the extent of the dead region along the bottom plate increases with increase in the separation distance between the lower end of the draft-tube and the orifice plate. This is expected from the flow dynamics.

Interestingly, the draft-tube spouted bed was found to compact after spouting commenced; in a

Table 3

A comparison of the peak pressure drops between full and half-column spouted beds.

No.	Material	Figure No.	Peak pressure drop (KN/m^2) Full column	Half column	percentage increase of full column ΔP max against half column ΔP max
1	Barley	4	2.81	2.22	27%
2	Oat	5	2.04	1.90	7.4%
3	Corn	6	1.85	1.84	0.5%

Table 4

Minimum spouting velocity - A comparison of experimental data with predictions using Mathur-Gishler correlation*

Material	Bed height (m)	D_i (m)	Minimum spouting velocity U_{ms} (m/s) Predicted	Experimental
Oat	0.152	0.051	0.503	0.52
	0.230	0.051	0.62	0.613
	0.305	0.051	0.711	0.867
	0.230	0.038	0.56	0.92
	0.305	0.038	0.646	0.917
	0.457	0.038	0.792	0.922
Corn	0.235	0.051	1.357	1.41
	0.152	0.051	1.093	1.12
	0.292	0.051	1.513	1.46
	0.152	0.038	0.993	1.058
	0.203	0.038	1.147	1.288
	0.254	0.038	1.282	1.371
Barley	0.305	0.051	0.793	0.998
	0.152	0.051	0.561	0.79
	0.229	0.051	0.687	0.93
	0.152	0.038	0.51	0.756
	0.254	0.038	0.658	1.01

$$* \; U_{ms} = \left(\frac{d_p}{D_c}\right)\left(\frac{D_i}{D_c}\right)^{1/3}\left(2gH\,\frac{\rho_s - \rho_f}{\rho_f}\right)^{1/2}$$

normal spouted bed, the bed expands after spouting commences. In the former case there was no air flaring into the annulus whereas in the latter an appreciable fraction (depending on the bed height) of the inlet air flares through the interface into the annulus. The presence of this air expands the bed. This behaviour was observed for all materials tested (viz; barley, oat and corn). Pressure drop curves obtained for draft-tube spouted beds of different materials are shown in Figs. 7 through 9.

For the draft-tube bed the entire ΔP-U curve for decreasing flow lies above the curve in the following cases (i) Barley - separation distance = 0.023 m and (ii) Corn- separation distance = 0.066 m. This behaviour is quite the opposite of what is found for the conventional spouted bed (conical or flat bottom). The curves for the remaining cases cross each other. It is also noteworthy that the separation between the increasing and decreasing flow curves is relatively small for draft-tube spouted beds.

Thus, the draft-tube spouted beds show minor hysterisis as compared to the normal spouted beds. Once again spouted beds of barley are an exception when spouted with large separation distances (0.06-0.10 m) between the draft-tube and the orifice plate. Significant hysterisis was observed for spouted beds of barley.

The pressure drop across a draft-tube spouted bed increases with increasing separation distance. Note that no direct comparison between the pressure drop across conventional and draft-tube beds could be made because the former results were confined to small bed heights (less than 0.35 m) while the bed heights in the latter case were upto 1.0 m. Presence of a draft-tube improved spontability of the bed with much lower pressure drops. The peak pressure drop across conventional spouted beds was calculated using an empirical correlation*. The ratios of the peak pressure drop across a normal spouted bed (predicted) and across a draft-tube spouted bed (measured) are summarized in Table 5. It is important to note that the pressure drop across the draft-tube beds at small separation distances is very small as compared to that for conventional spouted beds, or draft-tube spouted beds with large separation distances. This is, however, accompanied by a lower solids circulation rate. Thus the counteracting effects of draft-tube on bed pressure drop and solids circulation rate must be evaluated for each application. Operating a draft-tube spouted bed at an optimum separation distance (between the draft-tube and the orifice plate) may reduce the pressure requirements significantly.

Table 5

Comparison of peak pressure drops of normal and draft-tube spouted beds at same bed heights

Material	Separation distance (m)	Bed height (m)	Max ΔP for Normal spouted bed / Max ΔP for draft-tube spouted bed
Barley	0.023	0.913	12.8
"	0.06	0.95	8.3
"	0.10	0.99	2.0
Oat	0.06	0.95	8.6
"	0.10	0.99	2.6
Corn	0.06	0.95	4.1

CLOSURE

Experiments with half-column spouted beds were found to give representative results for full column beds only for relatively smooth solid particles, eg. wheat, corn etc; rough textured particles (eg. barley) behave differently in the two cases. Spouting along a smooth surface was found to be more stable than conventional spouting. This may explain partly the good spout stability found in draft-tube spouted beds which are a special case of spouting along a surface. This spout stability can be applied successfully in a slot spouted bed in which air enters through a slot at the bottom of a rectangular column, tangential to the inclined surface. It is felt that this spouting stability along a surface can be exploited in multiple spouting, which is known for its instabilities, by directing upwards the spouts along the corners of a rectangular column. Orcourse, in this case the stagnant region can be prevented by an improved choice of design for the bottom of the bed which allows for free flow of particles towards the corners. In a flat-bottom spouted bed the dead zone can be eliminated or decreased significantly by proper choice of parameters such as D_c/D_i, D_c/D_t and the tube separation distance. It was found that smaller the separation distances smaller the stagnant region and lower the pressure drop accompanied by reduced solids circulation rates.

Further work is needed with a wider assortment of rough particles and artificially "wet" model particles to clarify the differences between half-column and full-column results. Since the aerodynamics of particle motion determines the heat transfer and drying rates in such equipment, it is necessary to obtain further insight into this aspect as well.

ACKNOWLEDGEMENT

The authors are indebted to Purnima Mujumdar for typing this paper.

REFERENCES

1. Mathur, K.B. and Epstein, N., Spouted Beds, Academic Press, N.Y. (1974)
2. Gishler, P.E., Proc. 2nd Int. Symp. Spouted Beds, Vancouver, Canada, Oct. 3-6, 1982. pp.1-3
3. Khoe, G.K. and Van Brakel, J., A draft-tube Spouted Bed as Small Scale Grain Drier, I. Chem E. Symp. Series No. 59, paper presented at the Internat. Conf. on Solids Separation Processes, Dublin, April 1980
4. Buchanan, R.H. and Wilson, B., The fluid-lift solids recirculator. Mech. Chem. Eng. Trans. 1, 117 (1965)
5. Khoe, G.K., and Van Brakel, J., Drying characteristics of a Draft-tube Spouted Bed, Proceedings of the 2nd Int. Symp. on spouted beds, Vancouver, (1982).
6. Schwedes, J. and Otterbach, J., Design of pneumatic particle-mixers, Verfahrenstechsik 8(1974) Nr 2, 42-47
7. Krambroek, W., New pneumatic method for homogenizing free-flowing bulk goods, Verfahrenstechnik 8(1974) Nr 2, 48-53.

NOMENCLATURE

D_c	spouted bed diameter	(m)
D_t	diameter of draft-tube	(m)
D_i	inlet port diameter	(m)
H	Bed depth	(m)
d_p	particle diameter	(m)
p_s	particle density	(kg/m^3)
p_f	density of spouting fluid	(kg/m^3)
U_{ms}	Minimum spouting velocity	(m/s)
g	Acceleration due to gravity	(m/s)
ΔP	Pressure drop across the bed of solid particles	(KN/m^2)
s	Tube separation distance	(m)

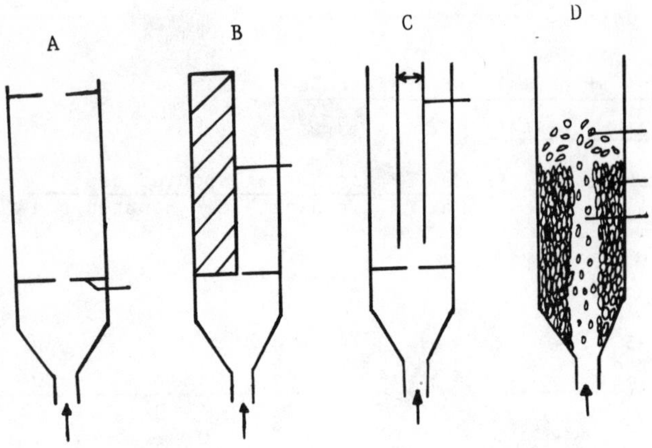

Fig. 1. Schematics of spouted beds studied.
(A) Flat-bottom spouted bed without internals
(B) Spouted bed with partition
(C) Spouted bed with draft-tube
(D) Conventional conical-bottom spouted bed

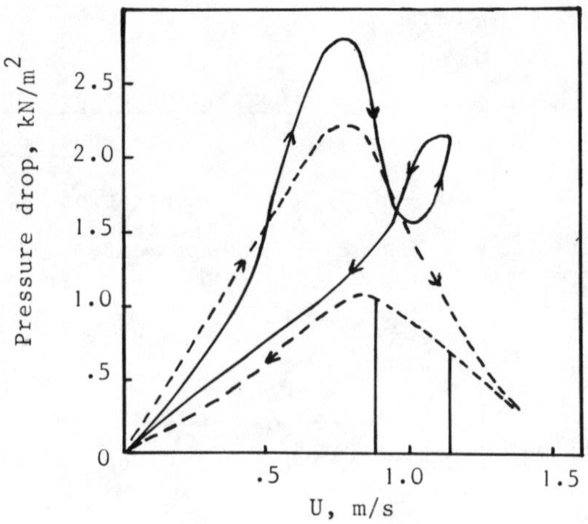

Fig. 3. A comparison of the pressure drop characteristics of full and half column beds of Barley
D_i= 0.038m, H = 0.254m, D_c= 0.23m
—— Full column, -------- Half column

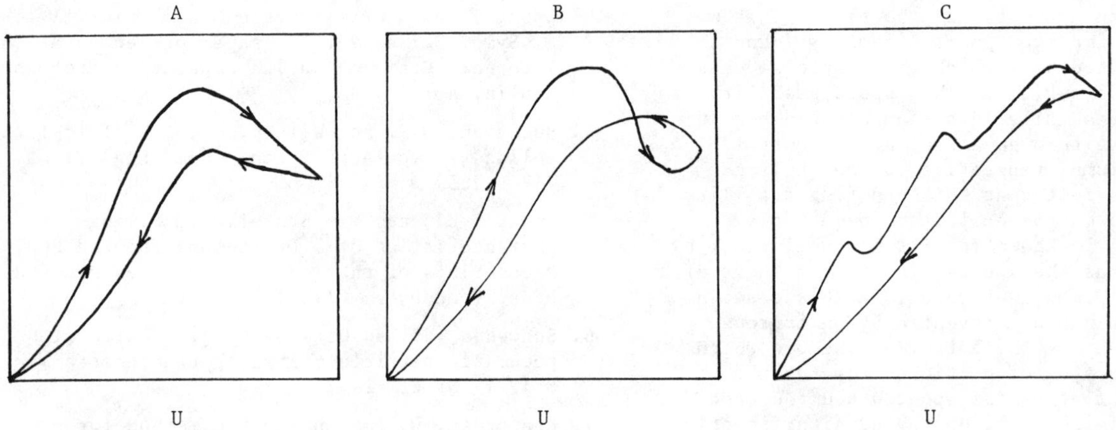

D_c= 0.23m D_c= 0.23m D_c= 0.23m
Corn D_i= 0.051m, 0.038m Barley D_i= 0.051m, 0.038m Oat D_i= 0.038m
Oat D_i= 0.051m

Fig 2. Typical shapes of ΔP.U curves obtained in present work with full column, flat-bottom spouted beds.

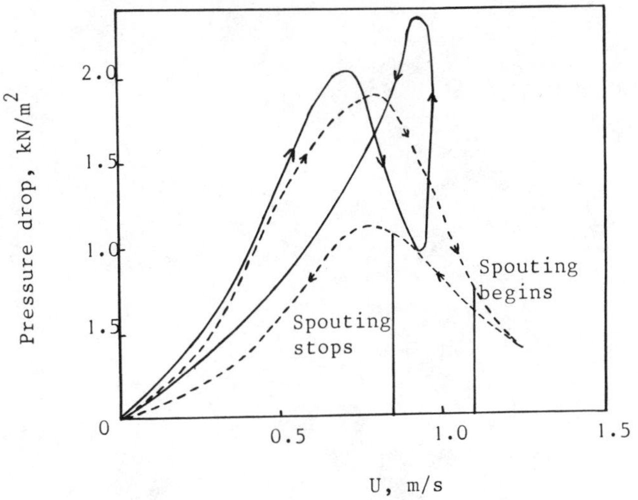

Fig. 4. Comparison of the pressure drop characteristics of full (solid line) and half column beds of Oat. (D_i= 0.038m, H=0.305m D_c= 0.23m.)

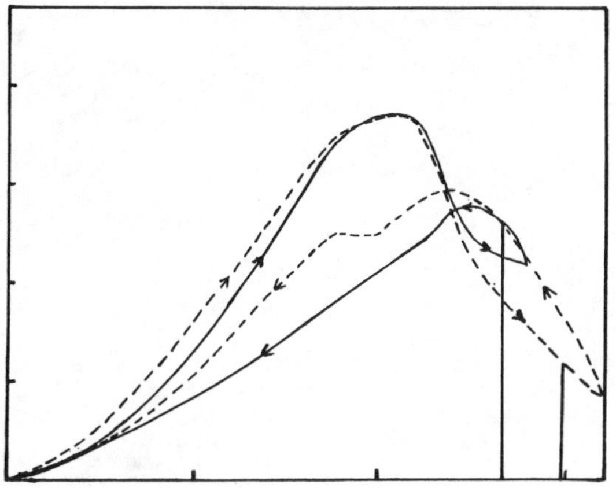

Fig. 5. Pressure drop characteristics of full and half column beds of Corn
D_i= 0.038m, H = 0.254m, D_c = 0.23m

—— Full column, ------- Half column

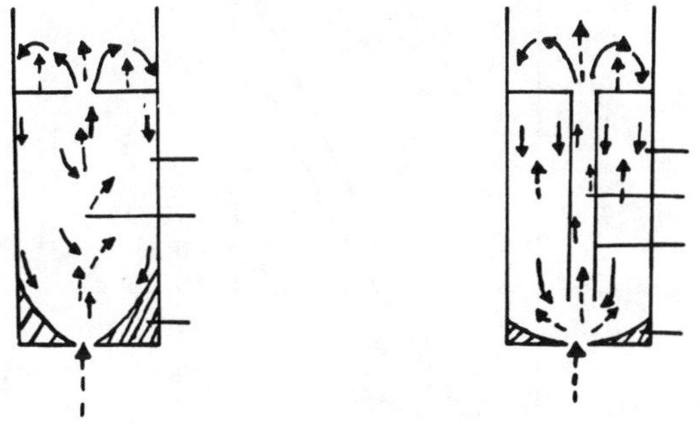

Fig. 6. Schematic representation of the flow pattern in a flat-bottom spouted bed
A. Without internals
B. With a draft tube
Solid line denotes particle path
Dotted line denotes air path

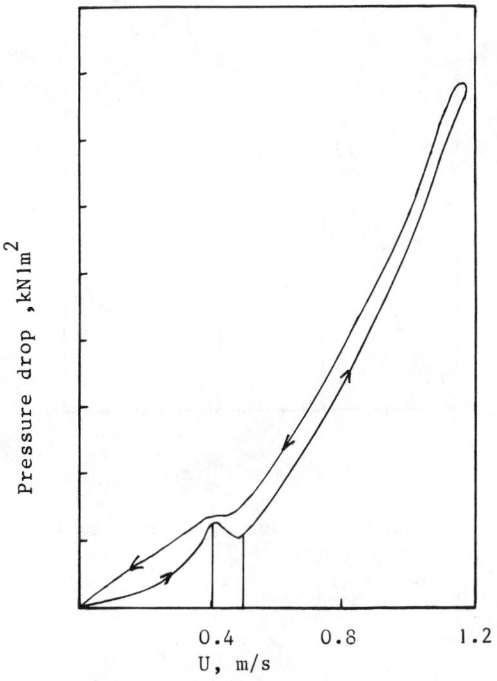

Fig. 7. Pressure drop characteristics of a draft-tube spouted bed of barley.

D_i = 0.038 m ; H = 0.91 m; S=0.023 m; D_c = 0.23 m.

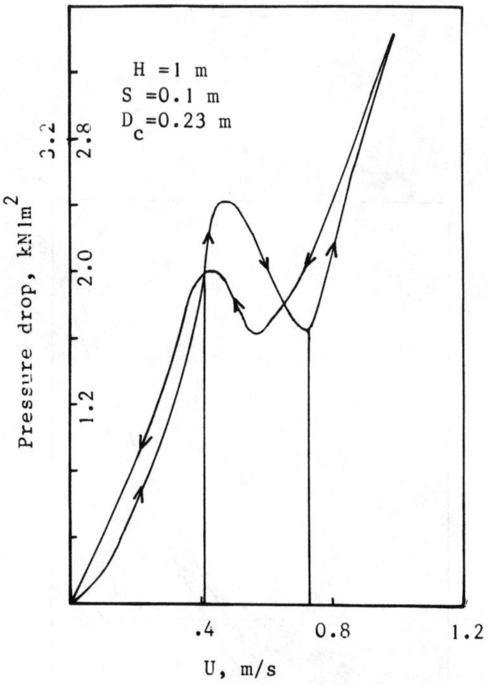

Figure 8. Pressure drop for draft-tube spouted bed of oat

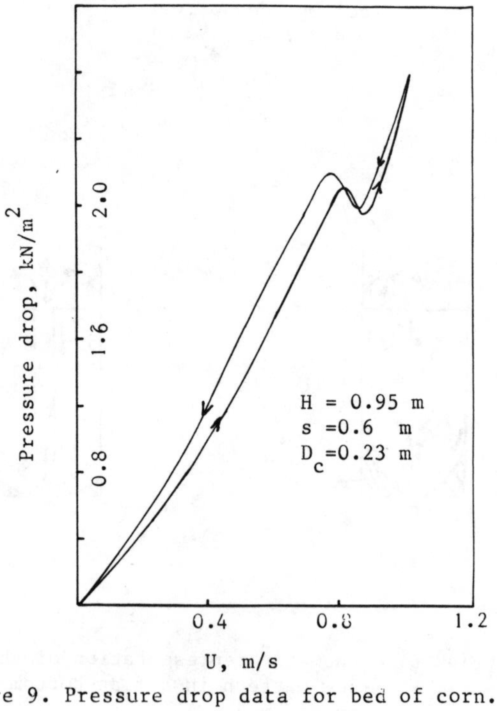

Figure 9. Pressure drop data for bed of corn.

CHARACTERISTICS OF ASYMMETRIC TWO-DIMENSIONAL SLOT SPOUTED BEDS FOR GRAINS

K. Anderson[1], G.S.V. Raghavan[1] and A.S. Mujumdar[2]

[1]Department of Agricultural Engineering, Macdonald Campus of McGill University, Ste. Anne de Bellevue, Que. H9X 1C0

[2]Department of Chemical Engineering, McGill University, Montreal, Que., Canada H3A 2K6

ABSTRACT

Pressure drop-velocity-drying data are presented for beds of wheat, corn, oat and barley grains spouted in a specially designed asymmetric two-dimensional apparatus. Slot spouted beds may be used to offset some of the fundamental scale-up and capacity limitations of the conventional tubular spouted bed for drying as well as other possible applications such as coating, granulation, heterogeneous chemical reactions, etc. The velocity-pressure drop curves displayed hysteresis typical of all spouted beds, but had shapes quite different from those reported earlier for conventional beds. Effects of grain type, bed height and moisture level are evaluated in the light of data obtained.

INTRODUCTION

Mujumdar (1) has presented a brief review of recent studies in spouted bed technology relevant to drying of spoutable particles. Abrahamson and Geldart (2) have characterized powders with respect to their fluidizability and spoutability. Most grains present formidable problems in fluidization but are readily amenable to spouting. Indeed, the first proposed application for the spouted bed concept was for drying of wheat (3). The conventional spouted beds are noted for a number of limitations which have discouraged their applications on the commercial scale. Among these one may cite: difficulty with scale-up because of strong nonlinearities, an inability to achieve good spouting in large-scale vessels (the maximum reported vessel is only one metre in diameter), high peak pressure drop (although for operation a lower pressure drop is needed) and a need to fix aeration rates dictated by the fluid dynamic requirement of spouting rather than that demanded by the heat/mass transfer operation being carried out. Mujumdar (1) has discussed some of the "modified" spouted bed designs which offset these limitations. He has also presented a new and more detailed classification for spouted bed equipment. No attempt will be made to repeat this material here. It should be noted that he has pointed out the advantages of a slotted two-dimensional spouted bed (2-D SB) viz. flexibility, simplicity of scale-up and fabrication, economic construction, ability to operate in combined direct-indirect drying mode for high thermal efficienty in drying, etc.

Prior work in this modified SB appears to be very limited. While Mujumdar (1) proposed a fully two-dimensional vessel (i.e., aspect ratio of cross-section of the order of 10), the earlier cited work of Miter and Volkov (see Mathur and Epstein (4)) was apparently done with three-dimensional slotted beds (small aspect ratios) which would retain all the limitations of the tubular spouted beds due to wall effects. Possibility of secondary flows in corners would further impede performance of such dryers.

The book of Rabinovich (5) in Russian presents only one set of pressure drop data for a two-dimensional spouted bed of peas with and without horizontal tube bundle. They noted that the 2-D bed can be operated at higher velocities than the conventional spouted bed. Spouting was stable at aspect ratios of 10 to 20. Larger aspect ratios apparently led to poor spouting, uneven air distribution along the slot and presence of dead zones. They observed that the minimum spouting velocities and maximum operating velocities were 50-100% higher for the 2-D SB compared to the axisymmetric case. The pressure drop was about the same for both geometries. Presence of a tube bundle (8 mm diameter with 24-mm pitch) in the bed increased the peak pressure drop and shifted it to higher air velocities. This was attributed to the damping of instabilities by the tube bundle. This stability should disappear at incipient fluidization. Indeed, the author noted the peak pressure drop for peas to occur at a velocity of 1.56 m/s while the minimum fluidization velocity was at 1.67 m/s.

Perhaps the most recent reference related to planar two-dimensional spouted beds is by Sawyer et al. (6) who studied the flow and combustion characteristics of such beds in a 12.7 mm x 100 mm (cross-section) x 300 mm (height) vessel. They chose the planar geometry solely to study optically the particle motion, assuming it to be similar to that in an axisymmetric spouted bed. Their objective was to obtain better understanding of the interaction of combustion and the bed fluid mechanics. Interestingly, they noted that the bed exhibited an unsteady, near-periodic motion both in cold and combusting flows. Particle to gas mass flow ratios of about ten were observed. Measurements of particle velocities in the moving bed region provided particle circulation rates in the bed. Their conclusions went on to point out that further optical probing is required to characterize the flow.

The objectives of this paper are primarily to justify the concept and feasibility of the 2-D SB and to present preliminary flow data for one variant of the 2-D SB viz. one with a slotted inlet along one wall. This is the analog of the half column of the tubular spouted bed. According to Swaminathan and Mujumdar (7) half-column spouted beds are more stable and present slightly reduced pressure drop especially for rough (and possibly wet) particles. Thus the asymmetric spouted bed chosen for this first study may be a better geometry than the symmetric slotted spouted bed.

MATERIALS AND METHODS

An asymmetric two-dimensional slot spouted bed of aspect ratio 50 was designed for the study. A schematic diagram of the apparatus shown in Fig. 1 presents essentially a vertical column of rectangular cross section into which air enters through a long slot.

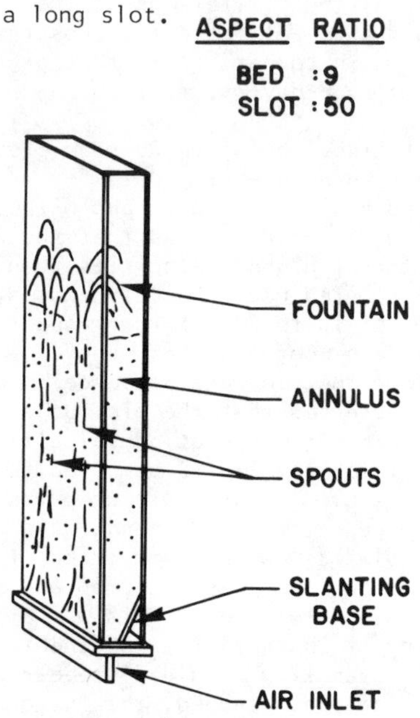

Fig. 1. Schematic of an asymmetric two-dimensional slot spouted bed.

For comparison, a schematic of a conventional cylindrical or circular column spouted bed with its main components is also presented (Fig. 2).

The experimental two-dimensional bed dryer under investigation was connected to a blower with proper duct work. Provision was made to measure air velocity and pressure drop during the study. Air flow into the bed was controlled by adjusting the orifice restrictor and bleed valve of the blower. Bed depths used were between 10 and 30 cm for the four types of grains, i.e. wheat, corn, oats and barley. The moisture contents selected for the grains varied between 11% and 30% wet basis. For each depth, the air flow was increased from very low values through those required for spouting to

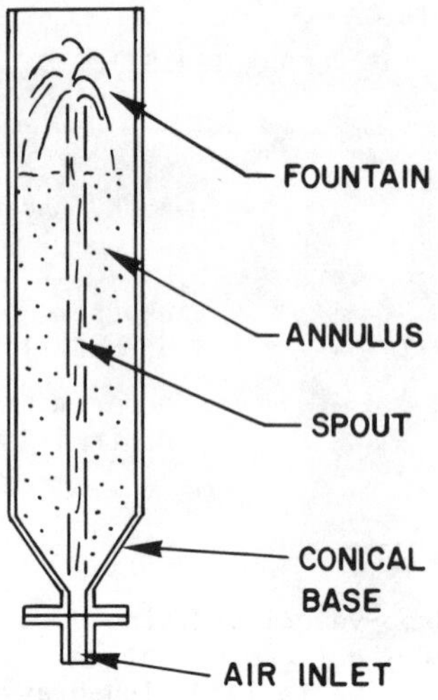

Fig. 2. Schematic of a conventional circular SB.

those at which aggregative fluidization began to occur, then decreased. Pressure drop-velocity readings were taken at various times during the run. Initial and final moisture contents were also noted.

RESULTS AND DISCUSSION

The data obtained from all the experiments were analyzed and the characteristic pressure drop vs. superficial velocity curve was determined. This is shown in Figure 3.

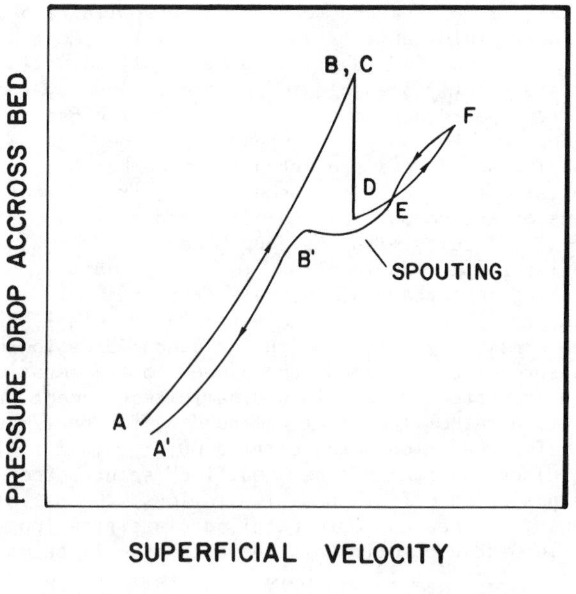

Fig. 3. Characteristic Δp vs. U curve for a 2-D SB.

The points ABDEF and FEB'A' correspond to the increasing and decreasing air flows respectively. Points B, D and B' are the locations representing the maximum pressure drop, the minimum spouting velocity and the velocity and pressure drop obtained just after the collapse of the spout, respectively. The point C shown is the point of incipient spouting. The characteristic curve shown in Fig. 3 for the asymmetric two-dimensional bed can be compared with the characteristic curve, Fig. 4 of a half section circular spouted bed dryer.

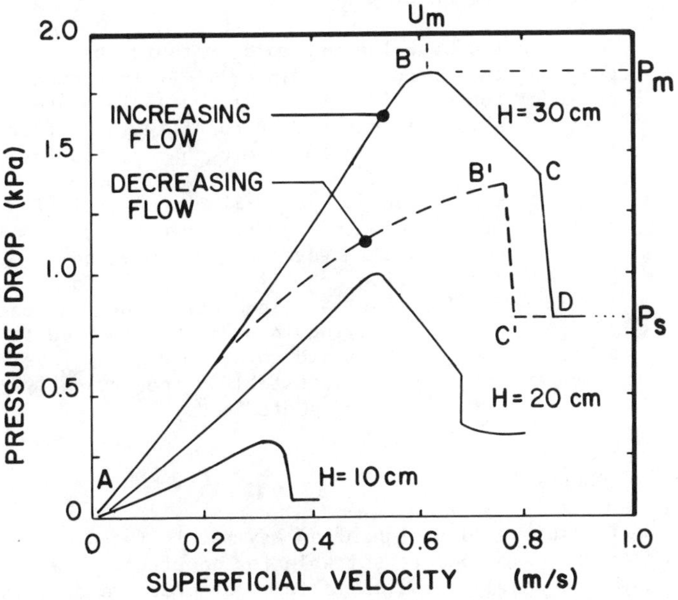

Fig. 4. Characteristic Δp vs. U curve for a 15.2 cm diameter half section circular spouted bed dryer (after Mandonna et al. (8)).

In circular spouted beds the points B and C are not the same as in the 2-D bed. There occurs a gradual pressure drop along BC followed by a sudden drop from C to D at the spouting velocity; whereas, in a 2-D bed the points B and C are coincident, that is to say, that there is no formation of the recognizable internal spout before the onset of spouting. This is due to the larger area ($48\ cm^2$) of the slot compared to the smaller area ($5\ cm^2$) of air inlet associated with the circular spouted bed of about the same grain volumes. A similar observation was made by Rabinovich (5) in his high aspect ratio spouted beds. The research results presented in the literature for the conventional circular bed cover the increasing flow range (ABCD, Fig. 4) up to spouting followed by decreasing flow (DC'B'A). In the present study, velocity was increased beyond spouting point as shown in Fig. 3.

A typical set of curves obtained for oats is shown in Figs. 5 to 7. They correspond to low and high depths (Figs. 5 and 6) and low and high moisture contents operating at the same depth (Figs. 5 and 7). Hysteresis is the main factor which can be observed in the sets of curves.

Spouting pressure drops for varying depths were obtained for the different grains at lower moisture contents (12 - 14% wb). These are shown in Fig. 8 along with their measured bulk density values. For most bed depths, the pressure drop is highest for wheat followed by corn, oats and barley. Variations in particle shape and grain bulk density

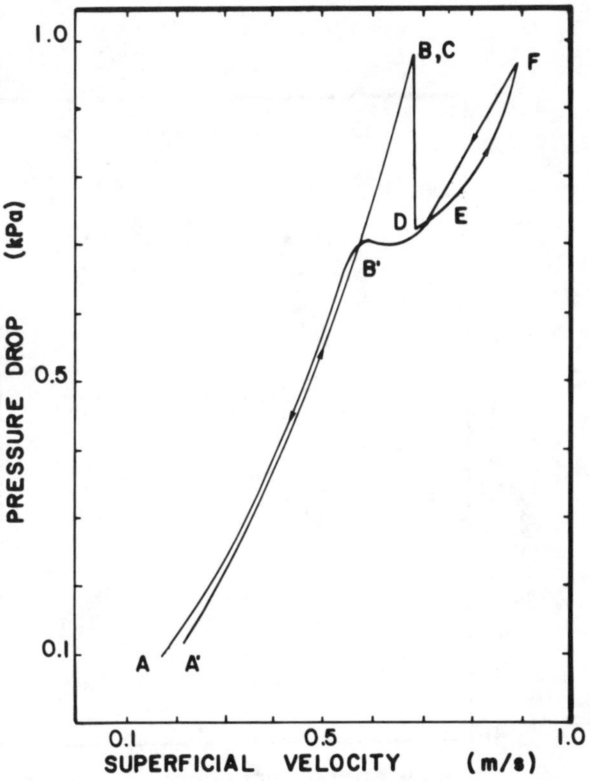

Fig. 5. Total pressure drop across the grain bed vs. superficial velocity for oats at 12% wb (bed depth 15.6 cm).

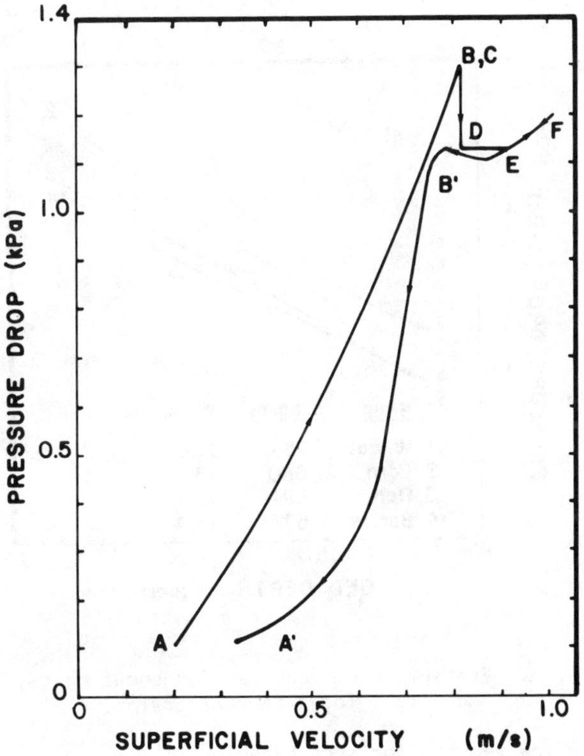

Fig. 6. Total pressure drop across the grain bed vs. superficial velocity for oats 12% wb (bed depth 23.6 cm).

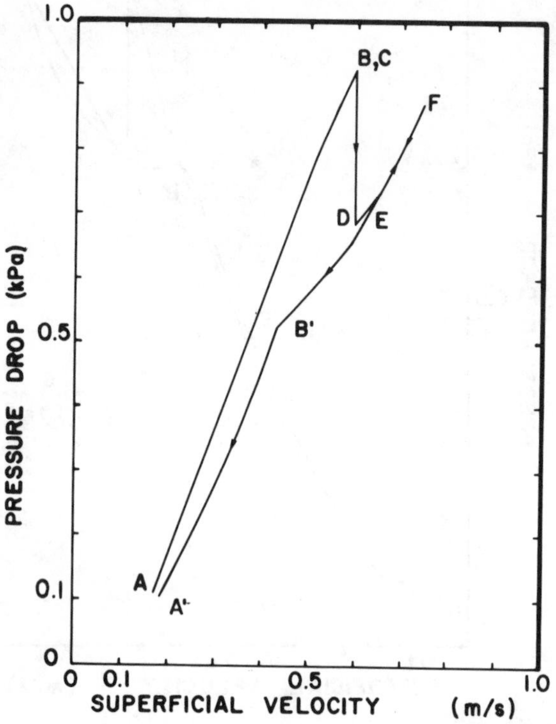

Fig. 7. Total pressure drop across the grain bed vs. superficial velocity for oats at 22.5% wb (bed depth 15.6 cm).

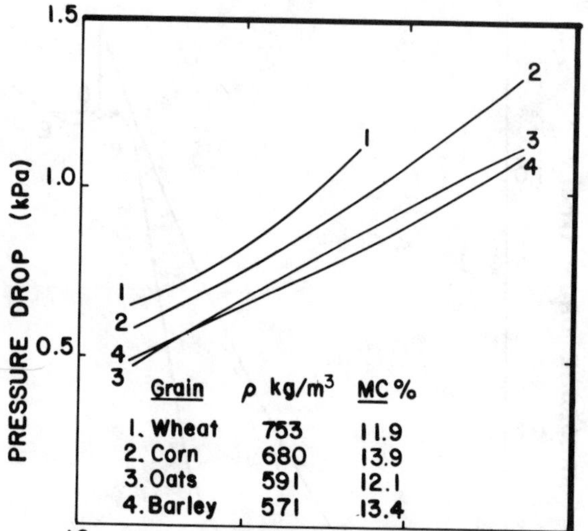

Fig. 8. Pressure drop required for spouting vs. bed height for different grains.

are the reasons for such behavior.

The overall results were analyzed and compiled for different grains. The characteristic curves for each grain are presented in Figure 9. These curves show the effect of low and high depth as well as low and high moisture contents of the grain. It may be seen that for all cases, the high depth and high moisture content plots are characterized by a greater amount of hysteresis between the increasing and decreasing rate curves. This is due to the air channels formed during the transition from a spouting to a static bed which remain open even after cessation of spouting.

The effect of varying moisture contents was studied only for the low depth (< 15 cm). At this depth moisture content did not affect the performance. Further tests are required to quantify the effect of this parameter on aerodynamic characteristics.

The 2-D SB was used for performing some preliminary drying tests. These tests were conducted to quantify the percentage improvement that one can obtain under spouting conditions compared to the static bed. The results using ambient air indicated 50% faster rate of drying in a 2-D SB compared to a static one for wheat and corn. Many more investigations are required to establish the drying characteristics in 2-D spouted beds.

CONCLUSIONS

The suggested design of an asymmetric two-dimensional slot bed is suitable for spouting. The basic characteristics in terms of pressure drop vs. superficial velocity were similar to an asymmetric circular bed. The velocity pressure drop curves displayed hysteresis typical of all spouted beds, but had shapes quite different from those reported earlier for conventional beds. The hysteresis was minimal at low depths. The pressure drop for a given depth was highest for wheat followed by corn, oats and barley. The moisture content did not affect the performance at low depths (< 15 cm). The drying studies indicated that the rate of drying in 2-D SB was 50% faster than in the static bed. Further studies are required in this area.

ACKNOWLEDGEMENT

The financial support of the NSERC through the NSERC summer award is gratefully acknowledged.

REFERENCES

1. Mujumdar, A.S. Proc. Brazilian Conf. on Transport Processes in Particulate System. São Carlos, Brazil (1982).
2. Geldart, D. and Abrahamsen, A.R., Powder Tech. 19: 133-136 (1978).
3. Gishler, P.E. Proc. 2nd Int. Symp. Spouted Beds. Vancouver, B.C. (1982).
4. Mathur, K.B. and Epstein, N., Spouted Beds, Academic Press, N.Y. (1974).
5. Rabinovich. Spouted Beds, Naukova Kumka, U.S.S.R. (1977).
6. Sawyer, R.F., Hart, J.R. and Ohtake, K. Lawrence Berkeley Lab. (LBL-142421),CA.(1982).
7. Swaminathan, R. and Mujumdar, A.S. Drying '84 Hemisphere, N.Y.(1984).

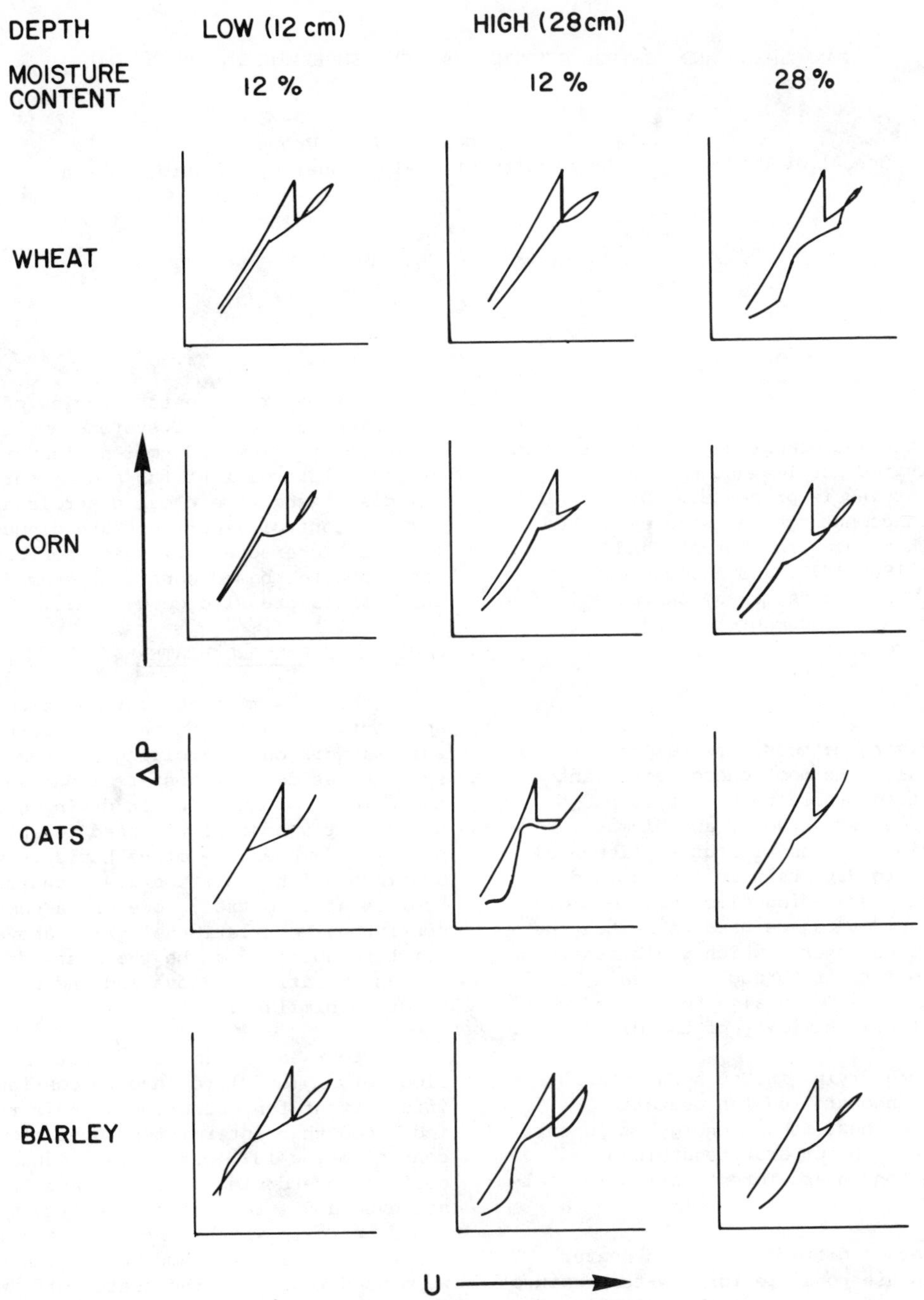

Fig. 9. Typical pressure drop vs. superficial velocity graphs for four grains at varying moisture contents and depths.

8. Mandonna, L.A., Lama, R.F. and Brisson, W.L. Brit. Chem. Eng. 42: 14 (1964).

MANAGEMENT AND CONTROL SYSTEMS FOR LOW TEMPERATURE CORN DRYING

G.S. Mittal, L. Otten and R.B. Brown
School of Engineering, University of Guelph, Guelph, Ontario, Canada

ABSTRACT

A review of the management and control systems employed and evaluated for low and ambient temperature corn drying is presented. Different fan and heater management schemes were evaluated using hourly weather data from Toronto and London for the period of 1965-1978, and a computer simulation model. Design features, characteristics and performance of a microcomputer based control system are also described.

INTRODUCTION

Of all the drying methods, low-temperature drying is potentially the most energy efficient technique and most adaptable to use of such alternate energy sources as solar and biomass. Continuous operation of fans without supplemental heat is sufficient to dry grain in favorable drying weather. Under adverse drying climates, control criteria and mechanisms can be used to switch the fan on and off. Thus, energy which would otherwise be used to force moist air through the grain is saved. Supplemental heat may also be applied to maintain a low relative humidity of the drying air.

Low temperature drying in Southern Ontario using continuous, uncontrolled fan operation without supplemental heat is not energy efficient in comparison with high temperature drying in 62% of the years for the London and Toronto areas (Mittal and Otten 1982).

A large number of different fan and heater management schemes are possible for low-temperature drying, but only those which meet the criteria of optimal energy use and performance are acceptable. Once a suitable set of management strategies is developed, it must be implemented using manual or automatic control of the drying process. In general, the control system should provide supplemental heat to the ambient drying air when the total amount of water removed from the grain bed is negative due to adverse weather conditions. When an excessive amount of heat is needed to produce a useful drying potential, the fan should be stopped. It will, however, be necessary to ensure that the wet grain remains sufficiently cool to prevent excessive spoilage and the fan may have to run periodically to cool the grain.

This paper presents a review of the management and control systems employed and evaluated for low and ambient temperature corn drying. Intermittent fan and heater operations, humidistat and time clock controls along with various control algorithms are discussed. Design features, characteristics and performance of a microcomputer based control system developed by the authors are also described.

CONTROL OF FAN AND HEATER

Table 1 summarizes automatic control applications in low-temperature corn drying. The initial work on controlling low temperature drying systems was done by Foster (1953) while testing the effects of airflows in drying corn. Humidistatic control was used to operate the fan only when ambient relative humidity was below 85%. Continuous fan operation was found superior to the humidistatic control; however, a combination of continuous ventilation of grain above 15.5% moisture appeared to be the most effective method of fan operation. Shove and Andrew (1969) reported similar results.

A ten-season simulation was conducted by Flood et al. (1972) to compare continuous and intermittent fan operation. Their results indicated that intermittent operation at moisture content below 18% (wet basis) reduced operating cost and overdrying. Arnholt and Tuite (1976) achieved a 50% reduction in energy consumption in field tests in which continuous and intermittent heater control were compared. Time clock controls were used to switch the heater off during the hours when the predicted equilibrium moisture content was below the desired final grain moisture content. A second field test involved manual heater control according to the same criterion. A 35% reduction in energy consumption over continuous heater operation was reported.

Simulation studies of Pfost et al. (1977) have shown that a 40% reduction in fan energy can be achieved using humidistat fan control. The scheme used a high airflow rate (2.77 m^3/min. tonne) when the relative humidity was <75%, otherwise, a lower airflow rate (0.14 m^3/min. tonne) was used. The cost of overdrying which can be significant, was not accounted for, so the energy savings by using a humidistat or clock

Table 1: Investigations of the automatic control applications in low-temperature grain drying

Year	Investigators	Type of Experiment	Control Type	Control for	Conditions for Fan Shut-off and Heater on	Energy Savings (%)	Recommended Fan Operation	Other Results and Remarks
1953	Foster	Field	Humidistat	Fan	RH > 85% at 3.2 m^3/min-t	-	Continuous above 15.5% MC	-------
1969	Shove and Andrew	Field	Thermostat and electric eye	Fan	>4.5°C, at night	30	Continuous	Total cost was more for intermittent.
1972	Flood et al.	Simulation	On-off	Fan	MC < 18%	-	Intermittent	Less overdrying and operating cost.
1976	Williams et al.	Field	Clock-timer	Collector-fan	5 pm to 9 am	3180 kWh for 4°C rise	Bin-fan Continuous	i) Prevented negative radiation on clear nights.
1976	Arnholt and Tuite	Field	Time-clock	Heater	EMC < MC	50	--	-------
1977	Pfost et al.	-	Humidistat	Fan	2.77 m^3/min-t for RH < 75% .14 m^3/min-t for RH > 75% fan off for RH > 95%	40	Intermittent	-------
1978 & 1979	Morey et al.	Simulation	Humidistat, thermostat, and time clock	Fan	1) MC at top > 19% continuous 17 to 19%, RH > 90% <17%, RH > 80% 2) RH > 90% at all MC	yes	Continuous with winter shut down	Increased overdrying and dry matter loss.
1978	McLendon and Allison	Field	Humidistat and thermostat (3 stages)	Fan and Heater	RH > 60% and RH > 50%, airflow 5.4 m^3/min-t	-	Intermittent with RH > 50% or continuous	DM loss not considered.
1979	Bauman and Finner	Field	Combination thermostat-humidistat in parallel	Heater	RH > 65% and Temp. < 1.7°C	50	Intermittent	-------
1979	Broder et al.	Field and Simulation	Two humidistats and a counter for heater switch	Heater and Fan	Wet at RH > 80%, Dry at RH < 65%, Temp. <-1.6°C, Switch on when the accumulated hours of wet weather exceeded the hours of dry weather	yes	Intermittent	Less overdrying.
1979	Colliver et al.	Simulation	On-off	Fan	1) RH < 70% -high airflow RH > 70% -low airflow 2) from 9 tc 24h-high airflow from 0 to 9h -low airflow	14 to 49 7 to 29	Intermittent	DM loss not predicted.
1979	Wilcke et al.	Field	Semi programmable, logic integrated circuits	Fan and Heater	Optimized algorithm	27	--	Increased overdrying not successful yet.
1979	Stone & Currelly	Field	Humidistat	Heater	High RH and No Sunshine	Yes	Intermittent	6.65 kWh/t/1% MC
1979	Stevenson	Field	Humidistat	Heater	RH > 75%	Yes	Intermittent	Dried from 21 to 17% MC

control may in practice be small or even non-existent.

Morey et al. (1979) found contrary to Pfost et al. (1977), that with a humidistat control, grain deterioration and over-drying was increased as a result of the delayed movement of the drying front. Continuous fan operation, under an appropriate fall shut off and spring start-up procedure, was considered preferable to management strategies which involve turning off the fan based on relative humidity, temperature, or a time clock.

McLendon and Allison (1978) used solar-heated air during the daylight hours as long as the relative humidity remained below 60%. When the relative humidity exceeded this value the fan was turned off. At night the drying air was heated if the relative humidity exceeded 60%. The energy efficiency of this system offered a significant improvement over conventional drying systems.

Bauman and Finner (1979), Stone and Currelly (1979) and Stevenson (1979) reported favourable results with humidistat control of supplemental heating of the drying air.

Various schemes have been used to implement different energy saving ideas. For example, Colliver et al. (1978 and 1979) developed an optimal management procedure for low temperature drying. A dynamic model first forecasted weather conditions and then chose the optimal mode of operation to minimize energy cost. The optimization was based on a critical path method and several modes of operation were tested; high airflow drying, low airflow aeration, electrical heat supplementation, and solar heat supplementation. Simulation results indicated that their optimal management scheme was superior to all conventional control methods.

Two simpler management schemes using relative humidity and time of the day as the switching parameters were also tested. Switching operation of drying and aeration fans based on relative humidity and time of day indicated 14 to 49% and 7 to 29% reduction in energy, respectively. These schemes were only slightly less effective than the optimal management.

A manual, closed loop control system consisting of an initialization control, fan control, heater control, exit from the drying mode, aeration control, exit from aeration, and fan alarm were developed by Kranzler (1977) using digital electronic circuits. Wilcke et al. (1979) used this system with some modifications, and found an average of 27% reduction in electrical energy consumption compared with a conventional, continuous, low-temperature drying system.

Our review of the literature has shown that the published results are not conclusive and that the recommendations are often conflicting, probably as the result of different climatic and harvest conditions. For example, although simulations have shown that humidistat and time clock control of dryer fans may reduce energy consumption, not all researchers agree that the resulting intermittent fan operation is a good practice. Reasons for this are the increased overdrying and dry matter losses associated with intermittent fan operation.

It is generally accepted that early in the drying season continuous air movement is required to prevent spontaneous heating of the wet grain. Kranzler (1977) suggested that late in the drying season, when grain temperatures are low and the main drying front has passed through the top, the fan and heater could be shut off during periods of high relative humidity, thus preventing rewetting and reducing energy consumption. He proposed shutting off the fan using humidistatic control when the average grain moisture is <18%, the moisture content of the top layer is <22%, the grain temperature of the top layer is <7°C, and the relative humidity is >90%. In addition, Colliver et al. (1978) recommended aeration of the grain when there had been no airflow in the previous 24 h. They also proposed that rewetting should not increase the average moisture content more than 0.025 percentage points per hour.

ALTERNATE MANAGEMENT SCHEMES

As a result of the variations in published results, the authors (Mittal and Otten, 1981) evaluated the following 12 fan and heater management schemes using hourly weather data from London and Toronto. The simulation model developed by Mittal and Otten (1982) was used to simulate the corresponding drying processes:

1. Continuous fan operation with no supplemental heat.
2. Intermittent fan operation (discussed below) without supplemental heat.
3. Continuous fan operation and 1.5^{o}C temperature rise of drying air.
4. Same as (3) except intermittent fan operation.
5. Continuous fan operation and controlled supplemental heat between 1800 h and 0800 h to raise the air temperature by 1.5^{o}C.
6. Same as (5) except intermittent fan operation.
7. Same as (3) except a 3°C temperature rise.
8. Same as (4) except a 3°C temperature rise.
9. Same as (5) except a 3°C temperature rise.
10. Same as (6) except a 3°C temperature rise.
11. Same as (4) except a 4°C temperature rise.
12. Fan and heater control.

Intermittent Fan Operation

The fan was operated continuously in the early part of the drying process and intermittent fan operation was started when the average grain moisture content reached 18%, the maximum grain moisture was 22% and the maximum grain temperature was 7°C. No airflow was provided when the total moisture removal in a time interval of 8 h was $\leqslant$0.001 kg/tonne of grain. The fan was restarted for at least one 8 h time interval when no airflow was provided in the last 24 h.

Fan and Heater Control

This scheme was used to control the fan and the heater when the moisture removed in a specified time interval was less then 0.02 kg/tonne of grain. Air temperature increases for 1.5, 3.0 and 4.0°C were assumed for the simulated heaters. When more than 4°C supplemental heat was required to achieve suitable drying conditions, the fan was turned off, provided that the intermittent fan operation criteria was satisfied.

The simulation results for the 12 management schemes are summarized in Table 2 for Toronto and London. The highest airflow rate is for scheme 1, continuous airflow without supplemental heat, and the lowest value is for scheme 12, fan and heater control. In general, the airflow rates at Toronto are lower than those at London because the ambient temperatures at Toronto tend to be lower at the start of the drying process.

Results from Duncan's multiple range test (Table 2) indicate that for Toronto scheme 5 requires the maximum energy input, while scheme 2, intermittent fan operation without supplemental heat, consumes the least energy. On the basis of energy consumption, schemes other than 2 and 5 do not differ significantly. For London, the most energy inefficient scheme is 9, and the most efficient scheme is 12, fan and heater control. However, the most energy efficient scheme cannot dry corn in unfavorable (humid and hot) years.

At Guelph, which is situated between London and Toronto, about 3.9 MJ/kg is required to dry corn from 27 to 14.5% moisture content employing an efficient high temperature dryer (Otten and Brown 1982). This indicates that significant energy is not saved in many years by using any of the low-temperature drying schemes for drying corn with an initial moisture content of 24% or more. However, many of the schemes were found to save a significant amount of energy for average and favorable years.

The following conclusions can be drawn from the discussion:

1. In favorable years, continuous fan operation without supplemental heat is sufficient to dry corn efficiently.
2. With better management of fan and heater operations, an airflow rate between 29.4 and 32.6 L/(s.m^3) is sufficient to dry corn from 24 to 15% moisture content in Southern Ontario in any year.
3. Addition of supplemental heat during unfavorable years significantly reduced the minimum airflow rates require to dry grain successfully.
4. The probability of getting corn dried in the fall increases with the addition of sufficient heat ($\geq$3°C) to the drying air.
5. For Southern Ontario, supplemental heat is necessary during unfavorable years for bins filled in one operation.
6. None of the management schemes was found to be energy efficient for all the years as compared to high-temperature drying.

MICROCOMPUTER BASED CONTROL SYSTEM

Application of microprocessors is expanding rapidly in agriculture due to the availability of more reliable and less expensive high speed computational devices. For the sensing and control of various processes a microcomputer is a powerful and versatile tool for initiating and processing on-line decisions. Advantages of a microprocessor-based control system over conventional systems include increased precision, increased flexibility of control algorithms and the ability to collect and process data.

For low-temperature grain drying systems, the ambient temperature and relative humidity can be monitored by the microcomputer as a measure of the input conditions. A model of the low-temperature drying process stored in the memory enables the microcomputer to calculate the grain moisture content at assigned locations in the bin. Control decisions can then be based on predicted moisture levels and on predicted or measured grain temperatures.

The authors (Mittal and Otten 1983) developed a microcomputer based low-temperature corn drying system (Figure 1). A North-Star-Horizon**, 48K, quad density, dual disk drive microcomputer was selected for this purpose. Two disc drives with floppy discs were used to store the programs (CP/M). Relative humidity and temperature sensors were interfaced through a TM-AD212-40-PGL analog to digital converter and timer-counter board made by TECMAR Inc., Cleveland, Ohio, USA.

Power to the outlets was switched by solid state relays used to control power input to such inductive loads as motors and such resistive loads as electric heaters. With these relays, the computer was able to control power input to the fans and heater of the low-temperature drying system. System operation required the computer to monitor input data every hour and perform a control operation based on a simulation model and a control algorithm.

The simulation model (Mittal and Otten, 1982) was used in the control system to predict the moisture content at various locations in the grain bin. The moisture distribution data were then used in the control algorithm to make control decisions.

Control Algorithm

Suitability of various algorithms

The following five methods of operating a low-temperature drying system were evaluated for five typical years using the simulation model:

**The mention of firm names or trade products does not imply that they are endorsed or recommended over other firms or similar products not mentioned.

TABLE 2: Average performance parameters and Duncan's multiple range test for different schemes

Scheme	Average airflow rate (L/(s.m^3))		Average specific energy consumption (MJ/kg)		Duncan's test for energy consumption	
	Toronto	London	Toronto	London	Toronto	London
1	23.5	26.1	3.4	4.2	A/B*	A/B
2	22.5	25.1	3.0	3.7	B	A/B
3	22.6	20.4	3.7	4.6	A/B	A/B
4	22.2	20.9	3.5	4.2	A/B	A/B
5	23.2	22.2	4.8	4.5	A	A/B
6	22.7	22.7	4.3	4.0	A/B	A/B
7	21.7	21.5	4.0	4.3	A/B	A/B
8	20.4	22.1	3.8	4.1	A/B	A/B
9	21.5	22.7	3.7	4.7	A/B	A
10	21.8	23.0	3.6	4.3	A/B	A/B
11	21.0	23.4	3.9	4.0	A/B	A/B
12	20.9	18.9	4.6	3.5	A/B	B

*If two groups are connected by a series of the same letters, then they are not significantly different at a confidence level of 0.95. The order of letters is in the direction of decreasing magnitude.

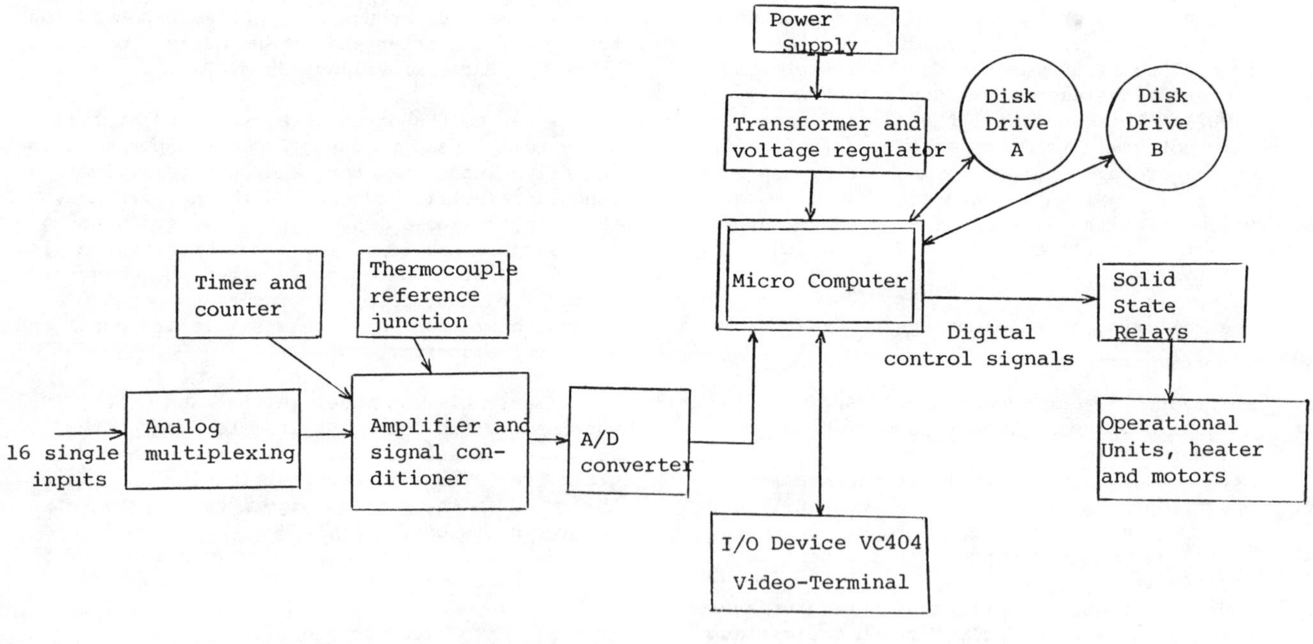

Figure 1: Schematic diagram of microcomputer control system

I Broder's criterion with a 2°C temperature rise of the drying air or aeration.
II Broder's criterion with a 1, 2 or 3°C temperature rise of the drying air or aeration.
III Continuous fan operation without supplemental heat.
IV Continuous fan operation with supplemental heat to give a 1.5°C temperature rise.
V Continuous fan operation with controlled supplemental heat (1800 to 0800) to give a 1.5°C temperature rise.

The first method involved increasing the temperature of the drying air by 2°C when supplemental heat was needed; however, if this temperature rise was insufficient to obtain a relative humidity of the drying air less than 80%, the drying fan was stopped and the aeration fan was started. The second method was the same except that three successive temperature increases were tried to obtain a relative humidity below 80%.

Simulation results show that the first method failed to dry grain in the average year without grain deterioration, while the second method was successful in every year. However, more energy was required in average and favorable drying years than with the uncontrolled method (III).

Method III failed to dry grain in unfavorable years due to excessive dry matter loss. On the other hand, methods IV and V dried grain successfully, but the corresponding energy consumptions were greater than those for high-temperature drying. Therefore, these five methods are not energy efficient under all drying conditions.

Colliver's control criterion was examined by assuming that the following five drying modes were available to the control system: (i) air-flow without supplemental heat; (ii) aeration at 2.5 L/(s.m^3); (iii) airflow with supplemental heat to raise the air temperature by 1.5°C; (iv) airflow with supplemental heat to raise the air temperature by 2.5°C; (v) airflow with supplemental heat to raise the air temperature by 4.0°C.

The simulation results indicated that grain can be dried without excessive deterioration in three out of five years. In the remaining two years the combination of Colliver's criterion and the five drying modes did not dry the grain successfully because the three modes involving supplemental heat were not used by the control scheme. Instead, frequent selection of the aeration mode resulted in excessive dry matter losses.

Simulation work using the Wilcke's criterion showed that no significant energy savings were obtained when compared with the uncontrolled drying method.

New control algorithm

A new control algorithm was developed after a detailed study of the various factors influencing the low-temperature drying process (Mittal and Otten, 1983). The control algorithm was based on five indices which should be specified before drying is started:

i) RHX: relative humidity index to control the fan.
ii) RHAX: relative humidity index to start searching for alternatives other than continuous fan operation without supplemental heat.
iii) RHXF: relative humidity index to control the heater.
iv) TRX: initial time period for which continuous fan operation is acceptable.
v) TLMCX: moisture content in the upper 10% of the bin which is used to control the aeration fan.

These indices are used in making the following decisions and control steps:

i) If the relative humidity of the air, RHA $\leqslant$ RHX, or the total drying time, TR < TRX, the drying fan is on but the heaters and aeration fan remain off.
ii) If RHX < RHA $\leqslant$ RHAX, the heaters are turned on to decrease RHA to, or below, RHXF.
iii) If RHA > RHAX or the heaters are unable to decrease RHA to, or below, RHXF, and TR $\geqslant$ TRX two alternate decisions are made depending on the moisture content of the grain (TLMC) in the upper 10% of the bin; namely,
    a) if TLMC > TLMCX, the heater and drying fan are shut off and the aeration fan is started.
    b) if TLMC $\leqslant$ TLMCX, all fans and heaters are shut off.

In addition to this, the new autumn shut off criterion used a shut-off date of December 31 when the grain temperature at the top 10% of the bin remained at about 0°C and the air temperature did not fall below -4°C (Mittal and Otten, 1983).

The optimum values of the indices were selected, after a sensitivity analysis of the control algorithm to variations in the indices, providing minimum energy consumption and over-drying, and an acceptable amount of dry matter deterioration. The optimum values of the indices for London conditions are: RHX = 0.8, RHXF = 0.7, RHAX = 0.9, TRX = 336 and TLMCX = 0.18. Similar values can be calculated from climatological data for other locations.

Simulation results of the new control algorithm indicate that the new algorithm results in the lowest energy consumption and a dry matter deterioration of less than 0.5% in all five typical drying years. Furthermore, the new control criterion requires less energy consumption than the 3.9 MJ/kg observed for high temperature drying (Otten and Brown, 1982).

Hence, the microcomputer control can save energy from 5 to 31% compared with high-temperature drying, and from 10 to 19% compared with uncontrolled low-temperature drying in favorable years. The results also indicate that providing controlled supplemental heat by increasing the air temperature by 2°C provided drying under adverse climatic conditions. The new control algorithm decreased the overdrying amount in comparison with the continuous fan operation with no supplemental heat.

The abovementioned system was field tested in 1981 at Arkell Research Station near Guelph. A dedicated, small and compact unit is at development stage.

CONCLUSIONS AND DISCUSSIONS

The performance of low-temperature drying depends on many factors such as ambient temperature and relative humidity, airflow rate and depth of grain bed, management of filling, and fan and supplemental heat management policy. All these factors are important, but poor management and control policies can result in serious grain spoilage or even total loss. A selection of optimum management policies will require results of field and simulation tests under varying drying conditions. Since management policies are weather dependent, no single policy is suitable under various drying conditions. Fan and supplemental heat are required to be controlled based on prevailing weather conditions. Because the calculations needed to determine the desired control actions are complex and must be performed frequently throughout the day, automatic control is more suitable.

Electromechanical or solid state controls can be designed to handle almost any specific control problem, but a minor change in control strategy will usually require changes in hardware. On the other hand, a microprocessor-based control system would allow many complex control strategies to be implemented simply by changes in software.

REFERENCES

1. Arnholt, D.J. and J. Tuite. 1976. Mold study results with electric drying - a progress report. Presented at the 1976 Farm Electrification Conference. Purdue Univ., West Lafayette, USA.

2. Bauman, B.S. and M.F. Finner. 1979. Reducing energy costs in a solar corn drying system. Pap. No. 79-3025. Amer. Soc. Agric. Eng., St. Joseph, MI., USA.

3. Broder, M.F., G.H. Foster and K.D. Baker. 1979. Management and control of solar collector-heat storage grain drying systems. Pap. No. 79-3529. Amer. Soc. Agric. Eng., St. Joseph, MI., USA.

4. Colliver, D.G., R.C. Brook and R.M. Peart. 1978. Optimal management procedures for solar grain drying. Pap. N. 78-3512. Amer. Soc. Agric. Eng., St. Joseph, MI., USA.

5. Colliver, D.G., R.M. Peart, R.C. Brook and J.R. Barrett. 1979. Comparison of minimal energy usage management procedures for low temperature grain drying. Pap. No. 79-3027, Amer. Soc. Agric. Eng., St. Joseph, MI., USA.

6. Flood, C.A., M.A. Sabbah, D. Mecker and R.M. Peart. 1972. Simulation of a natural air corn drying system. Trans. of Amer. Soc. Agric. Eng. 15:156.

7. Foster, G.H. 1953. Minimum airflow requirements for drying grain with unheated air. Agric. Eng. 34:681.

8. Kranzler, G.A. 1977. A digital electronic control system for low temperature drying of corn. Ph.D. Dissertation. Iowa State Univ., Ames, USA.

9. McLendon, B.D. and J.M. Allison. 1978. Solar energy utilization in alternate grain drying systems in the Southeast. Pap. No. 78-3013. Amer. Soc. Agric. Eng., St. Joseph, MI., USA.

10. Mittal, G.S. and L. Otten. 1981. Evaluation of various fan and heater management schemes for low temperature corn drying. Can. Agric. Eng. 23:97.

11. Mittal, G.S. and L. Otten. 1982. Simulation of low-temperature corn drying. Can. Agric. Eng. 24:111.

12. Mittal, G.S. and L. Otten. 1983. Microprocessor controlled low-temperature corn drying system. Agric. Systems 10:1.

13. Morey, R.V., H.M. Keener, T.L. Thompson, G.M. White and F.W. Bakker-Arkema. 1978. The present status of grain drying simulation. Pap. No. 78-3009. Amer. Soc. Agric. Eng., St. Joseph, MI., USA.

14. Morey, R.V., H.A. Cloud, R.J. Gustafson and and D.W. Petersen. 1979. Management of ambient air drying systems. Trans. of Amer. Soc. Agric. Eng. 22:1418.

15. Otten, L. and R.B. Brown. 1982. Low temperature and combination corn drying in Ontario. Can. Agric. Eng. 24(1):51.

16. Pfost, H.B., S.B. Maurer, L.E. Grosh, D.S. Chung and G.H. Foster. 1977. Fan management systems for natural air dryers. Pap. No. 77-3526, Amer. Soc. Agric. Eng., St. Joseph, MI., USA.

17. Shove, G.C. and F.W. Andrew. 1969. Cooling, chilling, dehydrating stored shelled corn. Agric. Eng. 50:360.

18. Stevenson, K. 1979. Performance measurements for low temperature drying. Presented at the On-Farm Grain Drying Conference, Univ. of Guelph, Guelph, Ontario, Canada.

19. Stone, R.P. and J.C. Currelly. 1979. Experiences with a solar assisted low temperature grain dryer. Pap. No. 79-3020, Amer. Soc. Agric. Eng., St. Joseph, MI., USA.

20. Wilcke, W.F., G.A. Kranzler and C.J. Bern. 1979. Digital electronic control system for a low temperature corn dryer. Pap. No. 79-3028, Amer. Soc. Agric. Eng., St. Joseph, MI., USA.

21. Williams, E.E., M.R. Okos, R.M. Peart and A.F. Badenhop. 1976. Solar grain drying and collector evaluation. Pap. No. 76-3512, Amer. Soc. Agric. Eng., St. Joseph, MI., USA.

CONVECTIVE HEAT TRANSFER BETWEEN A FLAT PLATE AND A JET OF HOT GAS IMPINGING ON IT

K. Kataoka, H. Shundoh, and H. Matsuo

Department of Chemical Engineering
Kobe University
Rokkodai, Kobe 657, Japan

ABSTRACT

The characteristics of free jet development and impingement heat transfer are experimentally studied for the case when a free jet of high-temperature gas issuing from a circular convergent nozzle into low-temperature quiescent air impinges normally on a circular flat plate cooled from the back. A new concept of jet development is introduced taking into account the effect of jet-to-ambient fluid density ratio. A generalized model is proposed, which permits local prediction of wall heat-fluxes over a wide range of the initial density ratio.

NOMENCLATURE

a, b, c	constants of Equation (1), K, K/m, K/m^2
D	nozzle exit diameter, m
H	nozzle-to-plate spacing, m
h	heat transfer coefficient, W/m^2K
Nu	Nusselt number, —
Nu_o	$= h_s D/\kappa_s$
Nu_s	$= 2h_s r_{1/2U}/\kappa_s$
p	static pressure, Pa
Δp	pressure difference $= p - p_\infty$, Pa
Δp^*	$= (p - p_\infty)/(p_s - p_\infty)$
Pr	Prandtl number, —
Pr_o	$= \nu_o/\alpha_o$
Pr_s	$= \nu_s/\alpha_s$
q	wall heat-flux, W/m^2
q^*	$= q/q_s$
r	radial distance from the jet axis, m
r^*	$= r/r_{1/2}$
$\overline{r}$	$= r/D$
$r_{1/2}$	jet half-radius, m
$\overline{r}_{1/2}$	$= r_{1/2}/D$
Re	Reynolds number, —
Re_o	$= U_o D/\nu_o$
Re_s	$= 2U_m r_{1/2U}/\nu_s$
T	temperature, K or °C
ΔT	temperature difference $= T - T_\infty$, K
ΔT^*	$= (T - T_\infty)/(T_m - T_\infty)$
$\overline{\Delta T}_m$	$= (T_m - T_\infty)/(T_o - T_\infty)$
$\overline{\Delta T}_{bs}$	$= (T_{bs} - T_\infty)/(T_o - T_\infty)$
Tp	temperature core length, m
$\overline{Tp}$	$= Tp/D$
U	axial velocity, m/s
U^*	$= U/U_m$
$\overline{U}_m$	$= U_m/U_o$
U_r	radial velocity outside the boundary layer along a flat plate, m/s
Vp	velocity core length, m
$\overline{Vp}$	$= Vp/D$
X	mole fraction of CO_2, —
y	distance in heat-flux gauge from the heat transfer surface, m
Z	axial distance, m
$\overline{Z}$	$= Z/D$

Greek Letters

α	thermal diffusivity, m^2/s
$\overline{\alpha}$	$= \alpha/\alpha_o$
δ	thickness of momentum boundary layer, m
κ	thermal conductivity, W/m K
$\overline{\kappa}$	$= \kappa/\kappa_o$
ν	kinematic viscosity, m^2/s
ρ	fluid density, kg/m^3
ρ_i	initial density ratio $= \rho_\infty/\rho_o$, —

Overline and Superscript

—	dimensionless quantity with respect to nozzle exit value
*	dimensionless quantity with respect to jet axis value or jet half-radius at an axial position

Subscripts

b	in the bulk fluid
c	copper plug
m	on jet axis
o	at nozzle exit
p	for pressure
q	for wall heat-flux
r	radial direction
s	at stagnation point
T	for temperature
U	for velocity
w	at the wall of flat plate
y	y-direction
1/2	jet half-radius
∞	in the ambient surroundings

INTRODUCTION

The research that will be described in this paper is a general study of the convective heat transfer that results when a free jet of gas much hotter than the surrounding quiescent gas impinges normally on a plane surface cooled from the back.
Impinging jets of this kind are of great importance because of their practical applications for heat and mass transport processes such as drying of paper and plastic film, glass toughening, and annealing of metal sheet.

When a submerged, round free jet issuing from a circular convergent nozzle impinges normally on a flat plate, the flow field consists of four separate regions shown in Figure 1.

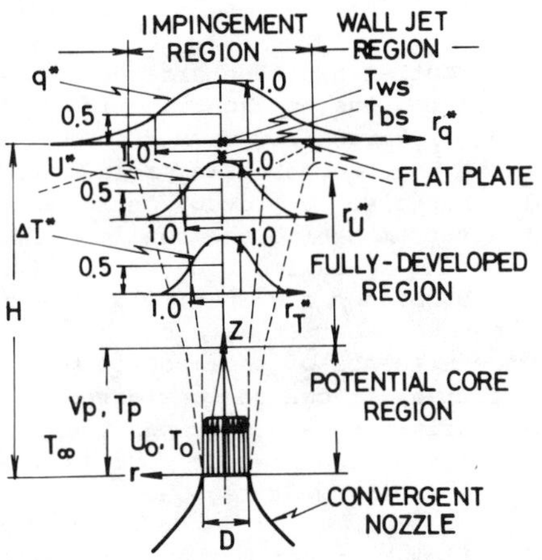

Figure 1 FLOW CONFIGURATION AND THE COORDINATES TO BE USED.

The first region encountered just below the nozzle exit is the potential core region in which the centerline velocity and temperature remain unchanged from their initial values. After a short transition region, the fully-developed free jet region is established, where the centerline quantities decrease and the jet width increases linearly with increasing axial distance. This fully-developed region ends at the entrance to the impingement region where the jet flow is deflected.
Heat transfer occurs between the deflected flow and the surface of the flat plate. The radial flow is accelerated there due to negative gradients of the surface pressure produced by the jet impingement. The wall jet region is formed downstream of the impingement region after the pressure gradients disappear.

The turbulence of axial velocity first increases from a small initial value with axial distance, and gradually decreases in the fully-developed region. In between, maximum values occur just below the end of the potential core. The turbulence is further amplified owing to the jet impingement.

The impingement heat transfer is significantly enhanced by the so-called free-stream turbulence. In order to attain a maximum rate of heat transfer, therefore, a heat transfer surface should be placed at the position where the centerline axial velocity remains almost unchanged from its initial value although its turbulence reaches a maximum.

For isothermal jets, the optimum nozzle-to-plate spacing lies between six and eight nozzle diameters (1 - 5). By using submerged impinging jets of aqueous electrolytic solution, Kataoka and Mizushina(6), Kataoka, Komai, and Nakamura(7), and Kataoka, Kamiyama, Hashimoto, and Komai(1) successfully observed local values of the velocity gradient and mass transfer rate on the surface of a flat plate. The stagnation-point mass transfer reaches a maximum when the flat plate is placed at six nozzle diameters from the convergent nozzle. Only the mass transfer is enhanced owing to the velocity turbulence in the wall region of the momentum boundary layer, whereas the momentum transfer is insensitive to such wall turbulence. It has been found that large-scale eddies produced owing to the disintegration of the ring-shaped vortices play a leading role in the production of the velocity turbulence in the wall region where the mass transfer takes place.

The jet development of high-temperature gas exhausting into quiescent low-temperature gas is greatly influenced by the initial density ratio $\bar{\rho}_i$; as the gas temperature at the nozzle exit is raised, the potential core does not only become short, but the decay and spreading rates of the jet are also increased owing to the compressible turbulent mixing. An important problem arises where the heat transfer surface should be placed for nonisothermal jets to attain the largest impingement heat transfer.

The problem should be considered by two steps: [1] the development of nonisothermal free jet and [2] the heat transfer in the impingement region. Regarding the compressible jet development, there are several turbulent mixing models:/1/ the turbulent mixing coefficient model(9, 10), /2/ the dynamic eddy transfer coefficient model(11 - 16), and /3/ the extended Reichardt free-mixing model (17). The present authors(15, 16) successfully extended the Prandtl's eddy diffusivity model to nonisothermal turbulent jets by taking into account the change in volume of traveling eddies due to large thermal gradients. Their generalized model indicates that all the eddy transfer coefficients for momentum, heat, and chemical species are proportional to the square root of the initial density ratio in the fully-developed region.

At the present stage, however, it is practically more fruitful to establish a simpler model of the jet development applicable over a wide range of the initial density ratio. The present authors (18) proposed a practically useful model for prediction of the velocity and temperature variations in nonisothermal free jets at various temperature levels.

There are some models proposed for prediction of stagnation-point heat/mass transfer for the case of isothermal jets(19 - 21). Donaldson, Snedeker, and Margolis(19) proposed a turbulence correction factor model, and Chia, Giralt, and Trass(21) improved its turbulence correction term by considering the turbulence intensity of the axial velocity at the beginning of the impingement region. Their models can also predict local heat/mass transfer coefficients by applying the turbulence correction factor to the laminar values theoretically obtained for turbulence-free condition. However, there is little work on nonisothermal jet problems, in which the impinging flow and heat transfer must be a function of the initial density ratio as well as of the nozzle-to-plate spacing and the jet Reynolds number.

EXPERIMENT

The general arrangement of experimental apparatus is shown in Figure 2. It consists of a stainless steel nozzle assembly, a horizontal circular flat plate, and a measuring system with a traversing device for thermocouple and total-head tube.

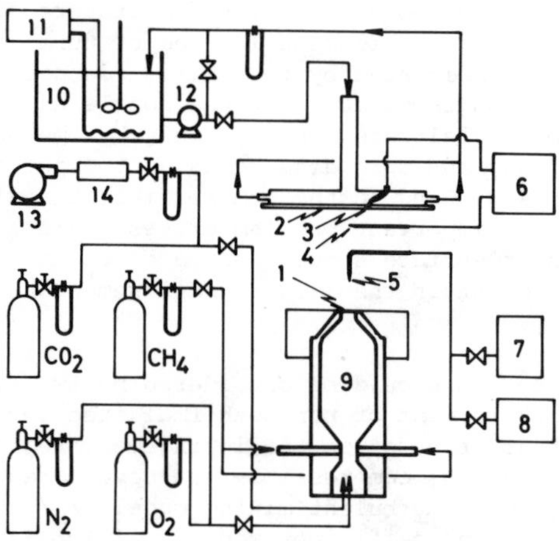

Figure 2 GENERAL ARRANGEMENT OF EXPERIMENTAL APPARATUS. (1: convergent nozzle, 2: flat plate, 3: heat-flux gauge, 4: Pt/PtRh thermocouple, 5: total-head tube, 6: multichannel mV recorder, 7: Goettingen manometer, 8: gas chromatograph, 9: combustion chamber, 10: constant-temperature water tank, 11: temperature controller, 12: pump, 13: compressor, 14: mist eliminator)

The nozzle assembly, shown in Figure 3, consists of a burner having a venturi throat for intimately mixing methane gas with an oxidizing stream, a 60 mm-ID combustion chamber with a straight section 135 mm long, and a 10 mm-ID convergent nozzle with a contraction ratio of 1/36. The inside surface of the nozzle has been carefully machined so as to make the exit velocity profile as uniform as possible.

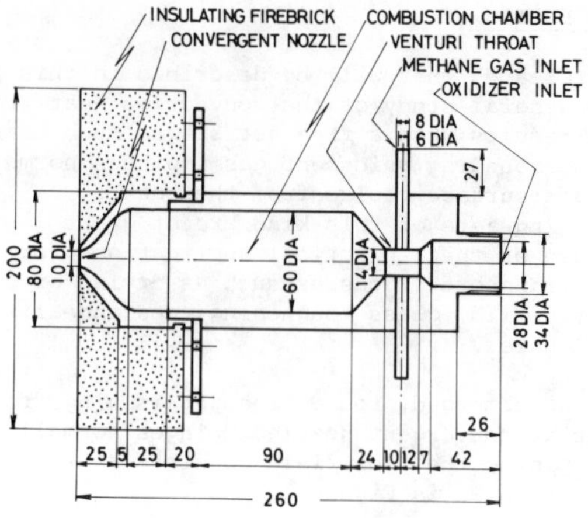

Figure 3 NOZZLE ASSEMBLY. DIMENSIONS GIVEN ARE IN MM.

The nozzle outlet has been arranged flush with, and central to, the horizontal surface of an insulating firebrick(200 mm x 200 mm) not only to provide the exit temperature profile as uniform as possible, but also to prevent entrainment of the ambient air from behind the nozzle. This nozzle belongs to the so-called blunt edged nozzle classified by Donaldson and Gray(10). The jet issuing from it can be regarded as a semi-confined jet(22, 23). The influence of semi-confinement on impingement heat transfer can be considered to be small for the nozzle-to-plate spacings examined because the attention of the present work is directed to the central impingement region.

The molar flow rates of methane, oxygen, and nitrogen are separately regulated at a ratio of $CH_4 : O_2 : N_2 = 1 : 3.3 : 5.2$ to make the nitrogen/oxygen ratio of the burned gas equal to that of the ambient air. For simplicity, the physical properties are determined from those of air since the average molecular weight of the burned gas is approximately equal to that of air. A hot jet of the burned gas produced in the combustion chamber is exhausted vertically upward from the nozzle into the quiescent air at room temperature and impinges normally on a horizontal flat plate. According to the previous work(15), the effects of free convection and thermal radiation are of negligible importance on the jet development as well as the thermal measurementation in the present experimental condition. It has also been confirmed by means of gas chromatograph that no burning mixture issues from the nozzle owing to the excess oxidizer supplied. Gas temperature T is measured directly by a 0.1 mm-diameter Pt/PtRh thermocouple. The error due to the thermal effect of radiation is confirmed to be within 2% by comparison with a suction thermometer used in the previous work(15). Axial velocity U is calculated from the dynamic pressure $(1/2)\rho U^2$ measured by a miniature total-head tube of 1.4 mm-OD quartz tube with the aid of the observed temperature distribution. These probes can be placed at any position (r, Z) by means of the traversing device with accurate lead screws.

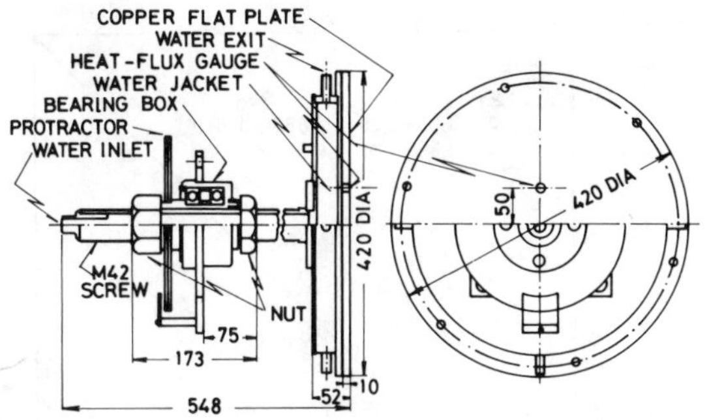

Figure 4 CIRCULAR FLAT PLATE FOR JET-IMPINGEMENT HEAT TRANSFER EXPERIMENT. DIMENSIONS GIVEN ARE IN MM.

Figure 4 shows the circular flat plate made of copper used as the heat transfer surface.
The flat plate has a water jacket on the back side to make heat flow from the front to the back.
The supporting axis of the flat plate having an accurate male screw outside to change the nozzle-to-plate spacing serves as a water inlet pipe.
Local measurement of wall heat-flux q and wall temperature T_w is accomplished on the surface of the circular flat plate by means of a plug-type heat-flux gauge mounted flush with the surface.
The flat plate can also be rotated at a fixed nozzle-to-plate distance by using the bearing box.
The axis of the convergent nozzle is fixed 50 mm away from the rotation axis of the flat plate, so that a 50 mm diameter circle can be formed on the flat plate as the locus of geometric stagnation point by rotation. A 10.20 mm-ID hole drilled at a point on the circle has been precisely finished by means of reamer. An about 10.21 mm-OD cylindrical plug-type heat-flux gauge made of copper, shown in Figure 5, has been mounted into the reamed hole by shrinkage fit, so that contact resistance around the cylindrical surface of the gauge can be neglected. Three 0.3 mm-diameter constantan leads, inserted in 0.9 mm-diameter drilled holes, independently form three hot junctions of C/C thermocouple within the copper plug.

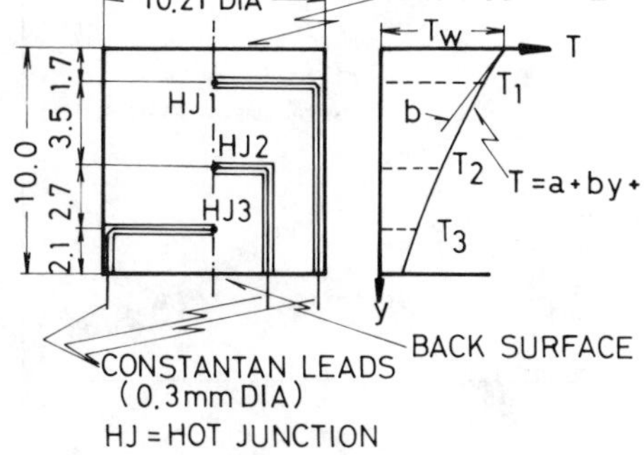

Figure 5 CIRCULAR PLUG HEAT-FLUX GAUGE. DIMENSIONS GIVEN ARE IN MM.

A quadratic approximate equation of temperature distribution is obtained by using three hot junction temperatures:

$$T = a + b\,y + c\,y^2 \tag{1}$$

where these constants a, b, c are a function of the radial position of the heat-flux gauge, respectively.
The wall heat-flux and wall temperature can be evaluated by the equations:

$$q = \kappa_c \left.\frac{\partial T}{\partial y}\right|_{y=0} = b\,\kappa_c \tag{2}$$

$$T_w = T\big|_{y=0} = a \tag{3}$$

The effect of lateral conduction in the plug can also be estimated changing the radial position of the gauge by rotation:

$$\frac{1}{r}\frac{\partial}{\partial r}(r\,q_r) + \frac{\partial q_y}{\partial y} = 0$$

or

$$(r\,q_r)\big|_{r+\Delta r} - (r\,q_r)\big|_{r} = 2\,c\,\kappa_c r\,\Delta r \tag{4}$$

As will be shown later, the surface of the flat plate is kept almost uniform in wall temperature in spite of the radial variation of the bulk gas temperature. Two supplementary experiments are conducted; the development of isothermal free jets is examined around $\bar{\rho}_i = 1$ by exhausting gas mixtures of CO_2 and air into air from the same nozzle. Another isothermal air jet is made striking an electrically heated Fe-Ni alloy foil stuck on an insulating flat plate for comparison with the isothermal jet data of the other investigators.

Experimental conditions are listed in Table 1. At high temperatures, the jet Reynolds number cannot be made very high owing to high viscosity and low density of the burned gas. As will be pointed out in the discussion, the present results are in satisfactory agreement with the previous investigations conducted at higher Reynolds numbers and much lower initial density ratios if the generalized model to be presented here is adopted.

Table 1 EXPERIMENTAL CONDITIONS

[I] FREE JET EXPERIMENT

I-1 HEATED JET

U_o	= 7.4 ~ 22.7	[m/s]
T_o	= 550 ~ 1260	[°C]
T_∞	= 15.0 ~ 29.0	[°C]
$\bar{\rho}_i$	= 2.73 ~ 5.26	[—]
Re_o	= 857 ~ 1810	[—]

I-2 ISOTHERMAL CO_2/AIR AND AIR/AIR JETS

U_o	= 5.76 ~ 12.55	[m/s]
$T_o = T_\infty$	= 13.5 ~ 24.0	[°C]
X_o	= 0 ~ 1.0	[—]
$\bar{\rho}_i$	= 0.66 ~ 1.0	[—]

Re_o = 4900 ~ 8300 [—]

[II] IMPINGING JET EXPERIMENT

II-1 HEATED JETS

U_o	= 7.4 ~ 45.1	[m/s]
T_o	= 550 ~ 1450	[°C]
T_∞	= 13.5 ~ 29	[°C]
$\bar{\rho}_i$	= 2.73 ~ 5.99	[—]
Re_o	= 857 ~ 2000	[—]
$\bar{H}$	= 2 ~ 20	[—]

II-2 UNHEATED JETS

U_o	= 7.4 ~ 14.4	[m/s]
$\bar{\rho}_i$	= 1.0	[—]
Re_o	= 4900 ~ 10000	[—]
$\bar{H}$	= 3 ~ 12	[—]

EXPERIMENTAL RESULTS AND DISCUSSION

[1] Development of nonisothermal free jet

The centerline quantities U_m, ΔT_m and jet widths $r_{1/2U}$, $r_{1/2T}$ can be adopted as the characteristic variables of this flow system.
The initial radial distributions of axial velocity and temperature have been confirmed to be uniform over about 90% of the cross section of the nozzle exit. The first step in analyzing the jet development is to correlate the core lengths measured at various temperature levels. The core lengths Vp, Tp for velocity and temperature are determined both from the centerline decay curves of U_m and ΔT_m and from the jet spreading curves of $r_{1/2U}$ and $r_{1/2T}$, respectively. As will be seen later from Equation (9), for example, the velocity core length Vp is evaluated as the axial length of the horizontal, core-region line of $1/\bar{U}_m = 1.0$ intercepted by the inclined straight line of $1/\bar{U}_m$ vs. $\bar{Z}$ in the fully-developed region.
The core lengths tend to decrease with the initial density ratio raised. Their correlative equations have been obtained by the method of least square:

$$\overline{Vp} = 2.82\ \bar{\rho}_i^{-0.29}\ Re_o^{0.07} \qquad (5)$$

$$\overline{Tp} = 3.80\ \bar{\rho}_i^{-0.45}\ Re_o^{0.03} \qquad (6)$$

The comparison between the best-fit line and experiment shown in Figures 6 and 7 indicates satisfactory agreement within 20% error over a wide range of initial density ratio, i.e. $\bar{\rho}_i$ = 0.66 ~ 5.3.
As the jet Reynolds number increases, the core lengths increase slightly. This trend is similar to that of Witze's model(14) for high-velocity compressible jet.

The normalized radial distributions of jet velocity and temperature, shown in Figure 8, are respectively similar at various axial positions in the fully-developed region.

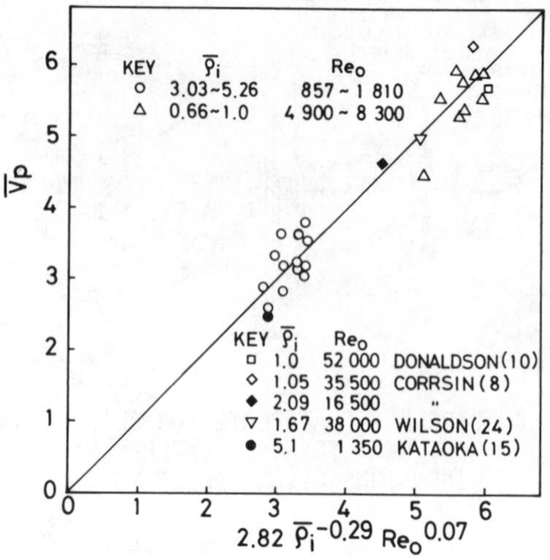

Figure 6 CORRELATION OF VELOCITY CORE LENGTHS.

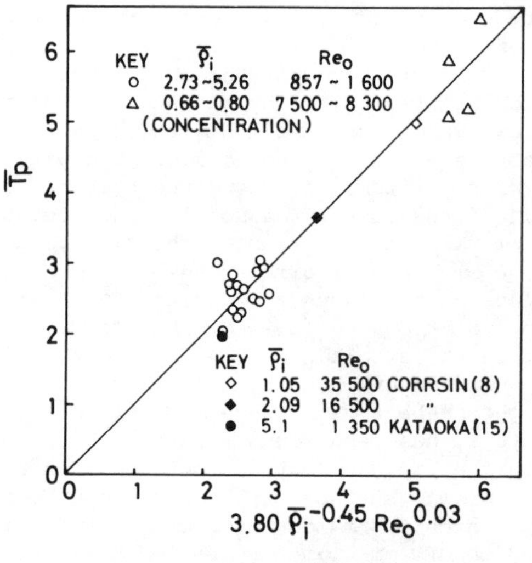

Figure 7 CORRELATION OF TEMPERATURE CORE LENGTHS.

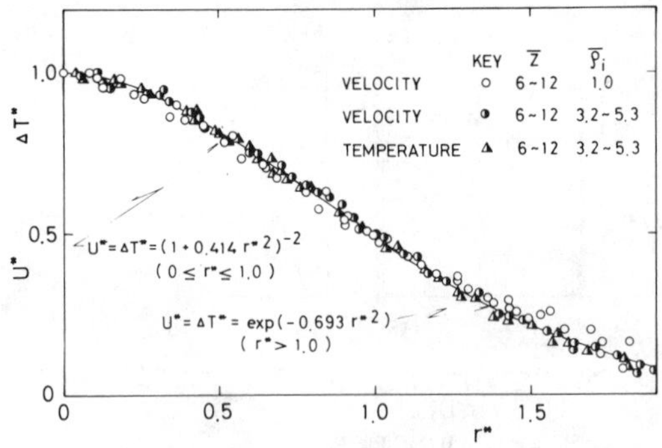

Figure 8 RADIAL DISTRIBUTIONS OF NORMALIZED VELOCITY AND TEMPERATURE IN THE FULLY-DEVELOPED REGION.

The best-fit curve connecting the experimental data points in Figure 8 can be expressed by Equation (7) in the inner jet region and Equation (8) in the outer intermittent region, respectively (15):

$$U^* = (1 + 0.414\, r_U^{*2})^{-2} \qquad (0 \le r^* \le 1) \qquad (7)$$
$$\Delta T^* = (1 + 0.414\, r_T^{*2})^{-2}$$

$$U^* = \exp(-0.693\, r_U^{*2}) \qquad (r^* > 1) \qquad (8)$$
$$\Delta T^* = \exp(-0.693\, r_T^{*2})$$

These equations are mathematically similar in form to those for unheated or isothermal jets.

The reciprocals of the centerline quantities $1/\overline{U}_m$ and $1/\Delta\overline{T}_m$ increase linearly with axial distance $\overline{Z}$. The decay rates of $\overline{U}_m$ and $\Delta\overline{T}_m$ tend to increase with $\overline{\rho}_i$. The previous investigators (12, 17, 24, 25) proposed $(\overline{\rho}_i)^{1/2}\,\overline{Z}$ as the characteristic streamwise coordinate. However, the effect of $\overline{\rho}_i$ on core lengths $\overline{Vp}$ and $\overline{Tp}$ should be taken into account in the definition of the jet development length.

Figures 9 and 10 show the axial decay of the centerline quantities with the jet development length. The best-fit equations are

$$1/\overline{U}_m = 0.16\, \overline{\rho}_i^{\,1/2}\, (\overline{Z} - \overline{Vp}) + 1.0 \qquad (9)$$

$$1/\Delta\overline{T}_m = 0.20\, \overline{\rho}_i^{\,1/2}\, (\overline{Z} - \overline{Tp}) + 1.0 \qquad (10)$$

The constant 1.0 comes from the definition of core lengths. The streamwise coordinate is similar to that used by Sforza(17). The only difference is that the present authors adopt the core length in place of virtual origin because the core length correlations are practically more useful.

The jet half-radii $r_{1/2U}$ and $r_{1/2T}$ increase linearly with axial distance and the spreading rate of the jet widths tends to increase very slightly with $\overline{\rho}_i$. Figures 11 and 12 show correlations of the jet half-radii for velocity and temperature, respectively.
The best-fit equations can be obtained by using similar streamwise coordinate:

$$\overline{r}_{1/2U} = 0.09\, \overline{\rho}_i^{\,1/10}\, (\overline{Z} - \overline{Vp}) + 0.5 \qquad (11)$$

$$\overline{r}_{1/2T} = 0.08\, \overline{\rho}_i^{\,1/3}\, (\overline{Z} - \overline{Tp}) + 0.5 \qquad (12)$$

The constant 0.5 also comes from the definition of core lengths.

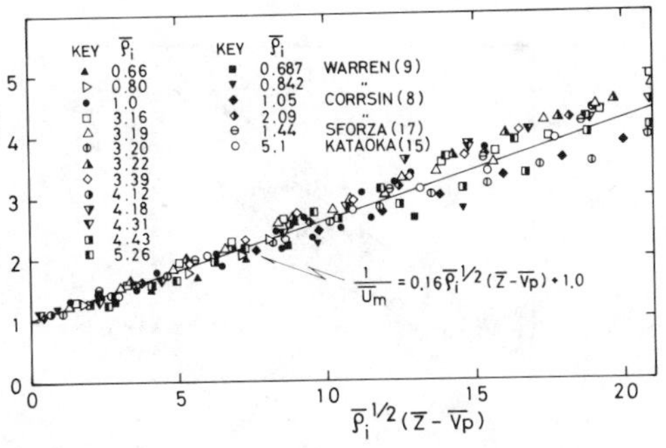

Figure 9 AXIAL DECAY OF CENTERLINE VELOCITY.

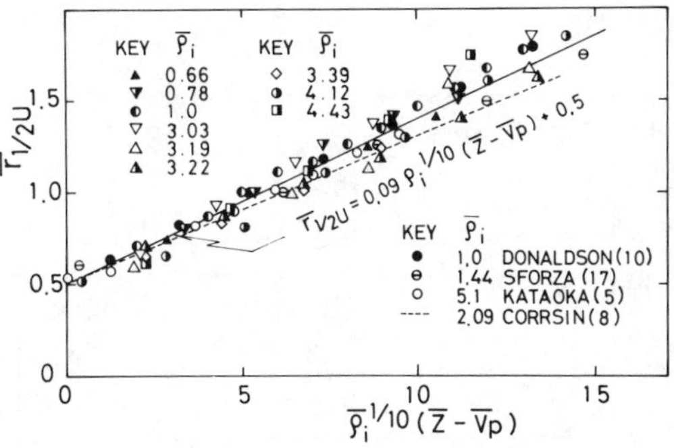

Figure 11 RADIAL SPREADING OF VELOCITY DISTRIBUTION.

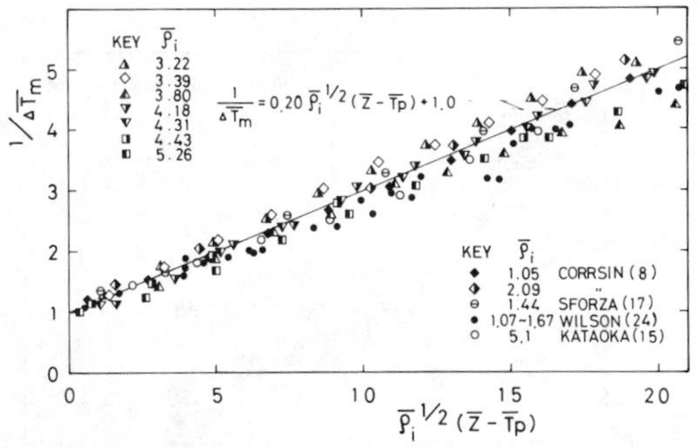

Figure 10 AXIAL DECAY OF CENTERLINE TEMPERATURE.

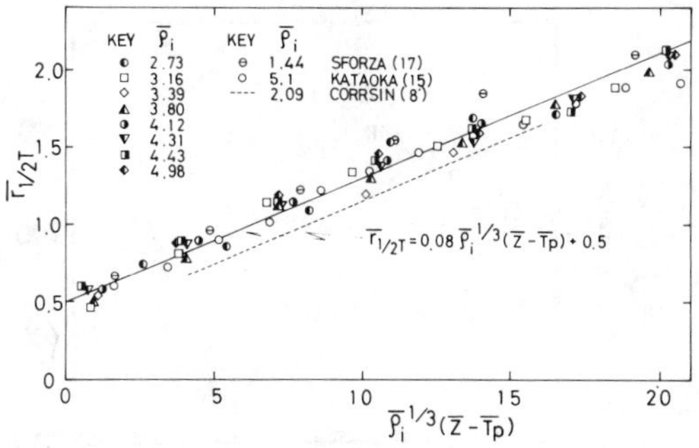

Figure 12 RADIAL SPREADING OF TEMPERATURE DISTRIBUTION.

The above set of correlative equations (Equations (5) to (12)) can describe the other investigations (8, 17, 19) done at higher Reynolds numbers within 20% error. Velocity and temperature profiles at any axial position in the fully-developed region of nonisothermal free jets can be predicted at various temperature levels by these equations.

[2] Impingement heat transfer

In a manner similar to Donaldson et al.(19), the centerline velocity U_m and velocity half-radius $r_{1/2U}$, that would exist at the plane of impingement in the absence of the flat plate, are adopted as the characteristic velocity and length scales to characterize convective heat transfer in the impingement region.

The convective heat transfer occurs between the wall temperature T_w and the bulk temperature T_b of non-isothermal impinging jet. Figure 13 shows axial variation of the centerline temperature in the presence of flat plate. When the jet approaches the flat plate, the centerline temperature deviates from the centerline decay curve of free jet and has a bulge near the plate. The bulge seems to be located near the beginning of the impingement region. The temperature corresponding to the bulge can be taken as the bulk temperature.

Figure 14 shows radial distributions of the wall and bulk temperatures. It can be seen from the figure that the surface of the flat plate used is kept at approximately constant temperature.
The bulk temperature also remains almost constant near the stagnation point, but goes down rapidly downstream with radial distance. It is not very significant to determine local heat transfer coefficients from the measured wall heat-fluxes except in the central impingement region, since not only the heat transfer coefficient h but also the temperature difference $T_b - T_w$ vary with radial distance r.

The stagnation-point heat transfer coefficient is defined as

$$q_s = h_s(T_{bs} - T_{ws}) \tag{13}$$

Figure 15 shows variation of the stagnation-point bulk temperature T_{bs} with characteristic jet development length $\overline{H} - \overline{Tp}$, which is similar to the axial decay of T_m for free jets.
The best-fit equation shown below is quite similar to Equation (10):

$$1/\Delta\overline{T}_{bs} = 0.146\,\overline{\rho}_i^{\,1/2}\,(\overline{H} - \overline{Tp}) + 0.83 \tag{14}$$

The constant 0.83 results from the fact that the bulk temperature T_{bs} exists near the beginning of

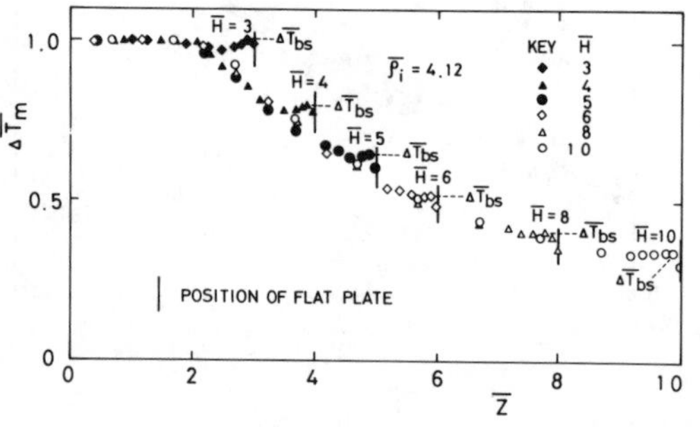

Figure 13 AXIAL VARIATION OF CENTERLINE TEMPERATURE IN THE PRESENCE OF FLAT PLATE.

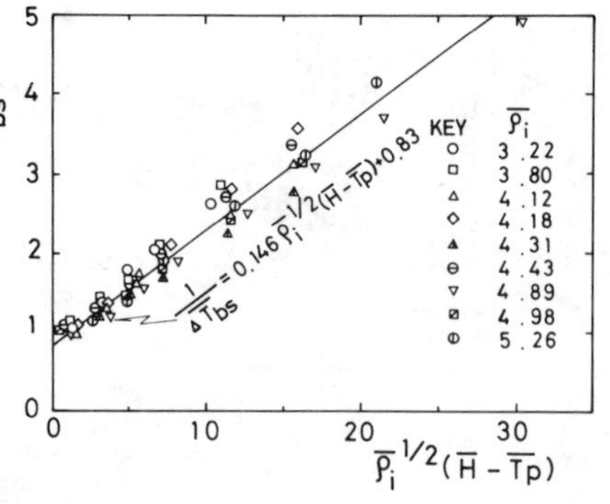

Figure 15 VARIATION OF STAGNATION-POINT BULK TEMPERATURE.

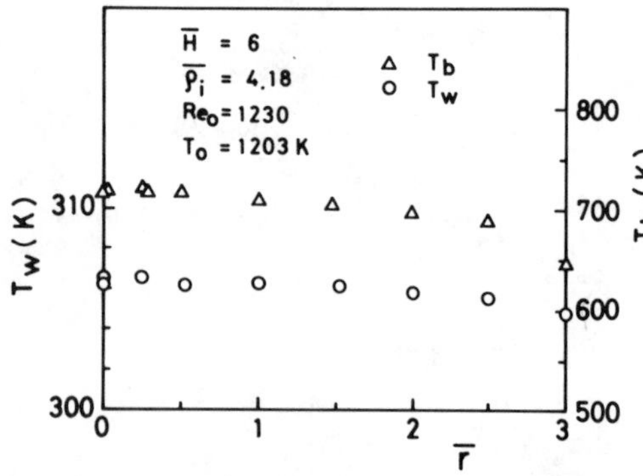

Figure 14 RADIAL DISTRIBUTIONS OF WALL AND BULK TEMPERATURES.

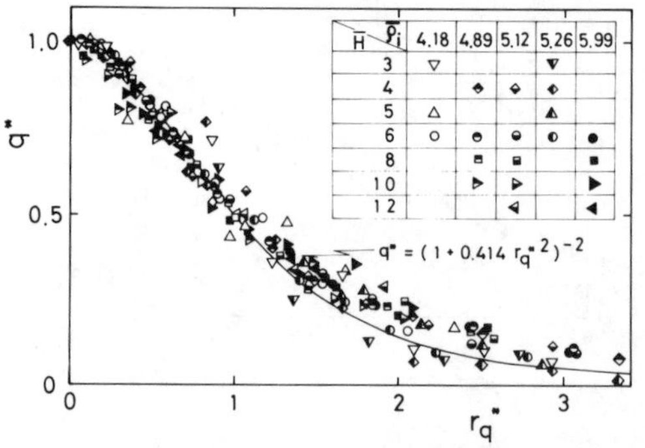

Figure 16 RADIAL DISTRIBUTION OF NORMALIZED WALL HEAT-FLUXES.

the impingement region or slightly upstream of the stagnation point.

Figure 16 indicates that the radial distribution of normalized wall heat-fluxes becomes similar at various nozzle-to-plate spacings if the radial coordinate is made dimensionless with respect to half-radius $r_{1/2q}$ of wall heat-flux. An off-stagnation maximum as reported by Obot et al.(23) cannot be discerned in the wall heat-fluxes near the stagnation point.
The following equation can be obtained in similarity to Equation (7):

$$q^* = (1 + 0.414\ r_q^{*2})^{-2} \qquad (15)$$

The wall heat-fluxes measured in the outer region are scattered above the curve of Equation (15) owing to the transition to the turbulent wall jet flow.

Similarly to the linear spreading of free jets, Figure 17 indicates linear variation of the heat-flux half-radius with characteristic jet development length.
The best-fit equation is given by

$$\overline{r}_{1/2q} = 0.21\ (\overline{H} - \overline{Vp}) + 1.0 \qquad (16)$$

This is similar in form to Equation (11) and (12), except in that the constant is 1.0 in place of 0.5. The half-radius $\overline{r}_{1/2q}$ is indirectly dependent on $\overline{\rho}_i$ through $\overline{Vp}$.

Although the present work has not measured the surface pressures on the wall of the flat plate, the lateral distribution of the surface pressure when $\overline{H} > \overline{Vp}$ can be assumed to be of the following form:

$$\Delta p^* = (1 + 0.414\ r_p^{*2})^{-2} \qquad (17)$$

The surface pressure at the stagnation point Δp_s is approximately given by

$$\Delta p_s = \frac{1}{2}\rho\ U_m^2 \qquad (18)$$

By applying Bernoulli's equation to the impingement region, the radial velocity outside the boundary layer along the flat plate is

$$U_r = U_m[1 - (1 + 0.414\ r_p^{*2})^{-2}]^{1/2} \qquad (19)$$

In the vicinity of the stagnation point ($0 \leq r_p^* << 1$), the radial velocity distribution can be approximated by the equation

$$U_r = [\sqrt{0.828}\ U_m/r_{1/2p}]\ r \qquad (20)$$

It should be noted that the coefficient in front of r is a function of jet half-radius $r_{1/2p}$ as well as U_m.
According to the laminar boundary layer solution of the axisymmetric Hiemenz flow(26), the boundary layer thickness at the stagnation point is

$$\delta/r_{1/2p} = 2.08\ (U_m r_{1/2p}/\nu)^{-1/2} \qquad (21)$$

The surface-pressure half-radius $r_{1/2p}$ can be assumed to be proportional to $r_{1/2U}$.
For moderate Prandtl numbers, the thichness ratio of thermal to momentum boundary layer is approximately proportional to $Pr^{1/2}$.
Therefore, the stagnation-point heat transfer coefficient should be of the form:

$$Nu_s/Re_s^{\ 1/2}\ Pr_s^{\ 1/2} = \text{constant} \qquad (22)$$

where the Nusselt and Reynolds numbers are based on U_m and $r_{1/2U}$.

Even for isothermal jets, however, the above dimensionless group of heat transfer varies greatly with nozzle-to-plate spacing because the turbulence of the centerline velocity is increased with the jet development and amplified by the jet impingement. Therefore, it can be expected that for nonisothermal jets the stagnation-point Nusselt number is a function of characteristic jet development length $(\overline{H} - \overline{Vp})$. The present work does not include the case when $\overline{H} < \overline{Vp}$, because of its small heat transfer coefficients.
Figure 18 indicates that the Nusselt numbers measured at the stagnation point are well correlated in the above functional form.

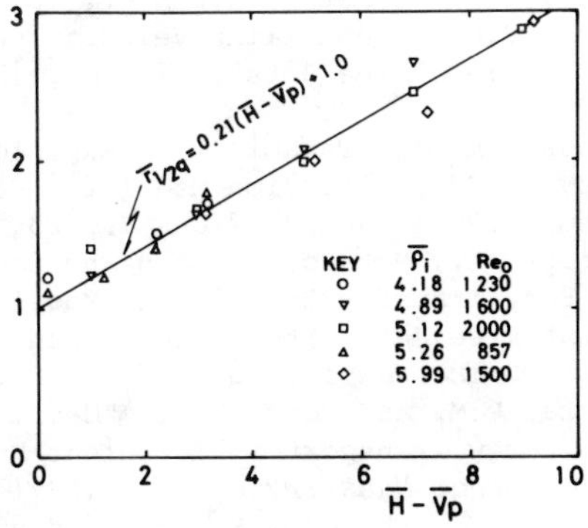

Figure 17 VARIATION OF HALF-RADIUS OF HEAT-FLUX.

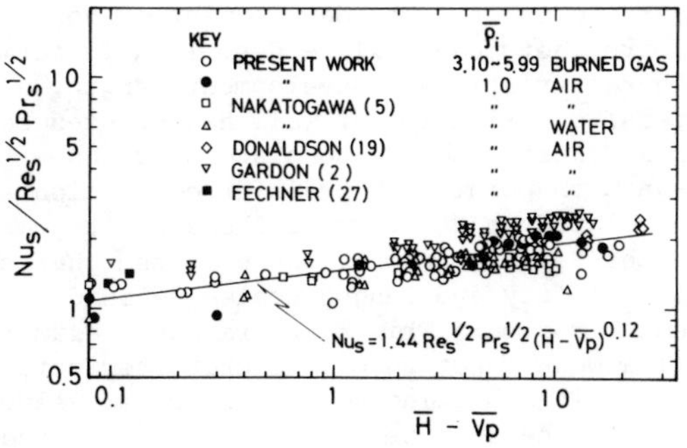

Figure 18 CORRELATION OF STAGNATION-POINT NUSSELT NUMBERS.

The following correlative equation can also describe the other isothermal jet experiment(2, 5, 19, 27) within average error of 30%:

$$Nu_s = 1.44\ Re_s^{1/2}\ Pr_s^{1/2}\ (\overline{H} - \overline{Vp})^{0.12} \tag{23}$$

Substitution of Equations (9) and (11) into the above equation gives the stagnation-point Nusselt number based on the nozzle-exit quantities:

$$Nu_o/Re_o^{1/2}\ Pr_o^{1/2} = 1.44\ \overline{\kappa}_s \overline{\alpha}_s^{-1/2}\ (\overline{H} - \overline{Vp})^{0.12} \times [0.16\ \overline{\rho}_i^{1/2}\ (\overline{H} - \overline{Vp}) + 1.0]^{-1/2} \times [0.09\ \overline{\rho}_i^{1/10}\ (\overline{H} - \overline{Vp}) + 0.5]^{-1/2} \tag{24}$$

This equation suggests that the stagnation-point heat transfer coefficient reaches a maximum when the flat plate is placed slightly downstream of the end of potential core. At the position, the turbulence of the centerline velocity becomes large enough to enhance heat transfer whereas the average velocity remains almost unchanged.

A generalized model has been established for prediction of local values of wall heat-flux in the impingement region. If the initial jet condition (U_o, T_o, T_∞, $\overline{\rho}_i$, Re_o) and the heat transfer requirement (T_w, $\overline{H}$) are specified, the design calculation can be performed in the following order:

(1) First calculate Vp and Tp from Equations (5) and (6).
(2) Estimate U_m by Equation (9) and $r_{1/2U}$ by Equation (11).
(3) Estimate T_{bs} by Equation (14) and evaluate Nu_s from Figure 18 or Equation (23).
(4) Calculate the stagnation-point heat-flux q_s by substituting the temperature difference ($T_{bs} - T_{ws}$) into Equation (13).
(5) Estimate $r_{1/2q}$ by Equation (16).
(6) Finally calculate local heat-fluxes q by using Equation (15).

CLOSURE

A generalized model for prediction of the impingement heat transfer has been established by experiment, taking into account the effect of the initial density ratio on it. The development of nonisothermal free jets can well be described in terms of a new characteristic jet development length.
The lateral distribution of wall heat-fluxes becomes similar at various nozzle-to-plate spacings.
The stagnation-point heat transfer coefficients are well correlated in the form of the Nusselt number against the Reynolds number based on the characteristic velocity and length scales defined in the impingement region. This heat transfer parameter is still a weak function of the characteristic jet development length owing to the effect of velocity turbulence. The bulk temperatures and half-radii for heat transfer are also correlated with the same jet development length.

REFERENCES

1. Kataoka, K., Kamiyama, Y., Hashimoto, S., and Komai, T., "Mass Transfer between a Plane Surface and an Impinging Turbulent Jet: the Influence of Surface-Pressure Fluctuations", *J. Fluid Mech.*, 119, 91 (1982).
2. Gardon, R. and Cobonpue, J., "Heat Transfer between a Flat Plate and Jets of Air Impinging on It", *Proc. of Int. Heat Transfer Conf.*, Part II, p.454 (1961).
3. Gardon, R. and Akfirat, J.C., "The Role of Turbulence in Determining the Heat-Transfer Characteristics of Impinging Jets", *Int. J. Heat Mass Transfer*, 8, 1261 (1965).
4. Gardon, R. and Akfirat, J.C., "Heat Transfer Characteristics of Impinging Two-Dimensional Air Jets", *J. Heat Transfer*, 88, 101 (1966).
5. Nakatogawa, T., D. Thesis, Univ. of Tokyo, Tokyo (1971).
6. Kataoka, K. and Mizushina, T., "Local Enhancement of Heat Transfer in an Impinging Round Jet by Free-Stream Turbulence", *Proc. of 5th Int. Heat Transfer Conf.*, Tokyo, Japan, Vol.2, FC8.3, p.305 (1974).
7. Kataoka, K., Komai, T., and Nakamura, G., "Enhancement Mechanism of Mass Transfer in a Turbulent Impinging Jet for High Schmidt Numbers", *The 2nd AIAA-ASME Thermophysics & Heat Transfer Conf.*, Palo Alto, Calif., May 24 - 26, 78-HT-5 (1978).
8. Corrsin, S. and Uberoi, M.S., "Further Experiments on the Flow and Heat Transfer in a Heated Turbulent Air Jet", *NACA TN* 1865 (1949).
9. Warren, W.R., Ph.D. Thesis, Princeton Univ., Princeton, N.J. (1957).
10. Donaldson, C.duP. and Gray, K.E., "Theoretical and Experimental Investigation of the Compressible Free Mixing of Two Dissimilar Gases", *AIAA J.*, 4, 2017 (1966).
11. Ferri, A., Libby, P.A., and Zakkay, V., "Theoretical and Experimental Investigation of Supersonic Combustion", *ARL 62-467*, Aeronautical Research Labs., Wright-Patterson Air Force Base, Ohio (1962).
12. Kleinstein, G., "Mixing in Turbulent Axially Symmetric Free Jets", *J. Spacecraft Rockets*, 1, 403 (1964).
13. Tomich, J.F. and Weger, E., "Some New Results on Momentum and Heat Transfer in Compressible Turbulent Free Jets", *AIChE J.*, 13, 948 (1967).
14. Witze, P.O., "Centerline Velocity Decay of Compressible Free Jets", *AIAA J.*, 12, 417 (1974).
15. Kataoka, K. and Takami, T., "Experimental Study of Eddy Diffusion Model for Heated Turbulent Free Jets", *AIChE J.*, 23, 889 (1977).
16. Kataoka, K., Matsuo, H., and Shundoh, H., "An Eddy Transfer Coefficient Model for Turbulent Free Jets with Variable Density", *J. Chem. Eng. Japan*, 15, 255 (1982).
17. Sforza, P.M. and Mons, R.F., "Mass, Momentum, and Energy Transport in Turbulent Free Jets", *Int. J. Heat Mass Transfer*, 21, 371 (1978).
18. Kataoka, K., Shundoh, H., and Matsuo, H., "A Generalized Model of the Development of Nonisothermal, Axisymmetric Free Jets", *J. Chem. Eng. Japan*, 15, 17 (1982).

19. Donaldson, C.duP., Snedeker, R.S., and Margolis, D.P., "A Study of Free Jet Impingement. Part II — Free Jet Turbulent Structure and Impingement Heat Transfer", *J. Fluid Mech.*, 45, 477 (1971).
20. Giralt, F., Chia, C.J., and Trass, O., "Characterization of the Impingement Region in an Axisymmetric Turbulent Jet", *Ind. Eng. Chem., Fundam.*, 16, 21 (1977).
21. Chia, C.J., Giralt, F., and Trass, O., "Mass Transfer in Axisymmetric Turbulent Impinging Jets", *Ind. Eng. Chem., Fundam.*, 16, 28 (1977).
22. Obot, N.T., "Effect of Suction on Impingement Heat Transfer", *Proc. of 7th Int. Heat Transfer Conf.*, Munich, Vol.3, FC67, p.389 (1982).
23. Obot, N.T., Douglas, W.J.M., and Mujumdar, A.S., "Effect of Semi-Confinement on Impingement Heat Transfer", *Proc. of 7th Int. Heat Transfer Conf.*, Munich, Vol.3, FC68, p.395 (1982).
24. Wilson, R.A.M. and Danckwerts, P.V., "Studies in Turbulent Mixing — II., A Hot-Air Jet", *Chem. Eng. Sci.*, 19, 885 (1964).
25. Thring, M.W. and Newby, M.P., "Combustion Length of Enclosed Turbulent Jet Flames", *4th Symposium on Combustion*, p.789 (1953).
26. Schlichting, H., *Boundary Layer Theory*, McGraw-Hill, New York (1966).
27. Fechner, G., D. Thesis, Technische Universität München, Munich (1971).
28. Kataoka, K., Takami, T., and Kawachi, Y., "Influence of the Surrounding Fluid on Impinging Jet Cooling", *Heat Transfer Japanese Research*, Scripta, 6, 22 (1977).
29. Kataoka, K., Shundoh, H., and Matsuo, H., "Characteristics of Convective Heat Transfer in Nonisothermal, Variable-Density Impinging Jets", to be presented at PACHEC'83, Souel (1983).

APPLICATION OF A THERMODYNAMIC THEORY TO DETERMINE CAPILLARY PRESSURE AND OTHER FUNDAMENTAL MATERIAL PROPERTIES AFFECTING THE DRYING PROCESS

M.J. LAMPINEN, DR
K. TOIVONEN, M.SC.

VALMET CORPORATION, PANSIO WORKS, TURKU, FINLAND

ABSTRACT

The paper presents a new fundamental capillary pressure equation, as well as the equations of diffusion and capillary flow resistant and conductivity coefficients. An application of capillary pressure equation is described.
A practical method of measuring and calculating the diffusion and capillary flow resistant and conductivity coefficients is presented. Finally the drying equations using the new material property equations are presented and the results are compared to measured ones.

NOMENCLATURE

a_1 = surface area of the 1-β interface per unit mass of constituent 1, m^2/kg

$a_{2'}$ = surface area of the 2-β interface per unit mass of constituent 2, m^2/kg

$a_{2''}$ = surface area of the 1-2 interface per unit mass of constituent 2, m^2/kg

a_2 = see $f_2^s a_2$, $h_2^s a_2$, $\varepsilon_2^s a_2$, $\mu_2^s a_2$.

c_α = ρ_α/ρ, mass fraction of constituent α

$c_{p\alpha}$ = specific heat of constituent α at constant pressure, J/kg K

c_p = mass fraction weighted average constant pressure heat capacity, J/kg K

D_{34} = diffusion coefficient, m^2/s

d_α = real density of constituent α, kg/m^3

f_1^s = Helmholtz free energy of surface a_1 J/m^2

$f_{2'}^s$ = Helmholtz free energy of surface $a_{2'}$ J/m^2

$f_{2''}^s$ = Helmholtz free energy of surface $a_{2''}$ J/m^2

$f_2^s a_2 = f_{2'}^s a_{2'} + f_{2''}^s a_{2''}$ J/kg

h_α = $\varepsilon_\alpha + p_\alpha v_\alpha$, enthalpy of constituent α, J/kg

$h_2^s a_2 = h_{2'}^s a_{2'} + h_{2''}^s a_{2''}$ J/kg

I_α = strength of the mass source or sink of constituent α, kg/m^3s

k = convective mass transfer coefficient, $kmol/m^2s$

k_2 = conductivity of water flow, m^2

k_β = permeability of moist solid, m^2

l_o = heat of vaporization of free water, J/kg

l = heat of vaporization of hygroscopically bound water, J/kg

M_α = molar mass of constituent α, kg/kmol

p_α = real pressure of constituent α, N/m^2

R = gas constant, $8.314 \cdot 10^3$ J/kmol K

s_1^s = entropy of surface a_1 J/m^2 K

$s_{2'}^s$ = entropy of surface $a_{2'}$, J/m^2 K

$s_{2''}^s$ = entropy of surface $a_{2''}$ J/m^2K

T = temperature K

T_{cr} = critical temperature of water, 647.3 K

t = time, s

u = ρ_2/ρ_1, moisture content

v_α = $1/d_\alpha$, specific volume of constituent α, m^3/kg

$\bar{w}_\alpha$ = velocity of constituent α, m/s

x = coordinate, m

α = convective heat transfer coefficient, W/m^2K

ε = specific internal energy of the mixture, J/kg

ε_α = specific internal energy of constituent α, J/kg

ε_1^s = h_1^s, internal energy of surface a_1 J/m^2

$\varepsilon_{2'}^s$ = $h_{2'}^s$, internal energy of surface $a_{2'}$ J/m^2

$\varepsilon_{2''}^s$ = $h_{2''}^s$ internal energy of surface $a_{2''}$ J/m^2

ε_b = effective diffusion resistant coefficient

λ = thermal conductivity, W/mk

λ_a = apparent thermal conductivity W/mK

γ_α = kinematic viscosity of constituent α, m^2/s

ρ_α = partial density of constituent α, kg/m^3
ρ = $\Sigma\, \rho_\alpha$ density of the mixture, kg/m^3
$\alpha = 1,2,\beta$
$\emptyset_\alpha$ = ρ_α/d_α, volume fraction of constituent α
ψ = relative vapor pressure

Sub- and superscripts

α = 1, solid phase
= 2, water
= 3, water vapor
= 4, dry air
= β, humid air

s denotes surface quantities

1. INTRODUCTION

Porous materials capable of binding water are termed hygroscopic. The hygroscopic material consists of four components; solid materials (subscript $\alpha = 1$), liquid water ($\alpha = 2$), water vapour (α=3) and dry air ($\alpha = 4$). Often, however, we prefer to regard the mixture as being composed of three components: solid material, liquid water and humid air ($\alpha = \beta$). The humid air is a mixture of the components $\alpha = 3$ and $\alpha = 4$.

The volume fraction of constituent α is defined by an equation $\emptyset_\alpha = \rho_\alpha/d_\alpha$, where ρ_α is the partial density of constituent α ($\rho_\alpha = dm_\alpha/dV$, where dm_α is the mass of constituent α in a volume dV of the mixture) and d_α is the real density of constituent α ($d_\alpha = dm_\alpha/dV_\alpha$, where the real volume occupied by dm_α is dV_α). It is immediately observed that $\emptyset_1 + \emptyset_2 + \emptyset_\beta = 1$ and $\emptyset_3 = \emptyset_4 = \emptyset_\beta$. Instead of the variable $\emptyset_2$ we may also use the moisture content u, which is defined by $u = \rho_2/\rho_1$. Hence the connection between the variables $\emptyset_2$ and u is $u = (d_2/d_1) \cdot (\emptyset_2/\emptyset_1)$. Mass fraction of constituent α is defined by $c_\alpha = \rho_\alpha/\rho$, where $\rho = \rho_1 + \rho_2 + \rho_\beta$.

When the moisture content of the material decreases, an increasing part of the water is located at the water-solid and water-air interfaces. Since this surface water is different from normal water, it follows that at lower moisture contents the water as a whole has exceptional properties which have to be taken into account. These exceptional properties are described by surface functions or by correction functions as we may also call them.

For instanse, let us considet the internal energy of the mixture (ε) in the equilibrium state. If the surfaces did not affect the internal energies, then $\varepsilon = c_1\varepsilon_1 + c_2\,\varepsilon_2 + c_\beta\varepsilon_\beta$ where ε_α(α = 1,2,β) are the usual "free state" internal energies of constituents. However, the surfaces affect the values of internal energies and, to get correct results, we have introduced the correction function $\varepsilon_1^s\, a_1$ and $\varepsilon_2^s\, a_2$ so that the equation
$\varepsilon = c_1(\varepsilon_1 + \varepsilon_1^s\, a_1) + c_2(\varepsilon_2 + \varepsilon_2^s\, a_2) + c_\beta\varepsilon_\beta$ is valid. In a similar way surface entropies have been instroduced.

Since the surfaces do not occupy any volume it is natural to define the surface internal energies and enthalpies as equal concepts i.e. $\varepsilon_\alpha^s\, a_\alpha = h_\alpha^s\, a_\alpha$. The definition of the surface chemical potential is $\mu_\alpha^s\, a_\alpha = h_\alpha\, a_\alpha - Ts_\alpha^s\, a_\alpha$ and, since $h_\alpha^s\, a_\alpha = \varepsilon_\alpha^s\, a_\alpha$, we may also write $\mu_\alpha^s\, a_\alpha = f_\alpha^s\, a_\alpha$.

The term $\varepsilon_2^s\, a_2$ (and $s_2^s\, a_2$ correspondingly) is composed of two parts

$\varepsilon_2^s\, a_2 = \varepsilon_{2'}^s a_{2'} + \varepsilon_{2''}^s\, a_{2''}$ where the subscript 2' denotes water-air interface and 2" water-solid interface. These two interfaces affect the internal state of the water, changing the internal energy and lowering the pressure of the water. In the following chapter we shall show how the surface function $f_2^s\, a_2$ can be determined by a thermodynamic method and also how the capillary pressure of water can then be calculated.

2. A NEW CAPILLARY PRESSURE EQUATION

2.1 A general theory

As has been derived in (1) in an equilibrium state the following general vapour pressure equation is valid:

$$v_3 dp_3 = \frac{h_3 - (h_2 + h_2^s\, a_2)}{T}\, dT + v_2 dp_2 + d(f_2^s\, a_2)_T \qquad (1)$$

When the temperature is kept constant, i.e. dT = 0, we get from (1)

$$v_3 dp_3 = v_2 dp_2 + d(f_2^s\, a_2)_T \qquad (2)$$

which describes the sorption isotherm of water in hygroscopic material in the equilibrium state. Using the ideal gas equation and by integrating (2) we obtain

$$\ln \psi = \frac{M_3}{RT}\,(v_2\,(p_2 - p_{2,0}) + f_2^s\, a_2) \qquad (3)$$

where $\psi = p_3/p_{3,0}$. Pressure $p_{3,0}(T)$ is the pressure of saturated vapour which is in equilibrium with free water at pressure $p_{2,0}$ (normally $p_{2,0}$ = 1 bar). A good numerical approximation for the saturated vapour pressure is

$p_{3,0}$ (T) = exp (11,78 · (T/K - 372,79)/ (T/K - 43.15) bar. This can be used in calculating p_3 (T,u) = ψ(T, u) · $p_{3,0}$(T).

From (3) we may see that the vapour pressure of bound water is changed for two reasons; 1) the pressure of water is lower than in the usual free state, and 2) the energetic state has also been changed, which is described by $f_2^s a_2$. Together the two terms on the r.h.s. of (3) present the total change of the chemical potential of hygroscopically bound water compared to free water.

The enthalpy of vaporization of hygroscopically bound water is

$$l = h_3 (T) - (h_2(T,p_2) + h_2^s a_2)$$

and correspondingly for free water it is

$$l_o = h_3(T) - h_2(T, p_{2,0})$$

so that the difference, the sorption heat is

$$l - l_o = h_2 (T, p_{2,0}) - (h_2 (T, p_2) + h_2^s a_2) \tag{4}$$

On the other hand, applying the relation $dh = c_p\, dT + (v - T\, \partial v/\partial T)\, dp$ (see e.g. (1), Eq (17.10)) we may write

$$h_2(T,p_2) = h_2 (T, p_{2,0}) + (v_2 - T \frac{\partial v_2}{\partial T}) \cdot (p_2 - p_{2,0}) \tag{5}$$

As has been shown in (1, Eq. (15.11)) the following equation holds:

$$h_2^s a_2 = \varepsilon_2^s a_2 = f_2^s a_2 - T \frac{\partial (f_2^s a_2)}{\partial T} \tag{6}$$

From Eq. (3) it follows that

$$p_2 - p_{2,0} = \frac{1}{v_2} \cdot \frac{RT}{M_3} \ln \psi - \frac{1}{v_2} \cdot f_2^s a_2 \tag{7}$$

Substituting (5), (6) and (7) into (4) we obtain

$$l - l_o = Tv_2 \frac{\partial}{\partial T} (\frac{1}{v_2} \cdot f_2^s a_2) - \frac{RT}{M_3} \ln \psi (1 - \frac{T}{v_2} \frac{\partial v_2}{\partial T})$$

and by intergating this we get

$$\int_{T_{cr}}^{T} d(\frac{f_2^s a_2}{v_2}) = \int_{T_{cr}}^{T} \frac{1}{v_2 T} \left[(l - l_o) + (1 + \frac{T}{v_2} \frac{\partial v_2}{\partial T}) \frac{RT}{M_3} \ln \psi\right] dT$$

If we choose T_{cr} = 647,3 K, the critical temperature of water, it is obvious that at this temperature, independent of the moisture content, the surface energies are zero, i.e. $f_2^s a_2 (T_{cr}) = 0$ and therefore also $\psi = 1$ and $l - l_o = 0$. Using this assumption we get from the above equation

$$\frac{1}{v_2} \cdot f_2^s a_2 = \int_{T_{cr}}^{T} \frac{1}{Tv_2} \left[(l - l_o) + (1 - \frac{T}{v_2} \frac{\partial v_2}{\partial T}) \frac{RT}{M_3} \ln \psi\right] dT \tag{8}$$

Consenquently from this assumption it follows that the term

$$\frac{RT}{M_3 v_2} \cdot \ln \psi$$

can be written into the form

$$\frac{RT}{M_3 v_2} \ln \psi = \int_{T_{cr}}^{T} \frac{\partial}{\partial T} (\frac{RT}{M_3 v_2} \ln \psi) dT$$

Substituting this and Equation (8) into Equation (7) we obtain after some manipulation

$$p_2 - p_{2,0} = \int_{T}^{T_{cr}} \frac{1}{Tv_2} \left[(l - l_o) - \frac{RT^2}{M_3} \frac{\partial (\ln \psi)}{\partial T}\right] dT \tag{9}$$

2.2 Laboratory experiments on a newsprint

In order to calculate the capillary pressure by using the thermodynamical equation (9) we need calorimetric measurements to determine sorption heat $l - l_o$ and vapour pressure measurements to determine the relative vapour pressure ψ at various moisture contents and temperatures.

The sorption heats have been calculated from the wetting heats achieved by a calorimeter shown in figure 1.

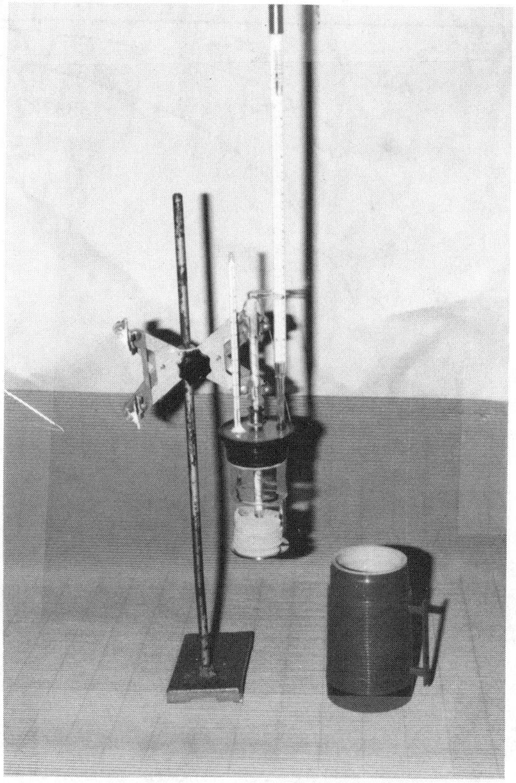

Fig. 1 - A calorimeter used to determine the sorption heats for a newsprint

The heat capacity of the calorimeter with 180 g water inside was 0,770 kJ/K and the heat capacity of the paper sample to be wetted is $c = m_1 (cp_1 + u \ cp_2)$ where m_1 is the mass of the dried paper sample. For instance in one measurement the following values were observed before the wetting process:

- the mass of the paper sample m_1 = 47.8 g
- initial temperature of 180 g water 19.70°C
- initial temperature of the paper sample 19,90°C
- initial moisture content of the paper sample u = 0,0
- the heat capacity c = 0,0478 · (1,40 + 0,0 · 4,186) = 0,0669 kJ/K.

The paper sample was then immersed in the water inside the calorimeter and the temperature of the water was carefully followed. After the paper sample had been completely wetted the temperature rose from 19,70°C to 22,26°C so that the energy released was E(T, u) = (0,770+0,0669)·22,26 - (0,0669·19,90+0,770·19,70) = 2,13 kJ. Such an amount of energy was released when 47,8 g newsprint was wetted from the initial moisture content u = 0 to its maximum value. These measurements were repeated at various initial moisture contents as well as temperatures, too.

After the curves $E(T, u)/M_1$ were constructed the sorption heats (in fact the adsorption heat, if we wish to emphasize the hysteresis) can be calculated immediately:

$$l - l_o = \frac{\partial}{\partial u} \left(\frac{E(T,u)}{m_1} \right)_T$$

Some results of these measurements and calculations are shown in figure 2.

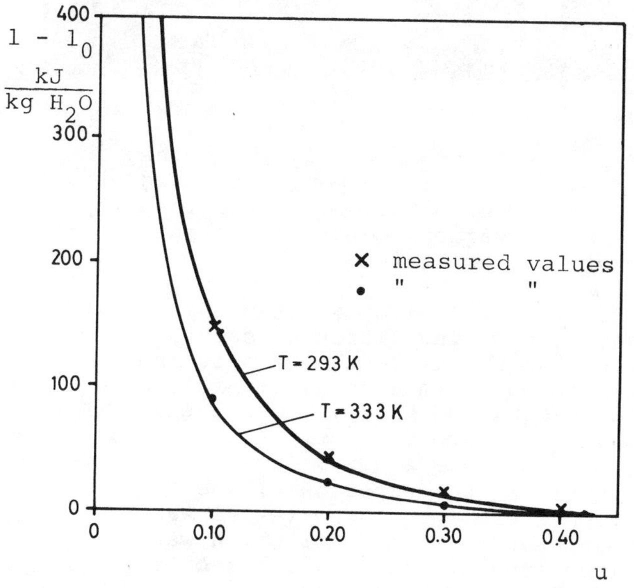

Fig. 2 - The sorption heats $l - l_o$ of a newsprint grade 45 g/m² (Enso-Gutzeit Co., Summa Paper Mill).

The relative vapour pressure ψ at various moisture contents and temperatures were measured by weighing the paper sample in the equilibrium state. A mechanical weighing system was placed inside a vessel where the paper sample was under controlled temperature and humidity conditions. The balance of the weighing system was read through a window (fig. 3).

Fig. 3 - An apparatus by which the relative vapour pressures of a newsprint were determined at various moisture contents and temperatures

Different relative vapour pressures were gained by using different saturated salt-water solutions in the vessel. The following salts were used: LiCl, K_2CO_3, NaBr, NaCl and K_2SO_4. The relative vapour pressures were respectively ψ = 0,1, 0,3, 0,55, 0,7 and 0,9 depending to some extent on the temperature. The temperature inside the vessel was regulated by heated air which was circulated around the jacket of the vessel. These measurements were made by 5^oC steps up to temperature level 150^oC. Some results of these measurements are shown in fig. 4.

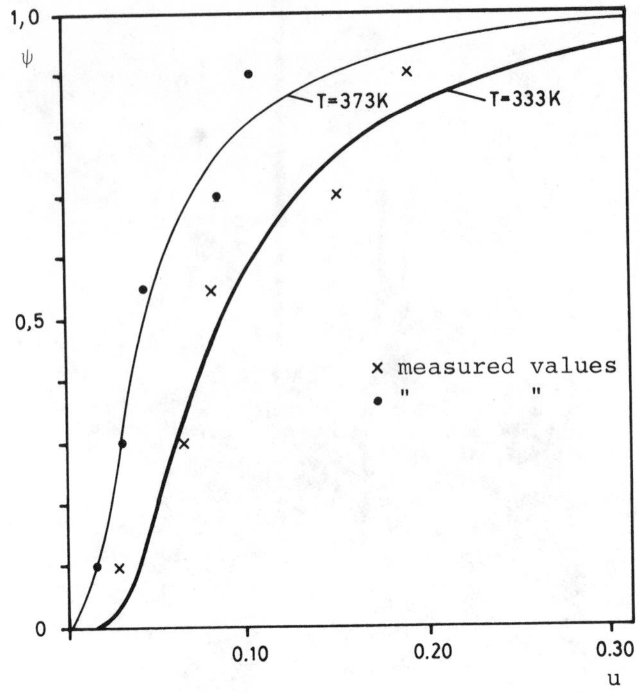

Fig. 4 - The relative vapour pressures of a newsprint quality 45 g/m^2

2.3 Application of the capillary pressure equation for a newsprint

To give a mathematical model for the quantities $1\text{-}l_o$ and ψ based on the measurements decribed above, two restrictions were introduced. The first restriction was the same as the assumption used in deriving equation (8), i.e. $1\text{-}l_o = 0$ and $\psi = 1$ as $T = T_{cr} = 647.3$ K at all moisture contents. The second restriction or assumption which has been made is that the functions $1\text{-}l_o$ and ln ψ have been sought in the forms

$$1 - l_o = f(u)\ g\ (T)$$

$$\ln \psi = h\ (u)\ k\ (T)$$

It turned out that it was very diffucult to find such a form of the function h (u) which would give a satisfactory correlation with measured values. Analyzing many different models, the form $h(u) = A \cdot u^B \cdot \exp(Cu)$ finally gave the best correlation.

The functions g(T) and k(T) have to disappear as T approches T_{cr}. To satisfy this requirement a model $(T_{cr} - T)^D$ was chosen.

Analyzing the measured values with the aid of statistical methods, the following approximate formulas have been found:

$$\ln \psi = A_1 \cdot u^{A_2} \cdot \exp (A_4 \cdot u) \cdot \left(\frac{T_{cr} - T}{A_5}\right)^{A_3} \tag{11}$$

$$l - l_o = A_6 \cdot u^{A_2} \cdot \exp (A_4 \cdot u) \cdot T^2 \cdot \left(\frac{T_{cr} - T}{A_7}\right)^{A_3 - 1} \tag{12}$$

where:

$A_1 = -3,433 \quad A_2 = -1,3820 \quad A_3 = 7,557$
$A_4 = -3,372 \quad A_5 = 583,5K \quad A_6 = 8,633 \cdot 10^{-3}$ kJ/kg K^2
$A_7 = 696,0K$

The correlation factors of (11) and (12) were 0,941 and 0,99 respectively.

Substituting (11) and (12) into (9) and keeping v_2 as constant ($v_2 = 10^{-3}$ m^3/kg), we can integrate it in a closed form and the following equation can be derived:

$$v_2 (p_{2,0} - p_2) = -\left(T + \frac{T_{cr} - T}{A_3 + 1}\right) \left(\frac{l - l_o}{A_3} \cdot \frac{T_{cr} - T}{T^2} + \frac{R \ln \psi}{M_3}\right) \tag{13}$$

Example 1

$u = 1,25 \quad T = 293$ K

$$\ln \psi = -3,433 \cdot 1,25^{-1,3820} \cdot \exp(-3,372 \cdot 1,25) \cdot \left(\frac{647,3-293}{583,5}\right)^{7,557} = -8,587 \cdot 10^{-4}$$

$$l - l_o = 8,633 \cdot 10^{-3} \cdot 1,25^{-1,3820} \cdot \exp(-3,372 \cdot 1,25) \cdot 293^2 \cdot \left(\frac{647,3-293}{696,0}\right)^{6,557}$$

$= 0,09608$ kJ/kg

$$v_2(p_{2,0} - p_2) = -\left(293 + \frac{647,3-293}{7,557+1}\right) \cdot \left(\frac{0,09608}{7,557} \cdot \frac{647,3-293}{293^2} + \frac{8,314 \cdot (-8,587 \cdot 10^{-4})}{18,0}\right)$$

$= 0,1151$ kJ/kg

$$p_2 = 100 \text{ kPa} - \frac{0,1151 \text{ kJ/kg}}{10^{-3} m^3/kg} = -15 \text{ kPa}$$

This means that the liquid water in the newsprint is already in these circumstances under tensile stress.

Calculating in the same way as in this example, the following curves, shown in figure 5, can be drawn:

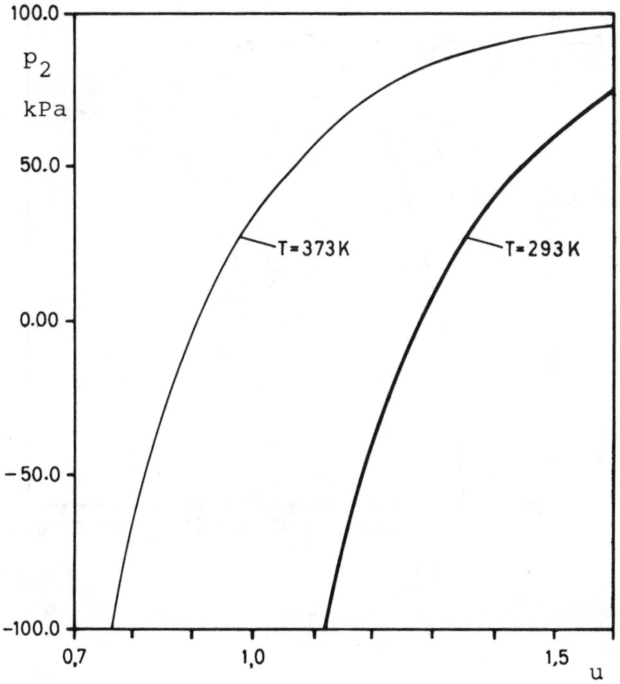

Fig. 5 - The pressure of water in a newsprint based on the equation (9) and the laboratory experiments

In our laboratory tests, as well as in others too, the total pressure of air has been 1 bar, if the temperature has been lower than 100°C. Therefore the pressure of the free water $p_{2,0}$ = 1 bar, when $T \leq 373$ K. In the experiments above 373 K the pressure of air has been increased according to the pressure of saturated free water, and then $p_{2,0}$ must be chosen according to this pressure.

From (7) it follows that

$$f_2^s a_2 = v_2(p_{2,0} - p_2) + \frac{RT}{M_3} \ln \Psi \tag{14}$$

Using the equations (13) and (11), the change of Helmholtz energy of water in a hygroscopic material (newsprint) can be calculated. The result of these calculations are shown in figure 6.

and does not consider other kinds of transformations in energetic contents and their influences on the vapour pressure.

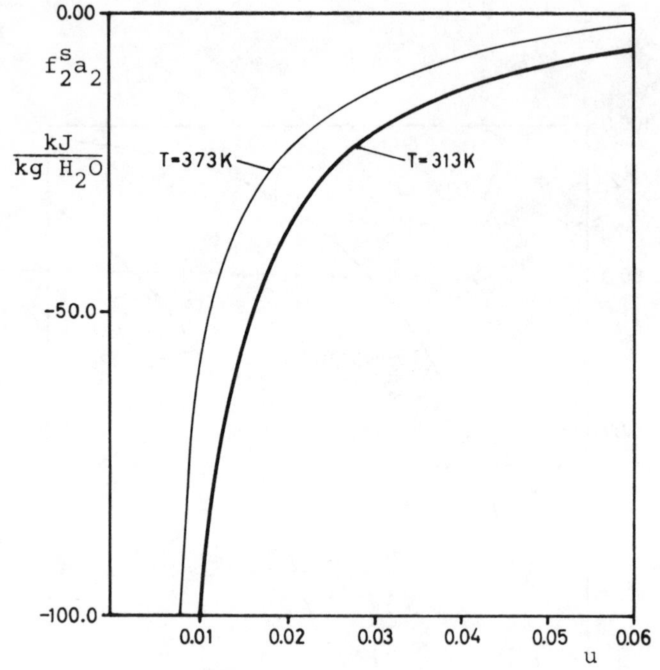

Fig 6 - The change of Helmholtz-energy of water in a newsprint

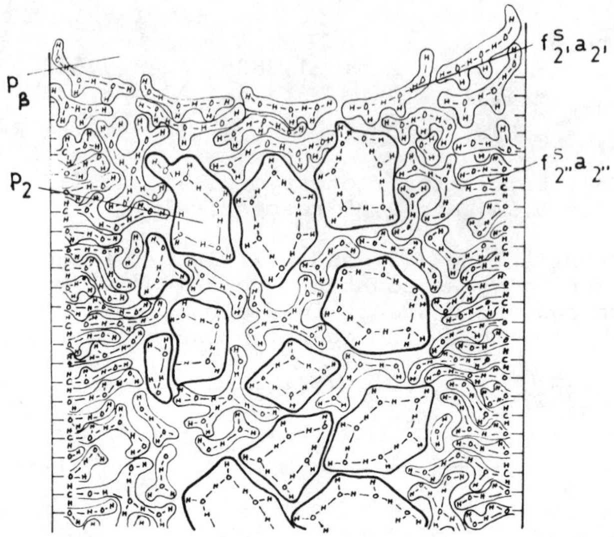

Fig 7 - Conceptual drawing of the various cluster species of liquid water in the cellulose surface and in the water-air interface.

2.4 Discussion

Perhaps one of the best-known capillary pressure equations is Kelvin's equation

$$p_2 - p_{2,0} = \frac{1}{v_2} \cdot \frac{RT}{M_3} \cdot \ln \psi \qquad (15)$$

Comparing this to equation (7) we may see that Kelvin's equation is correct only in the special case where $f_2^s\, a_2 = 0$.

To illustrate the difference between Kelvin's equation and the more general equation (7), let us consider the situation shown in figure 7. The water is in a small cavity and it is surrounded by two phase surfaces, the solid-water interface and the water-air interface. Due to the surface tension $f_{2'}^s$ and its concave form the pressure of the water is lower than the pressure of the surrounding air.

Besides the low pressure of the water, its internal structure has also been changed due to the interfaces (the lower moisture content, the more water resembles frozen water). Kelvin's equation takes into account only the change of pressure in explaining the relative vapour pressure

Since $f_2^s a_2 = f_{2'}^s a_{2'} + f_{2''}^s a_{2''}$ and since $a_{2'} > 0$ and $a_{2''} > 0$ we may conclude from figure 6 that in this case the Helmholtz surface-energy $f_{2''}^s$ is negative and that $|f_{2''}^s a_{2''}| > |f_{2'}^s a_{2'}|$. For a free water-air surface the function $f_{2'}^s$ is known separately, for instance at temperature T = 323 K, that is $f_{2'}^s = 63{,}1 \cdot 10^{-3}$ J/m^2.

The significance of the term $f_2^s a_2$ depends on the geometry of the material's porous structure and its hydrophilic properties. In this example (a newsprint) the influence of $f_2^s a_2$ on the pressure is rather slight and in this case Kelvin's equation gives quite good approximations. For instance at T=293 K and u = 1,25, Kelvin's equation gives p_2 =-16 kPa, whereas equation (7) gives p_2 =-15 kPa as has been calculated in example 1. Obviously in most cases we have $f_2^s a_2 < 0$, and therefore Kelvin's equation gives too low pressures of water.

These laboratory experiments descriped above took quite a long time and particularly the measurements of vapour pressures required great perseverance. The weakness of these measurements was that the sorption heats were determined only at three temperature levels because of the technical difficulties of the calorimeter.

3 CONDUCTIVITY COEFFICIENTS OF AIR AND WATER FLOWS

3.1 General theory

Ignoring inertia forces the flow of liquid water (capillary flow) can be written as follows in one-dimensional case:

$$\rho_2 w_2 = -\frac{k_2}{\gamma_2} \cdot \frac{\partial p_2}{\partial x} \quad (16)$$

and correspondingly the flow of air

$$\rho_\beta \, w_\beta = -\frac{k_\beta}{\gamma_\beta} \quad \frac{\partial p_\beta}{\partial x} \quad (17)$$

As has been shown in (1, page 78) the following approximation can be used to calculate the conductivity k_2 from the information of k_β:

$$\frac{(1-s)^3}{k_\beta / k_{\beta,\ max}} + \frac{s^3}{k_2 / k_{\beta,\ max}} = 1 \quad (18)$$

$$\text{where } s = \frac{\emptyset_2}{1-\emptyset_1} = \frac{\rho_1}{d_2(1-\emptyset_1)} \cdot u$$

3.2 Application for a newsprint

The permeability k_β of a newsprint (45 g/m^2) has been experimentally studied at various temperatures and moisture contents in our laboratory. On the basis of our measurements the following results were obtained:

$$k_{\beta,\ max} = \frac{B_1}{T} (1 - \exp (B_2 \cdot T \cdot u_o^{B_3})) \quad (19)$$

$$k_\beta = \frac{B_1}{T}(\exp (B_2 \cdot T \cdot u^{B_3}) - \exp$$

$$(B_2 \cdot T \cdot u_o^{B_3})) \quad (20)$$

where
$B_1 = 7{,}988 \cdot 10^{-12}\ m^2 \cdot K$
$B_2 = -2{,}601 \cdot 10^{-3}\ 1/K$
$B_3 = 1{,}378$
$u_o = 1{,}581$

Example 2

$T = 293\ K \qquad u = 1{,}25$

$$k_{\beta,\ max} = \frac{7{,}988 \cdot 10^{-12}}{293} \cdot (1 - \exp(-2{,}601 \cdot$$

$$10^{-3} \cdot 293 \cdot 1{,}581^{1{,}378})) = 2{,}08 \cdot 10^{-14} m^2$$

$$k_\beta = \frac{7{,}988 \cdot 10^{-12}}{293} \cdot (\exp (-2{,}601 \cdot 10^{-3} \cdot$$

$$293 \cdot 1{,}25^{1{,}378}) - \exp (-2{,}601 \cdot 10^{-3} \cdot$$

$$293 \cdot 1{,}581^{1{,}378})) = 3{,}16 \cdot 10^{-15} m^2$$

$$\emptyset_1 = \rho_1/d_1 = 450/1560 = 0{,}288$$

$$s = \frac{450}{1000 \cdot (1-0{,}288)} \cdot 1{,}25 = 0{,}790$$

$$k_2 = \frac{s^3 \cdot k_\beta}{\frac{k_\beta}{k_{\beta,max}} - (1-s)^3} =$$

$$\frac{0{,}790^3 \cdot 3{,}16 \cdot 10^{-15}}{\frac{3{,}16 \cdot 10^{-15}}{2{,}08 \cdot 10^{-14}} - (1-0{,}790)^3} = 1{,}092 \cdot 10^{-14} m^2$$

Proceeding according this example, the curves in figure 8 have been constructed.

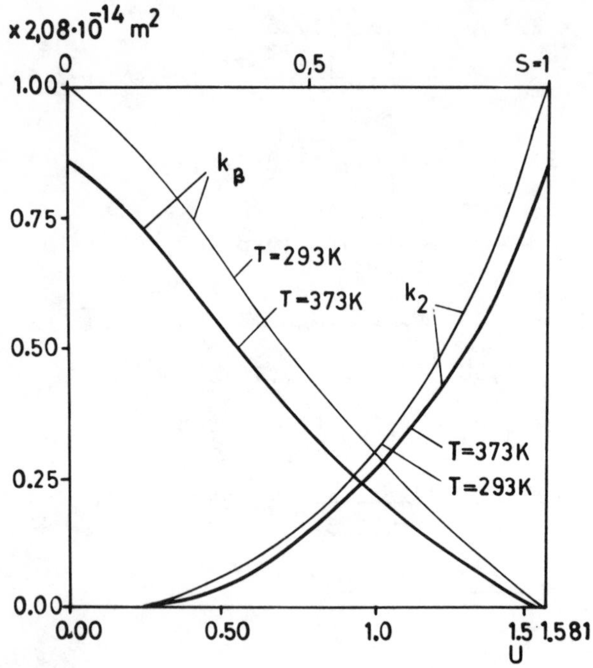

Fig 8 - Permeability (k_β) of a rewetted newsprint (45 g/m^2) based on our measurements and the conductivity coefficient k_2 based on the equation (18).

4 RESISTANT COEFFICIENT OF VAPOUR DIFFUSION

4.1 General theory

As has been shown in (2) the flow of water vapour can be expressed in the following way in a one-dimensional case:

$$\rho_3 w_3 = - \varepsilon_a \cdot \frac{p_\beta}{p_\beta - p_3} \cdot \frac{M_3}{RT} \cdot D_{34} \cdot \frac{\partial p_3}{\partial x} \quad (21)$$

where

$$\varepsilon_a = \frac{\phi_\beta}{1 + \frac{p_\beta}{p_\beta - p_3} \cdot \frac{M_3}{RT} \cdot D_{34} \cdot \gamma_\beta \cdot \frac{\phi_\beta}{k_\beta}} \quad (22)$$

which term can be called a diffusion resistant coefficient.

Let use a notation

$$\varepsilon_b = \frac{p_\beta}{p_\beta - p_3} \cdot \varepsilon_a$$

which can also be written into the form

$$\varepsilon_b = \frac{\phi_\beta}{\frac{p_\beta - p_3}{p_\beta} + \frac{M_3}{RT} \cdot D_{34} \cdot \gamma_\beta \cdot \frac{\phi_\beta}{k_\beta}} \quad (23)$$

This term can be called an effective diffusion resistant coefficient.

Using (23) the equation (21) can be expressed as follows:

$$\rho_3 w_3 = - \varepsilon_b \cdot \frac{M_3}{RT} \cdot D_{34} \cdot \frac{\partial p_3}{\partial x} \quad (24)$$

There is an essential difference between the diffusion of vapour in a boundary layer and in a solid material. To show this, let us first consider what form the diffusion equation (24) receives in the boundary layer.

By the boundary layer we mean the immediate region around the surface of the solid. In that region there exist only the constituents $\alpha = 3,4$ (water vapour, dry air), so that we may set $\phi_\beta = 1$ and $k_\beta = \infty$. Substituting these values into (23) and (24) we obtain:

$$\rho_3 w_3 = - \frac{p_\beta}{p_\beta - p_3} \cdot \frac{M_3}{RT} \cdot D_{34} \cdot \frac{\partial p_3}{\partial x} \quad (25)$$

which is the same as Fick's law combined with the factor of Stefan's flow. The factor is $p_\beta/(p_\beta - p_3)$.

From (25) we observe that the free surface of the moist solid cannot exceed 100°C (if $p_\beta = 1$ bar) unless the moisture content is so low that the relative vapour pressure $\psi < 1$. If $\psi < 1$ then $p_3 < p_\beta = 1$ bar, and the temperature may exceed 100°C.

Within the solid, on the contrary, the temperature may exceed 100°C independent of the moisture content. This is possible because the solid structure resists the flow of vapour so that the total pressure p_β can rise over the atmospheric pressure. If $\psi = 1$ and $T > 100$°C, then $p_3 > 1$ bar and at the same time p_β increases because p_β is the total pressure of the humid air ($p_\beta = p_3 + p_4$, Dalton's law). The term $p_\beta - p_3$ may go to zero and still $\varepsilon_b < \infty$ since $k_\beta < \infty$.(see eq.(23)).

4.2 Application for a newsprint

Example 3

$T = 293$ K $\qquad u = 1,25 \qquad p_\beta = 1$ bar

$\phi_1 = \rho_1/d_1 = 450/1560 = 0,288$

$\phi_2 = \rho_2/d_2 = u \cdot \rho_1/d_2 = 1,25 \cdot 450/1000 = 0,563$

$\phi_\beta = 1-(0,288 + 0,563) = 0,149$

$k_\beta = 3{,}16 \cdot 10^{-15} m^2$ (see example 2)

$\ln \psi = -8{,}587 \cdot 10^{-4}$ (see example 1)

$\psi = \exp(-8{,}587 \cdot 10^{-4}) = 0{,}9991$

$p_3 = 0{,}9991 \cdot 0{,}02337 \text{ bar} = 0{,}02335 \text{ bar}$

$\gamma_\beta = 15{,}08 \cdot 10^{-6} \text{ m}^2/\text{s}$

$D_{34} = 24{,}8 \cdot 10^{-6} \text{ m}^2/\text{s}$

$$\varepsilon_b = \frac{0{,}149}{\dfrac{\dfrac{1{,}0-0{,}02335}{1{,}0} + \dfrac{18}{8{,}314 \cdot 10^3 \cdot 293} \cdot 24{,}8 \cdot 10^{-6} \cdot 15{,}08 \cdot 10^{-6}}{} \cdot \dfrac{0{,}149}{3{,}16 \cdot 10^{-15}}} = 0{,}135$$

The curves in figure 9 have been calculated according to this example

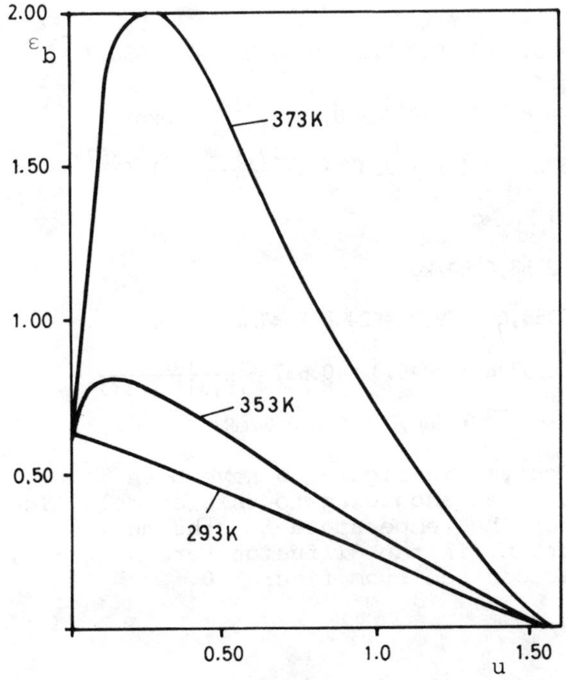

Fig 9 - The effective diffusion resistant coefficient ε_b for a newsprint (45 g/m²)

5 HEAT CONDUCTIVITY

5.1 General theory

For an isotropic material we have the following relation (Fourier´s law) between conductive heat flux and temperature gradient in a one-dimensional case:

$$q = -\lambda \frac{\partial T}{\partial x} \qquad (26)$$

where λ is the thermal conductivity of the material.

It is difficult to measure the heat conductivity of a moist material directly because of the diffusion of water vapour. Water evaporates from the hot surface and condenses on the cold surface, from which the condensed water flows back to the hot surface due to the capillary forces ("heat-pipe effect"). As has been shown in (1) the following equation holds for the measured heat flux density:

$$q_m = q + \rho_3 w_3 \cdot l$$

where $l = h_3 - (h_2 + h_2^s a_2)$

If we assume that the moisture gradients are small compared to the temperature gradients in λ measurements, then we can use an approximation

$$\frac{\partial p_3}{\partial x} = \frac{\partial p_3}{\partial u} \cdot \frac{\partial u}{\partial x} + \frac{\partial p_3}{\partial T} \cdot \frac{\partial T}{\partial x} \simeq \frac{\partial p_3}{\partial T} \cdot \frac{\partial T}{\partial x}$$

Using (24) together with this approximation we obtain

$$q_m = -(\lambda + l \cdot \varepsilon_b \cdot \frac{M_3}{RT} \cdot D_{34} \cdot \frac{\partial p_3}{\partial T}) \frac{\partial T}{\partial x} \qquad (27)$$

From equation (1) we get a formula for $\partial p_3/\partial T$

$$\frac{\partial p_3}{\partial T} = \frac{1}{Tv_3} + \frac{v_2}{v_3} \cdot \frac{\partial p_2}{\partial T} \qquad (28)$$

since $dp_2 = \frac{\partial p_2}{\partial T} dT + \frac{\partial p_2}{\partial u} du$. The term $\partial p_3/\partial T$ can also be calculated directly from the formula $p_3 = \psi(T,u)\, p_{3,0}(T)$.

The term

$$\lambda_a = \lambda + l \cdot \varepsilon_b \cdot \frac{M_3}{RT} \cdot D_{34} \cdot \frac{\partial p_3}{\partial T} \qquad (29)$$

can be called the apparent heat conductivity.

In the proper calculations, on the contrary, we need λ instead of λ_a and therefore it is important to find a model for λ.

Obviously, if measure λ_a at a moisture content $u = 0$, then $\rho_3 w_3 = 0$, and in that case $\lambda_a = \lambda$. We use a notation $\lambda(T,u=0) = \lambda_o$. A logical way to define λ_1 is

$$\lambda_o = \phi_1 \lambda_1 + \phi_\beta \lambda_\beta \qquad (30)$$

where λ_β is the heat conductivity of air.

The heat conductivity λ_1 defined by (27) is not necessarily the same as the heat conductivity of the solid substance itself because of the porous structure of the solid material. For instance the heat conductivity of pure cellulose is 0,21 W/mK (if T = 303 K), whereas for a newsprint we have λ_1 = 0,11 W/mK.

If we assume that water is located uniformly around or inside the solid material, then we may also assume that in the sense of conduction of heat, constituents α = 1,2 are parallelly coupled. According to this assumption we may write

$$\lambda = \Sigma \ \phi_\alpha \lambda_\alpha \tag{31}$$

α = 1,2,β

Since the heat conductivity λ_1 includes the information of the internal geometrical structure of the solid material and since in local analysis the parallel coupling assumption is obvious it is well justified to use the equation (31).

5.2 Application for a newsprint

From the relation $\ln p_3 = \ln \psi + \ln p_{3,0}$ we obtain

$$\frac{1}{p_3} \cdot \frac{p_3}{T} = \frac{\partial(\ln \psi)}{\partial T} + \frac{\partial(\ln p_{3,0})}{\partial T}$$

and from (11) and from the approximation for $p_{3,0}(T)$ (see ch. 2.1) we get

$$\frac{\partial \ln \psi}{\partial T} = A_1 \cdot u^{A_2} \cdot \exp(A_4 u) \cdot A_3 \cdot \left(\frac{T_{cr} - T}{A_5}\right)^{A_3 - 1} \cdot \left(-\frac{1}{A_5}\right)$$

$$\frac{\partial(\ln p_{3,0})}{\partial T} = 11,78 \cdot \frac{372,79 - 43,15}{(T - 43,15)^2}$$

Example 4

u = 0,10, T = 333 K
ϕ_1 = 0,288, ρ_1 = 450 kg/m^3
ρ_2 = $u \cdot \rho_1$ = 0,10 · 450 = 45 kg/m^3
ϕ_2 = ρ_2/d_2 = 45/1000 = 0,045
ϕ_β = 1 - (0,288 + 0,045) = 0,667
λ = 0,288 · 0,11 + 0,045 · 0,654 + 0,667 · 0,0250 = 0,0778 W/mK

$$k_\beta = \frac{7,988 \cdot 10^{-12}}{333} \cdot (\exp(-2,601 \cdot 10^{-3} \cdot 333 \cdot 0,10^{1,378}) - \exp(-2,601 \cdot 10^{-3} \cdot 333 \cdot 1,581)) = 1,703 \cdot 10^{-14} m^2$$

$$p_{3,0} = \exp(11,78\ (333 - 372,79)/(333 - 43,15)) = 0,1985 \text{ bar}$$

$$\ln \psi = -3,433 \cdot 0,10^{-1,3820} \cdot \exp(-3,372 \cdot 0,10) \cdot \left(\frac{647,3 - 333}{583,5}\right)^{7,557} = -0,550$$

ψ = exp (-0,550) = 0,577

p_3 = 0,577 · 0,1985 = 0,1145 bar

D_{34} = 33,4 · 10^{-6}m^2/s

γ_β = 18,99 · 10^{-6}m^2/s

$$\varepsilon_b = \frac{0,667}{\dfrac{\dfrac{1,0 - 0,1145}{1,0} + \dfrac{18,0}{8,314 \cdot 10^3 \cdot 333} \cdot 33,4 \cdot}{10^{-6} \cdot 18,99 \cdot 10^{-6} \cdot \dfrac{0,667}{1,703 \cdot 10^{-14}}}} = 0,637$$

$$\frac{\partial \ln \psi}{\partial T} = -3,433 \cdot 0,10^{-1,3820} \cdot \exp(-3,372 \cdot 0,10) \cdot 7,557 \cdot \left(\frac{647,3 - 333}{583,5}\right)^{6,557} \cdot \left(\frac{-1}{583,5}\right) = 0,01323$$

$$\frac{\partial(\ln p_{3,0})}{\partial T} = 11,78 \cdot \frac{372,79 \cdot 43,15}{(333 - 43,15)^2} = 0,04622$$

$$\frac{\partial p_3}{\partial T} = 0,1145\ (0,01323 + 0,04622) \frac{\text{bar}}{\text{K}} = 680,7 \frac{\text{Pa}}{\text{K}}$$

$$l - l_o = 8,633 \cdot 10^{-3} \cdot 0,10^{-1,3820} \cdot \exp(-3,372 \cdot 0,10) \cdot 333^2 \cdot \left(\frac{647,3 - 333}{696,0}\right)^{6,557} = 89,7 \text{ kJ/kg}$$

l_o = 2358,6 kJ/kg

l = 2358,6 + 89,7 = 2448,3 kJ/kg

$$\lambda_a = 0,0778 + 2448,3 \cdot 0,637 \cdot \frac{18,0}{8,314 \cdot 333} \cdot 33,4 \cdot 10^{-6} \cdot 680,7 = 0,308 \text{ W/mK}$$

The curves in figure 10 have been calculated according to this example. The higher the temperature is, the more important is the diffusion term $\rho_3 w_3 \cdot l$, as we can see from figure 10.

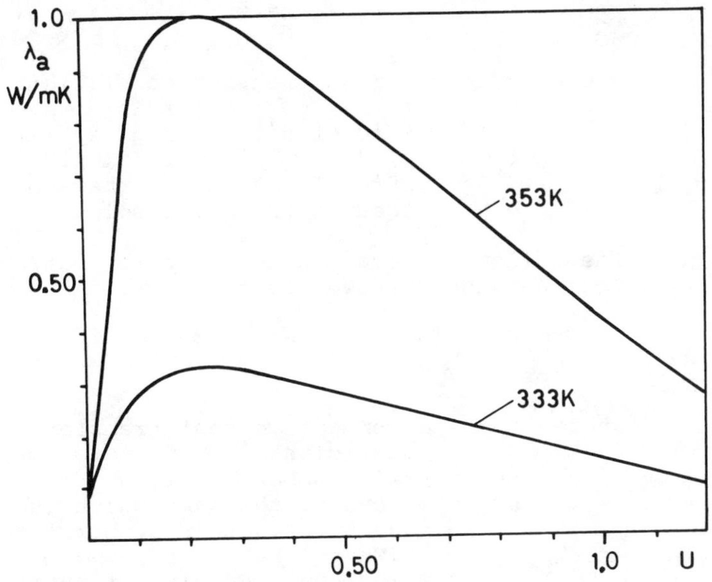

Fig 10 - The apparent thermal conductivity of a newsprint (45 g/m^2)

The results of a measurement shown in figure 11 can be regarded as an empirical proof for the heat pipe phenomena, and hence partly also for the curves in fig. 10. Instead of moisture content u, a relative volume ratio $\emptyset_2/(1-\emptyset_1)$ has been used.

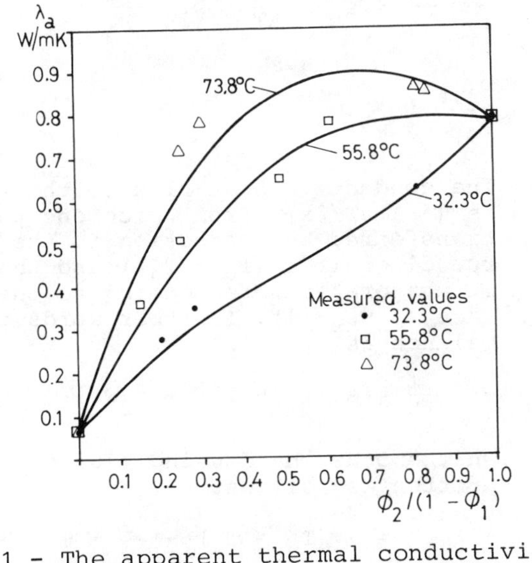

Fig 11 - The apparent thermal conductivity of a glass fiber mat based on measurements (3).

6 DRYING EQUATIONS

6.1 Balance equations

The mathematical solution of a drying problem requires the solution of the following differential equations: the mass balances, the momentum balances and the energy balance. The complete formulas of the balance equations are presented in (1).

In this simulation model we have considerably simplified some of these balance equations. For instance we have ignored the mass and momentum balances for the solid (α = 1) and we have simply regarded $\emptyset_1$ and ρ_1 as constant.

The mass balances for liquid water and water vapour are

$$\frac{\partial \rho_2}{\partial t} + \frac{\partial(\rho_2 w_2)}{\partial x} = I_2$$

$$\frac{\partial \rho_3}{\partial t} + \frac{\partial(\rho_3 w_3)}{\partial x} = I_3$$

By a numerial example it is easy to show that the term $\partial\rho_3/\partial t$ is small compared to the convection term $\partial(\rho_3 w_3)/\partial x$ so that this can be disregarded. Further taking into account that $I_2 + I_3 = 0$, we obtain by summing together these two equations

$$\frac{\partial \rho_2}{\partial t} = - \frac{\partial}{\partial x} (\rho_2 w_2 + \rho_3 w_3)$$

Disregarding the inertia forces, the momentum balances for liquid water and water vapour can be described by equations (16) and (24), respectively. Substituting these equations into the equation above and dividing that by ρ_1 (assumed to be constant) we get

$$\rho_1 \frac{\partial u}{\partial t} = \frac{\partial}{\partial x} \left(\frac{k_2}{\gamma_2} \frac{\partial p_2}{\partial x} + \varepsilon_b \cdot \frac{M_3}{RT} \cdot D_{34} \cdot \frac{\partial p_3}{\partial x}\right) \qquad (32)$$

The complete form of the energy balance is equation (11.2) in (1). Using equations (17.10) (1) and (18.2) (1) and then ignoring the kinetic energy and convection terms, deformation work and radiation term, the energy equation may be put into the following simplified form:

$$\rho\, c_p \frac{\partial T}{\partial t} + I_3 \cdot l = \frac{\partial}{\partial x} \left(\lambda \frac{\partial T}{\partial x}\right)$$

where $\rho\, c_p = \sum_{\alpha=1,2,\beta} \rho_\alpha c_{p\alpha}$

Again ignoring the term $\partial\rho_3/\partial t$, we can write $\partial(\rho_3 w_3)/\partial x = I_3$ and by substituting this into the equation above and by using (24), we get

$$\rho\, c_p \frac{\partial T}{\partial t} = l \cdot \frac{\partial}{\partial x}\left(\varepsilon_b \cdot \frac{M_3}{RT} \cdot D_{34} \cdot \frac{\partial p_3}{\partial x}\right) + \frac{\partial}{\partial x}\left(\lambda \frac{\partial T}{\partial x}\right) \qquad (33)$$

We have two unknown functions ($u(x,t)$, $T(x,t)$) and correspondingly two equations (32) and (33). All the coefficients depend on u and T; $k_2(T,u)$, $\varepsilon_b(T,u)$, $p_2(T,u)$, $p_3(T,u)$, $l(T,u)$, $\lambda(T,u)$. These questions we have analyzed in the previous chapters. Finally we give an example of the term $\rho\, c_p$.

Example 5

Newsprint, $u = 1{,}25$, $T = 293$ K
$\emptyset_1 = 0{,}288$, $\emptyset_2 = 0{,}563$, $\emptyset_\beta = 0{,}149$ (see example 3)
$\rho_1 = 450$ kg/m^3, $\rho_2 = 563$ kg/m^3, $\rho_\beta = \emptyset_\beta \cdot d_\beta = 0{,}149 \cdot 1{,}20 = 0{,}179$ kg/m^3
$c_{p1} = 1{,}40$ kJ/kg K, $c_{p2} = 4{,}19$ kJ/kg K, $c_{p\beta} = 1{,}05$ kJ/kg K
$\rho c_p = 450 \cdot 1{,}40 + 563 \cdot 4{,}19 + 0{,}179 \cdot 1{,}05 = 2990$ kJ/kg K

6.2 Boundary conditions in contact drying

In contact drying we have two different surfaces. The energy required for vaporization is transferred into the moist solid from the hot surface and from the opposite surface (upper face) vapour flows to the surrounding air.

On the contact surface water vaporizes and a corresponding amount of liquid water flows to the site. Hence the mass balance may be written as follows:

$$\rho_2 w_2 + \rho_3 w_3 = 0 \qquad (34)$$

Further, using equations (16) and (24), we get

$$\frac{k_2}{\gamma_2} \frac{\partial p_2}{\partial x} + \varepsilon_b \cdot \frac{M_3}{RT} \cdot D_{34} \cdot \frac{\partial p_3}{\partial x} = 0 \qquad (35)$$

The surface of the moist solid (against the hot plate) reaches the same temperature as the hot surface (see discussion in (2) of the correct mass disctributions) so that the energy boundary condition is simply

$$T\ (x = 0,t) = T_s\ (t) \qquad (36)$$

where T_s (t) is the temperature of the hot surface.

To the upper face, in contact with the surrounding air, flow vapour and liquid water (capillary flow) which evaporates on the surface. The sum of these two flows ($\rho_2 w_2 + \rho_3 w_3$) presents the total vapour outflow.

On the other hand the vapour outflow may be expressed by the formula $k\, M_3 \ln (p_{4,a}/p_{4,u})$ so that mass balance may be written as follows:

$$\rho_2 w_2 + \rho_3 w_3 = k\, M_3 \ln \frac{p_{4,a}}{p_{4,u}} \qquad (37)$$

where k = a mass transfer coefficient
$p_{4,a} = p_{\beta,a} - p_{3,a}$ = partial pressure of surrounding dry air
$p_{4,u} = p_{\beta,u} - p_{3,u}$ = partial pressure of dry air on the upper face of the moist solid

The following equation is an approximation for the mass transfer coefficient:

$$k = \frac{\alpha}{c_{pm}} \cdot Le^{1-n} \qquad (38)$$

where α = a convective heat transfer coefficient
Le = Lewis number
n = exponent, the same which is used in Prandtl's number ($Nu = A\, Re^m \cdot Pr^n$)
c_{pm} = molar heat capacity of humid air

The equation (38) as well as the formula $k\, M_3 \ln (p_{4,a}/p_{4,u})$ can be found in all standard textbooks on boundary layer theory.

Substituting (16) and (24) into (37) we obtain

$$\frac{k_2}{\gamma_2} \frac{\partial p_2}{\partial x} + \varepsilon_b \cdot \frac{M_3}{RT} \cdot D_{34} \frac{\partial p_3}{\partial x} = -\, k\, M_3 \ln \frac{p_{4,a}}{p_{4,u}} \qquad (39)$$

The conductive heat flux to the upper face is ($-\lambda\, \partial T/\partial x$), from which one part is transferred by convection to the surrounding air ($\alpha(T_u - T_a)$) and the rest of it evaporates water coming to the upper face ($\rho_2 w_2 \cdot l$). In other words the energy balance is

$$-\lambda \frac{\partial T}{\partial x} = \alpha\ (T_u - T_a) + \rho_2 w_2 \cdot l \qquad (40)$$

On the other hand using (16) this may be written as follows:

$$-\lambda \frac{\partial T}{\partial x} = \alpha\ (T_u - T_a) - \frac{k_2}{\gamma_2} \cdot l \cdot \frac{\partial p_2}{\partial x} \qquad (41)$$

The following example shows the calculation of total vapour outflow and convective heat flux:

Example 6

Newsprint, $u = 1{,}25$, $T_u = 353$ K
$\alpha = 40$ W/m^{2}K, $n = 0{,}4$, $T_a = 333$ K, $\psi_a = 0{,}50$

$$\ln \psi = -\,3{,}433 \cdot 1{,}25^{-1{,}3820} \cdot \exp(-3{,}372 \cdot 1{,}25) \cdot \left(\frac{647{,}3 - 353}{583{,}5}\right)^{7{,}557} = -\,2{,}113 \cdot 10^{-4}$$

$\psi = 0{,}9998$
$p_{3,u} = 0{,}9998 \cdot 0{,}4736 = 0{,}4735$ bar
$p_{4,u} = 1{,}0 - 0{,}4735 = 0{,}5265$ bar
$p_{3,a} = 0{,}50 \cdot 0{,}1992 = 0{,}0996$ bar
$p_{4,a} = 1{,}0 - 0{,}0996 = 0{,}9004$ bar

$$c_{pm} = \frac{p_3}{p_\beta} \cdot M_3\, c_{p3} + \frac{p_4}{p_\beta} \cdot M_4\, c_{p4} = \frac{0{,}4735}{1{,}0} \cdot 18{,}0 \cdot 1{,}85 + \frac{0{,}5265}{1{,}0} \cdot 29{,}0 \cdot 1{,}05 = 31{,}8 \text{ kJ/kmol K}$$

$Le = 1{,}65$

$$k = \frac{40}{31{,}8 \cdot 10^3} \cdot 1{,}65^{1 - 0{,}4} = 1{,}699 \cdot 10^{-3}\ \frac{\text{kmol}}{\text{m}^2\text{s}}$$

$$k\,M_3 \ln \frac{p_{4,a}}{p_{4,u}} = 1{,}699 \cdot 10^{-3} \cdot 18{,}0 \cdot \ln \frac{0{,}9004}{0{,}5265} = 0{,}01641 \text{ kg/m}^2\text{s}$$

$$\alpha\,(T_u - T_a) = 40 \cdot (353 - 333) = 800 \text{ W/m}^2$$

The boundary conditions are illustrated in figure 12.

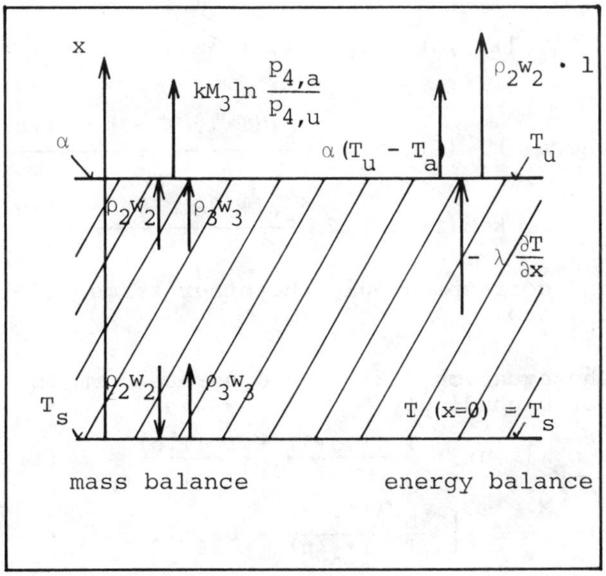

Fig. 12. Boundary conditions in contact drying.

Comment

In the numerical analysis presented in chapter 7 we allow mass for the surface elements also, and therefore the boundary conditions became a combination of these ideal boundary conditions and the balance equations.

7 APPLICATION OF THE DRYING THEORY FOR CONTACT DRYING OF PAPER

7.1 Description of an empirical drying simulation

In order to test the new material property and drying equations a drying simulator was built at Valmet Corporation Pansio Works. The principle of the simulator is shown in figure 13.

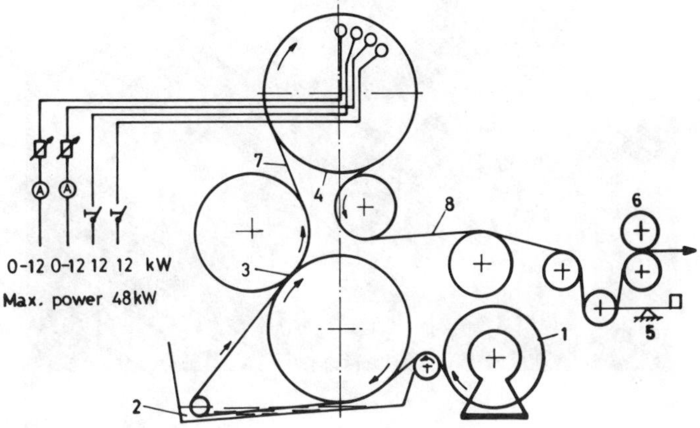

Fig. 13. The paper drying simulator.
1. Dry newsprint roll 2. Rewetting pool 3. Pressing nip 4. Electrically heated drying cylinder 5. Tension regulator 6. Waste paper roller 7-8. Locations of moisture content measurements

Some technical data on the drying simulator:

- diameter of the drying cylinder 414 mm and width 500 mm
- drying length (270/360 · π · 414) = 975 mm
- the width of the paper web 470 mm.

The drying curves of a rewetted newsprint have been experimentally determined with this simulator. The drying time was varied by changing the speed of the web. The shortest time achieved was about 0,3 seconds. The initial moisture content of the paper web was regulated by changing the force of the press. The surface temperature of the cylinder was controlled by electric heating.

After a stable running condition had been achieved the surface temperature of the cylinder was measured, as well as the moisture contents of the paper web before and after the drying cylinder. Moisture contents were determined by taking samples at the locations 7 and 8 shown in figure 13.

In order to obtain reliable results two men were needed to carry out the measurements. One of them took care of the drying condition and the other looked after the running of the paper web (fig. 14). The tension of the web was in most tests 106 N/m.

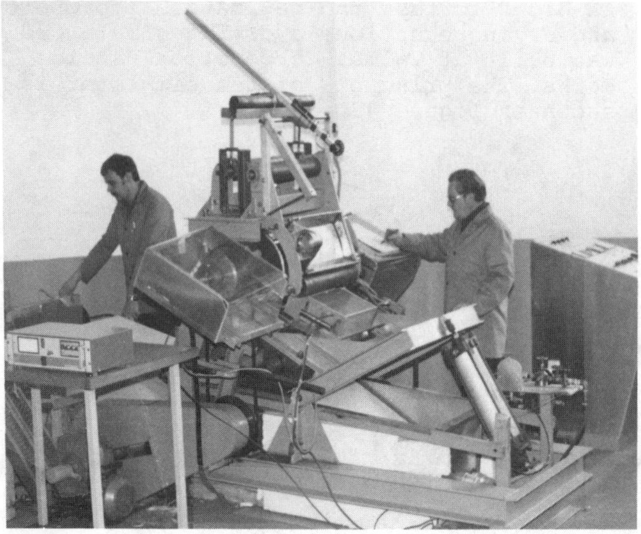

Fig. 14. The drying simulator in use.

Some results of these measurements are shown in figure 16.

7.2 Numerical solution of the drying equations

The region described by two independent variables is a part of a plane. The length variable, x, varies between 0 and Δs, where Δs is the thickness of the paper. The time variable, t, increases without limit from zero. When these two independent variables are replaced by discrete variables, the new variables are defined at points located as shown in figure 15.

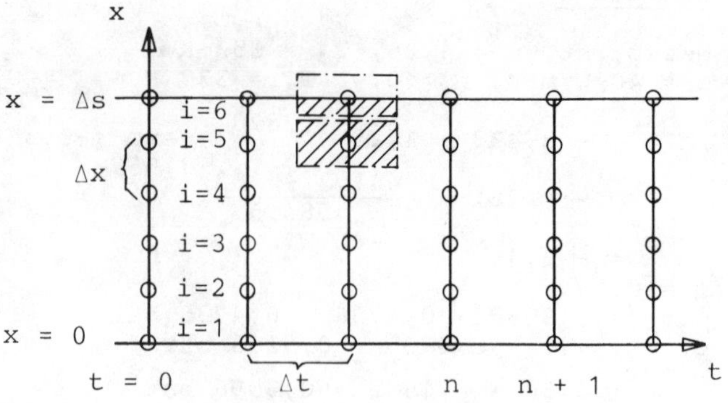

Fig. 15. Grid points for the problem of drying paper.

At the internal points (i = 2...5) the values of u and T are determined by using the equations (32) and (33). Using forward difference equations we obtain from (32)

$$\rho_1 \frac{u\,(i,n+1) - u\,(i,n)}{\Delta t} = \frac{1}{(\Delta x)^2} \cdot$$

$$\cdot \left[\frac{k_2\,(i+\tfrac{1}{2},n)}{\gamma_2\,(i+\tfrac{1}{2},n)} \cdot (p_2\,(i+1,n) - p_2\,(i,n)) - \right.$$

$$\left. - \frac{k_2\,(i-1,n)}{\gamma_2\,(1-\tfrac{1}{2},n)} \cdot (p_2\,(i,n) - p_2\,(i-1,n)) \right] +$$

$$+ \frac{M_3}{R(\Delta x)^2} \cdot \left[\frac{\varepsilon_b\,(i+\tfrac{1}{2},n)}{T\,(i+\tfrac{1}{2},n)} \cdot D_{34}\,(i+\tfrac{1}{2},n) \cdot \right.$$

$$\cdot (p_3\,(i+1,n) - p_3\,(i,n)) - \frac{\varepsilon_b\,(i-\tfrac{1}{2},n)}{T\,(i-\tfrac{1}{2},n)} \cdot$$

$$\left. \cdot D_{34}\,(i-\tfrac{1}{2},n) \cdot (p_3\,(i,n) - p_3\,(i-1,n)) \right] \qquad (42)$$

where $k_2\,(i+\tfrac{1}{2},n) \equiv \dfrac{k_2\,(i+1,n) + k_2\,(i,n)}{2}$

$$k_2\,(i-\tfrac{1}{2},n) \equiv \frac{k_2\,(i,n) + k_2\,(i-1,n)}{2}$$

and correspondingly the other terms γ_2, ε_b, etc.

The equation (33) in a discrete form is correspondingly

$$\rho\, c_p\,(i,n) \cdot \frac{T\,(i,n+1) - T\,(i,n)}{\Delta t} = l\,(i,n) \cdot$$

$$\cdot \frac{M_3}{R(\Delta x)^2} \cdot \left[\frac{\varepsilon_b\,(i+\tfrac{1}{2},n)}{T\,(i+\tfrac{1}{2},n)} \cdot D_{34}\,(i+\tfrac{1}{2},n) \cdot \right.$$

$$\cdot (p_3\,(i+1,n) - p\,(i,n)) - \frac{\varepsilon_b\,(i-\tfrac{1}{2},n)}{T\,(i-\tfrac{1}{2},n)} \cdot$$

$$\cdot\ D_{34}\ (i-\tfrac{1}{2},n)\ \cdot\ (p_3\ (i,n)\ -\ p_3\ (i-1,n))\Big]\ +$$

$$+\ \frac{1}{(\Delta x)^2}\Big[\lambda\ (i+\tfrac{1}{2},n)(T\ (i+1,n)\ -\ T\ (i,n))\ -$$

$$-\ \lambda\ (i-\tfrac{1}{2},n)(T\ (i,n)\ -\ T\ (i-1,n))\Big] \qquad (43)$$

which form is used at points i = 2...5.

According to the boundary condition (36) the temperature at points i = 1 is

$$T\ (1,n)\ =\ T_s\ (n) \qquad (44)$$

The boundary conditions (35), (39) and (41) are reformulated in the numerical solution so that the mass of the surface element (see fig. 15) is taken into account. According to this idea the boundary condition (35) takes the following form:

$$\frac{k_2\ (1+\frac{1}{2},n)}{\gamma_2\ (1+\frac{1}{2},n)}\ \cdot\ \frac{p_2\ (2,n)\ -\ p_2\ (1,n)}{\Delta x}\ +$$

$$+\ \frac{\varepsilon_b\ (1+\frac{1}{2},n)\ \cdot\ M_3\ \cdot\ D_{34}\ (1+\frac{1}{2},n)}{R\ \cdot\ T\ (1+\frac{1}{2},n)}\ \cdot$$

$$\cdot\ \frac{p_3\ (2,n)\ -\ p_3\ (1,n)}{\Delta x}\ =\ \rho_1\ \frac{\Delta x}{2}\ \cdot$$

$$\cdot\ \frac{u\ (1,n+1)\ -\ u\ (1,n)}{\Delta t} \qquad (45)$$

Reformulating equation (39) in a similar way we obtain

$$\frac{k_2\ (5+\frac{1}{2},n)}{\gamma_2\ (5+\frac{1}{2},n)}\ \cdot\ \frac{p_2\ (6,n)\ -\ p_2\ (5,n)}{\Delta x}\ +$$

$$+\ \frac{\varepsilon_b\ (5+\frac{1}{2},n)\ \cdot\ M_3\ \cdot\ D_{34}\ (5+\frac{1}{2},n)}{R\ \cdot\ T\ (5+\frac{1}{2},n)}\ \cdot$$

$$\cdot\ \frac{p_3\ (6,n)\ -\ p_3\ (5,n)}{\Delta x}\ +\ k\ M_3\ \ln\ \frac{p_{4,a}}{p_4\ (6,n)}\ =$$

$$=\ -\ \rho_1\ \frac{\Delta x}{2}\ \cdot\ \frac{u\ (6,n+1)\ -\ u\ (6,n)}{\Delta t} \qquad (46)$$

Finally the boundary condition (41) has to be transformed into a form where energy can be stored also on the surface element. Applying (33) together with the boundary conditions we get

$$-\ \rho\ c_p\ (6,n)\ \cdot\ \frac{T\ (6,n+1)\ -\ T\ (6,n)}{\Delta t}\ \cdot\ \frac{\Delta x}{2}\ =$$

$$=\ l\ (6,n)\ \cdot\ \frac{M_3}{R\ \cdot\ \Delta x}\ \cdot\Big[\frac{\varepsilon_b\ (5+\frac{1}{2},n)}{T\ (5+\frac{1}{2},n)}\ \cdot$$

$$\cdot\ D_{34}\ (5+\tfrac{1}{2},n)\ \cdot\ (p_3\ (6,n)\ -\ p_3\ (5,n))\Big]\ +$$

$$+\ l\ (6,n)\ \cdot\ k\ \cdot\ M_3\ \cdot\ \ln\ \frac{p_{4,a}}{p_4\ (6,n)}\ +$$

$$+\ \frac{1}{\Delta x}\Big[\lambda\ (5+\tfrac{1}{2},n)(T\ (6,n)\ -\ T\ (5,n))\Big]\ +$$

$$+\ \alpha\ (T\ (6,n)\ -\ T_a) \qquad (47)$$

After the initial state (u (i,0), T (i,0)) has been given, the new values (u (i,1), T (i,1)) can be calculated at each point i = 1...6 with the aid of equations (42) - (47) without any iteration. After this step the new values for n = 2 can be calculated correspondingly. Proceeding in the same way, the whole drying process can be computed.

The numerical solution was applied to the case of the drying simulator. The results of these tests are shown in figure 16.

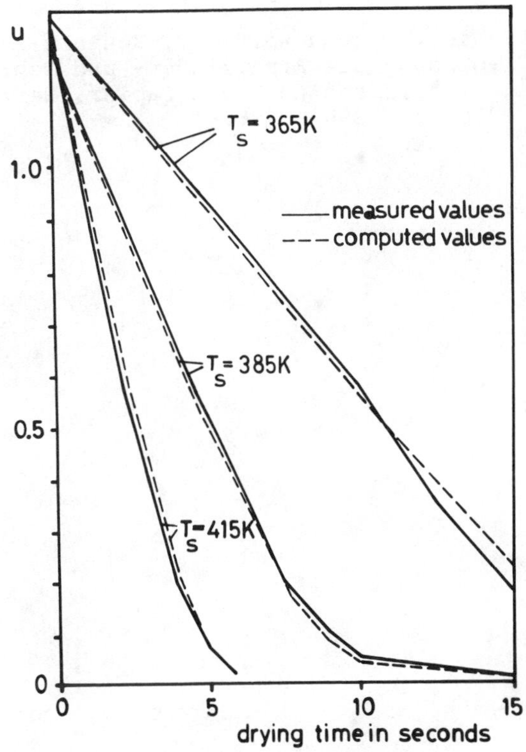

Fig. 16. A comparison of computed and measured values of drying curves achieved by the drying simulator at various surface temperatures.

The stability of the numerical solution method was studied experimentally. The result is

$$\Delta t\ \leq\ \frac{20\ (I\ -\ 1)}{I}\ \cdot\ \Delta x\ =\ \frac{20\ (6\ -\ 1)}{6}\ \cdot$$

$$\cdot\ 0{,}02\ \cdot\ 10^{-3}\ =\ 3{,}33\ \cdot\ 10^{-4}\ s$$

where $I = 6$, the number of grid points in x direction

$\Delta x = \Delta s/5 = 0{,}1 \cdot 10^{-3}/5 = 0{,}02 \cdot 10^{-3}$ m

In other words, if the time step is equal or less than $3{,}33 \cdot 10^{-4}$ s, the procedure is stable. Using $\Delta t = 3{,}33 \cdot 10^{-4}$ s the computing time was about 15 minutes in calculating a 15 seconds long drying process.

REFERENCES

1. Lampinen, M., Mechanics and Thermodynamics of Drying, Acta Polytechnica Scandinavica, Mechanical Engineering Series No. 77. Helsinki 1979. 104 pp.

2. Lampinen, M., Mechanics and Thermodynamics of Drying - A Summary, Drying 80 Volume 2: Proceedings of Second International Symposium, Montreal, McGill University. 1980.

3. Schiel. C. und Wolf, R., Kontakttrocknung auf Einzylinder- und Mehrzylindermaschinen, Das Papier, Heft 7, 1978, pp. 290 - 298.

EVAPORATION OF LIQUIDS WITH DIFFERENT SCHMIDT NUMBER FROM A SURFACE SUBJECT TO PERPENDICULAR STREAM FROM A SINGLE SLOTTED NOZZLE

F. Křížek Ing.

National Research Institute for Machine Design,
Praha 9 250 97 ČSSR

ABSTRACT

The results are given from the experimental research of the evaporation of different liquids (acetone, toluene, carbontetrachloride, ethylacetate etc.) from the plate subject to an impact stream from a slot nozzle. The processing of the results has shown us that the usual criterion functions of $Sh=k\ Re^m Sc^n$ and/or $j_d=k\ Re^{m-1}$ type cannot be used. On the basis of the study of the recent papers concerning the topic of the evaporation of various liquids from a plate along which the flow streams in the longitudinal direction the criterion of phase conversion K and simplex expressing the ratio of the total pressure in air stream to a partial pressure of liquid vapours on the phase boundary were used for the expression of the results. Such processing yielded satisfactory results so that the equation for the calculation of the mass transfer in the investigated case of mass transport during the evaporation of various liquids could be determined.

NOMENCLATURE

D = diffusion coefficient, m^2/h
F = surface , m^2
G = mass, kg, g
L = characteristical dimension, m
R = gas constant, J/kg deg
T = absolute temperature, K
$\bar{T}$ = absolute temperature for K
a = coefficient of heat conductivity, m^2/h
b = nozzle width, mm, m
c_k= specific heat of liquid, J/kg deg
P_p''= vapour pressure at saturation limit, N/m^2,
P = total atmospheric pressure, N/m^2
r = evaporation heat, J/kg
s = distance between plate and nozzle mouth, mm
t = temperature, ^oC
$\bar{t}$ = temperature as per relation (9), ^oC
w = flow velocity, m/s
w_o= flow velocity in nozzle mouth, m/s
α = coefficient of heat transfer, W/m^2deg, $kcal/m^2h^oC$
β = coefficient of mass transfer, m/h
λ = heat conductivity, W/m deg
ν = kinematic viscosity, m^2/s

SUBSCRIPTS

k = for liquid
v = for air
D = for vapour
m = for wet bulb thermometer
P = for surface

INTRODUCTION

The drying practice very often requires an evaporation from a material to be dried in cases when the liquid is not water. Mostly the liquids reaching higher evaporation rates than the evaporation rate of water are concerned - these are the highly volatile liquids. Such liquids are often evaporated during a nozzle drying which means during an impingement flow. For this reason we are interested in this matter in connection with evaporation problems - mass transfer during the air stream outlet from the nozzles normal to the evaporating (during the impingement flow).

In principle this is the verification of the validity of the relations for mass transfer at an impingement streaming which relations are obtained from water evaporation and/or from a sublimation of naphthalene (L 1, L 2) and which apply to cases of evaporation of different liquids.

ANALYSIS OF PAPERS DEALING WITH AN EVAPORATION OF DIFFERENT LIQUIDS

Many papers are dealing with the problems of the evaporation of various liquids which evaporate from a porous surface or from a free level. Only a single paper (L 3) shortly describes the evaporation of various liquids during the impingement flow i.e. in case of a single circular stream. The evaporated liquids were butanol (C_4H_{10}) and acetic acid ($C_2H_4O_2$) which saturated the evaporating surface formed by a layer of fine sand. The measurements proved that the dependence of Sh on Re maintains its character independently on the kind of evaporated liquid. Contrary to this the dependence of Sc number on the evaporated matter was clearly apparent although this number varied within a considerably small extent only. The dependence of mass transfer on Sc was determined by exponent 0.34 at Sc number.

Rather more papers are dealing with the problems of evaporation in case of heat and mass transfer which are investigated oftener (e.g. mass transfer from longitudinally by-passed plate, single ball etc.)

Very important paper in this branch is (L 4) endeavouring to summarize the results of the evaporation or sublimation of various matters in different cases of heat and mass transfer. The authors of this paper introduced the expression of the heat and mass transfer by means of j_H and j_D factors. In the occuring equations the number of exponent $n = n' = 0.33$ is used for Pr and Sc although various cases of mass transfer are compared. The authors have been aware of this inaccuracy and the paper states the comparison of the results of various authors for the different cases of mass and heat transfer and the occured errors are discussed.

In (L 5) the relations for j - factors are given in a different form which respects the differences of exponents n, n' for various cases of heat and mass transfer. For the rate of evaporation from the surface of balls the literature (L5) states the relation

$$m = \frac{\lambda_v \cdot \Delta t}{r} \pi D' \left[2 + 0,303 (Re_D \cdot S_c)^{0,6} \left(\frac{\lambda_v}{\lambda_D} \right)^{0,5} \right] \quad (1)$$

in which D' is the diameter of the ball, λ_v is thermal conductivity of air, λ_D is thermal conductivity of vapours and r is evaporation heat. We see that the effect of the physical properties of the evaporated matter is expressed in considerably more complicated form in dependence on several variables, mostly on Sc criterion. The latter has, nevertheless, a common exponent with Re_D. Another dependence is on quotient $\frac{\lambda_v \cdot \Delta t}{r}$ and finally on coefficient $\left(\frac{\lambda_v}{\lambda_D}\right)^{0,5}$ which, according to (L 5), considers approximately the simultaneous course of heat and mass transfer. This fact indicates that the mutual relation between the heat transfer and mass transfer may be different in certain cases for the evaporation of various liquids. The extent of mutual influence of the processes may vary.

The results of the investigations of the effect of Sc and Pr number on the course of the transport phenomena during the liquid flow inside the pipe are summarized in (L 7). It was ascertained that it was impossible to use one exponent equalling to 0.33 but that its value varied from 0.5 (for Pr, Sc numbers near one) to 0.25 (for Pr and Sc numbers higher than one).

Further processing of the analogy between the heat transfer and mass transfer resulted in supplementing of the criterion equations of the usual type

$$S_h = k \cdot R_e^m \cdot S_c^n \qquad (2)$$

with so called dimensionless pressure number $\frac{P - P_p''}{P}$ (L 8) so that the equation for mass transfer has the following form

$$S_h \cdot \frac{P - P_p''}{P} = k \cdot R_e^m \cdot S_c^n \qquad (3)$$

Such processing was used e.g. by Loos (L 9) for the results of the experiments with water evaporation from porous bodies of various shapes.

The introduction of the pressure number same as the quotient $\frac{\lambda_V}{\lambda_D}$ in the relation (1) show that the better knowledge of transport phenomena and the improved measuring methods enabled more accurate analysis which proved that such phenomena are effecting one another if occuring simultaneously and that often the simple criterion equations are not enough. The research is aimed at the solution of the question whether simultaneous mass transfer decreases or increases heat transfer and/or under what conditions the increase or decrease occurs. This question is being dealt with in many Soviet papers issued in the course of the past ten years.

The results as per (L 10) proved that the evaporation of various liquids caused an increased intensity of the heat transfer due to an increased intensity of evaporation. The processing of the results required the introduction of another criterion

$$Gu = \frac{t - t_m}{T}$$

which enabled the processing of the results by means of a single equation. Another knowledge gained from this investigation is that the coefficient of mass transfer β increases with the increase of molecular weight of the liquid.

The literature (L 11) gives the results of the experiments (longitudinally bypassed plate) at parameters which resulted in the high evaporation intensity. The high evaporation intensity is apparent by a higher transversal mass flow which is one of the factors effecting the relation of the transport phenomena. The transversal flow is expressed by the velocity of the motion of the vapours of the evaporated liquid from the evaporating surface usually in a ratio to the velocity of air streaming over the surface of evaporation. In this paper the mutual influence of the transport phenomena and thus also of the transversal flow is respected by criteria K or Gu. Criterion K is defined by relation

$$K = \frac{r}{c_k \cdot (t - t_p)} \qquad (4)$$

and is called the criterion of phase conversion.

This paper is linked up by with (L12) which also uses criterion K for the processing of the experiment results. Contrary to (L 11) it introduces also a simplex expressing the ratio of the total air pressure P to the partial pressure of liquid vapours on the phase boundary

P_p" in the criterion equations. The paper states that with the increasing of the transversal flow the mass transfer increases and the heat transfer decreases.

Common character of the papers (L 10, L 11, L 12) is the fact that they do not include the usually employed criteria S_c and P_r in the processing of the results. The given analysis of some papers dealing with the evaporation of different liquids shows us that the method of processing of the results and equations arising from it are subject to development in connection with the improved knowledge of the transport phenomena and with the more accurate measuring methods and it can be assumed that such development has not been terminated yet.

The following part of the paper is in ended for an elucidation of the results of the evaporation of various liquids (butanol $C_4H_{10}O$, ethyl alcohol C_2H_6O, methyl alcohol CH_4O, carbon tetrachloride CCl_4, acetone C_3H_6O, toluene C_7H_8, ethylacetate $C_4H_8O_2$) from a plate subject to the action of an impingement flow from a single slotted nozzle. When processing the results proper the knowledge gained from the study of the above mentioned literature has been taken into account. Below given data will show what results were received when using the individual methods of processing.

EXPERIMENTAL METHOD

The evaporating surface was made of a continuous frit plate 100 x 170 mm in size for the checking of the mass transfer during the evaporation of various liquids when it is sufficient to follow the mean coefficient of the mass transfer.The frit plate was cemented into a supporting plate of duralumin. The application of the frit plate for the model tests of termination of the coefficient of mass transfer and the device for the realization of this method are protected by Czechoslovak Patent No. 157384 (L 13).Liquid was supplied into the space formed under the frit plate by means of one mouth of the veseel placed on Sartorius electronic weighing machine. The quantity of the evaporated water was checked by the continuous weighing of the vessel. The holes for the placing of the thermocouples were bored on three places of the frit plate just under its active surface.General arrangement of the plate is shown in Fig. 1.

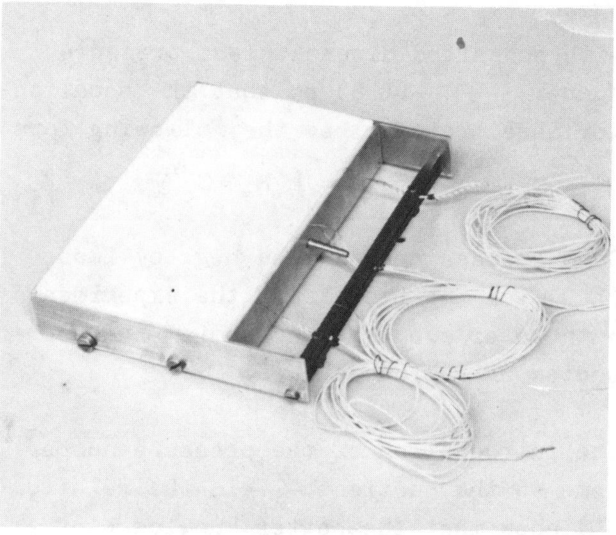

Fig. 1

In the course of the tests the testing plate was located under the slotted nozzle installed on the testing equipment for the investigation of impingement drying described in (L 1). The plate was positioned to the nozzle so that the flow centre line divided the plate into halvec(case of a uniform plate by-passing).

EVALUATION OF RESULTS

The used calculation relations are the following:

$$\bar{m} = \frac{\Delta G}{F \cdot \tau} \quad , \qquad (5)$$

$$\bar{m} = \frac{\beta}{R} \cdot \frac{P_p''}{\bar{T}} \quad , \qquad (6)$$

$$= \frac{GR\bar{T}}{F \cdot \tau \cdot P_p''} \quad , \qquad (7)$$

$$S_h = \frac{\beta \cdot L}{D} \quad , \qquad (8)$$

$$R_e = \frac{w_o \cdot L}{\sqrt{\frac{s}{b}} \cdot \nu} \quad , \qquad (9)$$

where $\frac{s}{b}$ is the relative distance between the nozzle and plate,

L is the width of active surface strip symmetrical to the flow (nozzle) centre line; in this case the length of the plate (m)

Vapour pressure at the saturation limit P_p'' in relations (6) and (7) is determined on the basis of mean temperature t_p of the plate surface measured by thermocouples. For the determination of in equation (9) there is decisive the air temperature in the nozzle mouth. As the definition temperature $\bar{t}$ for the determination of $\bar{T}$ and D the mean temperature of the limit layer was chosen according to the relation

$$\bar{t} = \frac{t_o + t_p}{2} \qquad (10)$$

For the evaluation of the results it is also necessary to know the pressure of vapours P_p'', diffusion coefficient D for an air-evaporated liquid system and gas constant R. For further processing of the results by means of the criteria it is also needed to know some thermophysical data of liquids used during the experiments. Many particulars exist in the literature for the functions and values of these constants which particulars differ slightly. Separate data were compared and critically evaluated. This formed the basis for the choice of such functions and data which seemed to be reliable. More details in this matter exceed the scope of the submitted paper.

EXTENT OF TESTS

The tests were performed with slot width b = 10 mm, distance between nozzle and plate s = 100 mm and flow rate in nozzle mouth w_o = 20, 30, 40 m/s. These parameters were chosen so as to lay within the domain of equation

$$S_h = 0{,}2\ b^{-0{,}155} R_e^{0{,}77} S_c^{1/3} =$$

$$= \bar{k} \cdot R_e^{0{,}77} \cdot S_c^{1/3} \qquad (11)$$

which was determined in the previous papers (L 1).

A variation of rate w_o brought the data S_h at various R_e which permitted the verification of the exponent value and this criterion. The influence of different $\frac{S}{b}$ and s was not investigated because if equation (11) is valid for $\frac{S}{b}$ = 10 and S = 10 mm and for different R_e for the evaporation of various liquids then there is no reason to doubt its suitability for $\frac{S}{b}$ and b differing from the previously mentioned values.

About 60 tests were performed in all. The evaporation intensity was always determined as the mean of 2 or 3 tests.

PROCESSING OF TEST RESULTS

The values of criteria S_h and R_e were yielded on the basis of the experiments. Such values were, first of all, used for the verification of validity of exponent at R_e number in the equation (11). For such purpose the values of simplex

$$\frac{S_h}{\bar{k} \cdot S_c^{1/3}}$$ (where $\bar{k} = 0.1399$) depending only on R_e were determined. The graph of the results gained in this way for CCl_4, CH_4O, $C_4H_8O_2$, C_2H_6O, $C_4H_{10}O$ and C_3H_6O is given in Fig. 2. The results show that the bellow condition is fulfilled with sufficient accuracy

$$\frac{S_h}{\bar{k} \cdot S_c^{1/3}} \quad R_e^{0,77}$$

If we express the results in the functional dependence

$$\frac{S_h}{\bar{k} \cdot R_e^{0,77}} = S_c^{1/3}$$

which represents an adapted relation (11) we ascertain that the influence of the thermophysical properties of the evaporating liquids as respected in the equation (11) $S_c^{1/3}$ is insufficient. Fig. 3 shows the sizes of the deviations from the assumed dependence for the individual liquids. It is apparent that in some cases the deviations are considerable (CH_4O, CCl_4, C_3H_6O). This figure also answers the question why such differences were not found when comparing the water evaporation tests and naphthalene sublimation tests.In this case the validity of relation (11) is fulfilled.

Further processing of the results investigated how the arrangement of the measured values would be effected by the introduction of the dimensionless pressure number $\frac{P - P_p''}{P}$ ($P = 10000$ kp/m^2). The results are given in Fig. 4. We see that still wider dispersion variance of the measured values occurs. More attention was, therefore, paid to the processing of the results by means of criterion K.

First of all we tried to express the results in the method similar to (L 11) which means that criterion S_c in the criterion equation was substituted by criterion K. K was determined for the individual tests according to relation (4) and values $\frac{S_h}{\bar{k} \cdot R_e^{0,77}}$ were plotted in dependence on its values. The graph of this is shown in Fig. 5. The yielded image of distribution of points did not show any notable tendency of their dependence on K.

Only the introduction of another factor used in (L 12) i.e. the ratio of total atmospheric pressure P to vapour pressure P_p'' led to a satisfactory processing of the measured results. The values $\frac{S_h}{\bar{k} \cdot R_e}$ were divided by this ratio and again plotted in dependence on K. The result is given in Fig. 6. In the given case the measured values already show a unique dependence on K. The dependence of values $\frac{S_h}{\bar{k}\, R_e^{0,77} \cdot \frac{P}{P_p''}}$ on K was elaborated in a linear regression method and the below equation was determined for the straight line passing through the measured points

$$\frac{S_h}{\bar{k} \cdot R_e^{0,77}} \cdot \frac{P_p''}{P} = 1,002\ k^{-1,184} \qquad (12)$$

and/or

$$S_h = 1,002\ \bar{k} \cdot R_e^{0,77} \cdot k^{-1,184} \cdot \frac{P}{P_p''} \qquad (13)$$

This means that the intensity of mass transfer decreases with the increase of the criterion of phase conversion. This is in agreement with the results as per (L 12) which deals with the longitudinally by-passed plate and concludes similarly with the only difference in K for which exponent -0.6 is stated. Fig.6

also includes the points yielded from the test with the method of naphthalene sublimation for which the criterion K is higher by one order (sublimation heat is taken as r). Even so if we extend the straight line according to equation (12) into this domain the deviation between the points and straight line is not too big.

The processing of the experimental data showed that a mere respecting of the thermophysical properties of the evaporating liquid and of the mutual effects of heat and mass transfer with the help of S_c number and ratio $\frac{P - P_p''}{P}$ is insufficient. On the other hand the application of phase conversion criterion K and ratio $\frac{P}{P_p''}$ enables the finding of the correlation of the result and the determination of the relation (13) which may be used for the determination of the mass transfer coefficients for the impingement flow from a slotted nozzle or a system of slotted nozzles i.e. within $S_c = 0.6 - 2.5$ and/or with $K = 7 - 500$. The ranges of the other parameters ($w_o = 10 \div 50$ m/s, $b = 5 \div 40$ mm, $\frac{L}{b} = 1{,}25 - 20$, $\frac{S}{b} = 8.5 - 80$) remain the same as in the original equation (10).

The problem of the exponent at K (eq.13) is still unsolved because its value for the case of the longitudinally by-passed plate and the investigated case of the impingement flow differs considerably which fact indicates its dependence on the case of the mass transfer. Future work will have to verify if the same exponent at K is reached as in equation (13) e.g. for the case of an impingement flow from a circular nozzle.

REFERENCES

L 1 Korger M., Křížek Fr.: Strojírenství, 1967, No. 7, p. 536 - 541

L 2 Křížek Fr.: Research Report SVÚSS 71-09001, Běchovice 1971

L 3 Schrader M.: VDI-Forschungsheft 484, Ausgabe B, 1961

L 4 Chilton T.H. Colburn A.P. Industrial and Engineering Chemistry, 1934, No.11, p. 1183 - 1187

L 5 Greber G.: Die Grundsetze der Wärmeübertragung (Springer, Berlin, 1955)

L 6 Maisel D.S. Sherwood T.K. Chemical Engineering Progress, 1950, No.3, p. 131 - 139

L 7 Spalding D.B. Convective Mass Transfer (Arnold, London, 1963)

L 8 Krischer O. Die wissenschaftlichen Grundlagen der Trocknentechnik (Springer, Berlin, 1956)

L 9 Loos G.: Diploma paper TH Darmstadt, D 17, 1957

L 10 Sergejev G.T. IFŽ, 1961, No.2, p. 77 - 81

L 11 Fedorov B.I. IFŽ, 1964, No. 1, p. 21 - 27

L 12 Vajnberg P.Š. IFŽ, 1967, No. 1, p. 51 - 58

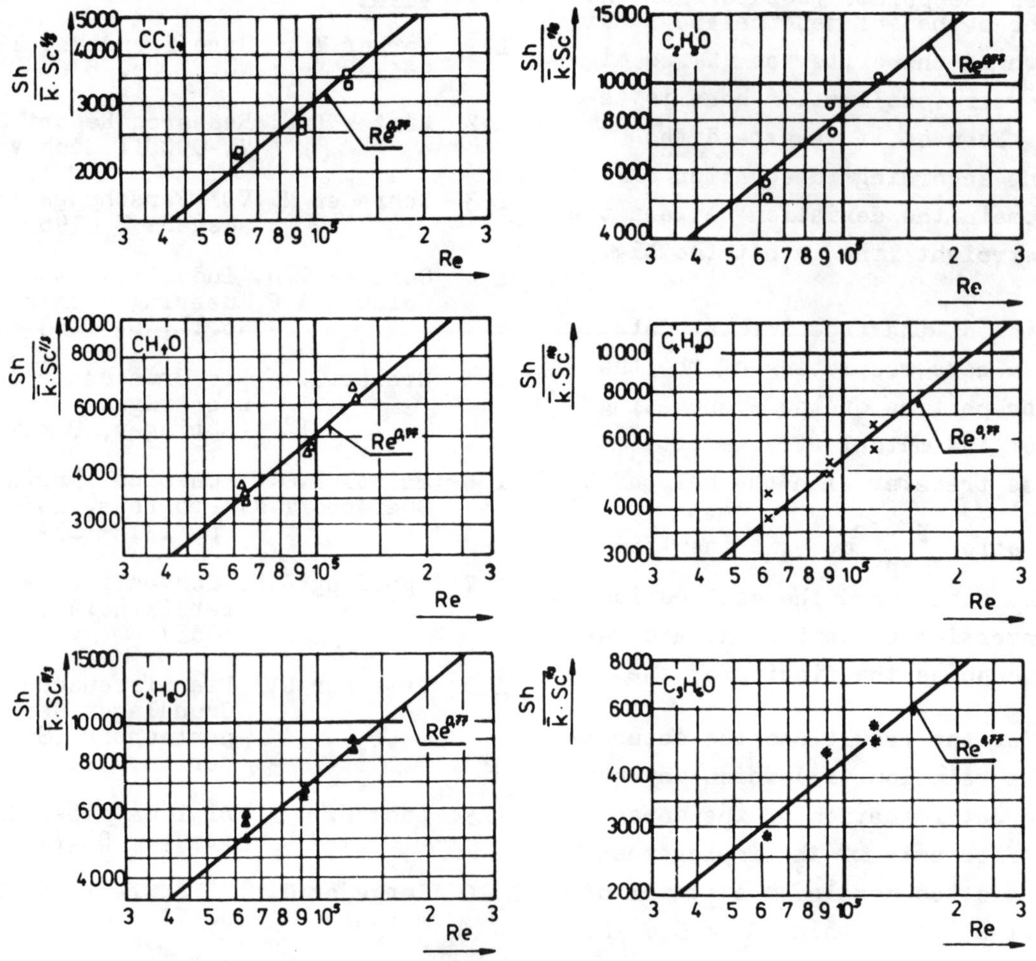

Fig. 2

Dependence of $\frac{Sh}{\bar{k}\ Sc^{1/3}}$ on Re number for $CCl_4, CH_4O, C_4H_8O_2, C_2H_6O, C_4H_{10}O, C_3H_6O$.

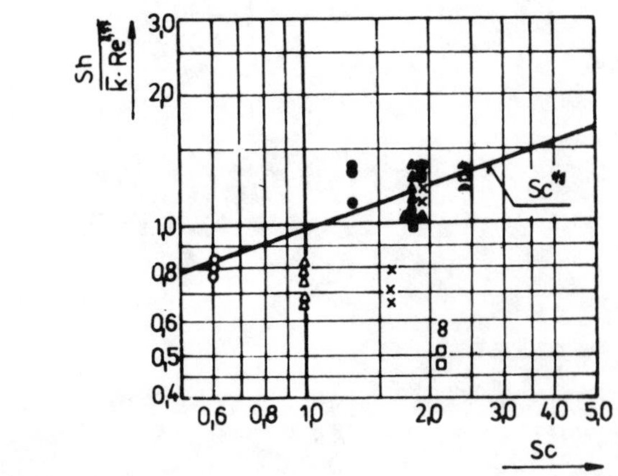

Fig.3

Dependence of $\frac{Sh}{\bar{k}\ Re^{0,77}}$ on Sc number

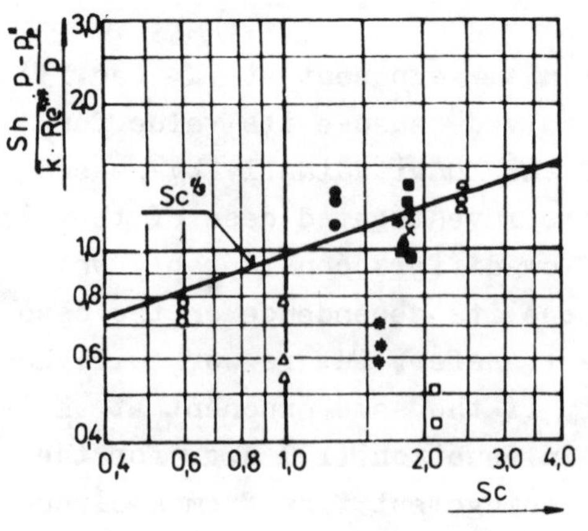

Fig. 4

Dependence of $\frac{Sh}{\bar{k}\ Re^{0,77}} \cdot \frac{P-P_p''}{P}$ on Sc Number

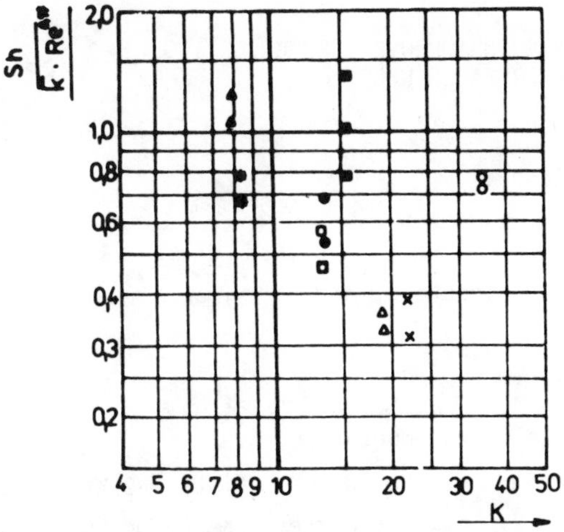

Fig.5
Dependence of $\frac{Sh}{\bar{k}\ Re^{0,77}}$ on phase conversion criterion K

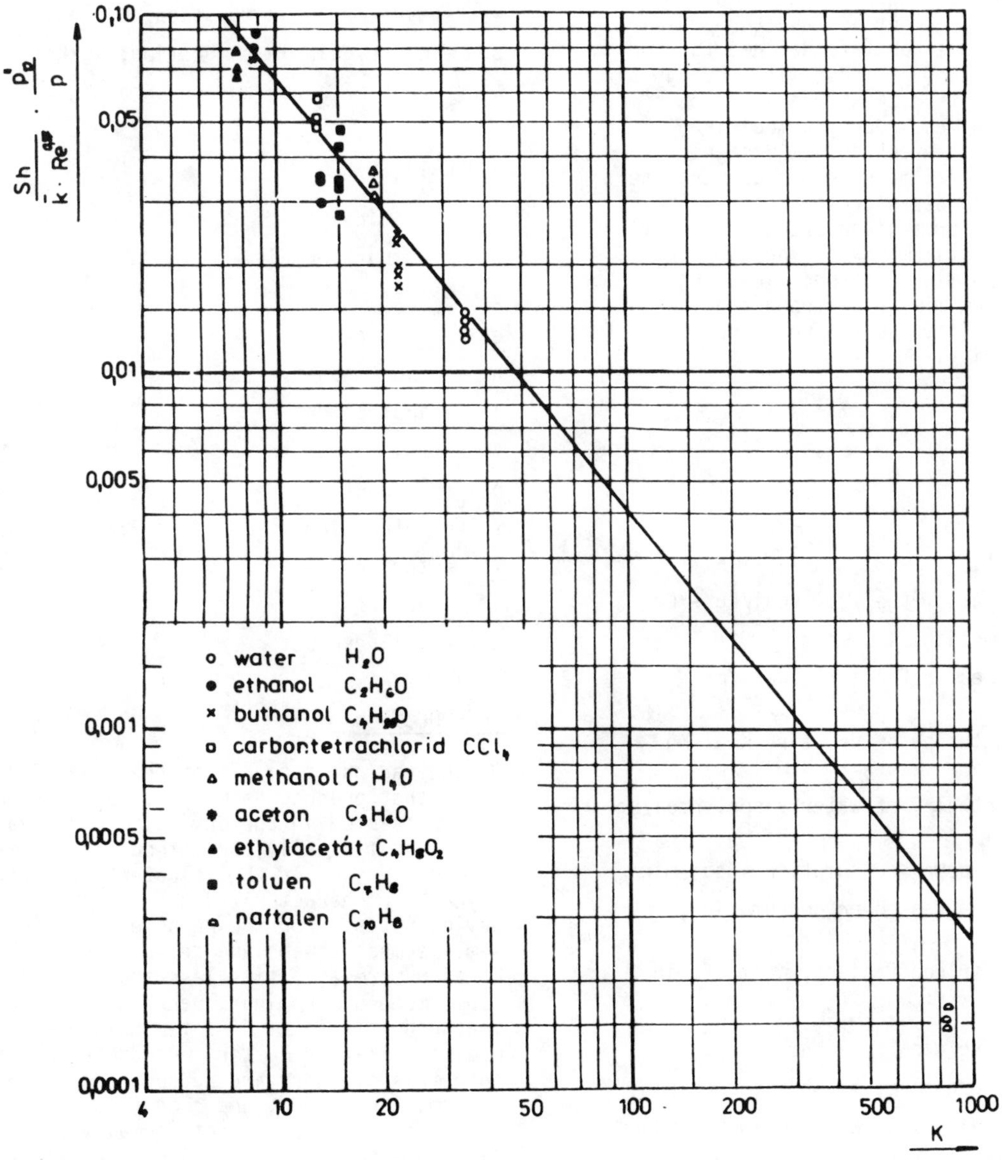

Fig. 6 Dependence of $\frac{Sh}{\bar{k}\ Re^{0,77}} \frac{P''}{P^p}$ on criterion K

USING HIGH VELOCITY IMPINGEMENT AIR TO IMPROVE THROUGH DRYING PERFORMANCE ON SEMI-PERMEABLE WEBS

Kenneth R. Randall

Honeycomb Systems, Inc., Biddeford, ME.

ABSTRACT

A mathematical model is developed to predict the heat and mass transfer rates for the process of drying a semi-permeable web by the simultaneous application of impingement air and through-air flow. Semi-permeable webs are defined for the purposes of this study as those with through-air velocities of less than 24 standard cubic meters/min-m^2 when measured by a modified Frazier Tester. Different system designs involving the critical machine design parameters are examined on the basis of the model. The results show that for a given product and drying load there is an optimum mix of through-air/impingement air to minimize specific energy cost. This optimum mix is a strong function of sheet air permeability and the relative costs of fuel and electricity.

NOMENCLATURE

a = constant

A = area, m^2

B = coefficient

C_p = specific heat of the web, kcal/kg - C

f = fuel consumption, Kcal/min

h = heat-transfer coefficient, Kcal/m^2 - min - C

h_a = specific enthalpy of dry air, Kcal/Kg dry air

h_v = specific enthalpy of water vapor, Kcal/Kg water

ΔH_x = exhaust fan power consumption, Kcal/min

ΔH_R = recirculation fan power consumption, Kcal/min

K = mass transfer coefficient, Kg/m^2 - min

m = mass, Kg

$\dot{m}$ = mass flow, Kg/min

P = barometric pressure, mm of water column

P_v = vapor partial pressure, mm of water column

Q = heat transfer rate, Kcal/min

t = time, min.

T = temperature, C

W_p = sheet water content, Kg water/Kg product

X = water content, Kg water/Kg dry air

α = air lost in combustion, Kg/Kcal

β = water gain in combustion, Kg/Kcal

λ = heat of vaporization, Kcal/Kg

SUBSCRIPTS

cr = critical

c = combustion air

d = system exhaust

e = hood exhaust

h = hood supply

i = impingement

ℓ = roll leak

p = product

s = saturation

t = through air

v = vapor

x = roll exhaust

INTRODUCTION

Of the various segments of the paper industry, those that produce "semi-permeable" products are restricted in the drying methods available. These are typically the methods of steam heated cylinder drying or the "floater" type convective dryer both characterized by relatively low drying rates. For the purposes of this paper semi-permeable webs are defined as those with Frazier permeabilities less than 24 standard cubic meters per minute per square meter when measured at 1270 mm of water column with a modified Frazier Tester. Products that fall into this category are newsprint, corrugating medium, linerboard, and some pulp products; those which one would not normally consider a candidate for through-drying.

Air through-drying has become an established

drying method for permeable products (e.g. textiles, tissue, towel, filter, nonwovens) in new and upgraded installations. As the name implies, a heated air/water vapor mixture is passed through a moving sheet causing heat transfer to the web fibers by convection and, consequently, mass transfer. In the conventional method of steam heated cylinder drying, heat must be transferred to the web interior by conduction from those fibers in contact with the surface. Water is then carried away by vapor diffusion from the interior and by the complex mechanisms of free and forced convection from the web exterior. In through drying, process air comes into intimate contact with the individual fibers thereby increasing the effective heat transfer area and reducing the vapor boundary layer buildup which inhibits mass transfer thereby yielding drying rates many times greater than these other methods.

Through-Dryers offer certain advantages over cylinder and Yankee dryers and "floater" dryers:

(1) Better thermal performance due to a higher degree of saturation of the drying medium. A rotary through air dryer consumes as little as 780 Kcal/Kg of evaporated water while Yankee dryers typically use 1200 to 1700 Kcal/Kg water and steam cylinders use 900 Kcal/Kg water when the energy required to generate steam is included.

(2) Reduction in space requirements for a given drying task. Higher drying rates give smaller space needs resulting in smaller building requirements.

(3) Product enhancement, where desired, through increased bulk/permeability. Bulk or permeability can, however, be controlled by pre or post treatment on other equipment when these qualities are not desired.

The rotary through-air design features a porous honeycomb cylinder which supports the web on the drying arc and permits passage of the air through the product and through the shell. The shell is of high open area construction thus permitting air passage with minimal pressure drop. The exhaust system draws the heated air through the web. A fraction of the circulating air is discharged from the system carrying with it the water evaporated from the web. The major portion is recirculated and reheated to supply air temperature by an indirect or direct fired heat source. The heated supply air then enters the hood which coincides with the vacuum arc of the roll. Pressurized air from the hood is metered through a perforated orifice plate to assure uniform distribution in the drying area.

The direct application of rotary through dryers to semi-permeable webs has been restricted to date by the structural limitations of Honeycomb Rolls[R] resulting in low through flow rates and consequently low drying rates. The introduction of a novel roll design in 1979 [1] capable of withstanding roll vacuums up to 3200 mm of water column (w.c.) has increased the prospects in this application. The marriage of the through-drying concept with other drying techniques, e.g. infrared and impingement, thereby substantially increasing water removal rate while retaining the thermodynamic efficiency inherent in the through-drying process is highly desirable.

The concept of using high velocity impingement air in conjunction with through air drying is not new. The concept was first proposed and subsequently patented by Burgess et. al [2] in 1965 at the Pulp and Paper Research Institute of Canada. A pilot facility was constructed to demonstrate the feasibility of the process and upon successful completion an experimental unit was installed on a newsprint machine at Canadian International Paper Company [3]. Drying rates of up to 142 Kg/m^2-hr were demonstrated at production speed with negligible change in product quality when compared to paper dried convectionally on the same machine. The demonstration "Papridryer" consisted of a 1.52 meter diameter drilled suction roll covered with a wire sleeve and surrounded by a high velocity hood with profiling capability.

The air system configuration differs markedly from that required in normal through air drying. The system is shown schematically in Figure 1. The supply air plenum is replaced by a high velocity hood. A portion of the supply air expended from the jets is drawn through the web by a differential pressure created by the roll exhaust fan. The remainder is exhausted from the hood by the main loop recirculation fan, mixed with fresh intake air to replenish that withdrawn through the roll, and reheated to supply temperature. The air exhausted from the roll when compared to the hood exhaust air is relatively cool (100 versus 300 degrees Celsius) and moist (90% versus 5% relative humidity); hence, the system operates most efficiently when all system exhaust air is expended through the roll.

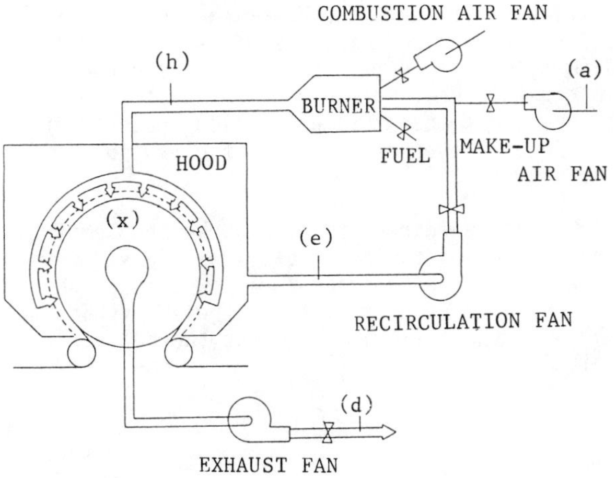

FIGURE 1. SYSTEM SCHEMATIC-IMPINGEMENT/THROUGH DRYER

The equipment technology for all components currently exists. High velocity hoods are used extensively in the paper industry and a wealth of information is available on the heat transfer theory of high velocity jets and their performance in conjunction with Yankee Dryers [4,5]. The use of a Honeycomb Roll instead of a drilled shell as the web transport surface offers certain unique advantages: (1) More flow area and therefore a larger percentage of the drying capacity by through drying and (2) less pressure drop and hence lower power requirements required to purge air through the shell. It is therefore the purpose of this paper to propose a model for computing the drying rate and specific energy consumption for simultaneous impingement and through-air heat and mass transfer as a function of the independent variables. As in all models simplifying engineering assumptions must be made to account for the unknown transfer phenomena and to make the mathematical solution manageable. It is then necessary to compare the analytical results to laboratory or field test data for accuracy and modify as necessary where incorrect assumptions are made. A model developed in this manner allows one to extract fundamental operating data from production observations, to study the basic machine response characteristics and inherently leads to optimum control of machine parameters.

MODEL DESCRIPTION

A summary of the early attempts to develop through-drying models is presented by Rohrer and Gardiner [6]. Much of the prior work has focused on the adiabatic saturation process taking place in the sheet. This has included both purely empirical results and simplified thermodynamic assumptions. These simplifying assumptions included: (1) air exiting the sheet is at the saturation temperature (2) equal mass flows through the sheet produce equal exit relative humidities and (3) constant moisture pickup by air in passing through the sheet. Many authors have also chosen to either neglect or simplify the transfer processes that occur in the warmup to the adiabatic saturation temperature and in the falling rate drying regions. At the high speeds typical of newsprint production, for example, the total dwell time on the dryer can be as little as 0.5 second and 30% of the surface of a 4.3 meter diameter rotary dryer is in the warmup region. To neglect sheet warm-up to the adiabatic saturation temperature and falling rate drying can lead to overestimation of the drying capacity.

Rohrer and Gardiner approached the problem from a purely mechanistic point of view. They derived a heat-transfer expression for flow through porous webs by working from classical relationships and obtained the following form for the heat-transfer relationship:

$$h = B\left(\frac{\dot{m}}{A}\right)^{0.8} \qquad (1)$$

The exponent was determined from numerous experimental tests. The B factor is a function of paper and vapor mix properties and is determined experimentally for each product type from through drying tests. The model has been used extensively to size through dryer systems and optimize process performance and has been checked with field test data from production machines.

Crotogino and Allenger [7] in their model of the Papridryer proposed that the individual contributions to the drying rate by pure impingement and pure through drying be combined linearly to compute the total drying rate. This assumes that the driving force for heat transfer by impingement and by through flow is the difference between the supply air temperature and the sheet temperature. They also assumed that the air passing through the web becomes fully saturated and its temperature is equal to the adiabatic saturation temperature in the constant rate zone. This is valid if the through air velocity is low or the heat-transfer coefficient is high. This assumption tends to overestimate the drying rate that will be realized in most applications and gives a model with limited applicability for a wide range of products. The analytical results were compared to the results of the pilot trials and the results from the mill trials. Agreement was within 11% of the measured data.

The thermodynamic model being proposed here for predicting drying rates, thermal efficiencies and machine system response for combination impingement and through drying is derived from the model proposed and used by Rohrer and Gardiner for through drying only. The mechanism equations have been altered, however, to include the effects of simultaneous heat transfer by impingement air and through air. The rate of heat transfer to a saturated web by combined impingement and through air flow is theoretically and experimentally difficult to determine. Theoretical studies of the impingement heat and mass transfer with wall suction have been accomplished [8]. The results of these studies are of limited value here because of (1) the assumptions of laminar flow and (2) the boundary conditions imposed upon the suction air flow.

The present method prefers to apply the two transfer mechanisms in true sequential fashion. It is assumed that first the impingement jet gives up its energy to the product and then the through air flow gives up its heat. In the first instance the driving potential is the difference between supply air and product temperature and in the second is local impingement exhaust air and product temperature. This assumption seems appropriate in view of the fact that, in most practical applications of the concept, the jet Reynolds numbers are sufficiently high such that the flow is fully turbulent and thus well mixed. The jet flow is also applied to a relatively small fraction (less than 2%) of the total drying surface and therefore sufficient time exists after impingement upon the web for the wedge flows to become fully mixed.

Sheet Model

The drying of porous webs where the mode of heat transfer is by convection alone may be divided into three clearly defined zones in which different rate controlling mechanisms govern: warmup period, constant rate period and falling rate period.

Constant Rate Drying

The constant rate drying period is classically defined as drying by vapor diffusion from a saturated surface the rate of which is controlled by heat transfer to the surface. It is assumed here that sufficient unbound moisture exists to keep the web and fiber surfaces saturated. In the absence of external heat sources (i.e. radiative and conductive surfaces which are neglected in this analysis) the web temperature is at the adiabatic saturation temperature and dynamic equilibrium is established between the rate of heat transfer to the surface and the rate of vapor removal from the web.

Because in this region the sheet is at the adiabatic saturation temperature, the energy conservation equates the rate of sensible heat input by impingement to the rate of evaporation:

$$\frac{dm_i}{dt} = Q_i/(\lambda + hv_e - hv_s) \qquad (2)$$

where the quantity $(hv_e - hv_s)$ is the energy required to raise water vapor from the saturation temperature to the exhaust temperature.

The heat input by impingement air, Q_i, is expressed by the following rate equation.

$$Q_i = h_i A\ (T_h - T_s) \qquad (3)$$

The impingement heat transfer coefficient, h_i, is calculated using the correlation developed by Chance [9] for round jets.

The heat transferred to the web by convection is equal to the sensible heat loss of the air stream in passing from the supply conditions of temperature and humidity, T_h and X_h respectively, to the local impingement exhaust temperature.

$$\dot{m}_h\ (ha_h + X_h hv_h) - \dot{m}_h\ (ha_e + X_h hv_e) = Q_i \qquad (4)$$

With a relationship between the dry air and water vapor enthalpy and temperature, Equation (4) can be solved explicitly for the local impingement exhaust temperature, T_e.

The mass conservation equation for water equates the rate of vapor removal by impingement air to the moisture gain by the air stream.

$$\frac{dm_i}{dt} = \dot{m}_h\ (X_e - X_h) \qquad (5)$$

Where the quantities X_e and X_h are the weights of vapor associated with a unit weight of dry air at the local impingement exhaust and supply conditions, respectively, and by the following equation bear a proportionality to the partial pressure of water vapor in the air.

$$X = 0.622\ \frac{P_v}{P - Pv} \qquad (6)$$

Implicit in the solution of Equation (3) is the assumption that the adiabatic saturation temperature, T_s, is known. Standard methods exist for computing this temperature, the simplest of which is an energy balance assuming that the exit air is fully saturated yielding

$$\lambda(X_s - X_h) = (ha_h + X_h hv_h) - (ha_s + X_h hv_s) \qquad (7)$$

With a relationship between the partial pressure of saturated steam and temperature and Equation (6), Equation (7) can be solved iteratively. It is important to note that for pure convection drying the evaporation temperature is independent of the drying method and dependent only upon supply air conditions.

By defining an impingement mass transfer coefficient, K_i, the rate of mass transfer can be alternately stated

$$\frac{dm_i}{dt} = K_i A\ \ln \frac{(1 + X_s)}{(1 + X_h)} \qquad (8)$$

Equations (2) thru (5) are identically repeated for the through drying contribution with the exception that the driving force for heat transfer is the difference between the local impingement exhaust air temperature and the saturation temperature.

$$\frac{dm_t}{dt} = h_t\ A\ (T_e - T_s) \qquad (9)$$

$$\frac{dm_t}{dt} = K_t\ A\ \ln \frac{(1 + X_s)}{(1 + X_e)} \qquad (10)$$

The total evaporation rate is expressed as the sum of the impingement contribution and the through drying contribution.

$$\frac{dm}{dt} = K_i\ A\ \ln \frac{(1 + X_s)}{(1 + X_h)} + K_t\ A\ \ln \frac{(1 + X_s)}{(1 + X_e)} \qquad (11)$$

Warmup Zone

The warmup period is characterized primarily by sensible heat convection to the web. The driving force for mass transfer is assumed to be the differential of the vapor partial pressure of water at the web temperature and the vapor partial pressure at the supply air temperature.

Using the rate equation established by Equation (8) the mass transfer rate becomes

$$\frac{dm_i}{dt} = K_i A \ln\frac{(1 + X_p)}{(1 + X_h)} \quad (13)$$

The local impingement exhaust humidity is

$$X_e = X_h + \frac{dm_i/dt}{\dot{m}_h} \quad (14)$$

The mass transfer rate for through air is obtained in a like manner from equation (10).

$$\frac{dm_t}{dt} = K_t A \ln\frac{(1 + X_p)}{(1 + X_e)} \quad (15)$$

The overall heat balance equates the energy transferred to the web by impingement and through air to the change in heat content of the web and water and the heat of vaporization.

$$\begin{aligned}\dot{m}_p (C_p + W_p C_w)\frac{dT}{dt} &= h_i A(T_h - T_p) \\ &+ h_t A(T_e - T_p) \\ &- \frac{dm_i}{dt}\left(\lambda + h_{v_e} - h_{v_p}\right) \\ &- \frac{dm_t}{dt}\left(\lambda + h_{v_x} - h_{v_p}\right) \quad (16)\end{aligned}$$

Falling Rate Zone

The falling rate drying period begins when the constant rate period ends at the critical water content. In this zone, drying is not controlled by heat input but by internal diffusion within the mat and fibers and is characterized by a decreasing drying rate. Falling rate can under some circumstances be on a macroscopic scale where non-uniform web permeabilities and basis weights result in uneven air flow distribution and hence drying but only the microscopic case is considered here. To classify the critical water content based upon a micro-analysis of web properties is highly desirable but has not been successful to date. The only recourse is to rely upon published data which is only approximate since the critical water content is somewhat dependent upon drying history or to rely upon laboratory trials using identical drying methods.

The rate of drying in the falling rate zone has been variously correlated in the literature with the drying rate in the constant rate zone. These correlations have taken the form of linear, logarithmic, hyperbolic, and elliptic relationships between drying rate and the local water content. The relationship preferred here was determined experimentally by Gottsching and Rhodus [10] in through drying tests and has been supported by the authors own observations. That relationship expresses the ratio of local falling rate drying rate to the constant rate drying rate as a function of local water content.

$$\frac{(dm/dt)}{(dm/dt)} = \sqrt{(1 + a)\frac{X_p}{X_{cr}} - a\left(\frac{X_p}{X_{cr}}\right)^2} \quad (17)$$

The coefficient a determines the shape of the normalized drying curve and is influenced by the product qualities.

System Model

Equations (1) thru (17) constitute the mathematical description of the sheet history on the dryer. The equations are a set of coupled nonlinear first order differential equations and are solved in full transient finite difference form. It only remains to present those equations which describe the components that constitute the remainder of the air system. Because the purpose here is to present a simplified model for the study of machine system performance, the air system has been idealized. Losses associated with convection from exposed ductwork and air system imbalance between the hood and the roll have been neglected.

Using the roll exhaust as a control volume, the energy conservation equation is given by

$$\dot{m}_t H_x + \dot{m}_\ell H_a + \Delta H_x = (\dot{m}_t + \dot{m}_\ell)H_d \quad (18)$$

Where $\dot{m}_\ell$ is the ambient roll leak into the system around the seals that baffle off the unwrapped portion of the roll and the subscript x implies the mixed sheet exhaust conditions. The roll exhaust fan power consumption, ΔH_x, is determined by the system losses which in this application are dominated by the sheet pressure drop. In order to determine the sheet pressure drop required to pass the required mixture flowrate, a permeability plot must be made on the web using a modified Frazier test. A typical plot for standard 49 gsm newsprint is shown in Figure 2. The test is performed over the operating vacuum range in question and the results placed into the model. The model then uses a linear interpolation between the wet and dry permeabilities to compute through flow at the specified roll vacuum. The advantage of the continuous relationship is the ability of the model to follow the permeability over the full operating range when examining various design parameters. The model corrects the pressure drop for the effects of temperature and humidity on viscosity and density.

The mass conservation equations for air and water using the recirculation loop consisting of the recirculation fan, makeup air and burner as a control volume are given by

$$\dot{m}_h = \dot{m}_e + \dot{m}_a + \dot{m}_c - \alpha f$$

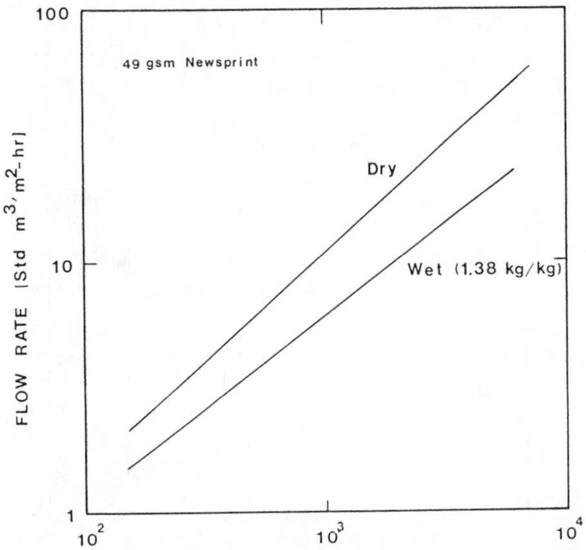

FIGURE 2. WEB PERMEABILITY

$$\dot{m}_e = \dot{m}_h - \dot{m}_t \tag{19}$$

$$\dot{m}_h X_h = \dot{m}_e X_e + \dot{m}_a X_a + \dot{m}_c X_a + \beta f \tag{20}$$

Where $\propto$ and β are the mass lost and water gained in the combustion process per unit of energy input, f, respectively. These quantities and the combustion air flowrate are computed based upon stoichiometric relationships. In an indirect fired system, $\dot{m}_c$, $\propto$ and β are naturally zero and the through air flowrate, $\dot{m}_t$ is equivalent to the makeup air flowrate, $\dot{m}_a$. The equation for the conservation of energy for the control volume is

$$\dot{m}_h(h_{ah} + X_h h_{vh}) = \dot{m}_e(h_{ae} + X_e h_{ve}) + (\dot{m}_a + \dot{m}_c)h_a + f + \Delta H_R \tag{21}$$

The recirculation fan power consumption is a function of the specified system losses and hood pressure required to pass the specified mass flow through the specified hood open area via the orifice equation.

Equations (18) thru (21) constitute three equations in the three unknowns $\dot{m}_a$, f and X_h, the solution of which is iterative because of the coupling of H_h and X_h. In addition, since the sheet model equations are coupled through supply air humidity to the system model equations the total problem solution, i.e. the sequential solution of the sheet model equations and the system model equations, is iterative also.

The specific energy consumption is defined by

$$E = (f + \Delta H_R + \Delta H_x) \,/\, (\frac{dm}{dt}i + \frac{dm}{dt}t) \tag{22}$$

It is not standard practice to include fan power consumptions in the above definition [4,6,7]. These are included in the present analysis because in some instances ΔH_R and ΔH_x can amount to as much as 50% of the total energy input which if neglected would lead to erroneous conclusions concerning the thermodynamic efficiency when compared to other drying techniques.

ANALYTICAL RESULTS

The governing equations are used to investigate the effect of various parameters on the drying rate and the specific energy consumption/cost. The specific case examined is that of drying standard newsprint stock in the critical range, from 58 percent water to 44 percent water. The advantages of applying through/impingement drying in this range are twofold: increased drying rate and hence a reduction in building space requirements and positive sheet control in the water content region where the web is weakest thereby reducing the number of sheet breaks and production downtime. In some cases the machine design will be fixed, i.e. roll and hood configurations will be selected and held constant. Except where otherwise noted, the following assumptions are made:

Physical

Roll size	4.27 m
Effective wrap	270^0
Hood nozzle diameter	5 mm
Hood open area	2%
Hood nozzle/roll spacing	15.9 mm
Supply/Exhaust duct losses	76 mm of w.c.

Process

Basis Weight	49 gm/m^2
Sheet Speed	1070 m/min
Width	7620 mm
Sheet Entrance Water	58% (1.38Kg/Kg)
Sheet Entrance Temperature	40^oC
Sheet Exit Water	44% (0.8 Kg/Kg)
Permeability	Figure 2
Supply Air Temperature	400^oC

Economic

Fan Power Cost	0.047 $/Kw-hr
Fuel Cost	13.9 $/$10^6$Kcal

Mass Flow

A given amount of drying can be accomplished with any mix of impingement mass flow and through air mass flow. At one extreme high impingement velocity coupled with low through air flow can be used to obtain the desired water removal rate while at the other extreme low impingement velocity coupled with 100% through air flow can accomplish the same task. This effect is shown in Figure 3. The inability to obtain the total drying rate (178Kg/hr-m^2) at 100% through air with fixed equipment is an anomaly inherent in the definition of drying rate. Through air water removal is defined here as the product of the mass flow of dry air passing through the web and the increase in the moisture content of the air in going from the local hood exhaust condition to the roll exhaust condition.

This definition thus depicts the true advantage obtained by adding roll suction. To attribute to through air drying an increase in moisture content from the supply air humidity (X_h) to the roll exhaust humidity (X_x) overstates the through air drying contribution to total drying rate. In the absence of roll suction, the through flow is exhausted from the hood at the hood exhaust humidity and inherently has an initial contribution to the total drying rate.

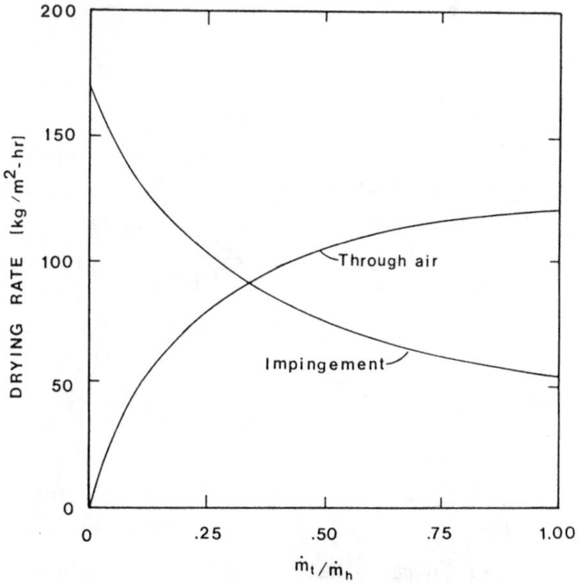

FIGURE 3. EFFECT OF THROUGH AIR MASS FLOW ON DRYING RATE

The decision on the proper mix of through air to supply air dry mass flowrate as the basis for design must be an ecomonic one. Figure 4 shows the effect of through drying water removal rate on the specific operating energy cost in units of US dollars per metric ton. The increase in operating cost to the left of the minimum is dominated by the power requirements of the recirculation fan while that to the right of the minimum is dominated by the requirements roll exhaust fan. High hood air flow rates result in high nozzle pressure drops while high through air flows result in high pressure differentials across the product. The optimum operation for this example occurs when through air is removing 33% of the total water load.

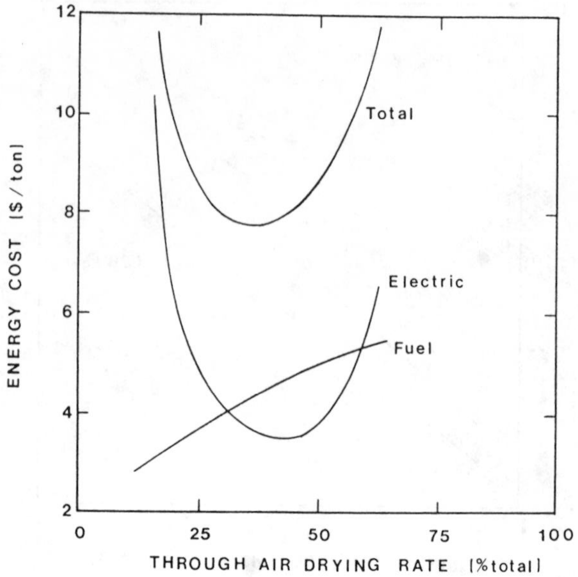

FIGURE 4. EFFECT OF THROUGH-AIR DRYING RATE ON OPERATING COST

The air exhausted from the system through the web must be made up by fresh intake air and air addition at the burner as combustion air. This therefore has a direct bearing on the equilibrium supply air humidity. This effect is shown graphically in Figure 5. No limit has been placed here on the obtainable supply air humidity at low through air flowrates which in practice is limited by air imbalance of the roll/hood interface and system leakage. At high through flowrates the minimum air humidity is controlled by the ambient make-up air humidity. The effect of increased water removal by through air is thus to decrease supply air humidity and to increase specific energy consumption.

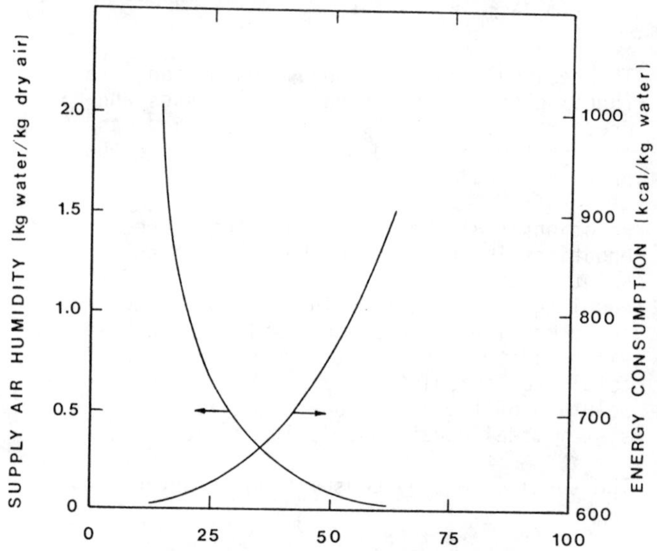

FIGURE 5. EFFECT OF THROUGH-AIR DRYING RATE ON SUPPLY HUMIDITY AND ENERGY CONSUMPTION

Increased supply air humidity increases the specific volume and results in increased power requirements while decreasing the supply humidity results in a large amount of ambient intake air resulting in relatively large fuel costs. The state diagram for the optimum operating point is shown in Figure 6. The state designations correspond to the locations in Figure 1. A comparison of the magnitude of the horizontal lines at states (d) and (e) representing exhaust and recirculation fan power input respectively with that at (h) representing the energy input at the air heater relates graphically the relative magnitudes of the energy inputs to the system. The figure also shows the degree of saturation attained by through air which is about 50% relative humidity in this example. To have assumed that the sheet exit air attained complete saturation would have overestimated the through air contribution to total drying rate by 12%.

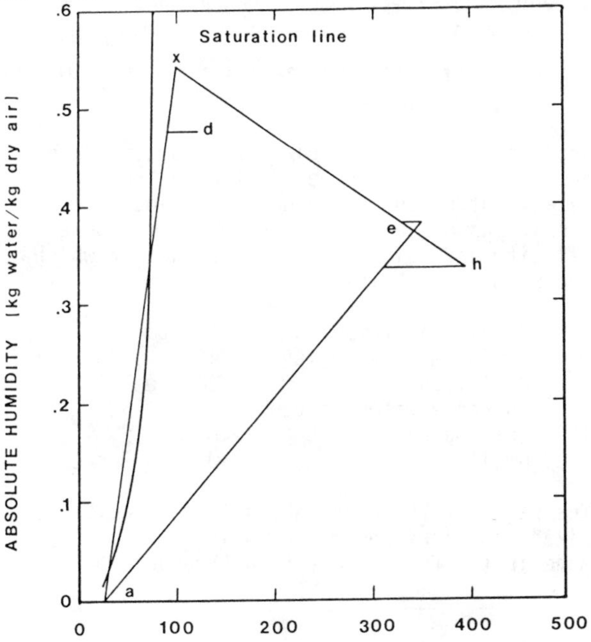

FIGURE 6. SYSTEM STATE DIAGRAM

The energy efficiencies obtained in this process (Figure 5) are quite attractive when compared to other convective drying schemes [4,6]. This can be directly attributed to the fact that the system dump air is relatively cool and is saturated to the fullest after passing through the sheet. In the traditional sense one could view through air flow as an energy recovery from the air which must be exhausted from the system to maintain a system mass balance.

Roll Size

At the fixed supply air condition of temperature, a given drying task can be accomplished with either high mass flows in small rolls or low mass flows in large rolls. The choice is then one of economics: that of weighing operating cost for electricity and fuel versus capital cost for drying equipment, associated air handling equipment and building space requirements. The optimum total operating cost and cost for electricity for the stated example are shown in Figure 7. As the roll size decreases the electrical cost approaches total cost asymptotically because the decrease in drying residence time must be offset by an increase in the impingement and through air velocity heat-transfer coefficients. This effect results in very large recirculation and roll exhaust fans at small roll diameters to achieve equivalent drying loads.

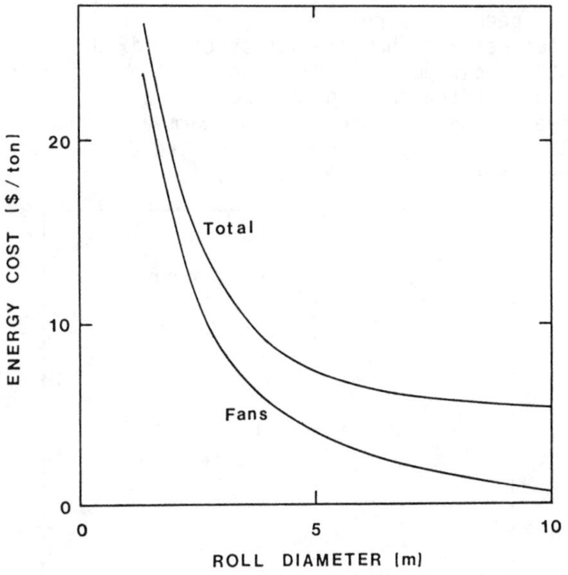

FIGURE 7. EFFECT OF ROLL DIAMETER ON OPERATING COST

Sheet Permeability

The sheet permeability has a direct effect on the power requirement of the roll exhaust fan, the supply air humidity and the economic optimum through air drying rate. Figure 8 shows the

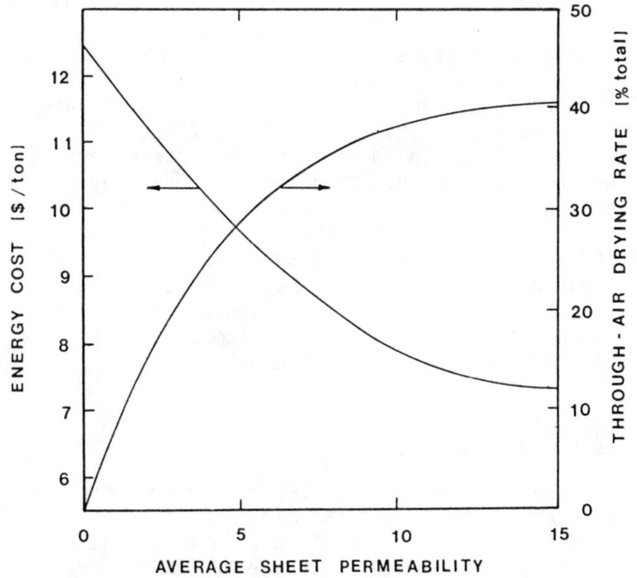

FIGURE 8. EFFECT OF SHEET PERMEABILITY ON OPERATING COST AND DRYING RATE

effect of average sheet permeability on drying cost and percent through air drying rate at the economic optimum operating point. Provided that the permeability is over 15 std $m^3/min-m^2$ it has little effect on system performance with fixed equipment for this example. A smaller operating cost in this range can be obtained by altering physical design characteristics and in some instances using 100% through air drying.

Sheet Speed

Web speed has a relatively minor effect on the water removal characteristics of a fixed machine with specified fans unless one is traversing into or out of the falling rate zone as a result of speed change (Figure 9). The warm-up period takes up a larger proportion of the total drying surface area at higher speeds.

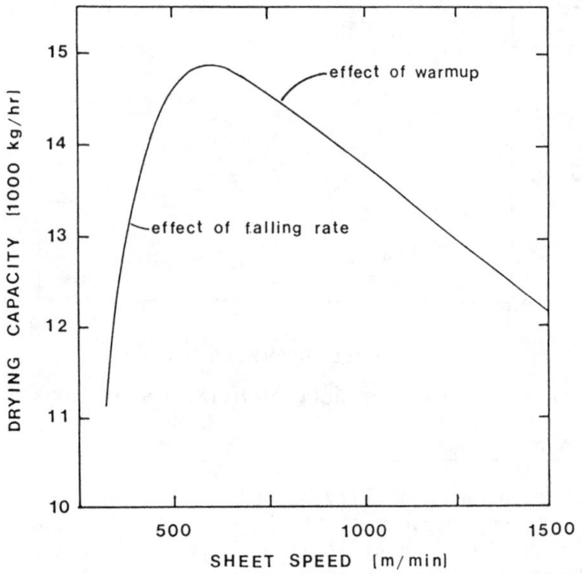

FIGURE 9. EFFECT OF SPEED ON DRYING CAPACITY

Supply Temperature

The effect of supply temperature on drying rate is primarily on the temperature differential which is the driving force for heat transfer. It has the greatest effect on water removal by impingement since the model assumes the driving force to be the differential supply temperature and web temperature and less of an effect on through air drying rate. Figure 10 shows this effect and stresses the importance of being able to operate dryers at elevated temperature to minimize capital equipment cost.

EXPERIMENTAL RESULTS

Preliminary experimental tests were conducted on a pilot dryer to test the model and the assumptions under which it was formed. The pilot dryer is a 1.83 meter diameter by 610 mm wide high-vacuum Honeycomb Roll . The roll is surrounded on 60 degrees of arc by a high velocity hood delivering air to the web through slots. The

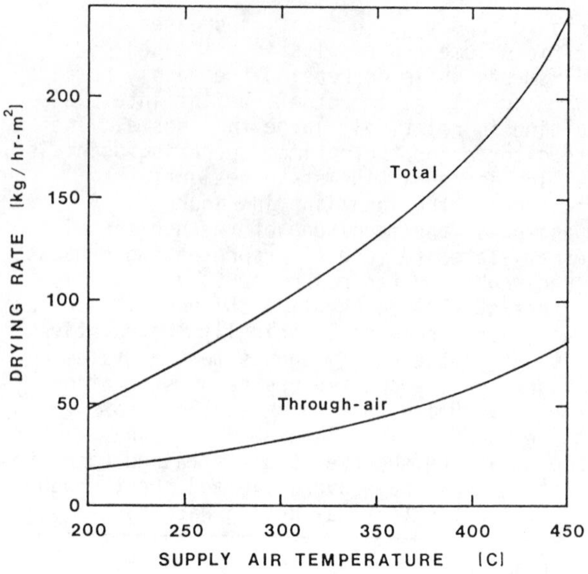

FIGURE 10. EFFECT OF SUPPLY AIR TEMPERATURE ON DRYING RATE

product used for the trials was 49 gsm standard newsprint. It was rewet to the desired ingoing water content and dried under varying process conditions of supply temperature, impingement velocity and roll vacuum, the ranges of which are summarized below.

Supply air temperature	177 to 260^{0}C
Impingement velocity	4000 to 4600 m/min
Roll Vacuum	380 to 1270 in w.c.
Entrance water content	30 to 75%
Exit water content	5 to 25%
Dwell Time	0.6 to 3 seconds

The range of these variables is limited by fixed fan sizes and minor modifications must be made in the future to extend the range of trials.

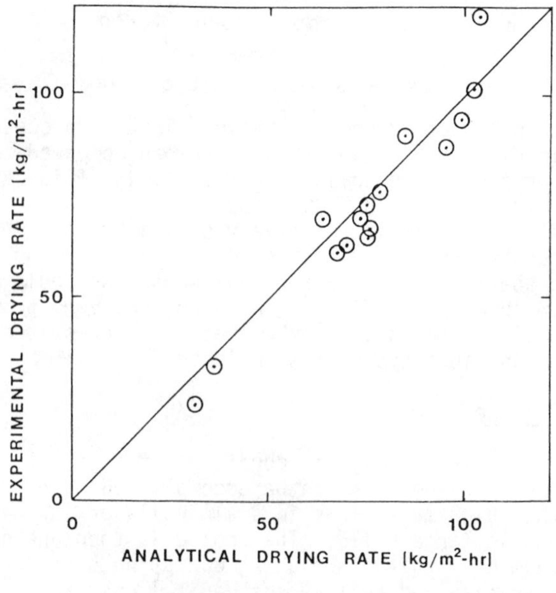

FIGURE 11. COMPARISON OF EXPERIMENTAL AND ANALYTICAL DRYING RATES

Figure 11 is a graphical comparison of the experimentally determined drying rate and the rate computed by the model with the independent process variables of supply temperature, impingement velocity, and roll vacuum as inputs. The agreement is very good (within 8% of the predicted result). More extensive tests must, however, be conducted over the full range of operation to fully test the model.

CONCLUSION

A mathematical model which simulates the drying process by combination impingement and through drying has been developed. The model was used to demonstrate its usefulness as a tool for predicting component sizes and performance for a selected example. The analytical results demonstrate that this dryer concept can be a viable alternative to other drying methods. The drying rates and specific energy consumptions are very attractive when compared to other drying alternatives.

The preliminary experimental results presented show that the mathematical model is a good simulation of the experimental data. This can be taken as an indication that the assumptions on which the process was modeled may be substantially true. This must be tempered, however, by the limited test data available and further tests must be undertaken to exercise the model over the full range of operating variables. To improve the model further, considerations must be given to:

(1) The heat and mass transfer phenomena in the falling rate zone.

(2) The combined effects of turbulent impingement heat transfer with wall suction.

(3) Variations of the sheet moisture and temperature through the thickness of the product.

ACKNOWLEDGEMENT

This paper was originally presented at the 1982 annual meeting of the American Institute of Chemical Engineers. It is with that body's kind permission that it is reproduced here.

REFERENCES

1. K. Randall, "High Vacuum Through Drying in a Roofing Felt Mill Cuts Energy Costs and Boosts Production", TAPPI, 64 (5), (1981).

2. B.W. Burgess, S.M. Chapman, U.S. Patent 3418723, (1968).

3. B.W. Burgess, W. Koller, E., Pye, "The Papridryer Process Part II - Mill Trials", Pulp and Paper Mag. Can, 73 (11) (1972).

4. D. Rounds, G. Wedel, "Drying Rate and Energy Consumption for an Air Cap Dryer System", First International Symposium on Drying, Science Press, (1978).

5. D. Kruska, H. Holik, A. Weinmann, "Energiesparender Einsatz von Hochleistungsduschauben in der Papiertrochnung", Wochenblatt for Papierfabrikation, 15, (1977).

6. J.W. Rohrer, F.J. Gardiner, "Through-Drying: Heat Transfer Mechanism and Machine System Response", TAPPI, 59 (4), (1977).

7. R.H. Crotogino and V. Allenger, "Mathematical Model of the Papridryer Process", Transactions of the CPPA, 5 (4), (1979).

8. Y. Li, A. Mujumdar, W. Douglas, "Coupled Heat and Mass Transfer Under a Laminar Impingement Jet", Proceedings of the first International Synposium or Drying, Science Press, (1978).

9. J. Chance, "Experimental Investigation of Air Impingement Heat Transfer Under an Array of Round Jets", TAPPI, 57 (6), (1974).

10. L. Gottsching and D. Rhodius, "Der Trocknungsverlauf Von Papier und Pappe in Abhangigkeit Von Trocknungstechnischer and Papier Technologischen Parametern", Das Papier, 10, (1977).

A SIMULATION MODEL FOR COMBINED IMPINGEMENT AND THROUGH DRYING USING SUPERHEATED STEAM AS THE DRYING MEDIUM

Esther Loo[1] and A.S. Mujumdar[2]

1. Texaco Ltd., Edmonton, Alberta, Canada
2. McGill University, Montreal, Canada.

The objective of this work is to evaluate the technical feasibility of using superheated steam for drying of permeable paper grades. A transient mathematical model was developed for combined impingement and through drying of newsprint using superheated steam as the drying medium. The simulation results as compared with those for air as the drying medium, indicate higher average drying rates and lower final paper moisture content. The most significant variables and parameters affecting average drying rates include the jet temperature, pressure differential across the web, percent heat loss and dryer geometry. Use of either slotted or round jet configuration shows no significant change in drying rates. At the drying conditions examined, a through drying contribution in the order of 65% was observed to be optimal in balancing energy expenditure versus moisture removal efficiency.

It is concluded that combined impingement and through drying feasible process for drying permeable grades of paper. However, experimental work is required to verify the mathematical model presented in this paper and also to study the effect of steam drying on paper properties.

NOMENCLATURE

A	Ratio of nozzle-to-web spacing to nozzle opening
A_{hoa}	Ratio of jet open area to drying area
a,b,c,K	Coefficients in Das' correlation for impingement heat transfer coefficient
B	Parameter in the correction factor for high evaporation rates
c_d	Heat capacity of paper fiber, kJ/kg.K
c_p	Heat capacity of steam, kJ/kg.K
c_v	Mean heat capacity of steam, kJ/kg.K
c_w	Heat capacity of water, kJ/kg.K
d	Nozzle opening, m
EC	Energy costs, \$/T
EM	Moisture removal efficiency, %
EP	Overall process efficiency, %
ET	Thermal efficiency, %
f	Ratio of nozzle opening to inter-nozzle spacing
H	Enthalpy of steam, kJ/kg
h	Nozzle-to-web spacing, m
Δh_A	Heat of adsorption of water on fiber, kJ/kg
Δh_o	Latent heat of vaporization of water at 0^0C, kJ/kg
Δh_p	Latent heat of vaporization of water at the paper temperature, kJ/kg
I_c	Cross-flow interference parameter in Chance's correlation
L	Inter-nozzle spacing, m
$\dot{M}$	Mass flux of steam, kg/m^2.s
m_p	Dry fiber basis weight of paper, g/m^2
Nu	Nusselt number ($\alpha d/\lambda$)
n	Exponent for Reynolds number in Chance's correlation
P	System pressure, psia
pr	Prandtl number ($c_p\mu/\lambda$)
P_{vp},P'_{vp}	Corrected vapor pressure, psia
P^*_{vp}	Saturation vapor pressure at Θ_p, psia
ΔP	Pressure differential across the web, Pa
q^*_i	Pure impingement heat transfer rate, kW/m^2
q_r	Radiant heat transfer, kW/m^2
q_t	Through drying heat transfer rate, kW/m^2
R	Total local drying rate, kg/m^2.s
$\bar{R}$	Average total drying rate, kg/m^2.hr
Re	Reynolds number ($\rho_j V_j d/\mu_j$)
R_i	Local Impingement drying rate, kg/m^2.s
R^*_i	Local pure impingement drying rate, kg/m^2.s
R_t	Local through drying rate, kg/m^2.s
r	Vapor pressure depression factor
T	Temperature, K
T_r	$0.5(T_j+T_e)$, K
T_{avg}	$0.5(T_j+T_p)$, K
t	Time, s
V_j	Jet velocity, m/s
V_{mach}	Machine speed, m/s
V_t	Percolation velocity, m/s

X_e	Fraction of original inlet steam in hood exhaust
X_t	Fraction of original inlet steam in vacuum exhaust
Y	Paper moisture content, kg water/kg dry fiber
Y_f	Final paper moisture content, kg water/kg dry fiber
Y_o	Initial paper moisture content, kg water/kg dry fiber
z, Z	Length of dryer, m

GREEK SYMBOLS

α	Heat transfer coefficient, $kW/m^2.K$
γ_r	Steam absorptance for radiation
ε	Emissivity
Θ_p	Paper temperature, ^0C
κ	Permeability of paper, m
λ	Thermal conductivity of steam, kW/m.K
μ	Viscosity of steam, kg/m.s
ρ	Density, kg/m^3
σ	Stefan-Boltzman constant, $kW/m^2.K^4$
ϕ	Fraction of heat input after heat loss
Ψ	Correction factor for high evaporation rates
$\omega_1 \ldots \omega_3$	Parameters in Chance's correlation

SUBSCRIPTS

j	Jet conditions
e,h	Hood exhaust conditions
t	Vacuum exhaust conditions
p	Paper conditions

INTRODUCTION

Conventional methods for drying paper such as steam-heated cylinder dryers are often plagued with low drying rates and limited paper drying capacity. This can be attributed to the inefficient process of heat transfer by conduction from the fibers in contact with the surface of the cylinder to the interior of the web as well as the low rate of water removal from the wet web.

Burgess et al. (1) have shown in their studies using air that drying rates as much as six to ten times those attainable by conventional methods, can be achieved by a combined impingement-through (CIT) drying process.

Mujumdar (2) proposed that significant improvement in drying rates and particularly the thermal efficiency be obtained by using superheated steam rather than hot air in impingement-percolation system. The concept of using superheated steam as a drying medium in drying of solids is not new. Several studies on drying granular solids indicated attainment of much higher drying rates in superheated steam than in air (3, 4, 5). With superheated steam, there is no mass transfer resistance and thus, the rate of evaporation is solely determined by the rate of heat transfer. Another advantage of using superheated steam is its inertness which eliminates explosion hazards. However, one disadvantage is that the temperature of the vapor must be kept above saturation conditions in order to avoid condensation on the drying surface. This type of drying is also limited to materials which are not heat-sensitive. Several technological difficulties will need to be overcome before the proposed system can be commercialized.

Based on above observations, it can be inferred that a impingement-percolation process using superheated steam as drying medium is viable for achieving higher drying rates. Thus, the primary objective of the current work was to investigate the technical feasibility of such drying scheme. A schematic of the proposed process using superheated steam is depicted in Figure 1. Essentially the dryer consists of a high velocity hood and a vacuum cylinder. Input steam is directed through an array of slots or round nozzles, and impinges on the paper web, thereby resulting in high heat and mass transfer rates. Vacuum is applied across the cylinder to remove the vapor formed within the sheet and between the sheet and cylinder. The steam from both hood and vacuum exhausts can be recovered, superheated, and recycled to the process.

In this paper, a mathematical model is developed for the combined impingement and through drying process using superheated steam as the drying medium, to calculate the drying rates, energy consumption and costs, and dryer thermal efficiencies as functions of the independent process variables and dryer geometry. The simulated results will be used to assess the performance of such dryer as compared to that using air. Heat transfer estimates were made assuming superheated steam to behave like air; appropriate fluid properties were used in the correlations used. Paper properties were likewise assumed to be the same (in the absence of data) in presence of steam.

MATHEMATICAL MODEL

Perhaps the first comprehensive work on mathematical modelling of the Papridryer process was carried out by Crotogino and Allenger (6). These investigators assumed that impingement and through drying occur simultaneously, and that the total drying rate can be expressed as a linear combination of pure impingement and pure through drying mechanisms.

Huang and Mujumdar (7) refined the Crotogino and Allenger model and overcame some of the limitations of the earlier model with a transient model. They incorporated variation of web temperature during the warm-up and falling rate rate periods, as well as presented calculations of local drying rates and web moisture content. Their unsteady state model for air was observed to yield more realistic results than the steady state model of Crotogino and Allenger. Different correlations were used to model the transfer rates in particular.

More recently, Randall (8) developed a sequential

model for the combined impingement and through drying for air. The concept is illustrated in the simplified mass flux diagram shown in Figure 2. Randall proposed that the impingement drying occurs first, thereby removing part of the web moisture. For this contribution, the driving potential is the difference between the supply air and paper temperature. While a portion of the resulting stream from impingement drying is exhausted through the hood, the remaining part passes through the web, where additional moisture is removed from the paper by the through drying contribution. The driving force in the latter process is the difference between impingement exhaust air and paper temperature. Randall's model is more simplistic as compared to those examined earlier. In his constant rate zone, all exhaust conditions can be determined based on the adiabatic saturation web temperature alone, even without knowing the paper moisture content. Randall's model does not allow for variation of the web temperature during the falling rate period. Furthermore, his computations for drying rates in the warm-up and falling rate zones were solely based on the values of mass transfer coefficients obtained earlier in the constant rate zone.

It is not possible to adopt Randall's sequential model to superheated steam drying in view of the difficulties encountered in determining the mass transfer rates during the warm-up period (since there is no mass transfer resistance in superheated steam drying), and also because of the lack of sufficient information required to relate the drying rates in the falling rate and constant rate periods.

The mathematical model developed here for superheated steam is adopted from Huang and Mujumdar's transient model. However, the assumption of thermal equilibrium established between the air and paper cannot be extended to the case of superheated steam since at the paper temperature, condensation of steam would inevitably occur. Instead, in the proposed model, an appropriate assumption is made to allow the vacuum exhaust temperature to be above saturation conditions and to vary with the local paper moisture content, thus giving a more realistic picture of the steam drying process. Various other assumptions implicit in the model are discussed by Huang and Mujumdar (7).

This model is comprised essentially of two transient water and heat balances on a section of paper moving through the dryer. The resulting differential equations must be solved simultaneously along with other supplementary expressions to yield the local paper moisture content and temperature along the length of the dryer. Figure 3 shows the geometry of the dryer which the mathematical model is based upon.

Water Balance

The overall water balance equates the inlet steam flux and the rate of water removal from the newsprint to the sum of the exhaust mass fluxes on the hood and vacuum sides, as follows:

$$\dot{M}_j - \frac{m_p}{1000}\frac{dY}{dt} = \dot{M}_e + \dot{M}_t \tag{1}$$

If $(1-X_e)$ and $(1-X_t)$ are defined as the fractions of the exhaust fluxes, $\dot{M}_e$ and $\dot{M}_t$ respectively, which represent the amount of moisture being removed from the paper by each stream, then a balance can also be written between the inlet steam mass flux and the exhaust mass fluxes; viz.,

$$\dot{M}_j = X_e\dot{M}_e + X_t\dot{M}_t \tag{2}$$

Using Equations (1) and (2), the overall drying rate can be expressed in terms of the moisture removed, $(1-X_e)$ and $(1-X_t)$, and the exhaust fluxes via,

$$R = \frac{m_p}{1000}\frac{dY}{dt} = \dot{M}_e(1-X_e) + \dot{M}_t(1-X_t) \tag{3}$$

In this equation, the drying rate attributed to the through drying mechanism is given by:

$$R_t = \dot{M}_t(1-X_t) \tag{4}$$

Hence, the drying rate due to impingement is:

$$R_i = \dot{M}_e(1-X_e) \tag{5}$$

The impingement drying rate can be related to the pure impingement drying rate through the linear model proposed by Crotogino and Allenger for combining pure impingement with pure through drying; viz., (6)

$$R = R_i + R_t = \frac{\dot{M}_e X_e}{\dot{M}_j} R_i^* + R_t \tag{6}$$

It follows from Equations (5) and (6):

$$\frac{(1-X_e)}{X_e} = \frac{R_i^*}{\dot{M}_j} \tag{7}$$

which can be used to solve explicitly for X_e.

Heat Balance

The overall heat balance is derived by equating the heat input both by convective and radiant heat transfer from the hood, to the sensible heat in the exhaust streams and the change in heat content of the paper web, taking into account the heat of adsorption. Thus,

$$\dot{M}_j H_j \phi + q_r - \dot{M}_e H_e - \dot{M}_t H_t = \frac{m_p}{1000}\frac{d}{dt}\{(c_d + c_w Y)\Theta_p - \Delta h_A Y\} \quad \ldots(8)$$

where ϕ is the heat loss fraction (e.g., for 5% heat loss, $\phi = 0.95$); c_d and c_w are the heat capacities of paper fibre and water (1.38 and 4.1868 kJ/kg.K, respectively); and Θ_p is the paper temperature, ^{0}C.

The enthalpy for each stream is determined by:

$$H = 1906.714 + 2.042\ T,\ \text{kJ/kg (T in K)}$$

This expression was found by performing a linear regression on superheated steam enthalpy values 1 atm (9).

The heat of adsorption as a function of paper moisture content is given by (10):

$$\Delta h_A = \exp(7.2-17.3\ Y) \tag{9}$$

The radiant heat transfer to the paper is estimated as follows:
q_r = Total irradiation from surroundings - Emmittance from paper
where the total irradiation from surroundings is equal to the irradiation from steam jet and the irradiation from the hood transmitted through the steam. Thus, q_r is given by:

$$q_r = \varepsilon_j\sigma T_j^4 + \varepsilon_h\sigma T_r^4 - \gamma_r\varepsilon_h\sigma T_r^4 - \varepsilon_p\sigma T_p^4 \tag{10}$$

where the ε's are the emissivities of steam, hood, and paper (0.24, 0.94, 0.74 respectively (7, 11)); γ_r is the steam absorptance for radiation, $\varepsilon_w'\ (T_j/T_r)^{0.45}$ with $\varepsilon_w' = 0.245$ (11); σ is the Stefan-Boltzman constant, 5.669×10^{-11}kJ/m^2.sec.K^4; T_p is the paper temperature in K; and $T_r = 0.5(T_j+T_e)$, K.

The overall heat balance, after expanding the differential term becomes:

$$\dot{M}_jH_j\phi + q_r - \dot{M}_eH_e - \dot{M}_tH_t = \frac{m_p}{1000}\{(c_d+c_wY)\frac{d\Theta_p}{dt} + c_w\Theta_p\frac{dY}{dt} - \Delta h_A\frac{dY}{dt} \tag{11}$$

Transformation of Time Differential to Length Differential

For a fixed machine speed, the time differential can be replaced by the length differential via,

$$\frac{d}{dt} = V_{mach}\frac{d}{dz}$$

Hence, the water balance becomes:

$$\frac{dY}{dz} = \frac{1000(\dot{M}_j-\dot{M}_e-\dot{M}_t)}{m_pV_{mach}}\ , \tag{12}$$

and the heat balance,

$$\frac{d\Theta_p}{dz} = \frac{1000(\dot{M}_jH_j\phi + q_r - \dot{M}_eH_e - \dot{M}_tH_t)}{m_pV_{mach}(c_d+c_wY)} + \frac{(\Delta h_A-c_w\Theta_p)}{(c_d+c_wY)}\frac{dY}{dz} \tag{13}$$

Individual Heat Balances

An individual heat balance for each exhaust stream can also be written by taking into consideration the heat entering through the hood, the heat added to the steam by vaporizing moisture from the web, the sensible heat in the exhaust stream, and the convective heat transfer from steam to web. For the hood exhaust side, the heat balance yields:

$$X_e\dot{M}_eH_j\phi + \dot{M}_e(1-X_e)(\Delta h_o+c_v\Theta_p) - \dot{M}_eH_e - X_e\dot{M}_eq_i^*/\dot{M}_j = 0 \tag{14}$$

and for the vacuum exhaust side:

$$X_t\dot{M}_tH_j\phi + \dot{M}_t(1-X_t)(\Delta h_o+c_v\Theta_p) - \dot{M}_tH_t - q_t = 0 \tag{15}$$

where Δh_o is the latent heat of vaporization at 0^0C, c_v is the mean heat capacity, and q_i^* and q_t are the convective heat transfer terms from steam to web due to pure impingement and through drying, respectively. The expression used for c_v is (9):

$$c_v = (3.55217 + 5.72196\times10^{-4}T_{avg})/\ 2.0$$

where $T_{avg} = 0.5(T_j+T_p)$.

Impingement Drying

In the absence of mass transfer resistance, pure impingement drying rate can be assumed to be controlled by the rate of heat transfer; viz.,

$$R_i^* = q_i^*/(\Delta h_p+\Delta h_A) \tag{16}$$

where Δh_p is the latent heat of vaporization at T_p. The heat of adsorption term is included in order to account for the effect of bound moisture in the paper beyond the critical moisture content. An expression for Δh_p as a function of T_p is given by Huang and Mujumdar (7):

$$\Delta h_p = 2.26\times10^3((1-T_p/647.3)/0.42353)^{0.38}$$

The pure impingement heat transfer rate can be determined by:

$$q_i^* = \alpha\,(T_j - T_p)\,(\Psi) \tag{17}$$

where α is the heat transfer coefficient and Ψ is a correction factor for reduction in the transfer rates due to high evaporation rates. For laminar flow, Ψ can be evaluated by correlations recommended by either Keey (12) or more recently by Kast (13), as follows:

$$\Psi = \frac{\ell n\,(1+B)}{B} \quad \text{(Ref. 12)} \tag{18}$$

or

$$\Psi = \frac{\ell n\,(1+B)^{1.4}}{(1+B)^{1.4} - 1} \quad \text{(Ref. 13)} \tag{19}$$

while for turbulent flow conditions, Ψ is given by Kast (13) as

$$\Psi = \frac{Bk}{\exp(Bk)-1} \tag{20}$$

with k = 1.5 for steam. Here B in all three cases is defined by

$$B = c_v(T_j - T_p)/\Delta h_p$$

Several correlations exist for the estimation of the average impingement heat transfer coefficient depending on the jet configuration. For slotted jets, these include correlations by Martin (14), Saad et al. (15), and Das (16). The Martin correlation will not be considered in this study since Huang and Mujumdar (17), have demonstrated in their paper that better results are obtained with Saad et al.'s correlation than with Martin's. Das' correlation accounts for the influence of temperature dependent physical properties and hence, is useful for comparison with Saad et al.'s correlation for slotted jets with central exhausts i.e. without crossflow interference effects. These two expressions are summarized below:

(a) Saad et al.,

$$Nu = 0.14\, Re^{0.775} A^{-0.286} f^{0.314}$$

where A is the ratio of nozzle-to-web spacing to nozzle opening (h/d), and f is the ratio of slot opening to inter-nozzle (slot) spacing (d/L). The ranges of validity for this correlation are:

$3000 \le Re \le 30{,}000,\ 8 \le A \le 24,\ 0.015 \le f \le 0.08$

(b) Das,

$$Nu_j = K\, Re_j^a\, (h/d)^b\, (T_j/T_p)^c\, Pr^{1/3}$$

where K, a, b, and c are all functions of L/2d. The ranges of validity are:

$5000 \le Re_j \le 20{,}000,\ 8 \le h/d \le 12,\ 1.18 \le T_j/T_p \le 2.06$

For multiple round jet configurations, the following correlation recommended by Chance (18) was used to estimate the heat transfer coefficient:

$$Nu = \omega_1\omega_2\omega_3\, Re^n\, Pr^{1/3}\, A_{hoa}^{1.0146}$$

where: $n = 0.561/A_{hoa}^{0.0835}$;

$\omega_1 = 2.06$;
$\omega_2 = 1.0 - 0.236\, I_c$;
$\omega_3 = 1.0 - (h/d)\,(0.023 + 0.182\, A_{hoa}^{0.71})$;
$I_c = A_{hoa}L/d$, a cross-flow interference parameter;

and A_{hoa} = ratio of jet open area to drying area.

The ranges of validity are:

$2 \le h/d \le 8,\ 0.012 \le A_{hoa} \le 0.07,\ I_c \le 1.8$

Except for Das' correlation in which Nu is evaluated at jet conditions, the steam properties are evaluated at film conditions, T_{avg}. The following expression for the thermal conductivity of steam was used (7):

$$\lambda x 10^3 = 0.86868x10^{-2} + 0.11318x10^{-4}T + 0.85793x10^{-7}T^2$$

The hood exhaust conditions can be determined from the knowledge of q_i^* and R_i^*. Equations (7), (16), and (17) yield the following expression for X_e

$$\frac{1-X_e}{X_e} = \frac{\alpha(T_j - T_p)\,\Psi}{\dot{M}_j(\Delta h_p + \Delta h_A)} \tag{21}$$

where $\dot{M}_j = A_{hoa}\,\rho_j\,V_j$

The hood exhaust temperature is determined from Equations (14 and (17) as follows:

$$H_e = X_e H_j \phi + (1-X_e)(\Delta h_o + c_v\Theta_p) - X_e q_i^*/\dot{M}_j \tag{22}$$

$$T_e = (H_e - 1906.714)/2.042 \tag{23}$$

Through Drying

Through drying rates greatly depend on the amount of steam which is able to pass through the sheet. This mass flux can be expressed in terms of the percolation velocity and X_t via,

$$\dot{M}_t = \rho_t V_t / X_t \tag{24}$$

The percolation velocity itself is a function of the structure of the paper as well as the pressure difference applied across it. Brundrett and Baines (19) presented an expression for calculating the percolation velocity:

$$V_t = \kappa\,\Delta P/\mu \tag{25}$$

where κ is the permeability of paper, μ is the steam viscosity, and ΔP is the pressure differential across the web. The permeability of newsprint as a function of moisture content has been experimentally determined and is given by (7):

$$\kappa \times 10^{10} = 1.726913 + 0.617884Y - 0.977268Y^2 + 0.312143Y^3$$

Thomas and Jackson (20) proposed the following equation for the dynamic viscosity of water vapor:

$$\mu = 0.99354 \times 10^{-6} + 0.27067 \times 10^{-7}T + 0.88906 \times 10^{-11}T^2$$

The through drying rate was computed as follows. It was assumed that the amount of moisture removed from the paper by the exhaust stream (i.e., $1-X_t$) is determined by the vapor pressure exerted by the moisture in the paper. Thus, X_t is given by:

$$X_t = (P-P_{cp})/P \tag{26}$$

where P is the system pressure and P_{vp} is the vapor pressure exerted by the moisture in the paper. In the constant rate zone where unbound moisture still prevails, P_{vp} is simply the saturation vapor pressure at the paper temperature. However, as the moisture content decreases below the critical point (as in the falling rate period), the vapor pressure exerted by the bound moisture in the paper must decrease. In order to account for this effect, the saturation vapor pressure is corrected by a vapor pressure depression factor r; viz.,

$$P_{vp} = P^*_{vp} r \tag{27}$$

where P^*_{vp} is the true saturation vapor pressure calculated from the Antoine's equation:

$$P^*_{vp} = 0.0193 \exp(18.58 - 3986/(\Theta_p + 233.7))$$

and the vapor pressure depression factor is given by:

$$r - 51.3\ Y^{1.86} \text{ fro } 0 \leq Y \leq 0.07$$
$$r = 1 - \exp(0.544-14.5\ Y) \text{ for } Y > 0.07 \tag{28}$$

Similarly, the vacuum exhaust temperature is assumed to be at the saturation temperature corresponding to the pressure of the system. However, as in the case of X_t, the effect of bound moisture in the falling rate period must be accounted for. Thus, the vacuum exhaust temperature is determined from the corrected vapor pressure which is given by:

$$P'_{vp} = P/r \tag{29}$$

while T_t can be solved from Antoine's equation:

$$T_t = 3986/(18.58 - \ell n(P'_{vp}/0.0193)) + 39.3 \tag{30}$$

SIMULATION OF THE PAPRIDRYER PROCESS

The two differential equations of water and heat balances along with the other related equations were numerically integrated using the fourth-order Runge-Kutta technique. A step size of 2 mm was chosen. A dryer length of 1 m per pass was used and the dryer was assumed to be a 4-pass dryer system (See Ref. 1). The initial conditions used for the simulation runs are listed below:

- Initial paper moisture content, Y_o = 1.92 (kg water/kg dry fiber)
- Initial paper temperature (^{0}C), Θ_p = 50.0
- Basis weight of paper (g/m^2), m_p = 52.08
- Dryer machine speed (m/s), V_{mach} = 2.032

It should be noted that in the simulation of the four-pass dryer, heat loss between passes is neglected (i.e., paper temperature at the beginning of the second pass is assumed to be equal to the paper temperature at the end of the first pass), and thus only one warm-up period for the first pass is considered.

Various jet configurations, jet velocities and temperatures, and pressure differentials across the web were studied. Aslo, the effects of heat loss on the dryer performance were examined. In all cases, the energy consumption and costs along with the thermal efficiency of the system were calculated for comparison purposes.

Energy Consumption and Costs Calculations

In order to render the steam drying process attractive, it is proposed to recycle the steam from both exhausts and to bleed off only the moisture removed from the paper. As shown in Figure 1, the whole hood exhaust stream is recycled since this stream is at a quite high temperature, and only a portion of the vacuum exhaust is used for making-up the total steam flux required for dryer input. Since both streams are at a lower temperature than the required inlet jet temperature, a heater is provided to superheat the mixed stream. The vacuum exhaust fan power consumption is mainly determined by the pressure drop established across the web, while the recirculation fan power consumption depends on the pressure required to pass the desired steam flux through the jet open area. This pressure can be calculated from the orifice equation.

The energy costs used for the two fans and the heater are estimated at \$0.047/KW-hr and \$0.012/KW-hr, respectively (8).

Efficiency Calculations

Three efficiencies were defined and calculated in the program:

(a) Dryer thermal efficiency which is defined as the ratio of the energy required for evaporation to the net energy input to the system. The latter is equal to the inlet steam enthalpy and paper sensible heat less the energy recovered from the steam that is recycled to the process.

(b) Overall process efficiency which is the ratio of the energy required for evaporation to the total energy input to the fans and heater.

(c) Overall moisture removal efficiency.

RESULTS AND DISCUSSION

Typical simulation results for the four-pass CIT dryer are shown in Figures 4 to 7. Figure 4 depicts the local drying rate and web moisture content along the length of the dryer. As shown in this figure, there is a quick warm-up period extending over 0.25 m dryer length followed by a constant increase in drying rate (3 m of dryer length), and then a falling rate period of 0.75 m. The moisture content decreases linearly along the dryer, and the critical moisture content is about 0.34 kg water/kg dry fiber, which is in agreement with Crotogino and Allenger's value of 0.30 kg water/kg dry fiber (6).

A breakdown of the drying rate contributions is presented in Figure 5. Clearly, the observed steady increase in drying rate during the middle two passes can be attributed to the through drying contribution which accounts for almost 67% of the total drying rate in this particular case. The impingement drying contribution is relatively constant throughout the dryer. This steady increase in through drying rate during the supposedly "constant rate" period can be explained in Figure 6 which shows the variations of the exhausts' conditions along the dryer. It can be seen that the vacuum exhaust mass flux increases steadily along the dryer up to the critical moisture point. This is because the paper becomes more permeable as moisture is being removed which is reflected in the drying rate R_t since R_t is equal to $\dot{M}_t(1-X_t)$. It should be noted that the units used in Figures 5 and 6 for mass fluxes and drying rates are quite different .

The warm-up period is primarily due to the increase in moisture removed by through drying, as reflected by the sharp drop of X_t during the first 0.25 m of the Papridryer system. On the other hand, beyond the critical moisture content, the decrease in vacuum exhaust mass flux can be explained by two factors in Equations (24) and (25): (1) an increase in vacuum exhaust temperature since less moisture is present in the paper, resulting in a decrease in steam density and an increase in steam viscosity; and (2) a decrease in moisture being removed, as reflected by the sharp rise of X_t curve. These two factors then together account for the falling rate period observed in Figures 4 and 5.

Also, as shown in Figure 6, much of the steam in the hood exhaust comes from the original steam jet (since X_e is large), and hence the impingement drying contribution is relatively small. The value for X_e is constant throughout the dryer except for a slight increase towards the end of the dryer, which is due to the decrease in moisture being removed during the falling rate period. Furthermore, the behavior of the hood exhaust mass flux is observed to be in contrary to that of the vacuum mass flux, which is as predicted from the mass balance, Equation (2).

Figure 7 depicts the temperature variations of the paper and the exhaust streams along the length of the dryer. As expected, the paper temperature rises initially during the warm-up zone, and then attains a constant value of 55^{0}C in the "constant rate" period. However, in the falling rate zone, the paper temperature again increases slightly to 65^{0}C due to heating-up of the paper which contains little bound moisture. The vacuum exhaust temperature is always above 100^{0}C and shows similar behavior as paper temperature towards the end of the Papridryer. This is expected since less and less moisture is being picked-up by the percolating steam. On the other hand, the hood exhaust temperature remains relatively constant throughout.

The effects of varying the pressure differential across the web on average total drying rate under turbulent jet conditions were also investigated, and the results are shown in Figure 8. The simulations were carried out for a slotted jet configuration using both Saad et al. and Das correlations for heat transfer coefficient for comparison purposes. In addition, Kast's (13) correction factor for high evaporation rates under turbulent conditions was applied to the above heat transfer coefficients. The results indicate an almost linear relationship between average drying rate and the pressure differential in all cases. At 580 K, Das' correlation predicted drying rates of 10 to 15% higher than Saad et al.'s.

The effects of pressure differential on drying rates were also studied under laminar jet conditions. It is noteworthy that the Papridryer (1) operated under apparently laminar conditions with jet Reynolds number of only about 1000. The dryer geometry used was similar to that of Huang and Mujumdar (7) in their study for hot air CIT dryer. Figure 9 shows the simulated results along with the results for air from Huang and Mujumdar. It should be noted that Saad et al.'s correlation was used in all cases (even for air) in the absence of better correlations for seemingly laminar flow conditions. Also, Kast's equation for laminar flow was used to correct the heat transfer coefficient for high evaporation rates effect in this case. However, if Keey's correction factor was used instead, the results would not be much different since the largest variation found was less than 2% (see Table I). It can be seen from Figure 9 that a much better rate of drying is attained using superheated steam than air at the same jet conditions. Nevertheless, for the case of superheated steam, lower drying rates were obtained under laminar conditions as compared to operating under turbulent conditions.

Figure 10 shows the efficiencies of the system, energy costs, and total dryer length as a function of the through drying rate contribution, which is a direct consequence of the applied pressure differential across the web. The zero point on the through drying contribution axis corresponds to pure impingement drying or zero pressure differential. It is interesting to see that the efficiency of moisture removal starts to reach a plateau beyond 65% through drying contribution while the process effeciency begins to drop off significantly. Although the length of dryer required to reach the final moisture content is shortened beyond this point, the associated energy costs increase significantly. This means that elevating the through drying contribution above 65% will result in greater and inefficient expenditure in energy with little improvement of final moisture content. Thus, it is reasonable to conclude that an optimum through drying contribution for these conditions would be in the order of 65%, which corresponds to a pressure differential of 22,500 Pa across the web of newsprint.

The jet velocity was vaired between 96.5 m/s to 110.0 m/s for a slotted jet configuration; the resulting average drying rate and final moisture content are shown in Figure 11. Only a 2% increase in drying rate was achieved by changing the jet velocity from 96.5 m/s to 104.5 m/s, and an even smaller increase from 104.5 m/s to 110.0 m/s, hence not affecting much the average drying rates. However, a notable decrease in final moisture content is observed, but only at the expense of higher energy costs (see Table II).

Figure 12 and Table III summarize the predicted effects of jet temperature on average drying rate, final moisture content, and total dryer length for a slotted jet configuration. These results were obtained for turbulent jet conditions, but are plotted together with those for air under laminar conditions (7) for comaprison. At increased jet temperatures, higher drying rates are achieved and fewer CIT dryers are required to reach the final moisture content. The appropriate jet temperature to be used depends greatly on the dryer material of construction, heat sensitivity of paper as well as the capital cost associated with the high temperature equipment. Effects of steam drying on paper quality need to be determined.

Assumed heat loss from the CIT dryer as a fraction of the heat input was varied (very conservatively) from zero to 7% as shown in Table IV. It can be seen that longer dryers are required to acieve the same final moisture content, and also a dramatic increase in energy costs as well as a decrease in thermal efficiency can occur if the dryer system is not well insulated. In actural practice the loss may be higher than 7%.

Simulation runs using different dryer geometry and jet configuration were performed. The dryer geometry was varied by altering the fraction of jet open area. These results for slotted jet configuration using the Saad et al. correlation are summarized in Table V. Higher drying rates and lower final paper moisture content with shorter dryer length are attainable (up to a point) by increasing the hole open area, but at the expense of higher energy consumption and lower dryer thermal efficiency. These observations apply equally to round jets. For very tightly packed jets the correlations used are not applicable and an optimal open area may indeed exist.

Also for a round jet configuration, the ratio of nozzle-to-web spacing to nozzle opening, h/d, was varied between 2.0 to 8.0 (see Table VI). As h/d decreases within this range (i.e., jet opening is closer to surface of paper), higher drying rates are expected, resulting in smaller dryers for reaching the same final moisture content.

Finally in Table VII, a comparison was made on the CIT dryer performance using round jets versus slotted jets. The results indicate that the dryer performance is similar for both configurations. Round jets are just as good as slotted jets as far as the drying rates, and energy costs are concerned. The choice between these may be made on the basis of other constraints, notably mechanical design.

CLOSURE

A transient model which incorporates the major assumption that the exhaust temperatures should be above saturation to avoid condensation, was developed to assess the technical feasibility of the CIT process with superheated steam as the drying medium. The predicted drying curve consists of an initial warm-up period followed by a zone of steady increase in drying rate, and then a falling rate period. The steady increase in drying rate is due to the higher through drying rates as the paper becomes more permeable with decreasing paper moisture content. The through drying contribution was found to be the key factor in determining the economics of the process. At the drying conditions examined, a through drying contribution in the order of 65%, was observed to be optimal in balancing energy expenditure versus moisture removal efficiency.

Among the various process variables and parameters studied in this work, jet temperature, pressure differential across the web, percent heat loss, and dryer geometry were found to have significant effects on average drying rates and thus must be considered in arriving at an optimum dryer design. Also, in order to achieve high heat transfer rates, the dryer should operate at turbulent jet conditions. A change in jet configuration from slotted jets to round jets did not result in any significant difference in dryer performance.

A comparison between superheated steam and air as drying media indicates that higher drying rates and lower final paper moisture content can be achieved with superheated steam. Furthermore, the exhaust streams must be recycled to recover the exhaust heat. The simulation results thus show the attractiveness and feasibility of using superheated steam, and justify further experimental work to verify the mathematical model presented in this paper. In particular, the paper equilibrium moisture content in the presence of steam should be determined experimentally. Also, the possibility of applying a sequential drying (SIT) process (i.e., impingement and/or through) to certain sections of the dryer in order to achieve better efficiency and performance should be further explored. Experimental investigations of heat transfer with superheated steam jets are in progress in the author's (ASM) laboratory. Better correlations are needed to account for effects of suction, web motion, cross-flow and multiplicity of jets.

Finally it is important to note that the model presented here can be improved to utilize time-averaged local heat/mass transfer rates (rather than the average Nusselt number correlations used here) for the specific jet configuration considered and including the effects of web motion, suction at the surface, improved modelling of the through drying process, influence of the exhaust port locations (which may induce undesirable cross-flow effects) etc. Augmentation of impingement heat transfer by introduction of turbulence promoters (e.g. perforated plates between the nozzle plate and the web) may also yield higher efficiencies at lower capital costs. Recently, Kasagi et al. (21, 22) have shown that presence of appropriately selected perforated plates to enhance stream turbulence can augment impingement heat transfer by a factor of up to two. It may be difficult, if not impossible, to incorporate this concept in a CIT process for paper. Oblique impingement may yield a better heat transfer performance than normal impingement for the case of a moving impingement target viz. paper web.

In view of the multiplicity of possible design parameters it is very important to develop practically usable simulation models for the CIT process. The results reported in this paper, like those of Huang and Mujumdar (7), represent preliminary attempts in this direction. Selection of the nozzle geometry and configurations to yield optimal transfer rates per unit pumping power are especially important design criteria. The results of Hardisty (23) and Obot, Mujumdar and Douglas (24) avoide useful guidelines in this regard. See Mujumdar (25) for additional design-oriented information.

ACKNOWLEDGEMENTS

The authors gratefully acknowledge the assistance of Purnima Mujumdar in the preparation of this paper and the typescript. The comments of Dr. Iva Filkova, Czech Technical University, Prague Czechoslovakia, have been very useful and are sincerely appreciated. Thanks are also due to Dr. Kasagi of University of Tokyo for providing ASM with a copy od Dr. Ali's Ph.D thesis.

REFERENCES

1. Burgess, B.W., Chapman, S.M., and Seto, W., Pulp & Paper Mag. Can., 73 (11), T314-T323 (1972).

2. Mujumdar, A.S., Plenary Lecture, 32nd Ind. Chem. Eng. Conf., Madras, India, December 15-17 (1981).

3. Chu, J.C., Finelt, S., Hoerrner, W., and Lin, M.S., Ind. Eng. Chem., 51, 275 (1959).

4. Lane, A.M. and Stern, S., Mech. Eng., 78, 423 (1956).

5. Chu, J.C., Lane, A.M., and Conklin, D., Ind. Eng. Chem., 45, 1586 (1953).

6. Crotogino, R.H. and Allenger, V., CPPA Trans. 5 (4). Tr 84 (1979).

7. Huang, P.G. and Mujumdar, A.S., Drying'82, Ed. Mujumdar, A.S., Hemisphere-McGraw-Hill, pp 106-114, N.Y. (1982).

8. Randall, K.R., Paper to be published in Drying'83, Ed. Mujumdar, A.S., Hemisphere-McGraw-Hill, N.Y. (1983).

9. Smith, J.M. and Van Ness, H.C., Introduction to Chemical Engineering Thermodynamics, McGraw-Hill, N.Y. (1975).

10. Prahl, J.M., Thermodynamics of Paper Fiber and Water Mixtures, PhD Thesis, Harvard (1968).

11. Holman, J.P., Heat Transfer, McGraw-Hill, N.Y. (1976).

12. Keey, R.B., Drying Principles and Practice, Pergamon Press, N.Y. (1972).

13. Kast, W., Heat Transfer Conf. (Munich), 2, pp 263-268 (FC47), F.R.G. (1982).

14. Martin, H., Advances in Heat Transfer, 13, pp 1-60, Academic Press (1977).

15. Saad, N.R., Mujumdar, A.S., and Douglas, W.J.M. Drying'80, 1 Ed. Mujumdar, A.S., pp422-430, Hemisphere Publishing Corporation, N.Y. (1980)

16. Das, D., M. Eng. Thesis, Chemical Engineering, McGill University, Canada (1982).

17. Mujumdar, A.S. and Huang, P.G., IPPTA, 19 (2), 65 (1982).

18. Chance, J.L., TAPPI, 57 (6), 108 (1974).

19. Brundrett, E. and Baines, W.D., TAPPI, 49 (3), 97 (1966).

20. Thomas, F.A. and Jackson, T.W., The Viscosity of Steam, Thermodynamic and Transport Properties of Gases, Liquids nad Solids, McGraw-Hill, N.Y. (1959).

21. Ali Khan, M.M., Ph.D. Thesis, University of Tokyo, Japan, (1980).

22. Ali Khan, M.M., N. Kasagi, M. Hirata and N. Nishiwaki, 7th International Heat Transfer

Conference, Munich, Vol. 3, Hemisphere, N.Y. (1982).

23. Hardisty, H., Proc. Instn. Mech. Eng. (London), Vol. 197C, March (1983).

24. Obot, N.T., Mujumdar, A.S. and Douglas, W.J.M., Drying'80, Vol. 1, Developments in Drying, Ed. A.S. Mujumdar, Hemisphere, N.Y. (1980).

25. Mujumdar, A.S., in Handbook of Industrial Drying, Marcel Dekker, N.Y. (1985) (In Press)

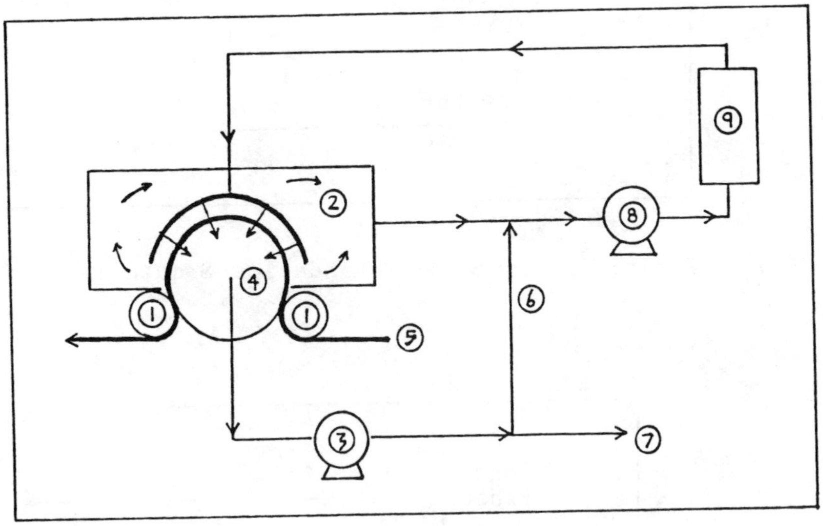

Figure 1. Flow Diagram of the Papridryer Process Using Superheated Steam as the Drying Medium.

(1) Paper Guide Rolls; (2) Hood Exhaust Chamber; (3) Vacuum Fan; (4) Vacuum Cylinder; (5) Paper Feed; (6) Make-up Steam; (7) Bleed Stream; (8) Recirculation Fan; and (9) Heater/Combustion Chamber.

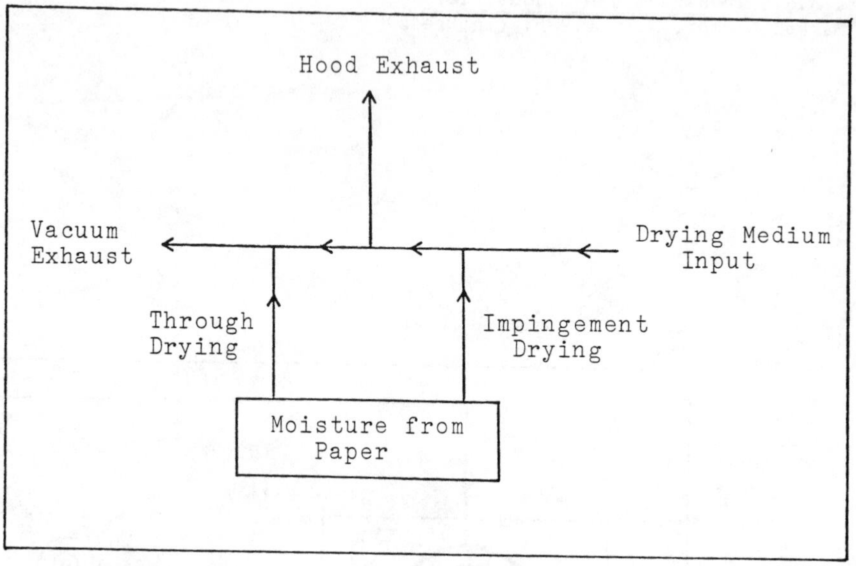

Figure 2. Concept of Randall's Sequential Model

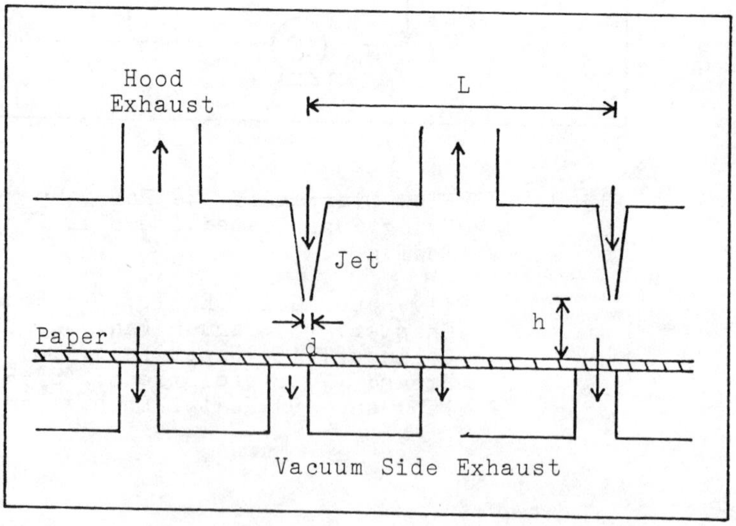

Figure 3. High Velocity Hood Configuration of the Papridryer

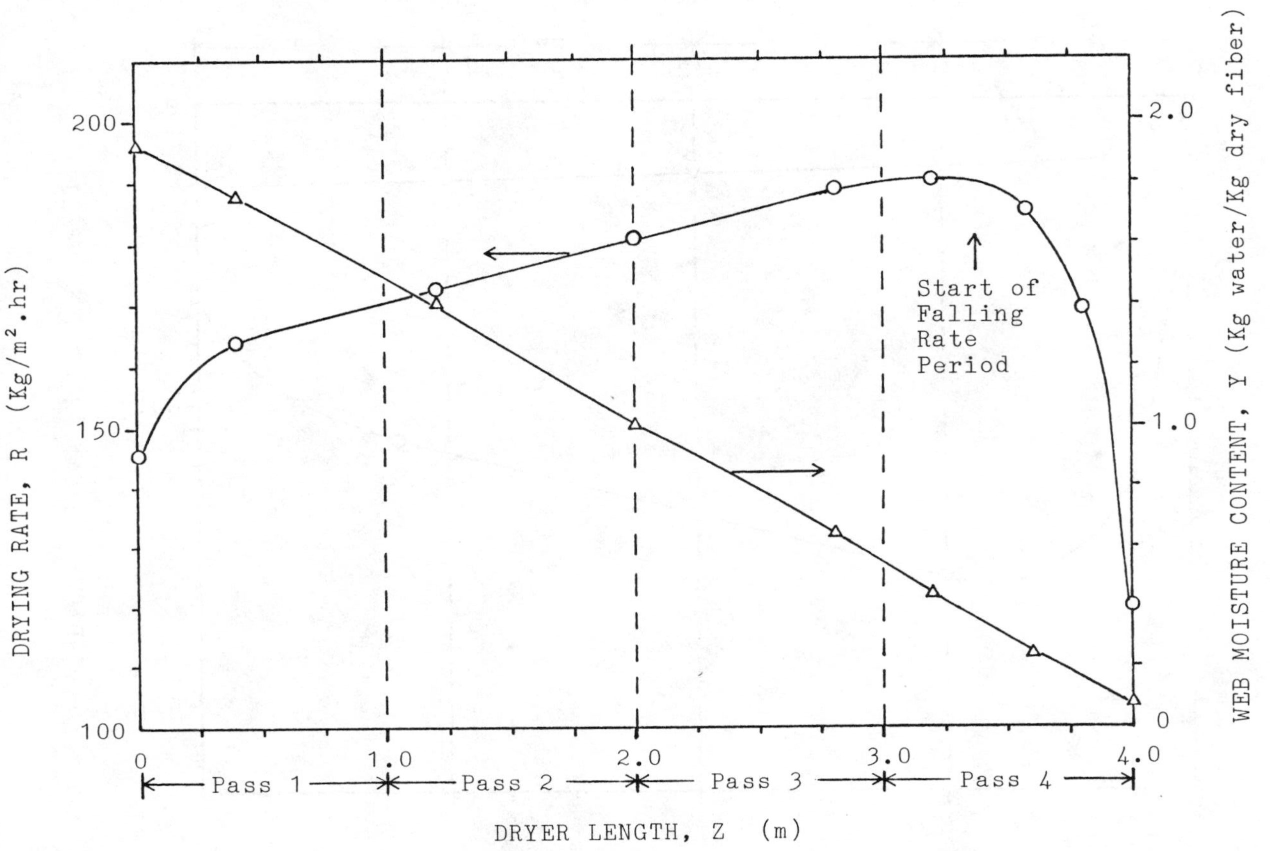

Figure 4. Drying Rate and Web Moisture Content Along the Length of the Four-pass Papridryer System (Base Case: T_j= 580 K, V_j=104.5 m/sec, ΔP=22500 Pa, A_{hoa}=0.0275)

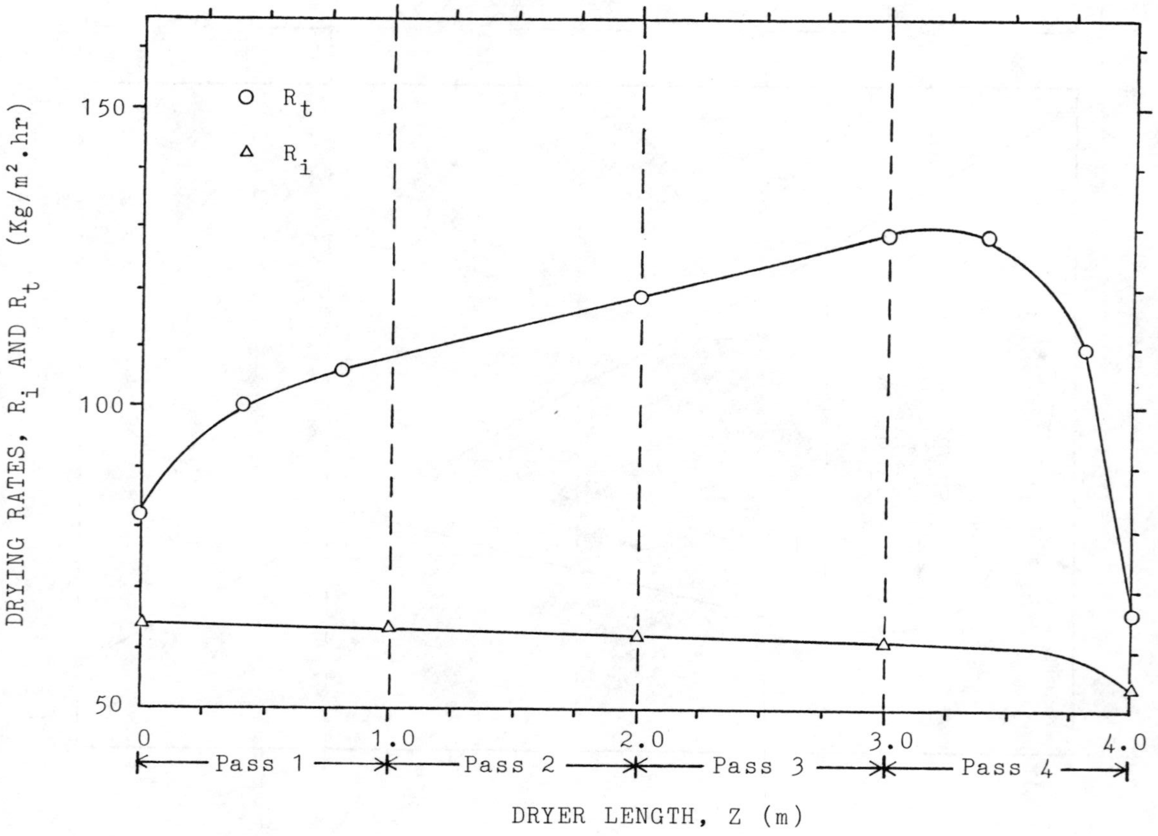

Figure 5. Impingement and Through Drying Contributions along the Length of the Dryer (Base Case, as Figure 4)

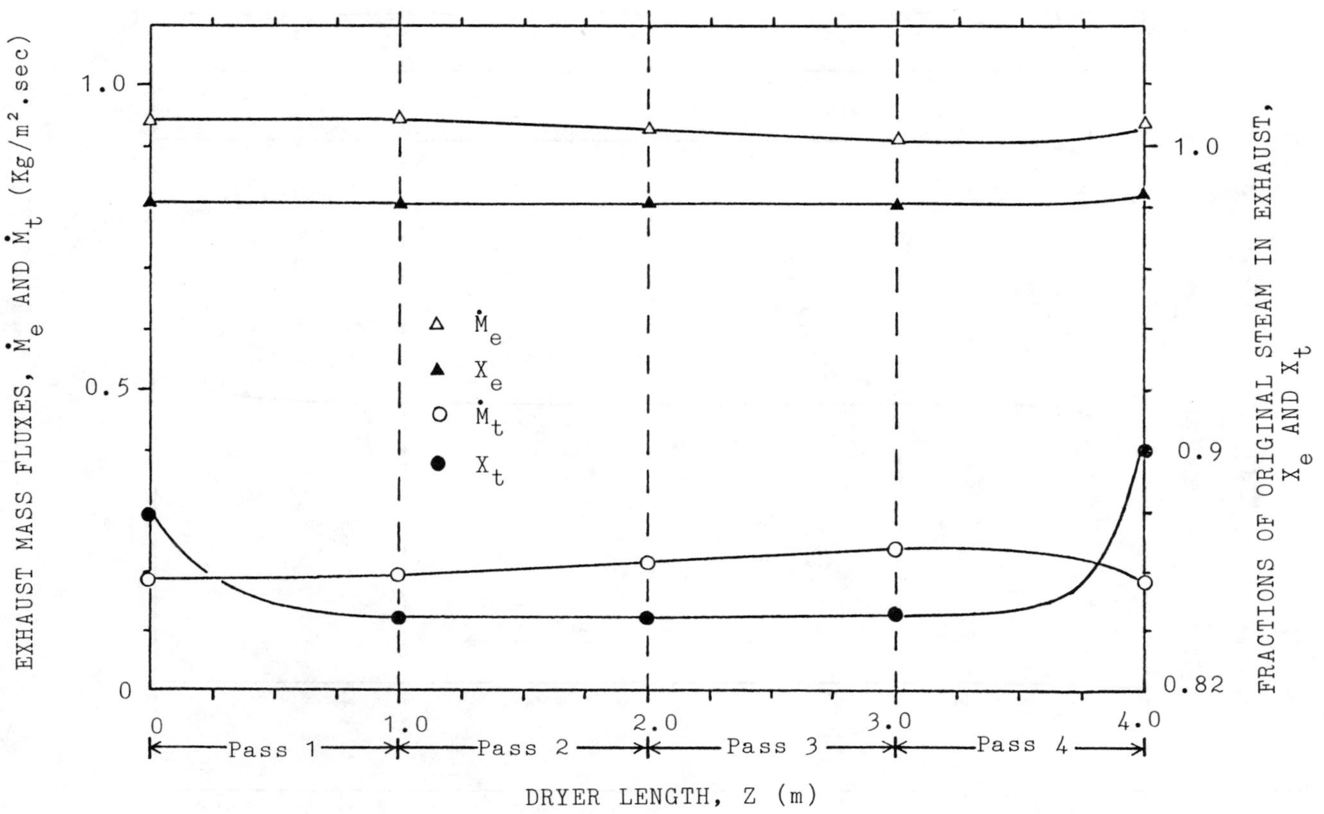

Figure 6. Hood and Vacuum Exhaust Conditions Along the Length of the Dryer
(Base Case, as Figure 4)

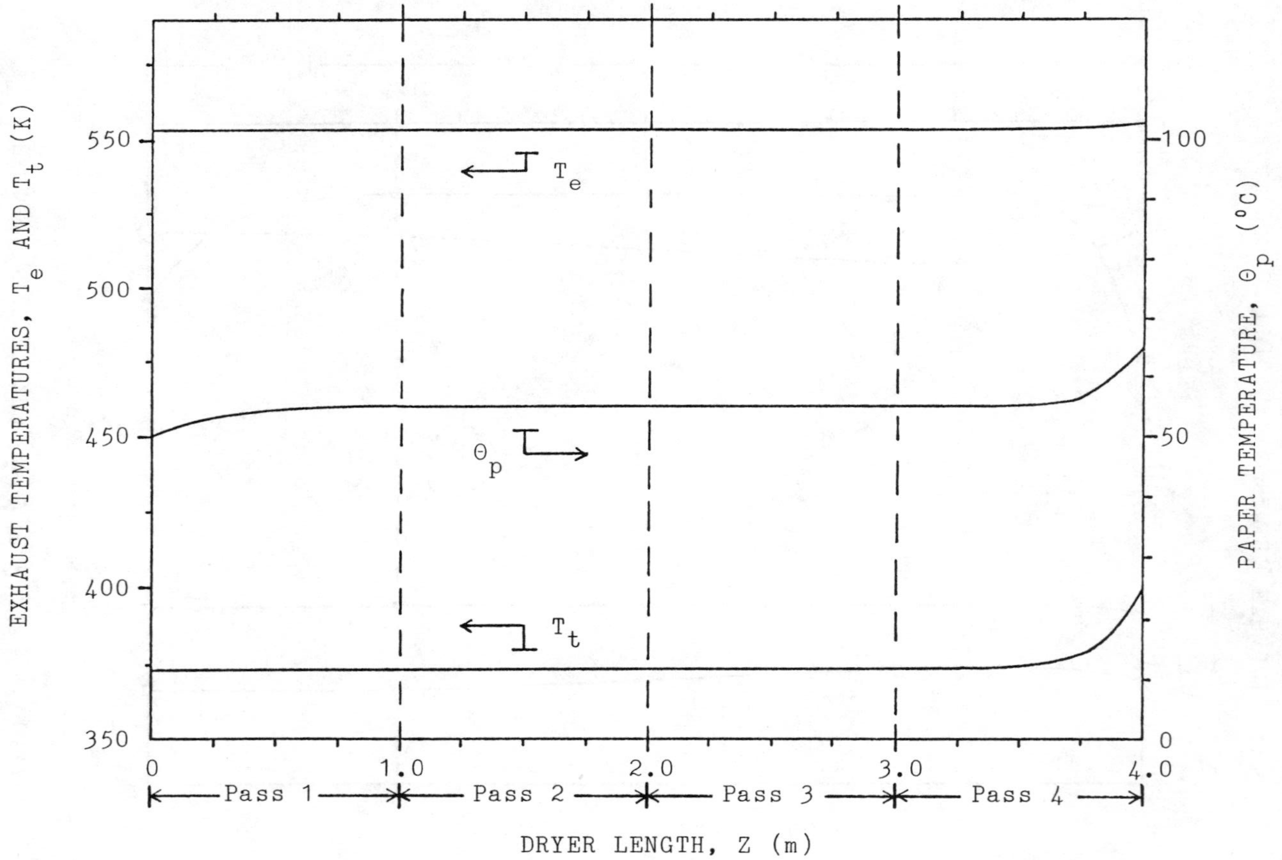

Figure 7. Hood and Vacuum Exhaust and Paper Temperatures Along the Length of the Dryer (Base Case, as Figure 4)

TABLE I

Comparison of Correction Factors for High Evaporation Rates using Saad's Correlation for Slotted Jets

(Laminar Conditions: T_j=580 K, V_j=96.5 m/sec, ΔP=22500 Pa, A_{hoa}=0.0135)

Correction Factor	$\overline{R}$	Y_f
Keey, laminar	159.15	0.2491
Kast, laminar	157.43	0.2670
Kast, turbulent	156.40	0.2779

TABLE II

Effect of Jet Velocity on Dryer Performance Using Saad et al's Correlation for Slotted Jets

(T_j=580 K, ΔP=22500 Pa, A_{hoa}=0.0275, Z=4.0 m)

V_j	EC	ET
96.5	23.69	51.98
104.5	25.54	51.76
110.0	26.92	51.58

TABLE III

Effect of Jet Temperature on Dryer Performance Using Saad et al's Correlation for Slotted Jets

(V_j=104.5 m/sec, ΔP=22500 Pa, A_{hoa}=0.0675)

T_j	Y_f	Z	EC	ET
500.0	0.3495	4.0	37.15	50.35
580.0	0.0115	3.658	35.57	49.97
650.0	0.0075	2.984	30.39	50.94

TABLE IV

Effect of Heat Loss on Dryer Performance Using Saad et al's Correlation for Slotted Jets

(T_j=650 K, V_j=104.5 m/sec, ΔP=22500 Pa, A_{hoa}=0.0675)

φ	$\overline{R}$	Y_f	Z	EC	ET
0.0	244.19	0.0075	2.984	30.39	50.94
0.95	206.93	0.0127	3.512	73.69	21.42
0.93	192.56	0.0165	3.774	95.49	16.57

TABLE V

Effect of Hole Open Area Fraction on Dryer Performance Using Saad et al's Correlation for Slotted Jets
(T_j=580 K, ΔP=22500 Pa)

V_j	A_{hoa}	$\bar{R}$	Y_f	Z	EC	ET
96.5(laminar)	0.0135	157.43	0.2670	4.0	19.40	52.46
104.5	0.0275	175.90	0.0732	4.0	25.54	51.76
104.5	0.0675	198.78	0.0115	3.658	35.57	49.97
104.5	0.1000	214.13	0.0114	3.396	41.81	49.30

TABLE VI

Effects of h/d and Hole Open Area on Dryer Performance for Round Jet Configuration
(T_j=580 K, V_j=104.5 m/sec, ΔP=22500 Pa)

A_{hoa}	h/d	$\bar{R}$	Y_f	Z	EC	ET
0.0645	8.0	188.48	0.0115	3.858	35.83	50.18
0.0645	4.0	214.86	0.0117	3.384	33.36	49.81
0.0645	2.0	228.05	0.0119	3.188	32.33	49.67
0.0271	8.0	174.21	0.0910	4.0	25.26	51.87

TABLE VII

Comparison of Round and Slotted Jet Configurations
(T_j=580 K, V_j=104.5 m/sec, ΔP=22500 Pa)

Configuration	Correlation	A_{hoa}	$\bar{R}$	Y_f	Z	EC	ET
Round	Chance	0.0645	188.48	0.0115	3.858	35.83	50.18
Slotted	Das	0.0675	217.80	0.0119	3.338	33.83	49.72
Slotted	Saad	0.0675	198.78	0.0115	3.658	35.57	49.97
Round	Chance	0.0271	174.21	0.0910	4.0	25.26	51.87
Slotted	Das	0.0275	187.75	0.0130	3.87	25.93	50.81
Slotted	Saad	0.0275	175.90	0.0732	4.0	25.54	51.76

AN EXPERIMENTAL AND ANALYTICAL INVESTIGATION OF A THERMALLY INDUCED VACUUM DRYING PROCESS FOR PERMEABLE MATS

F. Ahrens
I. Journeaux

The Institute of Paper Chemistry, Appleton, Wisconsin 54912

ABSTRACT

The hot-surface drying of an air-free permeable mat in a vacuum environment maintained by condensing vapor has been studied experimentally and analytically. Experimental data on paper drying rates are presented and shown to be several times greater than rates which occur under "conventional" paper drying conditions. A detailed mathematical model is presented which attempts to account for all the nonlinearities associated with the transport processes considered. This model and a simplified, linearized version of it are applied to prediction of the paper drying characteristics, with encouraging results.

NOMENCLATURE

C_L, C_s - specific heats of liquid and solid, $Jkg^{-1}k^{-1}$

C_p - specific heat capacity at constant pressure, $Jkg^{-1}K^{-1}$

k - permeability, m^2

L - thickness of permeable mat, m

P - pressure, Nm^{-2}

P_c - capillary pressure, Nm^{-2}

q_T - dimensionless time, -

Q - energy flux, Wm^{-2}

P_{rel} - relative vapor pressure, -

S_L - liquid saturation, -

t - time, s

T - temperature, K

u - dimensionless distance into solid (y/L), -

w - superficial velocity, ms^{-1}

X - solids moisture content (dry-basis)

y - cartesian coordinate perpendicular to solid surface, m

σ - surface tension, Nm^{-1}

ε - void ratio, -

ΔH - latent heat, Jkg^{-1}

Θ - dimensionless temperature, -

λ - thermal conductivity, $Wm^{-1}K^{-1}$

ρ - density, kgm^{-3}

μ - viscosity, Nsm^{-2}

Ω - dimensionless (modified) saturation variable, -

Subscripts

c - cold surface

H - hot surface

L - liquid

o - initial

ref - reference state

s - solid

v - vapor

Superscripts

‾ - average

* - saturation

INTRODUCTION

At the preceding International Drying Symposium ("Drying '80"), Lehtinen (1) presented laboratory data suggesting that the hot-surface drying of permeable mats under air-free conditions can result in drying rates nearly an order of magnitude greater than those typical of conventional dryer sections. Obviously, these results have generated interest within the pulp and paper industry.

The air-free drying process discussed by Lehtinen

and examined further in the present paper may be designated as a thermally induced vacuum (or "thermal/vacuum") drying process. Moisture evaporated in the mat contacting the hot surface is condensed on a nearby cooled surface, separated from the mat by a layer of permeable filler material to avoid rewetting (see Fig. 1). The vacuum level is thus controlled by the cooled surface temperature; the driving forces for drying are the temperature and vapor pressure gradients applied across the mat due to the presence of the heated and cooled surfaces.

Because thermal/vacuum drying has the potential for achieving both reduced equipment size and reduced energy cost for drying, it seems important to develop a further characterization and understanding of this process. Results of work toward this goal are reported here. Both experimental paper drying results and a detailed mathematical model of the thermal/vacuum drying of hygroscopic mats are included. Since the drying of paper and paperboard is the application of chief interest to the authors, results of applying the detailed mathematical model and a simplified version of it to the prediction of paper drying characteristics are presented.

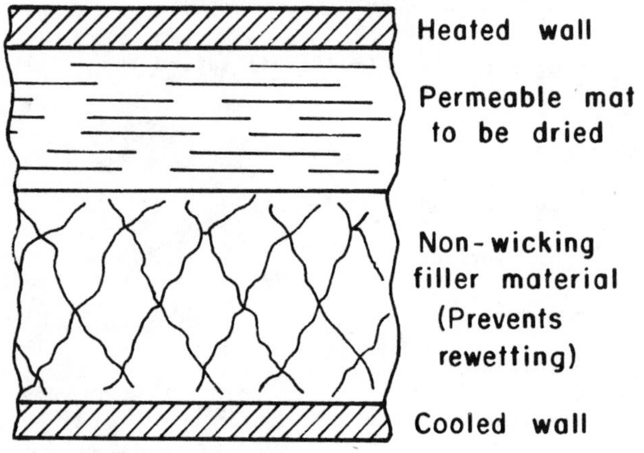

Fig. 1. Configuration for the Thermal/vacuum Drying Process

EXPERIMENTAL STUDY

The objectives of the experimental portion of this study were to corroborate and extend the thermal/vacuum drying of paper, to quantify the increase in drying rate via thermal/vacuum drying relative to "conventional" drying, and to provide data for comparison with the analytical results to be discussed later in this paper.

Apparatus and Procedure

A bench-scale drying apparatus capable of providing thermal/vacuum operating conditions was designed and constructed. A simplified schematic diagram of the device is shown in Fig. 2. The moist mat is dried within a chamber comprising a hot plate, a cylindrical housing, and a copper piston (heat sink). O-rings are used to provide vacuum seals, and the housing is clamped to the hot plate. The housing, machined from an epoxy/glass fiber composite, has an inside diameter of about 16 cm. The copper piston is approximately 5 cm thick and has a central hole to facilitate air removal and admission. It is usually operated at room temperature (24°C); it has a negligible temperature rise during a test, even though it serves as the site for condensation of the vapor removed from the mat. The filler material attached to the piston is actually a stack of plastic screens, approximately 2 cm thick. The hot plate is made of aluminum. It is about 1.3 cm thick and has a number of parallel, closely spaced, 0.65-cm-diameter holes through it, which act as flow channels for oil heated to a selected temperature and circulated at a high rate.

The paper fiber mats dried in this study were similar to and prepared in the same manner as the southern softwood fiber mats described in (2). A single basis weight, 0.205 kg/m^2 (typical of linerboard), was used in the present study, and the initial moisture content was always about 60% (wet basis). Each fiber mat was about 13 cm in diameter; prior to a drying test one was attached to the filler screens with fine threads. The initial thickness of these sheets was crudely measured to be about 0.7 mm, from which the overall porosity was calculated to be 0.81. From these values, the initial liquid saturation was calculated to be 0.53. This set of thickness, porosity, and initial saturation parameters was used in the predictions made with the mathematical models, to be described later.

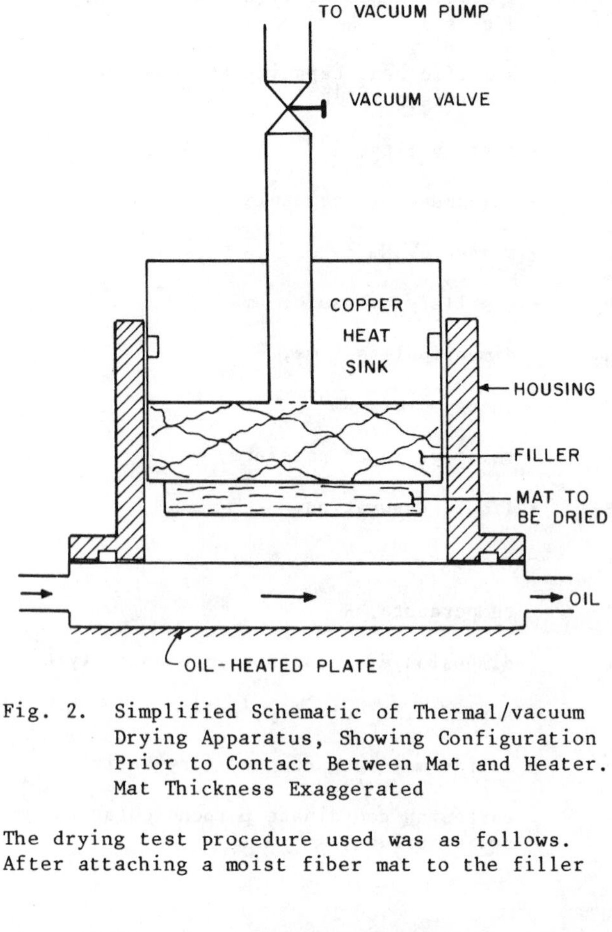

Fig. 2. Simplified Schematic of Thermal/vacuum Drying Apparatus, Showing Configuration Prior to Contact Between Mat and Heater. Mat Thickness Exaggerated

The drying test procedure used was as follows. After attaching a moist fiber mat to the filler screens, the piston/filler/mat assembly was partially inserted into the housing, such that the O-ring seal was engaged, but with the mat about 2 cm above the hot plate. The piston was held in this position for about 30 seconds, during which the vacuum pump removed the air from the system and the hot plate was heated to the desired temperature via the hot oil flow. Next, the vacuum valve was closed and the piston was released. The piston, exposed to a pressure difference of approximately 1 atmosphere, rapidly moved toward the hot plate, initiating the drying period. After a desired amount of drying time, the vacuum valve was opened to permit atmospheric air to reenter the chamber. The piston was then rapidly removed and the mat retrieved and weighed. By also determining the initial and oven-dry weights of the sample, the amount of water removal during the test could be found.

It was found that a small amount (e.g., about 5%) of the initial mat moisture evaporated during the chamber evacuation/hot plate heatup period. Since this did not constitute true drying, the test data were corrected for this effect.

Experimental Results

For a given hot surface temperature, a series of drying tests having different time durations were performed, so that a drying curve (moisture loss versus time) could be established. Data from a particular series (90°C hot plate temperature) are shown in Fig. 7, which appears later in this paper. The drying time, which was defined as the time required to reach a final moisture content of 6% (wet basis), was taken from the drying curve. It was noted upon inspection of the shape (slope) of the experimental thermal/vacuum drying curves obtained in this study (e.g., Fig. 7), that there is little, if any, indication of a "constant rate" drying period. Interestingly, the same type of mat does appear to have an appreciable region of relatively constant rate when dried on a hot surface in the atmosphere (2).

The drying times were used to compute average drying rates, and the results are presented in Fig. 3. For comparison purposes, some previous data (2), representing similar mats dried in the atmosphere with a mechanical pressure loading of 3.45 kPa (typical of that used in "conventional" paper machine dryer sections), are also shown. The dramatic improvements found for thermal/vacuum drying suggest potential applications ranging from significant reductions in dryer size via operation at relatively high temperature levels to the use of a less expensive energy source via operation at relatively low temperature while still matching or exceeding typical "conventional" drying rates. The effects of different operating conditions on paper properties was investigated by Lehtinen (1).

The increase in drying rate of the thermal/vacuum process compared with conventional drying seems to result from the combined effect of two differences in operating conditions. First, the thermal/vacuum process operates totally air-free, so that it is never impeded by diffusion effects. Second, the thermal/ vacuum system provides a mechanical pressure loading of about one atmosphere due to the pressure difference across the piston. This yields a reduction in thermal resistance between the heater and fiber mat. Data on the effect of higher than normal mechanical pressures (to 29.6 kPa) during drying in the atmosphere (2) suggest that pressure effects alone would typically account for less than half of the improvement shown in Fig. 3. The magnitudes of the drying rates found in the present study are generally compatible with the results of Lehtinen (1). The detailed variation with hot surface temperature, however, is not quite the same. Figure 3 suggests a linear increase in drying rate with temperature, whereas Lehtinen (1) suggests a linear decrease of drying time with temperature. It is felt that the representation in Fig. 3 is more physically plausible.

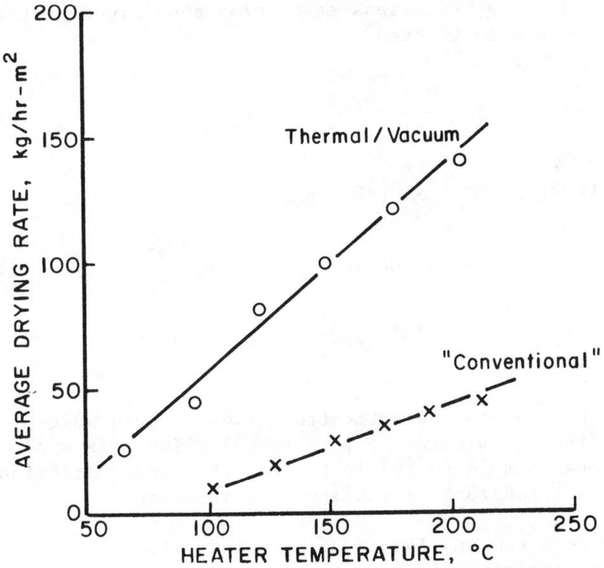

Fig. 3. Experimental Average Drying Rates for Southern Softwood Fiber Mats (0.205 kg/m^2 Dry Basis Weight, 60% Initial Moisture, 6% Final Moisture - Wet Basis). "Conventional" Refers to Hot-surface Drying in the Atmosphere with 3.45 kPa Mechanical Pressure Loading [Based on Data from (2)]

ANALYTICAL STUDY

The objectives of the analytical portion of this study were to develop a relatively detailed and general mathematical model of the thermal/vacuum drying of hygroscopic mats and to apply it to prediction of the details of paper drying. Although other analyses of vacuum drying have been published recently (e.g., 3,4), they do not appear to be general enough to meet the present objectives.

Analysis

The formulation is patterned after that given in (5) but invokes far fewer restrictions and simplifications and considers boundary conditions believed to better represent those occurring in the experiments described above. The analysis is based on the following assumptions:

1. There is no air in the mat
2. The solid structure of the mat does not deform during drying
3. Local thermal equilibrium exists at all points in the sheet
4. Pore Reynolds numbers are small enough that Darcy's law applies
5. Viscous dissipation and compression work are negligible
6. The filler material is such that there is no liquid return to the mat
7. Transient vapor storage is negligible compared with liquid storage

For the drying system under consideration, subject to the stated assumptions, the continuity and energy equations describing conditions in the mat may be written*:

$$\varepsilon\rho_L \frac{\partial S_L}{\partial t} = -\frac{\partial}{\partial y}(\rho_L w_L + \rho_v w_v) \qquad (1)$$

and

$$(\varepsilon S_L\rho_L C_L + \varepsilon(1-S_L)\rho_v C_{p_v} + (1-\varepsilon)\rho_s C_s)\frac{\partial T}{\partial t}$$

$$+ (\rho_L C_L w_L + \rho_v C_{p_v} w_v)\frac{\partial T}{\partial y} \qquad (2)$$

$$= -\frac{\partial Q}{\partial y} - \Delta H \frac{\partial}{\partial y}(\rho_v w_v)$$

where S_L is the saturation (i.e., liquid volume fraction within the pore space). The origin of the coordinate (y) in the mat thickness direction is taken to be the heater/mat interface.

The steam and liquid superficial velocities are represented by Darcy's law:

$$w_v = -\frac{k_v}{\mu_v}\left(\frac{\partial P_v}{\partial y} - \rho_v g\right) \qquad (3)$$

and

$$w_L = -\frac{k_L}{\mu_L}\left(\frac{\partial P_L}{\partial y} - \rho_L g\right) \qquad (4)$$

where g is the component of gravitational acceleration in the y-direction and k_v, k_L are the vapor and liquid permeabilities, represented as:

$$k_v = k_{abs}\, k_{rel_v} \text{ and, } k_L = k_{abs}\, k_{rel_L}$$

The absolute permeability (k_{abs}) for single phase flow is taken to be the same for both phases; the (dimensionless) relative permeabilities (k_{rel_v}, k_{rel_L}) are functions of S_L.

The liquid pressure can be expressed as:

$$P_L = P_v - P_c \qquad (5)$$

*When possible, the nomenclature used follows that recommended for the Third International Drying Symposium, Birmingham, September 1982.

where P_c is the capillary pressure, a function of S_L and, via the surface tension, of T.

The vapor pressure, P_v, depends on both T and S_L if the mat is hygroscopic (assumed here). For simplicity, the form of the relationship is taken to be:

$$P_v = P_v^*(T)\, P_{rel}\,(X) \qquad (6)$$

where P_v^* is the usual saturation pressure, and the (dimensionless) relative pressure (P_{rel}) depends on the dry-basis moisture content, X. The relation between X and S_L is:

$$S_L = \frac{\rho_s}{\rho_L}\left(\frac{1}{\varepsilon} - 1\right) X \qquad (7)$$

Since both P_v and P_L are functions of temperature and saturation, their gradients are written as [using Eq. (5)]:

$$\frac{\partial P_v}{\partial y} = \frac{\partial P_v}{\partial T}\frac{\partial T}{\partial y} + \frac{\partial P_v}{\partial S_L}\frac{\partial S_L}{\partial y} \qquad (8)$$

$$\frac{\partial P_L}{\partial y} = \left(\frac{\partial P_v}{\partial T} - \frac{\partial P_c}{\partial T}\right)\frac{\partial T}{\partial y} + \left(\frac{\partial P_v}{\partial S_L} - \frac{\partial P_c}{\partial S_L}\right)\frac{\partial S_L}{\partial y} \qquad (9)$$

The heat flux (Q) in Eq. (2) is due to conduction and can be expressed as:

$$Q = -\lambda\frac{\partial T}{\partial y} \qquad (10)$$

The thermal conductivity is assumed to be represented by the parallel-conductor model:

$$\lambda = (1-\varepsilon)\,\lambda_s + \varepsilon\, S_L\,\lambda_L + \varepsilon(1-S_L)\,\lambda_v$$

Two boundary conditions are needed at each mat surface. At the hot surface, which is solid, the net mass flux must be zero:

$$\rho_v w_v\,(0, t) + \rho_L w_L\,(0, t) = 0 \qquad (11)$$

For simplicity, the second condition at the hot surface is taken to be that the temperature of the mat is essentially that of the heater (T_H), assumed constant:

$$T\,(0, t) = T_H \qquad (12)$$

This assumes a negligible contact resistance between heater and mat and may have to be revised in the future.

At the mat/filler interface, the vapor pressure is considered to remain constant at a level dictated by the condensing (saturation) pressure corresponding to the temperature of the cooled wall, T_c, assumed constant:

$$P_v\,(L, t) = P_v^*\,(T_c) \qquad (13)$$

where L is the mat thickness. This assumes that both the vapor flow resistance of the filler and the thermal resistance of the condensate layer on the cooled wall are negligible.

The second condition at the mat/filler interface assumes that there is no heat loss due to conduction into the filler and that the interface is not a site of appreciable evaporation of liquid originally residing in the interior. These considerations imply:

$$\frac{\partial T}{\partial y}(L, t) = 0 \tag{14}$$

Taken together, these conditions applied at $y = L$ depict the interface as simultaneously heating and drying out, such that the hygroscopic relation [Eq. (6)] and Eq. (13) are continuously satisfied, under the action of energy input from the mat interior. The present boundary conditions are more difficult to implement than those assumed in an earlier, simplified analysis of thermal/vacuum drying (5), but are believed to more closely represent the conditions of the experiments discussed in the preceding section.

The initial conditions are taken to be that the mat has the same (uniform) temperature as the cooled wall and a known (uniform) moisture content:

$$T(y, 0) = T_c, \quad S_L(y, 0) = S_{L_o}$$

If the dependence on T and S_L is known for the various thermophysical and transport coefficients introduced so far, the model equations above can, in principle, be solved for cases of interest. Due to the nonlinearities which prevail, however, it is not possible to develop general (universal) solutions in terms of dimensionless variables and parameters. Nevertheless, it was felt to be useful to cast the equations into a dimensionless form prior to development of a solution methodology, in order to identify potentially significant terms and groupings of parameters and to facilitate comparison with the linearized model given in (5).

The final results of combining Eq. (1)-(10) and nondimensionalizing are the following continuity and energy equations:

$$\frac{\partial \Omega}{\partial q_T} = \frac{1}{\delta}\left\{-\frac{\overline{A}L}{\overline{B}(T_H-T_c)}\frac{\partial A'}{\partial u} + \frac{\partial}{\partial u}\left(B'\frac{\partial \theta}{\partial u}\right) + \frac{\partial}{\partial u}\left(C'\frac{\partial \Omega}{\partial u}\right)\right\} \tag{15}$$

and

$$K'\frac{\partial \theta}{\partial q_T} = \frac{\partial}{\partial u}\left(D'\frac{\partial \theta}{\partial u}\right) - \frac{\overline{E}L}{\overline{D}(T_H-T_c)}\frac{\partial E'}{\partial u} + \frac{\overline{G}\,\overline{B}}{\overline{D}\,\overline{C}}\frac{\partial}{\partial u}\left(G'\frac{\partial \Omega}{\partial u}\right) - \frac{\overline{H}L}{\overline{D}}H'\frac{\partial \theta}{\partial u} \tag{16}$$

$$+ \frac{\overline{I}(T_H-T_c)}{\overline{D}} I' \left(\frac{\partial \theta}{\partial u}\right)^2 + \frac{\overline{B}\,\overline{J}(T_H-T_c)}{\overline{C}\,\overline{D}} J' \frac{\partial \Omega}{\partial u}\frac{\partial \theta}{\partial u}$$

In these equations, the dimensionless time and position variables are defined as:

$$q_T = \frac{\overline{D}\,t}{\overline{K}\,L^2}, \quad u = y/L$$

The saturation and temperature variables are defined by:

$$\Omega = \frac{\overline{C}}{\overline{B}(T_H-T_c)}(S_L - S_{L_o}), \quad \theta = \frac{T-T_c}{T_H-T_c}$$

The notation for other quantities in Eq. (15) and (16) is that quantities with an overbar (e.g., $\overline{A}$) are constants representing values of property functions (to be defined below) evaluated at a certain reference state (T_{ref}, $S_{L_{ref}}$). Primed quantities (e.g., A') are ratios of the actual local values of the property functions to their corresponding reference values (e.g., $A' = A/\overline{A}$).

The appropriate definitions are:

$$A = \frac{k_L \rho_L g}{\varepsilon \mu_L} + \frac{k_v \rho_v^2 g}{\varepsilon \rho_L \mu_v}$$

$$B = \frac{k_L}{\varepsilon \mu_L}\left(\frac{\partial P_v}{\partial T} - \frac{\partial P_c}{\partial T}\right) + \frac{\rho_v k_v}{\varepsilon \rho_L \mu_v}\frac{\partial P_v}{\partial T}$$

$$C = \frac{k_L}{\varepsilon \mu_L}\left(\frac{\partial P_v}{\partial S_L} - \frac{\partial P_c}{\partial S_L}\right) + \frac{\rho_v k_v}{\varepsilon \rho_L \mu_v}\frac{\partial P_v}{\partial S_L}$$

$$D = \lambda + \frac{\Delta H \rho_v k_v}{\mu_v}\frac{\partial P_v}{\partial T}$$

$$E = \frac{\Delta H \rho_v^2 k_v g}{\mu_v}$$

$$G = \frac{\Delta H \rho_v k_v}{\mu_v}\frac{\partial P_v}{\partial S_L}$$

$$H = \frac{\rho_L^2 C_L k_L g}{\mu_L} + \frac{\rho_v^2 C_{P_v} k_v g}{\mu_v}$$

$$I = \frac{\rho_L C_L k_L}{\mu_L}\left(\frac{\partial P_v}{\partial T} - \frac{\partial P_c}{\partial T}\right) + \frac{\rho_v C_{P_v} k_v}{\mu_v}\frac{\partial P_v}{\partial T}$$

$$J = \frac{\rho_L C_L v_L}{\mu_L}\left(\frac{\partial P_v}{\partial S_L} - \frac{\partial P_c}{\partial S_L}\right) + \frac{\rho_v C_{P_v} k_v}{\mu_v}\frac{\partial P_v}{\partial S_L}$$

$$K = \varepsilon S_L \rho_L C_L + \varepsilon (1 - S_L) \rho_v C_{P_v} + (1 - \varepsilon) \rho_s C_s$$

The quantity δ appearing in Eq. (15) represents the ratio of two time scales, one implied by the continuity equation and the other by the energy equation. It is given by:

$$\delta = \overline{D/CK}$$

It is interesting to observe from the form of Eq. (16) that the quantity D defined above plays the role of an effective thermal conductivity; it represents the combined effect of conduction and latent heat transfer by bulk vapor flow. Whenever the mat is fairly moist, the value of D can far exceed that of λ alone.

In actuality, the reference state used in evaluating δ, $\overline{A}$, $\overline{B}$, etc., is arbitrary and has no effect on the values of the dimensional solution, T (y, t) and S_L (y, t), predicted by the variable-property model which has been presented. Of course, the dimensionless groups (e.g., $\overline{GB}/\overline{DC}$) appearing in Eq. (15) and (16) will be affected by the choice of reference state. In any case, the state used in the numerical solution of these equations was:

$$T_{ref} = (T_H + T_c)/2 \; ; \; S_{L_{ref}} = S_{L_o}/2$$

The counterparts of the previously given initial and boundary conditions appropriate to solving Eq. (15) and (16) become:

$$\theta\,(u, 0) = 0 \; ; \; \Omega\,(u, 0) = 0 \; ; \; \theta\,(0, q_T) = 1$$

$$-\frac{\overline{AL}}{\overline{B}(T_H - T_c)}\,A' + B'\,\frac{\partial\theta}{\partial u}(0, q_T) + C'\,\frac{\partial\Omega}{\partial u}(0, q_T) = 0$$

$$\frac{\partial\theta}{\partial u}(1, q_T) = 0$$

$$P_v\,(1, q_T)/P_v^*\,(T_c) = 1$$

Before discussing the numerical solution strategy for the nonlinear drying problem, it should be noted that Eq. (15) and (16) reduce to the same form as the linear ones solved in (5) if the following major approximations and assumptions are made: convection terms in energy equation negligible, gravity terms negligible, all primed quantities (property shape functions) constant with a value of unity. The linearized drying problem is then:

$$\frac{\partial\Omega}{\partial q_T} = \frac{1}{\delta}\left\{\frac{\partial^2\Omega}{\partial u^2} + \frac{\partial^2\theta}{\partial u^2}\right\} \tag{17}$$

$$\frac{\partial\theta}{\partial q_T} = \frac{\partial^2\theta}{\partial u^2} + \left(\frac{1}{1+\zeta}\right)\frac{\partial^2\Omega}{\partial u^2} \tag{18}$$

where:

$$\frac{1}{1+\zeta} = \frac{\overline{GB}}{\overline{DC}} \tag{19}$$

subject to:

$$\theta\,(u, 0) = \Omega\,(u, 0) = 0$$

$$\theta\,(0, q_T) = 1; \; \frac{\partial\theta}{\partial u}(0, q_T) + \frac{\partial\Omega}{\partial u}(0, q_T) = 0$$

$$\frac{\partial\theta}{\partial u}(1, q_T) = 0; \; \Omega\,(1, q_T) = -\,\theta\,(1, q_T)$$

It can be shown that the last boundary condition is equivalent to the constant vapor pressure condition stated earlier if the derivatives, $\partial P_v/\partial T$ and $\partial P_v/\partial S_L$, are constant (as they are assumed to be in this linearized form of the model).

Before either the nonlinear or the linear drying model can be used in predicting the details of drying for any specific situation of interest, the property functions appearing in the model equations must be specified. For the case of paper drying, a really complete and consistent set of such data is currently unavailable for representing any single pulp type. Therefore, the approach taken in developing the results to be presented in the next section was to simply use information from a variety of sources, in the hope of at least approximating the behavior of the drying system being studied. The relationships used for most of the paper-related properties will be indicated here; relationships for the widely available properties of steam and water will not be given.

The relative vapor pressure, P_{rel}, is a function of moisture ratio (X). Data by Prahl (6), represented in Fig. 4, relate relative vapor pressure to moisture ratio for a paper fiber and water mixture. The equation

$$P_{rel} = 1 - (1 + X/0.038)e^{-X/0.038}$$

was used to represent the curve. It should be mentioned that P_{rel} was used as a correction factor in calculating both the vapor pressure and the vapor density from their respective saturation values. The latter use assumes that the vapor behaves as an ideal gas.

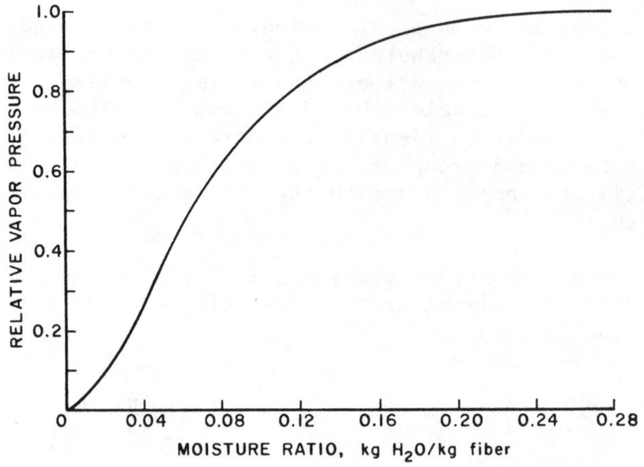

Fig. 4. Reduced Vapor Pressure Relationship Used in Modeling the Hygroscopic Behavior of Paper Fibers, Based on Data from (6)

Typical capillary pressure data for a kraft fiber mat are given by Cowan (7) and are shown in Fig. 5. These data were fit with the following equations:

$$P_c = 44.23\,e^{-5S_L}\left(\frac{\sigma}{\sigma_o}\right), \text{ cm Hg}; \; S_L \geq 0.24$$

$$P_c = 940.56\ e^{-17.74S_L} \left(\frac{\sigma}{\sigma_o}\right), \text{ cm Hg}; \ S_L \leq 0.24$$

The temperature dependence is introduced through the surface tension (σ), and σ_o is evaluated at 20°C.

Robertson (8) gives data for the relative permeability of a fiber mat to liquid flow. The liquid permeability in Fig. 6 is derived from these data and is represented by the following equations:

$$k_{rel_L} = 4.51 \times 10^{-12}\ e^{25.86S_L}, \ 0.82 < S_L \leq 1.0$$

$$k_{rel_L} = 6.93 \times 10^{-6}\ e^{8.54S_L}, \ 0.36 < S_L \leq 0.82$$

$$k_{rel_L} = 1.0 \times 10^{-11}\ e^{46.05S_L}, \ 0.15 < S_L \leq 0.36$$

$$k_{rel_L} = 0.0 \quad , \quad S_L \leq 0.15$$

Data on the relative permeability of fiber mats to vapor flow were not found in the literature. The relationships used here approximately reflect the expected behavior:

$$k_{rel_V} = 1.0 \quad , \quad S_L \leq 0.5$$

$$k_{rel_V} = 2 - 2S_L, \ 1.0 \geq S_L \geq 0.5$$

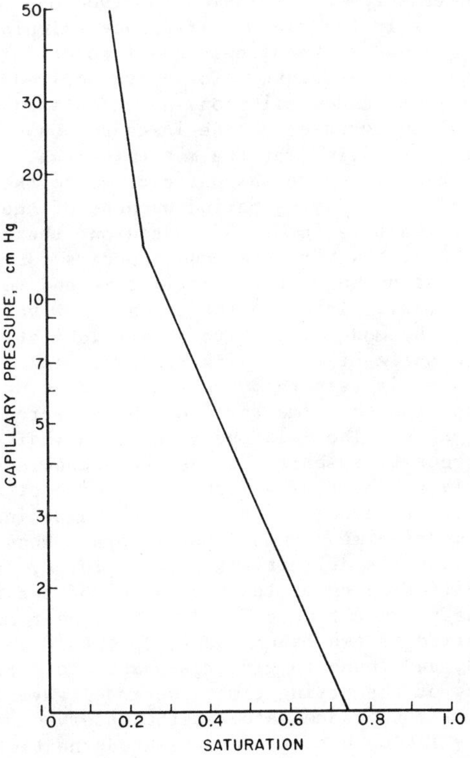

Fig. 5. Capillary Pressure Function used in Nonlinear Model, Based on Data From (7)

These equations are shown graphically in Fig. 6.

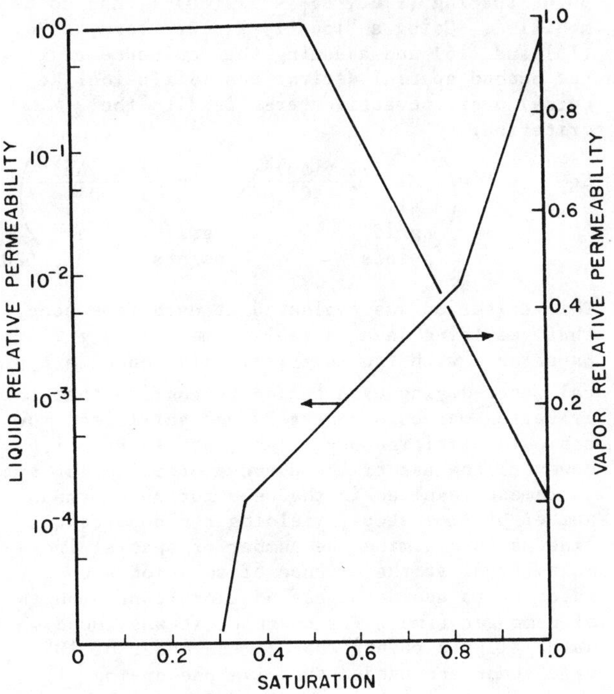

Fig. 6. Liquid and vapor relative permeability functions used in nonlinear model. Liquid relative permeability derived from data in (8)

Data on the increment in latent heat of vaporization (ΔH_H) of water in a paper fiber and water mixture due to hygroscopic effects were given by Prahl (6). The relationship used to describe the total latent heat of vaporization is:

$$\Delta H = \Delta H^* + \Delta H_H$$

where $\Delta H_H = 14.83\ e^{-10.09X}$ (cal/gm) and ΔH^* is the usual (saturation) value of the latent heat.

Finally, a reasonable value for the absolute permeability (9) was chosen as 10^{-12} m^2, and the values of the thermal conductivity and specific heat of the solid fiber material were taken as 0.24 w/m-°C and 2800 J/kg-°C, respectively.

The numerical solution procedure developed for the nonlinear drying problem which has been formulated above was based on an explicit (10,11) finite difference technique. That is, the time derivatives in Eq. (15) and (16) were approximated using forward differences, whereas all spatial derivatives were based on central differences in space, evaluated at the "old" time. In spite of the basic simplicity and ease of implementation of the explicit method, the strong nonlinearities produced by the variable properties resulted in certain difficulties being encountered during the development of an adequate solution procedure. For example, the usual stability criterion which, in the case of a single linear diffusion equation, leads to a global relation between time increment and grid point spacing [i.e., $\Delta q_T \leq 1/2(\Delta u)^2$], had to be modified. Using a "locally linear" form of Eq. (15) and (16) and assuming the dominance of the second spatial derivatives (diffusionlike terms) over convective terms lead to the criterion:

$$\Delta q_T = \leq \text{Min}\left(\underset{\substack{\text{all}\\ \text{grid}\\ \text{points}}}{\text{Min}} \left(\frac{1}{2}\frac{\delta(\Delta u)^2}{C'}\right), \underset{\substack{\text{all}\\ \text{grid}\\ \text{points}}}{\text{Min}} \left(\frac{1}{2}\frac{K'(\Delta u)^2}{D'}\right) \right) \tag{20}$$

This criterion was evaluated at each time step, thus resulting in a variable time step size. Experience with the numerical solution of a typical paper-drying case tended to confirm that this criterion was both necessary and sufficient for achieving satisfactory solutions. In practice, however, the use of the above expression for time increment resulted in the need for an enormous number of time steps, yielding a concomitant impetus to minimize the number of spatial increments used, at the expense of solution accuracy, in order to avoid the use of exorbitant amounts of computer time. For example, it was found when using 11 grid points that on the order of 10^6 time steps are needed to solve one drying problem, resulting in the use of several hours of computer time.

The other major difficulty presented by the extreme variability of some of the property functions is that the solution results are sensitive to the manner in which certain of the primed coefficients are calculated. This problem is compounded by the relatively large differences in dependent variable (Ω, θ) values (on which the properties depend) between neighboring grid points which can occur when a fairly coarse spatial grid is being used. The approach used was based on the recommendation of Patankar (10): coefficients representing "conductances" between two neighboring grid points in diffusionlike terms (B', C', D', and G' in the present case) are best represented by the harmonic mean of the values at the two grid points. For example, between grid points 1 and 2 the effective value of B' would be:

$$(B')_{eff} = \frac{2\ B_1'B_2'}{B_1' + B_2'} \tag{21}$$

This procedure would be exact if the properties at point 1 prevailed halfway to point 2 and those at point 2 prevailed halfway to point 1; it is derived on the basis of adding resistances in series.

The three boundary conditions on Eq. (15) and (16) that involve derivatives were approximated with finite difference equations. The fourth condition, that the vapor pressure at u = 1 remain constant, was used in conjunction with the Newton-Raphson method to determine the saturation at that boundary based on the calculated instantaneous temperature there.

Analytical Results

Computer programs for the solution of the linear and nonlinear thermal/vacuum drying model equations have been prepared and used to predict the characteristics of the process for paper-drying conditions typical of those investigated experimentally. In the case of the linear model, the results are dependent on the values of the dimensionless parameters δ and ζ, defined earlier. In turn, then, the linear model results depend on the choice of reference state (T_{ref}, $S_{L_{ref}}$) used to evaluate the property functions upon which δ and ζ are based. The state employed in the present study is:

$$T_{ref} = (T_H + T_c)/2 \text{ and, } S_{L_{ref}} = \sqrt{S_{L_o} S_{L_\infty}}$$

where S_{L_∞} is the final equilibrium saturation, which occurs when the mat reaches T_H while still at the vapor pressure P_v^* (T_c). The geometric mean was chosen for calculating $S_{L_{ref}}$ in the hope of better reflecting the highly nonlinear dependence of several of the properties on moisture content. It might be noted that for typical operating conditions the value of $S_{L_{ref}}$ is small enough that $k_{rel_L} \longrightarrow 0$, effectively reducing the linear model to the immobile liquid model discussed in (5), but with different boundary conditions.

Example predicted drying curves are shown in Fig. 7. The linear model solution involved use of 21 grid points in the finite difference calculations, whereas the nonlinear model solution was based on 11 grid points. The curve derived from the nonlinear model solution was calculated by numerical integration of the instantaneous saturation profile over the mat thickness. The nonlinear model curve was not carried to the full duration of the drying period because of the long computer run time involved. Based on the results shown, however, the model appears to overpredict the drying rate at early time and to underpredict it later in the process. Nevertheless, the model does give a fair indication of the time scale of the drying process. The linear model curve is seen to do well at predicting both the shape and the time scale of the measured drying curve. The relative success in indicating the correct time scale, it must be acknowledged, is partly a result of a fortunate choice of the reference state definition used in evaluating the linear model dimensionless parameters. When other plausible definitions were employed, drying times differing by at least a factor of two from the present result were found. The linear model was applied to two other cases, T_H = 66°C and T_H = 120°C, and found to give reasonably good predictions of the drying time; the model gave essentially the same value as the observed one for T_H = 120°C, but it overestimated the time by about 70% for T_H = 66°C.

The most detailed (and fundamental) information predicted by the models which have been presented are the distributions of temperature and moisture within the mat during the drying process. These give insight into the nature of the process. The results given by the linear model are shown in Fig. 8 and 9. Because the linear model does not account for the strong spatial and temporal variations of the property coefficients, one would expect the shapes of these curves to be only qualitatively correct. The corresponding nonlinear model results are given in Fig. 10 and 11; these are somewhat different in shape due to the variable property effects.

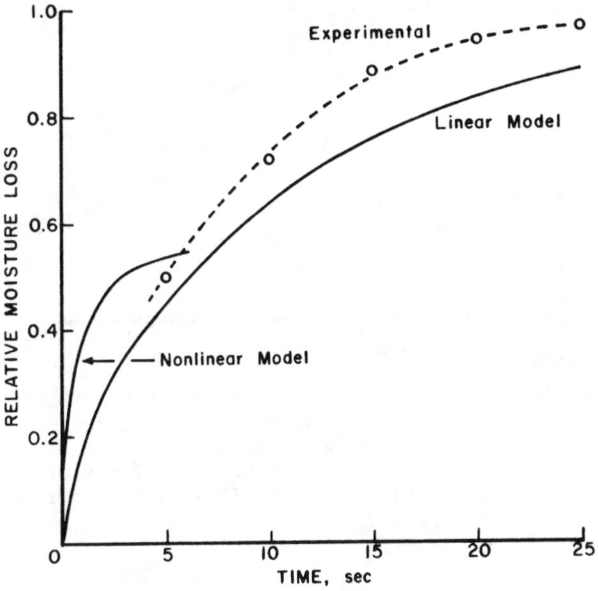

Fig. 7. Predicted and Measured Drying Behavior for T_H = 90°C, T_c = 24°C; Fiber Mat Described in Fig. 3. Linear Solution Based on δ = 5.08, ζ = 0.0057. Relative Moisture Loss = (Actual Loss)/(Moisture Loss Required to Achieve Pressure Equilibrium with Saturated Vapor at T_c)

Certain features of the drying process are exhibited by both models. For example, it is seen (Fig. 9 and 11) that very early in the process the saturation drops rapidly near the heater, due to evaporation there, while the saturation increases to values in excess of the initial level a little further into the mat, due to recondensation of the vapor. Concurrently, the temperature rises rapidly in the region of recondensation (Fig. 8 and 10). Both sets of results also show the dryout which occurs at the cooler boundary of the mat (u = 1) as the temperature rises there. Because of the nonlinearity of the vapor pressure function, this effect occurs more rapidly, according to the (presumably) more accurate nonlinear model, than is suggested by the linear model.

One aspect of the drying process is shown only by the nonlinear solution results (Fig. 10 and 11). That is, the dry zone which develops near the heater produces a large decrease in both the actual and the effective thermal conductivities (λ and D), resulting in a thermal barrier and a large temperature gradient there. This is followed by a region of nearly uniform temperature in the more moist region, except during the very early stages of mat heatup.

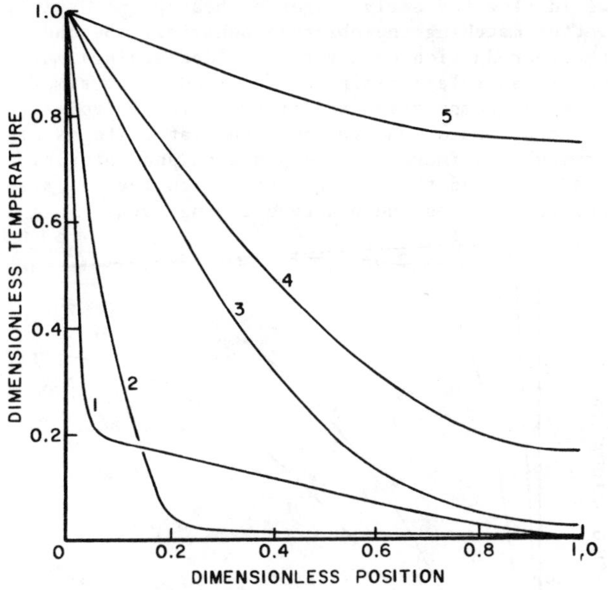

Fig. 8. Temperature Distributions Predicted by Linear Model for ζ = 0.0057, δ = 5.08 (Corresponds to T_H = 90°C, T_c = 24°C, Fiber Mat Described in Fig. 3). Drying Times: 1 = 0.003 s, 2 = 0.145 s, 3 = 2.5 s, 4 = 5.0 s, 5 = 20.0 s

DISCUSSION AND CONCLUSIONS

The thermal/vacuum drying process for permeable (hygroscopic) mats, with special reference to paper drying, has been investigated and analyzed. The experimental data reported certainly help to quantify the improvements to be expected from this process. However, from the viewpoint of aiding in the verification and further development of the mathematical model, it would be helpful to have additional experimental data available. This would include drying curves obtained for mats of other basis weights and possibly other materials, and more detailed data, such as the temperature history at several points within the mat or at its heated and cooled boundaries. Also needed are better data on the thermophysical/transport properties of the mat material used in the experiments. Some work toward meeting these needs is in progress.

In the analytical area, some needs also exist. Perhaps the major one is to adopt or develop a faster numerical solution procedure for the nonlinear model. This would enable solutions to be obtained more practically (i.e., with less computer time), making it possible to use more spatial grid points in order to overcome the difficulties presented by the highly variable property functions, and allowing comparison with experimental results to be easily made for a broad range of operating conditions.

The limited comparison between experimental and nonlinear model-predicted moisture loss vs. time (Fig. 7) also suggests that two extensions to the mathematical model might be needed. First, a contact resistance between the heater and mat would slow the early stages of heatup and drying, better matching the observed behavior. Second, the introduction of a variable (decreasing) mat thickness relationship into the model, to simulate mat compression and fiber shrinkage/collapse effects, would tend to speed the later stages of drying (via increasing temperature and saturation gradients and the average saturation level), also better matching the observed drying curve.

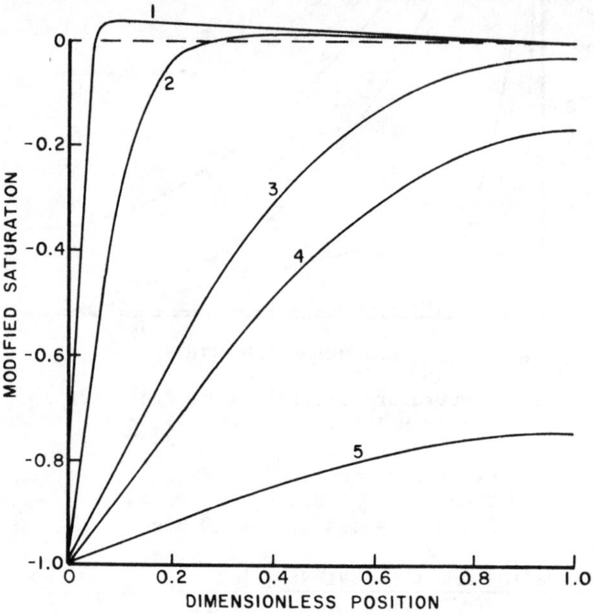

Fig. 9. Moisture Distributions Predicted by Linear Model for $\zeta = 0.0057$, $\delta = 5.08$ (Corresponds to $T_H = 90°C$, $T_c = 24°C$, Fiber Mat Described in Fig. 3). "Modified Saturation" Refers to Ω, Defined in Text. Drying Times: 1 = 0.003 s, 2 = 0.145 s, 3 = 2.5 s, 4 = 5.0 s, 5 = 20.0 s

A final (minor) comment on the nonlinear mathematical model is that, for the conditions to which it was applied here (i.e., rather thin mats), the terms involving gravity were found to be negligible.

On the basis of the information presented in this paper, the following overall conclusions are drawn:

1. The comparison between experimental data on the drying of similar paper fiber mats under thermal/vacuum and conventional atmospheric conditions suggests that the thermal/vacuum process has a significant potential benefit to the paper industry.

2. The mathematical models presented here yielded encouraging results when applied to the thermal/vacuum drying of paper, therefore recommending them as candidates for further application and development.

3. The mathematical analysis and the detailed numerical results thereof provide insight into the nature of the thermal/vacuum process.

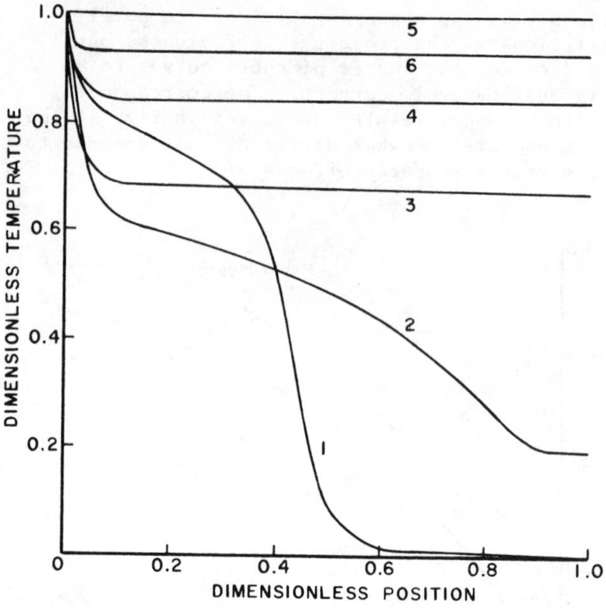

Fig. 10. Temperature Distributions Predicted by Nonlinear Model for $T_H = 90°C$, $T_c = 24°C$; Fiber Mat Described in Fig. 3. Drying Times: 1 = 0.035 s, 2 = 0.05 s, 3 = 0.104 s, 4 = 0.275 s, 5 = 2.25 s, 6 = 6.15 s

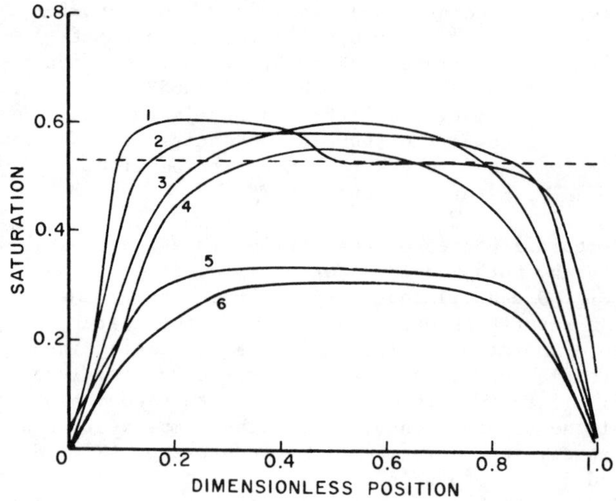

Fig. 11. Moisture Distributions (S_L) Predicted by Nonlinear Model for $T_H = 90°C$, $T_c = 24°C$; Fiber Mat Described in Fig. 3. Drying Times: 1 = 0.035 s, 2 = 0.005 s, 3 = 0.104 s, 4 = 0.275 s, 5 = 2.25 s, 6 = 6.15 s

ACKNOWLEDGMENT

The major role of J. Loughran in the design, construction, and operation of the experimental apparatus is gratefully acknowledged.

REFERENCES

1. Lehtinen, J. A., 1980, "A new vacuum-drying method for paper, board, and other permeable mats", Drying '80: Proceedings of the Second International Symposium, A. S. Mujumdar (Ed.), 2, pp. 347-354, Hemisphere, Washington.

2. Ahrens, F., Kartsounes, G., and Ruff, D., 1982, "A laboratory study of hot-surface drying at high temperature and mechanical loading", Proceedings of the CPPA Technical Section Annual Meeting, B., pp. 93-97, Montreal, Jan. 25-29.

3. Strek, F., and Nastaj, T., 1980, "Mathematical modelling and simulation of vacuum contact drying of porous media in the constant rate period", Drying '80: Proceedings of the Second International Symposium, A. S. Mujumdar (Ed.), 2, pp. 126-134, Hemisphere, Washington.

4. Strek, F., and Nastaj, J., 1980, "Mathematical modelling and simulation of vacuum contact drying of porous media in the falling rate period (boundary condition of the first kind)", Drying '80: Proceedings of the Second International Symposium, A. S. Mujumdar (Ed.), 2, pp. 135-143, Hemisphere, Washington.

5. Ahrens, F. W., 1982, "An analysis of a thermally-induced vacuum drying process for permeable mats", Proceedings VII International Heat Transfer Conference, Munich, Sept. 6-10.

6. Prahl, J. M., 1968, "Thermodynamics of paper fiber and water mixtures", Ph.D. Thesis, Harvard University, Cambridge.

7. Cowan, W. F., 1961, "An investigation of the hot surface drying of glass fiber beds", Doctoral Dissertation, The Institute of Paper Chemistry, Appleton, WI,

8. Robertson, A. A., 1963, "The physical properties of wet webs", Svensk Papperstidning, 66 (12), pp. 477-97.

9. Private Communication from J. Walsh, 1982, The Institute of Paper Chemistry, Appleton, WI, USA.

10. Patankar, S. V., 1980, "Numerical heat transfer and fluid flow", Hemisphere, Washington.

11. Myers, G. E., 1971, "Analytical methods in conduction heat transfer", McGraw-Hill, New York.

DRYING OF PAPER-A SUMMARY OF RECENT DEVELOPMENTS

W.-K. Cui[1] and A.S. Mujumdar[2]

1. Tianjin Paper Company, Tianjin, People's Republic of China

2. Chem. Eng. Dept., McGill University, Montreal, Quebec, Canada.

ABSTRACT

This brief communication is intended to summarize the various paper drying techniques currently employed or being investigated. Comparison between the various methods in terms of the modes of heat/mass transfer, drying rates attainable, advantages and disadvantages as well as the current status of the process, is presented in a concise tabular form. Key literature is cited for further examination by the interested reader.

Baker and Reay(1) have made estimates of energy consumption in the production of various products of the paper, printing and stationary industry in the U.K. Drying consumes about a third of the energy consumed by the industry. Following is an estimate of the energy use data for various products; drying constitutes a dominant part of the overall process.

TABLE 1

Packaging paper	9.95×10^3 MJ/t
Special purpose board	11.90×10^3 MJ/t
Toilet tissue	13.40×10^3 MJ/t
Corrugated sheet	8.56×10^3 MJ/t
Printing and writing paper	18.22×10^3 MJ/t
Newsprint	5.00×10^3 MJ/t

Baker and Reay (1) have also examined the potential energy savings in drying offered by different technological options available today e.g. heat recovery from dryer exhausts (with or without heat pump), better instrumentation and control, optimization of dryer design and operation, improved mechanical removal of water prior to drying etc. In general dryer retrofits for improved thermal efficiency lead to higher capital and sometimes higher operating costs. Depending on the specific prevailing conditions retrofits may or may not result in real cost savings.

In view of the above attempts have been made in recent years to develop and test new concepts for drying of paper. Clearly they are viable only if they yield a better product at a lower overall cost. Relative costs of fossil fuels and electricity constitute an important factor in determining the technoeconomic feasibility of most of the new concepts now being examined. Table 2 compares the various methods currently used or proposed to dry paper. The table is brief but self-explanatory. So, no detailed discussion is warranted. The interested reader is referred to the literature cited for details. Note that some of the techniques proposed are only at the laboratory or pilot stage as of now.

The vacuum drying techniques proposed by Lehtinen (7, 8) and Ahrens (10) appear to be very promising novel ideas for paper drying. Much remains to be done at laboratory and pilot scales before any conclusions can be drawn about their commercial viability. The superheated steam drying concept of Mujumdar (9) appears to be very attractive from the viewpoint of energy. No novel technology needs to be developed although laboratory scale experiments on drying rates and product quality need to be conducted. Problem of proper sealing of the equipment remains but it appears to be a problem that can be solved. The idea of substituting superheated steam for hot air can be extended to Yankee drying of tissue paper or simple through drying.

Finally, this summary does not include radiant or microwave/RF drying of paper as the potential for these essentially electrical drying methods is worth exploring only in areas where electricity is in excess supply and hence relatively cheap e.g. Quebec, Canada. In most parts of the world the potential for these drying techniques is almost nonexistent. RF drying has been applied commercially for correcting the final moisture profile; it is not used for the major part of drying to which this paper is confined.

ACKNOWLEDGEMENT

The authors are grateful to Purnima Mujumdar for the preparation of this typescript.

REFERENCES

1. Baker,C.G.J. and Reay, D., Proc. 3rd Int.

TABLE 2. COMPARATION OF THE VARIOUS TYPES OF METHODS FOR PAPER DRYING.

Drying Type	Mode of Heat & Mass Transfer	Drying Rate $kg/m^2 . h$	Advantages	Limitation	Comment
Multi-Cylinder (Conventional) (2)	Conduction heating. Mass transfer by diffusion	10 20 (average rate for drying Newsprint in U.S. & Canada is 12.7) (9)	It has been used for a long time. It is familiar to the papermaker. The temperature of each cylinder can be adjusted for different grades of paper.	Lower drying rate Because of high heat capacity, the control is insensitive, the moisture of the paper is not uniform. The equipment is very large.	It is hard to improve it further. Upper limit on speed due to wrinkling and/or sheet flutter.
Impingement Drying (2, 3)	Heating & mass transfer by convection	100 (For Vair=100 m/s. Tair=315^{0}C.)	High drying Rate. It is also suitable for drying coatings and pulp sheets. Fast response due to negligible thermal inertia.	May not give the sheet a smooth surface. High power consumption.	Extensively used in drying coatings, photographic film, pulp sheets etc.
Combined Impingement & through drying (4)	Convection heating Mass transfer by convection. Also some radiant heat transfer	145 (Typical of Papridryer)	Higher drying rate. Control is sensitive. A uniform moisture distribution in the sheet can be obtained. The stability of the paper is better. The length of the drying section is much shorten.	Higher power consumption; technical problems still to be resolved for optimization.	Trials in Papermills have been completed. Very promising new technology.
Yankee dryer combined impingement drying. (3)	Conduction and convection heating Mass transfer by convection (also radiation)	200 (For Vair=100 m/s Tair=315^{0}C, Psteam=8kg/cm^2 about half the heat is transferred from the air to the web)	Highest drying rate of all dryers. No sheet broken. Can run at high speeds. Very large units in operation (up to 5 m diameter)	It is only suited for drying tissue and MG paper.	70% tissue paper is dried in Yankee dryers.

TABLE 2. COMPARATION OF THE VARIOUS TYPES OF METHODS FOR PAPER DRYING.

Drying Type	Mode of Heat & Mass Transfer	Drying Rate $kg/m^2 . h$	Advantages	Limitation	Comment
Through Drying (5)	Convection heating & Mass Transfer by Convection	20 140 (8)	Higher drying rate. It is most suitable to be used in drying tissue paper . Gives high bulk product.	Higher power consumption. Good only for high permeability Paper.	It has become the dominant drying method for porous grades. Very promising.
Press Drying (6)	Conduction heating Mass transfer by diffusion	80 (For T=10^{0}C) 120 (For T=180^{0}C)	Increases strength of paper. A smooth surface of paper can be obtained. Poorer furnish can be used which is its main advantages.	It can not be operated at high speed. It requires a fabric that should be strong and porous enough and yet does not seriously mark the web. Web sticking to the cylinder is a problem.	The fabrics should be strong enough and porous enough . Technology still under development.
Convac Drying (7,8)	Conduction heating. Mass transfer by convection	(Using Lab static Convac dryer to test the board) 150 200	Shorten the length of the drying section, Saving in investment. A smooth surface of the sheet can be obtained.	High overall power consumption, higher steam pressure. Web sticking to the cylinder is a problem.	Much developmental work is needed before a definitive evaluation can be made on it.
Superheated Steam drying (Proposed by A.S. Mujumdar McGill Univ.) (9)	Convection heating. Mass transfer by convection	Drying rates of the order of hot air impingement at moderate temperatures. No experimental data available.	High drying rate may be obtained. All of waste Vapor can be reused to obtain high calorific efficiency. Already used commercially for drying of pulp.	It is complex to seal the equipment. Several technological problems remain to be identified and solved.	Laboratory Scale experimental work needed. May be extended to Yankee drying. Several alternatives for steam reuse possible.

Drying Symposium, Univ. Birmingham, U.K., Sept. 1982.

2. Pulp & Paper Manufacture, 2nd Ed., Vol. III, (Ed. R.G. MacDonald), McGraw-Hill, N.Y. 1970.

3. Gavelin, G., Drying of Paper, Lockwood, Lon., 1972.

4. Burgess, B.W. et al, Pulp & Paper Mag. Canada, 73(11), T314-323 (1972) and 73(11), T323-331 (1972).

5. Randall, K.R., DRYING'84, Ed. A.S. Mujumdar, Hemisphere, N.Y. (1984)

6. Settleholm, V.C., TAPPI, Vol. 62, No. 3, 1979, pp. 45-46

7. Lehtinen, J.A., DRYING'80, Vol. 2, Ed. A.S. Mujumdar, Hemisphere, N.Y. (1980).

8. Lehtinen, J.A., Proc. 3rd Int Drying Symposium, Vol. 2, Univ. Birmingham, Sept. 1982, pp. 382-394

9. Mujumdar, A.S. and Loo, E., DRYING'84, Hemisphere N.Y. (1984)

10. Ahrens, F.W., Proc. 7th Int. Heat Transfer Conf., Munich, F.R.G., Vol. 6, pp. 509-514, 1982.

SAFETY ASPECTS OF SPRAY DRYING

P. Filka

Potravinoprojekt, Prague

ABSTRACT

Main environmental risks of spray drying are fire hazard, explosion hazard and air pollution. The influence of these risks on the design and project engineering of spray drying plants is discussed. The spray dried substances are mostly inflammable and therefore also explosive in a dust cloud. Physical and chemical conditions of the environmental risks were studied and damage limitation methods were recommended. Automatic control system with an automatic fire detecting and fighting equipment should be a prominent part of a spray drying plant.

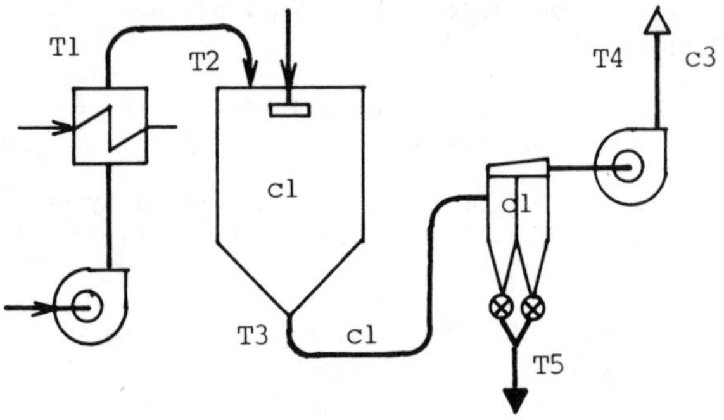

TYPE A DRYER Figure 1

INTRODUCTION

Every technological process carried out in large scale represents certain environmental risk. Not even spray drying is an exception.

This paper concerns main environmental risks of spray drying which are

1. Fire hazard
2. Explosion hazard
3. Air pollution.

We would like to evaluate the influence of these risks on the design and project engineering of spray drying plants. I will try to summarize the methods of solving these problems.

For the purpose of this paper we shall consider 3 basic types of spray dryers.

A - Indirectly heated spray dryer with pneumatic product discharge /Fig.1/

B - Indirectly heated spray dryer with mechanical product discharge /Fig.2/

C - Directly heated spray dryer with mechanical product discharge and multistage powder recovery /Fig.3/.

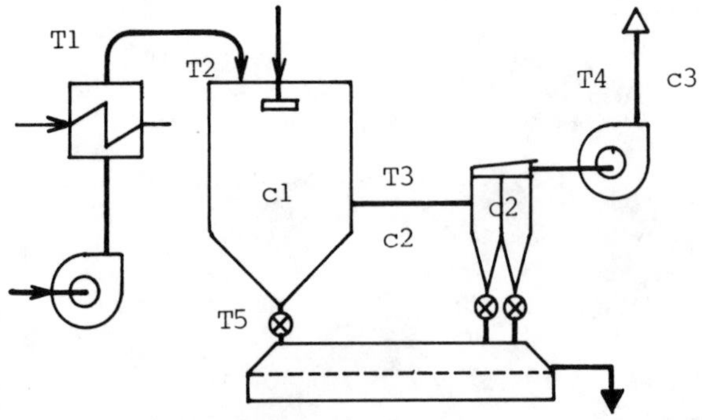

TYPE B DRYER Figure 2

1. Fire hazard

To analyze the fire hazard in a spray dryer following conditions must be studied :

- conditions of the combustion reaction,
- causes of ignition,
- possibilities of risk limitation.

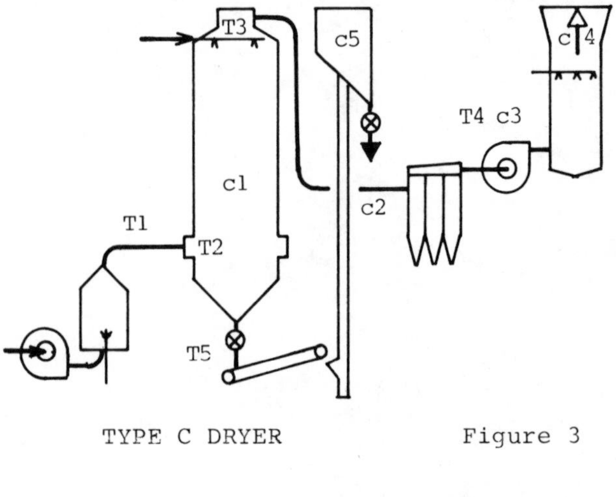

TYPE C DRYER Figure 3

1.1 Conditions of combustion are following.

a. Inflammability of a product

All food products as well as majority of chemicals, which could be spray-dried, are organic substances and therefore inflammable in dried form.
Inflammation temperatures of these powders measured in a layer or in a cloud differ substantially. The results of individual laboratories and separate measurements vary considerably as well. The example can be seen in Fig.4.

POWDER	INFLAMMATION TEMP. °C Layer	Cloud	EXPLOSION CONC.	EXPLOSION PRESSURE
Wheat starch	-	410-460	7-22	high
Pudding powder	-	-	20	high
Sugar powder	-	360-410	17-77	high
Cream topping	-	-	6.3	high
Monoglyceride	290	370	16	high
Monoglyceride + skim milk	282	435	32	high
Baby food	205	450	36	high
Milk concent.	190-203	440-450	22-32	high
Skim milk	134	460	52	high
Milk	142	420	54	high
Buttermilk	194	480	32	high
Whey	145	490	56	high
Skim milk + whey + fat	183-240	460-465	20-24	high
Coffee extract	160-170	450-460	50	
Cocoa	170	460-540	103	high
PVC	-	595	40	
Detergents	160-310	360-560	170-700	low

Fig.4.

Skim milk has the lowest inflammation temperature in a layer, while sugar and some detergents have the lowest inflammation temperature in a cloud.

b. Presence of oxygen

Oxygen content of drying air is 23.15 p.c. The O_2 content of the mixture of combustion gases and air in a directly heated spray dryer is about 18.2 p.c. Evaporated water decreases the oxygen content up to 22 p.c. or 16 p.c. respectively. These contents will be fully sufficient for the ignition and combustion.

c. Temperatures of product and drying gas were compared with the inflammation temperatures of products in layer and cloud, which can be seen in Fig.5.
This diagram which demonstrates the fire hazard is drawn for 3 typical dried product groups : milk products, coffee extracts, detergents.
We can see that the drying gas temperature nearly always exceeds the inflammation temperature in the layer. It proves that ignition is possible if deposits are formed or if drying conditions deviate from the regular ones.

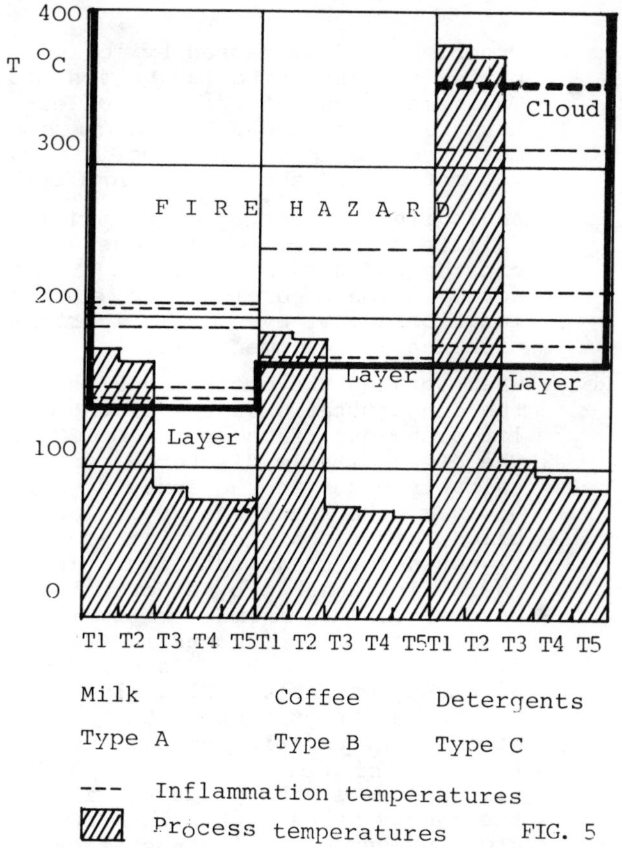

FIG. 5

In detergents, ignition is possible in the cloud as well.

1.2 Causes of ignition were described in detail by Masters /2/ and are as follows :

a. Spontaneous combustion in product deposits.
b. Hot solid particles entering the dryer with the drying gas.
c. Spark generation through friction.
d. Electrical failure.
e. Static electricity discharge.

The reasons of these phenomena are well known. We would like to add only a few comments.

Re a. Spontaneous combustion can occur if the oxidation heat is higher than the heat transferred to the environment. The layer therefore must be rather thick. However, such layer can be easily detected and has to be removed. This condition has to be paid attention to in designing the dryer, piping and cyclones. The lay-out of the plant should enable an easy transport of the removed powder to further processing.

Re b. The existence of red hot solid particles can be substantially limited by proper project and design engineering. Hot brick particles in directly heated dryers are particulary dangerous.

Re c. Trouble could be caused by moving parts, as fans, rotating valves etc. The temperature of the body of exhaust air fan should be continuously measured and any difference in electric motor load should be signalized.

Re d. An indirect effect of an electrical failure can be fire hazardous as well, causing undesirable friction or powder accumulation. Therefore the electric interlocking of even less important machines is recommendable.

Re e. When sprying the product in nozzles, the electrostatic charge of the droplet cloud can reach up to 20 000 V. The electrostatic discharge must inevitably occur in the spray dryer. However, the spark energy is low. Dried powder containing fat as monoglyceride or cream substitute is easily inflammable by electrostatic discharge. The minimal inflammation energy of these products is from 0.080 to 0.16 J, which is considered as highly dangerous. Slightly dangerous are products with the minimal inflammation energy from 1 to 10 J, which is usual with ordinary milk products. The products with the energy value higher than 10 J are considered as electrostatically safe.

1.3 Fire risk minimizing

Hazardous conditions exist during spray drying operations. How to minimize the existing fire hazard then ? There are 3 steps which must be taken in sequence :

a. Prevention of hazardous conditions
b. Detection of dangerous situation
c. Counteraction with damage limitation.

Basic fire prevention methods were described before /2/ and are based on proper operation and cleaning of the dryer. In addition to this attention should be drawn to the importance of an automatic control system of spray drying which could be an important safety factor. While the operator is able to control manually the drying temperatures within ± 10 °C, the ACS reaches easily ± 1 °C.
Present energy and economic pressures lead to drying at limit conditions, and therefore to maintain drying parameters precisely is extremely important. The ACS of dryer should include an automatic starting-up system as well. To start-up a dryer improperly is an easy way how to build dangerous product deposits.
Detection system should complement the ACS. The proper detection system should contain
- alarm of parameter deviation,
- deposit detection,
- drying gas analysis,
- continuous measurement of humidity, temperature and flow properties of final powdered product.

Some of these physical characteristics can be measured quite easily, but reliable continuous measurement of flow properties and humidity of powdered materials is not yet available. The necessity of the development of realiable measuring devices is evident.
Damage limitation in case of fire can be reached by an automatic system of counteraction. Such system provides
- operational activities as control of valves, air and product shut down,
- fire extinguishing action, which could be done by atomizing water in product nozzles or special nozzles and by steam injection,
- signalization of the counteraction effect.

The automatic counteraction system should be programmable including time delays determined for operator´s reaction and his possible intervention /3/.

2. Explosion hazard

The spray dried substances are mostly inflammable and therefore also explosive in a dust cloud.
Dust explosion hazard of typical products in succession is as follows:

1. sugar	6. skim milk
2. starch	7. PVC
3. cream substitute	8. whey
4. coffee extract	9. eggs
5. milk	10. detergents

The hazard depends on the value of dangerous concentration, on the oxidation velocity and on the explosion pressure. To bring a spray dryer to an explosion, it is necessary to meet certain physical and chemical conditions and to provide the initiation by ignition.

2.1 Physical and chemical conditions of a dust explosion are as follows :

a. The product forms a dust cloud.
b. The temperature, pressure and humidity of the cloud make the explosion possible.
c. The solid particles concentration is within dangerous limits.
d. The oxygen content in drying gas is sufficient.
e. The chemical reaction has a sufficient volume.

Re a. Dust cloud

The existence of a dust cloud is one of basic principles of spray drying. Dust clouds are formed by particles of 0-200 μm. Dangerous explosion pressures are produced by particles of about 0-100 μm. These are exactly powders which are dried in spray dryers, as can be seen in Fig.4. Very high explosion pressures were detected in all powdered materials containing either milk or fat, starch, sugar, skim milk, buttermilk and whey.

Re b. Temperature, pressure and humidity

The influence of temperature, pressure and humidity of the dust cloud can be important. Humidity of particles reduces the explosion hazard but cannot eliminate it. For example the increase of starch humidity from 1 p.c. to 8 p.c. causes the increase of the lower explosive concentration from 7 g/m^3 to 22 g/m^3. Therefore the parts of the drying equipment containing low humidity powder as cyclones are the most dangerous. Continuous product humidity measurement will be extremely important.

Re c. Solid particles concentration in a dust cloud

Examples of lower limits of explosive concentration are demonstrated on the right side of Fig.4.
On the contrary, the values of powder concentration in the course of the drying process are shown in Fig. 6:

Concentration g/m^3

Type	Product	C_1	C_2	C_3	C_4	C_5
A	milk	25	-	0.515	-	-
B	coffee	29.5	2.36	0.177	-	-
C	deterg.	66.6	4.0	0.255	0.005	250

The comparison of both explosive and process concentrations, evaluated for 3 typical cases, illustrates Fig.7. This figure demonstrates that under normal conditions dangerous concentration in the course of powder drying, transport and separation can exist.

The hatched field demonstrates powder concentrations in the course of the drying process /see Fig.1,2 and 3/ and is compared with minimal explosive concentrations, illustrated by the thick line. The thick line crossing the hatched field signalizes the explosion hazard in the corresponding part of the drying plant.
As mentioned before, the explosive concentrations depend on measurement conditions and can differ substantially. Therefore the powder concentration > 50 p.c. of lower explosive concentration is usually considered as dangerous and in such case, all safety measures should be taken /dotted line/.
Explosion hazard should be considered in the following parts of the drying plant :
starch products - spray dryer, piping, cyclones, exhaust fan,
milk products - spray dryer, piping, cyclones,

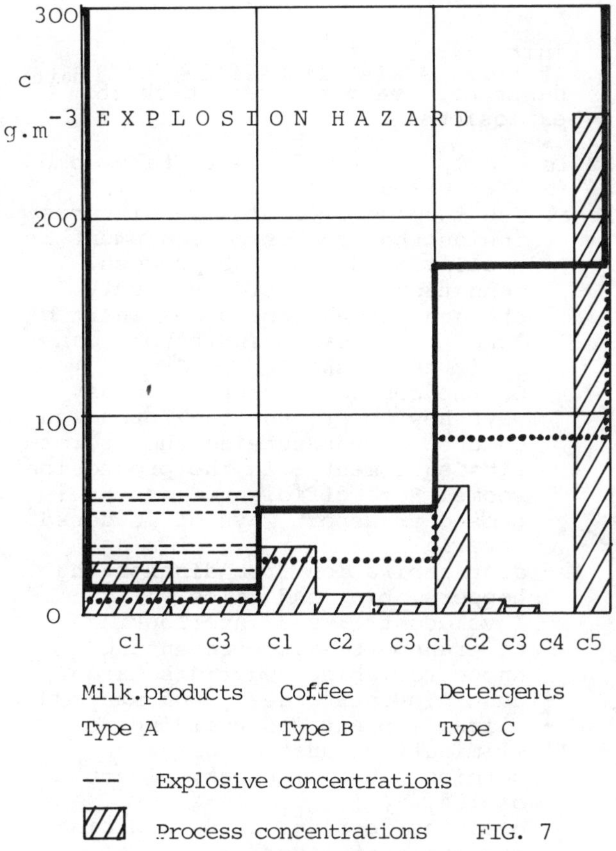

FIG. 7

coffee extracts - spray dryer, sometimes cyclones,
detergents - usually only pneumatic transport of powdered product.

2.2 Initiation by dust ignition is the necessary condition of an explosion. The dust cloud cannot explode by itself as for example the pyroforic substances. First it is necessary to heat a part of the cloud to inflammation temperature. Causes of ignition were described in part 1.2 concerning fire hazard.

The chain effect is typical for the initiation of dust explosion. The first ignition occurs usually in a small dust cloud with limited energy. The first explosion causes no damage but forms a much bigger secondary dust cloud with higher destructive capacity which shatters the plant and the building. If powder deposits or bins enable the formation of a tertiary dust cloud, the following explosion and fire could be fatal.

2.3 Risk and damage limitation methods

We have shown that spray drying includes explosion hazard. To limit the risk it is necessary to take the following steps :
a. To prevent the forming of explosive mixture.
b. If this is not possible, to prevent

initiation.

c. If this is also impossible, to limit destruction and to prevent chain explosions.

It is highly recommendable to take some preventive measures.

Re a. The forming of explosive mixtures inside the processing equipment is usually a part of the process conditions and could be hardly changed. In extreme cases inert gas and/or explosion inhibitors can limit the risk.
On the contrary, there are easy ways how to prevent the formation of dust clouds outside the processing equipment - in the production rooms. Some efficient measures :
- closed transport ways of powdered materials,
- dust aspiration from discharging hoppers, bins and silos,
- level control and interlocking of transport ways, preventing uncontrollable powder discharge,
- dust preventing design of equipment ,
- suitable plant and building design eliminating dust deposits on cornices, beams, pipes, electric cables,
- cleaning regulations determining maximum dust layer thickness /i.e. 1 mm/ and cleaning practice.

Re b. Prevention of the initiation is identical with fire prevention methods described in the par. 1.3.

Re c. To limit possible damage, the detection of dangerous conditions and primary explosion is essential. Temperature and pressure changes can be easily measured. Operators could be protected and preventive devices could be triggered.
Every dust processing equipment, where operating concentration is higher than 50 p.c. of the lower explosive concentration, has to be provided with safety device, certified by an authorized institution. Spray drying towers, pipes and cyclones should be usually equipped with safety doors, clappets or membranes. The necessary surface area and operating pressure is to be calculated using experimentally measured explosion pressure and velocity of pressure increase /4/.

It is evident that the laboratory measurement of dust explosion and fire properties is a necessary condition for the design and project engineering of a spray drying plant, taking into account all possible powdered products. In operating such a plant it is recommendable to measure product explosion properties periodically. Similarly to the equipment, it is necessary to protect all buildings, where dust explosion hazard exists or where an equipment with such a hazard is in operation. Therefore it is rather important how the environment in the building should be classified. Basically there are two separate questions to be considered :

1. The environment in the production rooms and buildings is dangerous if the dust concentration under usual operating conditions is higher than 50 p.c. of lower explosive concentration.
The decision must be taken by project engineers of the processing plant, building and electric installations. If the answer is positive, all necessary safety measures should be applied.
2. The interior of the processing equipment could be dangerous under similar conditions. The decision must be declared by the manufacturer of the equipment. If positive, certified safety devices should be applied.
Both safety characteristics of the rooms and of the equipment are decisive for the project engineering of the plant lay-out, emergency ways, explosion walls, air conditioning systems, electric and control installations.

3. Air pollution

Powder emission of a spray dryer is formed by powders which are frequently toxic, smelly and biologicaly active. Some organic powders react with the nitrogen oxides in the atmosphere and help to form smog.
From the environmental point of view the real powder emission is decisive. Undecisive are the values calculated from the theoretical values of separation efficiency. It has been proved that the true powder emission of spray dryer can be sometimes up to five times higher than the calculated emission values show. Luckily enough, usually nobody learns about it.
The reason of this difference can be :

- overloading
- higher content of fines
- insufficient cleaning
- oscillations of drying parameters.

Basic prerequisite for improving the situation mentioned is to have the possibility of continuous emission measurement. This is the way how to reveal the undesirable conditions and then corrective steps can be taken.
Further requirement on the dryer design is to equip the spray dryer with such separators which are effective even in unusual operating conditions.
Dust concentrations in the exhaust air of 4 different spray dryers are demonstrated in Fig. 8. Evidently, one stage separation is not sufficient any more.

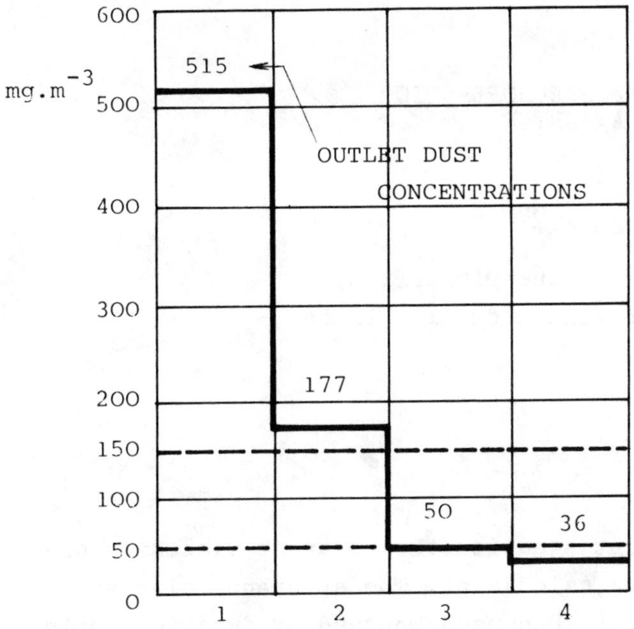

1,2 Cyclone separators

3 Scrubber

4 Bag filter

Environment saving limits

FIG. 8

Automatic control system of spray drying operation excludes excessive oscillations and therefore is favorable for the environment as well.

4. Summary

Leading ideas of safe and efficient spray drying :

- Automatic control of drying process is increasingly important.
- Automatic fire detecting and fighting equipment should be a prominent part of a drying control system.
- Continuous measurement of wall temperatures, product humidity and powder emission will be a healthy investment.
- Knowledge of dust properties including fire and explosion characteristics is essential for the design and project engineering of a spray drying plant.

Bibliography

/1/ Podlipný, V., Explosive media, /Czech/, SNTL Praha 1961

/2/ Masters, K., Spray Drying, Leonard Hill 1972

/3/ Filková, I., Automatic Control of Spray Dryer, Research report ČVUT Praha 218-6/1978

/4/ VDI Standard Nr. 3673

/5/ Filka, P., Dust explosions, causes and prevention, Informace PP, 5, 1975

/6/ Filka, P., Air Pollution of Spray Dryers, Informace PP, 1, 1978

DROP SIZE PREDICTION OF PSEUDOPLASTIC FLUIDS IN A SPRAY DRYER

I. Filková[1]
J. Weberschinke[2]

1. Faculty of Mechanical Engineering, Praha
2. Research Institute of Milk Industry, Praha

ABSTRACT

It is very important in the spray drying process to predict the quality of a spray, that means the characteristics size of droplets. Having known all parameters affecting the arising droplets, the optimum properties of dry product could be reached. However, the existing equations for drop size prediction do not take into account a non-Newtonian behaviour of sprayed liquids which are usually having high viscosity and mostly demonstrate pseudoplastic character. Thus our research work has been concerned with the development of suitable equations for the power-law fluids and the experimental verification.

Eight equations were selected for drop size prediction of Newtonian liquids, which were published in last years. All of them are valid in definite operating conditions. They contain dynamical viscosity as a fluid parameter, so that their application is restricted to Newtonian liquids only.

In order to generalize these equations also to non-Newtonian fluids it was necessary to solve a theoretical problem of a flow of power-law fluid through the vane of rotary atomizer and to derive the equation for radial velocity of fluid on the edge of an atomizer. Substituting this velocity into velocity gradient equation,the apparent viscosity can be calculated. The apparent viscosity was then used instead of dynamic one in the above mentioned equations.

After having some general equations for drop size prediction, their experimental verification has been done. An impact method of samples collecting was applied. All experiments have been carried out in the ANHYDRO Compact Spray Dryer. In accordance with the experimental results the most suitable equations are recommended in this paper.

NOMENCLATURE

$D_{3,2}$ = Sauter volume-surface diameter, μm
D_i = drop diameter, μm
K = consistency coefficient, $kg\ m^{-1}s^{n-2}$
M = mass feed rate, $kg\ s^{-1}$
N = revolutions, RPM
V_k = volume feed rate per vane, m^3s^{-1}
b = vane width, m
h = vane height, m
n = flow index, -
p_i = number of drops
r = wheel radius, m
u = mean velocity, ms^{-1}

$\dot{\gamma}$ = velocity gradient, s^{-1}
μ = dynamic viscosity, Pas
μ_a = apparent viscosity, Pas
ρ = density, kgm^{-3}
σ = surface tension, Nm^{-1}
ω = angular velocity, rad s^{-1}
τ = shear stress, Pa

INTRODUCTION

Desintegration of liquids into a spray is generally known and its use widely spread, mainly in food and chemical industries. It is usually applied in spray drying processes. Spray drying can be divided into three basic operations:

1. Liquid disintegration (atomization)
2. Drying (mass and heat transfer)
3. Dry powder collecting.

Drying process and dry powder properties are strongly affected by the first operation - an atomization. The quality of dry powder can be changed through changing the drop size and drop distribution. There is a direct dependence between drop size and particle size (1). Therefore our research work has concentrated on droplet size prediction and on the factors by which it is affected. The attention has been paid to wheel atomizetion and to non-Newtonian liquids. Whell atomizer was chosen because of its wide use in food and chemical industries. The following products are sprayed by means of wheel atomizer: milk and milk products, starch products, toppings,eggs, fruit juices etc. It was important to deal with non-Newtonian liquids too, as the sprayed materials, usually pre-evaporated, are highly consistent and the majority of them have pseudoplastic character. In fact, there have been no relations for droplet size prediction for such liquids. Rheological behaviour of mentioned liquids can be described by power law (see theoretical part). After a thorough consideration 8 various equations were chosen, all valid for drop size prediction of Newtonian liquids. Dynamic viscosity as a basic fluid parameter can be found in most of them. The question is, what parameter could substitute dynamic viscosity in case of non-Newtonian fluid. We tried to apply the so-called apparent viscosity. Our results is discussed further.

THEORETICAL PART

The surface-volume Sauter mean diameter for spray drying.

$$D_{3,2} = \frac{\sum_{i=1}^{j} D_i^3 p_i}{\sum_{i=1}^{j} D_i^2 p_i} \tag{1}$$

As mentioned above, 8 equations were selected, which enable to calculate Sauter mean diameter of any Newtonian liquid as a function of operating parameters, such as wheel geometry, revolutions, liquids properties, see table 1. Seven of these relations contain dynamic viscosity which will be substituted by apparent viscosity in case of pseudoplastic material.

Comparing Newton´s law

$$\tau = \mu \dot{\gamma} \tag{2}$$

and power law

$$\tau = K \dot{\gamma}^n = K \dot{\gamma}^{n-1} \cdot \dot{\gamma} \tag{3}$$

or

$$\tau = \mu_a \dot{\gamma} \tag{3A}$$

we obtain apparent viscosity as

$$\mu_a = K \dot{\gamma}^{n-1} \quad (4)$$

Equation (4) demonstrates that apparent viscosity depends on velocity gradient $\dot{\gamma}$, i.e. on wheel revolutions. Revolutions in an industrial atomizer are high enough so that the liquid disintegration occurs directly on the wheel edge (1). Thus, it was necessary to derive the mean outlet velocity of pseudoplastic fluid on the edge. By means of this velocity the egua- tion for apparent viscosity could be

Table 1

Equation
Friedman (8) $D_{3,2} = 0{,}4 \cdot r \left(\frac{M}{8b\rho N r^2}\right)^{0,6} \cdot \left(\frac{8\mu b}{M}\right)^{0,2} \cdot \left(\frac{6 b^3 \delta^3 \rho}{M^2}\right)^{0,1}$
Masters (9) $D_{3,2} = 0{,}37 r \left(\frac{M}{8b\rho N r^2}\right)^{0,6} \cdot \left(\frac{8\mu b}{M}\right)^{0,2} \cdot \left(\frac{6 b^3 \delta^3 \rho}{M^2}\right)^{0,1}$
Fraser (10) $D_{3,2} = 0{,}481 \left(\frac{1}{N}\right)^{0,6} \cdot \left(\frac{1}{\rho}\right)^{0,5} \cdot \left(\frac{\mu M}{2r}\right)^{0,2} \cdot \left(\frac{6}{8b}\right)^{0,1}$
Kremnev (11) $D_{3,2} = 0{,}241 \left(\frac{1}{N}\right)^{0,6} \cdot \left(\frac{1}{\rho}\right)^{0,3} \cdot \left(\frac{\mu M}{2r\rho}\right)^{0,2} \cdot \left(\frac{6}{8b}\right)^{0,1}$
Herring (12) $D_{3,2} = 2{,}74 \cdot 10^{-3} M^{0,24} (2Nr)^{-0,83} (8b)^{-0,12}$
Scott (13) $D_{3,2} = 6{,}3 \cdot 10^{-4} \left(\frac{M}{8b}\right)^{0,171} \cdot (2\pi r N)^{-0,537} \cdot \mu^{-0,017}$
Lastovcev (14) $D_{3,2} = 81 \cdot \frac{\delta^{0,46} h^{0,46} \mu^{0,08}}{r \ \rho^{0,54}}$
Hayashi (15) $D_{3,2} = 1{,}62 \cdot 10^{-3} \cdot N^{-0,53} M^{0,21} (2r)^{-0,39}$

determined. The outlet velocity equation was derived on the basis of the Navier- -Stokes equation and the Marshall's solu- tion of Newtonian fluid flow in the rota- ting blade (2, 3, 4). Resultant relation is as follows:

$$\bar{u} = \left[\left(\frac{r\omega^2\rho}{K}\right) \cdot \left(\frac{V_k}{b}\right)^{n+1} \cdot \left(\frac{n}{2n+1}\right)^n\right]^{\frac{1}{2n+1}} \quad (5)$$

Apparent viscosity in the case of flow of power law fluid in a wheel vane is following:

$$\mu_a = K\left[\frac{\bar{u}}{h}\left(\frac{2n+1}{n}\right)\right]^{n-1} \quad (6)$$

Substituting eq. (5) into eq. (6) we obtain the relation of apparent viscosi- ty and operating conditions:

$$\mu_a = K\left[\left(\frac{r\omega^2\rho}{K}\right)^2 \left(\frac{V_k}{b}\right)\left(\frac{2n+1}{n}\right)\right]^{\frac{n-1}{2n+1}} \quad (7)$$

From eq. (7) the apparent viscosity can be calculated by means of known wheel parameters and physical properties of disintegrated liquid. Apparent viscosi- ty of several pseudoplastic liquids was calculated in this way, substituted into equations in the table 1 and expe- rimentally tested.

EXPERIMENTAL PART

All experiments were carried out in ANHYDRO Compact Spray Dryer. Wheel atomi- zers were the standart type, D = 0,1 m with 8 straight vanes. Three different shapes of vanes were applied: circular, squar and rectangular. The research has been carried out at rotations from 17000 to 25000 RPM (3, 5). Three different aqueous solutions CMC were tested. The impact method has been used for spray analysis of all experiments. This method is suitable especially for sampling fine sprays and can also be applied in indu- strial dryers. Droplets were collected

on magnesium oxide coated slides where they left a well-defined craters. Slides were magnified 120 times by microscope and then photographed. Also correction of craters to actual drop size has been carried out. The slides were placed in the impactor and fixed in the horizontal position 0,1 m under the wheel. Five slides covered the distance 0,62 m between the chamber wall and the axis. Two or three photos were taken from each slide, see Figure 1.

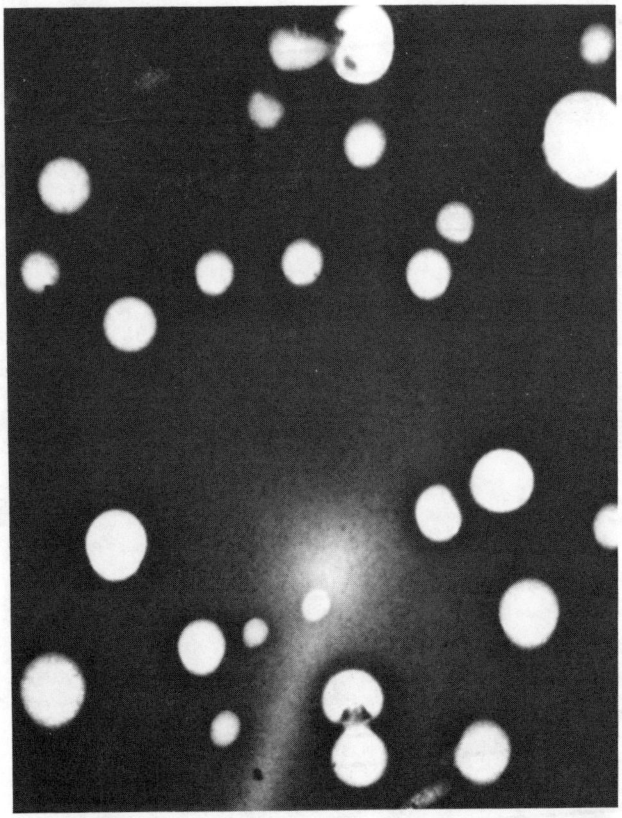

Figure 1 Sample of droplets

More then 500 droplets were evaluated for each revolution and liquid, so that about 3000 droplets had to be evaluated for three CMC concentrations. Each liquid was tested for 17, 19, 21, 23 and 25 000 RPM, for constant feed rate 0,51/min and co-current air flow. All details can be found in (3). The effect of vane shape on droplet size and size distribution was also studied (5). The results dealing with predicting of droplet diameter will be mentioned here. All results were drawn up into diagrams and compared with the results calculated from equations in table 1.
The experimentally obtained Sauter mean diameters for the highest CMC 3 concentration are shown in Figures 2, 3, 4 and compared with those predicted ones.

DISCUSSION

From all three figures it is evident, that eq. (15) and (11) are in the best agreement with experimental points. Taking into consideration the results of CMC 1 and CMC 2 as well, then the Kremnev equation (11) is the most suitable for our experimental conditions. The eq. (15), although not in full agreement, has another great advantage as it doesn´t contain any physical parameter of liquid. Therefore this equation can be applied to rough estimation of pseudoplastic droplets diameter. When the process of atomizetion calls for more accurate prediction of drop size, it is necessary to determine K and n, calculate μ_a and apply eq. (11). This equation can be recommended without any restrictions for pseudoplastic materials atomization under the conditions that are similar to those mentioned above, i.e. wheel diameter 0,1 m, 17 - 25000 RPM, feed rate 0,5 l/min and air velocity under 10 m/s.
We suppose that eq. (11) would be suitable for different conditions as well, but when exact results are required it is better to test this equation by the method described here.

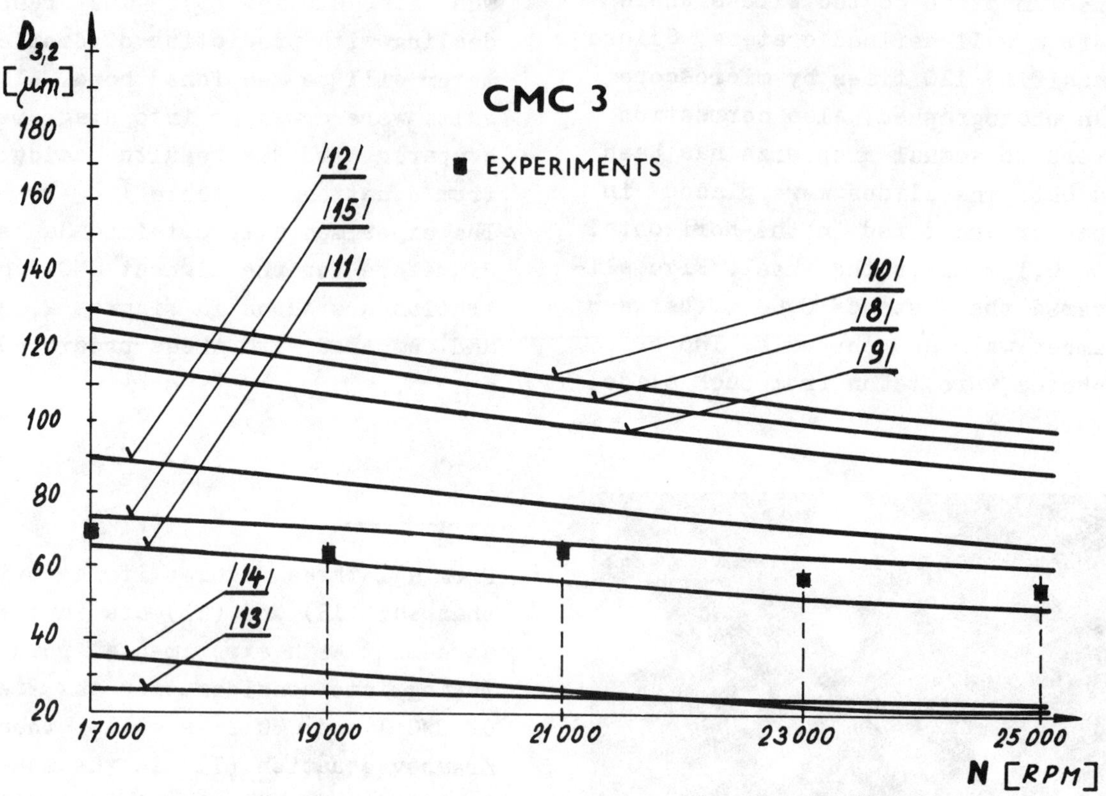

Figure 2 Predicted and experimentally obtained Sauter mean diameter; wheel with 8 rectangular vanes.

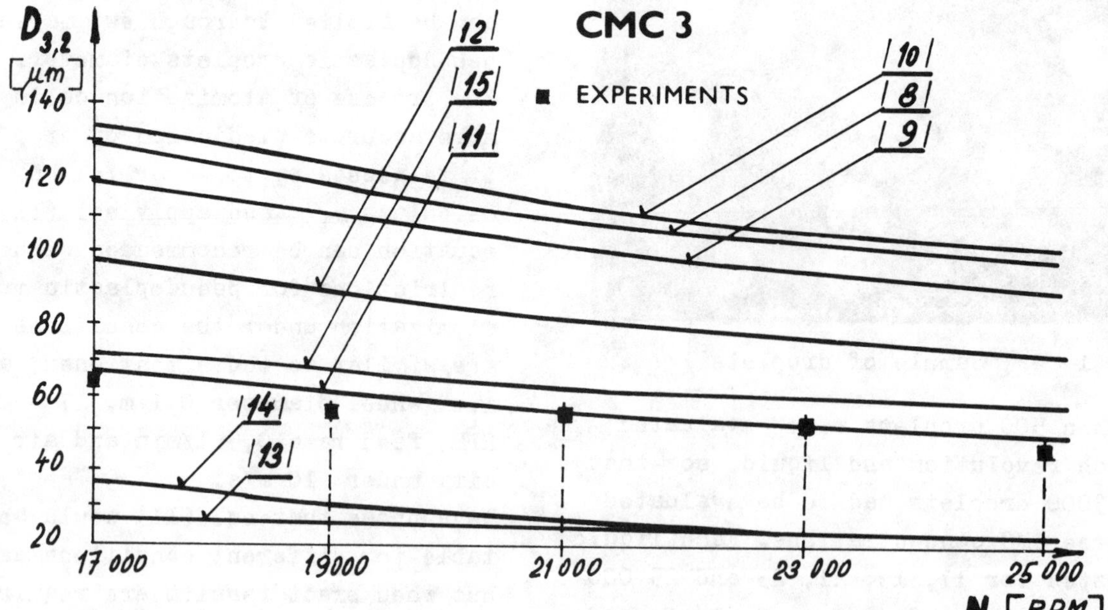

Figure 3 Predicted and experimentally obtained Sauter mean diameter; wheel with 8 square vanes.

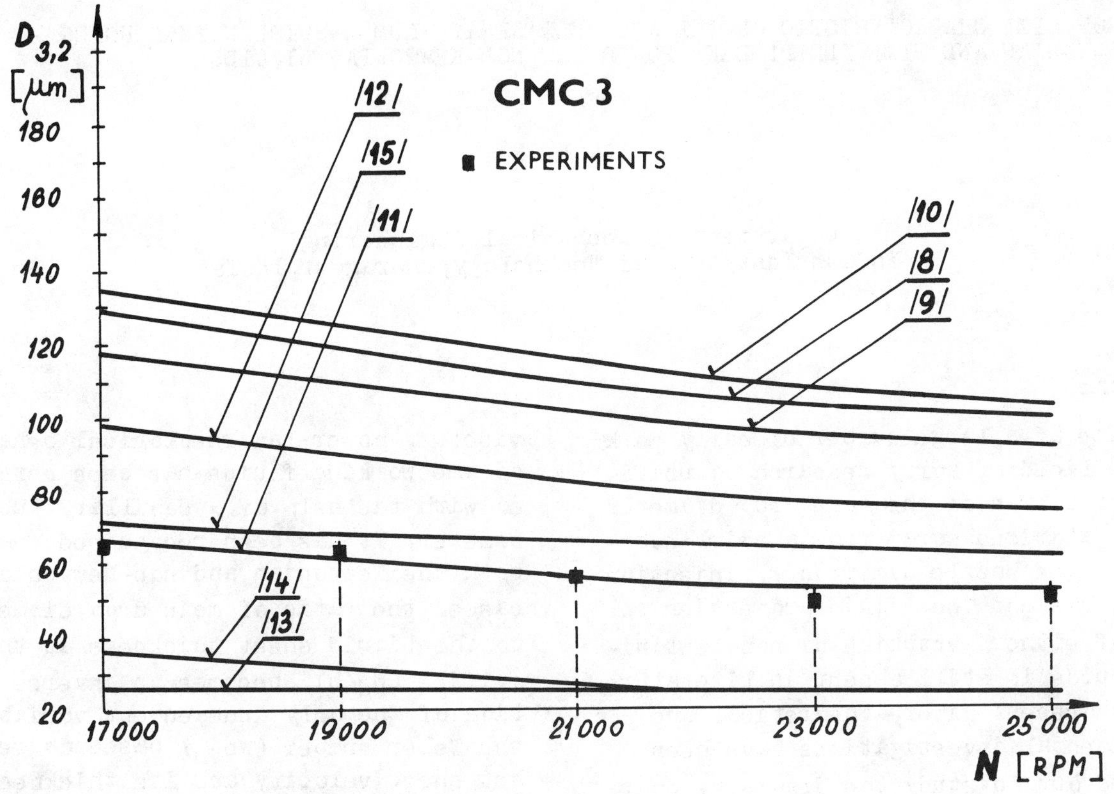

Figure 4 Predicted and experimentally obtained Sauter mean diameter; wheel with 8 circular vanes

CONCLUSION

Our research was aimed at drop size prediction in case of wheel atomization of pseudoplastic power law liquids in a spray dryer. Theoretical derivation of outlet velocity on the wheel edge was carried out and by the use of this velocity so-called apparent viscosity of power law liquid can be estimated. 8 equations for prediction of Sauter mean drop diameter were selected, which are valid for Newtonian liquids. Apparent viscosity was substituted instead of dynamical one in these equations so that their applicability increased for pseudoplastic materials as well. Theoretical results were compared with those experimentally obtained. The Kremnev equation proved to be the most suitable for the operating conditions described above.

REFERENCES

1. Masters, K., Spray Drying, John Willey Sons, N.Y. 1976
2. Weberschinke, J., Filková, I., Drying´81, McGraw - Hill, (Submitted)
3. Weberschinke, J., Doctorate Dissertation, ČVUT Praha, 1981
4. Trojan, Z., Doctorate Dissertation, ČVUT Praha, 1973
5. Filková, I., Weberschinke, J., Developments in Drying, 2nd International Symposium on Drying, Montreal, 1980

DROP SIZE CHARACTERISTIC OF THE ATOMIZED SPRAY FROM A SWIRL NOZZLE USING NEWTONIAN AND TIME INDEPENDENT POWER LAW NON-NEWTONIAN LIQUIDS

S. K. Som

Department of Mechanical Engineering
Indian Institute of Technology, Kharagpur, India

ABSTRACT

In spite of a large number of early works in the field of spray research, a unified correlation of the mean drop diameter of the atomized spray from a swirl nozzle with the nozzle dimensions, injection conditions and the liquid properties in case of either Newtonian or non-Newtonian liquids is still absent in literature. In the present paper, theoretical and experimental investigations have been carried out to study the drop size characteristics of the atomised spray of time independent power law liquids from swirl nozzles. In the first part of the theoretical analysis, an approximate expression has been developed with the help of the relevant earlier works of Squire (1) and Rayleigh (2) to predict the drop diameter in terms of the pertinent controlling parameters as $\frac{d_d}{h} = 3.78(Wb_{Vh})^{-0.50}$. In the subsequent part, an analysis of the hydrodynamics of flow inside a swirl nozzle has been presented from a brief review of the earlier works of the present author (3, 4, 5, 6) to recognise the influence of the nozzle geometry, the liquid properties and the injection conditions on the drop diameter. Experiments have been carried out to measure the mean drop diameter, the liquid sheet thickness and the sheet velocity at the nozzle orifice. Water and aqueous CMC (carboxymethyl cellulose sodium salt) solutions of different concentrations have been used as the working fluids. Typical time independent, pure viscous, power law rheological behaviour of the working fluids has been established with the help of a capillary tube viscometer. It has been recognised that in both the Newtonian and non-Newtonian cases, the ratio of mean drop diameter to the liquid sheet thickness at the orifice (d_d/h), becomes an inverse function of the only independent variable, the Weber number (Wb_{Vh}) based on resultant sheet velocity and its thickness at the orifice. In spite of correct qualitative trend predicted by the theory, large numerical discrepancies between the experiment and theory are arrived. This has been explained physically in the light of a few major limitations of the simplified theoretical analysis. An useful empirical equation is therefore developed as $\frac{d_{SMD}}{h} = 0.568(Wb_{Vh})^{-0.212}$. Similarly, another empirical relation regarding the dependence of liquid sheet thickness at the nozzle orifice with the pertinent parameters of influence has been established as, $\frac{h}{D_2} = 0.113(n)^{1.38} (Re_{G_1})^{-0.49}(D_2/D_1)^{-1.41}(2\alpha)^{-0.36}$. Finally, the two empirical equations are coupled into a single one as

$$\frac{d_{SMD}}{D_2} = 0.102(n)^{1.05}(Wb_{VD_2})^{-0.21}(Re_{G_1})^{-0.38}(D_2/D_1)^{-1.11}(2\alpha)^{-0.29}$$

NOMENCLATURE

- a - air core radius
- D_1 - swirl chamber diameter
- D_2 - orifice diameter
- d_d - mean drop diameter
- d_{SMD} - surface mean drop diameter
- E - a non-dimensional parameter defined by Eq. (15)
- E_R - a non-dimensional parameter defined by Eq. (31)
- h - liquid sheet thickness at the nozzle orifice
- h' - break-up thickness of the liquid sheet
- K - flow consistency index
- L_1 - length of the swirl chamber
- n - flow behaviour index
- P - static pressure
- P_b - back pressure of the nozzle
- P_i - inlet pressure to the nozzle
- P_δ - static pressure at the boundary layer - potential core interface at any section
- Q - volume flow rate
- R - longitudinal coordinate with respect to spherical polar coordinate system (Fig. 3)
- R_1 - radius of the swirl chamber
- R_2 - radius of the orifice
- R_o - a reference length in the longitudinal direction with respect to spherical polar coordinate system
- Re_{G_i} - generalised Reynolds number at inlet to the nozzle
- R_z - radius of the nozzle at any section
- r - radial distance from the nozzle axis
- U - longitudinal component of velocity with respect to spherical coordinate system
- V - resultant velocity of the liquid sheet at the nozzle orifice
- V_θ - component of velocity in axial plane perpendicular to R (Fig.3)
- V_r - radial velocity component
- V_z - axial velocity component
- $\bar{V}_{z_o}$ - average axial velocity at the orifice
- V_ϕ - tangential velocity component
- $\bar{V}_{\phi_o}$ - average tangential velocity at the orifice
- V_{ϕ_i} - tangential velocity at inlet to the nozzle
- W - component of velocity perpendicular to axial plane in spherical coordinate system
- Wb_{VD_2} - Weber number based on V and D_2
- Wb_{Vh} - Weber number based on V and h
- z - axial distance in the nozzle from its inlet plane
- α - half of the spin chamber angle
- δ - boundary layer thickness measured perpendicularly from the nozzle wall
- δ' - boundary layer thickness measured perpendicularly from the nozzle axis
- δ_2 - boundary layer thickness at the orifice measured perpendicularly from the nozzle axis
- θ - running coordinate in the transverse direction in the spherical polar coordinate system
- λ - wavelength of disturbance
- μ - dynamic viscosity
- σ - surface tension coefficient
- ν - kinematic viscosity
- ρ - density of the liquid
- ρ_a - density of air
- ϕ - running coordinate in the azimuthal direction with respect to the cylindrical polar coordinate system
- ψ - spray cone angle
- Ω - circulation constant

INTRODUCTION

One of the most important practical applications of a swirl nozzle lies in the field of spray drying in food, chemical and pharmaceutical industries. Most of the spray drying systems employ swirl spray pressure nozzles for the atomization of liquids. The different types of spray used in different industrial applications vary greatly in their performance characteristics, namely, the coefficient of discharge, spray cone angle and drop size distributions. The unified design approach of nozzles in those fields requires the interrelations between the different spray characteristics of the nozzles and the pertinent controlling parameters such as liquid properties, injection conditions and nozzle geometry. This needs a physical understanding of the hydrodynamics of flow inside the nozzle and of the mechanism of spray formation outside it. The different types of liquids that are normally used in the field of spray drying are either Newtonian or non-Newtonian in character, and the most important performance parameter of the nozzle considered in this field is the mean drop diameter of the atomized spray.

An exhaustive number of works (7 - 29) are already available in the field of spray research relating to the investigations on spray characteristics of a swirl nozzle. But most of the works (7 - 24) were carried out with Newtonian fluids only and gave empirical relations of injection pressure, liquid and ambient properties with the mean drop diameter of the spray. Quite a few number of works as those of Lee (7), Doble (8), Taylor (9) and Novikov (10) reported the effects of nozzle geometry, namely, the orifice length to diameter ratio on the different spray characteristics including the mean drop diameter. Even the author's earlier work (11) did not give any unified picture of drop diameter evaluated on the basis of both theory and experiment. On the other hand though the research in the field of non-Newtonian fluids has progressed considerably, only a limited number of works (5, 6, 25 - 29) in the area relevant to the present problem have been reported. Though Filkova (25) mentioned the effect of non-Newtonian character of a liquid on the drop diameter by considering the apparent viscosity, he did not evaluate finally a complete correlation of drop diameter with the exact rheological properties of the liquids. In fact there is no work so far which, being based on both theory and experiment, predicts the mean drop diameter with all the pertinent controlling parameters including mainly the different geometrical dimensions of the nozzle. Therefore the present paper reports such an attempt to investigate both theoretically and experimentally the effects of rheological properties in case of generalised time independent pure viscous power law liquids, injection condition and the basic geometrical dimensions of the nozzle, namely, the length and diameter of swirl chamber, angle of spin chamber and the orifice diameter on the mean drop diameter of the atomised spray.

THEORETICAL ANALYSIS

Mean Drop Diameter of an Atomized Spray

Liquid comes out from the orifice of a swirl spray atomizing nozzle as a high speed thin peripheral liquid film. This thin sheet due to its inherent instability with the interaction of the surrounding atmosphere breaks up into minute droplets. Mean drop size of the atomized spray has been computed after accepting the pioneering physical model of disintegration of thin sheet by Squire (1) and that of a jet by Rayleigh (2). A high speed liquid sheet is normally disturbed by aerodynamic waves. Fragments of liquids are broken off the wavy sheet and suffering continuous disintegration by air action tends to contract into unstable ligaments which ultimately disrupt into drops. Considering the system of forces acting on the slightly disturbed interface of a liquid sheet moving in air, it is found that surface tension force tends to draw the liquid back to its original position while the surrounding air acquiring an increased velocity and hence creating a local decrease in pressure, pulls the liquid outward. Under certain conditions i.e. for large values of the inertia force of the liquid sheet associated with its high velocity, aerodynamic forces exceed the interfacial tension and cause the formation of unstable waves which propagate at the same velocity as that of the sheet with exponentially increasing amplitude. Squire predicted this wavelength corresponding to maximum growth rate of the disturbance as,

$$\lambda = \frac{4 \pi \sigma}{\rho_a V^2} \qquad (1)$$

In an idealised break up mechanism, the most rapidly growing wave is detached at the leading edge in the form of a ribbon half a wavelength ($\lambda/2$) wide. This ribbon immediately contracts into a ligament of radius r_1 which subsequently disintegrates into drops of equal diameter. Equating the volumes of ribbon and ligament, one obtains

$$r_1 = \left(\frac{\lambda h'}{2\pi} \right)^{1/2} \qquad (2)$$

In accordance with Rayleigh's analysis (2), the collapse of a ligament of radius r_1 produces drops of diameter d_d given by,

$$d_d = 3.78\, r_1 \qquad (3)$$

With the help of Eqs. (1) and (2), the Eq. (3) is written as

$$d_d = 5.35 \left(\frac{\sigma h'}{\rho_a V^2} \right)^{1/2} \qquad (4)$$

Again, with the assumption that the disintegration process takes place within a very small distance from the orifice and the liquid sheet thickness remains more or less constant, the parameter h' in Eq. (4) can be replaced by h. Finally it becomes,

$$\frac{d_d}{h} = 3.78 (Wb_{Vh})^{-0.50} \qquad (5)$$

Where, the Weber number Wb_{Vh} is based on liquid sheet thickness h and sheet velocity V and is given by

$$Wb_{Vh} = \frac{\rho_a V^2 h}{2\sigma} \qquad (6)$$

It is apparently clear from this analysis that the instability phenomenon is guided by the kinematic conditions of the liquid sheet at the orifice, the surface tension of the liquid and the density of the ambient atmosphere. Hence the ratio of drop diameter to sheet thickness d_d/h becomes a function of the only pertinent variable parameter Wb_{Vh} as given by Eq. (5). The values of V and h are determined by the hydrodynamics of flow inside the nozzle and in turn depend on the injection condition, rheological properties of the liquid and the nozzle geometry. The Eq. (5) can be written in a more conventional and useful form as,

$$\frac{d_d}{D_2} = 3.78 \left(\frac{h}{D_2}\right)^{-0.50} (Wb_{V D_2})^{-0.50} \qquad (7)$$

Therefore to recognise the influence of liquid properties and nozzle geometries on drop diameter, an analysis of hydrodynamics of flow inside the nozzle is essential to predict the value of h/D_2 as a function of its independent pertinent parameters. A detailed analysis of the mechanics of flow in a swirl nozzle was already reported in the earlier works (3, 4, 5, 6) of the present author. It is briefly reproduced here for ready reference.

Hydrodynamics of Flow Inside the Nozzle

(A brief review of the earlier works (3, 4, 5, 6))

Perfect tangential entry of the fluid in a conventional swirl nozzle (Fig. 1) consisting of a cylindrical swirl chamber and a converging spin chamber was considered and the simplified model of the flow field was assumed to consist of the following three zones (Fig. 2) :

(i) Central air core near the axis

(ii) Potential core of only free vortex outside the air core.

(iii) Zone of boundary layer near the surface of the nozzle.

The assumption of only an irrotational vortex motion in the potential core was justified in the sense that for a perfect tangential entry, the axial and radial velocities in the potential core were small compared to those within the boundary layer. Moreover, the existence of a central hollow core was well explained by the free vortex motion. Separate coordinate systems and nomenclature were used in the mathematical analysis pertaining to the flow in the swirl chamber and in the spin chamber (Fig. 3).

Solution in the swirl chamber (Fig. 3)

The approximate laminar boundary layer equations for steady motion of an incompressible, purely viscous, time-independent power law non-Newtonian fluid flowing symmetrically with respect to the axis $r = 0$ in cylindrical polar coordinate system were considered as

$$\frac{\partial P}{\partial r} = \rho \frac{V_\phi^2}{r} \tag{8}$$

$$V_r \frac{\partial V_\phi}{\partial r} + V_z \frac{\partial V_\phi}{\partial z} = \frac{K}{\rho}\frac{\partial}{\partial r}\left[\left\{\left(\frac{\partial V_\phi}{\partial r}\right)^2 + \left(\frac{\partial V_z}{\partial r}\right)^2\right\}^{\frac{n-1}{2}} \frac{\partial V_\phi}{\partial r}\right] \tag{9}$$

$$V_r \frac{\partial V_z}{\partial r} + V_z \frac{\partial V_z}{\partial z} = -\frac{1}{\rho}\frac{\partial P}{\partial z} + \frac{K}{\rho}\frac{\partial}{\partial r}\left[\left\{\left(\frac{\partial V_\phi}{\partial r}\right)^2 + \left(\frac{\partial V_z}{\partial r}\right)^2\right\}^{\frac{n-1}{2}} \frac{\partial V_z}{\partial r}\right] \tag{10}$$

The approximate form of the continuity equation used was

$$\frac{\partial V_r}{\partial r} + \frac{\partial V_z}{\partial z} = 0 \tag{11}$$

The value of $\frac{\partial P}{\partial z}$ was evaluated by using Bernoulli's equation at the boundary layer interface with the potential core as

$$\frac{\partial P}{\partial z} = -\rho \frac{\Omega^2}{(R_1 - \delta)^3}\frac{d\delta}{dz} \tag{12}$$

The Eqs. (9) and (10) were solved by the momentum integral method. The value of $\partial P/\partial z$ was substituted from the Eq. (12) into Eq. (10) and the radial component of velocity V_r was eliminated with the help of Eq. (11). Finally, the transverse and axial momentum integrals became respectively

$$-\frac{\Omega}{(R_1-\delta)}\int_{R_1-\delta}^{R_1}\frac{\partial V_z}{\partial z}dr + \int_{R_1-\delta}^{R_1}\frac{\partial}{\partial z}(V_z V_\phi)dr = \frac{K}{\rho}\left[\left\{\left(\frac{\partial V_\phi}{\partial r}\right)^2 + \left(\frac{\partial V_z}{\partial r}\right)^2\right\}^{\frac{n-1}{2}}\frac{\partial V_\phi}{\partial r}\right]_{R_1-\delta}^{R_1} \tag{13}$$

$$2\int_{R_1-\delta}^{R_1} V_z\frac{\partial V_z}{\partial z}dr - \int_{R_1-\delta}^{R_1}\frac{\Omega^2}{(R_1-\delta)^3}\frac{d\delta}{dz}dr = \frac{K}{\rho}\left[\left\{\left(\frac{\partial V_\phi}{\partial r}\right)^2 + \left(\frac{\partial V_z}{\partial r}\right)^2\right\}^{\frac{n-1}{2}}\frac{\partial V_z}{\partial r}\right]_{R_1-\delta}^{R_1} \tag{14}$$

The polynomial distribution of velocity components were taken as

$$V_z = \frac{\Omega E}{R_1}(\eta - 2\eta^2 + \eta^3) \tag{15}$$

$$V_\phi = \frac{\Omega}{R_1}(2\eta - \eta^2) \tag{16}$$

$$\text{where,}\quad \eta = \frac{R_1 - r}{\delta} \tag{17}$$

The boundary conditions satisfied by the velocity components were

$V_z = V_\phi = 0$ at $r = R_1$

$V_z = 0$, $\frac{\partial V_z}{\partial r} = 0$, $V_\phi = \frac{\Omega}{(R_1-\delta)}$, $\partial V_\phi / \partial r = 0$ at $r = R_1 - \delta$

With the help of Eqs. (15),(16) and (17), the momentum integrals (13) and (14) were solved for E and δ as a function of z . The final expressions were developed as

$$\frac{d\delta_1}{dz_1} = \frac{225P(1 + 0.25E^2)^{\frac{n-1}{2}} E}{(E^2 + 105\,\delta_1)\,\delta_1^n} \quad (18)$$

$$\frac{dE}{dz_1} = \frac{60P(1 + 0.25E^2)^{\frac{n-1}{2}}}{\delta_1^{n+1}} - \frac{225P(1 + 0.25E^2)^{\frac{n-1}{2}} E^2}{(E^2 + 105\,\delta_1)\,\delta_1^{n+1}} \quad (19)$$

where the nondimensional variables were

$$\delta_1 = \frac{\delta}{R_1}, \qquad z_1 = \frac{z}{2R_1} \quad (20)$$

and the nondimensional parameter P was defined by

$$P = 8\left(\frac{2n}{3n + 1}\right)^n \frac{1}{Re_{G_1}} \quad (21)$$

The generalised Reynolds number Re_{G_1} at inlet to the nozzle was defined on the basis of the tangential velocity at inlet, the diameter of the swirl chamber, and the apparent viscosity of the fluid as

$$Re_{G_1} = \frac{\rho\, V_{\phi_1}^{2-n}\, D_1^n}{8^{n-1}\, K'} \quad (22)$$

with

$$K' = \left(\frac{3n + 1}{4n}\right)^n K, \qquad V_{\phi_1} = \frac{\Omega}{R_1} \quad (23)$$

Eqs. (18) and (19) were solved numerically with the initial conditions $\delta_1 = E = 0$ at $z_1 = 0$. The solutions were obtained for different values of the parameter P corresponding to the different values of the independent quantities n and Re_{G_1}.

Solution in the spin chamber

Spherical polar coordinate system (Fig.3) was used for the solution in this zone. The approximate boundary layer equations along with the equation of continuity were written respectively as

$$\frac{1}{R}\frac{\partial P}{\partial \theta} = \frac{\rho\, W^2 \cot\theta}{R} \quad (24)$$

$$U\frac{\partial U}{\partial R} + \frac{V}{R}\frac{\partial U}{\partial \theta} - \frac{W^2}{R} = -\frac{1}{\rho}\frac{\partial P}{\partial R} + \frac{K}{\rho\, R^{n+1}}\frac{\partial}{\partial \theta}\left[\left\{\left(\frac{\partial U}{\partial \theta}\right)^2 + \left(\frac{\partial W}{\partial \theta}\right)^2\right\}^{\frac{n-1}{2}} \frac{\partial U}{\partial \theta}\right] \quad (25)$$

$$U\frac{\partial W}{\partial R} + \frac{V}{R}\frac{\partial W}{\partial \theta} + \frac{WU}{R} = \frac{K}{\rho R^{n+1}}\frac{\partial}{\partial \theta}\left[\left\{\left(\frac{\partial U}{\partial \theta}\right)^2 + \left(\frac{\partial W}{\partial \theta}\right)^2\right\}^{\frac{n-1}{2}} \frac{\partial W}{\partial \theta}\right] \quad (26)$$

$$\frac{\partial U}{\partial R} + \frac{2U}{R} + \frac{1}{R}\frac{\partial V}{\partial \theta} = 0 \quad (27)$$

The term $\partial P/\partial R$ was evaluated from the equation in the potential core as

$$\frac{1}{\rho}\frac{\partial P}{\partial R} = \frac{\Omega^2}{R^3 \sin^2\theta} \quad (28)$$

Final forms of the longitudinal and transverse momentum integrals were

$$2\int_{\alpha-\delta/R}^{\alpha} U\frac{\partial U}{\partial R}\,d\theta + 2\int_{\alpha-\delta/R}^{\alpha} \frac{U^2}{R}\,d\theta + \int_{\alpha-\delta/R}^{\alpha} \frac{\Omega^2}{R^3 \sin^2\alpha}\,d\theta - \int_{\alpha-\delta/R}^{\alpha} \frac{W^2}{R}\,d\theta = \frac{K}{\rho R^{n+1}}\left[\left\{\left(\frac{\partial U}{\partial \theta}\right)^2 + \left(\frac{\partial W}{\partial \theta}\right)^2\right\}^{\frac{n-1}{2}} \frac{\partial U}{\partial \theta}\right]_{\alpha-\delta/R}^{\alpha} \quad (29)$$

$$- \frac{\Omega}{R^2 \sin\alpha} \int_{\alpha-\delta/R}^{\alpha} \left(R \frac{\partial U}{\partial R} + 2U\right) d\theta \; +$$

$$\int_{\alpha-\delta/R}^{\alpha} \frac{\partial}{\partial R}(UW)\, d\theta \; + \; 3 \int_{\alpha-\delta/R}^{\alpha} \frac{UW}{R}\, d\theta$$

$$= \frac{K}{\rho R^{n+1}} \left[\left\{ \left(\frac{\partial U}{\partial \theta}\right)^2 + \left(\frac{\partial W}{\partial \theta}\right)^2 \right\}^{\frac{n-1}{2}} \frac{\partial W}{\partial \theta} \right]_{\alpha-\delta/R}^{\alpha} \tag{30}$$

The arbitrary velocity profiles taken were

$$U = \frac{\Omega E_R}{R \sin\alpha} (\eta - 2\eta^2 + \eta^3) \tag{31}$$

$$W = \frac{\Omega}{R \sin\alpha} (2\eta - \eta^2) \tag{32}$$

$$\text{where } \eta = \frac{R(\alpha - \theta)}{\delta} \tag{33}$$

The usual boundary conditions satisfied by the velocity profiles were

$U = W = 0$ at $\theta = \alpha$

$U = 0,\; W = \Omega/R \sin\alpha ,\; \frac{\partial U}{\partial \theta} = \frac{\partial W}{\partial \theta} = 0$

at $\theta = \alpha - \delta/R$

With the help of equations (29) to (33), the final expressions relating the unknown quantities E_R and δ with the independent variable R were obtained as

$$\frac{d\delta_R}{dR'} = \frac{49}{R'} \frac{\delta_R}{E_R^2} - \frac{\delta_R}{R'} + \frac{15M(1+0.25E_R^2)^{\frac{n-1}{2}}}{R'^{\,n-2}\, \delta_R^{\,n}\, E_R} \tag{34}$$

$$\frac{dE_R}{dR'} = - \frac{49}{R' E_R} + \frac{E_R}{R'} - \frac{11M(1+0.25E_R^2)^{\frac{n-1}{2}}}{R'^{\,n-2}\, \delta_R^{\,n+1}} \tag{35}$$

Where, the non-dimensional variable parameters were defined as

$$\delta_R = \frac{\delta}{R_o}, \quad R' = \frac{R}{R_o}$$
$$M = 60(\sin\alpha)^n \left(\frac{2n}{3n+1}\right)^n \frac{1}{Re_{G_i}} \tag{36}$$

Eqs. (34) and (35) were solved numerically for δ_R and E_R. The initial conditions were matched from the end conditions of the solution for the swirl chamber. As the order of the radial velocity V_r was very small compared to that of the axial velocity V_z in the swirl chamber, the longitudinal component of velocity U at the entrance to the spin chamber satisfied the equation

$$U = -V_z \cos\alpha$$

Therefore the initial conditions for the solutions of (34) and (35) became

$$\delta_R = (\delta_1)_{z_1} \sin\alpha \cos\alpha$$
$$E_R = -(E)_{z_1} \cos\alpha \tag{37}$$

The values of $(\delta_1)_{z_1}$ and $(E)_{z_1}$ at any value of z_1 were found from the numerical solutions of the Eqs. (18) and (19). Finally, the solutions of (34) and (35) were obtained for different values of the independent parameters α, n and Re_{G_i}.

Liquid Sheet Thickness at the Nozzle Orifice

The most important and well known picture regarding the hydrodynamics of flow in a swirl nozzle is the formation of a central air core of uniform diameter within the nozzle. Due to the tangential entry, liquid attains a free vortex motion creating a very high velocity and low pressure near the nozzle axis. Above a certain injection rate, depending on the nozzle geometry and liquid properties, the strength of vortex motion inside nozzle becomes such that the static pressure falls below the back pressure of the nozzle creating a central air core throughout the nozzle. Under typical operating conditions, the formation of an air core takes place in almost all swirl nozzles. The liquid sheet thickness at the orifice therefore satisfies the relation

$$h = R_2 - a \qquad (38)$$

The air core diameter was predicted in the earlier works (3 , 6) with the use of Bernoulli's equation at the boundary layer-potential core interface and the air core-potential core interface at any section in the nozzle as

$$\frac{P_\delta}{\rho} + \frac{\Omega^2}{2(R_z - \delta')^2} = \frac{P_b}{\rho} + \frac{\Omega^2}{2a^2} \qquad (39)$$

Application of the Eq. (39) at the orifice where $P_\delta = P_b$, finally gave

$$a = R_2 - \delta_2 \qquad (40)$$

and hence, $h = \delta_2$ (41)

Therefore, the Eq. (7) can now be written as

$$\frac{d_d}{D_2} = 3.78\left(\frac{\delta_2}{D_2}\right)^{-0.50} (Wb_{VD_2})^{-0.50} \qquad (42)$$

Pertinent guiding parameters influencing the value of δ_2/D_2 as determined from the Eqs. (18), (19), (34) and (35) are as follows:

i) Re_{G_i}, generalised Reynolds number at inlet to the nozzle
ii) n, the flow behaviour index of the fluid
iii) $z_1 = L_1/D_1$, the length to diameter ratio of the swirl chamber
iv) $R' = D_2/D_1$, the orifice to swirl chamber diameter ratio
v) 2α, the spin chamber angle

Therefore, the Eq. (42) can be expressed in an implicit form as,

$$\frac{d_d}{D_2} = 3.78(Wb_{Vb_2})^{-0.50} F(Re_{G_i}, n, L_1/D_1, D_2/D_1, 2\alpha) \qquad (43)$$

The Eq. (43) finally shows all the pertinent non-dimensional parameters of influence on d_d / D_2 and guides the experiments accordingly.

EXPERIMENTAL INVESTIGATIONS

A number of nozzles having various geometrical dimensions were obtained from the different combinations of interchangeable swirl chambers and spin chambers (Fig. 4) fabricated with perspex. The different non-dimensional geometrical parameters which were varied during the range of experiments are shown in Table1. Water and aqueous CMC (carboxymethyl cellulose sodium salt) solutions with concentration values of 0.2, 0.4 and 0.6 % were used as the Newtonian and non-Newtonian working liquids to be injected into ordinary atmosphere(101 kN/m^2, 25°C). In the first part of the experimental work, the rheological behaviour of the working fluids was established with the help of a capillary tube viscometer.

Table 1. Catalogue of nozzles used in the experimental investigations.

Nozzle identification no.	Length to diameter ratio of swirl chamber (L_1/D_1)	Spin chamber angle 2α (degrees)	Ratio of orifice to swirl chamber diameter (D_2/D_1)
1	0.500	30	0.136
2	0.500	30	0.200
3	0.500	30	0.267
4	0.500	30	0.312
5	0.500	45	0.200
6	0.500	60	0.200
7	0.500	75	0.200

Determination of Rheological Properties of the Working Fluids

The experimental set-up (Fig. 5) in this

part consisted of a constant head tank T to supply the liquid to a capillary glass tube C of 12.7 mm internal diameter. Pressure drop readings across a length of 1.5 m in the developed region of the flow were taken with the help of a mercurry tube manometer M. The developed flow region was ensured by providing the necessary and sufficient straight entry length. The inlet end of the capillary tube was flared to ensure approximately a smooth and parallel flow at entry. A valve V was used at the downstream end to regulate the discharge rates which were measured by collecting the liquid in a measuring cylinder. Capillary tubes of different diameters were used and pressure drop readings over different lengths were measured to detect any time dependent and effective slip characteristics of the liquid. Finally it was recognised that the aqueous solutions of CMC powder behaved as a time-independent, purely viscous fluid obeying closely the Ostwald-de-Waele power law model for the constitutive equation. The values of the rheological properties, namely, the flow consistency index and the flow behaviour index, were determined from the measured values of pressure drops and flow rates. The values, being in fair agreement with those available in the literature, are given in Table 2 .

Table 2. Rheological properties of the working fluids

Concentration of the liquid(wt. % of CMC powder)	Flow behaviour index n	Flow consistency index $K(kg/ms^{2-n})$
0.2	0.90	0.010
0.4	0.82	0.025
0.6	0.75	0.150

Other properties such as density and surface tension of the liquids (aqueous CMC solution) were measured by the usual and standard laboratory methods and were found to be same as those of water. All the property values were measured at the room temperature (25°C).

Main Experimental Set-up and Measurement

In the main experimental set-up (Fig. 6), a centrifugal pump P was provided to supply high pressure liquid to the nozzle N fixed properly with the help of an adjustable clamp in the absolute vertical position facing downwards. One discharge valve V_1 and a by-pass valve V_2 were provided to control the flow through the nozzles. The rotameter R was used to measure the volumetric flow rates which were again checked by the direct method of collecting the nozzle flow in a volumetric tank during a known interval of time.

Determination of mean drop diameter

Surface mean drop diameters were found by taking the drop impressions on a glass slide coated with magnesium oxide and then counting an measuring the diameter of the drop impressions from the slide by means of a suitable microscope. To avoid coalescence of the drops, the time of spraying on the target was restricted by a shutter arrangement giving a correct expossure. The actual size of the drops were found out from the standard calibrated relationship given by May (30) that the ratio of impression size to droplet size is constant and equals to 1.16 for droplet sizes greater than 20 microns.

Determination of liquid sheet thickness at the nozzle orifice

The liquid sheet thickness at the orifice was determined by measuring the air core diameter and then using the relation given by the Eq. (38) . The air core diameter was measured with the help of a cathetometer. Due to the transparency of the perspex material and also of the liquid flowing inside, the central air core was distinctly visible from outside the nozzle. To facilitate a detailed observation, arrangements for proper illumination of the internal flow of the nozzle were made by placing a high power beam of light outside the nozzle. It was found that a uniform cylindrical air core was formed throught the nozzle. The diameter of the core was measured with the help of a cathetometer having a telescopic eye piece mounted on a scale bench with a vernier arrangement having a least count of 0.01 mm.

Determination of spray cone angle

To evaluate the Weber number Wb_{VD_2}, the liquid sheet velocity V at the nozzle orifice is required to be known. Due to the presence of both axial and tangential velocity, liquid sheet emerging from a swirl nozzle is conical in shape. Neglecting the order of radial velocity compared to those of the tangential and axial velocities at the orifice, the spray cone angle is usually expressed as

$$\psi = 2\tan^{-1}\left(\frac{\bar{V}_{\phi_o}}{\bar{V}_{z_o}}\right) \qquad (44)$$

With the help of the Eq.(44), the resultant velocity of the liquid sheet at the orifice can be written as.

$$V = \bar{V}_{z_o}\left(1 + \tan^2\frac{\psi}{2}\right)^{0.5} \qquad (45)$$

Hence to determined V from the Eq. (45) both $\bar{V}_{z_o}$ and ψ are required to be known. The average axial velocity was found out from the volumetric flow rates through the nozzle and the area of the orifice. Cone angle of spray was measured by obtaining a dark shadow of the spray on a ground glass placed behind the spray with frontal illumination. The ground glass fixed on a wooden frame (Fig.7) was placed behind the spray. An angular scale was drawn on the ground glass. The axis of the ground glass, i.e. the axis of the scale, was properly aligned with the axis of the nozzle to ensure the formation of a shadow symmetrical about the axis of the scale on the ground glass. The frontal illumination was obtained by using a special electric bulb and a built-in parabolic reflector arrangement. This was done to create a more or less parallel beam of light. Thus the direct shadow of the spray was formed on the ground glass and the cone angle was measured from the angular scale marked on the glass.

RESULTS AND DISCUSSION

It is recognised from both the theory and experiments (Fig. 8) that the ratio of mean drop diameter to the liquid sheet thickness at the nozzle orifice, d_d / h, can be expressed as an inverse function of the only independent non-dimensional parameter, Wb_{Vh}, the Weber number based on the liquid sheet thickness and the sheet velocity at the nozzle orifice. Although the theory predicts the correct qualitative trend, the theoretical values are always much higher than those of experimental. Moreover, the difference between theoretical and experimental values decreases with the increase in the Weber number Wb_{Vh}. An empirical equation has therefore been established from the experimental results as,

$$\frac{d_{SMD}}{h} = 0.568\,(Wb_{Vh})^{-0.212} \qquad (46)$$

The discrepancy between the theoretical and experimental results can be explained physically as follows :

Firstly, the liquid sheet emerging from a swirl nozzle is conical in shape and thus differs from the flat sheet, assumed in the theoretical analysis, both in curvature and uniformity of thickness. The reduction of sheet thickness from the orifice to the break-up section depends on the sheet velocity, the break-up distance and the spray cone angle. Typical values of this reduction ratio vary from 1/10 to 1/30 as reported from the experimental results of the earlier work (31). Secondly, the theory predicts uniform drops produced finally from an unstable ligament. But in practice, a second stage of atomization takes place with the further break-up of high speed coarser drops into fine droplets. Moreover, due to the induced motion of the ambient atmosphere, a complicated phenomenon of both the splitting and coalescence of drops take place simaltaneously in a very irregular manner ultimately resulting in a spectrum of drops. A complete theory taking account of all these practical factors is not yet found in literature. As the Weber number increases, the break-up occurs more close to the orifice and the difference between the liquid sheet thickness at the orifice and that at the break-up is reduced. Consequently the theoretical values of drop diameters become closer to the corresponding experimental values.

It was recognised earlier that to establish the effects of injection condition, liquid properties and the nozzle geometry on the drop size, an interrelation of the liquid sheet thickness at the orifice with those parameters was essential. The theoretical and experimental investigations relating to this aspect were described accordingly in the earlier sections. The results are shown in Figs. 9, 10 and 11. The ratio of liquid sheet thickness to the orifice diameter bears an inverse relationship with Re_{G_i}, D_2/D_1 and 2α and becomes a direct function of n. The physical explanation of these trends of variations lies in the mechanism of change of boundary layer thickness in the nozzle with those parameters. The theoretical work relating to the flow of Newtonian fluids is a special case of the solution of the generalised Eqs. (18), (19), (34) and (35) with $n = 1$ and K equals to μ (the Newtonian viscosity) and is corrborated by the experiments with water. Though the theoretical and experimental results show the same trend of variations, the theoretical values are always lower than the experimental values. This discrepancy occurs due to the simplified theoretical model of the flow field consisting of a laminar boundary layer along with a free vortex potential core. An analysis of turbulent boundary layer along with an inviscid core of combined free and forced vortex motions may be a closer approximation to the actual situation.

An empirical relation regarding the variation of the liquid sheet thickness with its independent parameters have been developed from the experimental results as

$$\frac{h}{D_2} = 0.113\,(Re_{G_i})^{-0.488}\,(n)^{1.376}\,(D_2/D_1)^{-1.410}\,(2\alpha)^{-0.364} \qquad (47)$$

Finally, the Eqs. (46) and (47) are coupled in a single and more useful form as

$$\frac{d_{SMD}}{D_2} = 0.1021(n)^{1.048}\,(Re_{G_i})^{-0.384}\,(Wb_{VD_2})^{-0.212}\,(D_2/D_1)^{-1.111}\,(2\alpha)^{-0.287} \qquad (48)$$

All the empirical equations are vallid within the range of values of the variables studied in the present experiments. The Eq. (48) shows in an explicit form the dependence of d_{SMD}/D_2 with all the pertinent independent parameters. Effects of injection conditions, namely, liquid velocity at inlet to the nozzle, and the liquid properties such as density, flow behaviour index, flow consistency index and the surface tension on drop diameter are shown by the non-dimensional terms Re_{G_i}, n and Wb_{VD_2} in the Eq. (48). The physical explanations for the typical trends of variation of drop size with these parameters are well known. The physical explanations of the effects of nozzle dimensions on the drop diameter are explained below:

Effect of Orifice to Swirl Chamber Diameter Ratio (D_2/D_1)

The ratio of surface mean drop diameter to the orifice diameter, d_{SMD}/D_2, is inversely proportional to D_2/D_1 at fixed values of other independent non-dimensional parameters. Actually, mean drop diameter decreases with the decrease in orifice diameter ; but this rate of decrease becomes less than the proportional decrease in the orifice diameter D_2 . A decrease in D_2 results in a higher value of discharge velocity of the liquid and a lower value of the liquid film thickness. Thus, a combined effect on reducing the droplet size is observed. The reduction in spray cone angle with decreasing orifice diameter plays to certain extent, a retarding role in reducing the droplet size. But the combined effect of discharge velocity and the liquid film thickness to reduce the drop diameter becomes more prominent than the adverse effect of spray cone angle, and finally results in a better atomization.

Effect of Spin Chamber Angle (2α)

As found from the Eq. (48), an increase in the value of 2α decreases the value of d_{SMD}/D_2 and vice versa. This is explained in a sense that with the increase in 2α, the liquid sheet thickness at the orifice is reduced and the spray cone angle is increased. Both these changes have a favourable influence in reducing the drop size. The increase in spray cone angle results in a better atomization due to the fact that more of the surrounding gases take part in the process of atomization.

CONCLUSION

The ultimate importance of the present work lies in the practical usefulness of the Eq. (48) in the field of designing a swirl nozzle. Starting with the values of liquid properties and pre-assigned spray characteristics, namely, the mean drop diameter and volumetric flow rate, the nozzle dimensions can be determined from the Eq. (48). The Eq. (48) has been developed in the present work from the empirical Eqs. (46) and (47) evaluated on the basis of theoretical concept. The Eq. (48) can be derived straightforward from the experiments after recognising all the pertinent input parameters from the theory. The experiments in that case should be designed in a way to generate different values of the input parameters independently. It was not possible to conduct the present experiment in a way to vary the Weber number Wb_{VD_2} at any fixed value of the Reynolds number Re_{G_1} and vice versa. But within the present experimental range, the Eq.(48) remains the same irrespective of the ways by which it is evaluated.

The limitations of the present work throws light on the scope of future works. A theoretical analysis of the hydrodynamic instability of a high speed conical liquid sheet with varying thickness becomes essential to predict the drop diameter more accurately. As the prediction of break-up distance which depends to a certain extent upon the size of initial disturbances becomes difficult, a semi-empirical approach can be made with the use of empirical values of the break-up distances regarding the final prediction of drop diameter. All the predictions of the present work is valid for a class of pure viscous, time independent, power law non-Newtonian liquids. The entire work can also be repeated with other different types of liquids after establishing their typical rheological characteristics before hand with any suitable rheometer.

REFERENCES

1. Squire, H.B., British J. Appl.Phys., Vol.4, June 1953, pp.167-169.
2. Rayleigh Lord., Proc.Lond.Math.Soc., Vol. 10, 1878, p. 4.
3. Som, S.K. and Mukherjee, S.G., Appl. Sci. Res., 36, 1980, pp. 173-196.
4. Som, S.K. and Mukherjee, S.G., Acta Mechanica, 36, 1980, pp. 79-102.
5. Som, S.K., J.N.N.Fl.Mech.(In Press)
6. Som, S.K., Appl. Sci. Res., (In Press)
7. Lee, W. Dana. and Spencer, R.C., NACA Technical Report, Note No. 424, 1932.
8. Doble, S.M., Engineering Vol.159, 1945, p. 21.
9. Taylor, G.I., Proc.Seventh Int. Congr. Appl. Mech., Vol. 2, pt.1, 1948, p. 280.

10. Novikov, I.I., Engineers' Digest, Vol. 10, No. 3, March 1949, p. 72.

11. Som, S.K. and Mukherjee, S.G., Inst. Engrs. (I), Vol. 60, pt. ME5, March 1980.

12. Schweitzer, P.H., J. Appl. Phys., Vol. 8, August 1937, p. 513.

13. Longwell, J.P., D.Sc. Thesis, M.I.T., 1943.

14. Merrington, A.C. and Richardson, E.G., Proc. Phys. Soc., Vol. 59, pt. 1, No. 331, January 1947, p. 1.

15. Larcombe, H.L.M., Chemical Age, Vol. 57, pt. 1, 1947, p. 563. pt.2.

16. Watson, E.A. and Clarke., J.S., J. Inst. Fuel, Vol. 21, No. 116, October 1947, p. 1.

17. Bown, J.C. and Joyce, J.R., Tech. Report, Shell Petroleum Co. Ltd., March 1948.

18. Giffen, E. and Massey, B.S., MIRA Report, No. 4, 1951, p. 11.

19. Giffen, E. and Lamb, T.A.T., MIRA Report, No. 5, 1953, p. 14.

20. Radcliffe, A., Inst. Mech. Engrs., London, 1954.

21. Giffen, E. and Neale, M.C., MIRA Report No. 6, 1954, p. 7.

22. Decorso, S.M. and Kemeny, G.A., Trans. ASME, Vol. 79, No. 3, April 1957, p. 607.

23. Mani, J.V.S. and Rao, M.N., J.Sci. and Engg. Res.(India), Vol. 1, No.1, January 1957, p. 113.

24. Decorso, S.M., Trans. ASME, J. Engg. Power, January 1960, p. 10.

25. Filkova, I., Drying 80, Vol. 1, p. 346.

26. Sears, J.T. and Ray, S., Drying 80, Vol. 1, p. 332.

27. Acrivos, A., Shah, M.J. and Peterson, E.E., J. Appl. Phys. 31, 963-968.

28. Bird, R. Byron, Stewart, Warren. E. and Lightfoot, Edwin. N., Transport Phenomena, John Willey and Sons, 1960.

29. Skelland, A.H.P., Non-Newtonian flow and Heat Transfer, John Willey and Sons, pp. 276-293, 1967,.

30. May, K.R., J. Sci. Instr. Vol. 27, May 1950.

31. York, J.L. and Stubbs, H.E., Trans. ASME, Vol. 74, 1952, p. 1157.

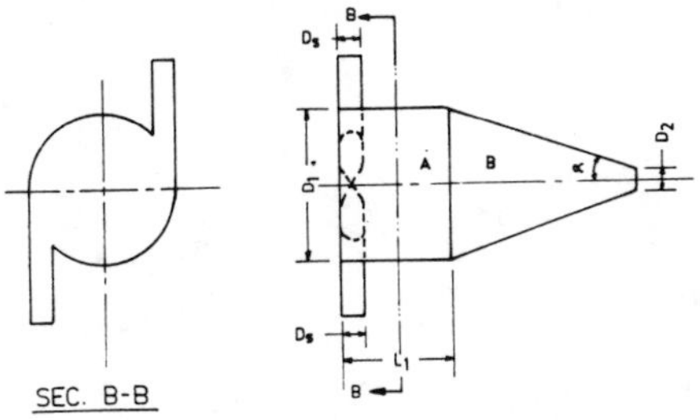

NOMENCLATURE OF THE SWIRL NOZZLE

A SWIRL CHAMBER
B SPIN CHAMBER
D_1 DIAMETER OF SWIRL CHAMBER
D_2 ORIFICE DIAMETER
D_s DIAMETER OF TANGENTIAL ENTRY PORTS
L LENGTH OF SWIRL CHAMBER

FIG. 1 CONVENTIONAL SWIRL NOZZLE.

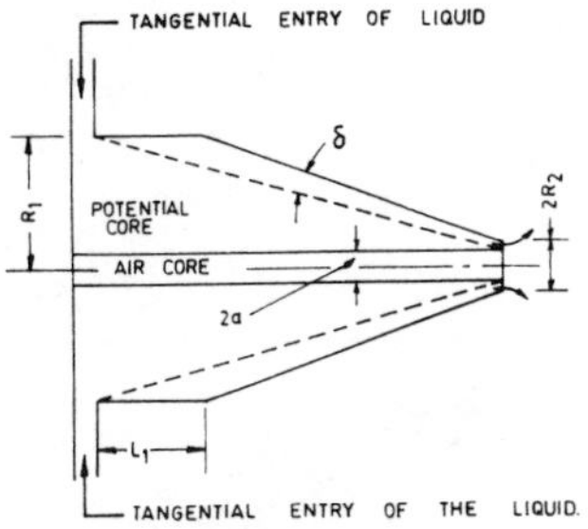

a RADIUS OF THE AIR CORE
L LENGTH OF THE SWIRL CHAMBER
R RADIUS OF THE SWIRL CHAMBER
R RADIUS OF THE ORIFICE
δ BOUNDARY LAYER THICKNESS

FIG. 2 HYDRODYNAMIC PICTURE INSIDE A SWRIL NOZZLE ACCORDING TO THE ASSUMED THEORETICAL MODEL.

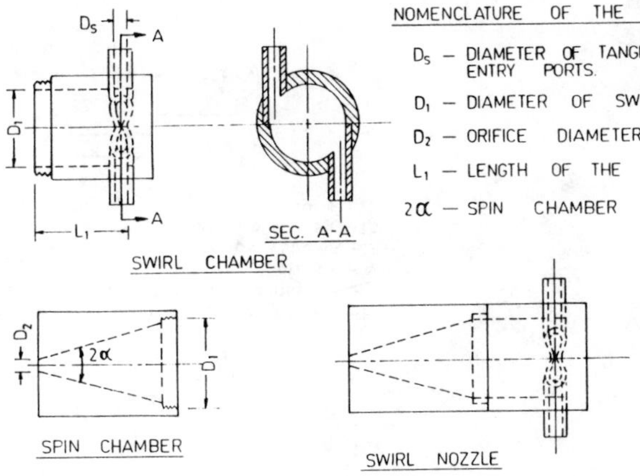

NOMENCLATURE OF THE SWIRL NOZZLE

D_s – DIAMETER OF TANGENTIAL ENTRY PORTS.
D_1 – DIAMETER OF SWIRL CHAMBER
D_2 – ORIFICE DIAMETER
L_1 – LENGTH OF THE SWIRL CHAMBER
2α – SPIN CHAMBER

FIG. 4 GEOMETRY OF CONVENTIONAL SWIRL NOZZLE FORMED FROM THE ASSEMBLY OF INTERCHANGEABLE SWIRL CHAMBERS AND SPIN CHAMBERS.

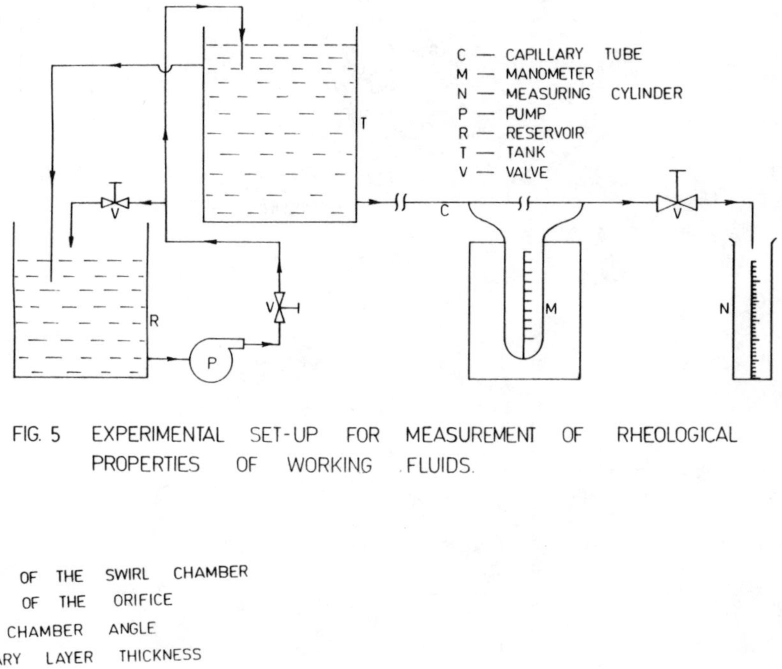

C — CAPILLARY TUBE
M — MANOMETER
N — MEASURING CYLINDER
P — PUMP
R — RESERVOIR
T — TANK
V — VALVE

FIG. 5 EXPERIMENTAL SET-UP FOR MEASUREMENT OF RHEOLOGICAL PROPERTIES OF WORKING FLUIDS.

A — SWIRL CHAMBER
B — SPIN CHAMBER

R_1 : RADIUS OF THE SWIRL CHAMBER
R_2 : RADIUS OF THE ORIFICE
2α : SPIN CHAMBER ANGLE
δ : BOUNDARY LAYER THICKNESS

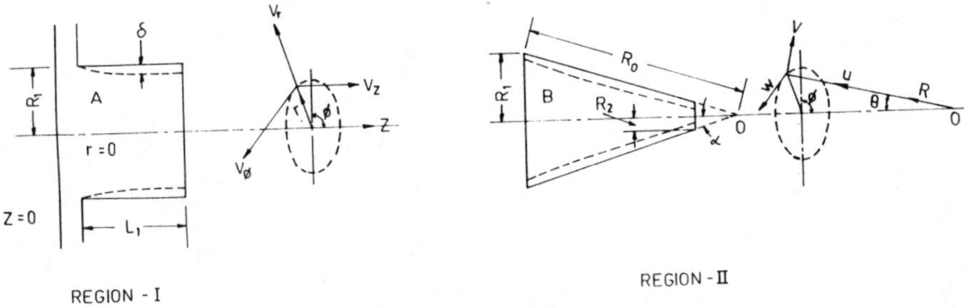

FIG. 3 SYSTEMS OF COORDINATES AND NOMENCLATURE USE IN THE THEORETICAL ANALYSIS

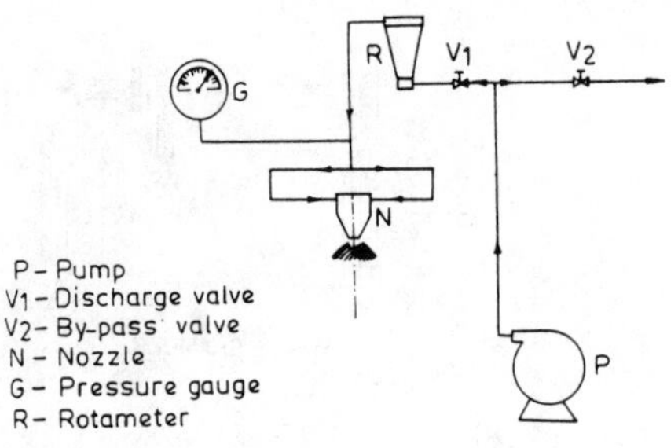

FIG 6 MAIN EXPERIMENTAL SET-UP

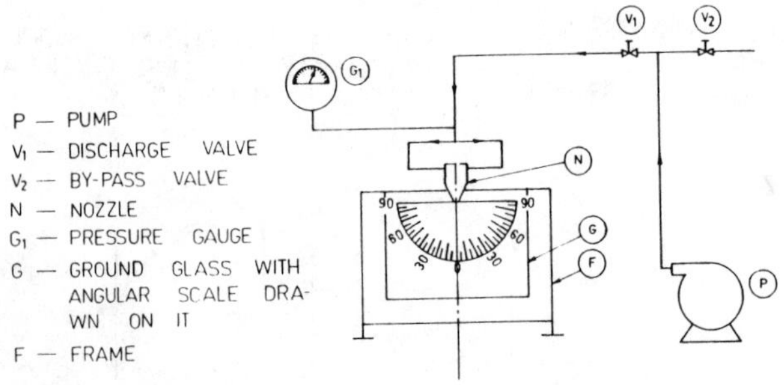

FIG. 7 LAYOUT OF THE EXPERIMENTAL SET-UP FOR MEASURING SPRAY CONE ANGLE.

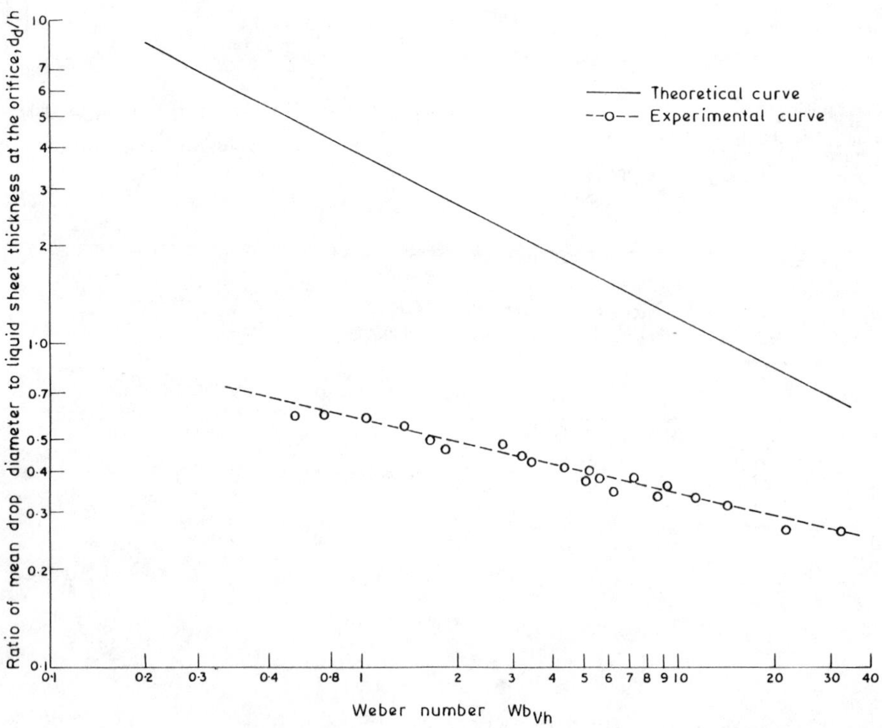

FIG. 8 VARIATION OF d_d/h WITH Wb_{Vh}

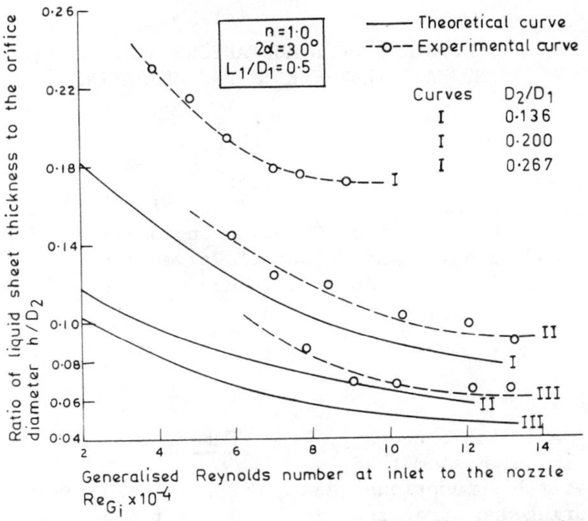

FIG. 9 VARIATIONS OF h/D_2 WITH Re_{G_i} AND D_2/D_1

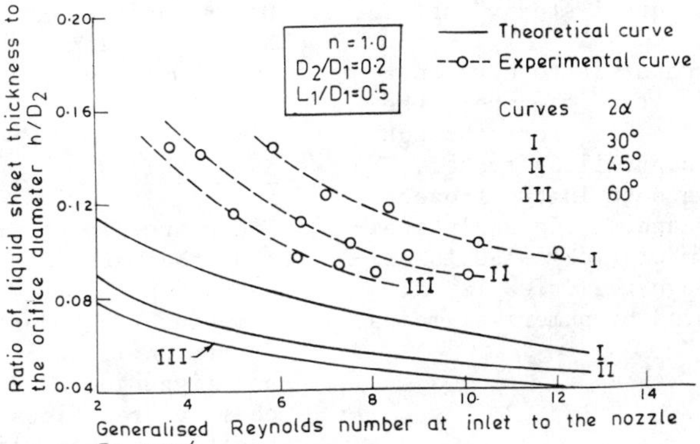

FIG 10 VARIATIONS OF h/D_2 WITH Re_{G_i} AND 2α

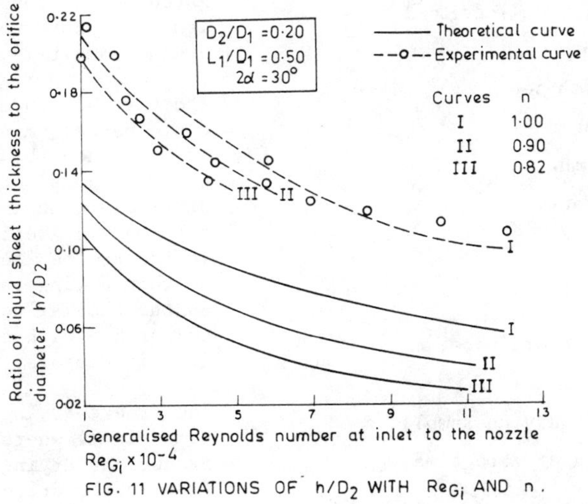

FIG. 11 VARIATIONS OF h/D_2 WITH Re_{G_i} AND n.

MODELING OF SIMULTANEOUS HEAT AND MASS TRANSFER IN FREEZE DRYING

B. Kısakürek

H. Çeliker

Department of Chemical Engineering
Middle East Technical University
Ankara, Turkey

ABSTRACT

Among the dehydration methods applicable to many food products, freeze drying is an ideal method for the maintenance of the desirable functional and structural characteristics during and after the drying process. Freeze drying is a very slow pro cess so that the transport equations presented in this paper are considered as "quasi-steady" states.

The primary purpose of this paper is to develop a retreating ice front model for the case where the removal of water vapor from the ice front through the dried layer is the rate controlling factor. The differential energy equations both in the frozen and dried layers are written and solved analytically. An expression in the form of Fick's law is utilized for the mass flux through the dried layer. The drying times are calculated by numerical means.

NOMENCLATURE

C_p = specific heat, cal/g -^oK

D = effective internal diffusion coefficient, m^2/sec

ΔH = heat of sublimation, cal/g

k = thermal conductivity, cal/sec-m-^oK

L = sample thickness, m

N = mass flow rate, g /m^2-sec

P = total pressure, mmHg

$\overline{P}$ = partial pressure, mmHg

R = universal gas constant, $3.462 * 10^{-3}$ m^3-mmHg/g -^oK

t = time, sec

T = temperature, ^oK

x = freeze drying model coordinate, m

X = interface position, m

W_s = weight of the freeze drying sample, g

W_{so} = weight of the sample at zero time, g

Subscripts

d = dried layer

Subscripts

d = dried layer
f = frozen layer
i = ice
L = position at frozen back face
0 = position at dried surface
X = position at interface
w = water vapor

INTRODUCTION

The principle of freeze-drying process is such that the material to be dried is frozen and the water vapor is removed by sublimation at low temperatures under vacuum. Keeping the material frozen until it is dry usually prevents shrinkage and migration of dissolved materials and inhibits chemical reactions to a very large extend. In addition, since moisture is removed at low temperatures the loss of flavor constituents is minimized. The dried product is porous and is rehydrated easily.

The mode of heat transfer that is to be investigated is the conduction heat transfer to the ice front both through the dried and frozen layers at the same time.

The thermal conductivities of the dried layers could be determined experimentally in the same way as it is done by Lusk, Karel and Goldblith (6) ussuming quasi steady-state conditions. They have performed experiments by making use of the radiation mode of heat transfer only. Heat is transferred to the ice front through the dried layer so that drying rates were heat transfer controlled as the thermal conductivities of dried layers are comparable with the thermal conductivities of many good insulators.

The freeze-drying process is assumed to be one-dimensional where the ice front receds into the product as drying occurs. Practically the vapor flow in the dried layer is in the transition region between Knudsen and Fickian diffusion. In this study, an effective diffusivity is used which eliminates the use of different momentum equations for different flow regimes. In this model the only limiting assumption is the fact that the diffusional flow is considered as the predominating flow regime so that the hydrodynamic flow is insignificant. On the other hand, heat transfer both through the dried and frozen layers and varying interface temperature and pressures are considered to increase the generality of the results.

EXPERIMENTAL

The freeze drier that is used is produced by the Virtis company. It mainly consists of two parts which are connected to each other.

1. Tray Drying Chamber,
2. Freeze Mobile Cabinet.

The drying medium of the tray drying chamber is cylindrical in shape and contains three shelves. There are electrically heated coils equipped under each shelf. There is a shelf heat control dial for each shelf and the temperature of each shelf could be adjusted to the desired temperature. The materials to be freeze dried are placed on the shelves in suitable containers. The heat is transferred to the sample both by conduction through the container of sample and also by radiation from the electrically heated coils of the upper shelf. During experiments the temperatures are measured by copper constantan thermocouples. One thermocouple is placed in the geometric center of the sample to measure the center temperature. Another thermocouple is attached to the drying surface of the sample to measure the surface temperature.

The Freeze Mobile Cabinet consists of four main parts:

i. Refrigeration - It is used to provide the low

Item	Description
1	Top Shelf Heat Control
2	Center Shelf Heat Control
3	Bottom Shelf Heat Control
4	Shelf Temperature
5	Refrigeration
6	Switch
7	Shelf Heat
8	Shelf (3)
9	Tray (3)
10	Drying Chamber
11	Door Lock
12	Lucite Door
13	Direct Heat/Control Switch
14	Thermistor Probe
15	Thermistor Jack
A	McLEOD GAUGE
B	VACUUM RELEASE-DEFROST
C	VACUUM
D	CONDENSER
E	DEFROST
F	CONDENSER TEMPERATURE

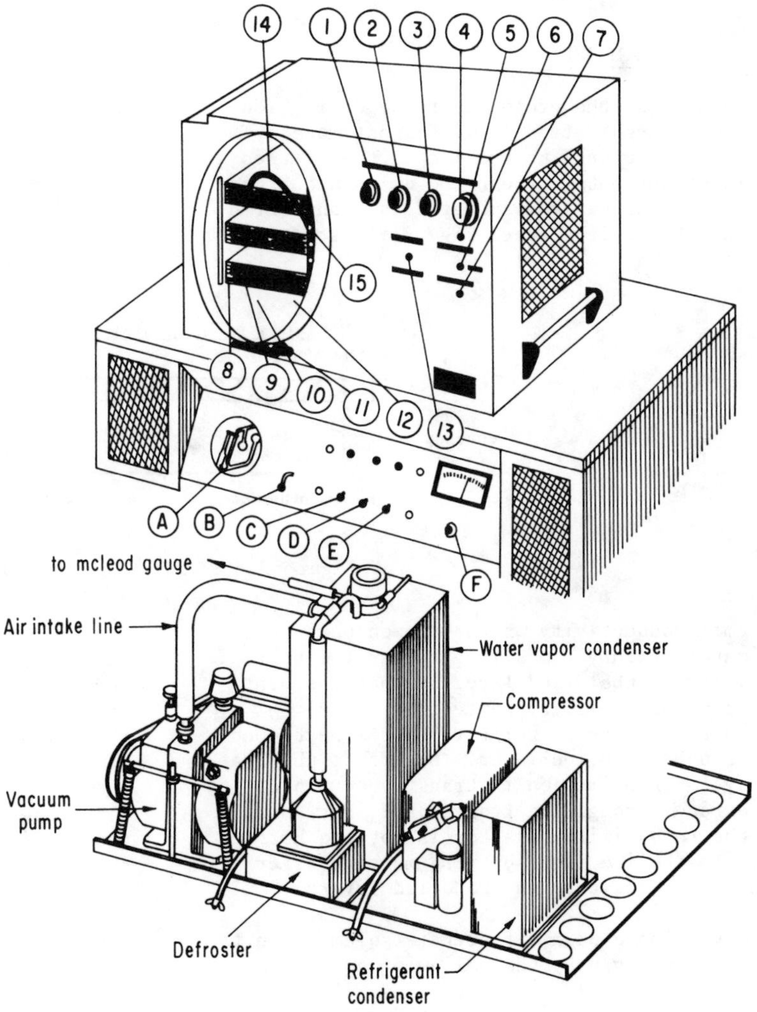

Figure 1. EXPERIMENTAL SET-UP.

temperatures required in the condenser. Condenser temperatures down to -55°C are provided by a 1/2 horsepower compressor. There is also a refrigerant condenser which is cooled by means of forced air. As refrigerant freon is employed.

ii. Condenser - There are two main connections to the condenser. The first one is to the drying chamber tube which has been inserted into the rubber vacuum hose on top of the condenser. The second connection to the condenser at the top has three branches. The first to the Mc Leod vacuum gauge, the second to the vacuum pump and the third to the defrost system.

iii. Vacuum Pump - It provides the vacuum in the drying chamber.

iv. Defrost system - It is used to melt the ice accumulated in the condenser. For this a 1/2 HP blower takes in ambient room air, adds the heat of its own drive motor and discharges it into the condenser.

THE MATHEMATICAL MODEL

In general terms, the proposed freeze drying model is the widely used retreating ice-front model. The heat necessary to sublime the ice is transferred to the ice-front both through the dried and frozen layers. The water vapor evaporated is transferred out through the dried layer. Fig. 2 illustrates the process.

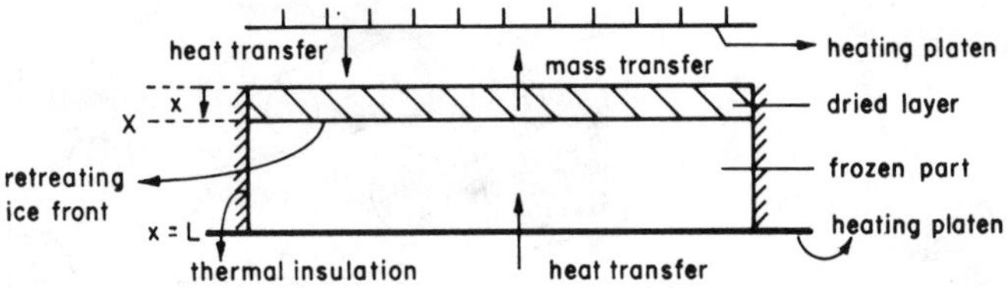

Figure 2. FREEZE DRYING MODEL.

The thermal conductivity of the frozen part is approximately twenty times larger than the thermal conductivity of the dried layer so that the heat through the non-subliming surface is conducted faster to the ice-front relative to the heat conducted through the dried layer. Therefore, the heat necessary for sublimation is transferred rapidly to the interface through the frozen part. For this reason the rate limiting factor is not the heat transfer to the ice front but the mass transfer from the ice front through the dried layer.

The energy balance on a differential element in the dried layer is given by the following equation.

$$\frac{d^2T_d}{dx^2} - \frac{C_p\rho_i\sigma}{k_d}\frac{dX}{dt}\frac{dT_d}{dx} = 0 \qquad (1)$$

where T_d is the temperature in the dried layer and X is the interface position.

The second term on the left hand side in Eqn.(1) is due to the mass transfer occuring in the dried layer which causes the temperature distribution to be non-linear.

The boundary conditions are as such,

At $x=0$; $T_d=T_o$

At $x=X$; $T_d=T_X$

In Eqn.(1) it is assumed that the solid matrix of the layer and the vapor phase in the capillaries of the solid matrix have the same temperature at any given position. The first and second terms represent the conductive and convective heat transfer respectively.

The solution of this differential equation can be obtained as

$$\frac{T_d-T_o}{T_X-T_o} = \frac{1-\exp(-\beta x)}{1-\exp(-\beta X)} \qquad (2.1)$$

where the coefficient of the exponential term, β, is defined as,

$$\beta = \frac{\rho_i\sigma C_p}{k_d}\frac{dX}{dt} \qquad (2.2)$$

As heat is conducted both through the dried and frozen layers, T_X, the interface temperature varies during drying. Since the interface moves very slowly, it is assumed that Eqn.(2.1) adequately represents the temperature distribution in the dried region at any instant during the drying process.

A differential energy balance in the frozen layer results in Eqn.(3).

$$\frac{d^2T_f}{dx^2} = 0 \qquad (3)$$

with the following boundary conditions.

At $x=L$; $T_f=T_L$

At $x=X$; $T_f=T_X$

The solution for this system is obtained as,

$$\frac{T_f-T_L}{T_X-T_L} = \frac{x-L}{X-L} \quad (4)$$

where L is the slab thickness and X is the interface position.

As it can easily be understood from Eqn.(4), since there is no mass transfer occuring in the frozen region, the temperature distribution is linear.

An energy balance would be in the form as given in Eqn.(5) if we consider the dried region as the system to be investigated.

$$N_w H = k_f \frac{dT_f}{dx}\bigg|_{x=X^+} - k_d \frac{dT_d}{dx}\bigg|_{x=0^+} - C_p N_w (T_o - T_X) \quad (5)$$

where N_w=flux of water vapor, H=heat of sublimation and C_p=specific heat.

The first two terms on the right-hand side of Eqn. (5) are differential terms. The first term could be obtained by differentiating Eqn.(4) and then evaluating it at $x=X^+$. The second term is obtained by differentiating Eqn.(2) and then evaluating it at $x=0^+$.

When Eqns.(2.1) and (4) are differentiated and the results obtained are substituted into Eqn.(5), we may obtain,

$$N_w \Delta H = \frac{k_f(T_X-T_L)}{X-L} - \frac{k_d(T_X-T_o)}{1-\exp(-\beta X)} - C_p N_w (T_o-T_X) \quad (6)$$

The argument value of the exponential term in Eqn. (6) is much less than one, thus it can be expanded in a Taylor series with the second- and higher-order terms being neglected. Then,

$$\exp(-\beta X) = 1-\beta X \quad (7)$$

can be obtained.

Upon combination of Eqn.(6) with Eqn.(7), Eqn.(8) results.

$$N_w = \frac{\dfrac{k_f(T_X-T_L)}{X-L} - \dfrac{k_d(T_X-T_o)}{X}}{\Delta H + C_p(T_o-T_X)} \quad (8)$$

On the other hand, the expression for the mass balance at the interface could be expressed in terms of the time rate of change of interface position as follows,

$$N_w = \rho_i \sigma \frac{dX}{dt} \quad (9)$$

Now considering that the flow regime is somewhere between the Knudsen flow and Fickian flow diffusion regimes, the mass flux expression can be written as,

$$N_w = \frac{D}{RT} \frac{(\bar{P}_{Xw} - \bar{P}_{ow})}{X} \quad (10)$$

where D is an effective internal diffusion coefficient which is a function of both the Knudsen diffusion and bulk diffusion coefficients. It is also assumed to be independent of position in the dried layer, i.e., dD/dx=0 .

The partial pressure of water vapor at the interface, $\bar{P}_{Xw}$, is a function of the interface temperature at any instant and it could be obtained from the data of the paper which is present by Dyer, Carpenter and Sunderland (2).

Equating Eqns.(8) and (9) with each other, the following equation is obtained.

$$\frac{dX}{dt} = \frac{\dfrac{k_f(T_X-T_L)}{X-L} - \dfrac{k_d(T_o-T_X)}{X}}{\rho_i \sigma \left| \Delta H + C_p(T_o-T_X) \right|} \quad (11)$$

with the initial condition,

At= t=0; X=0

Comparing Eqns.(8) and (10), we get,

$$\frac{D}{RT} \frac{(\bar{P}_{Xw}-\bar{P}_{ow})}{X} = \frac{\dfrac{k_f(T_X-T_L)}{X-L} - \dfrac{k_d(T_o-T_X)}{X}}{\rho_i \sigma (\Delta H + C_p(T_o-T_X))} \quad (12)$$

According to Ma and Peltre (7), the effective internal diffusion coefficient, D, in cm^2/sec, is given by the following expression at 293^oK.

$$D=78.5/(3.4+P, \text{mmHg}) \quad (13)$$

The temperature dependence of the effective internal diffusion coefficient could be neglected as the maximum variation of the effective internal diffusion coefficient due to the temperature change compares to be of the same nagnitude with the minimum uncertainity in the value of it. For this reason, this equation is used in estimating the effective internal diffusion coefficient.

In Eqn.(12) different values are assigned for X, the interface position, and T_X is solved implicitly. In this way T_X as a function of interface position is obtained. The best fitting curve for $T_X=f(X)$ is obtained by non-linear polynomial regression. It is then substituted into Eqn.(11) and upon numerical integration, drying times are calculated.

It is possible to check the estimated D which is obtained by Eqn.(13) with the experimental results. For this, the interface position is expressed in terms of the fraction of water remaining.

$$\frac{L-X}{L} = \frac{W_s - W_{se}}{W_{so} - W_{se}} \quad (14)$$

Differentiating Eqn.(14), Eqn.(15) is obtained.

$$\frac{dX}{dt} = -\frac{L}{(W_{so}-W_{se})}\frac{dW_s}{dt} \tag{15}$$

Substituting Eqn.(15) into Eqn.(9),

$$N_w = -\rho_i\sigma\frac{L}{(W_{so}-W_{se})}\frac{dW_s}{dt} \tag{16}$$

Equating Eqns.(16) and (10) the following expression is obtained.

$$W_{so}-W_s = \frac{D(W_{so}-W_{se})^2(\bar{P}_{Xw}-\bar{P}_{ow})}{RTL^2\rho_i\sigma\left(-\frac{dW_s}{dt}\right)} \tag{17}$$

If $(W_{so}-W_s)$ versus $\dfrac{\bar{P}_{Xw}-\bar{P}_{ow}}{\left(-\frac{dW_s}{dt}\right)}$ is plotted, a straight line is obtained from the slope of which the effective internal diffusion coefficient could be calculated and checked with the estimated value.

W_{so}, W_s and dW_s/dt are experimentally measurable quantities. $\bar{P}_{Xw}$ and $\bar{P}_{ow}$ would be calculated by means of the experimentally measured center and surface temperatures. Obtaining $\bar{P}_{Xw}$ from the measurable center temperature requires the assumption that the temperature distribution is uniform all over the frozen layer.

EXPERIMENTAL RESULTS AND DISCUSSION

A sample plot is given in Fig.5 for the experimental determination of the effective internal diffusion coefficient. The time rate of change of the weight of freeze drying sample, dW_s/dt, is obtained by evaluating the slopes at several points on the curve of Fig.4. The surface and center temperatures are obtained experimentally and are plotted in Fig.3 with respect to time.

The specifications for the sample which is freeze dried to get Figs.(3), (4) and (5) are as such:

Sample	= Beef meat
Thickness of Slab, L	= 1.7 cm
Surface area	= 10*11 cm
Initial weight of sample, W_{so}	= 242 gr
Grain orientation	= Vertical
Porosity,	= 0.75
Density of ice, $_i$	= $0.92*10^6$ gr/m^3
% moisture	= 72.7 %
Pressure in the drying chamber	= 0.8 mmHg

The value of the effective internal diffusion coefficient which is obtained from the slope of Fig.5 is $1.765*10^{-4}$ m^2/sec. On the other hand, choosing the average pressure within the dried layer during drying to be about 3 mmHg, it is estimated from Eqn.(13) that the effective internal diffusion coefficient is $1.226*10^{-3}$ m^2/sec.

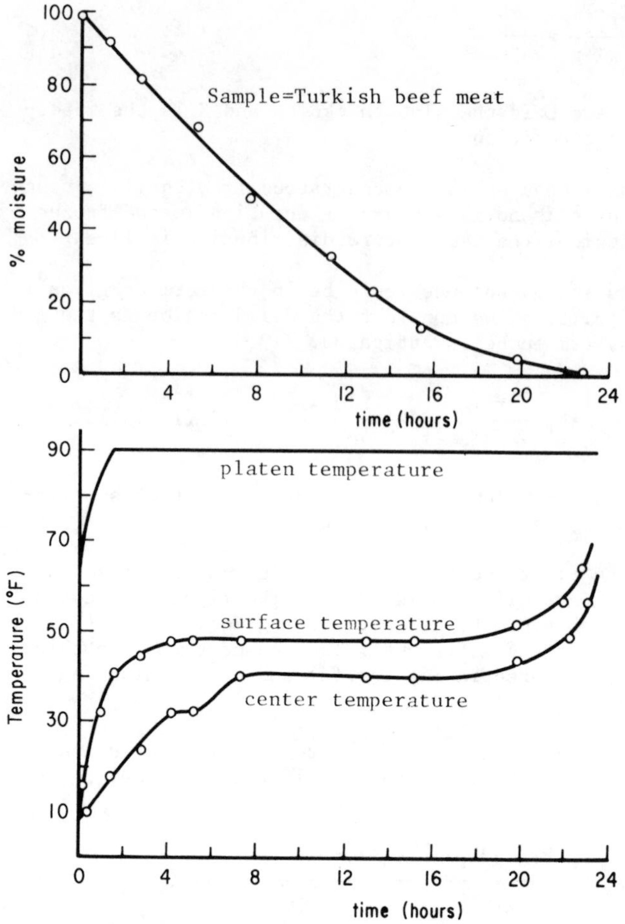

Figure 3. % MOISTURE AND PLATEN, SURFACE AND CENTER TEMPERATURES WITH RESPECT TO TIME.

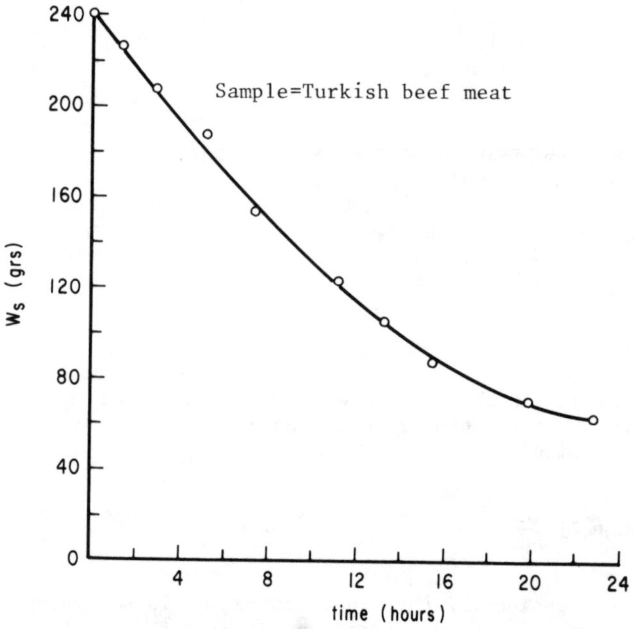

Figure 4. CHANGE OF SAMPLE WEIGHT WITH RESPECT TO TIME

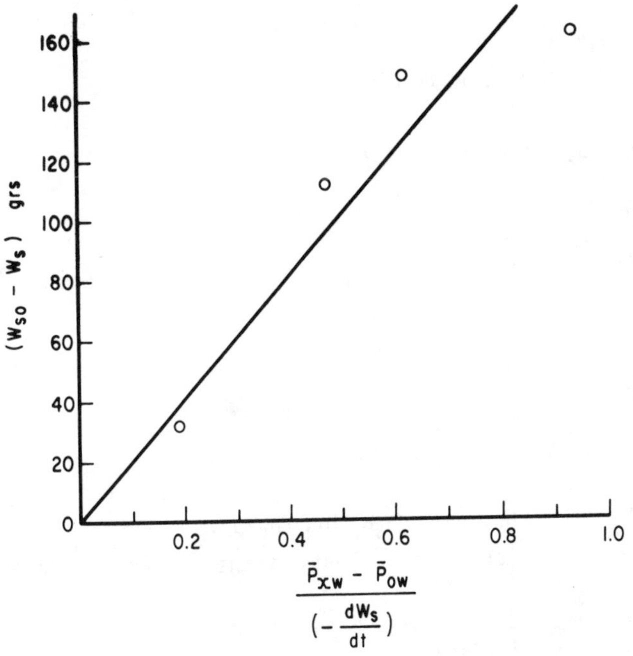

Figure 5. CALCULATION OF EFFECTIVE INTERNAL DIFFUSION COEFFICIENT.

The primary purpose of this work is to develop a uniformly retreating ice front model for freeze drying. In the mathematical model developed an internal diffusion coefficient, D, is dėfined which is infact both the functions of temperature and pressure. For this reason the temperature and vapor pressure of the frozen layer should be uniform at all points. Otherwise, D would be non-uniform which would result in a non-uniform retreat of the ice-front. So this means that great care should be taken during experimentation to achieve the necessary conditions for the uniform retreat of the ice-front so that the experimental and theoritical results will be simulated better.

In this study the variation of interface temperature with interface position is assumed to be an expression of order higher than one. This relation has to be determined by non-linear mathematical regression which is a more realistic approach for the solution of the problem.

It is required to estimate D and for this reason the expression which is proposed by Ma and Peltre (Eqn.13) is used. It is assumed that D does not change with respect to the distance in the dried layer as it is very difficult to get D in terms of x and it is beyond the scope of this work. Although the above assumption is used it is pointed out by Ma and Peltre that the variation of distance in the dried layer has considerable effect on D. For this reason as the best thing to do it is advised to get comprehensive experimental data and try to simulate the effect.

An experimental and easier way of determining the numerical value of thermal conductivity for the model is the one that is done by Lusk, Karel and Goldblith (6) and the errors that are associated with this type of determination are discussed in their paper.

It is not practical to have thermocouples situated very closely along the thickness of slabs to measure the temperatures at certain X values. Therefore when D is to be determined experimentally, $\bar{P}_{Xw}$ is calculated always at the temperature that is indicated by the thermocouple which would be situated at the geometrical center of the slab. $\bar{P}_{ow}$ is evaluated at the temperature indicated by the thermocouple that is situated at the drying surface of the slab.

The mathematical model proposed in this study will be tested with the experimental results which are obtained for beef meat and several marine products (fish). We are just at the stage of application of the mathematical model. The comparison of the mathematical and experimental results will be published in the future.

REFERENCES

1. Burke, R.F. and R.V.Decerau, "Advances in Food Research, Academic Press, N.Y.", V13, pp 1-88, 1964.
2. Dyer, D.F., D.K.Carpenter and J.E.Sunderland, J.Food Sci., V31, pp 196-201, 1966.
3. Dyer, D.F. and J.E.Sunderland, Journal of Heat Transfer, Trans ASME, November, pp 379-384, 1968.
4. Harper, J.C. and A.L.Tappel, "Advances in Food Research, Academic Press, N.Y.", V7, pp 171-234, 1957.
5. Hill, J.E. and J.E.Sunderland, Int.J.Heat Mass Transfer, V14, pp 625-637, 1970.
6. Lusk, G., M.Karel and S.A.Goldblith, Food Tech., October, pp 121-124, 1964.
7. Ma, Y.H. and P.R.Peltre, AIChEJ,No.2, V 21, pp 335-350, 1975.
8. Mujumdar, A.S., Y.K.Li and V.Jog, Drying '80, Developments in Drying, Edit. A.S. Mujumdar, V1, pp 233-241, 1980.
9. Sandall, O.C., C.J.King and C.R.Wilke, AIChE J, No.3, V13, pp 428-437, 1967.

WOOD PARTICLE DRYING
A MATHEMATICAL MODEL WITH EXPERIMENTAL EVALUATION

FERHAN KAYIHAN and MARK A. STANISH

Weyerhaeuser Technology Center
Tacoma WA 98477, USA

ABSTRACT

A two-dimensional wood drying model was developed using a three-phase diffusion and local equilibrium concept. The three moisture phases which are identified in the model are free water, bound water, and vapor. Partial differential equations of the model were transformed into ordinary differential equations using a discrete space compartment and continuous time approach. Equilibrium relations were represented through vapor pressure and sorption isotherm correlations. The differential and the nonlinear algebraic equations of the model were solved through a sequential iterative technique.

Experimental drying tests were run on sawdust and flake-like particles and the data were used to evaluate the model's performance. Simulations of sawdust drying behavior for different particle sizes, temperatures, and initial moisture contents were in close agreement with the measured data. Flake drying simulations were less effective at matching experimental results and consequently revealed important limitations of the current model when extended to larger or thicker materials. Needed improvements are identified in the modeling of external surface mass transfer and of internal free water migration.

NOMENCLATURE

Bi	Heat transfer Biot number
c_p	Specific heat of wet wood (J/kg K)
D_b	Bound water diffusivity (m²/sec)
D_{bf}	Bound water + free water (effective moisture) diffusivity (m²/sec)
$D_{H_2O,AIR}$	Bulk vapor diffusivity in air (m²/sec)
D_f	Free water diffusivity (m²/sec)
D_{rp}	Effective vapor random pore diffusivity (m²/sec)
D_v	Effective vapor diffusivity
d	Attenuation factor for vapor diffusivity
H	Humidity fraction (p_v/p_v^s)
h	Convective heat transfer coefficient (J/s m² K)
k_c	Convective mass transfer coefficient (m/sec)
M	Molecular weight
M	Number of discrete cuts along y-axis
N	Number of discrete cuts along x-axis
Nu	Nusselt number
p_v	Partial pressure of vapor (atm = 101.33 kPa)
p_v^s	Saturation partial pressure (atm)
Pr	Prandtl number
R	Gas constant
R	Local evaporation rate (1/sec)
Re	Reynolds number
Sc	Schmidt number
Sh	Sherwood number
T	Temperature (K)
u_b	Bound water moisture content (kg bound water/kg dry wood)
u_{be}	Equilibrium bound water content (kg bound water/kg dry wood)
u_{bf}	Combined bound and free water contents (kg water/kg dry wood)
u_f	Free water moisture content (kg free water/kg dry wood)
u_v	Vapor moisture content (kg vapor/kg dry wood)
v	Specific volume (m³/kg dry wood)
x	Distance along x-direction (m)
y	Distance along y-direction (m)
α_x	Wood thermal diffusivity in x-direction (m²/sec)

α_y Wood thermal diffusivity in y-direction (m^2/sec)

β Fraction of bound water

ε Void fraction of wet wood

ε_o Void fraction of dry wood

λ Heat of evaporation (J/kg)

ρ Density (kg/m^3)

INTRODUCTION

Drying theory has been examined by numerous researchers for a variety of reasons including process understanding, equipment design, and process control. Our goal in this study is to obtain a mechanistic wood drying model derived through fundamental relations which can be used for both interpretive (of data) and predictive applications.

Literature on drying models is extensive and will not be summarized here. However, the reader is referred to texts such as Luikov [1], Keey [2], Siau [3], and Skaar [4] for basic understanding of drying in porous bodies and for moisture behavior in wood. Recent developments in this area including modeling work can be found in proceedings and books edited by Mujumdar [5,6,7,8].

We interpret the drying process as the simultaneous phenomena of heat transfer, mass transfer, and local thermodynamic equilibrium. Because of the hygroscopic nature of wood we identify three moisture phases: free water, bound water, and vapor (mixed with air). Such an approach was previously used by Kayihan et al. [9] to describe the press drying of fiber and particle boards. During drying the heat and mass transfer processes and phase changes occur at specific rates depending on the size of the specimen, on the external heat and mass transfer driving forces, and on the changing local moisture content and porous structure. We use the continuum approach through diffusivities in describing the transport of moisture and heat in wood. Thermodynamic equilibrium between phases is then responsible for the interaction of the transport rates. Accordingly, phase changes can occur at any location and affect the heat and mass transfer potentials while continuously satisfying the equilibrium requirements. Thus, instead of assuming sequential drying rate periods (like constant, falling, etc.) or considering evaporation only at the surface or at a receding front we try to predict all expected phenomena from a unified model.

Published experimental data on wood particle drying are extremely rare. Malte et al. [10] report rate data for small Douglas-fir particles convectively dried at moderate temperatures below about 750°K. We needed data which would help us evaluate specific aspects of the model; therefore, carefully controlled, well characterized drying tests were designed to provide accurate data and to minimize the number of unknown operating parameters. Important simulation variables such as particle density and external heat and mass transfer coefficients were measured by independent means. The experimental program consisted of two major thrusts. In one set of tests, the particles were artificially made to conform to the geometry and boundary conditions imposed on the model equations. Since these physical constraints were satisfied, we were able to focus on the validity of the mechanistic assumptions built into the model. The other primary effort was to determine how well the model can simulate the behavior of typical materials such as sawdust and flakes under a variety of drying conditions.

In the following sections, we will give a brief account of the basic assumptions, develop the model equations and summarize the numerical approach. Then, the experimental testing and the results used to evaluate and analyze the model will be described. Additional details of the numerical problem will be discussed in the appendices.

DEVELOPMENT OF THE MODEL

Assumptions

1. Particles are two dimensional.

2. Wood is represented in terms of a rigid hygroscopic porous solid structure.

3. Moisture in wood occurs in three possible states: bound (or adsorbed) water phase, free (or condensed) liquid phase, and vapor phase.

4. There is local thermal and phase equilibrium at every point in the particle. Bound water and vapor phase equilibrium obeys sorption isotherms and vapor stays at saturated conditions while free water is present.

5. Drying occurs due to surface and internal evaporations supplemented by diffusion of moisture in liquid (free and bound) and vapor phases. Complete drying phenomena can be described by simultaneous heat and mass transfer equations coupled with phase change requirements according to the equilibrium relations.

6. Transport properties are direction (x and y) dependent and they can be locally computed as long as correlating relations are available. For practical reasons in the implementation of the model it was assumed that the properties change with average moisture and temperature while staying uniform across the particle at any given time. Density and apparent void fraction are point functions depending on local moisture.

7. There is no swelling or shrinkage with moisture. Adsorbed water does not occupy any accessible volume. Free water and vapor share the available void space.

8. Particle externals in both directions are exposed to the same ambient conditions. Heat and mass transfer boundary conditions are characterized by convective transfer coefficients.

9. Gas permeability is high and as a result internal pressures do not exceed the ambient atmospheric value. The excess of evaporated internal vapor instantly leaves the system to keep the total pressure constant; this form of migration accounts for bulk transport. Vapor and air in the void space are mixed and a proportionate amount of air leaves the system with vapor during bulk transport.

Geometry

We are considering an idealized two-dimensional cross section of a long particle with all four sides affected by the same ambient conditions. Therefore, there is a geometric symmetry that we can exploit in our modeling. As shown in Figure 1 we consider only the lower right quadrant of the particle cross section.

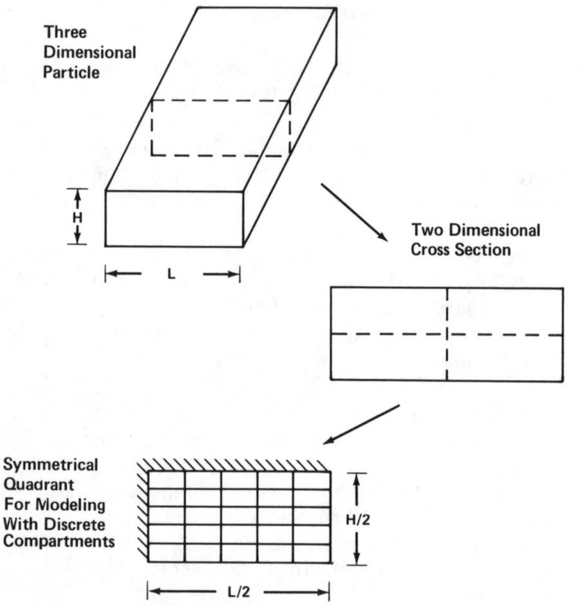

Figure 1.
Geometric aspects of two dimensional particles and the spacial approximation through discrete compartments

Characterization of Solid Structure

We assume that the following interpretations adequately describe the physical structure of wood for drying purposes.

Overall dry density of wood:

$$\rho_{dry} = \rho_o(1-\varepsilon_o) \tag{1}$$

where ρ_o is the cell wall density and ε_o is the dry porosity.

Local wet density:

$$\rho = \rho_o(1 - \varepsilon_o)\ (1 + u_{bf}) \tag{2}$$

where $u_{bf} = u_b + u_f$ (kg bound + free water/kg dry wood)

Local void fraction occupied by free water:

$$\rho_o(1-\varepsilon_o)\ u_f/\rho_w \tag{3}$$

where $\rho_w = \rho_{water}$

Local void fraction occupied by vapor (apparent porosity):

$$\varepsilon = \varepsilon_o - (\rho_o/\rho_w)\ (1-\varepsilon_o)\ u_f \tag{4}$$

Note that the parameters ρ_o, ε_o and the initial moisture content u_{bf} (t=0) must be consistent such that ε is always positive. Thus, there is a maximum u_{bf} for any given ρ_o and ε_o. Interpreted differently, we get an upper bound on initial moisture ($u_{vapor} \simeq 0$) as:

$$(u_{bf})_{max} = u_b + \frac{\varepsilon_o}{1-\varepsilon_o}\ (\rho_w/\rho_o) \tag{5}$$

where u_{bf} represents combined bound and free moistures in dimensionless units of (kg water/kg dry wood). With $u_b \simeq 0.3$ and $\rho_o \simeq 1500$ kg/m³

$$\text{we get}\quad (u_{bf})_{max} \simeq 0.3 + \frac{2}{3}\ \frac{\varepsilon_o}{1-\varepsilon_o} \tag{6}$$

Equation (6) becomes an important relation to be satisfied in specifying structural parameters and initial moisture content for the model. Figure 2 shows a graph of Equation (6).

Partial pressure of vapor in the void space and vapor moisture content are related through the ideal gas law as:

$$u_v = 219.4\ \frac{p_v}{T}\ v \qquad \text{kg vapor/kg dry wood} \tag{7}$$

where $v = \varepsilon/\rho_o(1-\varepsilon_o)$

$$\text{or}\quad v = \varepsilon_o/\rho_o(1-\varepsilon_o) - u_f/\rho_w \tag{8}$$

is the void volume per unit dry mass (m^3/kg dry wood) of wet wood.

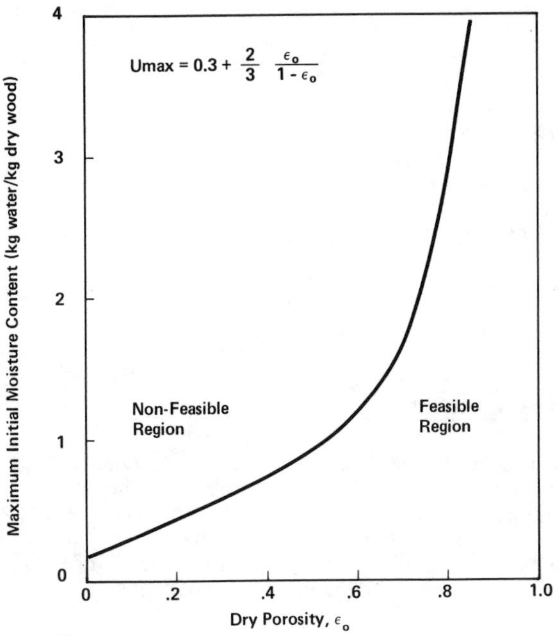

Figure 2.
Maximum possible moisture level for given dry porosity according to assumed solid structure

Diffusion Equations

Thermal Diffusion. A local two-dimensional transient enthalpy balance with possible phase change gives:

$$\frac{\partial T}{\partial t} = \alpha_x \frac{\partial^2 T}{\partial x^2} + \alpha_y \frac{\partial^2 T}{\partial y^2} - R \frac{\lambda}{C_p} \qquad (9)$$

Here α_x and α_y are the directional thermal diffusivities in (m^2/s). R represents the local rate of evaporation, the total phase change from bound and free water into vapor. As it will become apparent $R = R\ (T, u_v, u_{bf})$ is the link between the dependent variables of the model. Essentially it represents the simultaneous effects of heat and mass transfer through the phase equilibrium relations.

Liquid Diffusion. Bound and free water are assumed to migrate due to a potential difference defined in terms of $u_{bf} = u_b + u_f$. A local transient material balance gives:

$$\frac{\partial u_{bf}}{\partial t} = D_{bfx} \frac{\partial^2 u_{bf}}{\partial x^2} + D_{bfy} \frac{\partial^2 u_{bf}}{\partial y^2} - R \qquad (10)$$

It is expected that bound water will diffuse slower than free water and to reflect that an effective total moisture diffusivity is approximated by:

$$D_{bf} \simeq \beta D_b + (1-\beta) D_f \qquad (11)$$

where $\beta = u_b/u_{bf}$ is the fraction of bound water which varies with space and time. This approach was preferred to the alternative of writing individual diffusion equations for each of the free water and the bound water phases in the interest of saving computation time during numerical solution.

Vapor Diffusion. The driving force for vapor diffusion is the partial pressure gradient. In terms of p_v in (atm = 101.33 kPa) we get:

$$\rho_o(1-\varepsilon_o) \frac{\partial u_v}{\partial t} = \frac{\partial}{\partial x} \left(D_{vx} \frac{219.4}{T} \frac{\partial p_v}{\partial x}\right) + \frac{\partial}{\partial y} \left(D_{vy} \frac{219.4}{T} \frac{\partial p_v}{\partial y}\right) + R\ \rho_o(1-\varepsilon_o) \qquad (12)$$

Assuming quasi steady state with respect to temperature and using Eqs. (7) and (8) we express the diffusion equation in terms of the dimensionless vapor moisture content u_v (kg vapor/kg dry wood).

$$\varepsilon \frac{\partial u_v}{\partial t} = D_{vx} \frac{\partial^2 u_v}{\partial x^2} + D_{vy} \frac{\partial^2 u_v}{\partial y^2} + \varepsilon R \qquad (13)$$

or

$$\frac{\partial u_v}{\partial t} = \frac{1}{\varepsilon} \left(D_{vx} \frac{\partial^2 u_v}{\partial x^2} + D_{vy} \frac{\partial^2 u_v}{\partial y^2}\right) + R \qquad (14)$$

In the implementation of the model we assumed that the effective vapor diffusivity will depend on temperature and porosity through bulk diffusivity and random pore effects with an adjustable correction factor to account for the restriction of closed pores in wood. Thus, we estimate the effective vapor diffusivities as:

$$D_{vx} = d_1 D_{rp}\ , \qquad 0 < d_1 < 1 \qquad (15a)$$

and

$$D_{vy} = d_2 D_{rp}\ , \qquad 0 < d_2 < 1 \qquad (15b)$$

where $D_{rp} \simeq D_{H_2O,AIR}\ \varepsilon^2$ is the effective random pore diffusivity.

Now Equation (14) becomes:

$$\frac{\partial u_v}{\partial t} = D_{H_2O,AIR}\ \varepsilon \left(d_1 \frac{\partial^2 u_v}{\partial x^2} + d_2 \frac{\partial^2 u_v}{\partial y^2}\right) + R \qquad (16)$$

Equilibrium Relations

We identify three moisture phases in wood during drying: (1) condensed liquid phase, (2) bound water phase, and (3) vapor phase. Vapor and liquid phases are at equilibrium at the saturation pressure determined by local temperature. This applies to all locations of the particle (surface and internals) as long as liquid water is present. For the correlating function we used Yaws' [11] suggestion as:

$$p_v^s = \frac{1.0133 \times 10^5}{760} \; 10^{f(T)} \quad \text{Pa} \tag{17}$$

where

$$f(T) = 16.373 - 2818.6/T - 1.6908 \log_{10} T - 5.7546\text{x}10^{-3} T + 4.0073\text{x}10^{-6} T^2 \tag{18}$$

with T in °K.

In general the equilibrium bound water content is related to temperature and relative humidity. Therefore, for our purposes local vapor and bound water phases must be consistent with sorption isotherms at local temperature to satisfy the equilibrium requirements. This, of course, means that bound water equilibrium is a function of temperature at 100% saturation when free liquid is present but that it depends on both temperature and local relative humidity when there is no free water. We used the sorption isotherm correlation as suggested by Skaar [4]. Comparisons with actual data as reported by Rosen [12] show that the correlation holds for up to 140°C. The equilibrium bound water content is given by:

$$u_{be} = \frac{18}{W} \; a_2 H \left(\frac{1}{f_1} + \frac{a_1}{f_2} \right) \tag{19a}$$

where

$$W = 5.72\text{x}10^{-3} T_F^2 + 1.961\text{x}10^{-2} T_F + 216.9$$

$$a_1 = -1.547\text{x}10^{-4} T_F^2 + 3.642\text{x}10^{-2} T_F + 3.73$$

$$a_2 = -1.714\text{x}10^{-6} T_F^2 + 1.0533\text{x}10^{-3} T_F + 0.674$$

$$f_1 = 1 - a_2 H$$

$$f_2 = 1 + a_1 a_2 H$$

with u_{be} as the adsorbed equilibrium moisture (mass of water/mass of dry wood) and **H** as the relative humidity (fraction) = p_v/p_v^s. The correlation is in its original temperature units of degrees F.

For model implementation we also need the inverse function which predicts local relative humidity from equilibrium bound moisture and temperature. Algebraic manipulation of Equation (19a) gives:

$$H = \frac{1}{2} \left[z_1 + \left(z_1^2 + \frac{4}{a_1 a_2^2} \right)^{\frac{1}{2}} \right] \tag{19b}$$

where

$$z_1 = \frac{1-z_2}{a_2} - \frac{1+z_2}{a_1 a_2}$$

$$z_2 = \frac{18}{W u_{be}}$$

Boundary Conditions

The internal faces of the particle quadrant as we pictured in Figure 1 have reflective boundary conditions. There is no heat or mass transport through these surfaces. On the other hand, the boundary conditions for the exposed surfaces are described through convective conditions.

Thermal convection at the boundaries depends on the local temperature driving forces between the surface and ambient and on the convective transfer coefficient h (J/s m^2 K). Liquid and vapor diffusion equations have the convective boundary conditions defined in terms of the local partial pressure driving forces and the vapor mass transfer coefficient k_c (m/s) with the appropriate conversion (M/RT) to retain mass flux units. It is important to note that the surface water vapor partial pressure is determined through Equation (17) if surface wood has free water and that Equation (19b) must be used to compute the equilibrium vapor pressure if there is only bound water (i.e., $p_v = H \, p_v^s(T)$).

Physical Properties

Wood Handbook [13] suggests the following correlations for thermal conductivity of wet wood in (J/m s K).

$$k = (sg)\,(0.20 + 0.40\,u) + 0.024 \qquad u < 0.4 \tag{20a}$$

$$k = (sg)\,(0.20 + 0.55\,u) + 0.024 \qquad u > 0.4 \tag{20b}$$

where u (= $u_{bf} + u_v$) is the total moisture content in dry basis.

For convenience we chose:

$$k \simeq 1.5\,(1-\varepsilon_o)\,(0.20 + 0.50\,u) + 0.024 \tag{21}$$

as applicable for all moisture contents.

For specific heat of moist wood we used Siau's [3] suggestion as:

$$c_p = \frac{u + 0.324}{1 + u} (4184) \quad J/kg\ K \tag{22}$$

Water vapor bulk diffusivity in air is represented by a simple correlation in terms of absolute temperature (K) as proposed by Kanury [14].

$$D_{H_2O,AIR} \simeq 1.2 \times 10^{-9} \quad T^{1.75} \quad m^2/s \tag{23}$$

For heat of vaporization we used the simple temperature dependent correlation as suggested by Yaws [11]:

$$\lambda = 267\ (647 - T°K)^{0.38}\ kJ/kg \tag{24}$$

This expression holds when moisture content is above fiber saturation. To reflect the increased heat of vaporization of bound water we used the following approximation:

$$\lambda = 267(647 - T°K)^{0.38}\ [1 + 0.4\ (1 - \frac{u}{0.3})] \tag{25}$$

where $u < 0.3$ is the moisture content on dry basis.

NUMERICAL SOLUTION

Discrete-Space-Continuous-Time Approach

Three nonlinear partial differential equations, the equilibrium relations, the physical property correlations, and the convective boundary conditions constitute the model as described above. These equations were solved simultaneously through a numerical approach to simulate the drying process. We handled the numerical solution of the problem by converting the partial differential equations into ordinary differential equations. Assuming that the two-dimensional geometry of the particle can be represented by a discrete collection of uniform property compartments removes spacial variations from the equations. Thus, as shown in Figure 1, when we represent the particle segment in terms of NxM compartments we created NxM ODEs (ordinary differential equations) for each partial differential equation. This technique, in essence, uses the finite difference approach for the spatial variables but retains time as a continuous variable. The advantage is in the flexibility of using powerful ODE solvers which are commercially available rather than having to solve finite difference equations in time. This particular approach also guarantees numerical stability and accuracy. Then the approximation errors depend only on the assigned compartment sizes.

Algorithm

The equilibrium relations between the three phases keep the diffusion equations coupled. Therefore, the simultaneous solution of all the relations require that we continuously satisfy the algebraic equations of the phase equilibria as the differential equations are integrated. We devised an algorithm which handled algebraic and differential equations separately through small time increments while preserving the simultaneous integrity of the solution.

The basic steps of the algorithm can be summarized as:

Step 1. Initialize variables and constants. Set incremental integration time (1/50th of the thermal time constant of the particle was sufficiently small).

Step 2. Update physical properties.

Step 3. Integrate diffusion equations for each compartment through the incremental time.

Step 4. Solve equilibrium relations iteratively and correct individual phase moisture contents and temperatures at each compartment.

Step 5. Print results as necessary.

Step 6. Increment time. Go to Step 2 for more computations or stop if end of simulation.

The integrations of the ODEs required a stiff-solver with sparse matrix (banded-jacobian) capabilities. We used the LSODE package provided by Hindmarsh [15]. The development of the ordinary differential equations, the algebraic solution procedure of the equilibrium requirements and some other details of the numerical procedure used for this model will be described in Appendices A and B.

EXPERIMENTS

Materials and Equipment

Drying rates were experimentally determined for a variety of particle types and sizes. Douglas fir sawdust was separated by screening into two size fractions, 0.5-1.0 mm and 5-7 mm, and sufficient water was added to bring the moisture content to approximately 100% or 200% db. This material was then left to equilibrate for a minimum of three weeks before drying tests were performed. The particles in both size fractions were box-like in appearance. The dimensions of the 0.5-1.0 mm particles were in the approximate ratio 1:2:3 with the wood fibers oriented in the direction of the longest dimension. In the 5-7 mm particles, the dimension ratio was approximately 1:5:7 and the fibers were parallel to the shortest dimension.

The drying rates of flakes and of thin blocks of wood were also measured. The flakes were less than one millimeter thick and were from aspen with the grain oriented along the flake length. Blocks measuring approximately 3 x 15 x 20 mm on each side were cut from southern pine in specific orientations. Each was aligned such that the smallest dimension (thickness) lay in either the longitudinal or transverse direction relative to the wood grain. After cutting and trimming, the blocks were stored under water until tested.

The specimens were dried using the apparatus shown schematically in Figure 3. A Cahn 2000 electronic recording microbalance provided gravimetric information from which average moisture content data could be calculated. In tests with flakes and blocks, a single specimen was hung vertically on the balance hangdown wire; for sawdust, a number of particles (approximately 40 mg, dry weight) were distributed on a coarse steel mesh pan which was hung from the wire. In every case, the sample was dried with air at controlled temperature and flowrate. A Barnes model 12-8592 infrared thermometer calibrated for southern pine wood was used to measure the surface temperature of the flakes and blocks as they were dried. Sample weight and surface temperature and the temperature of the air immediately upstream of the sample were recorded at regular time intervals by an automated data acquisition device. At the conclusion of each test, the dry-basis moisture content for each reading was back-calculated using the final, bone-dry weight of the sample.

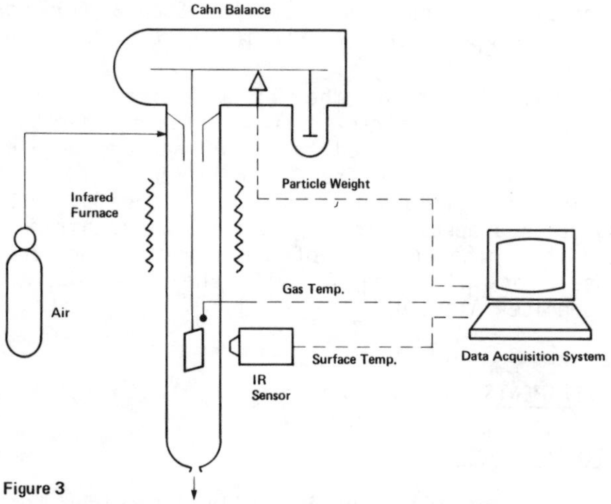

Figure 3
Particle drying unit

Convective Transport Parameters

Since external heat and mass transfer coefficients are required in the model simulation, an effort was made to characterize them for the conditions at which the drying tests were performed. Our approach was to measure the sublimation rate of naphthalene particles having sizes and shapes similar to those of the wood specimens studied. With the placement and orientation of these particles in the dryer apparatus replicated carefully, rates were measured at room temperature over a range of air flowrates (or relative velocities). Mass transfer coefficients were then calculated using the measured rate of weight loss and the total surface area of the particles.

Experimental mass transfer coefficients (in the form of Sherwood numbers) were correlated with the Reynolds and Schmidt numbers. For small sawdust-like particles, the functional form was assumed to follow that for spherical bodies, while for flakes and blocks, the form for flat plates was adopted [16]. Heat transfer coefficients were then available by analogy since for the conditions of this problem the Nusselt number is expected to follow an identical functional dependence on Reynolds and Prandtl numbers. The resulting correlating equations are:

for sawdust:

$$Sh = \begin{cases} 0.39 + 0.085\, Re^{.87}\, Sc^{.33} & 0.5\text{-}1.0 \text{ mm} \\ 1.39 + 0.37\, Re^{.65}\, Sc^{.33} & 5\text{-}7 \text{ mm} \end{cases} \quad (26)$$

$$Nu = \begin{cases} 0.39 + 0.085\, Re^{.87}\, Pr^{.33} & 0.5\text{-}1.0 \text{ mm} \\ 1.39 + 0.37\, Re^{.65}\, Pr^{.33} & 5\text{-}7 \text{ mm} \end{cases} \quad (27)$$

for flakes and blocks:

$$Sh = 3.074\, Re^{.335}\, Sc^{.33} \quad (28)$$

$$Nu = 3.074\, Re^{.335}\, Pr^{.33} \quad (29)$$

The equations for flakes and blocks are independent of thickness.

Results

The performance of the particle drying model was evaluated by comparing the results of model simulations with data from the experimental drying tests listed in Table 1. All of the input information required by the model was available from experimental data with the exception of the moisture diffusivities. The attenuation factor for vapor diffusivity in the longitudinal direction was taken to be unity; for transverse vapor diffusion the attenuation factor was assumed to be 0.005. A sensitivity analysis showed that this parameter had only a minor impact on the simulation results. As a first approximation, the effective moisture diffusivity was assumed to be independent of moisture content. This was implemented in the model simulations through Equation (11) by equating the diffusivities for bound and for free water in each direction. On the basis of reported bound-water diffusivities [17], transverse diffusivity was assumed to be one order of magnitude smaller than that in the longitudinal direction. Since diffusivity was in effect an adjustable parameter, one measure of the model's validity is the degree of consistency in the diffusivity values that are required to correctly simulate different drying experiments.

Table 1

Experimental Drying Tests

Test No.	Species	Size (mm)	Initial MC (%db)	Air Temp. (C)	Air Vel. (m/s)
1	Douglas fir sawdust	5-7 mesh	175	95	0.17
2	Douglas fir sawdust	0.5-1.0 mesh	180	95	0.17
3	Douglas fir sawdust	5-7 mesh	224	200	0.17
4	Douglas fir sawdust	0.5-1.0 mesh	181	200	0.17
5	Douglas fir sawdust	0.5-1.0 mesh	116	95	0.17
6	Douglas fir sawdust	5-7 mesh	146	200	0.17
7	Aspen flake	0.63 x 18.8 x 25.5	134	95	0.14
8	S. pine block	3.20 x 18.0 x 20.1	106	95	0.19
9	S. pine block	2.80 x 19.0 x 21.5	129	95	0.19
10	S. pine block	3.10 x 14.6 x 20.9	121	150	0.16

Table 2

Input Parameters for Sawdust Simulations
(5 x 5 compartments)

Test No.	Size (mm)	Particle Dimensions X x Y x Z Direction (mm)	Moisture Diffusivity D_{bf} (m^2/s) X-dir	Y-dir	Fig. Ref.
1	5-7	0.8 x 4.60 x 6.0	1×10^{-9}	1×10^{-10}	4
2	0.5-1.0	1.1 x 0.35 x 0.7	1×10^{-9}	1×10^{-10}	5,8
3	5-7	1.0 x 5.00 x 5.7	1×10^{-9}	1×10^{-10}	6
3	5-7	1.0 x 5.00 x 5.7	1×10^{-8}	1×10^{-9}	6
3	5-7	1.0 x 5.00 x 5.7	5×10^{-8}	5×10^{-9}	6,9
3	5-7	1.0 x 5.00 x 5.7	1×10^{-7}	1×10^{-8}	6
4	0.5-1.0	1.1 x 0.35 x 0.7	1×10^{-9}	1×10^{-10}	7
4	0.5-1.0	1.1 x 0.35 x 0.7	1×10^{-8}	1×10^{-9}	7
4	0.5-1.0	1.1 x 0.35 x 0.7	5×10^{-8}	5×10^{-9}	7
5	0.5-1.0	1.1 x 0.35 x 0.7	1×10^{-9}	1×10^{-10}	8
6	5-7	1.3 x 4.10 x 6.3	5×10^{-8}	5×10^{-9}	9

Sawdust:

The data from Test No. 1 were used to establish diffusivity values that are appropriate for use in the drying model. Good agreement between the simulation results and the test data (Figure 4) was obtained using diffusivity values of 1×10^{-9} m^2/s in the x-direction (longitudinal) and 1×10^{-10} m^2/s in the y-direction (transverse).

Simulation of Test No. 2 using the same diffusivities produced the results shown in Figure 5, demonstrating that the model accurately simulates the drying behavior of differently sized particles at 95°C.

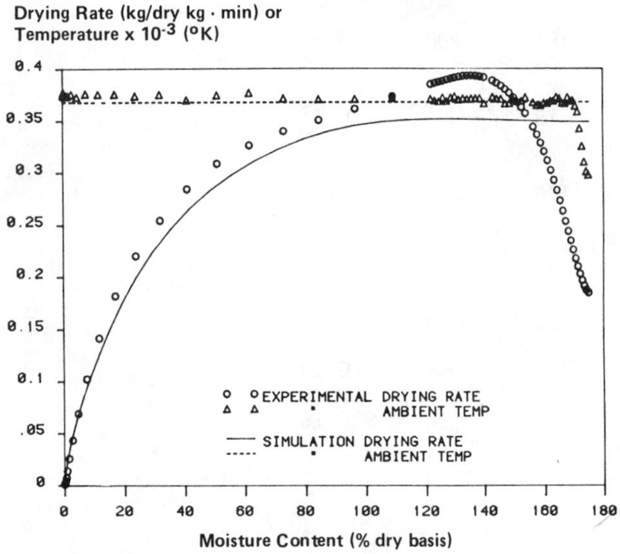

Figure 4a

Drying comparison for 5-7mm Douglas-fir sawdust at 95° C

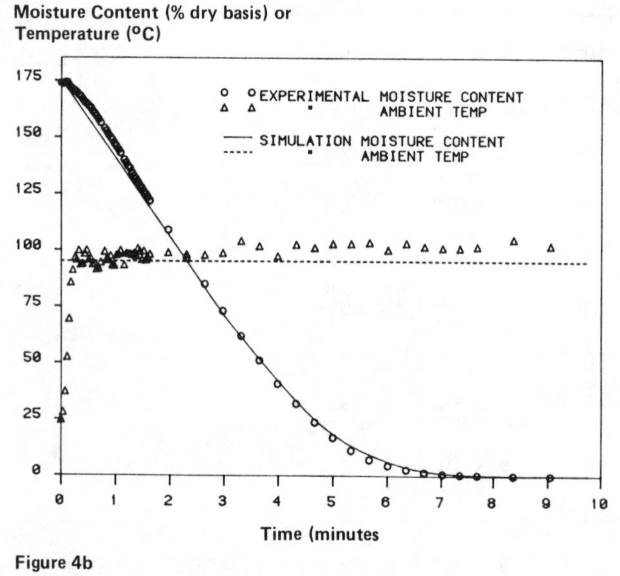

Figure 4b

Drying comparison for 5-7mm Douglas-fir sawdust at 95° C

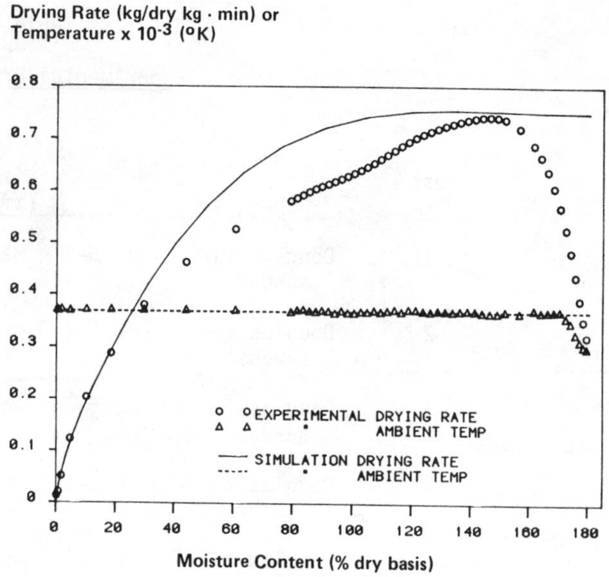

Figure 5a

Drying comparison for 0.5-1.0mm Douglas-fir sawdust at 95° C

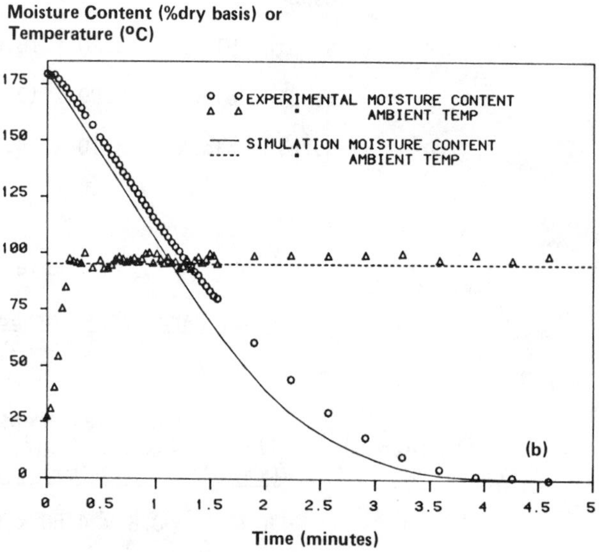

Figure 5b

Drying comparison for 0.5-1.0mm Douglas-fir sawdust at 95° C

Since moisture diffusivity is a function of temperature, it is unlikely that values applicable to drying at 95°C should be valid also for drying at 200°C. Therefore, simulations of Test No. 3 results were performed in order to establish the diffusivities appropriate for that temperature. Curves showing drying rate vs. moisture content were more useful in these comparisons and will, henceforth, be used exclusively. The representative results in Figure 6 show that the diffusivities applicable to drying experiments at 95°C were in fact inadequate. The diffusivities $D_{bf_x} = 5 \times 10^{-8}$ and $D_{bf_y} = 5 \times 10^{-9}$ m^2/s providing a good match with experiments at 200°C are greater than those for drying at 95°C by a factor of approximately 50. Although the agreement between

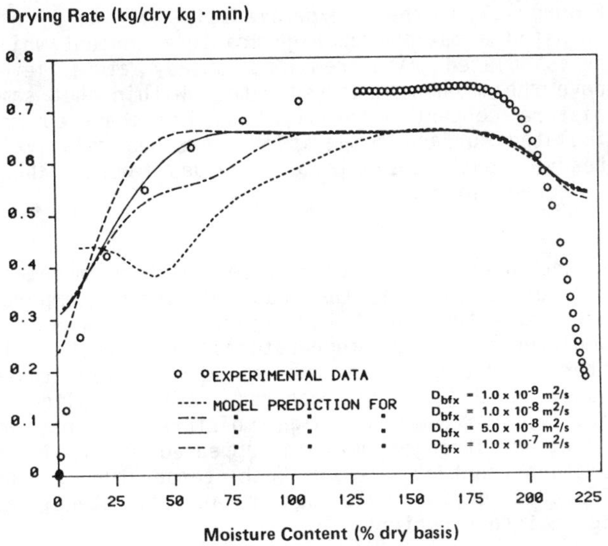

Figure 6

Rate comparison for 5-7mm Douglas-fir sawdust at 200° C

simulation and experiment is quite good for the 5-7 mm sawdust particles at 200°C, a simulation using the same diffusivity parameters for 0.5-1.0 mm particles at 200°C substantially over-predicts the rate of drying (Figure 7).

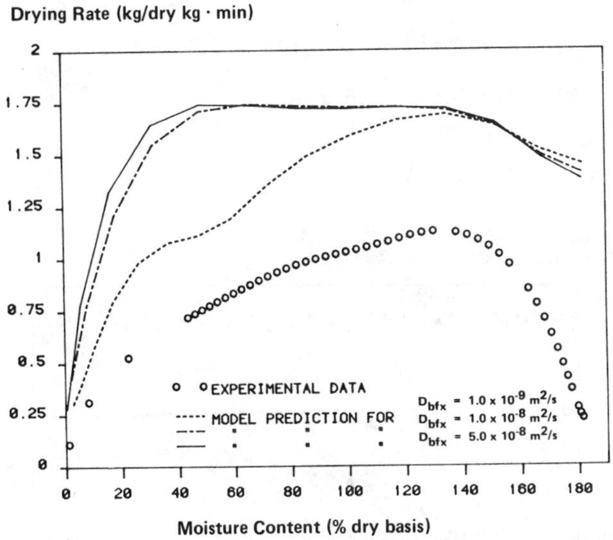

Figure 7

Rate comparison for 0.5-1.0mm Douglas-fir sawdust at 200° C

Finally, model simulations for sawdust particles having different initial moisture contents were evaluated by comparing those results with experimental data. Figure 8 demonstrates that the model is capable of simulating the drying behavior of 0.5-1.0 mm sawdust at 95°C using the diffusivity parameters $D_{bf_x} = 1 \times 10^{-9}$ and $D_{bf_y} = 1 \times 10^{-10}$ m^2/s previously established for that temperature.

Similarly, model simulations for 5-7 mm sawdust dried at 200°C using the diffusivities $D_{bf_x} = 5 \times 10^{-8}$ and $D_{bf_y} = 5 \times 10^{-9}$ m^2/s appropriate for that temperature agree reasonably well with experimental results (Figure 9).

A summary of the diffusivity values used in the sawdust test simulations is listed in Table 2.

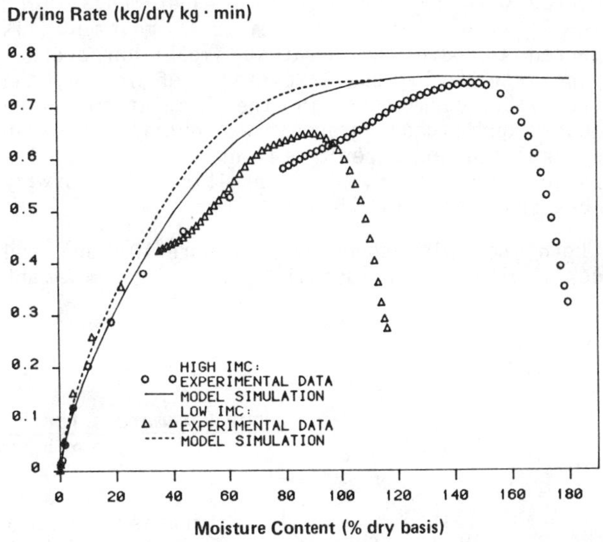

Figure 8

Rate comparison for 0.5-1.0mm Douglas-fir sawdust with different initial moisture content at 95° C

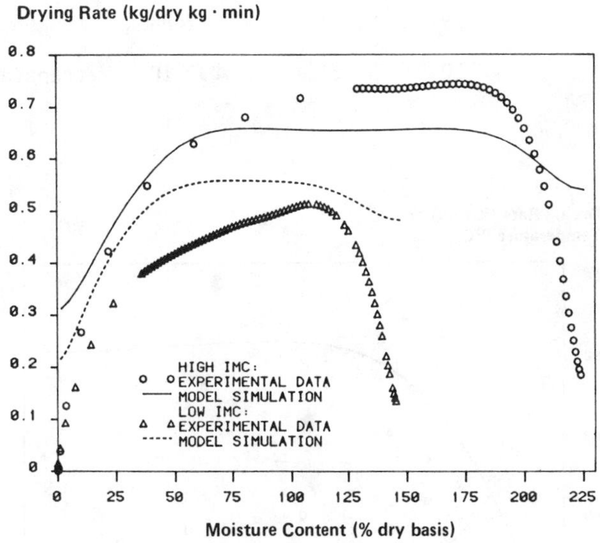

Figure 9

Rate comparison for 5-7mm Douglas-fir sawdust with different initial moisture content at 200° C

Flakes and Blocks:

Each specimen in Tests 7 and 10 was coated with epoxy on the two ends corresponding to longitudinal faces; thus, moisture movement was two-dimensional because it was restricted to the transverse (radial and tangential) directions only. The specimens in Tests 8 and 9 were coated on four sides to produce one-dimensional moisture movement. Due to the particular orientation of each specimen within the original wood material, moisture transport in Test No. 8 was in the longitudinal (along the grain) direction, while that in Test No. 9 was in the tangential direction.

Model simulations of flake and block drying were compared with the corresponding experimental test data. Since internal moisture movement is expected to have the greatest impact on overall drying rate during the last stages of drying, the diffusivity parameters in the simulations were adjusted until good agreement was obtained between the simulated and experimental drying rates at low moisture content. The resulting diffusivity values are listed in Table 3.

In both the thin aspen flake (Figure 10) and the block with longitudinal moisture movement (Figure 11), the experimental drying rate exhibited a maximum at high moisture content while the simulated rate remained steady at a level above the maximum measured rate. Within that same moisture content interval, both the measured and simulated surface temperatures remained relatively steady, but the latter values were about 10 degrees lower.

In the blocks with transverse moisture movement (Figures 12 and 13), the measured surface temperatures rose throughout the drying period while the simulated surface temperatures increased only after substantial amounts of drying had taken place. At 95°C, the measured rate exhibited a very weak maximum at high moisture content; at 150°C, a stronger maximum appeared at a lower moisture content. For both temperatures, the simulated drying rate was relatively steady at high moisture content.

Table 3

Input Parameters for Flake and Block Simulations
(5 x 5 compartments)

Test No.	Species	Thickness (mm)	(X-direction) Grain Orientation	Air Temp. (C)	Moisture Diffusivity D_{bf} (m^2/s) X-dir	Y-dir	Fig. Ref.
7	Aspen	0.63	transverse	95	5×10^{-10}	5×10^{-10}	10
8	S. pine	3.20	longitudinal	95	3×10^{-9}	0	11
9	S. pine	2.80	tangential	95	6×10^{-10}	0	12,14
10	S. pine	3.10	transverse	150	6×10^{-9}	6×10^{-9}	13

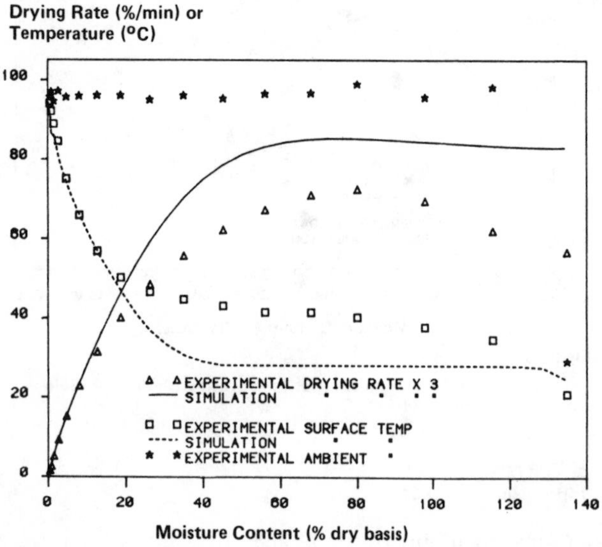

Figure 10

Drying comparison for Aspen flake at 95° C

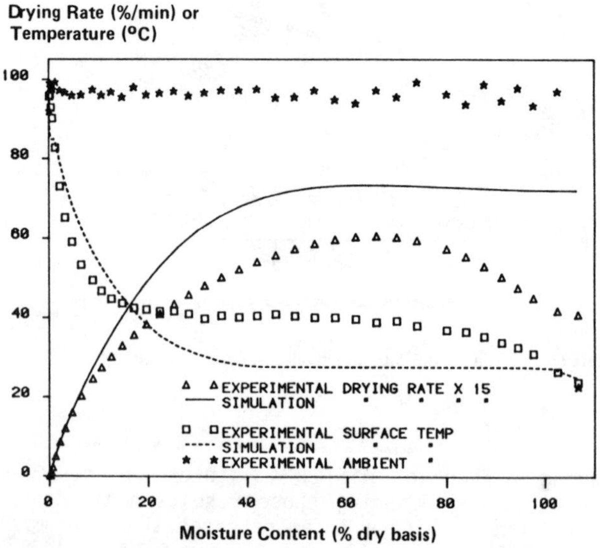

Figure 11

Drying comparison for southern pine block with one-dimensional longitudinal moisture movement at 95° C

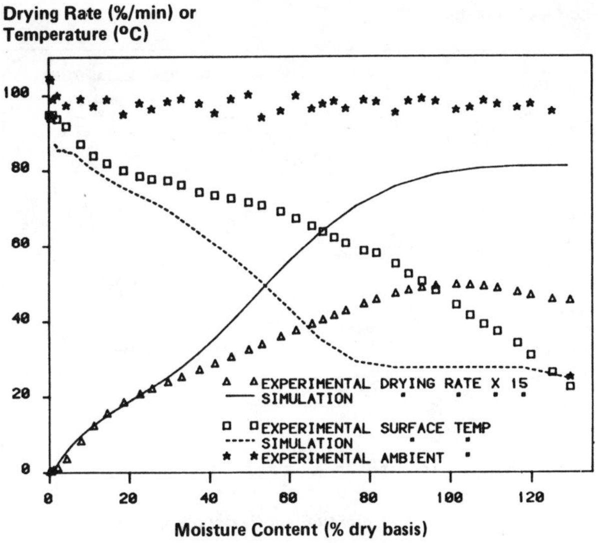

Figure 12

Drying comparison for southern pine block with one-dimensional tangential moisture movement at 95° C. (Test no. 9)

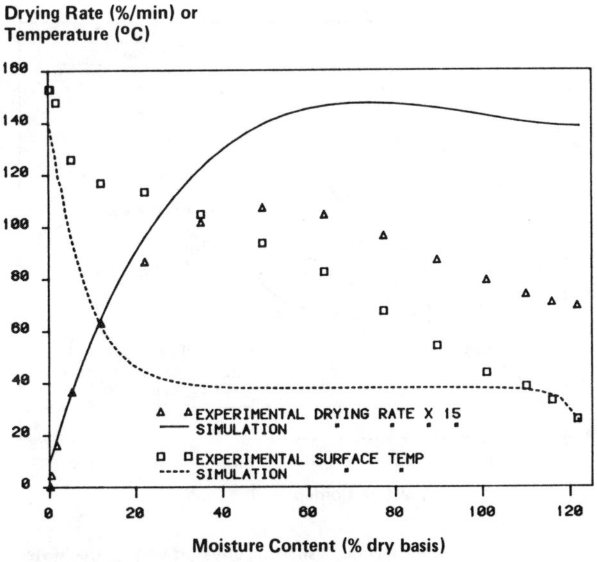

Figure 13

Drying comparison for southern pine block with two-dimensional transverse moisture movement at 150° C

ANALYSIS

Using consistent diffusivity parameters, the present model provided simulations which closely matched the drying behavior of small particles (i.e., those with a shortest dimension of one millimeter or less) at low temperature, including the variability of drying rate with particle size and with initial moisture content. For drying at higher temperatures (200°C), the specified diffusivities had to be increased in order to maintain good agreement between simulation and experiment.

For flakes and blocks, the simulated drying rates at high moisture content were significantly faster than those in the experimental drying tests. Since the external heat and mass transfer processes were characterized by an independent method, it is unlikely that the discrepancies between the experimental and simulated drying rates were due to errors in the specified transport coefficients. However, another explanation is suggested by the differences between the experimental and simulated surface temperatures at high moisture content in Tests 7 and 8. Since the effective surface areas for external heat transfer and mass transfer are assumed to be equivalent in the present model, the simulations indicated that the normal wet-bulb temperature was maintained during steady-state drying. However, if a flake or block surface is incompletely wetted, the area not covered by liquid water would be less effective for mass transfer because the local water vapor pressure could be below the saturation pressure of free water. Since the rate of heat transfer does not depend on complete wetting of the surface, the effective area for heat transfer would be greater than that for mass transfer. Therefore, the surface temperature during constant-rate drying would be higher than the normal wet-bulb temperature and the drying rate would be slower. Exactly these kinds of deviations were observed in flakes and in the block with longitudinal moisture movement, implying that the assumption of equivalent effective surface areas for heat and for mass transfer is incorrect.

With further drying, liquid water eventually vanishes from the surface and consequently the vapor pressure at the surface decreases. The rate of external mass transfer therefore decreases and the surface temperature rises. In tests with thin flakes and with the block with longitudinal moisture movement, this temperature rise did not occur until much of the total moisture was lost. Apparently, moisture migration to the surface was sufficiently rapid at high moisture content to replenish the water at the surface as it evaporated, thereby maintaining a low surface temperature. In blocks with transverse moisture movement, the surface temperatures started to rise almost as soon as drying began, implying that liquid water vanished from the surfaces almost immediately and therefore that moisture movement was considerably slower in these specimens.

In the model simulations, however, the surface remains in steady-state at the wet-bulb temperature until the moisture content of the entire surface compartment falls below the fiber saturation point. Since a certain length of time is required to evaporate the free water in the surface compartments, the duration of the steady-state drying period may be overestimated. The effect should diminish as the number of compartments is increased because the amount of mass in each surface compartment decreases. This phenomenon is illustrated in Figure 14 for simulations of Test No. 9 using 3x3, 5x5 and 7x7 compartment matrices, corresponding respectively to surface compartment thicknesses of 33, 20, and 14 percent

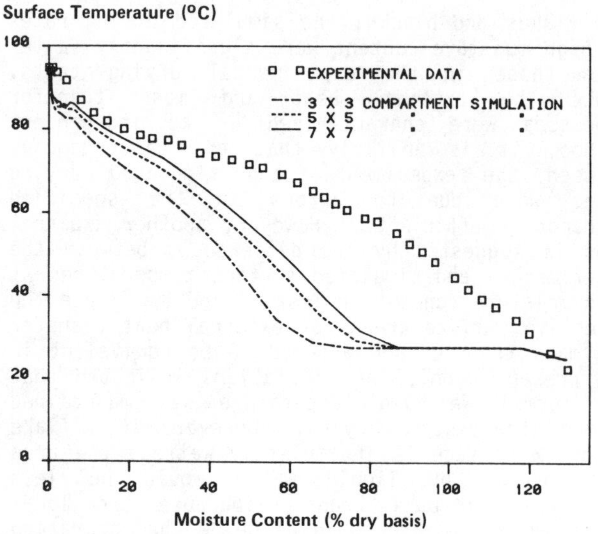

Figure 14a

Drying comparison for test no. 9 showing effect of compartment size on simulation results. (D_{bfx} = 6.0 x 10^{-10} m^2/s)

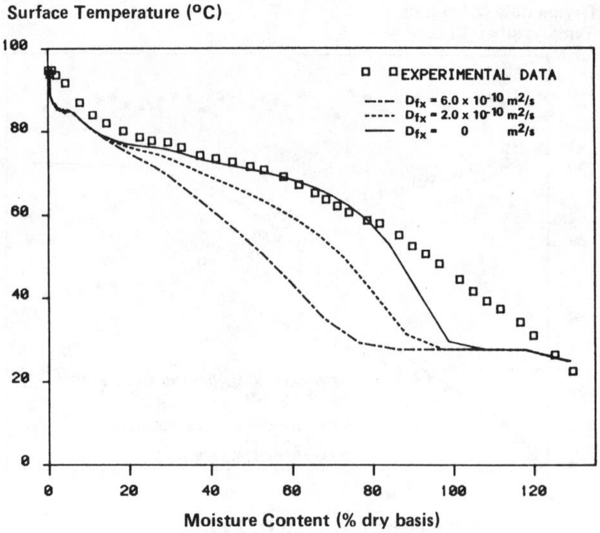

Figure 15a

Drying comparison for test no. 9 showing effect of varying free water diffusivity, D_{fx}, on simulation results with D_{bx} = 6.0 x 10^{-10} m^2/s. (5x5 compartment simulations)

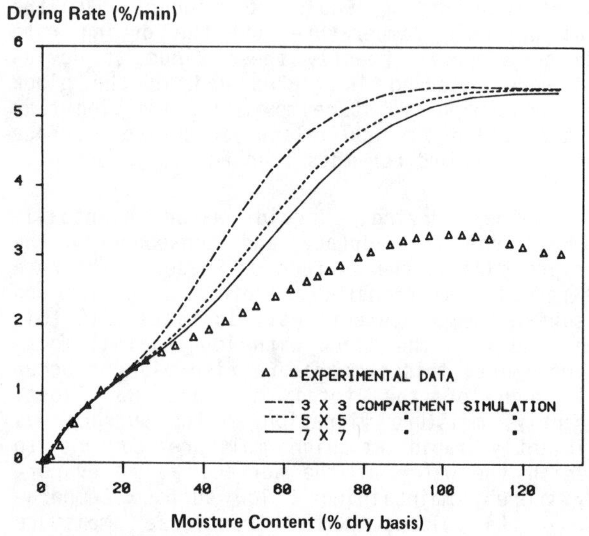

Figure 14b

Drying comparison for test no. 9 showing effect of compartment size on simulation results. (D_{bfx} = 6.0 x 10^{-10} m^2/s)

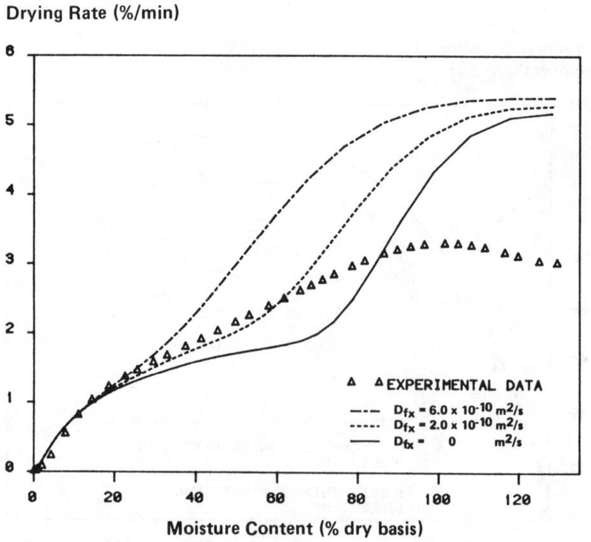

Figure 15b

Drying comparison for test no. 9 showing effect of varying free water diffusivity, D_{fx}, on simulation results with D_{bx} = 6.0 x 10^{-10} m^2/s. (5x5 compartment sumulations)

of the total particle thickness. The progressive shortening of the steady-state period suggests that this simulation could be further improved if the surface compartments were made very thin. This would allow the surface to dry and the rate to fall earlier in the simulation, providing better agreement with the actual drying behavior.

Finally, better agreement between simulation and test data was obtained when the diffusivity for bound and free water, Dbf, was allowed to vary with moisture content. In fact, Figure 15 shows that better matches of Test No. 9 resulted when the specified free water diffusivity, Df, was less than the bound water diffusivity, Db. Since the value of diffusivity used in the model computations is a weighted average of Df and Db given by Equation (11), these simulations show that the model provided better results when the diffusivity effectively increased with decreasing moisture content. This contradicts the moisture content dependence of diffusivity that is widely reported in the literature [3] and therefore some doubt is cast on the validity of the bound and free water migration mechanism represented by Equations (10) and (11). It is likely that a more realistic representation of free water migration would be helpful.

In summary, these results suggest that in the present model the fraction of surface area taken up by the wood substance should be accounted for in considering surface mass transport. This would help to remedy the discrepancies between simulated and experimental drying rates and surface temperatures at or near steady-state. Also, surface drying should be characterized in a way that better represents the decreasing mass-transfer rate at the surface. Finally, the mechanism of moisture migration, especially that of free water, needs to be modeled in a more realistic manner. These modifications would improve the performance of the model in applications involving thicker materials, in which internal moisture migration plays an important role in determining overall drying rate, and high temperatures, when surface drying is relatively fast.

CONCLUSIONS

A two-dimensional wood particle drying model has been developed which describes movement of free water, bound water and water vapor together with heat conduction under the constraint of local thermodynamic equilibrium. The general mathematical model consists of four partial differential equations - one for thermal diffusion and three for diffusion of each of the water phases - coupled with non-linear algebraic equations for the local equilibrium condition. The equations for free and for bound water diffusion could be combined and the resulting three partial differential equations transformed into a set of ordinary differential equations by converting the spatial coordinates into discrete compartments while retaining time as a continuous variable. The differential transport equations and coupled algebraic equilibrium constraints could then be solved in a sequential iterative manner.

Evaluation of this model using experimental drying rate measurements shows that for small particles such as sawdust, the simulations closely match the drying behavior observed for different particle sizes, temperatures, and initial moisture contents. For larger, thicker materials such as flakes, the agreement between model simulation and experimental data is less satisfactory, as the former overestimates the duration of the steady-state drying period at high moisture content. Moreover, experimental tests show that at steady-state, the drying rate is slower and the surface temperature is higher than the corresponding model predictions. Analysis of these results suggests that improvements are needed in the model descriptions of free water migration and of convective surface transport processes.

APPENDIX A: DISCRETE SPACE ORDINARY DIFFERENTIAL EQUATIONS

The partial differential equations are converted into ordinary differential equations through a spatial approximation which considers the two-dimensional particle as a collection of equivalent rectangular compartments. This was depicted in Figure 1. Here, we will show the derivation of the working equations for the model.

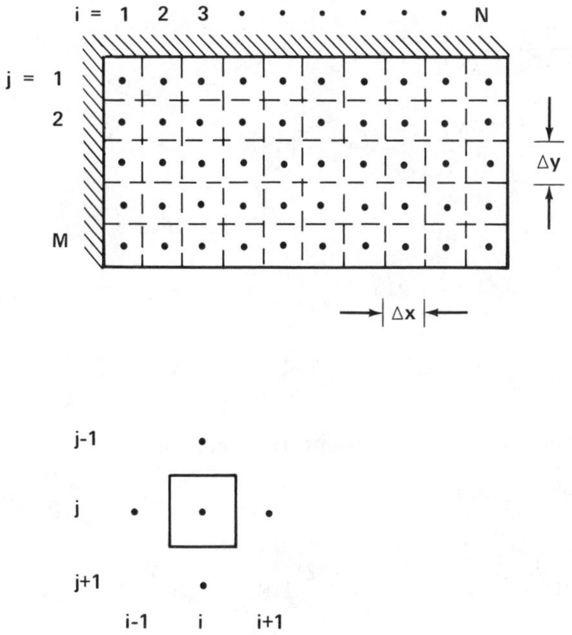

Figure A 1.

Indexing of compartments in the symetrical quadrant of the particle

Figure A1 shows the compartment arrangements in terms of x and y coordinate indexing through the integer variables "i" and "j" Considering variables as point functions (at the middle of the compartments) we get the following differential equation for the thermal transient of a general compartment.

$$\frac{dT_{i,j}}{dt} = \frac{\alpha_x}{\Delta x^2} (T_{i-i,j} - 2T_{i,j} + T_{i+1,j}) + \frac{\alpha_y}{\Delta y^2} (T_{i,j-1} - 2T_{i,j} + T_{i,j+1}) - (\frac{\lambda}{c_p} R)_{i,j} \qquad (A\ 1)$$

$i = 1,2, \ldots, N-1$
$j = 1,2, \ldots, M-1$

Boundary conditions are handled by specifying temperatures or by modifying the generalized equation (Equation A 1).

For reflective conditions we have:

$$T_{i-1,j} = T_{i,j} \qquad i=1, \text{ all } j$$

$$T_{i,j-1} = T_{i,j} \qquad \text{all } i, \ j=1$$

Convective boundary conditions at the exposed surface require a new set of differential equations for these compartments. We get:

$$\frac{dT_{N,j}}{dt} = \frac{\alpha_x}{\Delta x^2} [(T_{N-1,j} - T_{N,j}) + \frac{\Delta x}{L/2} Bi_x (T_\infty - T_{N,j})] + \frac{\alpha_y}{\Delta y^2} (T_{N,j-1} - 2T_{N,j} + T_{N,j+1}) - (\frac{\lambda}{c_p} R)_{N,j} \qquad (A\ 2)$$

$$j = 1,2, \ldots, M-1$$

where T_∞ is the ambient temperature.

We also get:

$$\frac{dT_{i,M}}{dt} = \frac{\alpha_x}{\Delta x^2} (T_{i-1,M} - 2T_{1,M} + T_{i,M}) + \frac{\alpha_y}{\Delta y^2} [(T_{i,M-1} - T_{i,M}) + \frac{\Delta y}{H/2} Bi_y (T_\infty - T_{i,M})] - (\frac{\lambda}{c_p} R)_{i,M} \qquad (A\ 3)$$

$$i = 1,2, \ldots, N-1$$

and

$$\frac{dT_{N,M}}{dt} = \frac{\alpha_x}{\Delta x^2} [(T_{N-1,M} - T_{N,M}) + \frac{\Delta x}{L/2} Bi_x (T_\infty - T_{N,M})] + \frac{\alpha_y}{\Delta y^2} [(T_{N,M-1} - T_{N,M}) + \frac{\Delta y}{H/2} Bi_y (T_\infty - T_{N,M})] - (\frac{\lambda}{c_p} R)_{N,M} \qquad (A\ 4)$$

The Biot numbers are defined as:

$$Bi_x = \frac{h(L/2)}{k_x} \text{ and } Bi_y = \frac{h(H/2)}{k_y} \qquad (A\ 5)$$

where k_x and k_y are the directional thermal conductivities of wet wood.

In the same manner we develop the liquid diffusion equations. For a general internal compartment we get:

$$\frac{du_{bf\ i,j}}{dt} = \frac{D_{bfx}}{\Delta x^2} (u_{bf\ i-1,j} - 2u_{bf\ i,j} + u_{bf\ i+1,j}) + \frac{D_{bfy}}{\Delta y^2} (u_{bf\ i,j-1} - 2u_{bf\ i,j} + u_{bf\ i,j+1}) - R_{i,j} \qquad (A\ 6)$$

$$i = 1,2, \ldots, N-1$$
$$j = 1,2, \ldots, M-1$$

Reflective boundary conditions require:

$$u_{bf\ i-1,j} = u_{bf\ 1,j} \qquad i=1, \text{ all } j$$

$$u_{bf\ i,j-1} = u_{bf\ i,j} \qquad \text{all } i, \ j=1$$

For convective boundary conditions at the exposed surfaces the equations are:

$$\frac{du_{bf\ N,j}}{dt} = \frac{D_{bfx}}{\Delta x^2} (u_{bf\ N-1,j} - u_{bf\ N,j}) - \frac{k_p}{\Delta x} [p_v^s (u_{bf\ N,j}) - p_{v\infty}] + \frac{D_{bfy}}{\Delta y^2} (u_{bf\ N,j-1} - 2u_{bf\ N,j} + u_{bf\ N,j+1}) + R_{N,j} \qquad (A\ 7)$$

$$j = 1,2, \ldots, M-1$$

$$\frac{du_{bf\ i,M}}{dt} = \frac{D_{bfx}}{\Delta x^2} (u_{bf\ i-1,M} - 2u_{bf\ i,M} + u_{bf\ i+1,M}) + \frac{D_{bfy}}{\Delta y^2} (u_{bf\ i,M-1} - u_{bf\ i,M}) - \frac{k_p}{\Delta y} [p_v^s (u_{bf\ i,M}) - p_{v\infty}] + R_{i,M} \qquad (A\ 8)$$

$$i = 1,2, \ldots, N-1$$

and

$$\frac{du_{bf\ N,M}}{dt} = \frac{D_{bfx}}{\Delta x^2} (u_{bf\ N-1,M} - u_{bf\ N,M})$$

$$+ \frac{D_{bfy}}{\Delta y^2} (u_{bf\ N,M-1} - u_{bf\ N,M})$$

$$- k_p \left(\frac{1}{\Delta x} + \frac{1}{\Delta y}\right) [p_v^s (u_{bf\ N,M}) - p_{v\infty}]$$

$$+ R_{N,M} \qquad (A\ 9)$$

In the above equations k_p is defined as:

$$k_p = \frac{k_p}{\rho_o (1-\varepsilon_o)} \qquad (A\ 10)$$

where k_p is the convective mass transfer coefficient around the particle in (kg/m^2 s atm) and $p_v^s(u_{bf\ i,j})$ is the saturation pressure of moisture in compartment (i,j). If free water is present (i.e., $\beta < 1$) then $p_v^s(u_{bf\ i,j}) = p_v^s(T_{i,j})$ is the vapor pressure of water at $T_{i,j}$. However, if $\beta = 1$ (i.e., no free water), then $p_v^s(u_{bf\ i,j})$ corresponds to the sorption isotherm partial pressure computed from $u_{bf\ i,j}$ and $T_{i,j}$. This value will be less than $p_v^s(T_{i,j})$ since the relative humidity, defined as $H = p_v/p_v^s$, is less than 1.

The vapor diffusion equations are written in terms of a modified vapor moisture content as:

$$\frac{du_{v\ i,j}}{dt} = \frac{D_{vx}}{\varepsilon \Delta x^2} (\tilde{u}_{v\ i-1,j} - 2\tilde{u}_{v\ i,j} + \tilde{u}_{v\ i+1,j})$$

$$+ \frac{D_{vy}}{\varepsilon \Delta y^2} (\tilde{u}_{v\ i,j-1} - 2\tilde{u}_{v\ i,j} + \tilde{u}_{v\ i,j+1})$$

$$+ R_{i,j} \qquad (A\ 11)$$

$i = 1,2, \ldots, N-1$
$j = 1,2, \ldots, M-1$

Here $\tilde{u}_{v\ i,j} = u_{v\ i,j}$

and $\tilde{u}_{v\ i-1,j} = \frac{T_{i-1,j}}{T_{i,j}} \frac{v_{i,j}}{v_{i-1,j}} u_{v\ i-1,j}$, etc.

where v is the available vapor volume (void space) as defined in Equation (8). These corrections are necessary because the actual vapor driving force is the partial pressures and not the vapor moisture content.

The reflective boundary conditions are similar to those defined for T and u_{bf}. The convective boundary conditions at the exposed surfaces lead to:

$$\frac{du_{v\ N,j}}{dt} = \frac{D_{vx}}{\varepsilon \Delta x^2} (\tilde{u}_{v\ N-1,j} - \tilde{u}_{v\ N,j})$$

$$- \frac{k_c}{\varepsilon \Delta x} (\tilde{u}_{v\ N,j} - \tilde{u}_{v\ N,j\infty})$$

$$+ \frac{D_{vy}}{\varepsilon \Delta y^2} (\tilde{u}_{v\ N,j-1} - 2\tilde{u}_{v\ N,j} + \tilde{u}_{v\ N,j+1})$$

$$+ R_{N,j} \qquad (A\ 12)$$

$j = 1,2, \ldots, M-1$

$$\frac{du_{v\ i,M}}{dt} = \frac{D_{vx}}{\varepsilon \Delta x^2} (\tilde{u}_{v\ i-1,M} - 2\tilde{u}_{v\ i,M} + \tilde{u}_{v\ i+1,M})$$

$$+ \frac{D_{vy}}{\varepsilon \Delta y^2} (\tilde{u}_{v\ i,M-1} - \tilde{u}_{v\ i,M})$$

$$- \frac{k_c}{\varepsilon \Delta y} (\tilde{u}_{v\ i,M} - \tilde{u}_{v\ i,M\infty})$$

$$+ R_{i,M} \qquad (A\ 13)$$

$i = 1,2, \ldots, N-1$

$$\frac{du_{v\ N,M}}{dt} = \frac{D_{vx}}{\varepsilon \Delta x^2} (\tilde{u}_{v\ N-1,M} - \tilde{u}_{v\ N,M})$$

$$+ \frac{D_{vy}}{\varepsilon \Delta y^2} (\tilde{u}_{v\ N,M-1} - \tilde{u}_{v\ N,M})$$

$$- \frac{k_c}{\varepsilon} \left(\frac{1}{\Delta x} + \frac{1}{\Delta y}\right) (\tilde{u}_{v\ N,M} - \tilde{u}_{v\ N,M_\infty})$$

$$+ R_{N,M} \qquad (A\ 14)$$

In these equations:

$$\tilde{u}_{v\ i,j\infty} = u_{v\infty} \frac{T_\infty}{T_{i,j}} \frac{v_{i,j}}{v_\infty} \qquad (A\ 15)$$

where u_{v_∞} and v_∞ reflect final steady-state conditions computed from ambient temperature and humidity.

APPENDIX B: NUMERICAL SOLUTION

The model equations need to be solved simultaneously. As we described before for a particle which is partitioned into NxM compartments we get 3xNxM ordinary differential equations describing the diffusion processes coupled with the NxM algebraic relations (the **R** terms) which enforce local phase equilibrium. If the terms could be written explicitly in terms of the dependent variables, then all we would need is to solve the differential equations using a reliable ODE solver. However, we assume that there is local thermodynamic equilibrium between the moisture phases and describe the interdependence through nonlinear algebraic equations. Therefore, we should be integrating the differential equations while simultaneously satisfying the algebraic equations as well. We approached the problem through a stepwise procedure as described below.

Consider the equivalent drying problem without the internal equilibrium restrictions. Here we would need to solve only the differential equations describing the transport phenomena with the associated boundary conditions. We consider this to be the base problem and modify our differential equations to reflect transport rates only and solve them for an incremental time period at a time. After each time interval we satisfy the algebraic requirements of the equilibrium conditions through iterative calculations and adjust the temperature and moisture contents (free, bound, vapor) of each compartment. The accuracy of the procedure depends on the frequency of the adjustments or the incremental time interval. We found 1/50th of the thermal time constant of the particle to be a sufficiently small time increment. The iterative procedure used for the equilibrium computations can be summarized as:

Step 1. Reset iteration counter.

Step 2. Compute vapor partial pressure and saturation pressures; if vapor pressure is greater than saturation, then go to Step 4; otherwise continue.

Step 3. If there is free water, then let free water evaporate to satisfy saturation and go to Step 5; otherwise go to Step 5 directly.

Step 4. Release excess air/vapor mixture to reduce partial pressure to saturation level.

Step 5. Compute equilibrium bound water content; if equilibrium content is greater than the present bound water content, then go to Step 7; otherwise continue.

Step 6. If there is free water, then the excess bound water transfers into free water; otherwise, the excess bound water evaporates; go to Step 8.

Step 7. If there is free water, then the free water provides the needed extra bound water to reach equilibrium; otherwise continue.

Step 8. Adjust compartment temperature to reflect the heat effects of the phase changes.

Step 9. Compute the stopping criteria in terms of the incremental temperature and partial pressure changes between iterations; if satisfied, then go to Step 1 and repeat the adjustments for the next compartment; otherwise, go to Step 2.

The procedure described above decouples the differential and algebraic problems during the integration time increment. With a judicial manipulation of the differential equations and the boundary conditions it is also possible to consider the transport equations as three groups of NxM equations rather than the original 3xNxM simultaneous equations. Note that heat and mass transfer equations are coupled through the boundary conditions. The reason this becomes an important factor is that the diffusion equations are "stiff" when algebraically coupled with the equilibrium relations and require special stiff solver ODE routines. These algorithms use implicit algebraic solution techniques and thus it is faster to solve three NxM systems than it is to solve a 3xNxM system.

	i = 1	2	3	4	5	6	7
j = 1	1	6	11	16	21	26	31
1	2	7	12	17	22	27	32
3	3	8	13	18	23	28	33
4	4	9	14	19	24	29	34
5	5	10	15	20	25	30	35

Figure B 1.

Single index referencing of compartments to simplify differential equation organization during numerical solution

For practical reasons the indexing of the variables in the differential equations are reduced from two to one. This required a special numbering system for the compartments as shown in Figure B1 for a 7x5 system. The index transformation is accomplished by:

$$C_{i,j} = C_n \qquad \text{(B 1)}$$

where

$$n = M(i-1) + j \qquad i = 1, \ldots, N \quad j = 1, \ldots, M$$

For the 7x5 system there are 35 differential equations to solve for each of the transport cases. These equations are of the form:

$$\frac{dx}{dt} = Ax + b \tag{B 2}$$

where x and b are vectors of size 35 and A is a 35x35 matrix. The coefficient matrix, A, represents the interdependence of neighboring variables according to the two-dimensional compartment arrangement. This results in a highly structured banded form for A. As shown in Figure B2 the only non-zero elements of A are in the main diagonal and in the three super and sub diagonals. The LSODE package which we used for the solution of the ODEs could take advantage of this special sparsity of the system and accelerate the solution.

$$A = \begin{bmatrix} x & x & x & & & & \\ x & x & x & x & & & \\ & x & x & x & & x & \\ x & & x & x & x & & x \\ & x & & x & x & x & & x \\ & & x & & x & x & x & & x \\ & & & x & & x & x & x & & x \\ & & & & x & & x & x & x \\ & & & & & x & & x & x & x \\ & & & & & & x & & x & x \end{bmatrix}$$

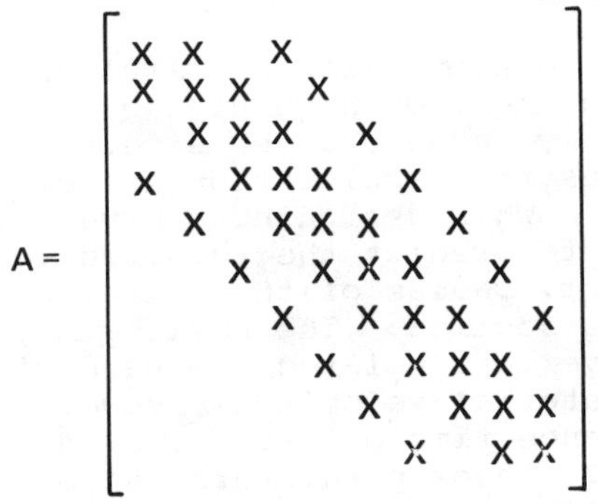

$$\dot{x} = Ax + b$$

Figure B 2.

Linear form of ordinary differential equations during the incremental integration time and the sparsity of the coefficient matrix

REFERENCES

1. Liukov, A. V., 1966, "Heat and Mass Transfer in Capillary-Porous Bodies," Pergammon Press, London.

2. Keey, R. B., 1972, "Drying Principles and Practice," Pergammon Press, London.

3. Siau, J. F., 1971, "Flow in Wood," Syracuse University Press, New York.

4. Skaar, C., 1972, "Water in Wood," Syracuse University Press, New York.

5. Mujumdar, A. S., 1980, "Drying '80, v1: Developments in Drying," Hemisphere Publishing, New York.

6. Mujumdar, A. S., 1980, "Drying '80, v2: Proceedings of the Second International Symposium," Hemisphere Publishing, New York.

7. Mujumdar, A. S., 1980, "Advances in Drying, v1," Hemisphere Publishing, New York.

8. Mujumdar, A. S., 1982, "Drying '82," Hemisphere Publishing, New York.

9. Kayihan, F., Johnson, J. A., and Lubon, C., 1981, "Preliminary Calculations of Heat and Mass Transfer During the Pressing Operation of Wood Composite Materials Manufacture," in "Numerical Methods in Thermal Problems," R. W. Lewis, K. Morgan, and B. A. Schafler, eds.; Proceedings of the Second International Conference held in Venice, Italy; Pineridge Press, Swansea, UK.

10. Malte, P. C., Robertus, R. J., Strickler, M. D., Cox, R. W., Messinger, G. R., Kennish, W. J. and Schmidt, S. C., 1976, "Experiments on the Kinetics and Mechanisms of Drying Small Wood Particles," Research Report TEL-76-8, Department of Mechanical Engineering, Washington State University, Pullman, Washington.

11. Yaws, C. L., 1977, "Physical Properties," McGraw-Hill, New York.

12. Rosen, H. N., 1979, "Psychrometric Relationships and Equilibrium Moisture Content of Wood at Temperatures Above 212°F," Proceedings of Wood Moisture Content and Humidity Relationships Symposium at Virginia Polytechnic Institute and State University, USDA Forest Service.

13. 1974, "Wood Handbook: Wood as an Engineering Material," Forest Products Laboratory.

14. Kanury, A. M., 1975, "Introduction to Combustion Phenomena," Gordon and Breach, New York.

15. Hindmarsh, A. C., 1980, Lawrence Livermore National Laboratory, California; private communications.

16. Bird, R. B., Stewart, W. E. and Lightfoot, E. N., 1960, "Transport Phenomena," Wiley, New York.

17. Choong, E. T., 1963, "Movement of Moisture Through a Softwood in the Hygroscopic Range," Forest Products J., 13 (11), p. 489.

Portions of this work were presented in the Third International Drying Symposium, Birmingham, England (1982) and in the Denver National AIChE Meeting, Separation Processing in the Forest Products Industry Session (1983).

A MATHEMATICAL MODEL OF A SPIRAL DRYER FOR FINE POLYDISPERSE MATERIALS

V.I. Mushtayev, A.S. Timonin, N.V. Tyrin,
A.V. Chernyakov, A.V. Levin, A.A. Pakhomov.

Moscow Technical University
Moscow, USSR.

Abstract

A mathematical description of the drying process of fine granular materials in a stream of a hot gas in spiral ducts is presented. A method of simultaneous numerical solution of the equations of particle motion and heat and mass transfer including material polydispersity is suggested. Polydispersity of the material is incorporated in the form of the particle size distribution function.

Nomenclature

- A, b - constants in eq. (2)
- b_d - effective duct perimeter
- c_{pg} - specific heat of gas
- d_p - particle diameter, m
- Δh_v - latent heat of vaporization
- K_m - average overall heat transfer coefficient
- K - step of the spiral
- ℓ - local duct length
- M - total mass of particles
- m - mass of size fraction
- $m_{p,d}$ - mass of bone dry particle
- Q - mass velocity of solids, kg/m^2s
- P - total pressure in the duct
- p_s - saturation pressure of vapor
- p - partial pressure of gas
- r_o - radius of the spiral
- R_v - gas constant of vapor
- S_d - duct cross-section area
- T - temperature, K
- u - velocity
- X - moisture content of fraction of size d_p
- λ - heat conductivity
- ρ - density
- ζ - coefficient of drag
- ζ_w - coefficient of wall friction
- τ - time
- Nu_p - particle Nussett number
- $\bar{X}$ - average moisture content of polydisperse material.

Subscripts

- a - ambient air acceleration zone
- g - gas
- 0 - initial
- s - saturation conditions
- p - particle

As pneumatic dryers with a spiral duct allow attainment of high drying rates with considerably reduced dryer size. Hence their industrial application is very promising. When designing a dryer it is necessary to predict the required length of the duct, radius of the spiral and the duct cross-section. The basic dimensions of the dryer are related to the drying gas and material velocities as well as to the residence time of particles in the apparatus. These parameters can be calculated by simultaneous integration of the governing particle motion and heat and mass transfer equations.

In this paper a method of numerical solution of the system of equations of particle motion and heat and mass transfer, including allowance for the general polydispersity of the material being dried, is reported.

The polydispersity of the material being dried is described in terms of the size distribution function, $f(d_p)=\frac{1}{M}(dm/dd_p)$, where M is the total mass of particles. The rate of moisture removal at a given point of the duct is then given by

$$\frac{dX}{d\tau} = \frac{d}{d\tau}\int_0^\infty f(d_p)\, X(d_p,\tau)\, dd_p$$

The following assumptions are made to formulate the system of equations:

1. The particle motion is defined by the balance of forces of inertia, drag and wall friction. Other forces of the full equation of particle motion [1] are negligible [2, 4].
2. Only constant drying-rate period is modelled during which the particle surface temperature is governed by the saturated vapor pressure according to the following equation [3]:

$$T_s = T_g - (p_s - p)\frac{10.270}{6.65}\frac{1 - \frac{0.56}{597}(T_s - 273)}{\left(1 + \frac{1}{7}\frac{p}{P}\right)\left(1 + \frac{p_s}{P}\right)} \qquad (1)$$

where p_s is calculated from

$$\ln p_s = A - B/T_s \qquad (2)$$

3. Pressure drop in the dryer is negligible.

With these assumptions the system of equations describing the drying process for a polydisperse material has the following form:

$$m_{p,d}(1+X)u_p\frac{du_p}{d\ell} = \zeta\frac{\pi d_p^2 \rho_g}{8}(u_g - u_p)^2 -$$

$$- \zeta_w \frac{m_{p,d}(1+X)u_p^2}{r_o(1 - 2K\ell/r_o^2)^{1/2}} n(\ell) \qquad (3)$$

where

$$n(\ell) = \begin{cases} 0 & \ell < \ell_a \\ 1 & \ell \geq \ell_a \end{cases}$$

$$\frac{dT_g}{d\ell} = \frac{K_m b_d}{u_g S_d c_{pg}\rho_g}(T_a - T_g) + \frac{\Delta h_v Q_0}{c_{pg}\rho_g u_g}\int_0^\infty \frac{\partial X}{\partial \ell} f(d_p)\,dd_p \qquad (4)$$

$$\frac{dp_s}{d\ell} = \frac{Q_0 R_v T_g}{u_g}\int_0^\infty \frac{\partial X}{\partial \ell} f(d_p)\,dd_p + \frac{p_s}{T_g}\frac{dT_g}{d\ell} \qquad (5)$$

$$\frac{dX}{d\ell} = \frac{6\,Nu_p \lambda_g}{\rho_p d_p \Delta h_v u_p}(T_s - T_g) \qquad (6)$$

Simultaneous solution of equations (3) through (6) with equations (1) and (2) was computed using a variable step Runge-Kutta method.

The computer program contains two options of possible computation procedures for the function $f(d_p)$: first directly from the experimental histograms and second using their forms corrected on the basis of normal or log-normal distributions.

Deviation of the results computed for particles in the range of diameter from 50 to 1000 μm from experimental results does not exceede 10-15%.

The results obtained can be applied to practical calculations of such dryers using basic technological data of the drying process provided the drying takes place entirely in the constnat rate period.

Acknowledgements

Translation by Dr. Z. Pakowski, Łódź Technical University. Edited by Dr. A.S. Mujumdar. Typescript prepared by Purnima Mujumdar, Montreal, Canada.

References

1. Soo S.L., Multiphase System Hydrodynamics, Mir, (1971)

2. Babukha G.L., Rabinovich N.I.- "Mechanics and heat transfer of polydisperse suspension stream", Naukova Dumka, (1969)

3. Krischer, O. - Scientific Principles of Drying Technology, Inostrannaya Literatura, (1961)

4. Timonin, A.S., Mushtayev, V.I.- Theoretical Principles of Chemical Technology, XIV, 3 (1980)

THE ® REMAFLAM PROCESS, A NEW METHOD FOR DRYING OF WOVEN FABRICS

Dr. Hans-Ulrich von der Eltz
Franz Schön

Hoechst Aktiengesellschaft
6230 Frankfurt am Main 80
F.R. Germany

ABSTRACTS

The energy consumption in the textile industry is extremely high, particularly for drying processes. Using conventional dryers as a basis for comparison, the Remaflam process and a corresponding drying range are described. Examples for application and details of practical experience with the process so far, enable a critical assessment by the finisher. Some considerations concerning the future prospects brings us back to the energy problem, whereby the Remaflam process is considered to have a good chance of success due to its low-pollution qualities.

NOMENCLATURE

- A = machine operators [numbers of workers]
- B = width of cloth [m]
- CH = capacity of trough [l]
- D = price of steam [DM 65/t]
- E_{ges} = total elec. power consumption [kWh]
- E_P = price of elec. power [DM 0.12/kWh]
- F = amount of dye used [g/l]
- F_P = price of dye [DM 20/kg]
- F_A = liquor pick-up in l liquor for 100 kg cloth [l/100 kg]
- f_1 = factor for volume contraction [3 % by vol.]
- f_2 = factor for the density of methanol [0.792 g/cm³]
- f_3 = preheating time for IR radiant heater [4 min]
- f_4 = factor for x % consumption of rated power [%]
- f_5 = factor for 20 % increase in consumption of steam (maintenance of temperature during dyeing and losses through radiation and in pipesystem) [%]
- f_6 = preheating time for thermocontact installation [10 min]
- f_7 = factor for water consumption [number of machine fillings]
- G = fabric weight [g/m²]
- G_P = price of gas [DM 0.30/m³]
- L = length of cloth batch [m]
- M = price of methanol [DM 0.50/kg]
- m_M = total amount of methanol [kg]
- m_W = total amount of water [m³]
- P_M = rated power of motors, etc [kW]
- P_{IR} = output of IR radiant heater [kW]
- R_W = cleaning water [l]
- S = wages [DM 28/h]
- t_{ges} = total working time [h]
- t_1 = operating time of installation [min]
- t_r = setting up time [min]
- v = production speed [m/min]
- V = volume of HT dyeing apparatus [3000 l]
- V_M = proportion of methanol in liquor [% by vol.]
- V_W = proportion of water in liquor [% by vol.]
- W = price of water [DM 2.30/m³]
- W_D = heat content of steam [2.1 GJ/t]
- W_G = heat content of gas [0.0335 GJ/m³]
- W_K = heat consumption of thermocontact installation [0.67 GJ/h]
- Δt = temperature difference [K]

Introduction

The main energy consumers are primarily the housedolds, followed closely by the industry and traffic. Whereas in the industry a great deal has been recently accomplished by energy-saving processes, the savings in other fields have been minimal up to now. The main task for the eighties is the substitution of oil by other energy carriers in order to lower the quota of oil in primary energy consumption (1).

If, for example, the energy consumption of the German textile industry (2) is considered, then the following observations will be made:

- The quota of brown and hard coal, which made up about 70 % only two centuries ago, has now shrivelled to below 10 %.
- Natural gas as a carrier of energy gained significance above all in the seventies and now makes up a proportion of more than 25 %.
- Up until 1973 there was a great increase in heating oil, but since 1973 this has dropped constantly, and in the course of the last ten years there has been a shift from light oil to heavy fuel oil. In 1978 fuel oil made up half of the energy consumption in the German textile industry, of which 75 % was for heavy fuel oil.
- In 1970 about 2 % of the turnover in textiles was consumed by the energy costs. Today this figure is already around 3.5 %.

The textile industry is still strongly dependent on oil. For this reason it is only understandable that the dependence on the crude oil producing countries led to appeals for energy conservation. An optimum solution would be to take advantage of sources such as the wind and the sun as, for example, illustrated in fig. 1, which shows a drying machine still operating in a milling range for heavy woollen "loden" cloth in the Steiermark (Austria) area. But this form of energy utilisation is probably an exception, and we must concentrate on improving the utilisation of energy on existing or new drying plant, e.g.

- by improved heat insulation
- by controlling the air flow rate by measuring the humidity of the exhaust air and/or the circulating air
- by measuring the residual moisture content, in order to avoid overdrying of the fabric.

and applying alternative energy where possible. In this connection, not the price but improved efficiency is important.

Fig. 1
Open-air dryer for "loden" cloth

Conventional drying methods

In the course of finishing and up to the final stages before completion, a textile fabric is subjected to numerous drying processes. From these let us select one - the drying of slop-pad dyed fabric in continuous dyeing processes. This can be carried out on different types of drying ranges, which include predryers, hotflues, perforated drum dryers, cylinder dryers and combined ranges. Let us take as an example the traditional machine for drying pad-dyeings: the hotflue combined with an infra-red predryer. With this machine alone many different types of heating systems are possible. The most common type is the infra-red predryer fitted with gas or electric radiators and the steam-heated hotflue. Now what is the energy utilisation like with this type of drying range?

According to HOUBEN (3) the total efficiency of a gas-heated infra-red predryer is approx. 55 %. This refers to absorbed radiation heat. The efficiency can be improved by recycling the heat from the exhaust air into the circulating air system by convection. With an electric radiator the total efficiency is approx. 68 %. This is higher than with gas radiators, because the convection losses are lower.

With a steam-heated hotflue the efficiency is approx. 78 % (4). It must be borne in mind, however, that there are substantial losses in the pipeline system, and this also applies to the steam generation itself. With the usual steam generators the efficiency is approx. 85 % (5). If the dissipation and utilisation of heat are considered as a whole, then this means that only half of the primary energy originally applied is, in fact, utilised (6).

The Remaflam drying technology

The "Remaflam" drying process (6, 7) is entirely different. Here there are no long energy-ducts with accompanying heat losses and no separate heat exchanger. Here the energy is used exactly where it is needed, that is on the textile fabric itself. The efficiency of the primary energy applied is, therefore, much greater.

Energy carriers in the "Remaflam" process are combustible solvents, which are conveyed by the fabric that is to be dried. In principle, for the Remaflam process all those solvents which burn without producing soot can be applied. In practice it has become evident that a mixture of solvents with water is the easiest and most suitable solution, but rather limits the selection. For the "Remaflam" process, therefore, solvents are used that can be mixed with water. The reasons for this are as follows:

- Solvents that cannot be mixed with water are subject to more stringent laws as regards transport, decanting, storage and handling than the water-miscible solvents (with the same flash points).
- If pure solvents were to be used, the energy released during combustion would be wasted. The only way to ensure the utilisation of a small proportion of the energy would be by heat recovery. By using solvent/water mixtures the combustion energy is used with a high efficiency to evaporate the water component.
- All the conventional dyestuffs as well as the auxiliaries and chemicals usually used in dyeing processes are suitable for application with water. Almost all these products can be applied in solvent/water mixtures, but not in pure solvent.
- Last but not least, the use of water has the added bonus of a distinct reduction in costs.

In spite of the restrictions of energy carrier for the "Remaflam" process in the form of solvents that can be mixed with water, there still remains a wide range of suitable solvents to choose from.

Which energy carriers are suitable?

Owing to the demands made, aliphatic alcohols have a special significance, but acetone and dioxane, too, seem to be specially suitable as far as the fundamental demands are concerned. At a very early stage it was realised that the properties as regards density, surface tension, wetting-out qualities, flash point and also threshold limit values (TLV) are very important. The following solvents can be characterised as regards their reaction in mixtures with water:

- Methanol
- Ethanol
- Isopropanol
- Acetone
- Dioxane

Let us commence with the threshold limit value, the so-called TLV or lower toxic limits. From table I we can see

Table I: TLV and flash points of solvents and chemicals

	TLV (ppm)	Flash points (°C)
Hydrogen peroxide	1	-
Hydrochloric acid	5	-
Formic acid	5	69
Acetic acid	10	40
Ammonia	50	-
Dioxane	50	11
Methanol	200	11
Isopropanol	400	13
Acetone	1000	- 20
Ethanol	1000	12

the TLV in parts per million (ppm) and the flash points for the five solvents mentioned as well as for some of the chemicals used daily in dyehouses.

From table I it can be clearly seen that the five named solvents are suitable with regard to the lower toxic limits or threshold limit values, and certainly cannot be described as extraordinary when compared with normal chemicals in dyeing. Acetone itself must be excluded due to its extremely low flash point.

Table II: Prices (including calorific values) and oxygen demand (situation: January 1983)

	DM/kg	kJ/kg	DM/MJ	Oxygen demand (Mol O_2/Mol)
Methanol	0.50	19929	0.025	1.5
Ethanol	1.56	26837	0.058	3
Isopropanol	1.23	30472	0.040	4.5
Dioxane	5.30	24587	0.216	5

In table II the costs are listed taking account of the calorific value and the oxygen requirement.

For cost reasons and owing to the large amount of oxygen required for combustion, dioxane, too, must be eliminated, so that with regard to density, surface tension and wetting-out properties we can limit ourselves to the first three products in table II.

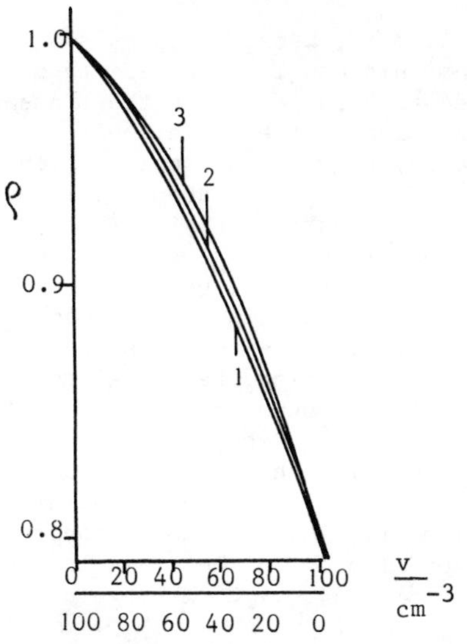

Fig. 2
Density $\frac{\rho}{[g \cdot cm^{-3}]}$ of alcohol (A) water (B) mixtures $\frac{V}{[cm^3]}$ at 20°C

1 Methanol/water mixtures
2 Ethanol/water mixtures
3 Isopropanol/water mixtures

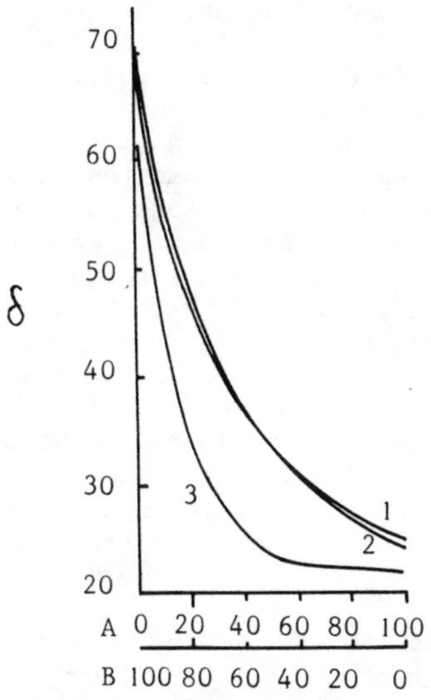

Fig. 3
Surface tension $\frac{\delta}{[dyn \cdot cm^{-1}]}$ of alcohol (A)/ Water (B) mixtures $\frac{V}{[cm^3]}$ at 20°C

1 Methanol/water mixtures
2 Ethanol/water mixtures
3 Isopropanol/water mixtures

In fig. 2 the densities of the respective mixtures with water at 20°C are given. The dependence of the surface tension on the respective mixtures with water can be seen in fig. 3, whereas the wetting-out properties of the mixtures with water are given in fig. 4. The flash points of the alcohol/water mixtures can be seen in fig. 5. As figs. 2 - 5 show, all three alcohols could be used for a combustion drying process. Although the longer chain alcohols ethanol and isopropanol have advantages with regard to the wetting-out

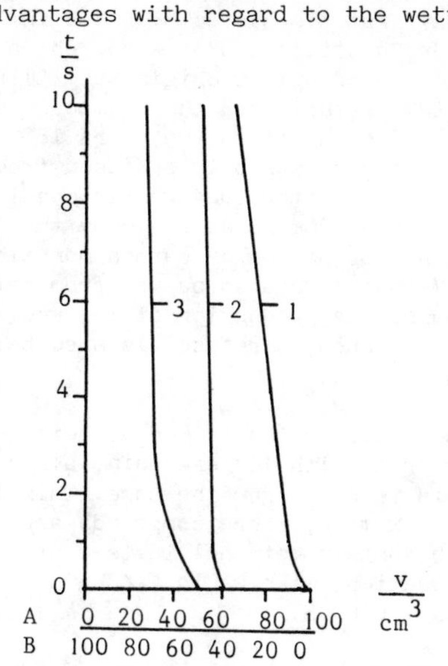

Fig. 4
Wetting-out values $\frac{t}{[s]}$ of alcohol (A)/water (B) mixtures $\frac{V}{[cm^3]}$ at 20°C

1 Methanol/water mixtures
2 Ethanol/water mixtures
3 Isopropanol/water mixtures

effect, if the cost situation according to table II and the oxygen requirement are taken into consideration, it will be seen that methanol has special priority as an energy carrier in drying. This is probably applicable to all countries in Europe. On the other hand, it must be considered that in some countries outside Europe the use of ethanol can have price advantages.

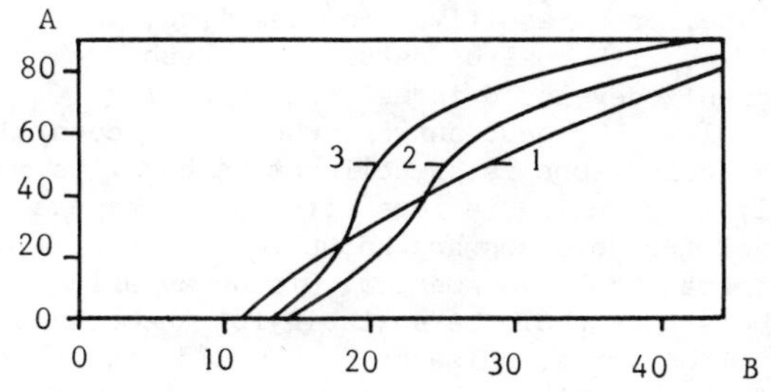

Fig. 5
Flash points of alcohol/water mixtures
A % by vol. water
B temperature (°C)

1 Methanol
2 Ethanol
3 Isopropanol

Methanol and ethanol

Methanol was first discovered by BOYLE in 1661 during the dry distillation of wood, which is where the term wood spirit originates. LIEBIG, DUMAS and PICTET elucidated the chemical constitution of methanol in the years 1835/36. At one time methanol was only produced from wood vinegar. Only after 1924 was it possible to win methanol on a commercial scale by the catalytical hydrogenation of carbon monoxide from coal. Methanol can also be won from natural gas or biomass. The production of methanol from waste products, such as refuse, is also being looked into.

Moreover, methanol can be termed as being similar to water, whereas with longer-chain, aliphatic alcohols this is no longer the case. This is obvious, for example, where compounds are concerned which contain methanol instead of water of crystallisation, like $CaCl_2 \cdot 4CH_3OH$, $CuSO_4 \cdot CH_3OH$, $MgCl_2 \cdot 6CH_3OH$.

Of the two alcohols, ethanol is most definitely the more pleasant and has been known a lot longer than methanol. At all events, ethanol was known in the form of alcoholic drinks in early history, as is proved by an old Sumerian tablet of cuneiform characters, about 9000 years old, describing the making of beer. The distillation of ethanol started between 1150 and 1250 in Southern Italy. Pure alcohol was first mentioned in 1554 and anhydrous ethanol was first prepared in 1796. Ethanol can be produced by alcoholic fermentation from biomass by catalytic hydrogenation from ethylene or by other processes. In the future the production of ethanol from agrarian products or even from agricultural waste promises to be the most interesting.

In recent years many countries have been endeavouring to use methanol and ethanol as gasoline, or as additives for gasoline. In countries with coal or natural gas resources and highly developed industrial technology, methanol is the best answer, whereas in countries where agriculture is predominant, ethanol is more likely to prevail. Serious efforts to produce ethanol from biomass are known to have been made in Africa and Latin America. Of the agrarian products that would be suitable for the ethanol production, rice, maize or corn would give the best yield. On the other hand, sugar millet, sugar cane or sugar beet could prove to be promising , because though the yield is not so high, a much smaller acreage is required (8).

Both methanol and ethanol are poisonous. If these alcohols are taken, the lethal concentration is about 250 g in the case of ethanol (equivalent to approx. 5 ‰ alcohol in the blood), and with methanol this is reckoned to be about 50 ... 75 g. The great danger in taking methanol is that this alcohol is only partly oxidised by the human organism, so that formaldehyde and formic acid can form. Through formaldehyde protein is precipitated and the oxidation processes in the body are obstructed. The retina in the eye is particularly affected, because it needs a great deal of oxygen. This can lead to defects of vision and even to blindness. For the transport, storage and handling of the two alcohols appropriate safety regulations apply, which, for example, in the FR of Germany are stipulated in the Ordinance on Combustible Fluids (VbF). Methanol and ethanol, being water-miscible fluids with flash points under 21°C, belong to group B. There are, therefore, certain regulations which must be observed. The alcohol/water mixture in the padder trough during the "Remaflam" process has a flash point of around 35°C and is no longer, therefore, subject to any special regulations. This becomes clear with the following comparison: a bottle of whisky with an alcohol content of 38 percent by volume is, in fact, more "dangerous" than the slop-pad liquor in the "Remaflam" process, which only contains 34 ... 35 % by volume of alcohol.

Description of the "Remaflam" process

As already mentioned, the most suitable energy source for the "Remaflam" drying process is methanol, in a quantity of approx. 35 % by volume. From fig. 2 it can be seen that the density of the methanol and water mixture depends on the methanol content. The greater the methanol content, the smaller the density, resulting in a difference between volume and weight percentage. To simplify matters we have decided to give the concentrations in percentages by volume. According to fig. 3 , as the methanol content increases, the surface tension decreases, too and this indicates that the wetting-out effect improves with the methanol content (fig. 4). It must be stressed, however, that the wetting-out effect only becomes interesting from approx. 50 % by volume upwards. Whether or not a wetting agent is required does, however, depend on the preparatory treatment of the fabric to be treated by the "Remaflam" process. If necessary, the addition of isopropanol to the methanol/water mixture would also result in a better wetting-out effect.

As can be seen from fig. 5 , in the case of a high methanol content the slop-pad liquor would have to be cooled for safety reasons. It has been proved, however, that a quantity of methanol of 35 % by volume can be safely used for the "Remaflam" drying process at room temperature, and must even be heated before ignition. For this reason, the impregnated fabric is heated by infra-red radiation.

During combustion only CO_2 and H_2O are formed, i. e. the process is not detrimental to the environment.

Methanol burns in three combustion stages:

1.	$2CH_3OH$	+	O_2 ———	$2CH_2O$	+	$2H_2O$
	Methanol		oxygen	formaldehyde		water

2.	$2CH_2O$	+	O_2 ______	$2CO$	+	$2H_2O$
	formaldehyde		oxygen	carbon monoxide		

3.	$2CO$	+	O_2 ______	$2CO_2$
	carbon monoxide		oxygen	carbon dioxide

According to table I the lower toxic limits or TLV are around 200 ppm, referring to 100 % methanol. With mixtures containing 35 % by volume of methanol, safe working conditions are guaranteed, as ranges in practical operation have proved.

The regulations concerning the storage of methanol in great quantities should be strictly oberserved. In practice, tanks with a content of 20 000 ... 30 000 litres of methanol are stored in the open. The methanol is transported to the machine in an enclosed ring pipe and is there mixed with the aqueous dyeliquor in the required ratio in an enclosed system. The working personnel does not come into contact with pure methanol at all. The methanol/water mixture, which is then fed to the padder trough, does not exceed the lower toxic limits in the vicinity of the padder.

The dyestuffs are first dispersed or dissolved in water in the usual way, whereby the methanol component is deducted from the total amount of the liquor. If, for example, 5000 litres of slop-pad liquor is to be prepared with 80 g/l of dyestuff, then the dyebath is set with the necessary 400 kilos of dyestuff in 3250 litres of water. Mixing with the methanol then takes place automatically in the mixing station. It should not at this point be concealed that this results in a minor error which can, however, be practically ignored. When mixing methanol with water a contraction in volume occurs which, with 35 % by volume of methanol, lies in the magnitude of 2.5 ... 3 %. When the fabric is padded with the methanol/water slop-pad liquor, you can expect practically the same liqour absorption as with water alone, if the liqour absorption is not expressed in percentage by weight but in percentage by volume. This means in practice that either the known values are applied, or that with new fabrics the liquor absorption can be determined in the usual way with water only, and by weighing before and after padding.

After padding the fabric first runs through an infra-red section, whereby the liquor on the fabric is heated above its flash point. A slop-pad liquor with 35 % by volume of methanol must, therefore, first be heated up to approx. 35°C. As soon as the flash point of the liquor on the textile fabric is attained, then adequate alcohol vapour will have formed on the surface of the fabric, which can then be ignited. The combustion of the alcoholic vapours causes the actual drying energy of the "Remaflam" process to be released and enables a further evaporation of the liquor on the fabric. The remaining alcohol component is thereby combusted still further and the water component is evaporated down to a controlled residual moisture. The fabric itself is only heated up to the marginal cooling temperature, and on no account can its temperature exceed 100°C. In practice, we have only ever measured max. 70°C on the fabric.

Description of the "remaflam" dryer

Some time ago in a joint venture, Brückner Trockentechnik GmbH & Co. KG and Hoechst AG, both in the FR of Germany, developed the first pilot plant, which was installed in the Technical Development Department (Dyestuffs) of Hoechst AG (fig. 6). There followed a trial period of approx. one year, in which time valuable knowledge was gained, which could be applied when constructing the first production plant. In the meantime further "Remaflam" ranges have been built and put into operation (fig. 7/8). The total set-up of the "Remaflam" range can be seen in the diagramm (fig. 9).

The fabric delivery and wind-up are accommodated on the operating side of the machine, so that only one operative is required to supervise the process.

The fabric to be dyed is fed via guide rollers an selvedge guiders to the dyeing mangle. It is advisable to maintain a constant fabric tension from the beginning to the end of a batch. The padder trough is situated as close to the two squeezing rollers as possible, so as to avoid colour streaking on the fabric.

Fig. 6
"Remaflam" pilot plant in Technical Development Department (Dyestuffs) of Hoechst AG, FR of Germany

Fig. 7
"Remaflam" production plant

After passing through the padder nip the fabric enters a dwell zone (guide rollers coated with polytetrafluorethylene) before entering the actual "Remaflam" chamber. In the drying oven the fabric first passes through an infra-red heating zone with six pairs of radiators, and then enters the combustion chamber. Then the fabric passes through a slotted diaphragm or partition into a section of the post-combustion chamber, and on leaving this passes through the counterflow air circulation system. The fabric is then taken up by a guide roller situated in the area of the

Fig. 8
"Remaflam" production plant

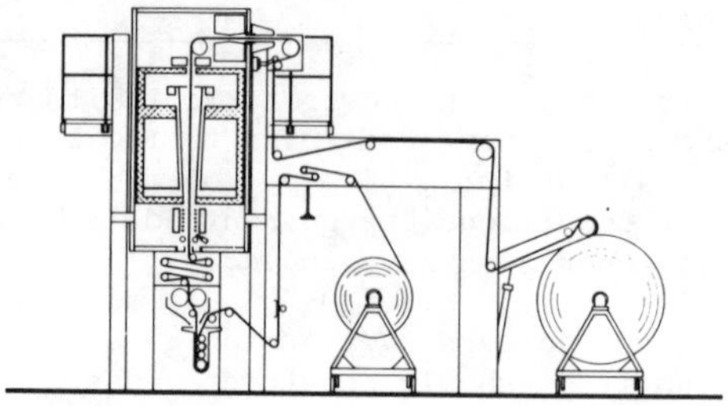

Fig. 9
Diagram of a "Remaflam" range

fresh air intake and is guided out of the machine. Fig. 10 diagrammatically illustrates the cold and hot air flow of the drying range. The light arrows mean cold air, which flows in from below and from above through the counterflow air circulation system. The dark arrows show the flow of hot air within the dryer.

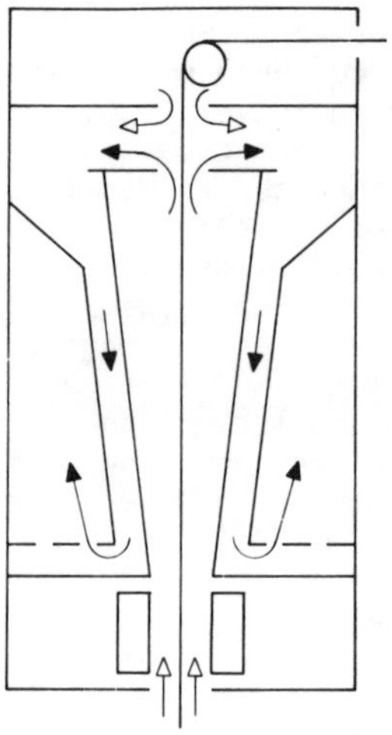

Fig. 10
Air flow on the "Remaflam" range

The hot air gases that form in the combustion chamber are extracted by a fan and conducted along the rear of the combustion chamber panels in counterflow. In this way heat energy is

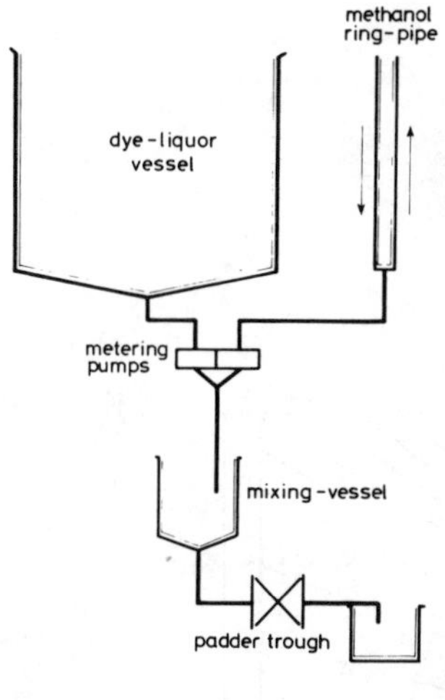

Fig. 11
Diagram of mixing and metering arrangement of a "Remaflam" range

1 Dyestuff preparation tank
2 Methanol ring-pipe
3 Metering pump
4 Mixing tank
5 Padder trough

transferred to the sheet steel panels, which then act as dark radiators and pass this energy on to the fabric in the combustion chamber, thereby accelerating the evaporation process. A large portion of the energy from the extracted hot air gases can eventually be recovered in a heat recovery plant.

A ring pipe leads to the liquor preparation station (fig. 11) and this conveys the methanol by means of a pump from the storage tank standing out in the open through the liquor preparation station and back to the storage tank. In a separate preparation tank the dye liquor is prepared in the usual way with water, except that the water component is reduced by the necessary amount of methanol.

Two metering pumps then mix the dye liquor with the methanol, thus producing the padding liquor. From a mixing tank this then passes via a solenoid valve to the pad mangle. This mixing unit is, by the way, also suitable for the slop-pad short dwell process with reactive dyestuffs, if for any reason the padder of the "Remaflam" plant is to be used by itself.

During mixing the temperature increases slightly. If the dye liquor has been set to 28°C and this temperature is maintained with the aid of a contact thermometer, then the slop-pad liquor can be fed to the padder trough with a temperature of approx. 30°C. At this temperature the paddingliqour is approx. 4 ... 5°C below its flash point and is, therefore, not combustible. The temperature is monitored in the padder by a contact thermometer, which initiates cooling of the trough.

Process sequence

After slop-padding the fabric first passes through the infra-red heating section, before entering the combustion chamber. The infra-red radiators are medium-wave radiators, the wave-length of which is in the main tuned in to substances containing hydroxyl groups.

Ignition takes place once only for each fabric batch on one side of the fabric, since the other side catches fire automatically. Fig. 12 shows the infra-red radiators, the moving fabric and the flame on both sides of the fabric. It can be clearly seen that the flash point of the methanol|water mixture is only reached after the second pair of radiators.

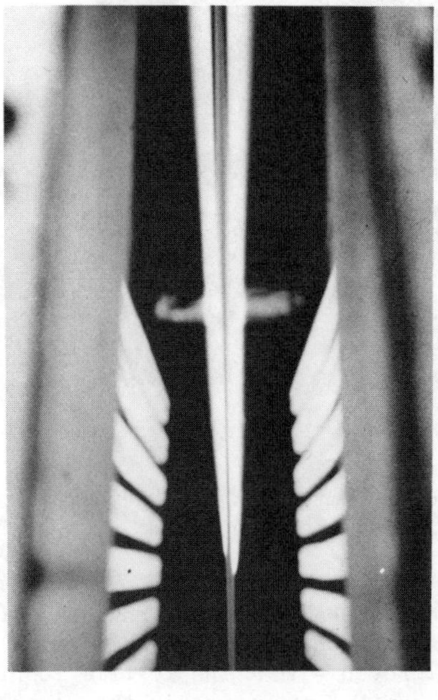

Fig. 12
Infra-red radiators with fabric web and enveloping flame

After passing through the infra-red zone the fabric surrounded by the flame enters the actual combustion chamber. The evaporation of the slop-pad liquor on the fabric which was initiated by the infra-red radiators is now continued with the aid of the thermal energy which is released when the methanol is combusted. The net calorific value (H_u = combustion heat minus condensation heat of the water quota) of the methanol is approx. 19 900 kJ/kg, thus the amount of energy involved is considerable. In addition to the energy available in the combustion chamber, heat is radiated from the steel panels of the combustion chamber which, as already mentioned, are heated by the hot air flowing behind them on the counterflow principle. This achieves optimum utilisation of the energy released during the combustion process within a very small area. Since fig. 13 was produced with an infra-red film, all cold areas appear dark on the picture. The higher the temperature of the photographed object, the lighter this area is on the picture. The flame is white, the radiators, the emission temperature of which is about 800°C, also appear very light, and the chamber panels are not so light (dark radiators), but the most important point about this illustration is that the fabric and its immediate vicinity appear dark, i. e. are cold.

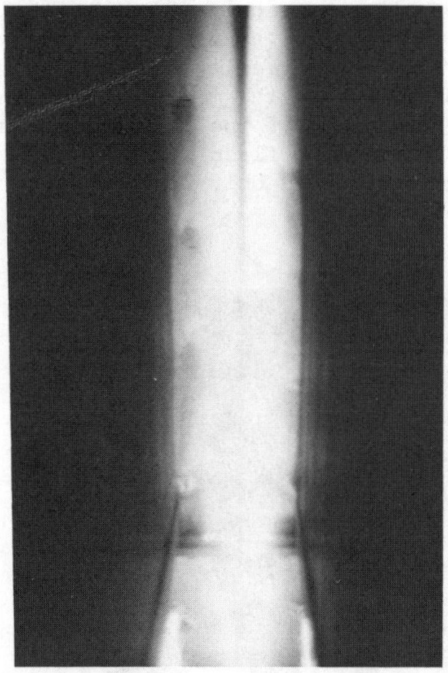

Fig. 13
Temperature distribution in the "Remaflam" chamber (infra-red photograph)

When a liquid evaporates it withdraws heat from its immediate surroundings. This means - as measurements proved - that the textile fabric in this case has a temperature in the combustion chamber of between 45 and 70°C, whereas the air in the upper part of the chamber achieves 600°C and in the post-combustion chamber over 800°C. The marginal cooling temperature of the fabric only exists as long as liquid is evaporating. The decisive point to ensure a correct functioning of the "Remaflam" plant is, therefore, the maintenance of the marginal cooling temperature of the fabric until it leaves the drying range.

The fabric speed in the "Remaflam" drying process demands as in the case of conventional drying methods, on the fabric weight, the amount of liquor applied and the required residual moisture. Thus, the lighter the fabric is, the quicker it can be dried. The greater the liquor absorption is, the longer the drying process takes. The length of the combustion chamber also plays an important role.

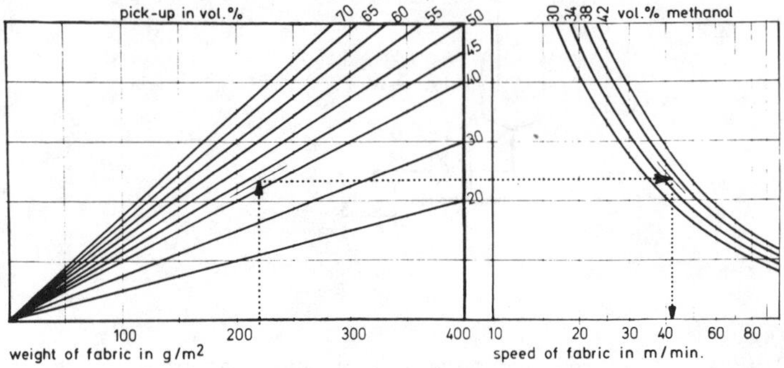

Fig. 14
Fabric speed graph for "Remaflam" process

Example:	Fabric weight	=	220 g/m^2
	Liquor pick-up	=	43 vol. %
	Methanol quota	=	36 vol. %
	Fabric speed	=	42 m/min.

Here there are similarities regarding conventional drying machines, because the more chambers a hotflue has, the faster is the passage of the fabric to be dried. An unrestricted fabric passage in the combustion chamber is, however, subject to limitations.

With the "Remaflam" drying method, an additional factor has to be considered, namely the methanol quota in the slop-pad liquor. Methanol requires much less energy to evaporate than water, i. e. the more methanol the slop-pad liquor contains, the faster the drying process is. These correlations are illustrated in fig. 14.

As the dotted arrows show, in the case of a polyester and cotton blend fabric with a weight of 220 g/m^2 and a liquor absorption of 43 % by volume using 36 % by volume of methanol, the fabric speed is 42 m/min.

In the "Remaflam" dryer the drying process begins with an initial speed of 10 m/min. After approx. 5 min the fabric speed reaches 70 % of the final speed, which is achieved after about 30 min. It ensues that the average drying speed also depends on the length of the fabric batch. There is, therefore, a direct correlation with the heating-up of the range mass. The differences in the fabric speed have, as trials with various fabrics have shown, practically no influence on the liquor absorption.

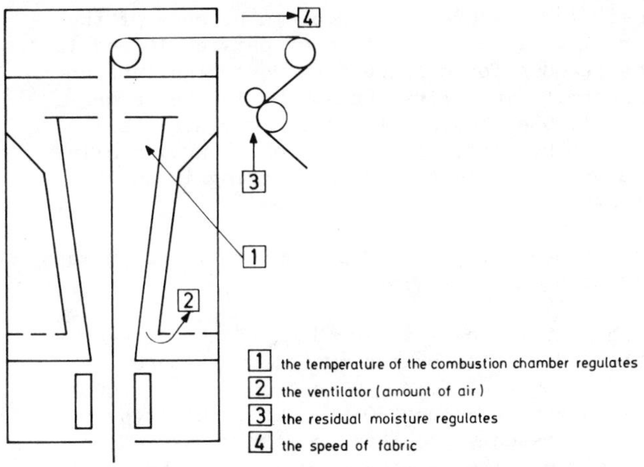

Fig. 15
Control diagram for a "Remaflam" range

It is understandable that the machine has to be automatically controlled. The operative has to prepare the machine. Then the infra-red radiators are heated up. When the machine is not in operation, the infra-red radiators are swung back by 180°. During the heating-up phase they radiate in the opposite direction to the fabric. After approx. 4 min the heating-up period is completed and the command "ready to start" lights up on the control console. After the start button has been pushed, controllers automatically take over all functions. The infra-red radiators swivel into position as soon as the wet fabric passes by and the pilot flame appears shortly afterwards. It only burns approx. 3s. As soon as this has taken place, the flame "feeds" itself.

A thermocouple in the combustion chamber takes over the control of the flue gas fan (fig. 15) and thereby meters the requisite air volume. The control is so designed that the preselected index temperature is always maintained. As a rule, a temperature of 600°C has been operated with up to now.

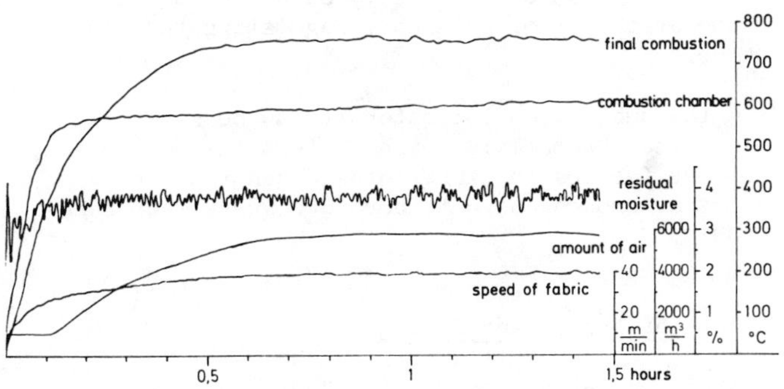

Fig. 16
Readings of a "Remaflam" long-range trial with polyester/cotton 320 g/m

A moisture measuring unit by Mahlo, FR of Germany, scans the fabric to determine the residual moisture, which can also be preselected. The measured values are passed on to the pad mangle and the batching unit, and the running speed of the machine is adjusted accordingly. From fig. 16 it can be seen that the optimum speed is achieved after about 30 min.

The optimum exhaust air volume of the fan is reached after approx. 45 min. The residual moisture of the fabric is already constant from the beginning of the process. In the combustion chamber the selected optimum temperature is attained after a few minutes, whereas the post-combustion does not achieve optimum conditions until after half an hour, owing to the heating-up of the metal mass.

All modern drying installations in the FR of Germany have to be inspected and accepted by the local authorities with regard to working safety and air pollution. The exhaust emission rates of the "Remaflam" range are in accordance with legal stipulations. Moreover, all technical safety appliances have been inspected and accepted by the relevant authorities.

Of course, such a highly automated plant as the "Remaflam" range is also equipped with automatically operating safety appliances.

During operation the flames in the combustion chamber are continuously monitored by a UV cell. If for any reason the flame should be extinguished, the UV cell immediately turns off the drive.

Another UV cell monitors the area between the padder and the combustion camber. If there is a power cut the padder is automatically sprinkled with water. This safety arrangement can, if required, be switched on by hand and used for cleaning the padder.

The exhaust fan provides adequate aeration of the plant. A 10 min preliminary running time ensures that at no time is there a build-up of an explosive methanol/air mixture in the dryer. The fan itself is controlled by an overspeed monitor.

Other safety equipment serves for the temperature control of the combustion chamber, the post-combustion chamber and the exhaust air.

Practical experience with the "Remaflam" drying process

The following are examples of the capacities that can be achieved when drying various fabrics in the approx. 1.60 m long drying zone of the "Remaflam" plant:

Cotton	(1.50 m wide, 230 g/m²)	
Liquor absorption		70 %
Max. fabric speed		23 m/min
Polyester/cotton 67/33	(1.50 m wide, 230 g/m²)	
Liquor absorption		42 %
Max. fabric speed		45 m/min
Polyester/cotton 50/50	(1.50 m wide, 110 g/m²)	
Liquor absorption		50 %
Max. fabric speed		60 m/min

These figures refer to a methanol application of 34 % by volume and a combustion chamber temperature of 600°C.

That an increase in the performance of the machine is possible is shown by some trials that were carried out on a newly installed machine.

The "Remaflam" plant was first put into operation under normal conditions, as are customary with other machines. Normal conditions are: 34 % by volume of methanol, 600°C combustion chamber temperature, 4 % residual moisture content.

Initially only cotton with a high liquor absorption was dried. The output achieved was around 25 m/min and it was intended to increase this slightly. According to our experience, and we have already carried out trials with many customers on all types of fabrics, as yet not one customer was interested in increasing the performance, but rather the contrary was true and it was requested that the plant should run slower.

To increase the output the amount of alcohol and the index temperature in the combustion chamber were increased:

Methanol	(% by vol.)	34	34	36
Liquor absorption	(% by vol.)	60	60	63
Temperature in combustion chamber	(°C)	600	700	750
Temperature in the post-combustion area	(°C)	800	950	1000
Fan capacity	(%)	15	20	35
Fabric speed	(m/min)	25	32	43

Up until now, in practical operation it has been proved that all blended fabrics of polyester fibre and cotton or rayon staple fibre can be fullyauto-matically dried on the "Remaflam" plant with successful results and without any problems. The same can be said of all-cotton and all-rayon goods. Good drying results have also been achieved with cotton corduroy and mixture corduroys of polyester/cotton. Also trials with jute fabric, used for pigment dyeing of textile wall coverings, were very successful.

The drying of fabrics of 100 % synthetic fibres will involve primarily polyester fibres.

Those polyester articles suitable for a continuous dyeing process must be divided into three groups:

Article made of filament yarns

These fabrics present problems because of the very low liquor absorption of between 10 and 12 %. The reasons for difficulties regarding fabric uniformity must first be sought in the padder itself. One should attempt to increase the liquor pick-up by means of auxiliaries. Fundamentally, the prerequisite for uniform drying is an even slop-padding process.

Fabric made of textured polyester fibres in warp and weft

Such articles have a very high liquor pick-up which, depending on the surface structure, is between 60 and 100 %. In most cases the liquor absorption is around 80 ... 85 %. This type of fabric presents problems in both slop-padding and drying. The surface structure of the fabric plays an important part in achieving uniform drying results. The problems with such articles are due to the fact that the liquor is not adequately absorbed by the fibres and practically remains just on the surface or "floats" between the fibres. This means that especially in the case of light-weight fabrics with a very smooth surface, there are levelling problems. Our previous trials with such fabrics have resulted in second-class quality goods. Only in the case of heavy fabric with a distinctive surface structure (e. g. twill weave) have good results been achieved so far. According to our information, on a "Remaflam" plant that was recently started up, heavy woven polyester fabrics are being dried with excellent results.

Lining materials made of polyester filament yarn in the warp and textured polyester in the weft

With these fabrics the liquor pick-up is around 30 %.

Consequently conditions are favourable both for the uniform impregnation of the fabrics on the padder and for subsequent drying.

A commercial unit has been giving troublefree service for some time in the drying of polyester lining material. It has proven advantageous for the unit to be operated at constant speed. This mode of operation can easily be achieved with automatic control by simply regulating the air flow.

On the subject of polyester it can be generally said that "Remaflam" drying offers the possibility to dry fabrics that are to be discharged in a further process at a temperature guaranteed below 100°C.

Costs of "Remaflam" drying

Points in favour of "Remaflam" drying which should be borne in mind when doing a calculation, are:

- independence form oil
- better energy utilisation
- smaller space requirement
- less time required for cleaning and maintenance
- no wasted energy through overdrying
- no energy consumption for preheating and during downtimes
- no increase in energy consumption when drying narrow fabric
- fully automatic control
- gentle treatment of the fabric with optimum drying results
- advantages of shock-drying
- no padding auxiliaries required
- heat recovery more interesting than with conventional ranges because of the high exhaust air temperature

Disadvantages could be:

- loss of jobs through fully automatic machine (one operative)
- uncertain supply situation with methanol (although the efforts in the motor vehicle sector signify that greater quantities of methanol and ethanol will be available in the coming years)
- no price guarantee (owing to the greater demand because of the potential as a petrol additive, the methanol price, though likely to increase, should be lower than for fuel oil in the long run)
- increased costs through the loss of the alcohol component in the residual liquor
- observation of certain regulations concerning the storage and handling of methanol
- unsuitable for drying knitted goods
- cut edges tend to melt or are singed, loose threads are burnt off
- limitation on the fabric speed

The following calculation is based on average prices. These were determined from price quotes obtained from various textile mills or from trade literature.

Water	2.30 DM/m³
Electricity	0.12 DM/kWh
Wages	28.00 DM/h
Methanol	0.50 DM/kg

The price for methanol refers to tank truck supply (20 000 ... 30 000 l).

Let us now determine the fabric finishing costs for a batch 5000 m long, padded and dried by the "Remaflam" process, taking as a basis a blended polyester/cotton fabric 67/33 with a width of 1.50 m and a weight of 220 g/m². The liquor pick-up of the fabric is 45 % by volume and is equivalent to the amount of fluid to be evaporated. The fabric is dried at an average speed of 40 m/min with a residual moisture content of approx. 4 % at the exit of the dryer. The methanol content of the slop-pad liquor is 34 % by volume.

The operating time is calculated from the running time of the fabric batch plus setting-up time to prepare the machine and to clean the padder on completion of the batch.

$$t_{ges} = t_l + t_r$$

$$t_l = \frac{L}{v}$$

$$t_{ges}\,[h] = \frac{\dfrac{L\,[m]}{v\,[m/min]} + t_r\,[min]}{60\,[min/h]}$$

$$t_{ges} = \frac{\dfrac{5000}{40} + 10}{60} = 2.25\ h$$

Practical experience has shown that a maximum setting-up time of 10 min between two batches is perfectly adequate. The preparation of the new fabric and the next padding liquor can take place during operation, which is fully automatic. On completion of a batch the machine merely has to be cleaned and the new fabric threaded in.

The quantity of methanol is calculated from the total liquor requirement, made up of the liquor absorption of the fabric and the fabric weight as well as the contents of the padder trough (residual liquor). When water and methanol are mixed, a contraction of approx. 3 % by volume takes place, which must be allowed for in the calculations.

The required water quantity is calculated in the same way. Another 400 ... 500 l of water must be added for cleaning the machine.

$$m_M[kg] = L[m] \times B[m] \times \frac{G[g/m^2]}{1000\ [g/kg]} \times FA[1/100\ kg] + CH[l] \times f_1[\%] \times V_M[\%\ by\ vol.] \times f_2[kg/l]$$

$$m_M = (5000 \times 1.5 \times 0.22 \times 0.45 + 50) \times 1.03 \times 0.34 \times 0.792 = 219.8\ kg$$

$$m_W[m^3] = \frac{\left(L[m] \times B[m] \times \frac{G[g/m^2]}{1000\ [g/kg]} \times FA[1/100\ kg] + CH[l]\right) \times f_1[\%] \times V_W[\%\ by\ vol.] + R_W[l]}{1000\ [l/m^3]}$$

$$m_W = \frac{(5000 \times 1.5 \times 0.22 \times 0.45 + 50) \times 1.03 \times 0.66 + 450}{1000} = 0.989\ m^3$$

The electricity requirement is made up of the actual power consumption for the infra-red radiators, the drive motors, the fans and the transport rollers.

The total installed power is fully effective for the infra-red radiators. The power consumption for the drive motors etc. is approx. 70 % of the installed power. The calculation allows for a heating-up time for the infra-red radiators of approx. 4 min before the machine is started up in order to attain their full capacity.

$$E_{ges}[kWh] = P_{IR}[kW] \times (t_1[h] + t_3[h]) + P_M[kW] \times f_4[\%] \times t_1[h]$$

$$E_{ges} = 48 \times (2.083 + 0.067) + 32 \times 0.7 \times 2.083 = 149.86\ kWh$$

Applying the values as determined, the following total drying costs emerge:

Wages					
DM/h x t_{ges}	=	28	x	2.25	= DM 63,--
Methanol costs					
DM/kg x m_M	=	0.50	x	219.8	= DM 109.90
Water costs					
DM/m^3 x m_W	=	2.30	x	0.989	= DM 2.27
Electricity costs					
DM/kWh x E_{ges}	=	0.12	x	149.86	= DM 17.98
Total costs					DM 193.15

From these total costs the following drying costs can be determined:

per running metre fabric	=	DM 0.0386
per m^2 fabric	=	DM 0.0258
per kg fabric	=	DM 0.1171

This calculation did not allow for the recovery of heating energy from the exhaust air of the "Remaflam" plant.

The following calculation, using polyester lining material as an example, is presented to show that for fabrics with low liquor pick-up the price of methanol exerts only a minor influence on the overall costs if not only the drying costs alone but the entire dyeing costs are considered. Furthermore, this example also shows how the costs compare with those of a customary dyeing method, HT exhaust dyeing (omitting capital expenditure and depreciation costs).

General data

4000 m polyeter lining fabric
60 g/m^2, 1.50 m wide

Exhaust method on the HT beam dyeing machine

Winding up at 60 m/min (1-man operation)

Total treatment time: 6 hours
Water consumption for dyeing, final scouring and rinsing:
4 x 3000 l (130°C, 60°C, 2 x cold), Δt = 150 K
Heating: steam at 7 bar (2.1 GJ/t)
Operators: 1 man for 3 machines
Installed power connection: 45 kW

Continuous method with Remaflam drying and thermosolling

(combined Remaflam/thermocontact unit)
Production speed: 50 m/min
Operators: 1 man
Liquor pick-up: 30 % by vol.
Methanol portion: 34 % by vol.
Setting up time: 10 min
Installed power connections:
Remaflam unit (with padder, without IR radiant heater): 32 kW
Thermocontact unit: 36 kW
IR radiant heater: 48 kW
Preheating time for IR heater: 4 min
Preheating time for thermocontact unit: 10 min
Heating for thermocontact unit
(natural gas): 0.67 GJ/h

Residual liquor: 50 l
Dye lost in residual liquor: 10 g/l
Cleaning water: 500 l

Exhaust dyeing in HT beam dyeing machine

Labour costs of winding up goods

$$\frac{L[m] \times A \times S\ [DM/h]}{v\ [m/min] \times 60\ [min/h]} =$$

$$\frac{4000 \times 1 \times 28}{60 \times 60} = DM\ 31.11$$

+ labour costs for dyeing

$$t_{ges}[h] \times A \times S\ [DM/h] =$$

$$6 \times 1/3 \times 28 = DM\ 56$$

+ water costs

$$\frac{V[l] \times f_7 \times W[DM/m^3]}{10^3 [l/m^3]} =$$

$$\frac{3000 \times 4 \times 2.3}{10^3} = DM\ 27.60$$

+ steam costs

$$\frac{\Delta t[K] \times f_5[\%] \times V[l] \times 4.187[KJ/l \times K] \times D[DM/t]}{2.1\ GH/t \times 10^6\ KJ/GJ} =$$

$$\frac{150 \times 1.2 \times 3000 \times 4.187 \times 65}{2.1 \times 10^6} = DM\ 69.98$$

+ power costs

$$P_M[kW] \times f_4[\%] \times t_{ges}[h] \times E_P\ DM/kWh =$$

$$45 \times 0.9 \times 6 \times 0.12 = DM\ 29.16$$

Continuous dyeing

$$t_1[min] = \frac{L[m]}{v\ [m/min]} = \frac{4000}{50} = 80\ min$$

Labour costs

$$\frac{(t_1[min] + t_r[min]) \times S[DM/h]}{60[min/h]} =$$

$$\frac{(80 + 10) \times 28}{60} = DM\ 42$$

+ power costs, motor drive

$$\frac{t_1[min] \times P_M[kW] \times f_4[\%] \times E_P[DM/kWh]}{60\ [min/h]} =$$

$$\frac{80 \times 68 \times 0.7 \times 0.12}{60} = DM\ 7.62$$

+ power costs, IR heater

$$\frac{(t_1[min] + f_3[min]) \times P_{IR}[kW] \times E_P[DM/kWh]}{60\ [min/h]} =$$

$$\frac{(80 + 4) \times 48 \times 0.12}{60} = DM\ 8.06$$

+ water costs

$$\frac{\left[\left(\frac{L[m] \times B[m] \times G[g/m^2] \times FA[1/100\ kg]}{10^3\ [g/kg]} + CH[l]\right) \times f_1[\%] \times V_W[\%] + RW[l]\right]}{\times\ (1/W(DM/m^3)) \times 10^3\ [l/m^3]} =$$

$$\frac{\left[\left(\frac{4000 \times 1.5 \times 60 \times 0.3}{10^3} + 50\right) \times 1.03 \times 0.66 + 500\right] \times 2.3}{10^3} = DM\ 1.40$$

+ methanol costs

$$\left(\frac{L[m] \times B[m] \times G[g/m^2] \times FA[1/100\ kg]}{10^3\ [g/kg]} + CH[l]\right) \times f_1[\%] \times V_M[\%] \times f_2[kg/l] \times M[DM/kg] =$$

$$\left(\frac{4000 \times 1.5 \times 60 \times 0.3}{10^3} + 50\right) \times 1.03 \times 0.34 \times 0.792 \times 0.5 = DM\ 21.91$$

+ heating cost, thermocontact unit

$$\frac{(t_1[min] + f_6[min]) \times W_K[GJ/h] \times G_P[DM/m^3]}{60\ [min/h] \times W_G[GJ/m^3]} =$$

$$\frac{(80 + 10) \times 0.67 \times 0.3}{60 \times 0.0335} = DM\ 9$$

+ dye costs (in residual liquor)

$$\frac{CH[l] \times F[g/l] \times F_P[DM/kg]}{10^3\ [g/kg]} =$$

$$\frac{50 \times 10 \times 20}{10^3} = DM\ 10$$

Total costs	Exhaust method	Continuous method
Labour	DM 87.11	DM 42.00
Water	DM 27.60	DM 1.40
Steam	DM 69.98	
Elec. power	DM 29.16	DM 15.68
Methanol		DM 21.91
Gas		DM 9.00
Dye		DM 10.00
Aftertreatment		DM 52.89
	DM 213.85	DM 152.88
	= 0.59 DM/kg	= 0.42 DM/kg

The aftertreatment costs for continuous dyeing (reductive cleaning on an open-width washing machine) is made up of the following costs that are known from commercial experience:

water consumption: 15 l per kg material, steam consumption: 1 MJ per kg material, power consumption (60 kW) and labour costs (1 workman) during an operating time of 50 minutes (= 80 m/min).

In the calculation the price of methanol comprises only 14.3 % of the overall costs, which means that even if the price of methanol doubled, the overall costs would rise by only 14.3 %.

The following calculation is also of interest:

The HT dyeing machine cannot always be operated with its optimum loading capacity of 4000 to 5000 m goods, for there are frequently smaller lots of 3000, 2000 or even 1000 m that must be dyed on the same machine.

As the following figures (DM per kg goods) show, the overal costs of continuous dyeing change only slightly while those for exhaust dyeing are markedly altered.

	Exhaust dyeing	Continuous dyeing
1 000 m	DM 2.12	DM 0.63
2 000 m	DM 1.10	DM 0.49
3 000 m	DM 0.76	DM 0.45
4 000 m	DM 0.59	DM 0.42
5 000 m	DM 0.49	DM 0.41
10 000 m	DM 0.49	DM 0.38
20 000 m	DM 0.49	DM 0.37

From the aforesaid it follows that the "Remaflam" drying method can be described as a cost-saving and fibresafe process and that alcohol can be used as energy carrier for drying processes. In the long run, methanol will be less expensive than oil. At present, the alcohol production all over the world is very sluggish, but in many countries there are earnest attempts to use alcohol as an alternative source of energy. Examples of this are South Africa and Brasil in their production of ethanol from biomass. In the first instance, of course, the idea is to substitute alcohol for motor fuel.

In the future alcohol should have a firm place among all the other energy carriers, if only because of its environmental protection qualities.

Literature

(1) Anonym, Energiekrise:vom Verbrauchs- zum Leistungsdenken, Ch/TI 29/81 (1979), 9, 639

(2) Anon., Energieprobleme in der Textilindustrie, Ch/TI 29/81 (1979), 9, 665

(3) HOUBEN, H., Wichtigste Trocknungsprinzipien für textile Flächengebilde, TV 9 (1974), 9, 394

(4) JOOS, W., Das Färben von Zellulosefasern mit Reaktivfarbstoffen nach dem Kaltlagerverfahren TV 10 (1975), 2, 72

(5) ERNST, H., STILLE, G., KURZ, J., Einsparungsmöglichkeiten bei der Wärmeenergie, R + W, 1 (1977), 30, 18

(6) PETERSOHN, G., Praktische Erfahrungen beim Färben und Trocknen nach dem Remaflam-Verfahren, TPI 34 (1979), 12, 1647

(7) VON DER ELTZ, H.-U., PETERSOHN, G., SCHÖN, F., Alkohole als Energieträger in der Textilveredlung: das Remaflam-Verfahren, ITB 2/81, 101

(8) BERNHARDT, W., MENRAD, H., KOENIG, A., Ethanol aus Biomasse als zukünftiger Kraftstoff für Automobile, Starch/Stärke 31 (1979), 8, 254

DEWATERING PROCESS ENHANCED BY ELECTROOSMOSIS

H. Yukawa
Gunma University, Kiryu, Japan

H. Yoshida
Oyama Technical College, Oyama, Japan

ABSTRACT

It is possible to apply electroosmosis to dewatering of sludge. Electroosmotic dewatering is very effective for gelatinous and colloidal sludges which are very difficult to dewater by mechanical methods.

The electroosmotic dewatering processes under conditions of constant electric current and constant voltage are explained in this chapter. The mechanisms of electroosmotic dewatering for compressible sludge under the conditions of constant electric current and constant voltage are discussed, based on a model of electroosmotic flow through a compressible-particle packed bed. The rate of electroosmotic dewatering and the electric power consumption are theoretically analysed for each operating condition. The equations obtained theoretically are experimentally confirmed by use of the compressible sludges such as white clay, magnesium hydroxide and bentonite. Bentonite is used as an example of a gelatinous sludge. The theoretical equations under both operating conditions are useful for the design of practical electroosmotic dewatering equipment. The electroosmotic dewatering is particularly effective for gelatinous sludge because the electroosmotic driving force of dewatering occurs in the internal sludge. The electroosmotic dewatering efficiency under the constant electric current is compared with that under the constant voltage, concerning dewatering rate, water content in dewatered sludge and dewatered volume per unit electric power consumption. In both operating conditions, a few characteristics of electroosmotic dewatering are also described.

NOMENCLATURE

A	= cross-sectional area of sludge	[m^2]
a	= experimental constant defined as Eq.(18)	[-]
D	= dielectric constant of liquid	[$F \cdot m^{-1}$]
d_p	= median diameter of particles	[μm]
E	= strength of electric field	[$V \cdot m^{-1}$]
E_{av}	= average strength of electric field	[$V \cdot m^{-1}$]
H	= height	[cm] or [m]
I	= electric current density	[$A \cdot m^{-2}$]
k	= Kozeny-Carman's constant	[-]
n	= experimental constant defined as Eq.(18)	[-]
p_E	= electroosmotic pressure	[kPa]
p_{EP}	= electroosmotic permeation pressure	[Pa]
p_{EP}'	= modified p_{EP} defined as Eq.(18)	[Pa]
Q_E	= volume dewatered by electroosmosis	[cm^3] or [m^3]
Q_G	= volume dewatered by gravitation	[cm^3]
Q_T	= total volume dewatered by electroosmosis and gravitation	[cm^3]
q_E	= electroosmotic flow rate	[$m^3 \cdot s^{-1}$]
q_E'	= flow rate of electroosmotic permeation	[$m^3 \cdot s^{-1}$]
S_v	= specific surface of particles	[$m^2 \cdot m^{-3}$]
t	= time	[s] or [min]
t_e	= time till end of dewatering	[s] or [min]
u_E	= electroosmotic velocity	[$m \cdot s^{-1}$]
u_{ES}	= superficial linear electroosmotic velocity	[$m \cdot s^{-1}$]
V	= applied voltage	[V]
W	= electric power consumption	[$W \cdot s$] or [$W \cdot min$]
α	= electroosmotic coefficient defined as Eq.(20)	[T^{-1}]
ϕ	= volume fraction of particles	[-]
ε_a	= porosity of air part	[-]
ε_m	= porosity	[-]
ε_w	= porosity of water part	[-]
λ	= equivalent specific conductivity	[$S \cdot m^{-1}$]
η	= dewatering efficiency defined as Eq.(32)	[-]
κ	= shape factor of particles	[-]
μ	= viscosity of liquid	[$Pa \cdot s$]
ρ_p	= density of particles	[$kg \cdot m^{-3}$]
ζ	= ζ-potential of particles	[V]

<Subscripts>

t = value at t

0 = initial value

1 = value for supernatant liquid layer or value for dewatering sludge bed

2 = value for thickening sludge layer or value for dewatered sludge bed

3 = value for thickened sludge layer

∞ = final value or terminal value

INTRODUCTION

Mechanical dewatering operations such as gravitational method, centrifugal method, expression, through flow method, and vibrational method have so far been studied theoretically and experimentally and have been practically used. But mechanical dewatering is very difficult for the precipitates of colloidal particles and the gelatinous sludge.

A particle dispersed in liquid has an electric potential, the so-called ζ(Zeta)-potential. The ζ-potential of particles dispersed in water is about 0.01~0.1 volt. Electrophoresis and electroosmosis are phenomena that the electric double layer occured at the interface between different phases is removed relatively by the action of a D.C. electric field. Accordingly, it is basically possible to apply them extensively to solid-liquid separation, liquid-liquid separation, dewatering of sludge, removal of ion, and the like. Interfacial electrokinetic phenomena such as electrophoresis, electroosmosis and electrodialysis have so far been applied to separation of fine particles from dispersed systems (refs.1-23). Application of these interfacial electrokinetic phenomena to separation techniques in chemical and other industries has many advantages.

An electroosmosis generated in sludge can be used for the dewatering of sludge (refs.24,25). Electroosmotic dewatering is caused by mass transfer in the electric double layer by the action of an electric field. Therefore, the mechanism of electroosmotic dewatering is qualitatively different from that of mechanical dewatering. Electroosmosis may be available to enhance the dewatering of sludge. According to author's experience, the electroosmotic dewatering is effective for the colloidal and the gelatinous sludges which are difficult to dewater by mechanical methods. If the water content of sludge can be decreased further by using the electroosmotic dewatering and the mechanical methods in the same time, the electroosmotic dewatering is more useful as pretreatment prior to drying. The above mentioned characteristics suggest that there are many possibilities of electroosmotic dewatering to the treatment of sludge.

Electroosmotic dewatering was originally performed with peat sludge by Lord Schwerin in an industrial scale (refs.24,25). Thereafter, electroosmosis has also been applied to the dewatering of sludge such as clay, soil, powder coal, magnesium hydroxide, nickel hydroxide, beating pulp, casein, and so forth (refs.24-28). However, engineering studies concerned with the electroosmotic dewatering of sludge have scarcely ever been made, namely the design equation for equipment that considers the physical properties of sludge and the operating conditions of equipment has not been proposed yet. Therefore, there is almost no development of practical electroosmotic dewatering equipment.

According to author's previous investigation, the electroosmotic dewatering of sludge is systematically explained in this chapter. The thickening process of sludge by electroosmotic dewatering carried out after the completion of gravitational settling of slurry is discussed. The electroosmotic dewatering of sludge can be performed under condition of either constant electric current or constant voltage. The fundamental design equations of electroosmotic dewatering for compressible sludge under both conditions of constant electric current and constant voltage are presented. To obtain them, the rate of electroosmotic dewatering and the electric power consumption are theoretically analysed based on a model of electroosmotic flow through a compressible-particle packed bed. These equations, obtained theoretically, are experimentally confirmed to be appropriate for the design of practical electroosmotic dewatering equipment. A comparison of electroosmotic dewatering efficiency under the constant electric current with that under the constant voltage is discussed here, involving dewatering rate, water content in dewatered sludge and dewatered volume per unit electric power consumption. It is experimentally proved that the electroosmotic dewatering is particularly effective for a gelatinous sludge such as bentonite which is difficult to dewater by mechanical methods. A few characteristics of electroosmotic dewatering are also described.

ELECTROOSMOTIC THICKENING PROCESS OF SLUDGE (refs.1,2)

Batch settling curves of slurries applied with a D.C. electric field are shown in **Figs.1** and 2 with E_{av} as a parameter, where E_{av} means the average strength of electric field V/H_0, V the applied voltage, H_0 the initial height of settling. These figures show that the settling rate and the final concentration of settled sludge are increased with increasing E_{av}. It is considered that an increase of the final concentration of the sludge is due

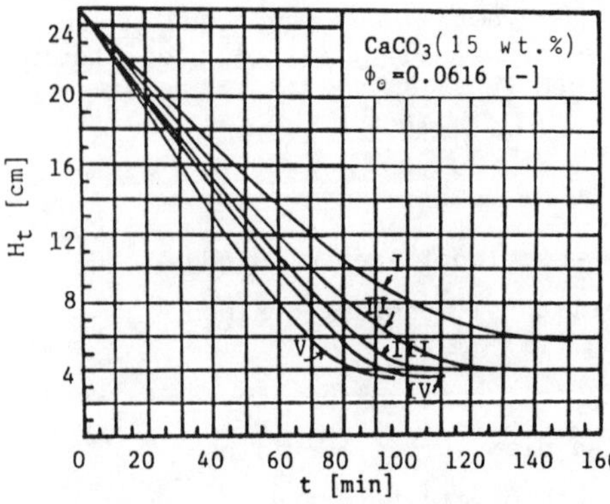

I : E_{av}= 0 [V·m^{-1}] , II : E_{av} =300, III: E_{av}=500, IV: E_{av}=700, V: E_{av}=900

Fig.1 Settling curves of $CaCO_3$ slurry in a D.C. electric field

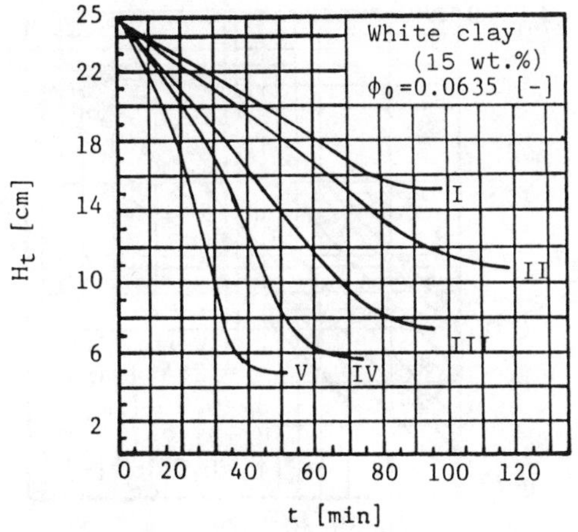

I: E_{av}=0[V·m^{-1}] , II: E_{av}=320, III: E_{av}=600, IV: E_{av}=960, V: E_{av}=1400

Fig.2 Settling curves of white clay slurry in a D.C. electric field

to electroosmotic dewatering occured in the sludge layer.

When a D.C. voltage is being applied to the slurry, the settling of slurry consists of three successive processes. In the first process, suspended particles settle with the resultant velocity of gravitational and electrophoretic components. In the second process, the sludge settled on the bottom is thickened by gravitational compression and electroosmosis. In the third process, the sludge is thickened only by electroosmotic dewatering after the completion of gravitational thickening. The state of these process is shown in **Fig.3** which shows that the electroosmotic pressure p_E increases abruptly as the settling height H_t approaches the final height of gravitational settling. This phenomenon attracts our attention and interest, suggesting that it is possible to dewater sludge by electroosmosis.

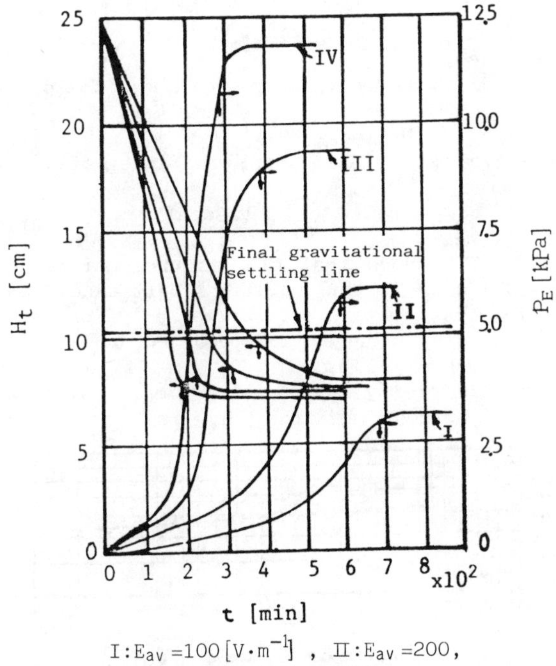

I: E_{av}=100[V·m^{-1}] , II: E_{av}=200, III: E_{av}=300, IV: E_{av}=400

Fig.3 H_t vs. t and p_E vs. t ($CaCO_3$ slurry)

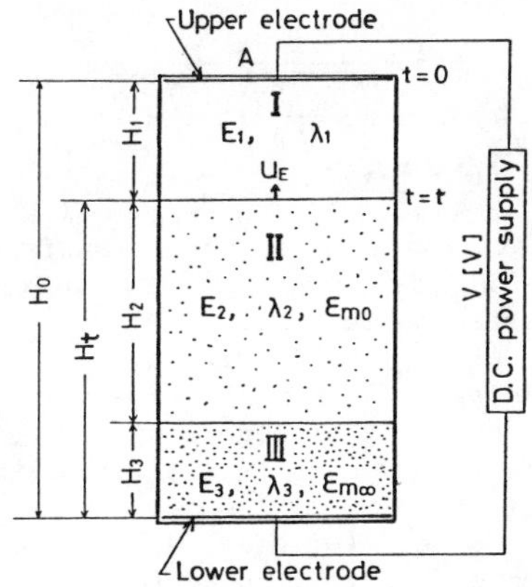

Fig.4 Schematic diagram of electroosmotic thickening process of sludge

The electroosmotic thickening process of sludge after the completion of gravitational settling of slurry is analysed by the following method. **Figure 4** shows the schematic diagram of electroosmotic thickening process of sludge. The water in the sludge is removed upwards with electroosmotic velocity u_E. In this figure, part I is the supernatant liquid layer formed by thickening, part II, the thickening sludge layer, which is being thickened, and part III, the thickened sludge layer, in which thickening is already finished. The notations of E_1, E_2, and E_3 are the strength of electric field in each layer, respectively, and λ_1, λ_2 and λ_3 are the equivalent specific conductivity of each layer, respectively. These values are assumed to be constant throughout each layer.

The electroosmotic velocity, u_E, in the sludge is represented by the following Debye-Hückel equation based on the electric double layer theory.

$$u_E = \frac{\zeta D}{\kappa \pi \mu} E_2 \qquad (1)$$

Where ζ, D, κ, and μ are the ζ-potential of particles, the dielectric constant of liquid, the shape factor of particles, and the viscosity of liquid, respectively. The strength of electric field, E_2, in the thickening sludge layer is expressed by Ohm's law as follows:

$$V = E_1 H_1 + E_2 H_2 + E_3 H_3 \;, \quad \lambda_1 E_1 = \lambda_2 E_2 = \lambda_3 E_3 \qquad (2)$$

and so that

$$E_2 = \frac{V}{(\lambda_2/\lambda_1)H_1 + H_2 + (\lambda_2/\lambda_3)H_3} \qquad (3)$$

In above equations, H_1, H_2 and H_3 are functions of time t. Considering the material balance on water in Fig.4, the following equation is obtained:

$$H_1 + H_2 \varepsilon_{m0} + H_3 \varepsilon_{m\infty} = H_0 \varepsilon_{m0} \qquad (4)$$

and the height of each layer is expressed by

$$\left.\begin{aligned} H_1 &= \varepsilon_{m0} \int_0^t u_E\, dt \,, \\ H_2 &= H_0 - \left(\frac{1-\varepsilon_{m\infty}}{\varepsilon_{m0}-\varepsilon_{m\infty}}\right) \varepsilon_{m\infty} \int_0^t u_E\, dt \,, \\ H_3 &= H_0 - (H_1 + H_2) \end{aligned}\right\} \qquad (5)$$

where ε_{m0} is the initial porosity, and $\varepsilon_{m\infty}$ the final porosity. By substituting Eqs.(1) and (5) into Eq.(3), E_2 is written as the following integral equation.

$$E_2 = \frac{V}{H_0 + B_2 \frac{\zeta D}{\kappa \pi \mu} \int_0^t E_2\, dt} \qquad (6)$$

As the solution of the equation (6), E_2 can be expressed by

$$E_2 = \frac{V}{\sqrt{\left(\frac{2B_2\zeta DV}{\kappa\pi\mu}\right)t + H_0{}^2}} \qquad (7)$$

in which $B_2 = \left(\frac{\lambda_2}{\lambda_1} - \frac{\lambda_2}{\lambda_3}\right)\varepsilon_{m0} - \left(\frac{1-\varepsilon_{m\infty}}{\varepsilon_{m0}-\varepsilon_{m\infty}}\right)\left(1-\frac{\lambda_2}{\lambda_3}\right)$

When the settling rate of sludge surface is indicated by $-(dH_t/dt)$, the electroosmotic flow rate q_E is expressed as follows:

$$q_E = A\varepsilon_{mt}u_E = -A(dH_t/dt) \qquad (8)$$

where A and ε_{mt} are the cross-sectional area and the average porosity of sludge at t, respectively, and ε_{mt} is written as Eq.(9) using the initial volume fraction of particles ϕ_0.

$$\varepsilon_{mt} = (H_t - H_0\phi_0)/H_t \qquad (9)$$

Substitution of Eq.(1) into Eq.(8) gives

$$-\frac{dH_t}{dt} = \varepsilon_{mt}\frac{\zeta D}{\kappa\pi\mu}E_2 \qquad (10)$$

Integrating the equation obtained by substitution of Eqs.(7) and (9) into Eq.(10), the following equation is obtained.

$$\psi(H_t) = \frac{H_0 - H_t}{H_0} + \phi_0 \ln\left(\frac{H_0 - H_0\phi_0}{H_t - H_0\phi_0}\right) = \frac{1}{B_2H_0}\left\{\sqrt{\left(\frac{2B_2\zeta DV}{\kappa\mu}\right)t + H_0{}^2} - H_0\right\} \qquad (11)$$

Equation (11) shows generally the relation between H_t and t. In the case where $\lambda_1=\lambda_2=\lambda_3$, E_2 becomes V/H_0 ($=E_{av}$) from Eq.(7). Therefore, $\psi(H_t)$ can be written as

$$\psi(H_t) = \frac{\zeta DV}{\kappa\pi\mu H_0}t \qquad (12)$$

Equation (12) shows that $\psi(H_t)$ is a linear function of t.

The experimental examinations of the relation between $\psi(H_t)$ and t were performed. In order to exclude the effect of gravitation on the thickening, the sludges settled completely by gravitation were used in this experiment. The relation between $\psi(H_t)$ and t for the sludge applied with a D.C. electric field is shown in **Figs.5** and **6** with E_{av} as a parameter. From the results obtained, it is found that $\psi(H_t)$ is proportional to t, and Eq.(12) is appropriate for the sludges of both $CaCO_3$ and white clay except the period in the neighborhood of the end of thickening. The relations between $\psi(H_t)$ and t for these sludges are expressed by Eq.(12) because these sludges contain some electrolytes, consequently, λ_1, λ_2, and λ_3 are nearly equal.

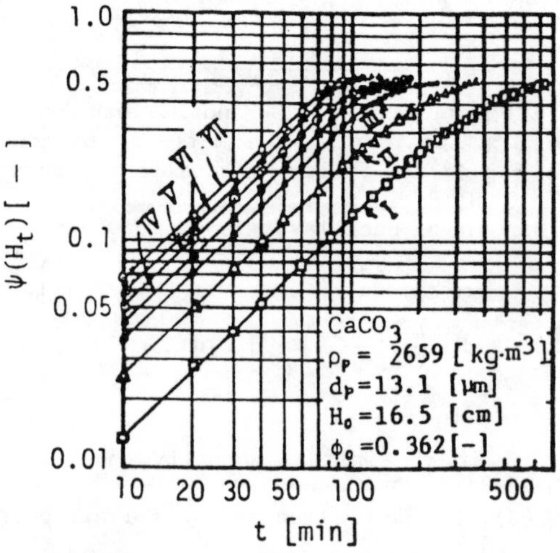

I: E_{av}=100[$V\cdot m^{-1}$], II: E_{av}=200, III: E_{av}=400, IV: E_{av}=600, V: E_{av}=800, VI: E_{av}=1000, VII: E_{av}=1200

Fig.5 Relation between $\psi(H_t)$ and t for $CaCO_3$ sludge

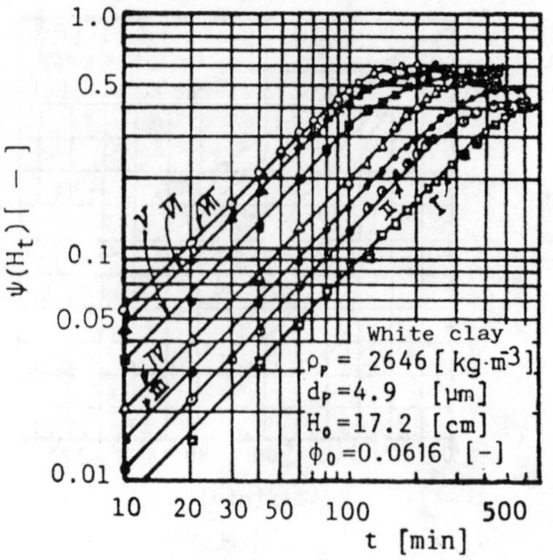

I: E_{av}=100[$V\cdot m^{-1}$], II: E_{av}=200, III: E_{av}=300, IV: E_{av}=400, V: E_{av}=600, VI: E_{av}=800, VII: E_{av}=1000

Fig.6 Relation between $\psi(H_t)$ and t for white clay sludge

After the completion of electroosmotic thickening, the superficial electroosmotic velocity u_{ES} is expressed as follows:

$$u_{ES} = \varepsilon_{m\infty}\frac{\zeta D}{\kappa\pi\mu}E_3 \qquad (13)$$

On the other hand, this velocity is also expressed by the following Kozeny-Carman equation:

$$u_{ES} = \frac{\varepsilon_{m\infty}{}^3}{k\mu S_v{}^2(1-\varepsilon_{m\infty})^2}\cdot\frac{dp_{EP}}{dH_3} \qquad (14)$$

where p_{EP} is the electroosmotic permeation pressure, k the Kozeny-Carman's constant, and S_v the specific surface of particles. Combination of Eqs.(13) and (14) leads to

$$\frac{dp_{EP}}{dH_3} = \frac{\zeta DkS_v{}^2}{\kappa\pi}E_3\left(\frac{1-\varepsilon_{m\infty}}{\varepsilon_{m\infty}}\right)^2 \qquad (15)$$

If H_3 is denoted by H_∞ when $t=t_\infty$, the following equation is obtained:

$$H_\infty(1-\varepsilon_{m\infty}) = H_0\phi_0 \qquad (16)$$

Using above the relationship, the integration of Eq.(15) gives

$$p_{EP} = \frac{\zeta DkS_v{}^2}{\kappa\pi}H_0\phi_0E_3\frac{(1-\varepsilon_{m\infty})}{\varepsilon_{m\infty}{}^2} \qquad (17)$$

Equation (17) shows that p_{EP} is proportional to $E_3(1-\varepsilon_{m\infty})/\varepsilon_{m\infty}{}^2$. The relations between $\varepsilon_{m\infty}$ and $E_{av}(1-\varepsilon_{m\infty})/\varepsilon_{m\infty}{}^2$ obtained experimentally for $CaCO_3$ and white clay sludges are plotted on log-log paper as shown in **Fig.7**. In this figure, the value of $\varepsilon_{m\infty}$ for any E_{av} can be estimated by trial and error method, and then H_∞ is obtained by

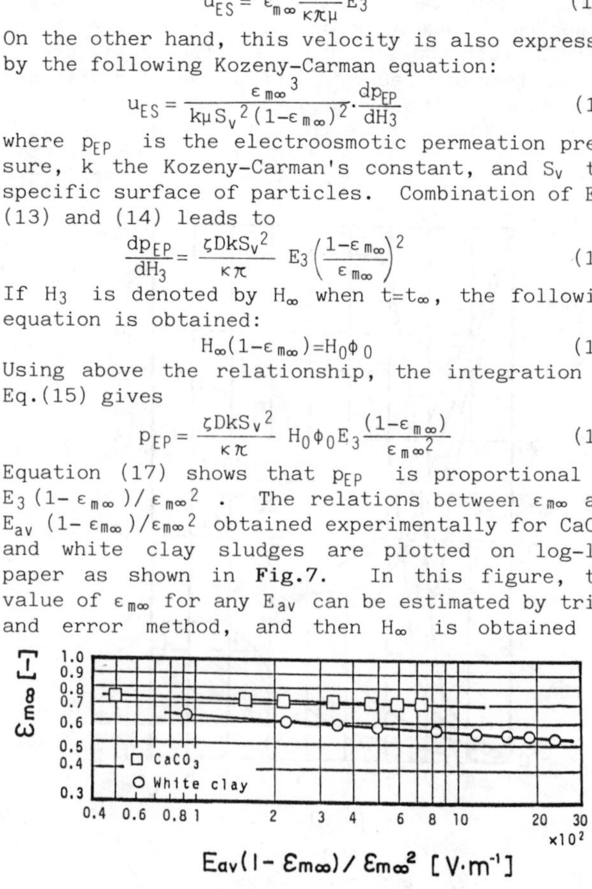

Fig.7 $\varepsilon_{m\infty}$ vs. $E_{av}(1-\varepsilon_{m\infty})/\varepsilon_{m\infty}{}^2$

Eq.(16). It is considered that the experimental result shown in Fig.7 is similar to the relation between equilibrium porosity and mechanically exerted pressure, which were reported by Ruth (ref. 29), Tiller (ref.30) and Yagi et al. (ref.31). If p_{EP} is replaced with p_{EP}' by modifying Eq.(17) with E_{av}, the relation between p_{EP}' and $\varepsilon_{m\infty}$ is expressed by the following empirical formula based on the result shown in Fig.7:

$$\varepsilon_{m\infty} = a(p_{EP}')^{-n} \qquad (18)$$

where a and n are the experimental constants depending on the property of sludge.

ELECTROOSMOTIC DEWATERING OF SLUDGE UNDER CONDITION OF CONSTANT ELECTRIC CURRENT (ref.32)

The electroosmotic dewatering of sludge can be performed by applying a D.C. electric field under the condition of either constant electric current or constant voltage.

In this section, the fundamental design equation of electroosmotic dewatering under the condition of constant electric current is described. The mechanism of elctroosmotic dewatering for compressible sludge under the condition of constant electric current is discussed based on a model of electroosmotic flow through a compressible-particle packed bed. Then the effects of the operating conditions (i.e. electric current density, sludge concentration, height of sludge bed, etc.) on the rate of electroosmotic dewatering and the electric power consumption are theoretically analysed.

The electroosmotic dewatering model for the compressible sludge under the condition of constant electric current is shown in **Fig.8**. A constant electric current density I_0 passes through the the cross-section of sludge bed. The polarity of

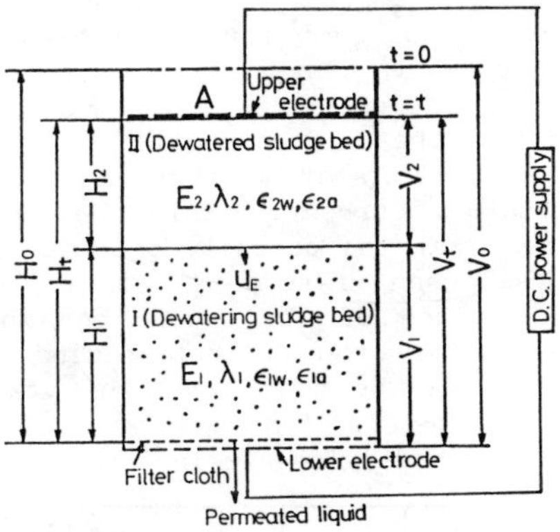

Fig.8 Schematic diagram of electroosmotic dewatering under condition of constant electric current

electrode is selected according to the polarity of ζ-potential of particles in order to remove downwards the water in the sludge. In the case of compressible sludge, the height of sludge bed decreases with the proceeding of dewatering and the initial height of sludge bed H_0 changes to H_t at the dewatering time t. In Fig.8, part I is the dewatering sludge bed, and part II the dewatered sludge bed. The notations of ε_{1w} and ε_{2w} mean the porosity of water part in each bed and ε_{1a}, ε_{2a} the porosity of air part in each bed, and λ_1, λ_2 the equivalent specific conductivity of each bed, respectively. These values are assumed to be constant throughout each bed.

As shown in Fig.8, the electroosmotic velocity, u_E, in the dewatering sludge bed is represented by Eq.(1) replacing E_2 with E_1. That is, Eq.(1) is rewritten as:

$$u_E = \frac{\zeta D}{\kappa \pi \mu} E_1 \qquad (1')$$

The strength of electric field, E_1, in the dewatering sludge bed is expressed by Ohm's law as follows:

$$E_1 = I_0/\lambda_1 \qquad (19)$$

By substituting Eq.(19) into Eq.(1'), u_E is rewritten as follows:

$$u_E = \alpha(I_0/\lambda_1) \quad , \quad \alpha = \zeta D/\kappa\pi\mu \qquad (20)$$

where α is the electroosmotic coefficient which can be obtained experimentally. As the experimental conditions are defined, α, I_0, and λ_1 are constant. Therefore u_E becomes also constant. The electroosmotic flow rate q_E is expressed by Eq.(21).

$$q_E = A\varepsilon_{1w} u_E \qquad (21)$$

Substituting Eq.(20) into Eq.(21), the volume dewatered by electroosmosis Q_E can be got by the integral of q_E with respect to t.

$$Q_E = \int_0^t q_E dt = A\varepsilon_{1w} u_E \int_0^t dt = A\varepsilon_{1w} \alpha \frac{I_0}{\lambda_1} t \qquad (22)$$

Equation (22) shows that Q_E is proportional to I_0 and t. The relation between the electric power consumption W and t is obtained by the following method. The voltage, V_t, applied on the electrodes at t is shown as

$$V_t = V_1 + V_2 = E_1 H_1 + E_2 H_2 = I_0(H_1/\lambda_1 + H_2/\lambda_2) \qquad (23)$$

where V_1 and V_2 are the voltage applied to each bed I, II in this model. Considering the material balances on liquid and solid in the sludge as shown in Fig.8,

$$\left.\begin{aligned} &AH_1\varepsilon_{1w} + AH_2\varepsilon_{2w} + A\varepsilon_{1w}\int_0^t u_E dt = AH_0\varepsilon_{1w} \quad \text{(liquid)},\\ &A(H_0-H_1)\{1-(\varepsilon_{1w}+\varepsilon_{1a})\} = AH_2\{1-(\varepsilon_{2w}+\varepsilon_{2a})\} \quad \text{(solid)} \end{aligned}\right\} \qquad (24)$$

From above equations, H_1 and H_2 are obtained as follows:

$$\left.\begin{aligned} H_1 &= H_0 - \frac{\varepsilon_{1w}\{1-(\varepsilon_{2w}+\varepsilon_{2a})\}}{\varepsilon_{1w}(1-\varepsilon_{2a})-\varepsilon_{2w}(1-\varepsilon_{1a})}\int_0^t u_E dt ,\\ H_2 &= \frac{\varepsilon_{1w}\{1-(\varepsilon_{1w}+\varepsilon_{1a})\}}{\varepsilon_{1w}(1-\varepsilon_{2a})-\varepsilon_{2w}(1-\varepsilon_{1a})}\int_0^t u_E dt \end{aligned}\right\} \qquad (25)$$

Substitution of Eq.(25) into Eq.(23) gives

$$V_t = \frac{I_0}{\lambda_1}\left[\frac{\alpha I_0 \varepsilon_{1w}}{\varepsilon_{1w}(1-\varepsilon_{2a})-\varepsilon_{2w}(1-\varepsilon_{1a})}\left\{\frac{1-(\varepsilon_{1w}+\varepsilon_{1a})}{\lambda_2} - \frac{1-(\varepsilon_{2w}+\varepsilon_{2a})}{\lambda_1}\right\} t + H_0\right] \qquad (26)$$

Hence, W is written as Eq.(28) from the following equation(27).

$$W = \int_0^t AI_0 V_t dt = AI_0 \int_0^t V_t dt \qquad (27)$$

$$W = \frac{AI_0^2}{\lambda_1}\left[\frac{\varepsilon_{1w}}{\varepsilon_{1w}(1-\varepsilon_{2a})-\varepsilon_{2w}(1-\varepsilon_{1a})}\left\{\frac{1-(\varepsilon_{1w}+\varepsilon_{1a})}{\lambda_2} - \frac{1-(\varepsilon_{2w}+\varepsilon_{2a})}{\lambda_1}\right\}\frac{\alpha I_0}{2}t^2 + H_0 t\right] \qquad (28)$$

If the porosity of air part in each bed is much smaller than that of water part, ε_{1a} and ε_{2a} can be neglected. Therefore, Eq.(28) is approximately written as Eq.(29).

$$W = \frac{AI_0^2}{\lambda_1}\left\{\frac{\varepsilon_{1w}}{\varepsilon_{1w}-\varepsilon_{2w}}\left(\frac{1-\varepsilon_{1w}}{\lambda_2} - \frac{1-\varepsilon_{2w}}{\lambda_1}\right)\frac{\alpha I_0}{2}t^2 + H_0 t\right\} \qquad (29)$$

From Eqs.(22) and (28), the electric power consumption per unit volume dewatered by electroosmosis W/Q_E is expressed as follows:

$$\frac{W}{Q_E}=\frac{1}{\varepsilon_{1w}}\left[\frac{\varepsilon_{1w}}{\varepsilon_{1w}(1-\varepsilon_{2a})-\varepsilon_{2w}(1-\varepsilon_{1a})}\left\{\frac{1-(\varepsilon_{1w}+\varepsilon_{1a})}{\lambda_2}-\frac{1-(\varepsilon_{2w}+\varepsilon_{2a})}{\lambda_1}\right\}\frac{I_0^2}{2}\,t+\frac{H_0}{\alpha}\,I_0\right] \quad (30)$$

Using Eqs.(22) and (29), the following approximate equation concerning W/Q_E can be obtained in the case of sludge with large water content.

$$\frac{W}{Q_E}=\frac{1}{(\varepsilon_{1w}-\varepsilon_{2w})}\left(\frac{1-\varepsilon_{1w}}{\lambda_2}-\frac{1-\varepsilon_{2w}}{\lambda_1}\right)\frac{I_0^2}{2}t+\frac{H_0}{\alpha\varepsilon_{1w}}I_0 \quad (31)$$

Also, defining the dewatering efficiency η as the ratio of Q_E to the initial total water content of sludge, η is expressed as

$$\eta=\frac{Q_E}{AH_0\varepsilon_{1w}}=\frac{\alpha I_0}{\lambda_1 H_0}t \quad (32)$$

And the time, t_e, till the end of electroosmotic dewatering is shown as Eq.(33) from the material balances on liquid and solid contained in the sludge bed at the end of dewatering.

$$t_e=\left[1-\frac{\{1-(\varepsilon_{1w}+\varepsilon_{1a})\}}{\{1-(\varepsilon_{2w}+\varepsilon_{2a})\}}\left(\frac{\varepsilon_{2w}}{\varepsilon_{1w}}\right)\right]\frac{\lambda_1 H_0}{\alpha I_0} \quad (33)$$

Neglecting ε_{1a} and ε_{2a}, t_e is approximately got as Eq.(34).

$$t_e=\left\{1-\frac{(1-\varepsilon_{1w})}{(1-\varepsilon_{2w})}\left(\frac{\varepsilon_{2w}}{\varepsilon_{1w}}\right)\right\}\frac{\lambda_1 H_0}{\alpha I_0} \quad (34)$$

The equations obtained theoretically were confirmed experimentally. The outline of batch-type electroosmotic dewatering apparatus used in this experiment is shown in **Fig.9**. The lower electrode

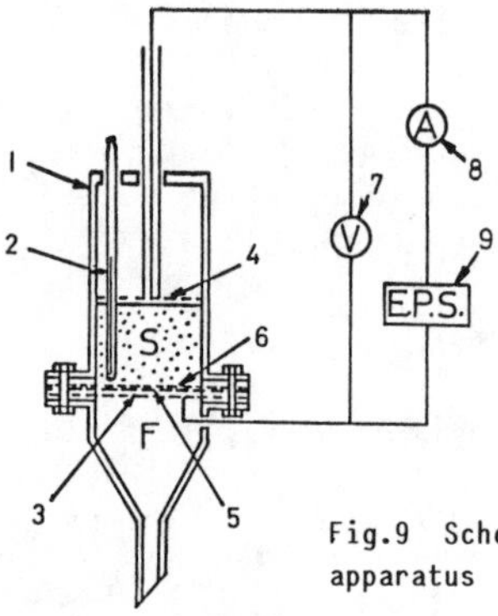

1. Acrylic resin cylinder
2. Thermometer
3. Electrode supporter
4. Upper electrode
5. Lower electrode
6. Filter cloth
7. D.C. voltmeter
8. D.C. amperemeter
9. Regulated D.C. power supply

Fig.9 Schematic diagram of experimental apparatus

5 was set at the bottom of an acrylic resin cylinder (90 or 110 mm inside diameter) **1**. The electrode made of a wire net of 20 mesh (opening of 840 μm) or a perforated plate was used at the lower side. The filter cloth (SP#7) **6** was set up in contact with the upper surface of lower electrode. The sludge of which powder was mixed adequately with deionized water was packed in part **S** in Fig.9 and the upper electrode **4** was set up in contact with the surface of this sludge bed. The upper electrode was made of a perforated plate with hole diameter of 3 mm in order to facilitate the remove of gas produced by electrolysis. Including the experimental examinations described in the following sections, copper, stainless steel, carbon, titanium, and platinum were variously used as the electrode material. Titanium and platinum were used as the electrodes in order to avoid anodic dissolution of electrode material by electrolysis, but copper, stainless steel and carbon may be used practically as the electrodes. As the height of sludge bed decreases with proceeding the dewatering of sludge, the position of the upper electrode was always adjusted so as to contact with the upper surface of sludge bed, but the sludge bed was not compressed by the upper electrode to avoid the dewatering by compression. The polarity of both electrodes is determined in consideration of the polarity of ζ-potential of particles, and then the electric fields under the condition of constant electric current are applied to the sludge bed by a regulated D.C. power supply **9**. As the water flows down into part **F** in Fig.9, the total volume, Q_T, dewatered by electroosmosis and gravitation was measured, and the voltage, V_t, applied to the sludge bed was also measured with an automatic recorder at the same time. The value of Q_E was calculated as the difference between Q_T and the gravitationally dewatered volume Q_G, and α was determined from the equation $\alpha=q_E'/A\varepsilon_{1w}E_{av}$. Here q_E' is the flow rate of electroosmotic permeation obtained in the early period of measurement and E_{av} is the average strength of electric field in this period.

The sludges used in this experiment were the compressible sludges such as white clay, bentonite and $Mg(OH)_2$. Bentonite was used as an example of a gelatinous sludge which was difficult to dewater by mechanical methods. Both white clay and bentonite particles have negative ζ-potential, and $Mg(OH)_2$ particle positive ζ-potential. The initial concentration of solid (particle) in the sludge was adjusted to be nearly equal to the concentration of solid thickened completely by gravitational sedimentation.

For both white clay and bentonite sludges, the relation between Q_T and t is shown with I_0 as a parameter in **Fig.10**. Q_T is the sum of Q_E and Q_G. It is clear that the total dewatering flow rate (dQ_T/dt) increases considerably with increasing I_0. It is noticed that electroosmotic dewatering is very effective for bentonite sludge which is difficult to dewater by gravitation. The experimental result for $Mg(OH)_2$ sludge also shows a similar tendency to that for white clay sludge. In the case of white clay, the bending points are found out at Q_T of about 50 cm^3 in Fig.10, and it is observed that the voltage applied to the sludge bed increases rapidly at those points under the constant electric current. This phenomenon may be considered that the porosity of water part in the sludge bed changes from ε_{1w} to ε_{2w} and the electric resistance of the sludge bed increases with decreasing the water content of sludge. That is to say, it may be considered that the first dewatering process is finished at the bending point and the secondary dewatering process follows

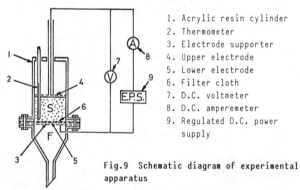

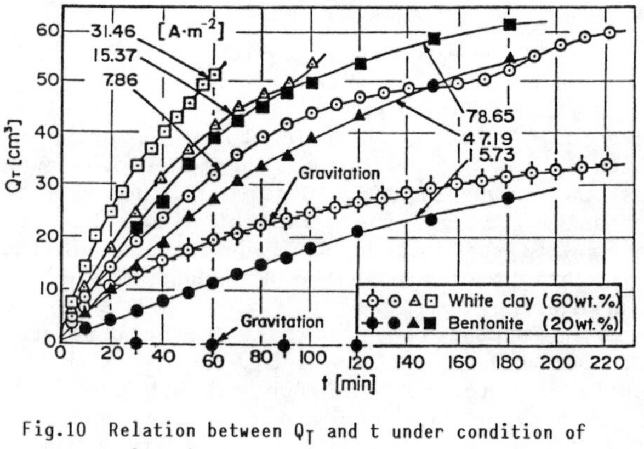

Fig.10 Relation between Q_T and t under condition of constant electric current

immediately. It is observed that large electric power consumption is needed in the secondary dewatering process. The secondary dewatering process shows several complex phenomena in the sludge bed, such as crack creation, temperature rising and drying by Joule's heat.

Figures 11 and **12** show the relation between Q_E and t on log-log paper for each sludge with I_0 as a parameter, and also show the comparison of the experimental results with the relationship expressed by Eq.(22). In these figures, each key shows the observed values and the solid lines express the theoretical results, respectively. The values calculated by Eq.(22) show good agreement with the observed values within ±20% range except the period in the neighborhood of the end of dewatering. The result for $Mg(OH)_2$ sludge also shows to be similar to that for the sludges above mentioned.

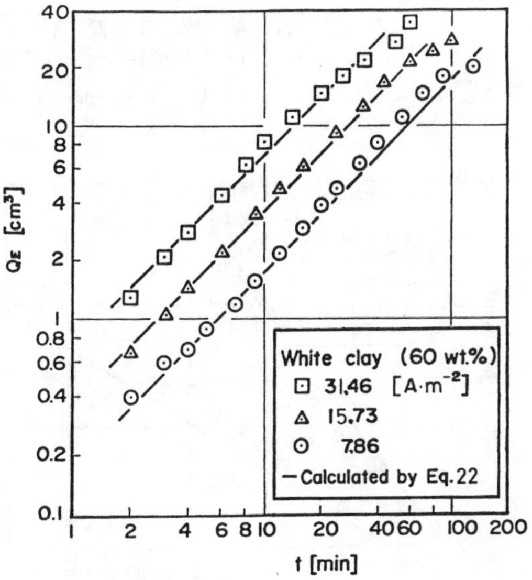

Fig.11 Relation between Q_E and t under condition of constant electric current

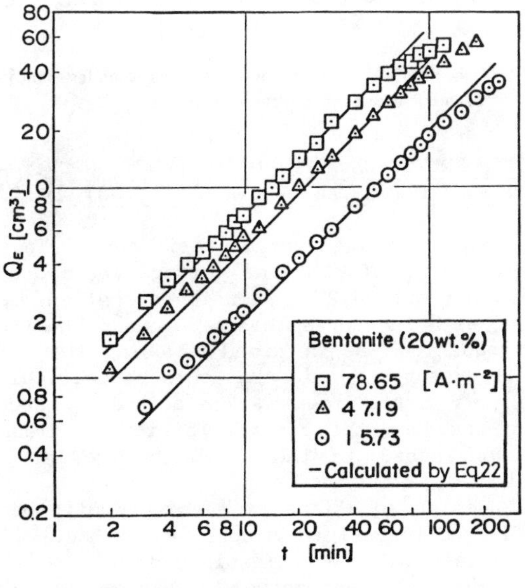

Fig.12 Relation between Q_E and t under condition of constant electric current

The relation between W and t is expressed by Eq.(29). **Figures 13** and **14** show the comparison of the experimental results with the values calculated by Eq.(29) with I_0 as a parameter. The values of ε_{2W} and λ_2 used for the calculation of W by Eq.(29) are, respectively, obtained by measurement of porosity of water part and by calculation with Ohm's law using the voltage at the time when the first dewatering process has finished. Equation (29) shows that W is a quadratic equation with respect to t. In the case of $Mg(OH)_2$ sludge, the experimental results coincide approximately with the theoretical relation with I_0 as

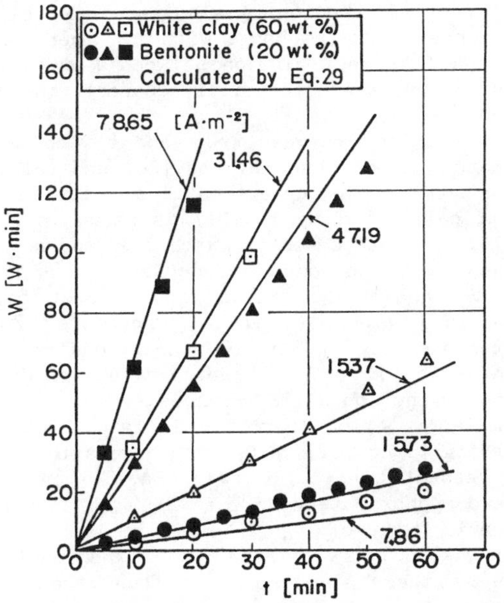

Fig.13 Relation between W and t under condition of constant electric current

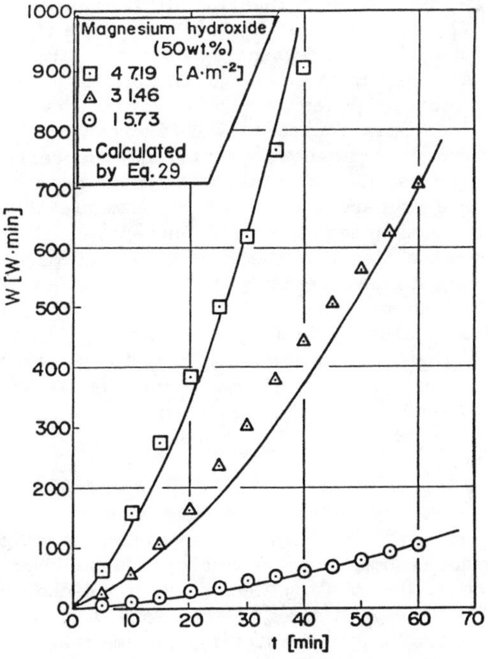

Fig.14 Relation between W and t under condition of constant electric current

a parameter as shown in Fig.14. In the cases of white clay and bentonite sludges, the observed relations between W and t are fairly linear as shown in Fig.13 and the values calculated by Eq.(29) are in agreement with the observed values. It is shown that these relations are approximately expressed as $W=(AI_0{}^2H_0/\lambda_1)t$ because the quadratic term of t is much smaller than the linear term of t in Eq.(29). The difference between the relation shown in Fig.13 and that in Fig.14 is due to the following experimental results. In the case of $Mg(OH)_2$ sludge, the electric resistance of sludge bed increases gradually with t under the condition of constant electric current. In the cases of white clay and bentonite sludges, their electric resistance scarcely change with t. Therefore, it is found that the relation between W and t is influenced by the electrical characteristics of sludge.

The relation between W/Q_E and t with I_0 as a parameter is expressed by Eqs.(30) and (31). The comparison of the experimental results with the values calculated by Eq.(31) is shown in **Figs.15** and **16**. In both cases of white clay and bentonite sludges, W/Q_E becomes independent of t because both Q_E and W are linear with respect to t as shown in Figs.11, 12 and 13, therefore W/Q_E is related to only I_0 and is proportional to I_0 as shown in Fig.15. The observed values of W/Q_E are a linear function of I_0 irrespective of t and show a similar relation to that calculated by Eq.(31). According to Eq.(31), W/Q_E is a quadratic equation with respect to I_0. However, Eq.(31) can be expressed approximately as $W/Q_E=(H_0/\varepsilon_{1w}\alpha)I_0$ when the quadratic term of I_0 is much smaller than the linear term of I_0. As to the experimental results of these sludges, it is found that Eq.(31) is nearly expressed by the linear relationship in respect of I_0. When the quadratic term of I_0 in Eq.(31) can not be neglected, W/Q_E is linear in respect of t. For $Mg(OH)_2$ sludge, W/Q_E is expressed as a linear function of t as shown in Fig. 16. It is found that the slopes of straight lines increase with increasing I_0 and that the value of W/Q_E at t=0 shows $H_0I_0/\varepsilon_{1w}\alpha$ in Eq.(31). Therefore, the electric power per unit volume dewatered by electroosmosis increases with increasing I_0, namely the electric power efficiency decreases with an increase of I_0. This fact must be noticed in the case of carrying out electroosmotic dewatering under the constant electric current.

The time, t_e, till the end of electroosmotic dewatering is expressed by Eq.(34). As an example, the values of t_e calculated by Eq.(34) under the constant electric current density of 15.73 $A\cdot m^{-2}$ are 113 min for white clay and 264 min for bentonite, respectively. These calculated values coincide nearly with the observed values in Figs.11 and 12. Therefore Eq.(34) is regarded as appropriate.

From above results, the electroosmotic dewatering mechanism of compressible sludge can be explained by the model of electroosmotic flow through a compressible-particle packed bed and it was confirmed that the equations obtained by theoretical analysis were appropriate for the design of electroosmotic dewatering equipment.

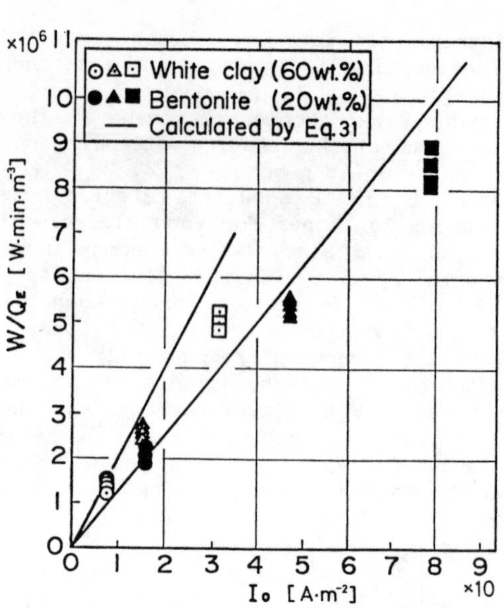

Fig.15 Relation between W/Q_E and I_0 under condition of constant electric current

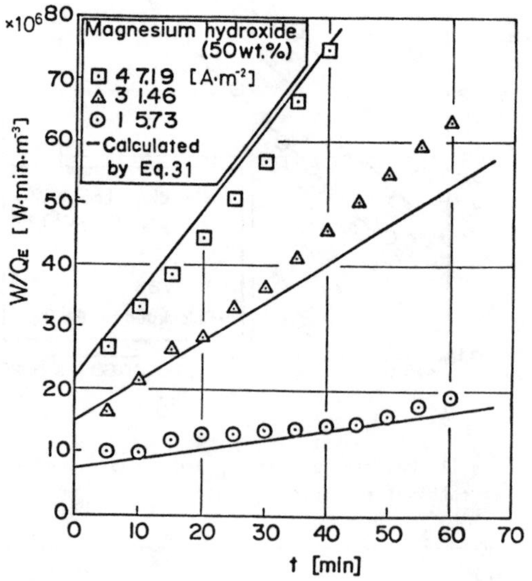

Fig.16 Relation between W/Q_E and t under condition of constant electric current

ELECTROOSMOTIC DEWATERING OF SLUDGE UNDER CONDITION OF CONSTANT VOLTAGE (ref.33)

The equation of electroosmotic dewatering under the condition of constant voltage was proposed by Komagata (refs.24,27). His investigation was concerned with incompressible sludge and the theoretical equation was obtained by using the straight capillary tubes model. But this theoretical analysis was not adequate for actual compressible sludge because the compressibility of sludge was not considered in the straight capillary tubes model.

The design equations of electroosmotic dewatering for compressible sludge under the condition of constant voltage are explained in this section. The rate of electroosmotic dewatering and the electric power consumption are theoretically analysed by using the model of electroosmotic flow through a compressible-particle packed bed, in the same manner as used under the condition of constant electric current in the previous section. The reliability of equations obtained theoretically was ascertained experimentally by use of the compressible sludges such as white clay sludge and gelatinous bentonite sludge.

The electroosmotic dewatering model for the compressible sludge under the constant voltage can be represented with almost the same as that under the constant electric current, namely V_t at t=t in Fig.8 has to be replaced with V_0. When a constant voltage V_0 is applied to the sludge bed, the water in the sludge is removed downwards by electroosmosis with the velocity u_E.

As shown in Eqs.(19) and (20), u_E is written as the following equation with α and E_1.

$$u_E=\alpha E_1 \tag{35}$$

Under the constant voltage, E_1 is expressed by Eq. (37) from the following Eq.(36).

$$V_0=V_1+V_2=E_1H_1+E_2H_2\ ,\ \lambda_1E_1=\lambda_2E_2 \tag{36}$$

$$E_1=\frac{V_1}{H_1}=\frac{V_0}{H_1+(\lambda_1/\lambda_2)H_2} \tag{37}$$

By substituting Eq.(37) into Eq.(35), u_E is rewritten as:

$$u_E=\frac{\alpha V_0}{H_1+(\lambda_1/\lambda_2)H_2} \tag{38}$$

In Fig.8, H_1 and H_2 under the constant voltage are also expressed by Eq.(25) as same equations obtained under the constant electric current, based on the material balances on liquid and solid in the dewatered and dewatering sludge bed. Substitution of Eq.(25) into Eq.(38) gives

$$\frac{\alpha V_0}{u_E}=\left\{\frac{\varepsilon_{1w}}{\varepsilon_{1w}(1-\varepsilon_{2a})-\varepsilon_{2w}(1-\varepsilon_{1a})}\right\}\left[\frac{\lambda_1}{\lambda_2}\{1-(\varepsilon_{1w}+\varepsilon_{1a})\}-\{1-(\varepsilon_{2w}+\varepsilon_{2a})\}\right]\int_0^t u_E dt + H_0 \tag{39}$$

Equation (39) is an integral equation with respect to u_E. Solving Eq.(39) by using the initial condition, $u_E=\alpha V_0/H_0$ at t=0, u_E is finally obtained as follows:

$$u_E=\alpha V_0\left[2\alpha V_0\left\{\frac{\varepsilon_{1w}}{\varepsilon_{1w}(1-\varepsilon_{2a})-\varepsilon_{2w}(1-\varepsilon_{1a})}\right\}\times\left[\frac{\lambda_1}{\lambda_2}\{1-(\varepsilon_{1w}+\varepsilon_{1a})\}-\{1-(\varepsilon_{2w}+\varepsilon_{2a})\}\right]t+H_0^2\right]^{-\frac{1}{2}} \tag{40}$$

Under the constant electric current, u_E is independent of t and becomes constant. However, under the condition of constant voltage, u_E is a function of t as shown in Eq.(40). Substituting Eq. (40) into Eq.(21) and integrating the substituted equation with respect to t, Q_E is written as Eq. (41).

$$Q_E=\int_0^t q_E dt=\frac{A\{\varepsilon_{1w}(1-\varepsilon_{2a})-\varepsilon_{2w}(1-\varepsilon_{1a})\}}{[(\lambda_1/\lambda_2)\{1-(\varepsilon_{1w}+\varepsilon_{1a})\}-\{1-(\varepsilon_{2w}+\varepsilon_{2a})\}]}\times\left[\left[2\alpha V_0\left\{\frac{\varepsilon_{1w}}{\varepsilon_{1w}(1-\varepsilon_{2a})-\varepsilon_{2w}(1-\varepsilon_{1a})}\right\}\left[\frac{\lambda_1}{\lambda_2}\{1-(\varepsilon_{1w}+\varepsilon_{1a})\}-\{1-(\varepsilon_{2w}+\varepsilon_{2a})\}\right]t+H_0^2\right]^{1/2}-H_0\right] \tag{41}$$

If the water content in the sludge is large, the porosity of air part, ε_a, in each sludge bed is much smaller than the porosity of water part, ε_w. Therefore, ε_{1a} and ε_{2a} can be neglected. In the case of above condition, Eq.(41) is approximately rewritten as Eq.(42).

$$Q_E=\frac{A(\varepsilon_{1w}-\varepsilon_{2w})}{\{(\lambda_1/\lambda_2)(1-\varepsilon_{1w})-(1-\varepsilon_{2w})\}}\times\left[\sqrt{2\alpha V_0\left(\frac{\varepsilon_{1w}}{\varepsilon_{1w}-\varepsilon_{2w}}\right)\left\{\frac{\lambda_1}{\lambda_2}(1-\varepsilon_{1w})-(1-\varepsilon_{2w})\right\}t+H_0^2}-H_0\right] \tag{42}$$

The relation between W and t is obtained by the following procedure. The electric current density, I_t, passing through the cross-section of sludge bed at t is shown by Ohm's law as follows:

$$I_t=\lambda_1E_1\ (=\lambda_2E_2) \tag{43}$$

From Eqs.(35) and (40), E_1 is got and then I_t is expressed as follows:

$$I_t=\lambda_1V_0\left[2\alpha V_0\left\{\frac{\varepsilon_{1w}}{\varepsilon_{1w}(1-\varepsilon_{2a})-\varepsilon_{2w}(1-\varepsilon_{1a})}\right\}\times\left[\frac{\lambda_1}{\lambda_2}\{1-(\varepsilon_{1w}+\varepsilon_{1a})\}-\{1-(\varepsilon_{2w}+\varepsilon_{2a})\}\right]t+H_0^2\right]^{-\frac{1}{2}} \tag{44}$$

Hence, W is written as Eq.(46) from the following Eq.(45).

$$W=\int_0^t AV_0I_tdt=AV_0\int_0^t I_tdt \tag{45}$$

$$W=\frac{A\lambda_1V_0\{\varepsilon_{1w}(1-\varepsilon_{2a})-\varepsilon_{2w}(1-\varepsilon_{1a})\}}{\alpha\varepsilon_{1w}[(\lambda_1/\lambda_2)\{1-(\varepsilon_{1w}+\varepsilon_{1a})\}-\{1-(\varepsilon_{2w}+\varepsilon_{2a})\}]}\times\left[\left[2\alpha V_0\left\{\frac{\varepsilon_{1w}}{\varepsilon_{1w}(1-\varepsilon_{2a})-\varepsilon_{2w}(1-\varepsilon_{1a})}\right\}\times\left[\frac{\lambda_1}{\lambda_2}\{1-(\varepsilon_{1w}+\varepsilon_{1a})\}-\{1-(\varepsilon_{2w}+\varepsilon_{2a})\}\right]t+H_0^2\right]^{\frac{1}{2}}-H_0\right] \tag{46}$$

By neglecting ε_{1a} and ε_{2a}, W is approximately rewritten as Eq.(47).

$$W=\frac{A\lambda_1V_0(\varepsilon_{1w}-\varepsilon_{2w})}{\alpha\varepsilon_{1w}\{(\lambda_1/\lambda_2)(1-\varepsilon_{1w})-(1-\varepsilon_{2w})\}}\times\left[\sqrt{2\alpha V_0\left(\frac{\varepsilon_{1w}}{\varepsilon_{1w}-\varepsilon_{2w}}\right)\left\{\frac{\lambda_1}{\lambda_2}(1-\varepsilon_{1w})-(1-\varepsilon_{2w})\right\}t+H_0^2}-H_0\right] \tag{47}$$

From Eqs.(42) and (46), W/Q_E is expressed as the following equation.

$$W/Q_E=(\lambda_1/\alpha\varepsilon_{1w})V_0 \tag{48}$$

As the experimental conditions are determined, α, ε_{1w} and λ_1 become constant. Therefore, the relation between W and Q_E is linear with V_0 as a parameter. Also, neglecting ε_{1a} and ε_{2a}, t_e is approximately shown as Eq.(49) from the material balances on liquid and solid in the sludge at the end of dewatering.

$$t_e=\frac{(\varepsilon_{1w}-\varepsilon_{2w})}{2\varepsilon_{1w}(1-\varepsilon_{2w})}\left\{\frac{\lambda_1(1-\varepsilon_{1w})}{\lambda_2(1-\varepsilon_{2w})}+1\right\}\frac{H_0^2}{\alpha V_0} \tag{49}$$

Furthermore, these theoretical equations were examined experimentally. The experimental apparatus and procedures in this examinations were almost the same as those under the condition of constant electric current in the previous section. As both white clay and bentonite particles have negative ζ-potential, the polarity of electrodes was determined as follows: the upper electrode was anode and the lower one was cathode. The electric fields under the constant voltage are applied to the sludge bed by a regulated D.C. power supply and I_t was measured with an automatic recorder.

For the sludges of both white clay and bentonite, the relation between Q_T and t is shown with V_0 as a parameter in **Fig.17**. This figure shows that the electroosmotic dewatering under the constant voltage is also very effective for the gelatinous bentonite sludge which is hardly dewatered by gravitation. In the case of bentonite sludge, the dewatering efficiency (ratio of terminal Q_T to the initial water content of sludge) was 28.5% under V_0 of 12 volt. In the experiments of vacuum dewatering of bentonite sludge, as an example, the dewatering efficiency was about 7.3% under a vacuum of 600 mmHg. That is, the volume dewatered by electroosmosis was about 4 times as much as the volume dewatered by vacuum method. From this result, it was found that electroosmotic dewatering was very useful for the gelatinous sludge which is difficult to dewater.

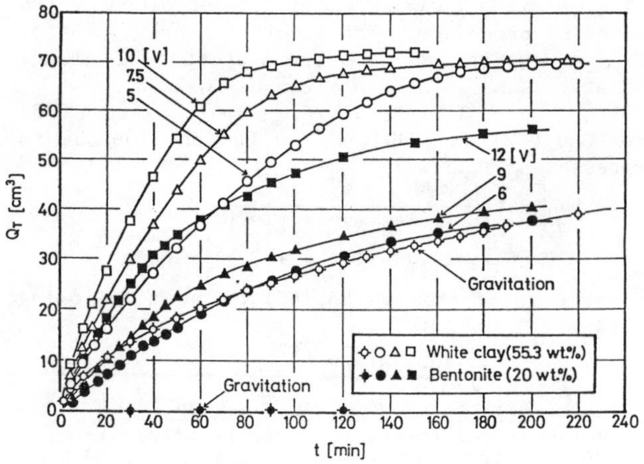

Fig.17 Relation between Q_T and t under condition of constant voltage

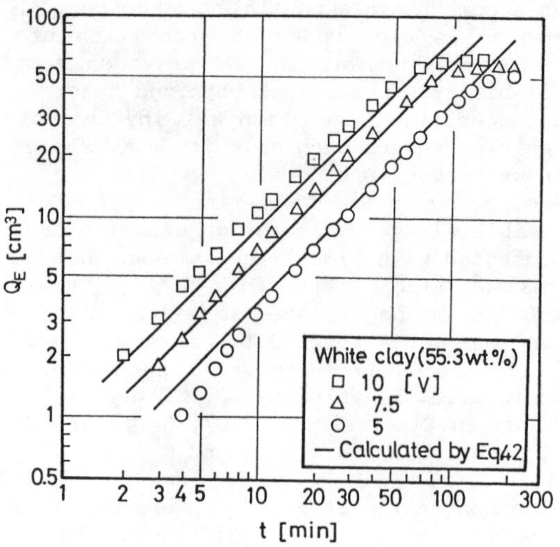

Fig.18 Relation between Q_E and t under condition of constant voltage

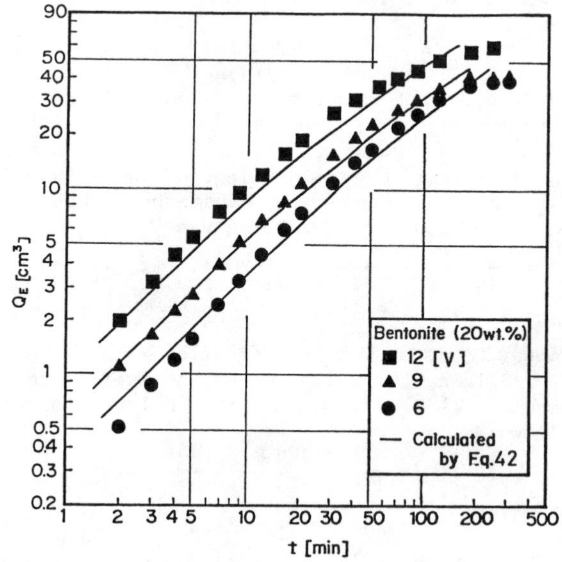

Fig.19 Relation between Q_E and t under condition of constant voltage

The time till the end of dewatering from the initiation was generally reduced by electroosmotic dewatering, but the final dewatered volume approached a nearly constant terminal value irrespective of V_0 in the case of white clay sludge, as shown in Fig.17. It is considered that there is a terminal water content in the case of electroosmotic dewatering under the constant voltage. Under the constant electric current, it was observed that Q_T amounted to the first terminal water content and that the secondary dewatering process followed consecutively. Under the constant voltage, however, it was observed that I_t decreased gradually because the electric resistance of the dewatered sludge bed increased with the proceeding of dewatering and then the dewatered volume finally reached a terminal value. Consequently, it may be discussed that the secondary dewatering process does not appear under the condition of constant voltage.

Figures 18 and 19 show the relation between Q_E and t on log-log paper for each sludge with V_0 as a parameter, and these figures show the comparison of the experimental results with the values calculated by Eq.(42). The values of ε_{2w} and λ_2 used for the calculation of Q_E by Eq.(42) were got by the measurements when the dewatering had finished. In the case of white clay sludge, the values calculated by Eq.(42) show good agreement with the observed values except the period in the neighborhood of the end of dewatering, as shown in Fig.18. In this case, it was observed that I_t almost never changed with t in spite of the constant voltage condition. Consequently, the relation between Q_E and t was nearly similar to the linear relationship under the constant electric current. When the electric resistance of sludge almost never changes in the dewatering, E_1 is approximately expressed as V_0/H_0 and becomes constant. Therefore, u_E also becomes constant from Eq.(35) and Q_E is given as the following approximate equation.

$$Q_E=A\varepsilon_{1w}\alpha(V_0/H_0)t \qquad (50)$$

In the case of bentonite sludge in Fig.19, the values calculated by Eq.(42) are in agreement with the observed values within ±20% range except the period in the neighborhood of the end of dewatering.

The value of t_e is obtained by Eq.(49) and t_e is inversely proportional to V_0. For white clay sludge, the values of t_e calculated by Eq.(49) under the conditions of constant voltage of 5, 7.5 and 10 volts are 184, 93 and 72 min, respectively, and the calculated values of t_e in the case of bentonite sludge are 221, 181 and 124 min for 6, 9 12 volts, respectively. These calculated values nearly coincide with the observed values in Figs. 18 and 19.

The relation between W and t is theoretically expressed by Eq.(47). The experimental results and the values calculated by Eq.(47) are shown with V_0 as a parameter in **Fig.20**. In the case of white clay sludge, the calculated values show good agreement with the observed values and W is nearly with respect to t. This is due to above-mentioned electrical characteristics that the electric resistance of sludge scarcely changes in the dewatering process. Therefore, I_t in Eq.(43) becomes constant and the relation between W and t for such a sludge can be approximately given as the following linear equation from Eqs.(43) and (45).

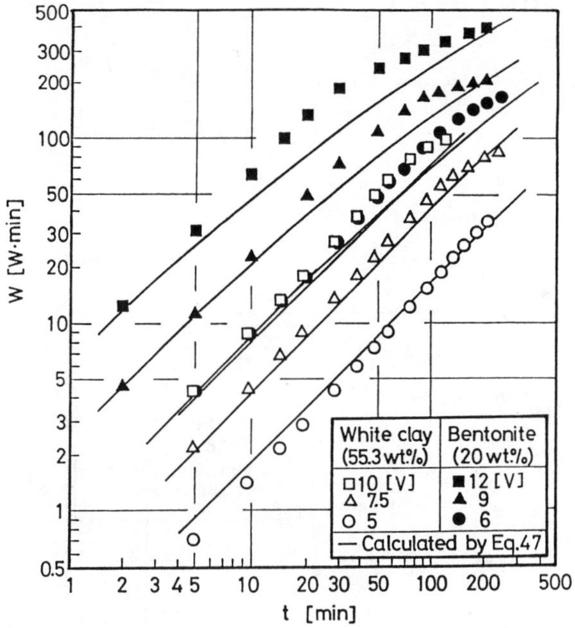

Fig.20 Relation between W and t under condition of constant voltage

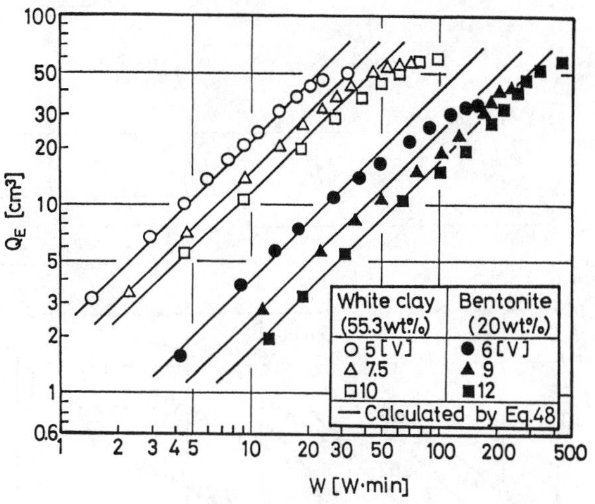

Fig.21 Relation between Q_E and W under condition of constant voltage

$$W=(A\lambda_1 V_0{}^2/H_0)t \quad (51)$$

For bentonite sludge shown in Fig.20, the observed values are slightly large compared with the values calculated by Eq.(47) and the difference between them increases with increasing V_0. It is thought that this difference is caused by the following reason. It was observed that I_t increased because of ionization from the surface of upper electrode by anodic oxidation and I_t decreased soon because of increasing the electric resistance of sludge bed with the proceeding of dewatering. However, I_t in Eq.(44) and W in Eq.(47) are obtained without considering the change of the electrical conductivity of sludge bed by the ionization of electrode material.

The relation between Q_E and W is expressed by a linear function with V_0 as a parameter as shown in Eq.(48). **Figure 21** shows the comparison of the experimental results with the values calculated by Eq.(48). For both white clay and bentonite sludges, the experimental results nearly coincide with the calculated ones except near part of the end of dewatering. The values of Q_E for the same value of W increase with decreasing V_0 for both sludges, as shown in Fig.21. That is to say, it is found that Q_E/W is inversely proportional to V_0 as expressed by Eq.(48). Accordingly, in electroosmotic dewatering under the constant voltage, there is an optimum operating condition with regard to dewatering rate and electric power consumption. Therefore, an optimum value of V_0 must be generally decided by the consideration of dewatering rate and electric power consumption. Under the constant electric current in the previous section, W/Q_E is expressed as a function of electric current density and dewatering time. However, W/Q_E under the constant voltage is proportional to only V_0, as shown in Fig.21.

As described above, the model used in the electroosmotic dewatering under the constant electric current is also applicable to the electroosmotic dewatering process under the constant voltage, and it was experimentally confirmed that the theoretical equations based on the model were available as the design equations of electroosmotic dewatering equipment.

COMPARISON BETWEEN ELECTROOSMOTIC DEWATERING EFFICIENCIES UNDER CONDITIONS OF CONSTANT ELECTRIC CURRENT AND CONSTANT VOLTAGE (ref.34)

In this section, a comparison of electroosmotic dewatering efficiency under the constant electric current with that under the constant voltage is made from the point of view of dewatering rate, water content in dewatered sludge and dewatered volume per unit electric power consumption.

As shown in the previous sections, the design equations of electroosmotic dewatering for compressible sludge under the conditions of constant electric current and constant voltage are listed in **Table 1**. In this table, the electroosmotically dewatered volume per unit electric power consumption Q_E /W under both operating conditins are written as Eqs.(52) and (53), respectively. These equations are obtained by transforming Eqs.(31) and (48) in the previous sections.

In order to compare the electroosmotic dewatering efficiency under the constant electric current with that under the constant voltage, the initial strength of electric field E_0 was adjusted so as to have the same value under both operating conditions. Experiments were performed in which the values of E_0 and H_0 were changed under the conditions of both constant electric current and constant voltage. $Mg(OH)_2$ sludge having positive ζ-potential was used in this experimental examinations. The initial concentration of solid and the porosity of water part ε_{1w} in the sludge were 26.9 wt.% and 0.861, respectively.

Figure 22 shows the relations between Q_E and t under the constant electric current and the constant voltage respectively, with E_0 as a parameter. At each value of E_0, the electroosmotic dewatering rates (dQ_E/dt) under both operating conditions are nearly equal in the early dewatering period. As shown in Eqs.(20) and (40) in Table 1, u_E under

Table 1 Equations obtained by theoretical analysis

Constant electric current condition

$$u_E = \alpha\frac{I_0}{\lambda_1} \qquad (20) \qquad\qquad Q_E = A\varepsilon_{1w}\alpha\frac{I_0}{\lambda_1}t \qquad (22) \qquad\qquad \frac{Q_E}{W} = 1\Bigg/\left\{\frac{{I_0}^2}{2(\varepsilon_{1w}-\varepsilon_{2w})}\left(\frac{1-\varepsilon_{1w}}{\lambda_2}-\frac{1-\varepsilon_{2w}}{\lambda_1}\right)t+\frac{H_0I_0}{\alpha\varepsilon_{1w}}\right\} \qquad (52)$$

Constant voltage condition

$$u_E = \alpha V_0\Bigg/\sqrt{2\alpha V_0\left(\frac{\varepsilon_{1w}}{\varepsilon_{1w}-\varepsilon_{2w}}\right)\left\{\frac{\lambda_1}{\lambda_2}(1-\varepsilon_{1w})-(1-\varepsilon_{2w})\right\}t+{H_0}^2} \qquad (40)$$

$$Q_E = \frac{A(\varepsilon_{1w}-\varepsilon_{2w})}{\{(\lambda_1/\lambda_2)(1-\varepsilon_{1w})-(1-\varepsilon_{2w})\}}\left[\sqrt{2\alpha V_0\left(\frac{\varepsilon_{1w}}{\varepsilon_{1w}-\varepsilon_{2w}}\right)\left\{\frac{\lambda_1}{\lambda_2}(1-\varepsilon_{1w})-(1-\varepsilon_{2w})\right\}t+{H_0}^2}-H_0\right] \qquad (42) \qquad\qquad \frac{Q_E}{W} = \frac{\alpha\varepsilon_{1w}}{\lambda_1 V_0} \qquad (53)$$

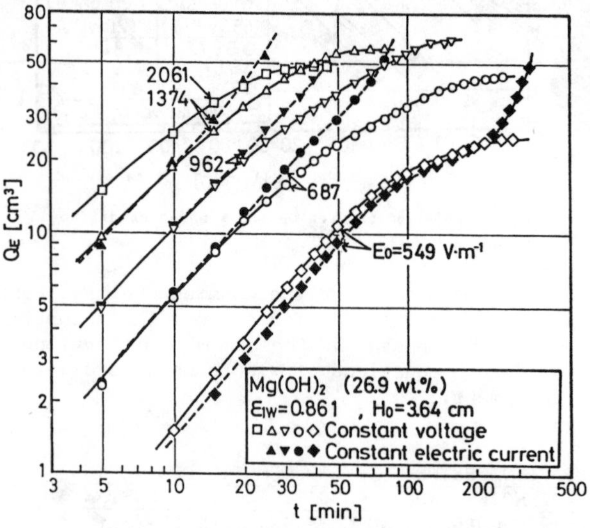

Fig.22 Relations between Q_E and t under conditions of constant electric current and constant voltage

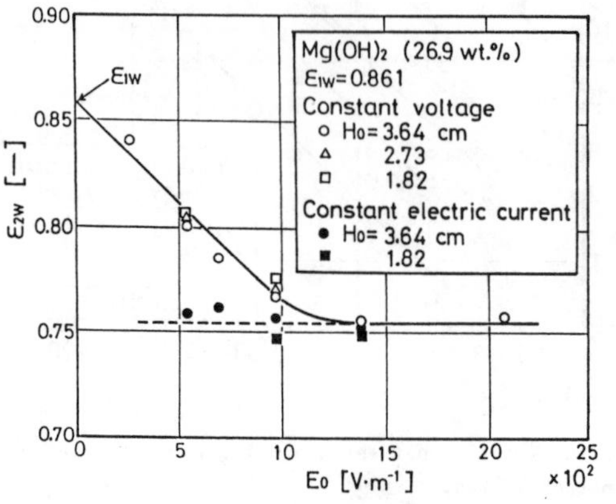

Fig.23 Relations between ε_{2w} and E_0 under conditions of constant electric current and constant voltage

the constant electric current is proportional to $I_0/\lambda_1(=E_0)$ irrespective of t; u_E under the constant voltage is approximately expressed as $\alpha V_0/H_0$), that is, proportional to $V_0/H_0(=E_0)$ when t is small. Therefore, in the case of the same value of E_0, it is found that u_E under both operating conditions is approximately the same in the early dewatering period. Under the constant voltage, Q_E gradually approaches a constant terminal value $Q_{E\infty}$ for each value of E_0, as shown in Fig.22. The value of $Q_{E\infty}$ becomes maximum at a certain E_0 with increasing E_0. This phenomenon may be explained as follows. It was observed that the sludge of small water content was formed near the upper electrode when E_0 was large. Accordingly, the electrical conductivity of the sludge of small water content decreases very much and the applied voltage almost drops in this part of sludge. Under the constant electric current, the relation between Q_E and t is expressed by a linear relationship as shown in Eq.(22). In Fig.22, Q_E under the constant electric current is approximately linear with respect to t for each value of E_0 and then dQ_E/dt increses further as Q_E approaches $Q_{E\infty}$ under the constant voltage. It is considered that this phenomenon is due to the secondary dewatering process after the completion of the first dewatering process, as described in the previous section. Judging from these results, the constant electric current condition is effective to increase Q_E.

When the dewatering process is finished, the sludge near the upper electrode is regarded as completely dewatered sludge of which the porosity of water part is ε_{2w}. **Figure 23** shows the relation between ε_{2w} and E_0 with H_0 as a parameter. The broken line and the solid line express the relations under the constant electric current and the constant voltage respectively. In this figure, it is regarded that the relation between ε_{2w} and E_0 is independent of H_0 under both operating conditions. In the case of constant voltage, ε_{2w} decreases linearly and then maintains a constant value over a certain value of E_0. The smallest value of E_0 minimizing ε_{2w} nearly coincides with E_0 at which a maximum value of $Q_{E\infty}$ has been obtained under the constant voltage in Fig.22. Accordingly, there is an optimum value of E_0 with regard to $Q_{E\infty}$ and ε_{2w}. In the case of constant electric current, ε_{2w} shows a constant value irrespective of E_0 and agrees with the minimum value under the constant voltage. This result is explained by the fact that the voltage applied to the sludge bed under the constant electric current increases gradually due to the increasing electrical resistance of sludge with the proceeding of dewatering. Consequently, it has become clear that the minimum values of ε_{2w} under both operating conditions are nearly equal.

The relations between Q_E and W under both operating conditions are shown with E_0 as a parameter in **Fig.24.** The broken lines and the solid lines express the results calculated by Eqs.(52) and (53), respectively. The experimental results nearly coincide with the calculated values except near part of the end of dewatering. The values of Q_E for the same value of W increase when E_0 is set to be small under both operating conditions. For the same value of E_0, Q_E/W under the constant voltage is slightly larger than that under the

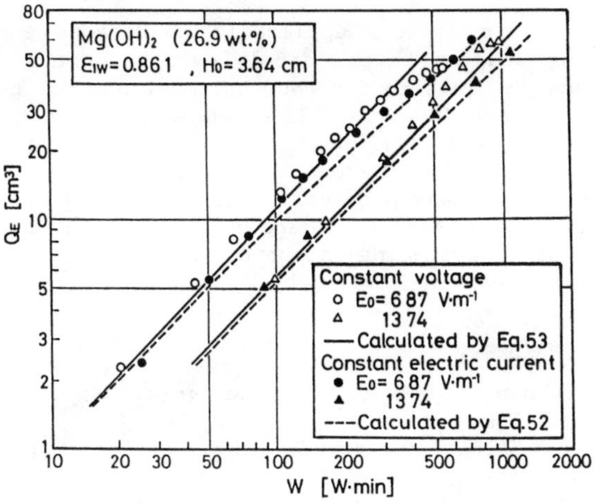

Fig.24 Comparison between Q_E/W values under conditions of constant electric current and constant voltage

constant electric current. It is considered that these results are caused by the following facts. In both operating conditions, the formation rate of completely dewatered sludge of small water content is enhanced with increasing E_0 because u_E is augmented with increasing E_0. As the electrical resistance of this dewatered sludge is large, the electric power loss in this part of sludge becomes greater with increasing E_0. Therefore, Q_E/W under both operating conditions increases when E_0 is small. In the case of the same value of E_0, u_E under both operating conditions is approximately equal in the early dewatering period, but u_E under the constant voltage decreases gradually with t compared with that under the constant electric current, as shown in Eqs.(20) and (40) in Table 1. Therefore, the foregoing electric power loss in the dewatered sludge under the constant electric current becomes greater with t compared with that under the constant voltage. Accordingly, Q_E/W under the constant voltage is larger with increasing W than that under the constant electric current except near part of the end of dewatering. **Figure** 25 shows the effect of H_0 on the relations between Q_E and W under both operating conditions. In this figure, the same value of E_0 is

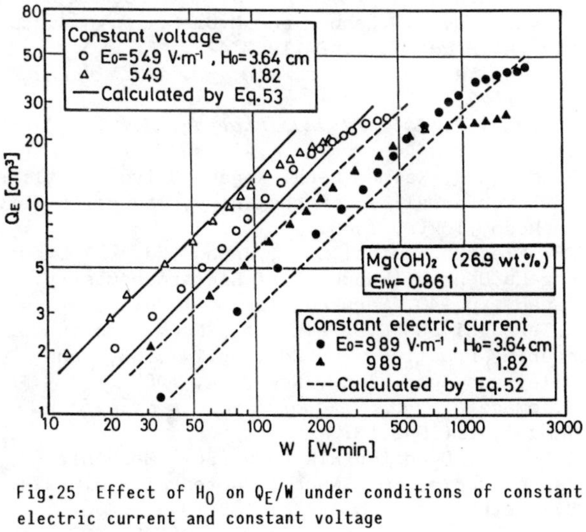

Fig.25 Effect of H_0 on Q_E/W under conditions of constant electric current and constant voltage

used for each operating condition. According to Eq.(52), Q_E/W under the constant electric current is inversely proportional to H_0 when the quadratic term of I_0 is much smaller than the linear term of I_0. Equation (53) is rewritten as $\alpha\varepsilon_{1w}/\lambda_1 E_0 H_0$, so that Q_E/W under the constant voltage is also inversely proportional to H_0. The experimental results nearly coincide with the calculated values except near part of the end of dewatering. In this range, it is found that Q_E for the same value of W under both operating conditions increases when H_0 is set to be small, because the applied voltage can be set to be small when E_0 is the same and H_0 is small. From the results in Figs.24 and 25, Q_E/W under both operating conditions increase when E_0 and H_0 are set to be small. In the cases of the same E_0 and H_0, Q_E/W under the constant electric current is smaller than that under the constant voltage.

SUMMARY

The electroosmotic phenomenon can be used effectively to dewater sludge. The thickening process by electroosmotic dewatering of sludge after the completion of gravitational thickening was theoretically analysed and was examined experimentally. The electroosmotic dewatering processes of compressible sludge under the constant electric current and the constant voltage were investigated experimentally and theoretically by using a model of electroosmotic flow through a compressible-particle packed bed. A comparison between electroosmotic dewatering efficiencies under the constant electric current and the constant voltage was discussed. The following results were obtained.

1) In the thickening process of sludge applied with a D.C. electric field after the completion of gravitational settling of slurry, an increase of the maximum concentration of particle is caused by the electroosmotic dewatering.

2) In the electroosmotic thickening of sludge, the relation between H_t and t is theoretically expressed by Eqs.(11) and (12). It was experimentally confirmed that $\Psi(H_t)$ was proportional to t in the case of uniform electrical conductivity throughout the sludge, as shown in Eq.(12).

3) The final porosity of sludge in the electroosmotic thickening is expressed by the empirical equation (18), which is similar to the relation between equilibrium porosity and mechanically exerted pressure. The maximum concentrations of particle obtained experimentally were about 80 wt.% for $CaCO_3$, and about 60 wt.% for white clay, respectively. These results are almost the same or over as those obtained by filter press and vacuum filter. According to author's investigation, enhanced settling rate and sludge concentration were obtained by application of electric field on slurry at the same time.

4) The electroosmotic dewatering mechanisms of compressible sludge under the constant electric current and the constant voltage can be explained by the model of electroosmotic flow through a compressible-particle packed bed. It was experimentally confirmed that the equations obtained by theoretical analysis were appropriate for the design of electroosmotic dewatering equipment.

5) Electroosmosis has remarkable effects on the decrease of water content of sludge. Especially, electroosmotic dewatering is very effective for a gelatinous sludge which is difficult to dewater mechanically.

6) In the electroosmotic dewatering under the constant electric current, dQ_E/dt and Q_E increase in proportion to I_0 during the first dewatering period. After the first dewatering period, the secondary dewatering process immediately follows. On the other hand, under the constant voltage, dQ_E/dt is proportional to V_0 but there is a terminal water content of Q_E independent of V_0. The values of W/Q_E under the constant voltage increase in proportion to V_0, as shown in Eq.(48). Therefore, an optimum operating condition must be decided by the general consideration concerning dQ_E/dt and W.

7) In the electroosmotic dewatering under the constant electric current, the secondary dewatering process appears after the completion of the first dewatering process. But the electric power consumption is large in this process because the electrical resistance of sludge in this process becomes gradually large. Accordingly, the operating condition of the constant electric current is effective to increase Q_E compared with that of the constant voltage. However, for the same E_0 and H_0, Q_E/W under the constant electric current is smaller than that under the constant voltage.

8) In the electroosmotic dewatering under the conditions of constant electric current and constant voltage, Q_E/W becomes more when E_0 and H_0 are set to be small. The terminal water content in the dewatered sludge is nearly equal under both operating conditions.

In conclusion, the characteristics of the electroosmotic dewatering method are described as follows; 1) The dewatering rate and the dewatering efficiency are easily controlled by regulating the electric current and the applied voltage. 2) The driving force of the dewatering occurs throughout sludge bed, consequently the dewatering rate does not decrease so much even though the thickness of sludge bed increases. 3) The driving force owing to electroosmosis is different from the mechanical filtration pressure. Therefore, the filter medium is not damaged and blocked so much. 4) Electroosmotic dewatering is very effective for colloidal and gelatinous sludges which are very difficult to dewater mechanically because of the reason described in 2). 5) Electroosmotic dewatering can be jointly used with mechanical dewatering methods such as pressure filtration and vacuum filtration. 6) Carrying out electroosmotic dewatering, mechanical dewatering equipment such as filter press and vacuum filter may be reduced and then may be unnecessary. 7) If the corrosion of the anodic electrode occurs, the contamination of sludge may be produced. 8) When the electrical conductivity of sludge is very large, the electric power loss by Joule's heat becomes large. On the contrary, when it is small, the electric power consumption increases since large applying voltage is necessary. In such a case, a suitable electrolyte is added to the sludge.

As stated above, although electroosmotic dewatering method has several advantages compared with mechanical dewatering one, it has hardly been used as a practical application. Because engineering studies have scarcely ever been made and the design equation for the equipment has not been proposed previously. Electroosmotic dewatering techniques are still being developed. Therefore, in the stage for the practical application of it, there may be many problems. According to author's experience, however, we are confident that electroosmotic dewatering can be expected to be a very useful dewatering method by applying it properly.

REFERENCES

1 H.Yukawa, *Kagaku Kogaku* (*Chem. Eng., Japan*), 28 (1964)732.
2 *idem, ibid.*, 29(1965)579.
3 *idem, ibid.*, 31(1967)151.
4 H.Yukawa and S.Egawa, *ibid.*, 32(1968)901.
5 H.Yukawa and Y.Yushina, *ibid.*, 32(1968)909.
6 H.Yukawa S.Kanai, O.Shimoyama and T.Karino, *ibid.*, 35(1971)657.
7 H.Yukawa, H.Chigira, T.Hoshino and M.Iwata, *J. Chem. Eng. Japan*, 4(1971)370.
8 H.Yukawa, in The Soc. of Chem. Engrs.,Japan(Ed.), Modern Chemical Engineering, p.111, Maruzen, Tokyo, 1972.
9 H.Yukawa, K.Kobayashi, Y.Tsukui, S.Yamano and M.Iwata, *J. Chem. Eng. Japan*, 9(1976)396.
10 H.Yukawa, K.Kobayashi, H.Yoshida and M.Iwata,in R.J.Wakeman(Ed.), Progress in Filtration and Separation, Vol.1, p.83, Elsevier, Amsterdam , 1979.
11 H.Yukawa, H.Obuchi and K.Kobayashi, *Kagaku Kogaku Ronbunshu* (*Chem. Eng.,Japan*), 6(1980)288.
12 *idem, ibid.*, 6(1980)323.
13 H.Yukawa, K.Kobayashi and M.Hakoda, *J. Chem. Eng. Japan*, 13(1980)397.
14 H.Yukawa, *MEMBRANE* (*Japan*), 6(1981)253.
15 J.D.Henry,Jr., Lee F.Lawler and C.H. Alex Kuo , *AIChE. Journal*, 23(1977)851.
16 K.Kobayashi, M.Iwata, Y.Hosoda and H.Yukawa, *J. Chem. Eng. Japan*, 12(1979)466.
17 K.Kobayashi, M.Hakoda, Y.Hosoda, M.Iwata and H. Yukawa, *ibid.*, 12(1979)492.
18 K.Matsumoto, O.Kutowy and C.E.Capes , *Powder Technology*, 28(1981)205.
19 M.C.Porter, *AIChE. Symposium Ser.*, 68(1972)120.
20 R.M.Jorden, *J. Am. Water Works Assoc.*, 55(1963)771.
21 S.P.Moulik, F.C.Cooper and M.Bier, *J. Colloid and Interface Sci.*, 24(1967)427.
22 S.P.Moulik, *Environmental Science and Technology*, 5(1971)771.
23 S.Stotz, *J. Colloid and Interface Sci.*, 65(1977) 118.
24 S.Komagata, Kaimen Denki Kagaku Gaiyo (Outline of Interfacial Electrochemistry, Japan), p.79 , Shokodo, Tokyo, 1969.
25 The Electrochemical Soc., Japan (Ed.), Denki Kagaku Benran (Handbook of Electrochemistry, Japan), p.940, Maruzen, Tokyo, 1974.
26 K.Terzaghi and R.B.Peck, Soil Mechanics in Engineering Practice, 2nd Ed., pp.147,383, John Wiley and Sons, Inc., New York, 1967.
27 S.Komagata, *Denki Kagaku* (*Electrochemistry* , *Japan*), 11(1943)13.
28 T.Mogami, Doshitsu Rikigaku (Soil Mechanics , Japan), pp.52,86, Iwanami Shoten, Tokyo, 1974.
29 B.F.Ruth, *Ind. Eng. Prog.*, 38(1946)564.
30 F.M.Tiller, *Chem. Eng. Prog.*, 49(1953)467.

31 S.Yagi and Y.Yamazaki, *Kagaku Kogaku* (*Chem. Eng., Japan*), 24(1960)81.
32 H.Yukawa, H.Yoshida, K.Kobayashi and M.Hakoda, *J. Chem. Eng. Japan*, 9(1976)402.
33 *idem*, *ibid*., 11(1978)475.
34 H.Yoshida, T.Shinkawa and H.Yukawa, *ibid*., 13 (1980)414.

DRYING HEAT SENSITIVE AND/OR HAZARDOUS MATERIALS UTILIZING A FULLY CONTINUOUS VACUUM SYSTEM

DIETER FORTHUBER
PATRICK MC CARTY

KRAUSS MAFFEI CORPORATION
WICHITA, KANSAS

Abstract

The paper presents a newly developed, continuously operating vacuum drying system. It utilizes a Vacuum Plate Dryer with Rotary Vacuum Locks for feeding and discharging material. The system is extremely vacuum tight and overcomes the problems usually encountered when processing heat sensitive and/or hazardous materials.

Introduction

Thermal separation of fine crystalline or powdery solids from H_2O or organic solvents in applications where

- solids are heat sensitive
  and/or
- solids and liquids have toxic and hazardous properties

requires a drying system with precise temperature control as well as strictly controlled environment.

These applications include processing of a variety of products such as pesticides (suspended in toluene, methanol, hexane, etc.), peroxides, surfactants, explosives, resins, pharmaceutical products and others.

The chemical and physiological properties of such materials often mandate the need for very low drying temperatures in the range of 20°-80^oC in order to avoid the risk of thermal degradation. Operating convection dryers at these temperatures requires large equipment with large quantities of drying gas flowing through the dryer and necessitates recycling the (inert) gas if the presence of oxygen is not permitted during the drying process. This inevitably leads to the risk of dust formation, high energy consumption and difficult recovery of evaporated solvents.

Dryers with conductive heat transfer operating under vacuum would eliminate the aforesaid problems. Additionally, the lower boiling temperature of volatiles (achieved by reducing their vapor pressure) results in a reduction of both solids temperature and duration of the drying process.

Vacuum operation also simplifies the solvent recovery. The often toxic and hazardous properties of solvents and solids create the need for a completely enclosed environment to minimize the danger to operating personnel and surrounding equipment. The selected dryer should therefore be of rigid construction and extremely gastight.

The tendency in the CP-Industry is going more and more towards continuous processes. Only the continuous operation provides the advantages of labor savings, consistent product quality and high production rates at reduced installation and operating costs.

Following is a description of a truly continuous drying system.

Operating Principle of the Vacuum Plate Dryer

The main component of the described continuous vacuum drying system is the Vacuum Plate Dryer, the vacuum version of the atmospheric Plate Dryer. This unit consists of a vacuum tight shell containing staionary vertically stacked plates, each of which are heated by a flowing medium such as water, steam or oil. The heat transfer in the dryer is solely made by contact of the product with the heated plates (conduction drying), eliminating the need for large quantities of drying gas.

Figure 1 is a schematic of the operation of the Plate Dryer.

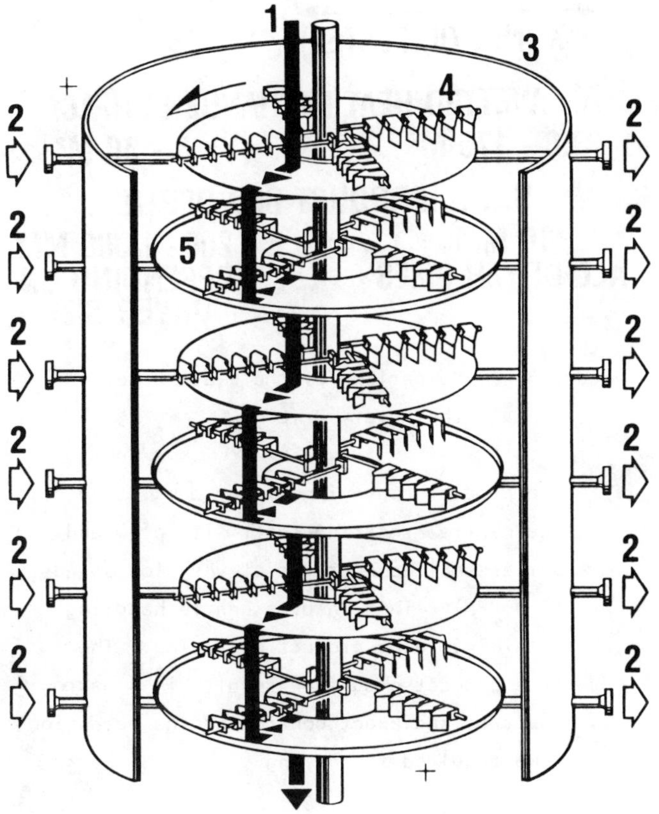

1. Product
2. Heating or Cooling Medium
3. Shell, housing
4. Conveying System
5. Plate

Figure 1 - Operation of the Plate Dryer

The wet product enters the top of the dryer and is gently conveyed in a spiral pattern across the plates by means of a vertical rotating shaft provided with radial arms and self-aligning plows. The alternating arrangement of small and large plates allows the product to drop from one plate to the next until it is discharged at the bottom of the dryer. As the product flows through the dryer it is continually turned over and over exposing each particle to the heated surface of the plate.

The evaporated volatiles are removed from the dryer by evacuation and subsequently condensed.

Advantages of the Vacuum Plate Dryer

The Vacuum Plate Dryer offers the following advantages:

a) Conduction drying under vacuum is much more efficient as compared to convection drying where a drying gas (usually nitrogen) has to be cooled, heated and recycled in order to remove the solvents evaporated in the dryer. Thus the overall operating costs of a vacuum dryer are significantly reduced.

b) The ability to operate at much lower drying temperature and drastically reduced retention time minimizes the risk of thermal degradation of the product. Figure 2 shows a comparison of an atmospheric plate dryer vs. a vacuum plate dryer. As can be seen, vacuum operation reduces the retention time by 40-50%. In addition, the atmospheric plate dryer must be operated at a 60% higher plate temperature to achieve the same drying time as the vacuum operation yields.

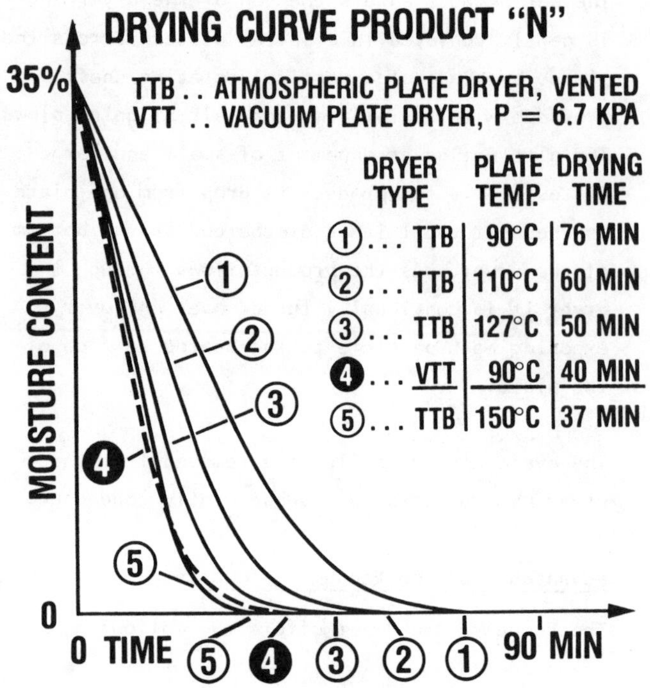

Figure 2 - Comparison of Atmospheric Plate Drying vs. Vacuum Plate Drying

c) Precise temperature control can be realized and a temperature profile can be established by adjusting the temperature of individual plates or plate groups.

d) Thin product layer (approximately 10 mm) on large heat exchange surface coupled with high product turnover improves both heat and mass transfer. Figure 3 shows a comparison of the plate dryer to rotary dryers in regard to product bed depth and wetted heat exchange surface.

In the Plate Dryer, the product layer is kept extremely low. The entire heat exchange surface is utilized for heat transfer. The product surface exposed to the surrounding atmosphere is even larger than the actual wetted heat exchange surface. The design of the product conveying system insures product turnover numbers in the range of 200-1500. In contrast, rotary type dryers have high bed depths coupled with small surface exposed to the surrounding atmosphere - both factors which limit mass transfer coefficients. In addition, only 50-65% of the total heat exchange surface is utilized for heat transfer.

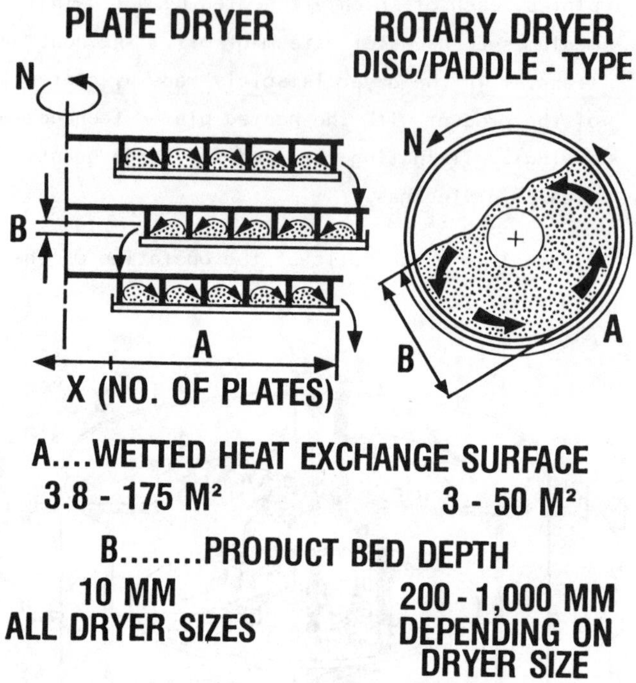

Figure 3 - Comparison of the Plate Dryer vs. the Rotary Dryer

e) The product conveying system consisting of the central rotating shaft with plow and arm assemblies operates at very low speed (1-4 RPM). This insures gentle handling thus minimizing product attrition. The true product plug flow results in a narrow spectrum and exact control of the retention time required for drying.

f) Solvent recovery is simplified by the use of a condenser to handle volatiles only.

g) Heavy gauge metal thicknesses and precise manufacturing procedures of the entire drying system insure extremely low leakage rates and ability to contain system pressure surges.

h) The low density of vacuum vapors limits their ability to carry dust thus eliminating the need for a dust collection system and the risk of dust explosions.

i) Modular design yields a wide range of dryer sizes with heat exchange surface between $3.8\ m^2$ to $175\ m^2$.

Vacuum Drying Systems

Due to the design of the equipment available, vacuum drying has been limited to batch-type operation. Various batch vacuum dryers are shown in Figure 4.

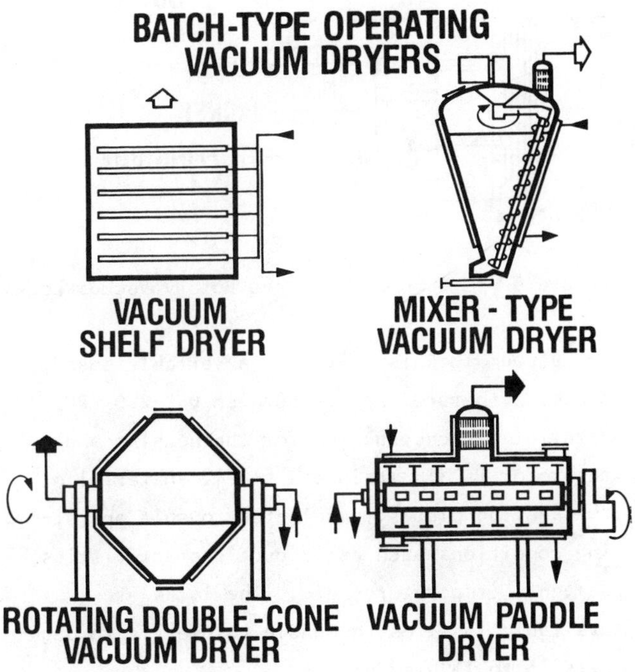

Figure 4 - Batch-Type Operating Vacuum Dryers

Several disadvantages arise from the usage of batch vacuum dryers:

a) Labor intensive product handling.
b) Product holding tanks necessary both upstream and downstream.
c) Risk of nonconsistent product quality.
d) Long drying time.
e) Normally limited to small production rates.

The Vacuum Plate Dryer coupled with the newly developed Rotary Vacuum Locks overcomes these disadvantages and creates a fully continuous drying train. The lock can serve to both feed and discharge the product continuously under full vacuum. In the past these functions had to be accomplished via alternating feeders and receivers provided with vacuum gate valves. Such a system actually operates on a semi-continuous basis. A comparison of both systems is shown in Figure 5.

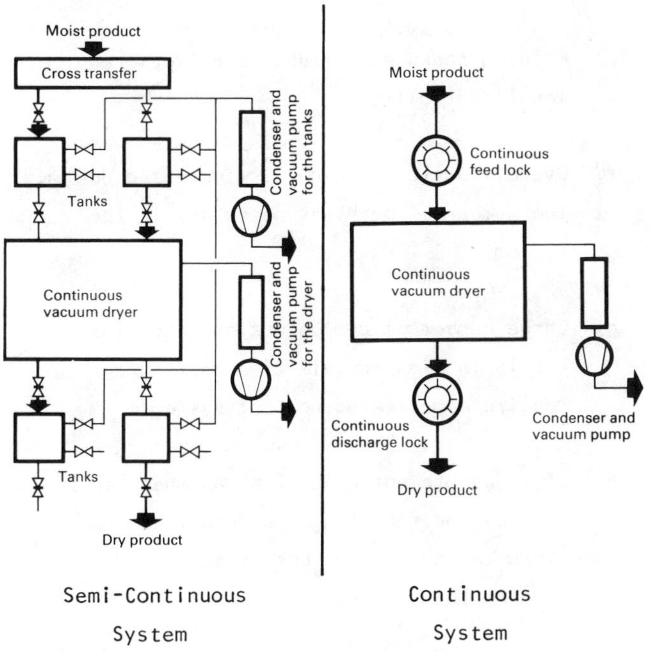

Figure 5 - Comparison of Vacuum Systems

The semi-continuous system shown in Figure 5 is a complex operation requiring intermittent switching between feeders and receivers. The wet product is fed via a reversible conveyor into the vented table feeder. At the same time, the adjacent feeder is under vacuum and feeds the material continuously into the vacuum dryer. After this feeder empties, it must be vented and re-filled, vacuumized and then begins feeding the dryer with product. The same alternating procedure occurs on the discharge end of the dryer.

Many disadvantages arise with the use of the semi-continuous system:

a) The time interval required for switching from one feeder to the other is lost drying time.

b) An additional vacuum pump and condenser is required for the feeders and receivers.

c) Vacuum valves require extensive maintenance (due to complicated mechanics and high number of cycle times).

d) Holding tanks and product receivers require level controlling.

f) Overall building height is increased due to the number of machines necessary in the system.

g) Large number of components and ancillaries result in high installation costs and require complicated control systems.

All of the aforementioned disadvantages can be overcome by incorporating the Rotary Vacuum Lock into the system for both feeding and discharging the product.

Rotary Vacuum Lock

The patented Rotary Vacuum Lock (shown in the Continuous System, Figure 5) is basically a vacuum tight star valve. The chambered rotor rotates in a pressure tight housing. Axial sealing between rotor and housing is accomplished by stuffing boxes. Radial sealing against the differential pressure between inlet and outlet is achieved by several consecutively arranged sealing elements. Each sealing element consists of a rectangular plastic sealing plate mounted on a metal support. This support is pneumatically pressed against the rotor by a pressure plate and an elastic membrane which transmits the pressure superposed by pressurized air. Figure 6 shows the schematic of the Rotary Vacuum Lock.

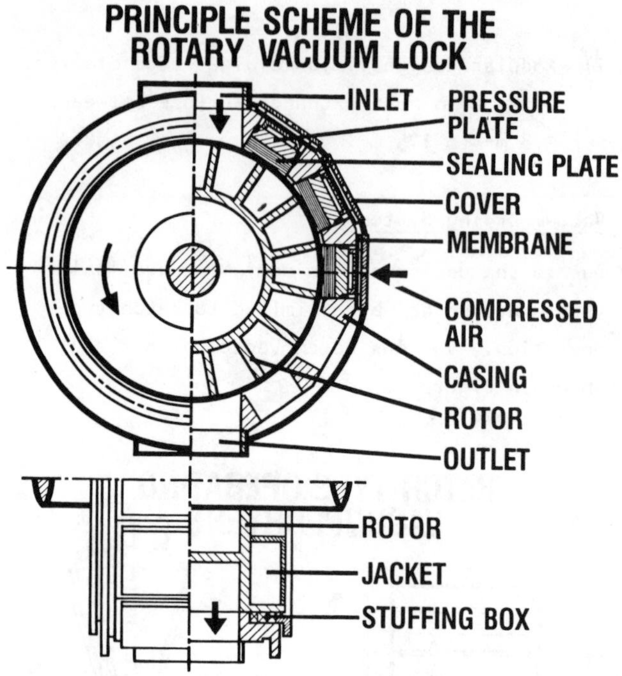

Figure 6 - Schematic of the Rotary Vacuum Lock

The vacuum lock is driven by a variable speed drive, with rotor speeds between 0.2-2.0 RPM. Material of construction for the housing and the rotor is 316 stainless steel. Material of the sealing plates is selected to suit operating conditions such as product characteristics, solvent, temperature, etc. The locks are available in sizes to handle volumetric throughputs up to 14.0 m^3/hr.

Continuous Vacuum Plate Drying System

The following benefits can be realized with the drying system incorporating both Vacuum Plate Dryer and Rotary Vacuum Locks:

a) The system operates fully continuously - resulting in greatly reduced installation costs and much more compact design.

b) The continuous drying process maintains consistent high product quality.

c) The system is completely contained under vacuum eliminating the risk of exposure to the hazardous product.

d) Standardized equipment and simple controls reduces both maintenance and operator surveillance.

Conclusion

Various drying systems utilizing the fully continuous vacuum process as described in this paper are in operation worldwide for drying a variety of heat sensitive and/or potentially dangerous products. Risk to personnel and environment has substantially been reduced. Energy consumption of the continually operating vacuum dryer is significantly lower as compared to common convection dryers with closed circuit systems. In addition, the fully continuous system versus the semi-continuous system (reference Figure 5) shows an overall reduction in installation costs of approximately 20%. At the same time, due to the fewer components involved in the continuous system, the reliability of the entire drying process is considerably improved.

PRESSURE DROP AND PARTICLE CIRCULATION STUDIES IN MODIFIED SLOT SPOUTED BEDS

Law, L.K.C., Ziaie, B., Malhotra, K. and Mujumdar, A.S.

McGill University, Montreal, Canada

ABSTRACT

Experimental results are presented on the pressure drop and solids circulation rates for symmetric (full) as well as asymmetric (half) slotted spouted beds fitted with vertical baffles. Tests were carried out with dry corn, peas, molecular sieve particles and binary mixtures thereof. Vertical baffles placed over the slot nozzle inlet were observed to stabilize spouting and induce a smooth, nearly plug flow of particles in the dense downcomer zone. Effect of positioning additional vertical baffles in the dense phase on the solids circulation rate, particle residence time and segregation phenomena in binary mixtures, was also examined.

INTRODUCTION

This brief communication summarizes a part of our exploratory experimental work to evaluate the concept of a fully two-dimensional slotted spouted bed (SSB) first proposed by one of the authors (ASM). The principal advantages of a SSB are the elimination of the need to scale-up, modular design and ability to obtain stable spouting, controlled solids circulation time and indirect heating/cooling by immersing vertical flat-plate heat exchangers. Use of vertical plates or baffles allows use of combined direct-indirect drying of large particulates (e.g. grains) in a spouted bed device. Appropriately placed vertical plates (over the inlet air slot nozzle) act in a manner similar to that of a draft tube in the conventional round spouted bed (see Ref. (1)). Although no experiments exist it is safe to conclude that use of such baffles eliminates the restriction of a maximum spoutable bed height. Anderson et al (2) have presented some of our earlier work on SSB's of grains. The results presented in this brief communication were obtained using a different apparatus.

EXPERIMENTAL SET-UP AND PROCEDURE

All experiments were carried out in a plexiglas vessel of rectangular cross-section, 20.2 cm x 5 cm. The inlet slot width was fixed at 2 cm giving an aspect ratio of nearly 10. A large mesh screen was placed over the slot to prevent the particles falling into the plenum chamber. The bed pressure drop was measured with a manometer; the static pressure taps were located just above the nozzle and well above the bed height. The air was exhausted from a slot on one side of the bed (narrow side, to maintain two-dimensionality). Exhaust from the top was found to make no discernible effect on the observed flow pattern or bed pressure drop.

Experiments were carried out with peas, corn and molecular sieve particles. The bed charge was always 3.20 kg. The bulk densities of peas, corn molecular sieve used are 684, 770 and 873 kg/m^3 respectively while the average particle dimensions were 5, 8 and 2 mm respectively. (For corn, 8 mm is the average largest dimension of the individual grain.) The terminal velocities for peas, corn and molecular sieve particles were estimated to be 13.1, 14.8 and 7.8 m/s respectively. These are only approximate values.

Table 1 summarizes the five different variants of SSB's used in this study. FSB (full spouted bed) refers to the case where the spouting takes place centrally as in a conventional bed. HSB refers to half spouted bed, since this case is analogous to the half-columns used commonly in conventional spouted bed studies. Here the nozzle is located on one side of the base.

Because of a limitation on the air source and associated ductwork spouting could not be achieved in the plain SSB i.e. without the help of two vertical plates (analogous to the draft tube in round SB) located centrally 5 cm above the slot and 3.5 cm apart. The plates were about 0.30 cm thick and 45 cm high.

In the case of a HSB only one plate was needed to effect stable spouting. Without these plates spouting could be obtained only for 15 cm high beds; even in this case the spout was unstable and oscillated significantly. (This was a limitation of the blower capacity.)

Since one of the potential commercial applications of the SSB is in drying of grains and since indirect drying is inherently more energy-efficient than direct drying, effect of immersing vertical plates (to simulate heat exchange surfaces) on the flow behaviour and solids circulation rate was studied for peas and binary mixtures of peas, corn and molecular sieve particles.

The solids circulation rate was calculated by simply timing with a stop watch the transit time of a tagged particle in the dense phase downcomer (each compartment in the case of multiple verticle plates).

The solids circulation rate is computed from the average particle velocity in each compartment, the width of the compartment and the bulk density of the dense phase (assumed equal to that in a static packed bed). The circulation rates given are for the apparatus studied i.e. they are in kg of material recirculated by the air jet per second per 5 cm bed width (z-direction). This number can be readily converted to any width of the bed providing the spouting is uniform across the width and no three-dimensional effects are present.

RESULTS AND DISCUSSION

Pressure drop data

Figures 1 through 4 present typical pressure drop (ΔP, N/m^2) versus superficial bed velocity (U, m/s) data. Note that the dotted lines are for decreasing air flow while the solid lines are for increasing flow. Since all data were taken well over the minimum spouting velocity (estimated to be in the order of 2 m/s) the hysteresis is minor and not significant. It is interesting that the pressure drop (N/m^2) is higher at lower slot velocities. This is possibly due to greater confinement of the jet air to the spout compartment i.e. less percolation through the outer dense phase. This effect is less pronounced in full beds as compared to the half beds. The effect of additional vertical partitions is also relatively unimportant as far as engineering design is concerned. The effect of particle type was also minor over the range of flow and geometric parameters studied. The half columns (HSB's) typically displayed lower pressure drops as compared to FSB's. However, this was accompanied by reduced solids circulation rate in the bed as noted later.

Solids circulation rates

Solids circulation rates were computed by timing the transit time for a number of tagged (colored) particles in each compartment (dense phase only) to travel from the top of the bed to the bottom. The reported rates are averages for 4-6 particles at each lateral distance. Transit times were obtained at various locations within each compartment. It was observed that the bed moved in a near plug flow in each compartment albeit with different mean flow rates. The flow rate dropped significantly in compartments away from the lean spouting zone.

Following is a summary of data obtained with peas, corn and molecular sieve particles. More data are needed to substantiate the general observations reported here. Caution should be exercised in applying these conclusions to other systems.

TABLE 2

Solids Circulation data for FSB-1 and FSB-2

H = 45 cm (approx.)

U, m/s	Particles	S, kg/s	Bed
3.5	Peas	0.27	FSB-1
4.3	"	0.34	"
3.5	Corn-Peas	0.25	"
4.3	"	0.43	"
3.5	Peas	0.28	FSB-2
"	M.S. + Peas	0.30	"
"	Corn + Peas	0.25	"
4.3	Peas	0.37	"
"	M.S. + Peas	0.33	"
"	Corn + Peas	0.37	"

The effect of inlet velocity is quite significant in determining the solids circulation rate i.e. the solids entrainment rate for the jet (Table 2). The overall rate is little affected by presence of additional partitions although the solids flow rate in each compartment is significantly different (but uniform within each compartment). The rate is also insensitive to the particle characteristics within the range studied. No segregation seems to appear in the dense phase. Most of the classification occurs in the fountain region.

TABLE 3

Solids circulation data: HSB-1, HSB-2, HSB-3

U, m/s	Particles	S kg/s HSB1	HSB2	HSB3
3.5	Peas	.073	.085	.084
"	Corn+Peas	.11	.10	.11
"	M.S.+Peas	.15	.14	.12
4.3	Peas	.15	.15	.26
"	Corn+Peas	.15	-	-

The effect of air flow rate on S is even more pronounced in the case of half-beds as seen in TAble 3. The HSB circulation rate is only about one-half to one-third of that in FSB's for the same slot width and air velocity. This is expected since solids entrainment takes place only along the edge of the half-jet in the case of HSB. As the number of partitions increase the data showed that progressively increasing proportion of the solids circulation rate shifts to compartments closer to the spout region. For HSB-2 data showed that 85% of the solids circulation occurs in the first compartment (next to spout). In the case of HSB-3, the solids circulation rate was 65-75% in the first compartment while it was only 7% in the outermost compartment.

It should be noted that these data are valid only for batch beds. If the solids discharge is located below the outermost compartment the dense bed velocity these will increase to a point; excessive discharge rate may cause the compartment to run "dry" (devoid of particles).

CONCLUDING REMARKS

Some preliminary flow data and observations on solids circulation rates are presented for some novel slotted spouted bed configurations. Work is currently in progress using a wider variety of particles. Parenthetically, it is interesting to note the effect of mechanical vibration applied to spouted beds. For shallow fluid beds the effect is generally beneficial. For spouted beds vibration tended to compact the dense phase and also counter the downward motion of the particles under certain vibrational acceleration levels the particles could be made nearly stationary in the downcomer zone. The solids circulation rate (or jet entrainment rate) is also minimal which makes application of vibration undesirable for spouted beds.

ACKNOWLEDGEMENT

The authors are indebted to Purnima Mujumdar for cheerfully typing this paper in a remarkably short time.

REFERENCES

1. Swaminathan, R. and Mujumdar, A.S., DRYING'84, Ed. A.S. Mujumdar, Hemisphere, N.Y., (1984).

2. Anderson, K., Raghavan, G.S.V. and Mujumdar, A.S., ibid, (1984).

TABLE 1

Spouted bed configurations investigated

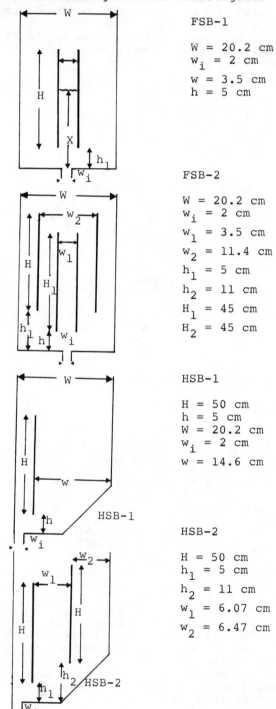

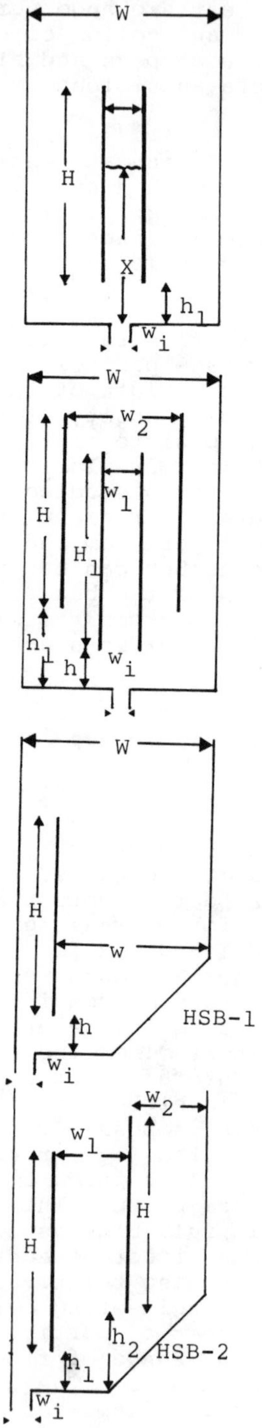

FSB-1

$W = 20.2$ cm
$w_i = 2$ cm
$w = 3.5$ cm
$h = 5$ cm

FSB-2

$W = 20.2$ cm
$w_i = 2$ cm
$w_1 = 3.5$ cm
$w_2 = 11.4$ cm
$h_1 = 5$ cm
$h_2 = 11$ cm
$H_1 = 45$ cm
$H_2 = 45$ cm

HSB-1

$H = 50$ cm
$h = 5$ cm
$W = 20.2$ cm
$w_i = 2$ cm
$w = 14.6$ cm

HSB-2

$H = 50$ cm
$h_1 = 5$ cm
$h_2 = 11$ cm
$w_1 = 6.07$ cm
$w_2 = 6.47$ cm

Table 1 continued

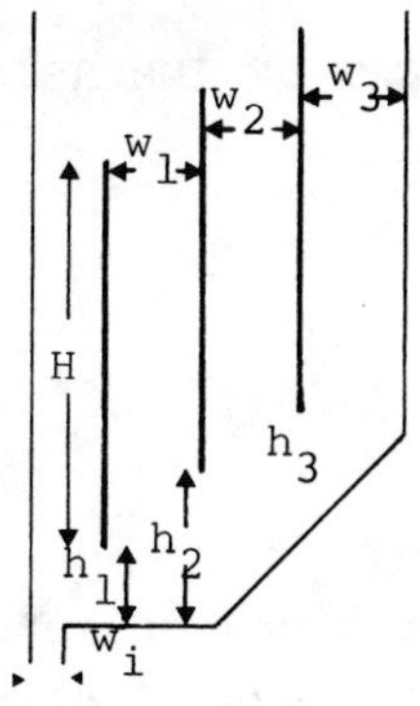

HSB-3

$H = 50$ cm

$w_1 = 4.4$ cm

$w_2 = 4.4$ cm

$w_3 = 4.8$ cm

$h_1 = 5$ cm

$h_2 = 11$ cm

$h_3 = 14$ cm

$w_i = 2$ cm

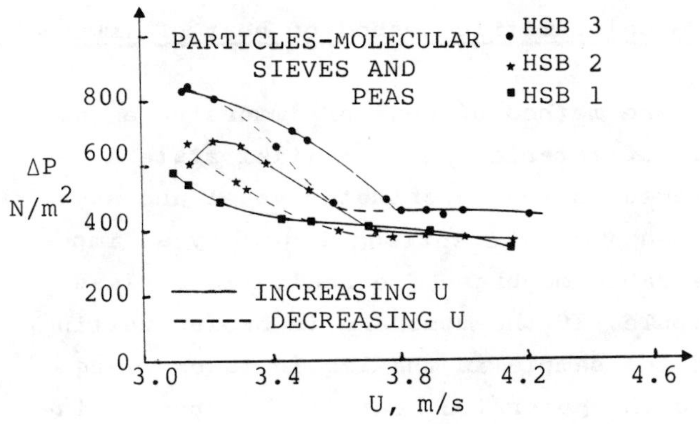

Fig. 1 Pressure drop data above minimum spouting velocity for half-beds

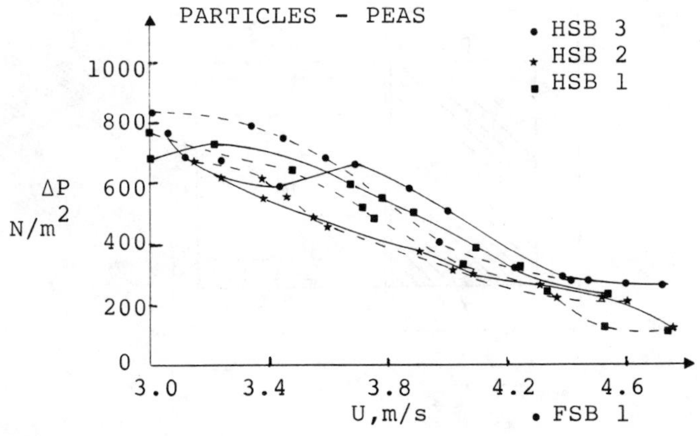

Fig. 2 Pressure drop data above minimum spouting velocity for half-beds of Peas.

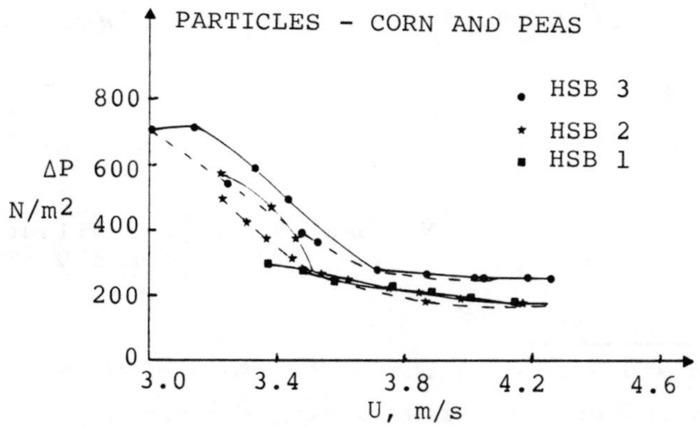

Fig. 3 Pressure drop data for half-spouted beds of Corn and Peas mixture.

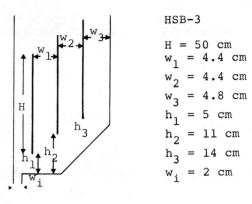

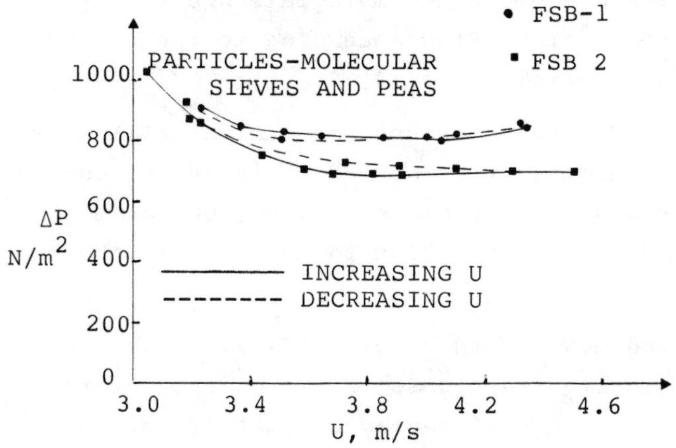

Fig. 4 Pressure drop data for full spouted beds of molecular sieve particles and Peas mixture.

MOISTURE BOND IN MATERIALS STUDIED BY THE CALORIMETRY OF HEAT OF IMMERSION

B. Čermák

National Research Institute for Machine Design
Praha 9 250 97

Introduction

Moisture bond is the basic cause of moistening of materials. During the process of drying this bond is destroyed while the liberated moisture is transported to the surface and from there into the drying medium. Dried materials are therefore often classified according to their bond with moisture /1, 2, 3/.

So far the moisture bond in materials has been classified according to drying curves drying thermograms and energograms /2, 4/ or according to sorption isotherms /5/.

The new method of classification according to the course of the characteristic curve in the diagram of state functions /6/ is based on a complex thermodynamic description of moisture bond with material using energy functions of potential, enthalpy and entropy. For a particular material the curve in the diagram is given by the experimentally determined values of potential (or chemical potential) $\mu_W^{MR}(u)_{T,p}$ and specific enthalpy of bound moisture $h_W^{MR}(u)_{T,p}$ for different specific moisture content u at a constant temperature T and pressure p.

The potencial μ_W^{MR} can be easily determined on the basis of sorption isotherms. For the determination of the enthalpy the most reliable among the available methods is the one of heat of immersion. Moreover, this method enables to investigate the character of moisture bond by means of comparison of various types of interaction of material with liquids of different molecular nature.

The calorimetric method of heat of immersion

In the method of heat of immersion a sample of material in initial state A is placed in the calorimeter vesel and separated from the wetting liquid by an impenetrable membrane f.e sealed in a glass ampule. If the membrane is broken wetting of the sample in the liquid takes place and the generated heat is transported into the surroundings (fig 1). The final state B is defined by the equilibrium when the sample is completely wetted by liquid.

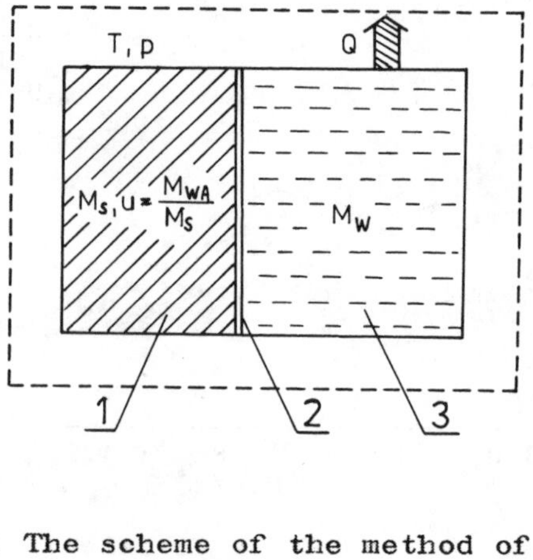

Fig.1: The scheme of the method of heat of immersion
1-sample of material, 2-membrane, 3-wetting liquid

If the initial and the final values of the temperature T and pressure p are the same, the energy effect of immersion Q is then determined according to the 1st law of thermodynamics by the change in enthalpy of the system $H_B - H_A$

$$H_A = M_S h_S + M_W h_{W0} + M_{WA} h_{W0} + \int_0^{M_{WA}} h_W^{MR}(u)\, dM_W$$

$$H_B = M_S h_S + M_{WB} h_{W0} + \int_0^{M_{WB}} h_W^{MR}(u)\, dM_W$$

$$H_B - H_A = \int_{M_{WA}}^{M_{WB}} h_W^{MR}(u)_{T,p}\, dM_W$$

The integral heat if immersion H_W, corresponding to the initial moisture content is defined by the relation

$$H_W(u_A)_{T,p} = \frac{Q(u_A)}{M_S} = \frac{1}{M_S}\int_{M_{WA}}^{\infty} h_W^{MR}(u_A)_{T,p}\, dM_W$$

from which follows

$$h_W^{MR}(u_A)_{T,p} = \left.\frac{\partial H_W(u_A)}{\partial u_A}\right|_{T,p}$$

In practice as the upper integral limit such M_{WB} is chosen that

$$\int_{M_{WB}}^{\infty} h_W^{MR}(u)\, dM_W \doteq 0$$

for example

$$M_{WB} = (10 \div 100)\, M_S \ .$$

For measurements of heat of immersion quasi-isothermic calorimeters are mostly used, either the type of a simple calorimeter with mixing of the liquid /9/ or the type of a double symmetric calorimeter with a thermobaterie /10/ or with a thermistor bridge (fig. 2).

The energy effect Q is in a quasi-isothermic calorimeter determined by the time integral of heat flux Q from the vessel into surroundings

$$Q = \int_0^{\infty} \dot{Q}(\tau)\, d\tau \ .$$

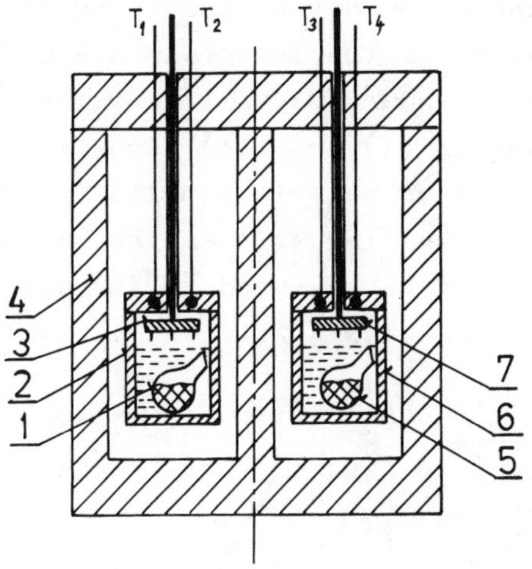

Fig.2: The scheme of the quasiisothermal calorimeter
1,5 - ampule with the sample
2,6 - calorimetric vessel with the wetting liquid
3,7 - mechanical braker
3,7 - mechanical breaker
4 - calorimeter block

For small deviations of the temperature of the vessel from the surroundings $\Delta T < 1K$ there is

$$\dot{Q}(\tau) = konst \,.\, \Delta T(\tau)$$

while ΔT is expressed by the bridge output signal $\Delta U(\tau)$

$$\Delta T = A . \Delta U(\tau) \ ,$$

where A is given by the parameters of the bridge. For instance, for a bridge of the Wheatson type with thermistors NRZ (produced by Pramet Šumperk Corp.) which is supplied by the Ni-Fe type baterie has been, at the temperature of $T \doteq 298K$, determined the value

$$A \doteq 19\ K.V^{-1} \ .$$

The heat Q given by the integral

$$Q = konst \int_0^{\infty} \Delta U(\tau)\, d\tau$$

is in practice evaluated indirectly by means of comparison with the calibration heat Q_K with the use of so-called corrected temperature or bridge output signal deviations ΔU, ΔU_K

$$Q = \frac{\Delta U^*}{\Delta U_K^*}\, Q_K \ .$$

The corrected deviations, proportional to the values of the integrals are determined in the standard way /11/from the time courses $\Delta T(\tau)$ or $\Delta U(\tau)$ (see fig. 3). The callibration heat is supplied by preciss electric heating so that the energy effect Q may be measured to the accuracy better than 1 % /8/.

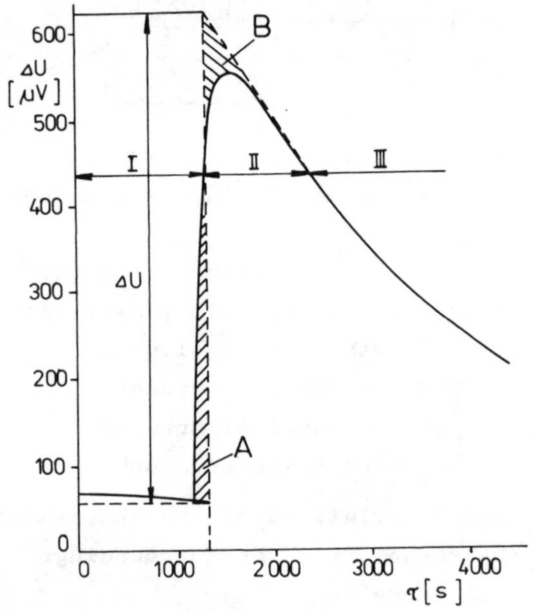

Fig.3: A typical time course of a bridge signal and a corrected deviation

The accuracy of the integral heat of immersion determination is influenced, besides the mechanical and thermodynamic effects connected with ampule breaking (whose value in the present case does not exceed 0.1 kJ.Kg^{-1}), especially by the non-homogenity of pre-wetting of the samples in the initial state A.

Fig. 4 shows a typical dependence $H_w(u)$ for silicagel. For homogenous sample there is for example at moisture content $u = 0.1$ kgkg^{-1} $H_w(0.1) = 39$ kJ.kg^{-1}. In an extreme case of a sample of which one half is dry and the other half has double moisture content ($u = 0.2$ kJ.kg^{-1}) the value of $H_w^*(0.1)$ would be

$$H_w^*(0.1) = \frac{H_w(0)+H_w(0.2)}{2} \doteq \frac{107+16.6}{2} = 61.8 \text{ kJ.kg}^{-1}$$

the absolute error would be

$$\delta H_w = H_w^*(0.1) - H_w(0.1) \doteq 22.8 \text{ kJ.kg}^{-1}$$

while the relative error would be

$$\frac{\delta H_w}{H_w(0.1)} \cdot 100 \doteq 58 \% .$$

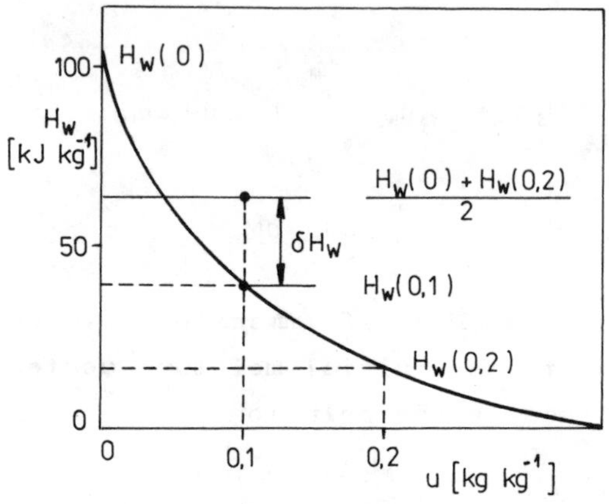

Fig.4: Influence of non-uniform pre-wetting for the case of silica gel at $T \doteq 298$K and $p \doteq 10^5$Pa

Bond character determination by means of heat of immersion

From the integral heats of immersion $H_w(u)$ moisture bond enthalpy may be derived:

$$h_w^{MR}(u) = \left| \frac{\partial H_w(u)}{\partial u} \right|_{T,p}$$

which is, together with the value of the potential $\mu_w^{MR}(u,T,p)$ decisive for the characteristic curve in the state functions diagram.

Fig. 5 shows the characteristic curves for the molecular sieve Nalsit 4 A (product of Lachema Brno Corp.) and for sweetened egg melange which were determined from heats of immersion, respectively sorption isotherms.

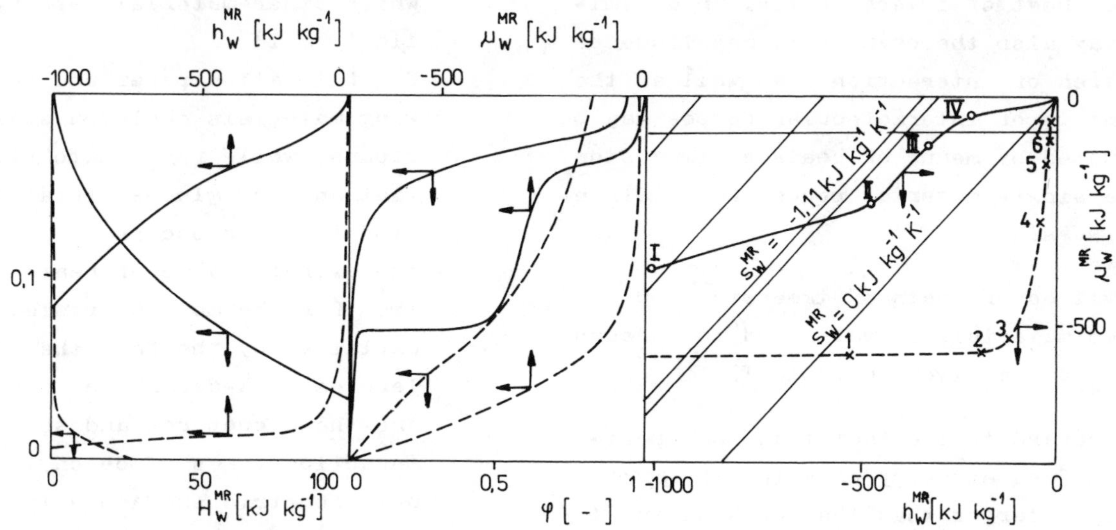

Fig.5: Heat of immersion $H_W(u)$ / enthalpy $h_W^{MR}(u)$, sorption isotherms $u(\varphi)$ moisture potential $\mu_W^{MR}(u)$ and resultant curves in the state function diagram at T =298K and $p = 10^5$ Pa for:
molecular sieve Nalsit 4A ———
/values of specific moisture content $u[kg.kg^{-1}]$: I - 0.074, II - 0.155, III - 0.164, IV - 0.170/
sweetened egg melange ----
/values of specific moisture content $u[kg.kg^{-1}]$: 1 - 0.016, 2 - 0.018, 3 - 0.020, 4 - 0.037, 5 - 0.060, 6 - 0.080, 7 - 0.160/

With regard to the general relationship

$$\mu_W^{MR} = h_W^{MR} - T s_W^{MR}$$

the prevailing enthalpy effect of the bond $|h_W^{MR}| > |\mu_W^{MR}|$ in the molecular sieve may be explained by a higher degree of organization and a lower degree of mobility of the molecules ($s_W^{MR} < 0$) of the moisture bound in the micropores. On the other hand, smaller enthalpy effect $|h_W^{MR}| < |\mu_W^{MR}|$ of the sweetened egg melange corresponds to a higher degree of mobility and a lower degree of organization of the molecules ($s_W^{MR} > 0$) following from partial swelling or solving of the material in moisture.

When classifying materials from the point of view of drying the molecular sieve may be considered as an inert material while sweetened egg melange must be regarded as a material sensitive to the drying regime /12/.

The interaction between material and moisture may be characterized also by ration of heats of immersion in polar and non-polar liquid. The heat of immersion will be proportional to the specific area of mutual interaction a and to the magnitude of the mutual surface enthalpy following from the nature of the intermolecular forces.

According to the character of the intermolecular forces we distinguish hydrophilic surface (high-energy surface) and hydrophobic (low-energy surfaces) /13/. For inert materials whose area a is the same when immersed in both polar and non-polar fluids the ratio of heats of immersion indicate whether the surface is hydrophilic or hydrophobic. If a given material swells or dissolves in one type of liquid, the area of interaction increases by order and so does the value of heat of immersion too. In a similar way also the changes in magnitude of the area of interaction a as well as the intensity of intermolecular forces may be evaluated by means of heats of immersion for a single material processed in different ways.

The values of heats of immersion for selected materials in water and in N-decan ($C_{10}H_{22}$) are given in table I.

With regard to the fact that the specific surface enthalpy of water is approximately 3 larger than that of N-decan /13/ we may suppose that:

- soot is strongly hydrophobic (the ratio of heat of immersion $\varkappa \doteq 0.2$) while other materials are hydrophilic ($\varkappa > 1$)
- for filtration paper and the following materials skeleton swelling occurs, while for potato flour partial and for glucose total dissolution in water occurs
- the extreme ratio of heats of immersion for the molecular sieve may be explained by the fact that the molecules of N-decan are larger than are the micropores and so the interaction takes place on the external geometric surface only whose area is negligible in comparison with the effective surface of the micropores.

Table I: Integral heat of immersion of dry materials in Water and N-decan

Material	$H_{H_2O}(0)$ [kJ.kg^{-1}]	$H_{C_{10}H_{22}}(0)$ [kJ.kg^{-1}]	$\varkappa = \frac{H_{H_2O}(0)}{H_{C_{10}H_{22}}(0)}$ [-]
soot "Chezacarb"	9.2	51	0.2
insulation material "Kryzolit"	41	3.1	13
gas concrete	35	2.4	15
filtration paper	21	1.2	18
bentonite	87	4.0	22
silica gel	107	4.9	22
potato flour	113	1.8	63
beech wood	71	1.1	65
glucose	27	0.05	510
molecular sieve "Nalsit 4A"	172	0.2	860

Conclusion

Calorimetry of heat of immersion is an easily available and reliable means for determination of the enthalpic effect of moisture bond in materials.

On the basis of heat of immersion and moisture potential the characteristic curve in the diagram of state functions can be constructed, enabling to objectively classify the moisture bond in materials from the point of view of thermodynamics especially with regard to the sensitivity to drying regime.

The comparison of the heats of immersion in liquids of different polar character may supply us with data concerning the degree of hydrophility or hydrophobity of the material in question, its swelling or solubility in moisture, or general information about the magnitude and quality of surface area under various conditions.

References

L 1 Rebindĕr, P.A.:
O formach svjazi vlagi s materialami v processach suški. Proc. conference Vsesojuznoje naučno-techničeskoje povĕščanije po suške, Profizdat, Moskow 1958

L 2 Lykov, A.V.:
Teorija suški, Energija, Moskow 1968

L 3 Keey, R.B.:
Drying principles and practice, Pergamon Press, Oxford 1972

L 4 Kazanskij, V.M.:
Opredelenije teploty isparenija vlagi, zaključennoj v poristom tele, IFŽ, č.8, IV, 1961

L 5 Gál, S.:
Die Methodik der Wasserdampf-Sorptionmessungen, Springer Verlag, Berlin 1967

L 6 Čermák, B.:
Characteristic moisture bond in materials with respect to drying using the state function diagram, Proc. Drying 80, Vol. 2 Hemisphere Pub. Corp. N.Y., 1980

L 7 Globus, A.M.:
Sovremennyje experimentalnyje metody opredĕlenije potenciala perenosa v trojfaznych poristych systemach, Proc. of conference in Minsk 1967, Tom 6 par. 1

L 8 Čermák B.:
Termodynamická metoda hodnocení charakteru vazby vlhkosti v materiálech, Thesis, SVÚSS, Prague 1977

L 9 Topic, M., Micale F.J., Leidheiser, H., Zettlemoyer, A.C.:
Calorimeter for measuring of heats of wetting of solids in organic media, Rewiew Scientific Instruments, 45, 4, 487-490 (1974)

L 10 Calvet, E., Prat, H.:
Recent progress in microcalorimetry, Pergamon Press, Oxford 1963

L 11 Gun, S.R.:
On calculation the corrected temperature rise in isoperibol calorimetry, Journal of the Chemical Thermodynamics, 3, 19-34, (1971)

L 12 Čermák, B.:
Použití diagramu stavových funkcí pro charakterizaci vazby vlhkosti s vysoušeným materiálem, ZTV 24, No. 2, 79 (1981)

L 13 Zettlemoyer, A.C.:
Hydrophobic surfaces, Proc. 155th meeting of ACHS, San Francisko, Academic Press 1969

THIN LAYER DRYING CHARACTERISTICS OF CASHEW NUTS AND CASHEW KERNELS

A. Chakraverty
M.L. Jain

Post Harvest Technology Centre
Indian Institute of Technology
Kharagpur, India

ABSTRACT

The effects of air temperatures and air velocities on thin layer drying characteristics of cashew nuts and cashew kernels have been studied. Empirical drying equations relating moisture ratio to drying constant and drying time have been developed and found to be in good agreement with the experimental data. The drying air temperatures of 75°C and 70°C have been considered to be optimum for drying of cashew nuts and cashew kernels respectively.

NOMENCLATURE

k = drying constant, 1/min
M = moisture content, per cent (d.b.)
MR = moisture ratio, fraction,
θ = drying time, min

Subscripts

1 = initial
2 = final
e = equilibrium

INTRODUCTION

The cashew kernel is a delicious and neutritious food because of its high fat and protein contents (Shivanna, 1972). It is the second largest doller earning crop in India. The processing of cashew nut consists of several operations namely, drying, cleaning, conditioning, roasting, centrifuging, shelling, peeling, grading and packaging. Of them, drying, conditioning and roasting operations have pronounced effect on the quality and quantity of the products of cashew nuts. But little information is available on drying of cashew nut and cashew kernels. Therefore, this project was undertaken with the following major objectives :

i) To study the effects of various air temperatures of 65°, 75°, 85°, and 95°C on drying characteristics of cashew nuts.

ii) To study the effects of various air temperatures of 40°, 50°, 60°, 70° and 80°C on drying characteristics of unpeeled cashew kernels.

iii) To study the effects air velocities of 18 and 30 m/min on drying characteristics of unpeeled cashew kernels.

MATERIALS AND METHODS

Freshly harvested cashew nuts as well as unpeeled cashew kernels were collected from Midnapore, West Bengal, India. The cashew kernels and cashew nuts were kept in polyethylene bags, sealed and stored safely in a refrigerator at 5°C.

Thin layers of cashew samples of 6 cm depth were dried in a laboratory drying set up equipped with weighing device and temperature and air flow rate controls which were described elsewhere (Chakraverty and Kausal, 1982). The temperature was controlled with ± 2°C for all experiments.

Unpeeled cashew kernel samples of 2000 g each were dried at each of the air temperatures of 40°, 50°, 60°, 70° and 80°C at the air velocities of 18 and 30 m/min. Freshly harvested cashew nut samples of 2000 g each were dried at each of the air temperatures of 65°, 75°, 85° and 95°C at an air velocity of 30 m/min. Relative humidity of the inlet ambient air was not controlled. However, variation in relative humidity in the closed room where the drying experiments were conducted was within ± 5 per cent only. Thin layers of cashew kernels and cashew nuts were dried at the above drying conditions upto equilibrium moisture content to study the drying characteristics. A few samples of cashew kernels were dried to 3.5 per cent moisture content to study the ease of peeling and other quality of the cashew products. A few samples of cashew nuts were dried to 7 per cent to study the quality of the cashew nuts.

RESULTS AND DISCUSSION

The equilibrium moisture contents of cashew kernels were found to be 4.095, 2.859, 2.102, 1.779 and 1.240 per cent at 40°, 50°, 60°, 70° and 80°C respectively at both the air velocities of 18 and 30 m/min. These show that equilibrium moisture content (M_e) of cashew kernels decreased with the increase of air temperature. The M_e of cashew nuts could not be determined as the shell oil came out of the shell after a certain period of drying. The drying characteristics of cashew kernels and cashew nuts are shown in Figures 1 through 3.

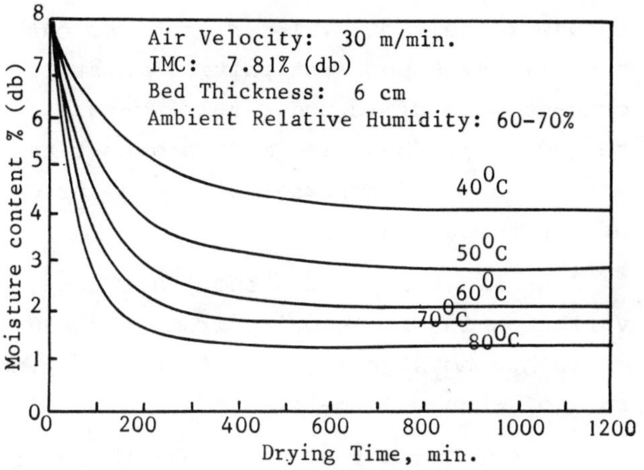

Fig. 1 Relation between Moisture Content and Drying Time for Cashew Kernels at 30m/min. Air Velocity.

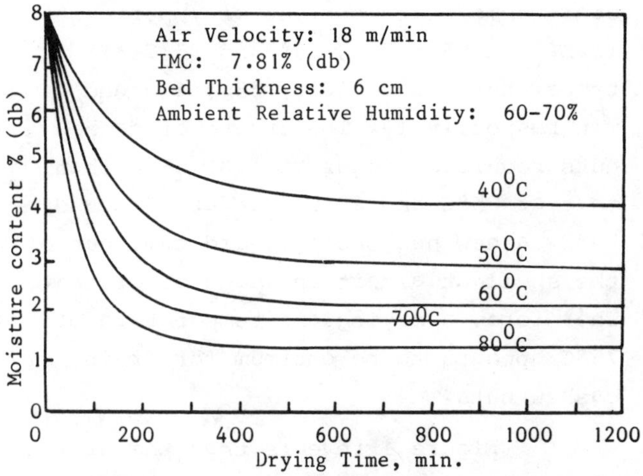

Fig. 2 Relation between Moisture Content and Drying Time for Cashew Kernels at 18m/min. Air Velocity.

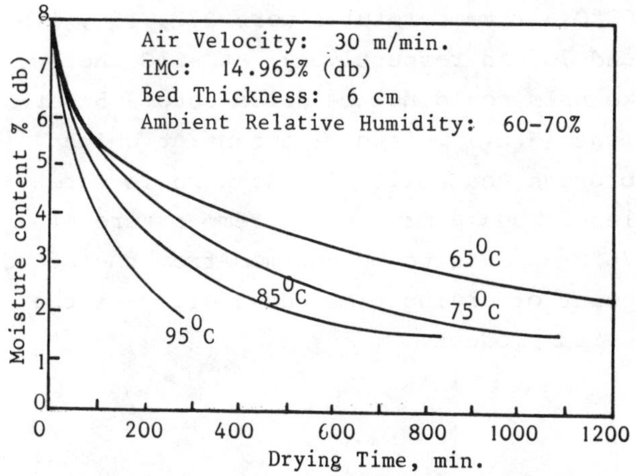

Fig. 3 Relation between Moisture Content and Drying Time for Freshly Harvested Cashew-nuts at 30m/min Air Velocity

Figures 4 through 6 show that both cashew kernels of 7.81 per cent and cashew nuts of 14.96 per cent initial moisture content were dried under falling-rate period only. They also indicate that the rate of drying increased with the increase of air temperature.

Figures 4 and 5 show that the variation in air velocity from 18 to 30 m/min had no significant effect on drying rate of cashew kernels.

Figure 3 indicates that the drying times required to dry the freshly harvested cashew nuts from an initial moisture content of 14.96 per cent (d.b.) to safe storage moisture content of 7 per cent (d.b.) at 65°, 75°, 85° and 95°C air temperatures were 1020, 650, 420 and 240 min respectively. The drying of cashew nuts required longer time at 65°C than that at 85°C and 95°C. But at 85°C and 95°C cashew nut shell liquid came out of the shell at a certain stage of drying. Therefore, a drying air temperature of 75°C appears to be optimum for drying of cashew nuts.

Figure 1 reveals that the drying times required to dry the unpeeled cashew kernels from 7.81 per cent to 3.5 per cent (d.b.) moisture content at 50°,60°,70° and 80°C air temperatures were 270, 160, 105 and 70 min respectively. At 40°C the kernels could not be dried upto 3.5 per cent (d.b.) moisture content at which brokens and splits in the kernels were less. Thus a drying air temperature of 70°C appears to be optimum from the stand point of drying time and quality of the dried product.

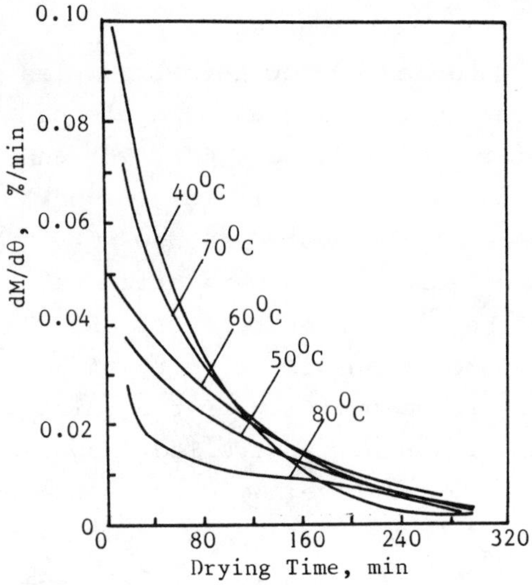

Fig. 4 Effect of Drying Air Temperatures on Drying Rate of Unpeeled Cashew Kernels at 30 m/min Air Velocity

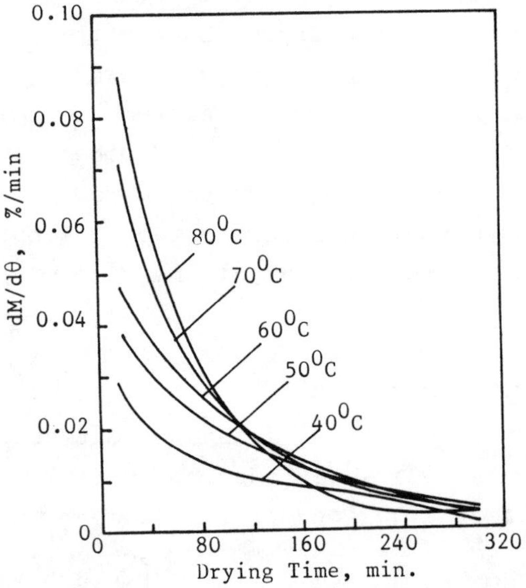

Fig.5 Effect of Drying Air Temperatures on Drying Rate of Unpeeled Cashew Kernels at 18m/min Air Velocity.

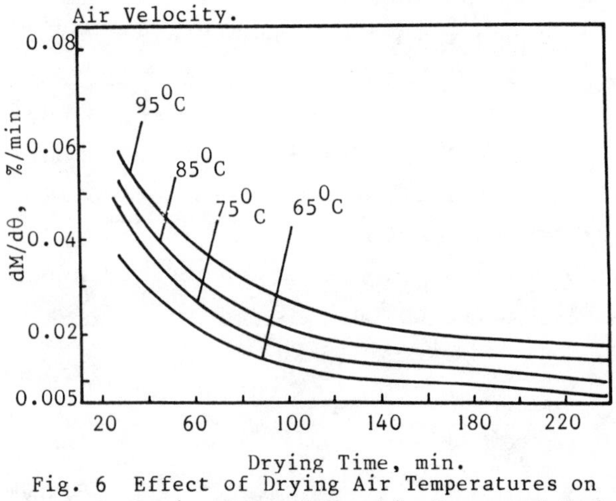

Fig. 6 Effect of Drying Air Temperatures on Drying Rate of Freshly Harvested Cashew-nut at 30 m/min Air Velocity.

It may be seen from Figure 7 that the drying constant, k increased as drying air temperature increased.

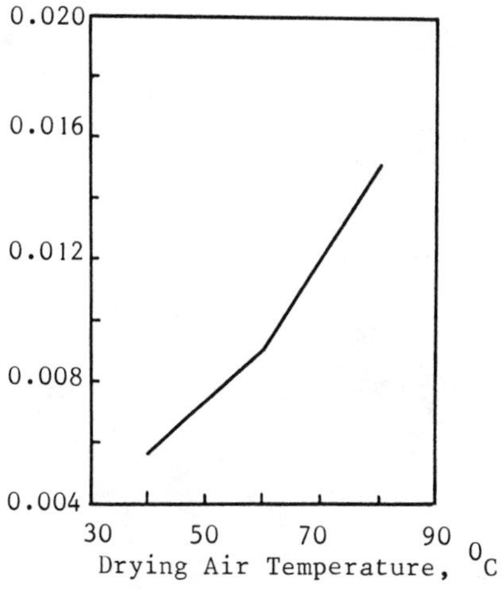

Fig. 7 Relation between Drying Constant K and Drying Air Temperature at 18 and 30 m/min Air Velocities for Cashew Kernels

It also shows that the variation in air velocity from 18 to 30 m/min had no effect on drying constant. The nature of the curves in Figures 1 through 3 is exponential. Therefore the plot of $(M_2-M_e)/(M_1-M_e)$ versus drying time, θ on semilog graph paper yields straight line as shown in Figure 8.

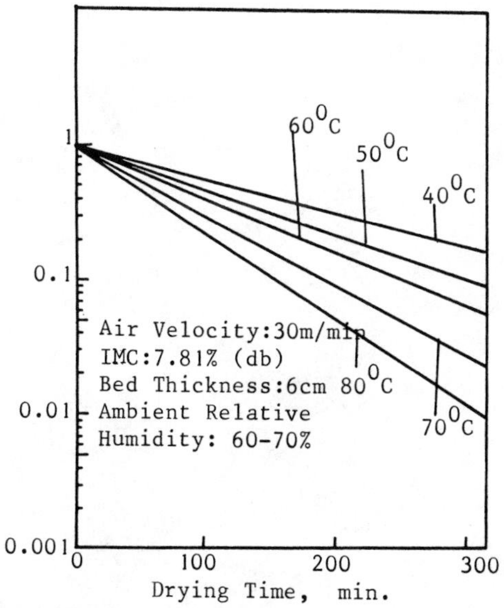

Fig. 8 Moisture Ratio vs. Drying Time for 30 m/min Air Velocity.

Accordingly the slope of the straight line is the drying constant k. The drying constant, k was thus calculated by the graphical method (Chakraverty and De, 1981). The drying equations developed by the graphical method are tabulated in Table 1.

Table 1. Empirical drying equations for cashew kernels of 7.81 per cent (d.b.) initial moisture content for both the air velocities of 18 and 30 m/min.

Air temp. °C	Equation	Corr. Coeff.
40	*$MR = e^{-5.694\times10^{-3}\theta}$	0.973
50	$MR = e^{-7.5\times10^{-3}\theta}$	1.000
60	$MR = e^{-9.02\times10^{-3}\theta}$	0.877
70	$MR = e^{-1.2\times10^{-2}\theta}$	0.928
80	$MR = e^{-1.469\times10^{-2}\theta}$	0.649

$$* \ MR = \frac{M_2 - M_e}{M_1 - M_e} = e^{-k\theta}$$

Table 1 shows that except at 80°C the correlation coefficients of the thin layer drying equations for cashew kernels vary from 0.877 to 1.000 which show their goodness of fit.

CONCLUSIONS

i) The equilibrium moisture content of cashew kernels decreases with the increase of drying air temperature but it is independent of drying air velocity.

ii) The drying rates of cashew nuts and cashew kernels increase as drying air temperature increases.

iii) The drying of cashew nuts and cashew kernels of 14.96 per cent and 7.81 per cent (d.b.) initial moisture contents takes place under falling-rate period only.

iv) The variation in air velocity between 18 and 30 m/min has no significant effect on rate of drying of cashew kernels.

v) The air temperatures of 75°C and 70°C are taken as optimum for thin layer drying of cashew nuts and cashew kernels respectively.

REFERENCES

1. Chakraverty, A. and Kausal, R.T., 1982. Determination of optimum drying conditions and development of drying equations for thin layer drying of parboiled wheat, AMA, Vol.XIII, No.1.

2. Chakraverty, A. and Dey, D.S. 1981, Post Harvest Technology of cereals and pulses, Oxford and IBH Pub.Co.

3. Shivanna, C.S. and Govindarajan, V.S. 1972, Processing of cashew nut, Indian Food Packer, Vol.XXVII, No.5.

EVALUATION OF A MULTI-PURPOSE DRYER USING NON-CONVENTIONAL ENERGY SOURCES

Jeon, Yong Woon
Leonides S. Halos
Clarence W. Bockhop

Agricultural Engineering Department
The International Rice Research Institute
Los Baños, Laguna, Philippines

ABSTRACT

The design of the warehouse dryer was intended to introduce the concept of intermediate drying and storing technologies in the context of non-conventional energy utilization and to provide a groundwork for the implementation of the concept.

The furnace presents an additional approach to broaden utilization of agricultural waste residues, while the vortex wind machine is an attempt to eliminate the dependence of drying on the operation of fossil-fueled engines or electric motors to drive blowers.

Evaluation of the system gave encouraging results to warrant further investigation and development.

NOMENCLATURES

δ integration interval

ε combustion efficiency

H latent heat of vaporization [$H = 2.501 + 1.775 \times 10^{-3}T$, Wilhelm (1976)], *MJ/kg*

m moisture contained in the material

Q_c actual drying heat energy

Q_{cm} theoretical minimum drying heat energy, *MJ*

Q_f heat energy from consumption of rice hull, *MJ*

Q_m heat energy to elevate and maintain the temperature of drying paddy, *MJ*

Q_r heat energy released by paddy, *MJ*

Q_t total drying heat energy, *MJ*

Q_x heat energy in exchanged air, *MJ*

INTRODUCTION

As a result of energy crisis during the last couple of decades, attentions have been focused on the need, not only to conserve conventional sources of energy but also to explore non-conventional energy sources such as biomass, wind, solar radiation and others.

Drying is an energy intensive agricultural operation which is greatly affected by the growing fossil fuel shortage.

Sundrying had been considered a most practical way to deal with the problem. However, because of the high humidity conditions during harvest, it is necessary that there be an added source of heat energy provided to supply the heat of vaporization. In temperate regions, dryers have been developed thoroughly but had not been successfully introduced into tropical and sub-tropical regions not only because of climatic differences but also because of the technological gap existing between temperate and tropical and sub-tropical regions and the very small land holdings that tend to be typical of tropical farmers. Therefore, the main challenge for these nations in this field is the ability to adapt known technology in their specific situations.

Recent researches had been geared toward the development of on-farm drying facilities. There had been attempts of utilizing available sources of energy like rice hull and solar energy but the use of mechanical or electrical power for the blower or fan is always coupled to the system. This is one of the reasons why these dryers are often not used in the rural areas; electricity is not always found in smaller towns and farms.

Countries in the tropical region have greater potential for the use of non-conventional sources of energy as an alternative energy source for drying. A well-designed drying system using non-conventional sources of energy could reduce drying costs as it can be developed along the concept of an intermediate technology wherein the system is simple in construction, low cost, makes use of materials available in the region and can be used for multi-purpose end.

This paper, therefore, discusses the performance of a multi-purpose dryer designed to make use of combustible agricultural waste materials as fuel sources and a vortex wind machine as power source.

DEVELOPMENT OF THE MODEL/PROTOTYPE

Of prime consideration in the development of the dryer is the analysis of the weather condition and the general trend in agriculture of the locality where the dryer will serve. It was assumed that the conditions studied were normal mean values for the tropical regions.

Typical weather conditions at UPLB and IRRI (1970-1980 records) and the general trends in agriculture (Appendix) were analysed and resulted to the conceptualization of a drying system which can be described as simple, easy to operate, versatile and economically feasible. The design was therefore kept to meet these standards and a dryer which makes use of agricultural waste materials as fuel and wind energy as power, was developed.

The dryer is a warehouse type which consisted of a center-tube furnace, a vortex wind machine and a material holding chamber. Each component were designed separately and then fused together for drying and storing application.

Center-tube Furnace

Basic experiments conducted had been very useful in the development of the center-tube furnace. It was provided with primary and secondary air inlets which can be adjusted to provide the proper amount of air for combustion. This observation entails maintenance of a furnace bed temperature to at least 350°C for sustained combustion. Different kinds of fuel material have different kindling temperature, therefore, bed temperature will vary based on this chemical property. Rice hull, for instance, will burn continuously at 90% efficiency in a furnace with bed temperature approximately 420°C. Similarly, coconut husks and sawdust will burn at the same efficiency at temperatures of 350 and 380°C, respectively. Fuel moisture contents below 14% will not materially affect burning efficiency (Jeon, et al, 1982).

To keep the furnace operational at satisfactory level, feed rate was predetermined based on burning time and volume change which affected temperature of the fuel bed. Estimates had been choosen to feed at a time when the fuel volume in the chamber has not yet been reduced to cause a drastic temperature drop below the kindling temperature.

A T-chimney design was also adopted to give sufficient pressure drop with minimum effect of wind direction.

A cut-out view of the furnace is shown in Figure 1.

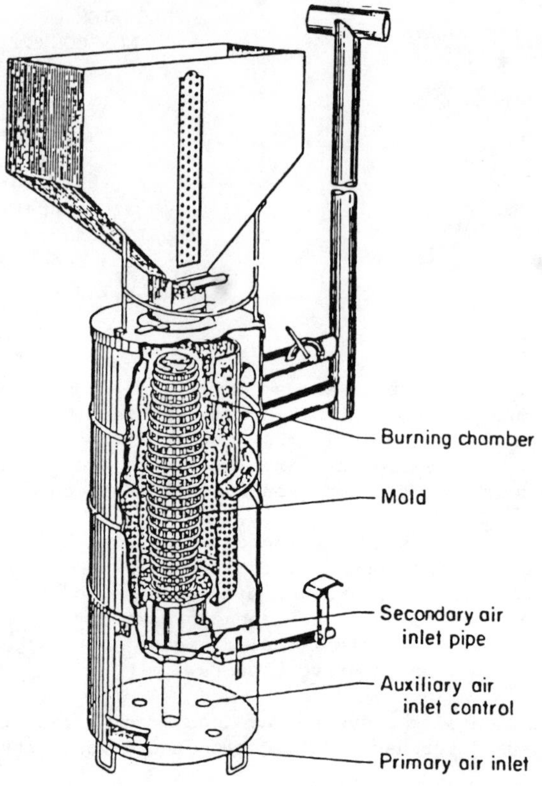

Fig. 1. Cut-out view of the center-tube furnace.

Vortex Wind Machine

The concept of aerodynamic circulation velocity was used to investigate the wind flow interaction in a hollowed tower with alternate air inlets. The swirling motion of the wind entering the system creates a low pressure zone at the core of the tower rendering it capable to admit a rush of air beneath the structure.

The vortex wind machine (Fig. 2) was designed and is perhaps the first innovative approach to reap power from the wind for aeration in drying and storage. The system consisted of a hollow cylinder with equally spaced opened- and closed-vanes oriented vertically and fixed to the circular holders at the top and bottom of the structure. A conical frustum (venturi) is located underneath the structure to provide a more effective low-pressure exhaust.

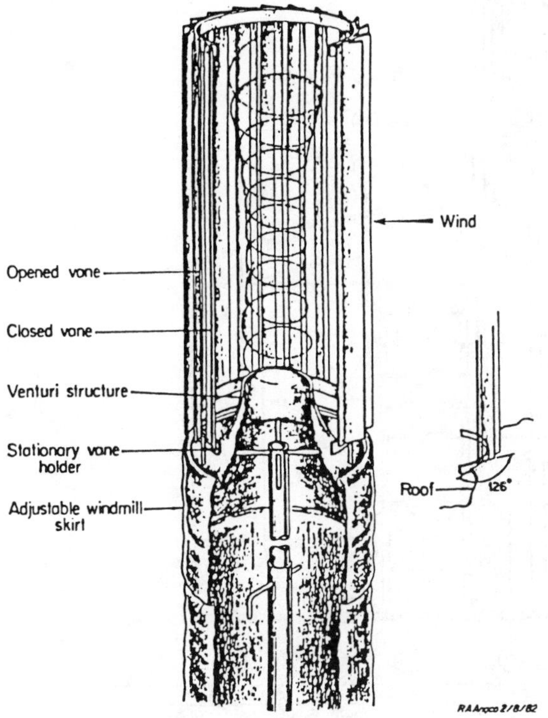

Fig. 2. Cut-out view of the vortex wind machine.

Analysis of wind distribution, in general, gives a wide range of gust variations of wind velocity. Observations in a vortex wind machine, however, showed that air suction velocity is more or less stable with maximum gust variation only about 0.17 m/sec occurring in 0.16 sec (Fig. 3).

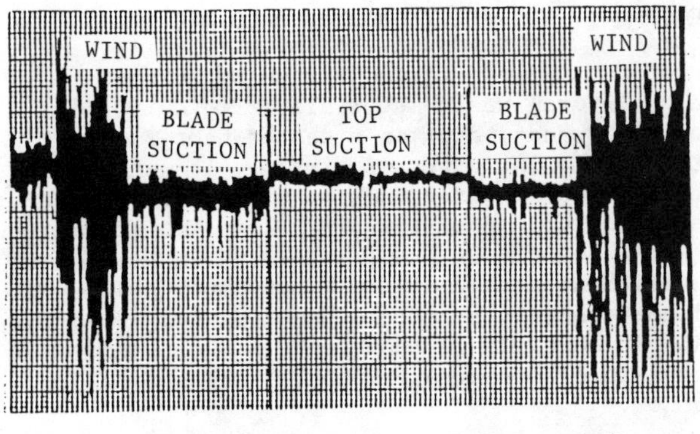

	Wind velocity	Suction velocity	
		Top suction	Blade suction
a. Gust amplitude	0.4 m/sec	0.07 m/sec	0.09 m/sec
b. Gust formation time	0.6 sec	2.7 sec	1.8 sec
c. Max. gust variation	1.15 m/sec	0.08 m/sec	0.17 m/sec
d. Gust lapse time	0.77 sec	0.11 sec	0.16 sec

Fig. 3. Gust variations of wind and suction velocity as observed from the vortex wind machine.

The observed actual air suction rate (Fig. 4) showed a trend similar to the theoretical air suction curve. The top suction velocity was approximately two times that of wind velocity (cross-sectional area of cone top is 0.031 m^2). Side suction, that is measured at the clearance between the venturi and bottom of tower wall revealed a possibility of controlled air movement which can be applicable for aeration in drying and storage. The vortex wind machine can generate a total suction of 17 m^3/min at the average wind velocity of 1.4 m/sec.

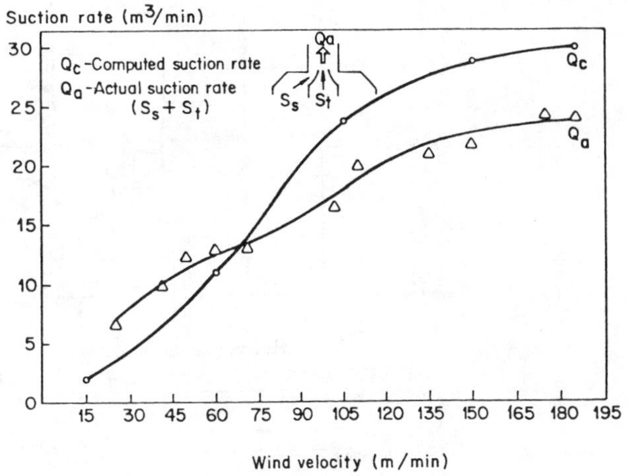

Fig. 4. Theoretical and actual air suction rate as a function of wind velocity.

Materials Holding Chamber

The material holding chamber was designed to allow materials to dry in a manner similar to thin layer drying and to simulate a tobacco drying model (Cundiff and Dodd, 1979). Thus, the main drying chamber consisted of several trays (bamboo flat baskets) one (1) meter in diameter, arranged in seven layers around a hexagonal frame to contain atmost 10 cm of granular drying material. A cabinet type pre-drying chamber was also provided.

TEST MODEL/PROTOTYPE AND THE DRYING PROCESS

A warehouse dryer (Fig. 5) was designed where a material holding chamber is provided. The center-tube furnace and the vortex wind machine were adapted to the structure as heat and power sources, respectively.

Two circular flue ducts were connected to the furnace and routed around and below the drying chamber and diverged at the opposite end of the warehouse (reference is made to furnace location), forming a circular heat exchanger.

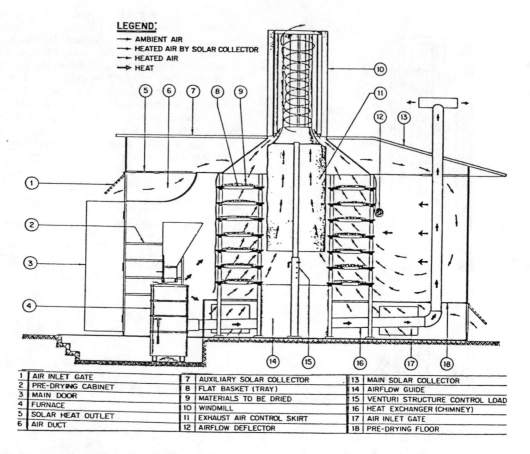

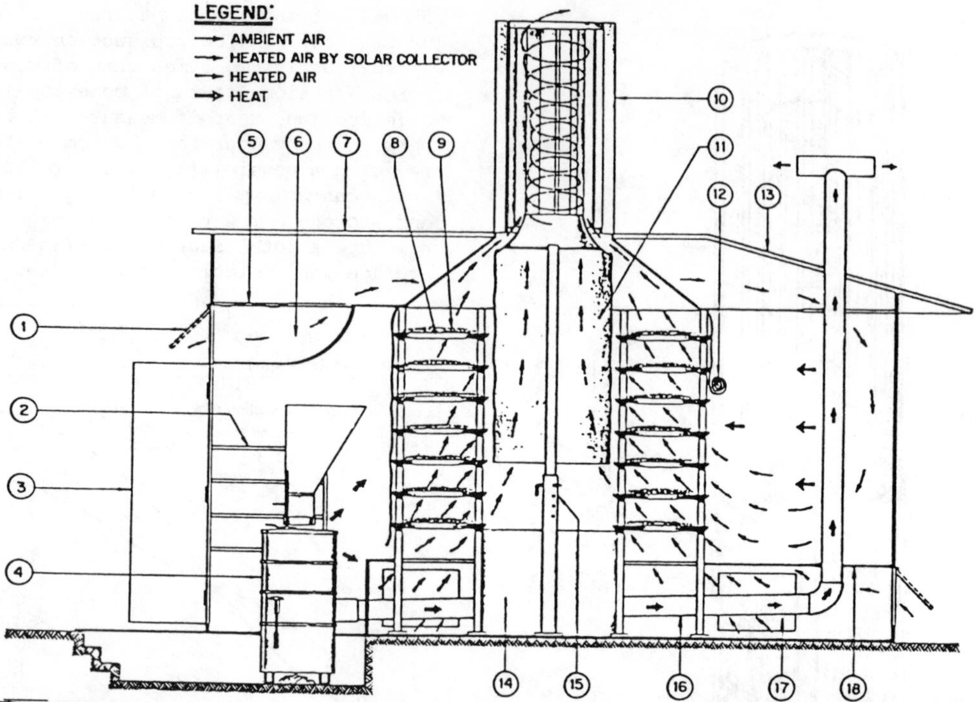

1	AIR INLET GATE	7	AUXILIARY SOLAR COLLECTOR	13	MAIN SOLAR COLLECTOR
2	PRE-DRYING CABINET	8	FLAT BASKET (TRAY)	14	AIRFLOW GUIDE
3	MAIN DOOR	9	MATERIALS TO BE DRIED	15	VENTURI STRUCTURE CONTROL LOAD
4	FURNACE	10	WINDMILL	16	HEAT EXCHANGER (CHIMNEY)
5	SOLAR HEAT OUTLET	11	EXHAUST AIR CONTROL SKIRT	17	AIR INLET GATE
6	AIR DUCT	12	AIRFLOW DEFLECTOR	18	PRE-DRYING FLOOR

Fig. 5. Cut-out view of the warehouse drying system.

The flue ducts were inclined at 4 degrees and connected to the vertical exhaust chimney. Adjacent to these ducts are air inlet doors where ambient air passes and heated by the hot ducts.

The vortex wind machine is positioned at the top center of the hexagonal drying chamber, with the venturi extending farther down to the middle of the tray layers by means of an adjustable canvass skirt. Jute sacks were also suspended around the drying chamber which served to control air-flow rates and drying air temperatures.

Drying proceeds in a manner such that the heated air is moved upwards from the heat exchanger through the layer of the drying materials or at the top or bottom surfaces. Saturated drying air is exhausted at the wind machine tower. The pressure drop resulting from the circulating motion of wind entering through the tower blades causes this upward movement of air.

EVALUATION PARAMETERS

The inter-relationship of the drying parameters were analyzed in the warehouse drying system. These included evaluation of the ambient and drying air conditions, heat flow and air movement, and the material moisture contents.

The main objective of the experimentation was to examine the feasibility of drying materials in a warehouse where air movement is effected only by natural draft of air and by the suction pressure created by the swirling motion of wind striking the vanes of the vortex wind machine. Heat is supplied to the system by burning agricultural waste materials (rice hull, corn cobs, sawdust, etc) in a center-tube furnace. Due to the low rates of airflow, it was assumed that the drying air and the grain will reach equilibrium before the air leaves the grain.

Sufficient amount of newly harvested paddy were loaded on the trays (flatbaskets) and dried to 16 per cent moisture content (d.b.). Each tray contained approximately 25 kgs which was 7 cm thick.

The variables considered in the analysis were the fuel material feed rate, airflow rates and weather conditions. These variables control the drying rate of the material and the uniformity and quality of materials dried.

DRYING CONDITIONS

Ambient air conditions observed for the duration of the drying tests were recorded for a gloomy weather with intermitent rain showers and an average relative humidity of 84%. Observations during a fair weather gave an average relative humidity of 77%. Wind velocity was observed within the range of 80 to 120 m per min with a gust variation of 1.0 to 1.25 m per sec.

The drying air temperature was maintained at 40-45°C with furnace feed rate of 30 kg per hr. Venturi blade and tower wall bottom clearance was fixed at 6.5 cm to ensure a more uniform airflow rate, thus, a uniform drying.

In general, temperature distribution inside the drying chamber had a standard deviation of 3.42, whereas suction velocity had standard deviation of 0.35. These observations showed that a uniform drying air condition can be maintained in the system despite the fact that air suction was very much dependent on the weather condition. It was also noted that with still air, suction at the wind machine was recorded at 5 m^3/ min which maybe due to the temperate gradient between the exhausting air and the swirling air.

The minimum wind velocity for safe drying using the developed wind machine was recorded at 0.75 m per sec, thus, an effective total drying time of 18 hours per day can be expected; 6 hours having velocities lower than 0.75 m per sec.

ENERGY REQUIREMENTS

Heat energy required for moisture evaporation from the drying material was accounted based on the energy model for forced air tobacco curing developed by Cundiff and Dodd (1979). The model describes the utilization of heat energy during the curing process. Heat energy is used to: (a) change the enthalphy of the exchanged air; (b) elevate and maintain the temperature of the drying material; and (c) elevate and maintain the temperature of the drying structure. The heat energy inputs are: (a) the combustion of fuel material (rice hull, for this determination), and (b) energy released due to respiration and as a by-product of the complex chemical reactions which takes place during drying.

The theoretical minimum energy required to dry the paddy is the sum of the heat energy required to raise the temperature of the drying paddy (sensible heat) plus the heat energy required to evaporate the total moisture removed (latent heat). Therefore,

$$Q_{cm} = Q_m + \delta\left(H_{xo}\dot{m}_o/2 + \sum_{j=1}^{n-1} H_{xj}\dot{m}_j + H_{xn}\dot{m}_n/2\right)$$

The actual drying energy was calculated for comparison from the relation given by:

$$Q_c = Q_m + Q_x$$

Where, at any time t, the following energy balance during paddy drying was satisfied:

$$Q_t = Q_f + Q_r = Q_m + Q_x + Q_s$$

About 95 percent of the total heat, Q_t required for drying was derived from the combustion of fuel materials, Q_f, the rest being the heat released by the paddy, Q_r. The exchanged air absorbed approximately 2/3 of this total energy while 1/3 was utilized to maintain the temperature of the structure. An excess heat energy of 4 percent is required to elevate and maintain the temperature of the drying materials (Table 1).

Table 1

Heat Energy for Paddy Drying in a Warehouse Type Dryer, kcal/kg of Paddy Dried

	Q_f Fuel	Q_r Released	Q_t Total	Q_x Exhausted	Q_s Structure	Q_m Material
Test A						
kcal/kg	1.16×10^5 (104 kg-hull)	6.12×10^3	1.20×10^5	8.16×10^4	4.32×10^4	4.8×10^3
%	95.8	4.2	100	68	36	-4
Test B						
kcal/kg	1.17×10^5 (102 kg-hull)	6.64×10^3	1.23×10^5	3.11×10^4	4.67×10^4	4.92×10^3
%	94.6	5.4	100	66	38	-4
Test C						
kcal/kg	1.09×10^5 (102 kg-hull)	5.75×10^3	1.15×10^5	7.70×10^4	4.25×10^4	4.6×10^3
%	95.0	5.0	100	67	37	-4

Test A: 1120 kcal/kg; average burning temperature = 740°C
Test B: 970 kcal/kg; average burning temperature = 670°C
Test C: 1070 kcal/kg; average burning temperature = 690°C

Dryer capacity for all tests: 2 tons of paddy

MOISTURE REMOVAL RATES

The suction air rate was greatly affected by the weather condition, consequently the drying rates.

Generally during a gloomy weather, suction air rate was observed to be lower than during a fair weather. This can be attributed to the high relative humidity of the spinning wind in the vortex wind machine. The drying air temperature can also be hardly maintained at the required level (40-45°C) without increasing fuel energy input. Thus, a longer drying time during a gloomy weather was observed.

Figure 6 represents the drying curve of paddy in the system during two different conditions of drying. It can be noted that the drying curve did not resemble that obtained from other drying systems. The variable airflow rate at any time during the drying operation effected this behavior. The material was also only partly heated at the on-set of the operation since an initial heat is required to elevate and maintain the temperature of the structure. In a forced air convection dryer, heat is directly utilized in elevating the temperature of the material.

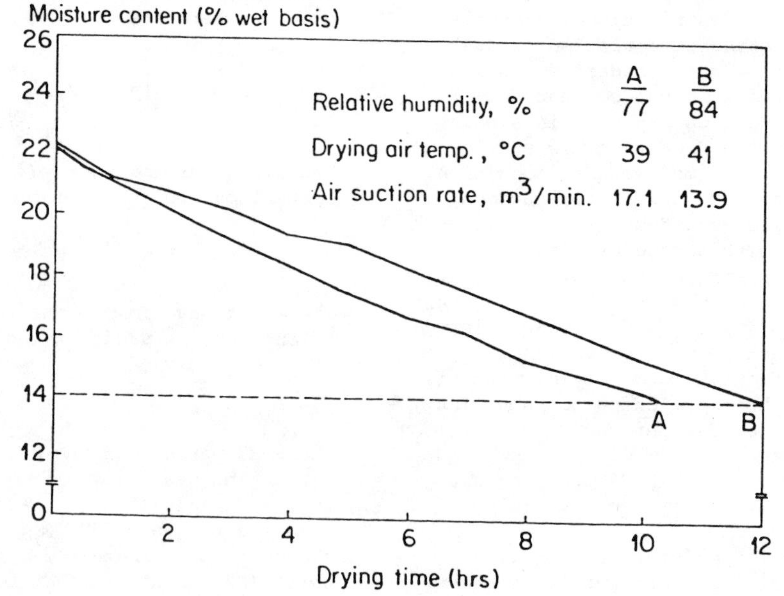

Fig. 6. Moisture content-time relationship in a warehouse dryer.

LIMITATIONS OF THE DESIGN

- An intensive aerodynamic analysis of airflow rate in the system had not yet been established due to lack of instrumentation. Results discussed were based mainly on measurements of air velocities and suction rates but the exact behavior of the direction of air movement had not been predicted.
- Quality assessment was limited to visual observation of the physical appearances of dried material. For paddy, the determination of broken grain; for some other materials, which were tested to a little extent due to time constraint, observation of mold growth and color changes.
- The concept of this dryer design would obviously be acceptable as intermediate technology but the context of application vary widely as weather condition as well as agronomic practices differ from one region to another; thus, may hinder the full adaptation of the dryer design in some strategic locations where the need for dryers and store houses are even more serious.

SUMMARY AND CONCLUSIONS

In general, the venting rate or airflow rate during drying has more influence on the drying efficiency of the system than any other management variables. The airflow rate control was found more critical than temperature control especially so that the drying operation is highly weather dependent. One should consider the fact that dryers are indispensable during the wet season harvest and therefore should adapt a system which can be economically sound. The system performed equally well even during a bad weather condition with a reasonable increase in total drying time.

Heat energy accounting revealed that paddy drying in the warehouse dryer is an energy efficient process as values of actual, theoretical and model calculated energy requirements (Table 2) paralleled each other.

Table 2

Comparison of Theoretical (Q_{cm}), Actual (Q_c) and Model Calculated (Q_f-Q_s) Heat Energy for Drying, x 10^4 kcal

	Q_{cm}	Q_c	$Q_f - Q_s$ ($\varepsilon = 0.95$)
Test A	6.79	8.64	7.18
Test B	6.60	8.60	7.03
Test C	5.97	9.16	6.65

It should be noted however, that the amount of energy used to elevate and maintain the temperature of the structure (1/3 of total energy) appears to be more potential for improvement in the energy efficiency through insulation of the structure. Further refinement of other system control as with the exchanged air control can only be done satisfactorily after this modification.

REFERENCES

1. Akyurt, M. and M. K. Selcuk. 1973. A solar drier supplemented with auxiliary heating systems for continuous operation. Solar Energy 14(3), 313-320.

2. Calderwood, D. L. 1970. Use of aeration to aid rice drying. Trans. ASAE 9(6), 893-895.

3. Cundiff, J. S. and J. W. Dodd. 1979. Energy model for forced air tobacco curing. Trans. ASAE 24(1), 211-215.

4. Dachtler, W. C. 1959. Temperature gradients in grain drying. Trans. ASAE 10, 150-153, 156.

5. De Padua, D. B. 1976. Grain drying principles, practices and systems. University of the Philippines at Los Baños, No. 35.

6. Eldridge, F. R. 1975. Wind machine. Division of Advanced Energy and Resources Research and Technology. USA Grant No. AER-75-12937. pp. 9-17.

7. Golding, E. W. 1976. The generation of electricity by wind power. Spon Ltd., London.

8. Herderson, S. M. and R. L. Perry. 1976. Agricultural process engineering. 3rd Edition. AVI Publishing Co., Westport Conn.

9. Hukill, W. V. 1955. Grain drying. In storage of cereal grains. Am. Assoc. Cereal Chem., St. Paul. Minn.

10. Hukill, W. V. and C. K. Shedd. 1955. Non-linear air flow in grain drying. Agr. Eng. 36, 462-466.

11. Jeon, Y. W., L. S. Halos and C. W. Bockhop. 1982. Furnace and vortex wind machine: A non-conventional approach to drying and storing technology. Paper presented during the 32nd PSAE Convention, Manila, Philippines, April 28-29.

12. Justus, C. G., W. R. Hargraves and A.Yalcin. 1976. Nationwide assessment of potential output from wind powered generators. J. Applied Met., 15, 673-678.

13. Khan, A. U. 1973. Rice drying and processing equipment for Southeast Asia. Trans. ASAE (16):1131-1135.

14. Ritzua, Y. 1975. Report on drying, storing and milling in the Philippines. Sci. Rept. L. Agr. Process Eng. No. 11 62-84.

15. Rajvir, Singh, T. P. Ojha and R. C. Maheshwari. 1980. Efficient use of agricultural wastes for energy production. AMA Vol. XI, No. 4, 34-37.

16. Stevens, M. J. M. 1979. The usefulness of weibull functions in determining the output of wind energy conversion systems, Dept. of Physics, Un. of Technology, Eindhoven, Netherlands, R-370-A.

APPENDIX: Typical weather conditions and general trends in agriculture in the Philippines (UPLB and IRRI, 1970-1980).

- Mean ambient air temperature : 29.4°C
- Mean relative humidity : 80 percent
- Average wind speed : 1.44 m/s
- Mean annual percent possible sunshine : 50 percent
- Daily insolation rate
  - Highest (April) : 22.97 mJ/m .day
  - Lowest (January) : 10.25 mJ/m .day
- Major crops grown : Rice and corn, with rice as the dominant and staple crop
- Average grain yields : 2.64 tons/ha (rice)
  2.94 tons/ha (corn)
- Harvest seasons
  - Wet season : September to December
  - Dry season : May to June
- Grain moisture content at harvest : 25 to 33 per cent, dry basis
- Farm size range : 1 to 10 hectares
- Agricultural wastes (i.e. rice hull, coconut husk and shell, sawdust, etc) are available energy sources
- Majority of farmers still dry their grains by spreading them on grounds or pavements under the sun
- Electricity is not available in most smaller towns and farms
- Most rural houses and farm storage buildings are constructed with corrugated galvanized iron (Cor. G.I.) sheets roofing materials.

THE THERMO-ECONOMIC MODEL OF CONVECTIVE DRYERS AND ITS OPTIMIZATION

K.E. Militzer

Technische Universität Dresden,
Dresden, German Democratic Republic

ABSTRACT

Cost and process model may be combined to give an thermo-economic model of convection dryers. Its solution depends on the definition of a suitable reference dryer. If j is the number of the types of costs accounted for, the model contains j-1 dimensionless cost factors (PAUER-numbers) and j simplexes of process parameters. A thermo-economic model of convection dryers is derived in a general form containing heat-, capital-, and electrical energy costs. As an example, the continuous single stage fluid bed dryer is used. A dryer with minimal air flow rate, i. e. with effluent air saturated serves as reference dryer. The model thus defined contains the number of transfer units in air flow direction as the only unknown. The resulting condition of optimum bed height cannot be solved explicitly but has to be treated numerically.

NOMENCLATURE

A	cross sectional area, m^2
C_P, C_V	PAUER-number, equ. (21), (22)
c_p	specific heat, kJ/kg
H	height, m
h	specific enthalpy, kJ/kg
k_P	price of electrical energy, M/J
k_Q	price of heat energy, M/J
k_V	specific annual dryer costs, M/m^3a
M	mass, kg
$\dot{M}$	mass flow rate, kg/s
Δp	pressure drop, N/m^2
P	fan power demand, W
$\dot{Q}$	heat flux, J/s
t_{op}	operation time, s/a
S	surface area, m^2
V	volume, m^3
$\dot{V}$	flow rate, m^3/s
w	velocity, m/s
$\overline{X}$	material moisture content, kg/kg
Y	air moisture content, kg/kg
Z	parameter defined by equ. (17)
α	heat transfer coefficient, W/m^2K
α^*	volumetric heat transfer coefficient, W/m^3K
η	air moisture content
η_{fan}	overall efficiency of fan
$\varkappa$	overall costs
ξ	dryer height
ϱ	density, kg/m^3
ψ	true drying rate of material

Superscripts

'	dryer inlet
"	dryer exit

Subscripts

g	gas
l	liquid
Pl	disperser plate
ref	reference dryer
sat	saturation
s	solid
var	variable
o	ambient state

INTRODUCTION

Increasing energy costs in the last years have stimulated an intensive search for optimal conditions of convective dryers. For heat exchangers detailed optimization procedures were published many years ago. But for optimization of convective dryers there are only few publications in the literature covering special topics. Even in recent editions of standard text books the problems of dryer economy are dealt with only in a very small scale. Therefore in this paper a general approach for dryer optimization based on both cost and process models is proposed.

COST MODEL

As main annual costs of convection dryers which may be influenced by dryer heat consumption, capital costs, and power requirements of fans are considered:

$$K_{var} = K_Q + K_V + K_P \tag{1}$$

Using simple statements for these items one gets

$$K_{var} = k_Q t_{op} \dot{Q} + k_V V + k_P t_{op} P \tag{2}$$

k_Q and k_P are specific energy prices, k_V are the annual capital costs per unit volume (or per unit area) of dryer and t_{op} the opteration time per year.

For convenience a reference dryer is defined with the same cost factors and annual operation time but different process parameters:

$$(K_{var})_{ref} = k_Q t_{op} \dot{Q}_{ref} + k_V V + k_P t_{op} P_{ref} \tag{3}$$

Dividing equ. (2) by the heating costs of the reference dryer one gets the dimensionless overall cost function

$$\frac{K_{var}}{k_Q t_{op} \dot{Q}_{ref}} = \frac{\dot{Q}}{\dot{Q}_{ref}} + \frac{k_V V_{ref}}{k_Q t_{op} \dot{Q}_{ref}} \frac{V}{V_{ref}} + \frac{k_P P_{ref}}{k_Q \dot{Q}_{ref}} \frac{P}{P_{ref}} \tag{4}$$

or

$$\varkappa = \frac{\dot{Q}}{\dot{Q}_{ref}} + c_V \frac{V}{V_{ref}} + c_P \frac{P}{P_{ref}} \tag{5}$$

Equ. (5) holds for any installations consuming energy and capital costs and may also be used for heat exchangers. For convective dryers it ought to be specified by calculating heat flux $\dot{Q}$ and fan power requirements P, assuming the same enthalpy differences $h'-h_o$, equal overall fan efficiency η_{fan}, and constant air density $\rho_g = \rho_{g,ref}$ for the model and the reference dryer:

$$\frac{\dot{Q}}{\dot{Q}_{ref}} = \frac{\dot{M}_g (h'-h_o)}{\dot{M}_{g,ref}(h'-h_o)} = \frac{\dot{M}_g}{\dot{M}_{g,ref}} \tag{6}$$

$$\frac{P}{P_{ref}} = \frac{\dot{M}_g \Delta p}{\rho_g \eta_{fan}} \left(\frac{\rho_g \eta_{fan}}{\dot{M}_g \Delta p} \right)_{ref} \approx \frac{\dot{M}_g}{\dot{M}_{g,ref}} \frac{\Delta p}{\Delta p_{ref}} \tag{7}$$

Thus the special form of cost model of convection dryers is derived form the general equation (5)

$$\varkappa = \frac{\dot{M}_g}{\dot{M}_{g,ref}} + C_P \frac{A}{A_{ref}} \frac{H}{H_{ref}} + \frac{\dot{M}_g}{\dot{M}_{g,ref}} \frac{\Delta p}{\Delta p_{ref}} \qquad (8)$$

The dimensionless constants C_V and C_P (sometimes called PAUER-numbers in honour of W. PAUER (1887 - 1971) nestor of power economics in Germany /2/) are to be calculated from given prices and from the parameters of the reference dryer. The simplexes Q/Q_{ref}, V/V_{ref} and $\Delta p/\Delta p_{ref}$ may be derived from suitable process models, depending on dryer type and operation and characteristics of the material to be dried.

EXAMPLE: CONTINUOUS SINGLE STAGE FLUID BED DRYER

The volume of a fluid bed dryer may be realized in many combinations of cross-sectional area and height of bed with different overall costs. On the one hand, air flow rate and heat requirements increase with cross-sectional area, resulting in higher heating costs. On the other hand, pressure drop and fan power demand grow with bed height, hence higher power costs are obtained. There exists a certain combination of area and height for which allover costs have a minimum. This optimal configuration will be calculated in the following section by a combination of cost and process models.

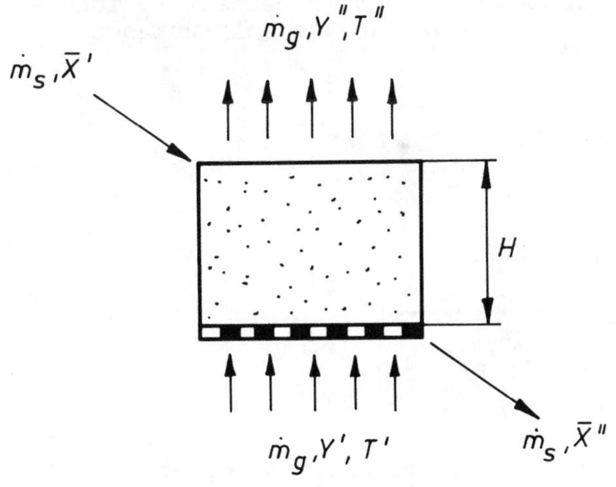

Figure 1 Fluid bed dryer

Assumptions of the process model

- The monodispersed solids are ideally mixed and have the same initial moisture content.
- Flow rate and inlet state of air are constant.
- Plug flow of air within the bed is assumed.
- Heat and mass transfer are allowed only by convection.
- Coefficients of heat and mass transfer are constant in time and locus.
- Drying process is adiabatic.

Exit gas concentration

As reference dryer the same kind of dryer is chosen with minimum gas flow rate, i. e. with water vapor saturation of gas reached at the exit.

A water balance gives

$$\dot{M}_l = \dot{M}_g (Y''-Y') = \dot{M}_{g,ref} (Y_{sat}-Y') \qquad (9)$$

Therefore the dimensionless water content of the exit gas is obtained for the simplex of gas flow rates:

$$\frac{\dot{M}_g}{\dot{M}_{g,ref}} = \frac{Y_{sat}-Y'}{Y''-Y'} = \frac{1}{\eta''} \qquad (10)$$

Variation of η with bed height ξ may be calculated according to VAN MEEL /3/ by

$$\frac{d\eta}{d\xi} = \psi (1-\eta) \qquad (11)$$

ξ is the dimensionless bed height expressed as its number of transfer units (NTU):

$$\xi = \frac{H}{H_{ref}} = \frac{H}{HTU} = \frac{\alpha (S/V)}{(w\, c_p)_g} H \qquad (12)$$

ψ is the true drying rate of the material to be dried, measured in an classic kinetic experiment with outer conditions kept constant /1/, sometimes referred to as characteristic drying curve. For convenience ψ is related to the moisture content $\bar{X}$ of the solid:

$\psi = \psi(\bar{X})$

If ideal mixing of solids is assumed, moisture content and true drying rate of the material remain constant all over the bed and equ. (11) may be solved to give

$$\eta'' = 1 - \exp(-\psi\zeta'') \tag{13}$$

Cross-sectional area of the bed

The cross-sectional area of the bed is calculated with the continuity equation $\dot{V} = A\, w_g$. The fluidizing gas velocity can be considered constant: $w_g \approx w_{g,ref}$. Hence

$$\frac{A}{A_{ref}} = \frac{\dot{V}_g/w_g}{(\dot{V}_g/w_g)_{ref}} \approx \frac{\dot{M}_g}{\dot{M}_{g,ref}} = \frac{1}{\eta''} \tag{14}$$

follows as simplex of areas.

Pressure drop of the dryer

Pressure drop of fluidized bed dryers include both pressure drop of the bed itself and pressure drop of the disperser plate. The latter may be expressed as an equivalent bed-height H_{Pl}. As reference system a dryer with the same grid but a different height is taken. Following equ. (12), H_{ref} = HTU.

$$\frac{\Delta p}{\Delta p_{ref}} = \frac{\rho_s g(H+H_{Pl})}{\rho_s g(H_{ref}+H_{Pl})} = \frac{\zeta + \zeta_{Pl}}{1 + \zeta_{Pl}} \tag{15}$$

The thermo-economic model and its optimization

Equations (10), (13), (14), and (15) contain all the required information about the single stage fluid bed dryer to solve the general cost model (8) of convective dryers.

Thus, as thermo-economic model, one gets in this case

$$\varkappa = \frac{1 + C_A\zeta'' + C_P \dfrac{\zeta'' + \zeta_{Pl}}{1 + \zeta_{Pl}}}{1 - \exp(-\psi\zeta'')} \tag{16}$$

This relationship after some transformation is shown in figure 2.

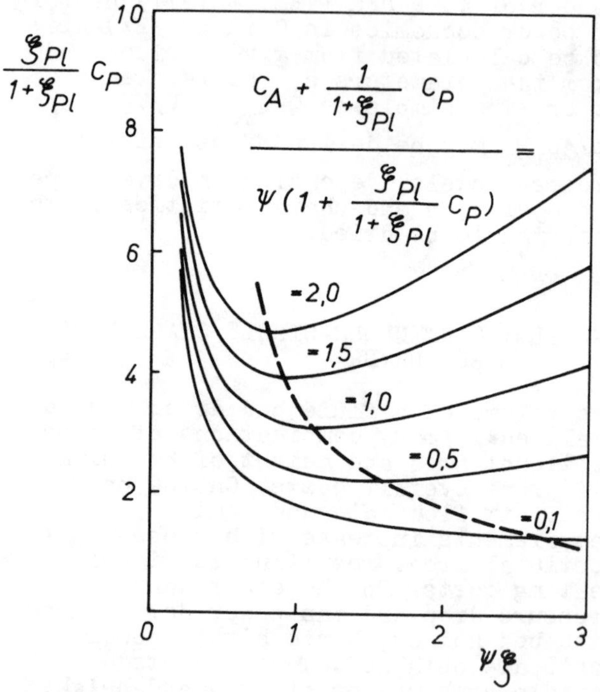

Figure 2 Overall costs of fluid bed dryer

Differentiation with respect to the bed height ζ'' as the only unknown gives

$$\exp(\psi\zeta''_{opt}) - \psi\zeta''_{opt} - 1 = \frac{1 + \dfrac{\zeta_{Pl}}{1+\zeta_{Pl}} C_P}{C_V + \dfrac{\zeta_{Pl}}{1+\zeta_{Pl}} C_P} = Z \tag{17}$$

Graphical representation of equ. (17) is given in figure 3.

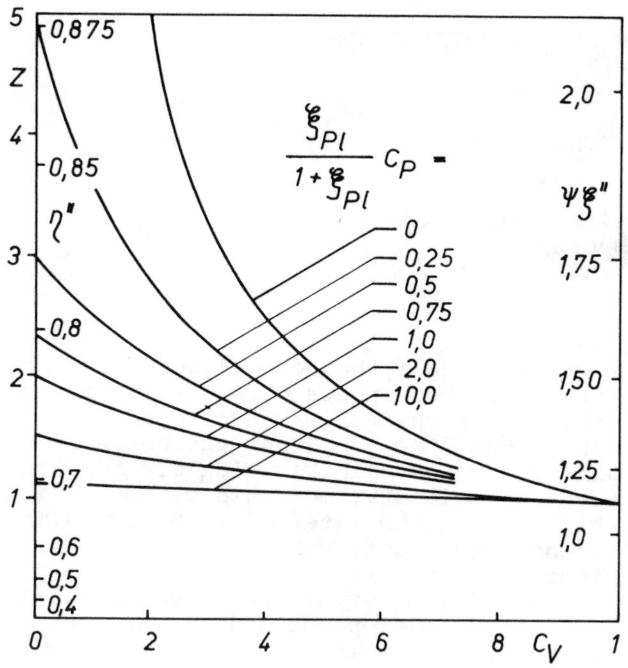

Figure 3a Optimal height of fluid bed dryer (according to the left side of equ. (17))

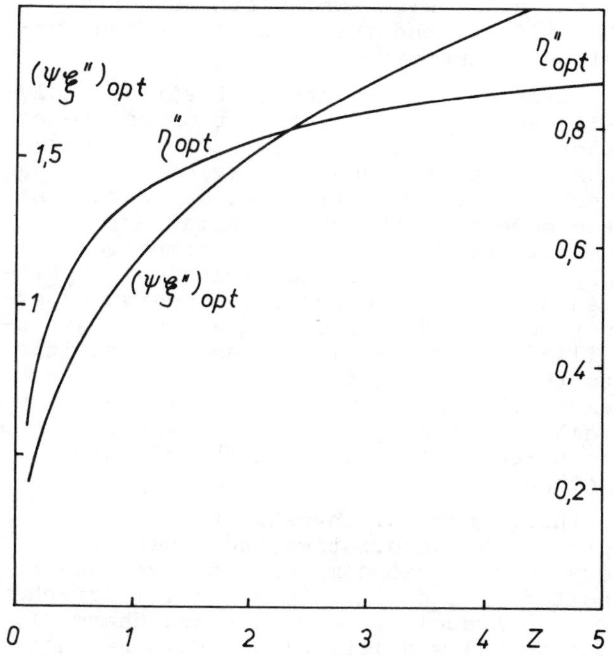

Figure 3b Optimal parameters of fluid bed dryer (according to the right side of equ. (17))

It can be shown that optimal bed height ξ''_{opt} increases with falling true drying rate ψ of the material. The influence of the complex $P = C_P \xi_{Pl}/(1+\xi_{Pl})$ depends on the value of the PAUER-number C_V:

for $C_V > 1$ $\psi\xi''_{opt}$ increases with P,

for $C_V < 1$ $\psi\xi''_{opt}$ diminishes with increasing P.

Equ. (17) cannot be solved explicitly and has to be treated numerically. But two empirical equations may derived, reflecting the relationship between Z and $\psi\xi''_{opt}$ with a maximum error of 1,2 % at intersection point Z = 1,4:

$$Z \leq 1.4: \psi\xi''_{opt} = \sqrt{\frac{4Z}{2+\sqrt{Z}}} \tag{18}$$

$$Z > 1.4: \psi\xi''_{opt} = \ln\left(1 + Z + \frac{5}{4}\sqrt[4]{Z}\right) \tag{19}$$

Minimal variable costs become

$$\varkappa = \left(C_V + \frac{C_P}{1+\xi_{Pl}}\right)\frac{\exp(\psi\xi''_{opt})}{\psi} \tag{20}$$

Resubstituting into definitions of C_V, C_P, and ξ_{Pl} results in the simple calculation equations:

$$C_V = \frac{k_V}{k_Q t_{op}\alpha^*(T'-T_o)} \tag{21}$$

$$C_P = \frac{k_P\,\Delta p_{Pl}}{k_Q(\ \ c_P)_{g,ref}\,(T'-T_o)\,\eta_{fan}} \tag{22}$$

REFERENCES

1. Ginsburg, A.S., Krasnikow, W.W., and Militzer, K.E. Durchführung und Auswertung von Trocknungsversuchen, Wiss. Z. TU Dresden, Vol. 31. no. 3, p. 19-24, 1982
2. Pauer, W., and Munser, H., Grundlagen der Kraft- und Wärmewirtschaft, Steinkopff, Dresden 1970
3. van Meel, D.A., Adiabatic Convection Batch Drying with Recirculation of Air, Chem. Eng. Sci., Vol. 9, p. 36-44, 1958

STATISTICAL ANALYSIS OF DRYING EXPERIMENTS

G.S. Raouzeos and G.D. Saravacos

Dept. of Chemical Engineering, National Technical University
Athens 147, GREECE

ABSTRACT

The statistical design of experiments and the analysis of variance provide a useful method for the evaluation of the effects and interactions of the important parameters on the drying time of various materials.

The factorial design of experiments was applied to the air-drying of slabs of starch gels, which simulate several food and polymeric materials. Two factorial designs were used· (1) The 3^3 design with the following parameters at three levels: Dry-bulb temperature T (30, 40, 50 ^{0}C), initial moisture content W^o (81, 86, 91%), and sample thickness L (2, 4, 6 mm). (2) The 3^4 design with the additional parameter of glucose content C (0, 25, 50%).

The analysis of variance of the experimental data demonstrated the significance of the main effects (L, T, C, W^o, in diminishing order), it revealed the significance of the two-factor interactions (TL, W^oL, CL, CT, and TW^o), and it showed that higher order interactions (3X and 4X) are insignificant. Further analysis of the data with the method of orthogonal polynomials showed the importance of the linear (L) and the quadratic (Q) effects and it revealed the significance of the following main effects and interactions for the two experimental designs: (1) For the 3^3 design: T_L, T_Q, W^o_Q, L_L, T_LL_L, $T_LW^o_L$, $T_QW^o_L$, $W^o_QL_L$, $W^o_LL_Q$, and $T_QW^o_LL_L$. (2) For the 3^4 design: T_L, T_Q, L_L, C_L, C_Q, W^o_Q, T_LL_L, C_LL_L, $C_LW^o_L$, and C_LT_L.

Application of regression (least-square) analysis resulted in two polynomials for each design, which can be used for the estimation of drying time as a function of the corresponding 10 significant main effects and interactions, for the range of values of the parameters used in the present work.

INTRODUCTION

The engineering design and the efficient operation of drying processes and equipment are usually based on laboratory or pilot-plant experiments on the specific material at a certain range of operating conditions. The important variables in the air-drying of solids include air temperature and velocity, moisture content, and thickness of the material to be dried. The drying experiments can give valuable indications about the mechanism of drying of the specific material, which can be utilized in the development and improvement of the theory of drying.

In the classical method of experimentation the parameters (independent variables) of the process are varied one at a time, the rest remaining constant. The corresponding response (dependent variable) is estimated by some method of measurement. This approach has some serious disadvantages, such as the large number of experimental tests for multivariable experiments, and the limited conclusions which can be derived on the effects and interactions of the various parameters.

Statistical design and statistical analysis can yield more information on the effect of the various parameters on the measured response (usually drying time) than the classical experimental approach. At the same time the experimental error can be evaluated. Of particular importance in physical, chemical and biological experimentations is the factorial design of experiments. In this method all data are utilized in the analysis, and, in addition to the main effects, the interactions among the independent variables can be evaluated. The number of experimental tests required is less than in the classical method.

Although several investigations on the drying characteristics and kinetics of drying of various materials have been reported in the literature (Keey 1, Krischer 2, van Arsdel 3, Saravacos and Charm 4), the formal statistical analysis has not been applied to drying experiments. Preliminary work by the authors of this paper (Raouzeos and Saravacos, 5), has indicated the importance of statistical analysis to the air-drying of starch gels.

In the present paper the statistical method of factorial design is presented in a thorough analysis of the air-drying of starch gels. The statistical data are used for the development of mathematical models which can predict quantitatively the effect of the important parameters and their interactions on the drying time. Starch gels were chosen as experimental materials because they simulate several food and polymeric materials, and they can be prepared as experimental samples of the desired dimensions and composition.

EXPERIMENTAL DESIGN

Statistical experiments are performed in random order, and they provide for the estimation of the experimental error, which is required for testing the significance of the various effects of the parameters on the measured response. Among the various statistical methods, the factorial design is well suited for drying experiments because of its advantages (Montgomery 6, Davies 7). Factorial designs are expressed in a power form, e.g. a 3^2 design indicates two factors (parameters) at three levels, resulting in 9 tests (measurements). In order to estimate the experimental error, replication of the tests is required, e.g. 2 replicates of the 3^2 design result in a total of 18 experiments. In higher-order designs, e.g. 3^4, replication would result in a very large number of experiments. This can be avoided by estimating the experimental error by a limited number of replicates at a certain point of the design table. Factorial designs of higher order than 3^4 are not practical, because the number of experiments becomes too high. In such cases the fractional factorial and other statistical designs are recommended, because they require less experimental tests (Hicks 8).

From previous experimental work and from the practice of drying, the following 4 parameters have been found important in the air-drying of hygroscopic materials, similar to starch gels: Dry-bulb temperature (T), initial moisture content (W^0), thickness (L), and composition (C). For practical reasons it was decided to limit the experiments to these 4 factors, excluding other parameters, which might have an important effect, such as air velocity and air humidity.

In the following mathematical analysis three parameters (factors) are used, but the same treatment can be extended to four or more parameters. The parameters A, B, C can take the corresponding values a, b, c, i.e. their values can vary at the a, b, c levels. All experiments are performed at random with n replications ($n \geqslant 2$), and the experimental data (parameters and responses), can be tabulated as shown in Table (4).

The response (dependent variable) y can be expressed as a function of the three parameters and their interactions in the following linear statistical model:

$$y_{ijkr} = \mu + \tau_i + \beta_i + \gamma_k + (\tau\beta)_{ij} + (\tau\gamma)_{ik} + (\beta\gamma)_{jk} + (\tau\beta\gamma)_{ijk} + \varepsilon_{ijkr} \quad (1)$$

where, $i = 1,2,\ldots,a$ $\quad k = 1,2,\ldots,c$
$\quad j = 1,2,\ldots,b$ $\quad r = 2,2,\ldots,n$

The term μ of (1) is the mean of the effects of the parameters A, B, C, while the terms τ_i, β_j, γ_k represent the main effects of parameters A, B, C, at the corresponding levels of i, j, k. The terms $(\tau\beta)_{ij}$, $(\tau\gamma)_{jk}$, and $(\tau\beta\gamma)_{ijk}$ express the two- and three- factor interactions at the corresponding levels of values. The last term ε_{ijkr} represents the experimental error.

The significance of terms of equation (1) is tested by statistical analysis of variance (ANOVA), (Montgomery, 6). The ANOVA tests are based on the test of the following hypotheses of H_0: $\tau_i = 0$, $\beta_j = 0$, $\gamma_k = 0$, $(\tau\beta)_{ij} = 0$, $(\tau\gamma)_{ik} = 0$, $(\beta\gamma) = 0$, and $(\tau\beta\gamma)_{ijk} = 0$. This means that it is assumed that all three parameters and their interactions are not significant. If the statistical analysis of the data shows that the H_0 hypothesis is not valid, the H_1 hypothesis must be accepted, which means that the corresponding parameters and their interactions have a significant effect on the experimental response (y).

The test of the H_0 hypothesis is carried out by an analysis of variance, which is summarized in Table (1). The derived values of Table (1) are calculated by the following expressions:

1) Sum of squares (SS) for the main effects of parameters A, B, C:

$$SS_A = \sum_{i=1}^{a} \frac{y_{i\ldots}^2}{bcn} - \frac{y_{\ldots\ldots}^2}{abcn} \quad (2)$$

$$SS_B = \sum_{j=1}^{b} \frac{y_{.j..}^2}{acn} - \frac{y_{\ldots\ldots}^2}{abcn} \quad (3)$$

$$SS_C = \sum_{k=1}^{c} \frac{y_{..k.}^2}{abn} - \frac{y_{\ldots\ldots}^2}{abcn} \quad (4)$$

2) Sum of squares for the two-factor interactions (AB, AC, BC):

$$SS_{AB} = \sum_{i=1}^{a}\sum_{j=1}^{b} \frac{y_{ij..}^2}{cn} - \frac{y_{\ldots\ldots}^2}{abcn} - SS_A - SS_B \quad (5)$$

$$SS_{AC} = \sum_{i=1}^{a}\sum_{k=1}^{c} \frac{y_{i.k.}^2}{bn} - \frac{y_{\ldots\ldots}^2}{abcn} - SS_A - SS_B \quad (6)$$

$$SS_{BC} = \sum_{j=1}^{b}\sum_{k=1}^{c} \frac{y_{.jk.}^2}{an} - \frac{y_{\ldots\ldots}^2}{abcn} - SS_B - SS_C \quad (7)$$

Table 1. Analysis of Variance for the Factorial Design of Three Parameters (A,B,C)

Parameter	Sums of Squares	Degrees of Freedom	Mean Squares	$F_{calculated}$	F^o
A	SS_A	a-1	MS_A	MS_A/MS_E	$F^o\lvert(a-1),abc(n-1),\alpha\rvert$
B	SS_B	b-1	MS_B	MS_B/MS_E	$F^o\lvert(b-1),abc(n-1),\alpha\rvert$
C	SS_C	c-1	MS_C	MS_C/MS_E	$F^o\lvert(c-1),abc(n-1),\alpha\rvert$
AB	SS_{AB}	(a-1)(b-1)	MS_{AB}	MS_{AB}/MS_E	$F^o\lvert(a-1)(b-1),abc(n-1),\alpha\rvert$
AC	SS_{AC}	(a-1)(c-1)	MS_{AC}	MS_{AC}/MS_E	$F^o\lvert(a-1)(c-1),abc(n-1),\alpha\rvert$
BC	SS_{BC}	(b-1)c-1)	MS_{BC}	MS_{BC}/MS_E	$F^o\lvert(b-1)(c-1),abc(n-1),\alpha\rvert$
ABC	SS_{ABC}	(a-1)(b-1)(c-1)	MS_{ABC}	MS_{ABC}/MS_E	$F^o\lvert(a-1)(b-1)(c-1),abc(n-1),\alpha\rvert$
Error	SS_E	abc(n-1)	MS_E		
Total	SS_T	abcn-1			

3) Sum of squares for the three-factor interaction (ABC):

$$SS_{ABC} = \sum_{i=1}^{a}\sum_{j=1}^{b}\sum_{k=1}^{c}\frac{y^2_{ijk.}}{n} - \frac{y^2_{....}}{abcn} - (SS_A + SS_B + SS_C + SS_{AB} + SS_{AC} + SS_{BC}) \quad (8)$$

4) Total sum of squares (T):

$$SS_T = \sum_{i=1}^{a}\sum_{j=1}^{b}\sum_{k=1}^{c}\sum_{r=1}^{n} y^2_{ijkr} - \frac{y^2_{....}}{abcn} \quad (9)$$

5) Error sum of squares (E):

$$SS_E = SS_T - (SS_{ABC} + SS_A + SS_B + SS_C + SS_{AB} + SS_{AC} + SS_{BC}) \quad (10)$$

In equation (2) to (10), $y_{....}$ is the sum of all experimental values of the response, $y_{i...}$ is the sum of the y values for factor A, $y_{.j..}$ is the sum of the y values for factor B, etc.

The mean squares (MS) of Table (1) are obtained by dividing the sum of squares by the corresponding degree of freedom. The ratio of the mean squares of the various effects and interactions to the error mean squares yields the corresponding value of F_{calc}, which is a measure of the statistical significance and it follows the F distribution. F_{calc} is compared to the corresponding value F^o of the Statistical Tables at a given significance level, e.g. $\alpha = 0,05$ (95% confidence level). If $F_{calc} > F^o$, the corresponding effect or interaction is significant. Otherwise, the H_0 hypothesis holds and any observed differences are due to experimental error.

REGRESSION ANALYSIS

Regression analysis can be used for the development of a mathematical model of the main effects and interactions of the parameters (x_i) on the experimental response (y). In statistical analysis linear polynomials are the preferred models, because of their advantages. In complex processes, like air-drying, with more than one parameters, multiple linear models can be applied.

A multiple linear regression model is expressed by the following equation (Montgomery 6):

$$y_j = b_o + b_1x_{1j} + b_2x_{2j} + \ldots + b_px_{pj} + \varepsilon_j \quad (11)$$

or

$$y_i = b_o + \sum b_ix_{ij} + \varepsilon_j$$

where, y_j = response value of the j experimental point
x_{ij} = independent variable (i) at the j experimental point
b_i = regression coefficient of i independent variable
ε_j = experimental error at the j experimental point

$i = 1,2,\ldots\ldots,p$
$j = 1,2,\ldots\ldots,n, \quad n > p$

The polynomial model is considered linear

if the regression coefficients (b_i) are linear, although the independent variable x_{ij} may be non-linear.

The regression coefficients (b_i) can be estimated by the method of least squares, utilizing the experimental data and statistical calculations of the analysis of variance. The polynomial model of equation (11) with k parameters requires n experimental points (measurements), where n > p.

Equation (11) can be written in a matrix form as follows:

$$Y = XB + \varepsilon \qquad (12)$$

where, Y = (nx1) column vector of the experimental responses
X = nx(p+1) matrix of the parameters
B = (p+1)x1 column vector of the regression coefficients
ε = (n+1) column vector of random errors

Solution of equation (12) by the least squares method yields:

$$\hat{B} = (X'X)^{-1}(X'y) \qquad (13)$$

where, $\hat{B}$ = estimate of the regression coefficients (b_i)
$(X'X)^{-1}$ = inverse of the matrix of sums of squares of the parameters x_{ij}
$(X'y)$ = px1 matrix of y

(X'X) is a (p+1)(p+1) symmetric matrix, which becomes simplified, if it is diagonal.

The polynomial model is tested for statistical significance by an analysis of variance (Montgomery 6). The H_0 hypothesis is that the model fits adequately the experimental data. The H_0 hypothesis is rejected if the following relationship is true:

$$\frac{C}{n-p} \Big/ \frac{E}{\sum(r_i-1)} > F^{o}(n-p,\sum(r_j-1),\alpha) \qquad (14)$$

where,

$$E = \sum_{j=1}^{n}\sum_{k=1}^{r_j}(y_{jk}-\bar{y}_j)^2 \qquad (15)$$

$$R = \sum_{j=1}^{n}\sum_{k=1}^{r_j}y^2_{jk} - \sum_{i=1}^{p}(\hat{b}^2_i\sum_{j=1}^{n}\sum_{k=1}^{r_j}x^2_{ijk}) \qquad (16)$$

$$C = R - E \qquad (17)$$

E = the sum of squares of the error (due to replication)
R = the sum of squares of the residuals between the predicted values by the model and the experimental values
C = the sum of squares of differences due to the lack of fit (inappropriate model).

The degrees of freedom of E, R, and C are $\sum_{j=1}^{n}(r_j-1)$, $(\sum_{j=1}^{n}r_j-p)$, and (n-p) respectively.

If the regression model is found acceptable the confidence limits of its coefficients ($\hat{b}_i$) can be estimated by the following relationship:

$$\hat{b}_i-\sqrt{V(\hat{b}_i)}\,t(\sum_{j=1}^{n}(r_j-1),\alpha/2) < \hat{b}_i < \hat{b}_i + \sqrt{V(\hat{b}_i)}\,t(\sum_{j=1}^{n}(r_j-1),\alpha/2) \qquad (18)$$

where,

$$V(\hat{b}_i) = \frac{E}{r_j\sum_{j=1}^{n}(r_j-1)\sum_{j=1}^{n}X^2_{ij}} \qquad (19)$$

r_j is the number of replications at the j^j experimental point.

In (18), the term

$$t(\sum_{j=1}^{n}(r_j-1),\alpha/2)$$

is the Student's "t" value taken from statistical tables for

$$\sum_{j=1}^{n}(r_j-1)$$

degrees of freedom at (α) significance level (two-tail value).

EXPERIMENTAL PROCEDURE

The experimental materials and methods used in this work were described by Raouzeos and Saravacos (5). Starch gels were prepared from corn starch/glucose, and they were dried in Plexiglass dishes as slabs of 6 cm diameter. A Laboratory tunnel dryer was used, operating at an air velocity of 2 m/s and at various dry-bulb temperatures. The loss of weight was measured continuously with a recording balance. The drying time (measured response) was taken as the time (in minutes) required for drying from the initial moisture content (W^o) of the sample to a final moisture of 30% wet basis, corresponding to approximately 95% loss of the initial moisture.

The following four parameters were considered in the design and analysis of the drying experiments: Dry-bulb temperature (T), initial moisture content (W^o), sample thickness (L), and glucose content of the sample (C). All these parameters are quantitative, which facilitates the statistical analysis of the data. Two factorial designs were selected, 3^3 and 3^4, which are presented in Tables (2) and (3):

Table 2. The 3^3 Factorial Design of Drying Experiments

Parameters	Units	Physical Values			Coded Values		
Dry-bulb temperature T	°C	30	40	50	-1	0	1
Initial moisture W°	% by wt	91	86	81	-1	0	1
Sample thickness L	mm	2	4	6	-1	0	1

Table 3. The 3^4 Factorial Design of Drying Experiments

Parameters	Units	Physical Values			Coded Values		
Dry-bulb temperature T	°C	30	40	50	-1	0	1
Initial moisture W°	% by wt	91	86	81	-1	0	1
Sample thickness L	mm	2	4	6	-1	0	1
Glucose content C	% solids	0	25	50	-1	0	1

The coded values x_i are derived from the physical values $\tilde{x}_i$ by the following relationship:

$$x_i = \frac{\tilde{x}_i - \tilde{x}_{io}}{\delta_i} \qquad (20)$$

where, $\tilde{x}_{io}$ = the physical value of parameter i at the central point of its definition

δ_i = the interval of variation of the physical value of parameter i around its central point of definition.

In the 3^3 design, 2 replicates were performed at each point, resulting in 54 experimental measurements. Replication at all points was considered not practical in the 3^4 design, and the experimental error was based on 12 replications at the central point of the design table. All computations were carried out with a desk calculator.

STATISTICAL ANALYSIS

Analysis of Variance of the Experimental data.

Table (4) and (5) show the experimental drying data for the two factorial designs 3^3 and 3^4. The experimental response is drying time in minutes. The following two linear statistical models, based on equation (1), can be used to represent the values of Tables (4) and (5).

$$t_{S,ijk} = \mu + T_i + W^o_j + L_k + TL_{ik} + W^oL_{jk} + TW^oL_{ijk} + \varepsilon_{ijk} \qquad (21)$$

$$t_{SG,ijkm} = \mu + C_i + T_j + W^o_k + L_m + CT_{ij} + CW^o_{ik} + CL_{im} + TW^o_{jk} + TL_{jm} + W^oL_{km} + CTW^o_{ijk} + CTL_{ijm} + CW^oL_{ikm} + TW^oL_{jkm} + CTW^oL_{ijkm} + \varepsilon_{ijkm} \qquad (22)$$

where, t=time (S=starch, SG = starch/glucose) μ, is the mean value of the drying time, T, W°, L, C are linear functions of the main factors (parameters), TW°, TL,..., the two- factor interactions, TW°L,... the three-factor interactions, and CTW°L is the four-factor interaction.

The significance of the main effects and interactions of the parameters of equations (21) and (22) is evaluated by statistical tests of the H_0 hypothesis, i.e. by assuming that all the parameters are equal to zero. This leads to 7 tests for eq. (21) and 15 tests for eq. (22). Equations (2) to (10) are used to calculate the various statistical sums of squares from the experimental data of Tables (4) and (5). The degrees of freedom, the mean squares and the F_{calc} are calculated according to Table (1). Finally, the corresponding F^o values are taken from the F-distribution Table, using a significance level $\alpha = 0{,}05$. The calculated data are shown in Tables 6 (3^3 design) and Table 7 (3^4 design).

Table (6) indicates that all main effects (T, W, L) and all two-factor interactions (TL, W°L, TW°) are significant ($F_{calc} > F^o$), while the three-factor interaction (TW°L) is insignificant. Table (7) indicates also the significance of the main effects C,T, L, and the two-factor interactions TL,CL, CT, while all other two-, three-,and four-factor interactions are insignificant.

The analysis of variance of the original experimental data confirms the known strong dependence of the sample thickness and the air temperature on the drying time. It also shows a strong effect of the glucose content, while the initial moisture content has a relatively weak effect. Strong two-factor interactions were revealed between temperature/thickness, initial moisture content/thickness, and glucose content/thickness, while the other significant two-factor interactions were weaker.

Orthogonal Polynomials and Orthogonal Coefficients

The conclusions drawn from the analysis of variance are limited to the significance (or no significance) of the main effects and interactions of the parameters of the drying experiments. This method can not compare and predict the effects of the various factors and interactions on the neasured response. More information can be obtained by an analysis of the experimental data by the method of ortho-

Table 4. Experimental Data for the 3^3 Factorial Design. Measured Response, t (min)

	T_{-1}			T_o			T_1		
	W^o_{-1}	W^o_o	W^o_1	W^o_{-1}	W^o_o	W^o_1	W^o_{-1}	W^o_o	W^o_1
L_{-1}	111	143	123	93	111	84	86	80	69
	122.5	129	128	90	111.5	90	77	76.5	71
	134	115	133	87	112	96	68	73	73
	245	258	256	180	223	180	154	153	142
L_o	288	365	413	193	212	293	177	148	152
	312.5	360	416.5	218	201	256.5	152	149.5	175
	337	355	420	243	190	220	127	151	198
	625	720	833	436	402	513	304	299	350
L_1	570	490	620	508	275	343	289	204	224
	536	478	581.5	464	290	360.5	269.5	219.5	222.5
	502	466	543	420	305	378	250	235	221
	1072	956	1163	928	580	721	539	439	445

Cell of Data:

First measurement
Mean
Second measurement
Sum of two measurements

Table 5. Experimental Data for the 3^4 Factorial Design. Measured Response, t(min)

		C_{-1}			C_o			C_1		
		T_{-1}	T_o	T_1	T_{-1}	T_o	T_1	T_{-1}	T_o	T_1
L_{-1}	W^o_{-1}	122	84	66	147	82	53	220	82	62
	W^o_o	96	104	68	176	150	78	192	73	56
	W^o_1	124	86	66	173	124	64	98	82	60
L_o	W^o_{-1}	330	237	122	362	310	262	592	209	211
	W^o_o	341	183	136	302	200	191	362	290	166
	W^o_1	375	203	179	344	351	170	515	315	156
L_o	W^o_{-1}	461	218	239	615	530	266	880	502	350
	W^o_o	431	284	221	695	374	327	604	408	248
	W^o_1	499	353	201	704	513	364	735	400	221

Table 6. Analysis of Variance for the 3^3 Factorial Design (Data of Table 4)

Parameter	Degrees of Freedom	Sums of Squares	Mean Squares	$F_{calc.}$	$F^o(0.05)$
T	2	306690	153345	189.6	3.35
W^o	2	10147	5074	6.3	3.35
TW^o	4	11505	2876	3.6	2.73
L	2	709972	354986	439.0	3.35
TL	4	92693	27173	33.6	2.73
W^oL	4	27742	6935	8.6	2.73
TW^oL	8	10732	1342	1.7	2.30
Error	27	21835	809		
Total	53	1191316			

$F^o(2,27,\ 0.05) = 3.35$

$F^o(4,27,\ 0.05) = 2.73$

$F^o(8,27,\ 0.05) = 2.30$

Table 7. Analysis of Variance for the 3^4 Factorial Design (Data of Table 5)

Parameter	Degrees of Freedom	Sums of Squares	Mean Squares	$F_{cal.}$	$F^o(0.05)$
C	2	117675	58838	20.2	3.89
T	2	658467	329234	113.3	3.89
CT	4	52179	13045	4.5	3.26
W^o	2	15708	7854	2.7	3.89
CW^o	4	23318	5830	2.0	3.26
TW^o	4	9483	2371	0.8	3.26
L	2	1452739	726370	250.0	3.89
CL	4	71528	17882	6.2	3.26
TL	4	122059	30515	10.5	3.26
W^oL	4	14292	3573	1.2	3.26
CTW^o	8	26196	3275	1.1	2.85
CTL	8	25671	3209	1.1	2.85
CW^oL	8	23220	2903	1.0	2.85
TW^oL	8	6126	766	0.3	2.86
CTW^oL	16	46344	2897	1.0	2.62
Error	12	34877	2906		
Total	92*	2699882			

$F(2,12,\ 0.05) = 3.89$
$F(4,12,\ 0.05) = 3.26$
$F(8,12,\ 0.05) = 2.85$
$F(16,12,\ 0.05) = 2.62$

*81 experimental points of the factorial design plus 12 replications at the center of the design table.

gonal polynomials, which can predict the effect of each level of the parameters on the response. The parameters are analyzed into linear and quadratic components for the three-level experimental design.

The orthogonal polynomials utilize the orthogonal contrast coefficients, which simplify considerably the regression analysis of the experimental data. The orthogonal coefficients can be used in the analysis of the drying data, presented in this paper, since the levels of all factors are equally spaced.

Two contrast coefficients c_i and d_i are orthogonal if

$$\sum_{i=1}^{a} c_i d_i = 0$$

where a is the number of treatments,which result in (a-1) contrasts. The orthogolal coefficients are given in Statistical Tables as functions of the number of levels of the factors. For the 3-level designs of this work, the orthogonal coefficients of the linear component (C_i^L) are -1, 0, 1 for the corresponding three coded levels -1, 0, 1. The respective coefficients of the quadratic component (C_i^Q) are 1, -2, 1.

The analysis of the statistical effects shown in Table (6) and (7) into linear and quadratic components requires complete tables of the orthogonal coefficients. Two complete tables of orthogonal coefficients of the 3^3 and 3^4 design were prepared,which are not given here because they occupy large space. The following example illustrates the procedure of calculation of the orthogonal coefficients: Considering the experimental data of the 3^3 design (Table 4), the linear component of temperature (T_L) will have orthogonal coefficients equal to -1 for the values of the measured response y_1- y_9, since the temperature is at its lowest level. The coefficients of T_L will be equal to 0 at y_{10}- y_{18} and 1 at y_{19}- y_{27}. The coefficients of the quadratic component of initial moisture (W_Q^o) will be 1 at y_1- y_3, -2 at y_4- y_6, 1 at y_7- y_9, etc. The orthogonal coefficients of the interactions are derived by multiplying horizontally the coefficients of the components of the parameter (e.g. $T_L \times W_Q^o$).

Analysis of Variance of Linear and Quadratic Effects

The significance of the linear (L) and quadratic (Q) effects were tested by the same procedure used in the statistical analysis of the original drying data (Tables 4-7). The sums of squares (SS) of the main effects (A_L, A_Q,....) and the various interactions (A_LB_L, A_LB_Q, ...A_LB_Q.. D_L), required for the analysis of variance, were calculated from the following relationships:

$$SS_{A_L} = \frac{\left(\sum_{i=1}^{a} C_i^L y_{i\ldots}\right)^2}{n\sum_{i=1}^{a} (C_i^L)^2} \qquad (23)$$

$$SS_{A_Q} = \frac{\left(\sum_{i=1}^{a} C_i^Q y_{i\ldots}\right)^2}{n\sum_{i=1}^{a} (C_i^Q)^2} \qquad (24)$$

$$SS_{A_LB_L} = \frac{\left(\sum_{i=1}^{a}\sum_{j=1}^{b} C_{ij}^{LxL} y_{ij..}\right)^2}{n\sum_{i=1}^{a}\sum_{j=1}^{b} (C_{ij}^{LxL})^2} \qquad (25)$$

$$SS_{A_LB_Q} = \frac{\left(\sum_{i=1}^{a}\sum_{j=1}^{b} C_{ij}^{LxQ} y_{ij..}\right)^2}{n\sum_{i=1}^{a}\sum_{j=1}^{b} (C_{ij}^{LxQ})^2} \qquad (26)$$

$$SS_{A_LB_Q\ldots D_L} = \frac{\left(\sum_{i=1}^{a}\sum_{j=1}^{b}\cdots\sum_{k=1}^{d} C_{ij\ldots k}^{LxQ..XL} y_{ij\ldots k}\right)^2}{n\sum_{i=1}^{a}\sum_{j=1}^{b}\cdots\sum_{k=1}^{d} (C_{ij\ldots k})^2} \qquad (27)$$

where, C_i^L, C_{ij}^{LxL}, ...= orthogonal coefficients, calculated as described in the previous section

$y_{i...}$, $y_{ij..}$, ... = corresponding sums of the experimental response (time in min, Tables 4,5)

The analysis of variance of the linear and the quadratic effects of the parameters for the experimental designs 3^3 and 3^4 is summarized in Tables (8) and (9). In Table (9), only the main effects and the two-factor interactions are given, since all three-factor interactions were found to be insignificant. It should be noted that each main effect and each interaction has one degree of freedom and the error has 27 degrees of freedom in the 3^3 design, and 12 in the 3^4 design.

Table (8) shows the statistical significance of the following components of the parameters on the drying time of the 3^3 design: Linear components of temperature and sample thickness (T_L, L_L), quadratic components of temperature and initial moisture content (T_Q, W_Q^o), the two-factor interactions $T_L L_L$, $T_L W_L^o$, $T_Q W_L^o$, $W_Q^o L_L$, $W_L L_Q$, and the three-factor interaction $T_Q W_L^o L_L$. These results are in general agreement with the results of the simple analysis of variance of the original data (Table 6), except that here the three-factor interaction was found to be significant. The conclusions derived from Table (8) are specific on the effects of the linear and quadratic components, while the data of Table (6) refer to the mean effects of the parameters.

Similar results were obtained with the data of the 3^4 design (Table 9): Significant effects were found with the linear components of the parameters C_L, T_L, L_L, the quadratic components C_Q, T_Q, W_Q^o, and the two-factor interactions $C_L T_L$, $C_L W_L^o$, $C_L L_L$, $T_L L_L$. All three-factor and four-factor interactions were insignificant. The analysis of variance with orthogonal coefficients has revealed the significance of the quadratic component of the initial moisture content (W_Q^o) and the interaction $C_L W_L^o$, which were found insignificant in the simple analysis of the experimental data (Table 7).

MATHEMATICAL MODEL FOR THE DRYING TIME

The complete factorial designs of 3^3 and 3^4, used in this work, require two linear polynomials theoretically with 27 and 81 terms respectively (equation 11). However, the statistical analysis of the main effects and interactions of the parameter components (Tables 8 and 9) show that only 10 main effects and interactions are significant in each experimental design.

Table 8. Analysis of Variance of the L[illegible] near and Quadratic Effects of [illegible] Parameters for the 3^3 Factoria[illegible] Design (Data of Table 4)

Parameter	Degrees of Freedom	Sums of Squares	Mean Squares	F_{calc}
T_L	1	303050	303050	374.7
T_Q	1	3640	3640	4.5
W_L^o	1	400	400	0.5
W_Q^o	1	9747	9747	12.0
$T_L W_L^o$	1	5704	5704	7.0
$T_L W_Q^o$	1	421	421	0.5
$T_Q W_L^o$	1	3612	3612	4.5
$T_Q W_Q^o$	1	1768	1768	2.2
L_L	1	708964	708964	876.7
L_Q	1	1008	1008	1.2
$T_L L_L$	1	88574	88574	109.5
$T_L T_Q$	1	1922	1922	2.4
$T_Q L_L$	1	180	180	0.2
$T_Q L_Q$	1	2017	2017	2.5
$W_L^o L_L$	1	1820	1820	2.2
$W_L^o L_Q$	1	10585	10585	13.1
$W_Q^o L_L$	1	14706	14706	18.2
$W_Q^o L_Q$	1	631	631	0.[illegible]
$T_L W_L^o L_L$	1	1640	1640	2.0
$T_L W_L^o L_Q$	1	281	281	0.4
$T_L W_Q^o L_L$	1	1027	1027	1.3
$T_L W_Q^o L_Q$	1	576	576	0.7
$T_Q W_L^o L_L$	1	3536	3536	4.4
$T_Q W_L^o L_Q$	1	306	306	0.4
$T_Q W_Q^o L_L$	1	3364	3364	4.2
$T_Q W_Q^o L_Q$	1	2	2	0.0
Error	27	21835	809	
Total	53	1191316		

$F^o(1,27,0.05) = 4.21$

For the 3^3 design the mathematical model becomes:

$$t_S = a_o z_o + a_1 z_1 + a_2 z_3 + a_{11} z_{11} + a_{22} z_{22} + a_{12} z_1 z_2 + a_{13} z_1 z_3 + a_{112} z_{11} z_2 + a_{223} z_{22} z_3 + a_{233} z_2 z_{33} + a_{1123} z_{11} z_2 z_3 \quad (28)$$

and for the 3^4 design:

$$t_{SG} = b_o x_o + b_1 x_1 + b_2 x_2 + b_4 x_4 + b_{11} x_{11} + b_{22} x_{22} + b_{33} x_{33} + b_{12} x_1 x_2 + b_{13} x_1 x_3 + b_{14} x_1 x_4 + b_{24} x_2 x_4 \quad (29)$$

where z_o, x_o = 1
x_1 = function of glucose content (C)
z_1, x_2 = function of air temperature (T)
z_2, x_3 = function of initial moisture content (W^o)
z_3, x_4 = function of initial sample thickness (L)

$$x_2 = \frac{T-\bar{T}}{\delta_T}, \quad z_2 = x_3 = \frac{W^o-\bar{W}^o}{\delta_{W^o}}$$

= mean values of C, T, W^o, and L
δ_L = respective intervals of variation of their values.

. Analysis of Variance of the Linear and Quadratic Effects of the Parameters for the 3^4 Factorial Design (Data of Table 5)

Parameter	Degrees of Freedom	Sums of Squares	Mean Squares	F_{cal}
C_L	1	94627.0	94627.0	32.6
C_Q	1	23048.0	23048.0	7.9
T_L	1	642577.0	642577.0	221.1
T_Q	1	15890.0	15890.0	5.5
C_LT_L	1	39171.0	39171.0	13.5
C_LT_Q	1	1716.0	1716.0	0.6
C_QT_L	1	4013.0	4013.0	1.4
C_QT_Q	1	7279.0	7279.0	2.5
W^o_L	1	361.0	361.0	0.1
W^o_Q	1	15348.0	15348.0	5.3
$C_LW^o_L$	1	14957.0	14957.0	5.1
$C_LW^o_Q$	1	4004.0	4004.0	1.4
$C_QW^o_L$	1	4186.0	4186.0	1.4
$C_QW^o_Q$	1	170.0	170.0	0.0
$T_LW^o_L$	1	2.4	2.4	0.0
$T_LW^o_Q$	1	5450.0	5450.0	1.9
$T_QW^o_L$	1	4016.0	4016.0	1.4
$T_QW^o_Q$	1	14.0	14.0	0.0
L_L	1	1451761.0	1451761.0	499.6
L_Q	1	978.3	978.3	0.3
C_LL_L	1	49203.0	49203.0	16.9
C_LL_Q	1	165.0	165.0	0.0
C_QL_L	1	12716.0	12716.0	4.4
C_QL_Q	1	9445.0	9445.0	3.2
T_LL_L	1	117295.0	117295.0	40.4
T_LL_Q	1	102.3	102.3	0.0
T_QL_L	1	4656.8	4656.8	1.6
T_QL_Q	1	6.5	6.5	0.0
$W^o_LL_L$	1	26.9	26.9	0.0
$W^o_LL_Q$	1	26.5	26.5	0.0
$W^o_QL_L$	1	10347.0	10347.0	3.6
$W^o_QL_L$	1	3392.0	3892.0	1.3
...	..	.....	...	...
Error	12	35887.0	2906.0	
Total	92	2699882.0		

$F^o(1,12,\ 0.05) = 4.75$

The regression coefficients a_i and b_i of equations (28) and (29) express the effects of parameters z_i and x_i on the experimental response (drying time t), while the coefficients a_{ij} and b_{ij} express the effects of the interactions z_iz_j and x_ix_j. The coefficients a_i and b_i are estimated by the least squares method (equation 13). The calculations are simplified if the matrix (X'X) is diagonal (i.e. all the non-diagonal terms are zero). A diagonal matrix (X'X) can be obtained if the corresponding design table of the experimental design is orthogonal

$$\left(\sum_{i=1}^{n} x_{\lambda i}x_{\mu i} = 0,\ \lambda \neq \mu\right).$$

Both the 3^3 and 3^4 design tables were transformed into orthogonal tables by simple calculations (Davies, 7).

For a diagonal matrix (X'X), the estimators of the regression coefficients are calculated by the following simplified relationship:

$$\hat{b}_k = \frac{\sum_{i=1}^{n} x_{ki}y_i}{\sum_{i=1}^{n} x_{ki}^2} \tag{30}$$

The calculated estimators of the regression coefficients ($\hat{a}_k$ and $\hat{b}_k$) are given in Table (10). The two linear models (equations 28 and 29) were tested for significance of their coefficients using the statistical analysis, described by the relationships 14, 15, 16 and 17. The results of the analysis are shown in Table (11).

Table 10. Estimated Regression Coefficients $\hat{a}_i$ and $\hat{b}_i$ and Confidence Limits for the 3^3 and 3^4 Designs

Coefficient $\hat{a}_i$	Coefficient $\hat{b}_i$
$\hat{a}_o = 242.9 \pm 7.9$	$\hat{b}_o = 269.7 \pm 13.1$
$\hat{a}_1 = -91.7 \pm 9.7$	$\hat{b}_1 = 41.9 \pm 16.0$
$\hat{a}_3 = 140.3 \pm 9.7$	$\hat{b}_{11} = -35.8 \pm 27.7$
$\hat{a}_{11} = 17.4 \pm 16.8$	$\hat{b}_2 = -109.1 \pm 16.0$
$\hat{a}_{22} = 28.5 \pm 16.8$	$\hat{b}_{22} = 29.7 \pm 27.7$
$\hat{a}_{12} = -15.4 \pm 11.9$	$\hat{b}_{12} = -33.0 \pm 19.6$
$\hat{a}_{13} = -60.7 \pm 11.9$	$\hat{b}_{33} = 29.2 \pm 27.7$
$\hat{a}_{223} = 42.9 \pm 20.9$	$\hat{b}_{13} = -20.4 \pm 19.6$
$\hat{a}_{112} = 21.2 \pm 20.6$	$\hat{b}_4 = 154.0 \pm 16.0$
$\hat{a}_{223} = -36.4 \pm 20.6$	$\hat{b}_{14} = 37.0 \pm 19.6$
$\hat{a}_{1123} = 25.7 \pm 25.2$	$\hat{b}_{24} = -57.1 \pm 19.6$

Table 11. Statistical Tests for Fit of Experimental Data to the Linear Models (Eq. 31, 32).

I. Regression Coefficients $\hat{a}_i$ (3^3 Design)

Variance	Degrees of Freedom	Sums of Squares	Mean Squares
R	43	39147	912
E	27	21835	809
C	16	17362	1085

F^o (16,27, 0.05) = 2.06
Mean Squares C/E = 1.34 < F^o (2.06)

II. Regression Coefficients $\hat{b}_i$ (3^4 Design)

Variance	Degrees of Freedom	Sums of Squares	Mean Squares
R	69	240786	3490
E	12	34877	2906
C	57	205903	3612

F^o (57,12, 0.05) = 2.38
Mean Squares C/E = 1.2 < F^o (2.38)

The data of Table (11) suggest that the H_0 hypothesis must be accepted at the 5% significance level, i.e. the linear model fits the experimental data adequately ($F_{calc} < F^o$ (Tables)). Therefore, both models are acceptable for the experimental drying data.

The confidence limits of the regression coefficients of both models were calculated according to the relationships (18) and (19), and they are given in Table (10). It is shown that the confidence limits of the coefficients of the mean drying time and the main effects T_L, C_L and L_L are close to the estimated values, while the variation in the interactions is much wider.

The two linear models for estimating the drying time (t_S and t_{SG}) for the 3^3 and 3^4 designs are represented by the following equations:

$$t_S = 242.9-91.7z_1+140.3z_3+17.4z_{11}+28.5z_{22}- - 15.4z_1z_2-60.7z_1z_3+21.2z_{11}z_2+ + 42.9z_{22}z_3-36.4z_2z_{33}+25.7z_{11}z_2z_3 \quad (31)$$

$$t_{SG} = 269.7+41.9x_1-109.1x_2+164.0x_4-35.8x_{11}+ + 29.7x_{22}+29.2x_{33}-33.0x_1x_2-20.4x_1x_3 + + 37.0x_1x_4-57.1x_2x_4 \quad (32)$$

DISCUSSION AND CONCLUSIONS

Starch gels were found to be convenient experimental materials for the numerous experimental measurements needed in a statistical factorial design. Samples of uniform composition and dimensions can be prepared, so that changes in the measured response t (drying time) can be attributed to changes of the process parameters and their interactions. The statistical method can be applied to more complex agricultural, food, and industrial products, provided that the experimental samples are prepared as uniform as possible.

Both experimental designs (3^3 and 3^4) proved their usefulness in the analysis of a complex physical process, such as air-drying of hygroscopic materials. The 3^3 design with two replications provided all the data meeded for: (1) an analysis for significance of the 3 main parameters, temperature (T), moisture content (W^o), and sample thickness (L), and their interactions; (2) an evaluation of the effect of the linear and quadratic components of these parameters on the drying time, and (3) the development of a linear mathematical model for the drying time as a function of 10 significant effects and interactions. The 3^4 design required a larger number of experimental measurements, and it provided additional information on the effect and interactions of glucose content of the gel (C). A fractional factorial design could perhaps provide similar information with fewer experiments and less computations than the full 3^4 design.

The strong linear effects of sample thickness L_L (positive) and temperature T_L (negative) are in agreement with the mechanism and the practice of drying. Here, the positive effect means an increase in drying time, while the negative effect has the opposite meaning. The significant positive effect of glucose content (C_L) has been observed in drying experiments and applications. The presence of glucose in a starch gel may change the physical structure of the polymeric material, resulting in a decrease of the diffusivity of water, especially at lower temperatures (Raouzeos and Saravacos, 5).

Significant quadratic effects were detected with T_Q (positive), C_Q (negative), and W_Q^o (positive). The presence of quadratic terms in the two regression models means that the relationship between drying time and the parameters of the process is more complex than linear. The linear effect of W_L^o and the quadratic effect of L_Q were found insignificant, contrary to drying theory. This may be explained by the strong two-factor interactions of these parameters, which may overshadow the main effects. The weak effect of W_L^o may be attributed to the high initial moisture contents of the gels used in this work. Some samples (91% moisture) developed cracks during drying, resulting in shorter drying times, and thus reducing the significance of W_L^o.

Significant two-factor interactions were detected between temperature/thickness ($T_L L_L$), moisture content/thickness ($W^o L_L$, $W_L^o L_Q$), glucose content/thickness ($C_L L_L$), and glucose/moisture content ($C_L W_L^o$), suggesting that diffusion of moisture is a controlling factor in the drying of gels.

The two mathematical models derived from the experimental data (equations 31 and 32) give close estimations of the drying time (min) of starch gels, within the range of values of the parameters used in this work. A limitation of the statistical models, compared to physical models, is that they cannot be used for extrapolation of the experimental data.

REFERENCES

1. Keey, R.,1972, "Drying Principles and Practice", Pergamon Press, Oxford.
2. Krischer,O.,1963, "Die Wissenschaftlichen Grundlagen der Trocknungstechnik", Springer Verlag, Berlin.
3. van Arsdel, W., Copley M., and Morgan A., 1973, "Food Dehydration", Avi.Publ. Co., Westport, Connecticut.
4. Saravacos, G.,and Charm,S.,1962, " A study of the mechanism of fruit and vegetable dehydration", Food Technology, 16, 78-81.
5. Raouzeos,G., and Saravacos, G., 1982, "Air-drying characteristics of starch gels", Proceedings of the 3rd International Drying Symposium, Vol. 1, 91-99, Birmingham.
6. Montgomery,D.,1976, "Design and Analysis of Experiments", John Wiley and Sons, New York.
7. Davies,O., 1967, "The Design and Analysis of Industrial Experiments", Oliver and Boyd, London.
8. Hicks, C., 1964, "Fundamental Concepts in the Design of Experiments",Rinehart and Winston, U.S.A.

DRYING CHARACTERISTICS OF LUCERNE

L. Imre, K. Molnár, S. Szentgyörgyi

Technical University Budapest
Hungary

ABSTRACT

For the purposes of the numerical analysis of the drying process of lucerne under different initial and boundary conditions the adsorption and desorption isotherms, the heat transfer coefficient, the mass transfer coefficient have been determined by experimental measurements. The leaf-separating force in the function of the moisture content and the pressure drop on a static lucerne layer with different thicknesses have also been investigated. The results have been compared with that of other authors.

NOMENCLATURE

- a = specific surface, m^2/m^3
- A = "empty" cross section, m^2
- B,C = constants
- c = specific heat, J/kg K
- F_e = phase-contact surface, m^2
- Δh = heat of evaporation, J/kg
- h = heat transfer coefficient W/m^2 K
- H = height, m
- k = heat conductivity, W/m K
- $\dot{m}$ = mass flow rate, kg/s
- M = mass of a mol, kg/kmol
- N = rate of drying, kg/m^2s
- P = pressure, Pa
- p = partial pressure, Pa
- q = heat flux density, W/m^2
- t = time, s
- T = temperature, K
- V = volume, m^3
- X = water content of the material /dry basis/; mass-ratio, kg/kg
- Y = absolut water content of the air, dry basis, kg/kg
- w = gas velocity based on "empty" cross section, m/s

Greek Symbols

- Δ = sign of difference
- φ = relative humidity,
- μ = dynamic viscosity, Ns/m^2
- ρ = density, kg/m^3
- σ = evaporation coefficient, kg/m^2s

Dimensionless numbers

- Le = Lewis number
- Nu = Nusselt number
- Pr = Prandtl number
- Re = Reynolds number

Subscripts

- a = material
- f = surface
- G = gas
- h = heating up ratio
- l = leaf
- lg = logaritmic mean
- n = wet
- s = stalk
- v = volumetric
- T = for volume
- $*$ = for equilibrium
- wb = wet bulb

INTRODUCTION

The hay produced from lucerne by drying is an important fodder of stock-breeding farms. The value of the fodder is determined by its initial nutritive content /carotine, proteine etc./.

The leaves and the stalks of lucerne contain the nutritive components in differing quantities. The nutritive content of the leaves are higher than that of the stalks. The high fibre-content of the stalks is necessary for the digestion of multi-maws animals.

If the drying process is not adequately controlled an overdrying of the leaves may occur resulting in falling-off of the leaves. Such a hay is a fodder of reduced value. To avoid the overdrying of leaves it is necessary to analyse the drying process of the leaves and of the stalks separately inside the dryer.

The drying characteristics of the leaves and of the stalks differ from each other. For the numerical analysis of the drying process of lucerne a two-component model has been proposed [1,2]. To the computation the values of the drying characteristics and the leafseparating forces in the function of the moisture content of the material have been determined by experiments.

DETERMINATION OF THE LEAFSEPARATING FORCES

After having mowed, lucerne starts drying inmidiately on the meadow. For energy reasons, on one hand, it may be desirable to dry on the swath to a certain extent until the moisture content is descreased by about 15-20 per cent [7,10].On the other hand, the control of the drying on the swath would be very difficult /e.g. by turning over the swath/, taking into consideration the far-reaching estates harvested during a day and the uncertainties of the weather forcast in some places of the world. In the case of natural drying on the swath, the top layer of the heap is exposed to direct radiation of the Sun and the effect of the wind. In bright and windy weather conditions the drying of the top layer is very quick and, after some hours may reach the water content critical in the respect of leaf-separating. In the case when the critical moisture content according to the leaf separating occurs in the top layer a large amount of the leaves fall down during the gathering up and get lost reducing the value of the fodder [9,11]. Some attempts have been made to avoid the leaf separating, e.g. by atomizing some water on the top layer of the swath before gathering up [3].

Leaf separating can also be occur in the bottom layers of a static bed dryer, when the moisture content of the drying air is very low.

The leafseparating forces in the function of the moisture content of the leaves have been investigated by laboratory experiments for different species of lucerne planted in Hungary and taking samples from the aceretions 1, 2, 3 resp.

The lucerne used for the measurements was slowly dried on trays. The stem of each lucerne plant was fixed in a press and its upper leaves in a device joined to a weight-loaded dynamometer. Then loading was increased until breaking. 5-10 leaves from each plant were torn off. Moisture content of the separated leaves was determined in vacuum dryer cabin, in well known way.

The results obtained showed fairly wide scattering even for the same plant [8]. Reason for that is, in the first line, that the size of the peduncles is different. Results gained for different sorts of lucerne /1st. aceretion/ are shown in Fig.1.

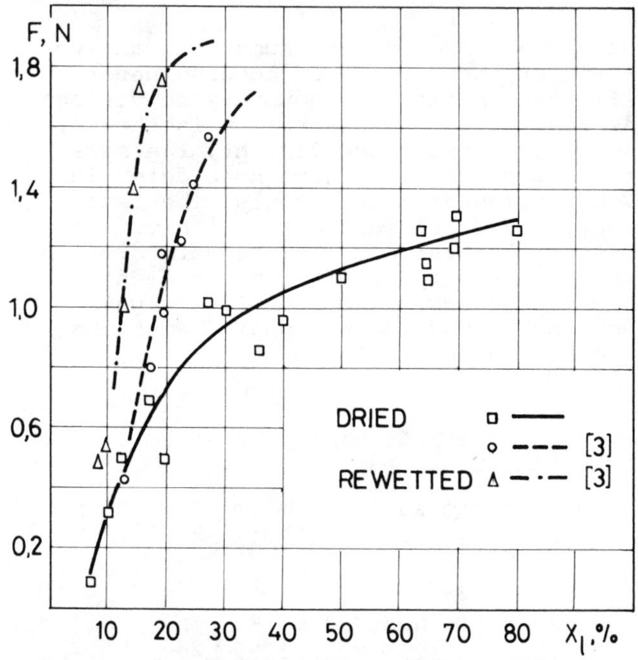

Figure 1 Effect of drying and rewetting on the leafseparating force of lucerne

The figure presents also results measured in the course of drying as well as after re-wetting, gained by other authors [3]. It can be stated that after re-wetting a higher separating force was measured with the same moisture content than in the course of drying /hysteresis/. A probable reason for that is that during drying the outer surface layer of the leaf peduncle is the driest, while after re-wetting the outer layer has the highest moisture content. When breaking off the leaves a greater portion of the loaded cross section falls on the outer surface layer. It can be stated from the figure that the value of the leaf separating force under X^{*} = 0,2 kg/kg /dry basis/ begins to decrease strongly.Therefore it is advisable to take this limit value into consideration on the basis of the desorption isotherm of the leaf, when controlling the state of the drying air.

DETERMINATION OF SORPTION ISOTHERMS

The isotherms for adsorption and desorption of lucerne have been separately determined for leaf and stem. A static method applied in constant space tension was used for the determination [4, 5]. For the purpose of control, some measurements were taken with dynamic method. Sorption isotherms of the stem are shown in Fig.2; those of the leaf in Fig.3.

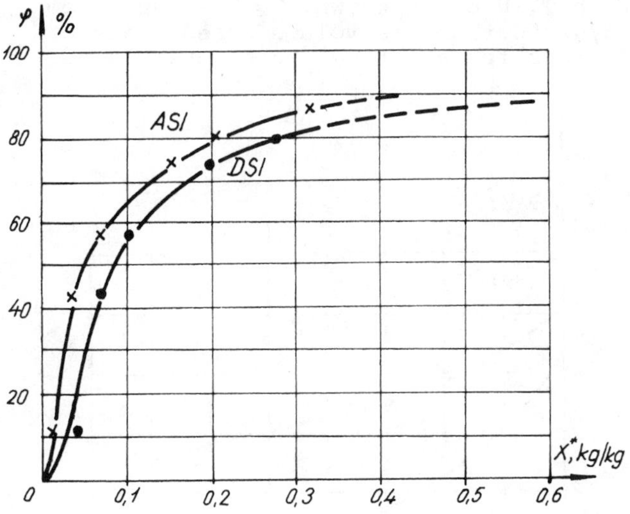

Figure 2 Adsorption /ASI/ and desorption /DSI/ isotherms of lucerne stalk at 25 °C.

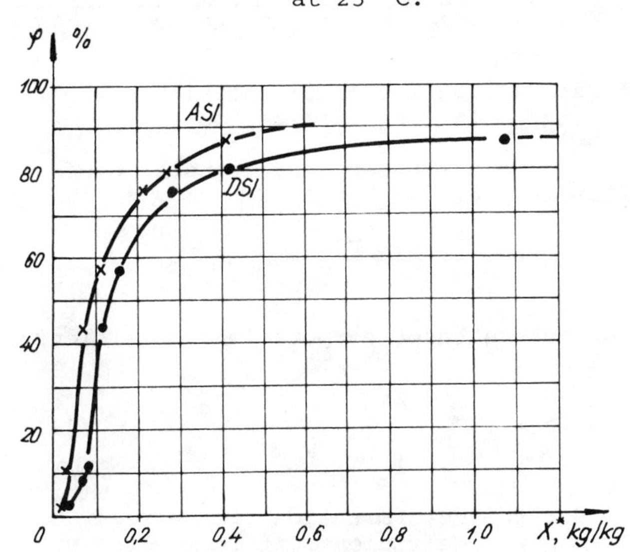

Figure 3 Adsorption /ASI/ and desorption /DSI/ isotherms of lucerne leaf at 25°C

For mathematical simulation of the drying process the stages of the sorption isotherms can be approximated with the following power functions

for the leaf: $0{,}14 \leq X_l \leq /X_l/_{cr} = 1{,}985$

$$\varphi_l = 0{,}89\, X_l^{0{,}17}$$

$$0{,}06 \leq X_l \leq 0{,}14 \qquad /1/$$

$$\varphi_l = 59\, X_l^{2{,}3}$$

for the stem: $0{,}14 \leq X_s \leq /X_s/_{cr} = 1{,}29$

$$\varphi_s = 0{,}96\, X_s^{0{,}16}$$

$$0{,}06 \leq X_s \leq 0{,}14 \qquad /2/$$

$$\varphi_s = 3{,}16\, X_s^{0{,}766}$$

DETERMINATION OF TRANSFER CHARACTERISTICS

Drying operation of lucerne in barns proceeds in high static bed, with state determinants changing in time and place.

For the simulation of different dryint strategies it is necessary to know the evaporation coefficient, the heat transfer coefficient and the pertinent value of the so called specific phase contact surface, separately for the stem and for the leaf. Moisture conductivity within the stem is not dealt with here [12].

From the point of view heat and mass transport phenomena the surfaces of the solid overlapping each other in the bed cannot be taken into consideration. In the static bed of lucerne with initial moisture content an initial specific phase-contact surface is forming, its value depends on the mass density of the bed. In the course of drying the bed collapses thus changing its mass density /porosity/ and the specific phase contact surface as well. In a drying barn the superposing after repeated mawings will also condense the bed. In view of the above reasons it is expedient to elaborate an experimental examination method which renders the determination of real transport factors and the volume transport factors belonging to the same air condition also possible.

PRINCIPLE OF DETERMINING VOLUME BASED TRANSPORT COEFFICIENTS

If a short lucerne bed is constructed so that $\dot{m}_G$, H, $Y_{G,in}$ and $T_{G,in}$ are set at constant value, then, after a short time, $Y_{G,out}$ will also be constant and will stay constant as long as the moisture content of the lucerne reaches the critical value at the spot in the bed where air enters. In this interval the volume evaporation coefficient can be calculated from the measured steady state parameters:

$$\sigma a = \frac{\dot{m}_G}{A_q H} \frac{Y_G - Y_{G,in}}{\Delta Y_{\ell g}} \qquad /3/$$

where

$$\Delta Y_{\ell g} = \frac{/Y_{w,G} - Y_{G,in}/ - /Y_{w,G} - Y_{G,out}/}{\ln \frac{Y_{w,G} - Y_{G,in}}{Y_{w,G} - Y_{G,out}}} . \qquad /4/$$

The determination of the volume based heat transfer coefficient can be realised in the same way, noting that in the case of adiabatic saturation $T = T_{wb}$ is constant

$$ha = \frac{\dot{m}_G c_{w,G}}{A_q H} \frac{T_{G,in} - T_{G,out}}{\Delta T_{\ell g}} \qquad /5/$$

where

$$\Delta T_{\ell g} = \frac{/T_{G,in} - T_{w,G}/ - /T_{G,out} - T_{w,G}/}{\ln \frac{T_{G,in} - T_{w,G}}{T_{G,out} - T_{w,G}}} . \qquad /6/$$

In the stage after adiabatic saturation the heat transferred from the air to the material is used for the evaporation of humidity and for warming the wet material:

$$q = \bar{N} \Delta h + q_{sens.} \; ,$$

$$\frac{\Delta Y_G}{\Delta T_G} = - \frac{\bar{N} c_{w,G}}{\bar{N} c_{w,G} + q_{sens}} . \qquad /7/$$

From the above relationship it can be seen that when free water runs out, $\Delta Y_G / \Delta T_G$ decreases. In this stage $\bar{T}_f > T_{wb,in}$ and $\varphi_f < 1$; the above statements are illustrated in Figure 4.

It can be supposed that in this stage of drying

$$\bar{X} \cong \bar{X}_f = X^* \quad \text{and} \quad \bar{T}_a = \bar{T}_f .$$

With the above, the absolute humidity content of the air is

$$Y_f = \frac{M_v}{M_G} \frac{p_{vt}/\bar{T}_f/}{P} \varphi /\bar{T}_f, X^*/ \qquad /8/$$

where $\varphi /\bar{T}_f, X^*/$ is the function of the desorption isotherm.

Thus

$$\frac{\Delta Y_G}{\Delta T_G} = \frac{\frac{M_v}{M_G} \frac{p_{vt}/\bar{T}_f/}{P} \varphi /\bar{T}_f, X^*/ - \bar{Y}_G}{\bar{T}_f - \bar{T}_G} \qquad /9/$$

In the above equation only $\bar{T}_f$ is unknown. Since the right and the left side are identical, T_f can be determined, by repeated trial.
When $\bar{T}_f$ becomes known, Y_f is given from /8/, and thus the volume based transport factors for any time can be calculated.

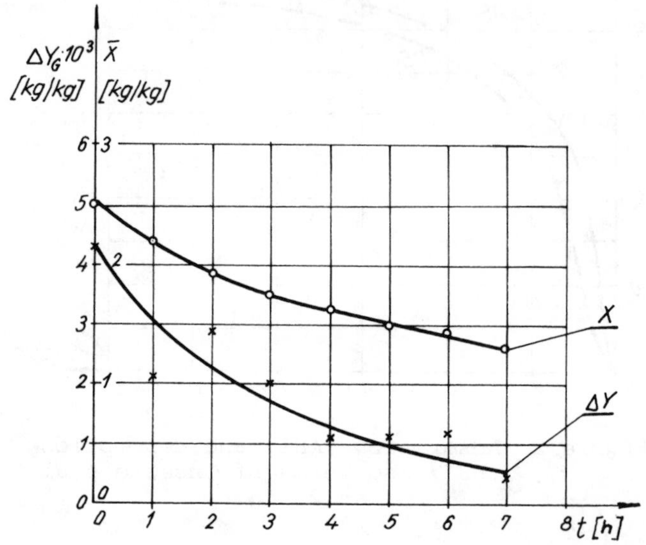

Figure 4 Saturation of the air in the falling rate period

The volume based heat transfer coefficient /ha/ is:

$$ha = \frac{\dot{m}_G c_{w,G}}{A_q H} \frac{\Delta T_G}{\bar{T}_f - \bar{T}_G} . \qquad /10/$$

The volume based evaporation coefficient /σa/ is:

$$\sigma a = \frac{\dot{m}_G}{A_q H} \frac{\Delta Y_G}{\bar{Y}_f - \bar{Y}_G} . \qquad /11/$$

For the determination of the real transport coefficients it is necessary to carry out through-flow drying investigations where the size of the specific phase contact surface is known exactly. E.g. 4-5 pieces of lucerne plants whose surface has been previously determined are tied on a rod and dried so that no covered solid surface exists. Following the method written above the "real" transport coefficients can be determined.

Determination of specific transport surfaces

From the point of view of heat transfer, the transport surface can be calculated from the following relationship [13]:

$$a = \left[\frac{h.a}{k_G} /1- \frac{\Delta V_a}{\Delta HA_q}/ \frac{1}{A\ Pr^C} /\frac{\mu_G\ A_q}{\dot{m}_G}/^B\right]^{\frac{1}{2-B}}, \quad /12/$$

where the volume based heat transfer coefficient is known from the foregoing as a function of H. In relationship /12/ A, B, C are the constants of the constitutional equation serving for the calculation "real" heat transfer coefficient:

$$Nu = ARe^B Pr^C \quad /13/$$

where

$$Nu = \frac{hV}{k_G F_e} \quad /14/$$

$$Re = \frac{m_G V}{A_q \mu_G F_e} \quad /15/$$

Constants A, B, C can be determined with the 4-5 pieces examination described above.

For a two-component simulation model of drying [1] the contact surface referred to volume unit a must be divided into two parts in the following way

$$a = \frac{F_{es}+F_{e\ell}}{V} = \frac{F_{es}}{V} + \frac{F_{e\ell}}{V} = a_s + a_\ell . \quad /16/$$

The specific geometrical surface of lucerne leaves are considerably larger than that of the stem; therefore the effect of mistake made on determining a_s is of minor importance. Further, it can be supposed that the phase boundary surface of the stem does not differ considerably from the geometrical one, while the leaves, owing to bed collapsing, are overlapping considerably. Thus the phase boundary surface of the leaves may be much smaller than their geometrical surface:

$$a_\ell = a - /a_s/_{geometrical} \quad /17/$$

The changing of the mass transfer coefficient and that of the specific surface can be followed with the changing of the psychrometric ratio

$$\frac{ha}{\sigma a c_{w,G}} = Le^{1-C} . \quad /18/$$

Results of experimental measurements

An experimental lucerne dryer was constructed where all the state characteristics of the air entering and leaving the bed could be continuously measured and registered. Mass density of the bed and height were varied. According to the description above first the constants A, B, C of the constitutive equation for calculating the real transport coefficients were determined.
The constitutive equation for calculating heat transfer coefficient is the following, Re < 200:

$$Nu = 6{,}26.10^{-4}\ Re^{0{,}84}\ Pr^{1/3}. \quad /19/$$

Further on, in the full drying cycle of the lucerne bed the volume based heat transfer coefficients were determined; then knowing /19/ and with consideration to relationship /17/ the specific phase contact surface a was also determined. Results are presented in Figure 5.

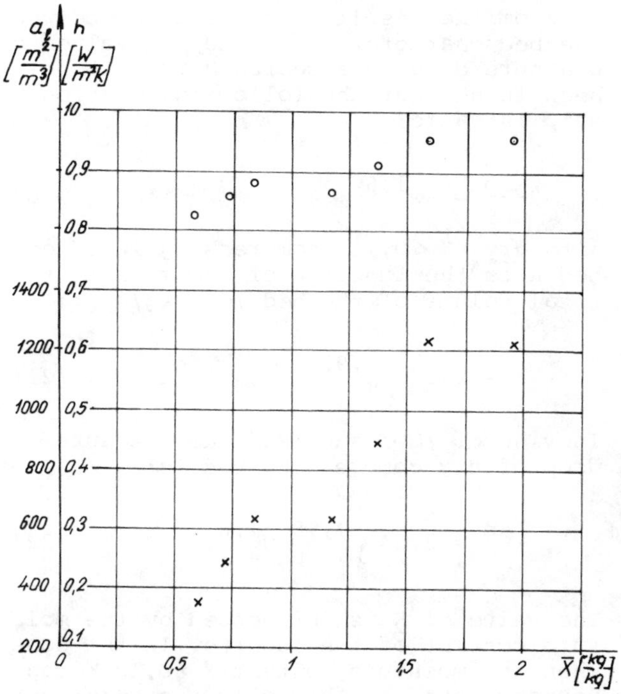

Figure 5 Heat transfer and specific phase contact surfaces in the function of the moisture content /O: h, x: a /

It can be seen from the figure that the heat transfer coefficient can be taken as practically constant during drying, while the specific phase boundary surface decreases considerably.

In the course of measurement the psychrometrical ratio was also checked with the measurement of volume based evaporation coefficient and volume based heat transfer coefficient, and according to /18/ values in the neighbourhood of unit have been received. Consequently the heat transfer surface and the evaporation phase contact surface are practically equal.

AIR PRESSURE DROP ON THE BED

For the design and simulation of lucerne dryers it is necessary to known the hydrodynamic characteristics of the bed, first of all function of the pressure drop. As has been pointed out, in the process of drying there are changes in the porosity of the bed and superposings may occur because of the periodic mowing.

To investigate the above phenomena an experimental equipment was built where the dried lucerne bed was undergoing an ever increasing weight load and the compression of the bed was measured as well as the pressure drop of the air on the bed.

From the results a correlation between the bed characteristics and the values of pressure drop were searched for. It has been found that the following relationship is valid:

$$\frac{\Delta p}{H} = Kw^{1,6} \quad . \qquad /20/$$

With dry /X < 0,12/ compressing lucerne bed K is the function of the mass in the total volume of the bed /ϱ_T, kg/m^3/:

$$K = 2,34\,\varrho_T^{1,98} . \qquad /21/$$

In view of /20/ and /21/ the pressure drop of dry compressing bed is:

$$\frac{\Delta p}{H} = 2,34\,\varrho_T^{1,98}\, w^{1,6} . \qquad /22/$$

The value of K is influenced by the moisture content of the lucerne. In the case when the moisture content X > 0,12, K can be calculated by the following relationship:

$$K = 1718\, e^{0,51X} \quad . \qquad /23/$$

Typical pressure drop characteristics are presented in Figure 6, and compared with the results of Scheuermann [14]. The parameter is the surface load of the bed.

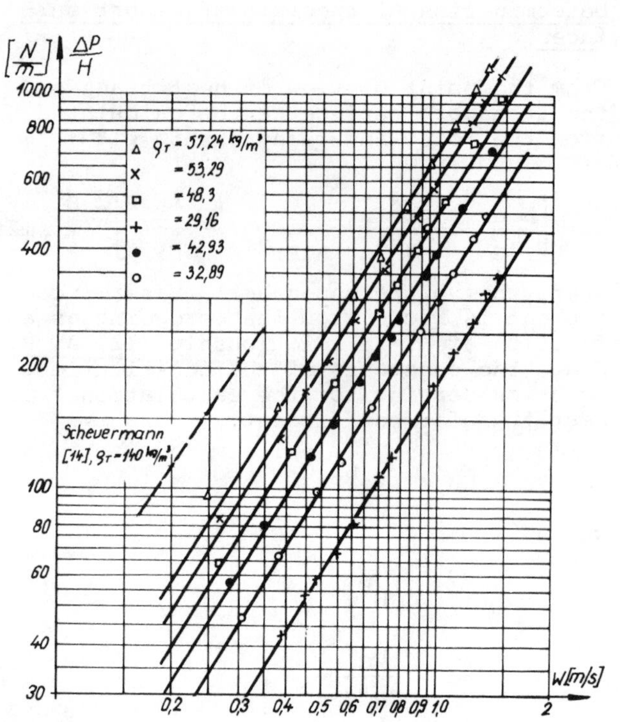

Figure 6 Pressure drop of a static bed of lucerne

REFERENCES

1. Imre,L.; Molnár,K.: Two component model of a lucerne convective bed dryer Proc.3rd Int. Drying Symposium /ed.by J.C. Ashworth/, Drying R. Ltd, Wolverhampton, Vol.2. pp.73-78, 1982.
2. Imre,L.; Farkas,I.; Kiss,I.L.; Molnár, K.: Numerical analysis of solar convective dryers, 3rd Int. Conference on Num. Methods in Thermal Problems, Seattle, USA, 1983.
3. Lack,A.; Fenyvesi,L.: On the mechanical effects causing leaf-separation of lucerne /in Hungarian/. Mez.Gép. Tan. MÉM Publ. pp. 1-8, 1981.
4. Imre,L.: The measurement of equilibrium relative humidity, Periodica Polytechnica El.Eng. Part 1. 8. pp. 181-191, Part 2. 8. pp.227-242, 1964.
5. Imre,L.: Handbook of Drying /in Hungarian/, Chapter 20.1., Müszaki Könyvkiadó, Budapest, 1974.
6. Likov,A.V.: Theory of Drying /in Hungarian/, Nehézipari Könyv- és Folyóiratkiadó Vállalat, Budapest, 1952.
7. Tullberg,J.N.; Angus,D.E.: Increasing the Drying Rate of Lucerne by the use of chemicals, Journ.Austr.Inst. Agric. Science, pp.214-215, 1972.

8. Shepherd,W.: Experimental Methods in Haymaking Trials, Aust.J.Agric.Res., pp. 27-38, 1958.

9. Shepherd,W.: The susceptibility of hay species to mechanical damage, Aust.J. Agric. Res., pp. 783-796, 1961.

10. Barrington,G.P.; Bruhn,H.D.: Effect of the Mechanical Forage-Harvesting Devices on Field-Curing Rates and Relative Harvesting Losses, Transactions of the ASAE, pp. 874-878, 1970.

11. Pedersen,T.T.; Buchele,W.F.: Drying Rate of Alfalfa Hay, Agricultural Engineering, pp. 86-89, 107, 1960.

12. Bagnall,L.O.; Millier,W.F.; Scott,N.R: Drying the Alfalfa Stem, Transactions of the ASAE, pp. 232-245, 1970.

13. Szentgyörgyi,S.; Molnár,K.: Drying Characteristics of Lucerne /Report, ed. by L.Imre/, ITES TECHN. UNIV. Budapest, 1982.

14. Scheuermann,A.: Der Strömungswiderstand bei der Belüftungstrocknung von blattreichem, dicht lagerndem Heu Grundl.Landtechn.Bd.16. /1966/ Nr.4.

Experience in drying coal slurries

F. Baunack

Deutsche Babcock Anlagen AG, Krefeld, Deutschland

Abstract

Experience in drying coal slurries

Coal slurry is obtained as a residue from the preparation of raw coal; it is thermally dried after mechanical water removal. The slurry varies both in water content, which may be between 10 and 40 %, and in ash content, which may range from 25 to more than 60 %, and its consistency varies accordingly.

The paper describes a direct-heated rotary drying plant for such slurries which handles a throughput of approx. 200 tons/hr and evaporates approx. 35 tons/hr of water. The operating results are discussed, and it is shown that such plant can be built for most difficult slurries and high capacities. It is particularly suited for slurries that are to be transformed into a non-dusting dried product.

Other concepts which include the fluidized-bed technology might be considered but have not yet been realized. They could be furnished with a second drying stage for the production of coal dust if required. If used for the production of hot gas, the fluidized-bed combustor would have the great advantage of burning high-ash slurries.

Introduction

For about two years now, a drying plant has been in operation in the French mining area, which delivers its dried fuel containing 7 % water to a power station. The coal slurries dried in this unit are coming from washeries and contain between 10 and 35 % water and between 25 and 60 % ash. As a consequence, these slurries differ very much in their consistency causing problems in the feeding equipment, but also in the dryer proper, due to coarse and hard agglomerates that may develop.

For this reason special types of internals and distributing facilities are required.
The required throughput is high, ranging from 120 to more than 200 tons/hr. To cover this duty, the plant had to be of considerable size and is hence slowly regulating. Since on the other hand, variety and rate of the coal slurry handled are subject to frequent changes for internal operational reasons, a highly efficient regulating and control system is required to cover both the changing throughput and meet the required narrow limits of residual moisture.

It is very likely that a fluidized bed dryer would also do the job if provided with a feeding system specially suited for slurries. In a second stage that could be added, such a plant might also produce coal dust if required. A plant of this kind will be described.

Coal slurries

The ash content of run-of-mine coal has increased contininonsly during the last two decades. For technical and economical reasons it is no more possible to separate the different coal qualities in the mine and to mine only the high quality coal.

On the other hand the mechanisation of coal excavation does not allow to follow correctly the boundary of the coal deposit. Such, also inert rock will be mined.

The part of coal which has to be washed is increasing steadily, in Central Europe nearly all run-of-mine coal goes to a preparation plant.

Mining mechanicly produces more fines, and the fines are separated from the coarser size by wet screening. Great amounts of slurry are produced, a part of it concentrated with froth flotation, thus producing a coal concentrate and a slurry with high ash content, another part used as solid fuel without further washing.

Those slurries have an ash content from 20 to 75 %, the moisture content after mechanical drying may be between 12 to 40 % and the granulometry is very different. Very fine granulated slurries may be found, but also slurries with a lower percentage of finest material. The consistence of this slurries is very different. It is influenced by the amount and the nature of minerals in the slurry and the moisture content. Slurries can be found with the consistence of wet sand, slurries like a paste, but also slurries which have the character of a viscous fluid.

I would like to make some remarks about an installation, which is able to dry several kinds of slurry.

Problem and solution

The problem was to dry coal slurries from some washeries in order to use it as combustible for the steam generators of a great power plant. No modification of the coalmills should be necessary when running with dried slurry. Most of the slurries have high ash content, therefore drying is needed, but the dryproduct shall not be dusty, in order to avoid difficulties when handling it.
The second requirement was to produce as much pulverized coal as needed for heating the dryer. The capacity of the unit was fixed on 100 to 230 tons/hr.

Fig. 1 shows the characteristics of the slurries. They are very different in its water and ash content. It can be seen on Fig. 2 that the granulometry is very different too. As a result of these facts the consistence was extremly different, like wet sand ore like a viscous paste as extremes.

It is experience that those slurries produce dust only when the moisture content is below 7 %. Difficulties with the coalmills are created if it is higher than 10 %. Therefore the final moisture content was limited to 7 - 9 %. 35 tons/hr. water evaporation was to be guaranteed.

A directly heated rotary drum dryer was choosen. In order to get sufficient pulverized coal for the hot gas generator, the gas velocity schould be above 5 - 6,5 m/s. This means a great length in comparison to the diameter, in figures 4 m diameter, 33 m length, equipped with lifter plates of special contruction to avoid sticking on and in order to reduce big balls of slurry falling down during rotation of the drum to small pieces.

As a support for making small pieces we installed heavy chains, 5 m long, on 15 m length of the dryer.
Fig. 3 shows the dryer with the view to the end, Fig. 4 is the view in direction of the feed.

The slurry is fed to the dryer through a plate conveyor, Fig. 5;
the dried product falls on a belt conveyor, Fig.6.

The rotation of the dryer can be regulated continionsly between 0 - 6 rpm by means of a hydraulic gear unit.
Normally it is running between 4 - 6 rpm.

Dust precipitation of the stack gases is done by a electrostatic precipitator, Fig. 7. The dust is used as combustible for the hotgas-generator.

Results of operation

Fig. 8 shows some of the results of operation. Slurry No. 1 can be dried easily. It forms only small agglomorates within the dryer and it is transported quickly through the dryer. The power input to drive is relativly low.

Slurry No. 2, the ash content is nearly 40 %, agglomorates more, the retention time is greater, the power input too.

Slurry No. 3 has an extreme pasty consistence. It forms big balls of considerable strength, the surface dry and hard, the interior wet. This caused clogging of the dryer outlet. A normal operation was not possible. Only with a gas temperature at the outlet of more than 200 ^oC continuous operation could be realized, but this is not acceptable for safety reasons.

A normal operation was possible - with the normal and guaranteed capacity of slurry No. 3 by adding another kind of slurry of coarser granulometry, and in that case, higher ash content. A mixture of 1/3 slurry No. 4 and 2/3 slurry No. 3, only by feeding both slurries to the bunker, allowed the production of 165 tons/hr. dried product without difficulties.

As you may remark on Fig. 8 the final water content was at the lower limit in spite of the maximal ratation of 6 rpm and a gas velocity at the end of the dryer of 6 - 6,5 m/s. In all probability the capacity of that installation can be increased by increasing the rotation velocity of the drum and by increasing the massflow of the gas, in order to ensure not only the heat demand of the dryer, but also the demand on combustible for the oven.

The consumption of combustible was 2 - 5 % of the dried product. Nearly all the dust separated from the stackgas was used as pulverized coal. Therefore the dust load of the gas entering the precipitator can be estimated to 20 - 40 $g/m^3{}_n$. In the gas leaving the electrostatic precipitator max. 40 $mg/m^3{}_n$ have been found. The dust-precipitator was operating excellent, Fig. 9.

The old experience was confirmed, that the ash content of the dust is higher, sometimes more than 30 % higher than the ash content of the wet slurry. This may arise problems in a pulverized coal fired hotgas generator with slagging.

Most of the ash of a pulverized coal fired hot-gas generator is blown as fly ash through the dryer and will be found in the dust separated by the electrofilter.
To avoid 100 % recirculation of this ash it is useful to have more dust from the electrostatic precipitor than needed for firing, and thus to eliminiate the flyash.

Sometimes, decomposition of coal as a result of drying has been feared. As a matter of fact no lost of volatile matter will happen, if the coal is fed well distributed into the stream of hot gases; the temperature of the gas will be dropped very quickly. We could not find, that the dried product had less volatile matter than the raw slurry.

The heat consumption per kg evaporation, shown in Fig. 8 includes all heat losses of the system: the losses of the dryer itself, the losses of the hotgas generator and the sensible heat of the final product.
Fig. 10 shows, that the efficency of the system with 71 % is good.

The safety regulations for coal dryers are different from country to country, namely in respect of the allowed maximal oxygen content of the gases.

The installation which is presented in this paper was planned and operated without recirculation of stack gas. The oxygen content, calculated on dry-gas basis, was 15 - 17 %, calculated on wet gas basis 12 - 13,5 %. If the slurry is dried to 7 % and deposits of product are avoided, safe operation is possible. No difficulties or incinerations happend in the electrostatic precipitator with a very dry dust of only 1 - 2 % water content.

A rotary drum dryer surely is suitable for drying of coal slurries. Very different slurries can be handled in the same installation, and only very pasty and viscous slurries need to be mixed with recirculated dry product or with a slurry of more favourable consistence or the feed has to be done by a sling.

Control

It is difficult up to now to run such an installation with an automatic control, such a control is very expensiv and complicated. We have choosen a semi-automatic control: the mass of slurry, fed to the dryer, and the outlet temperature are held automaticly on preselected values by means of a balance resp. variation of the gas temperature at the dryer inlet controled by variation of pulverized coal stream to the burners. If given limitations are exceeded, the operator has to take action.

It is recognized that Semi-automatic control has long reaction time as a result of the inertia of the dryer system. It takes more than two hours to bring the system from irregulate operation back to regulate values of temperature ore final water content. This is a disadvantage.

On the other hand, it is able to control important variations of the feed, if the velocity of variation is not too great. Anyway, the variation of temperature at dryer inlet is considerable.

I think, that modern computerized control systems may offer in the next future more effective and cheap systems.

Fluidized bed dryer for slurries

Rotary drum drying is, as we have seen, suitable for coal slurries. But investment costs are high, as well as space requirement. Flashdryers are cheaper, automatic control is less complicate. But they need recirculation of dry product and mixing equipment for the abrasive product. Moreover it is not easy to produce a final non-dusting product with some surface humidity.

Possibly, that a combination of flashdryer and fluidized bed is a better solution.

The slurry is fed to the fluidized bed by a sling where a part of the hot gases is running through. There it is loosened, despersed and dried at the surface. The product bounces against a wall of movable suspended iron bars and falls into the fluidized bed.

The fluidized bed is operated with high gas velocity, between 5 - 10 m/s, in order to prevent layers on the bottom.
It can be expected that weak granules are formed within the fluidized bed.

A part of the slurry will be entrained by the gas. It can be separated mechanicly be means of cyclones from the gas.

It is possible to cool and evaporate the product at the end of the fluidized bed, ore on a screen and to realize grinding and final drying to pulverized coal in a following system of hammermill and flashdryer.

That system has the advantage, that no recirculation of product is necessary, the gas-temperature at the inlet of sling, fluidized-bed and, if pulverized coal is to be produced, flashdryer is equal. Thus one hot gas generator is sufficient.

It can be a fluidized bed combustor, running with the product of the fluidized bed dryer as combustible and able to accept coal slurries with a very high ash content.

A fluidized bed dryer for the same capacity of the discussed rotary drumdryer 4 x 33 m, needs a bed of only 20 m^2, this means a diameter of 5 m.

I think, that such a system does not exist up to now. It would be of interest to test it, might be that it is less complicate and easier to operate than existing systems and is suitable for grat capacities.

Fig. 1 Characteristics of Slurry

Slurry No.	1	2	3	4	4+3
Moisture content%	21.9	18.4	29.0	9.5	19.0
ash%	25.6	38.3	32.6	50.9	45.3
Volatile matter%	26.7	26.0	26.7	22.0	23.5

Fig. 2 Size Distribution of Slurry (dry)

Slurry	Overflow%						
mm	2.4	1.2	0.9	0.63	0.4	0.1	0.075
1	1.3	4.4	7.8	14.5	23.9	52.5	57.0
2	11.8	20.5	25.1	31.2	34.0	66.3	70.3
3	0.6	1.8	3.2	5.6	9.6	28.2	33.2
4	35.1	50.5	56.9	63.7	66.7	90.9	93.5

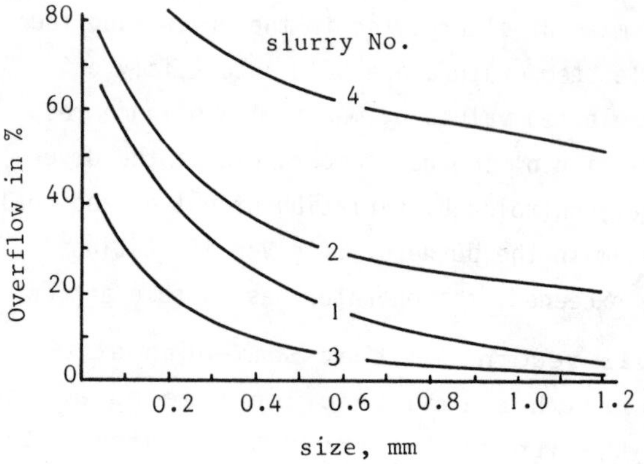

Fig. 3 Overflow as a Function of Particle Size.

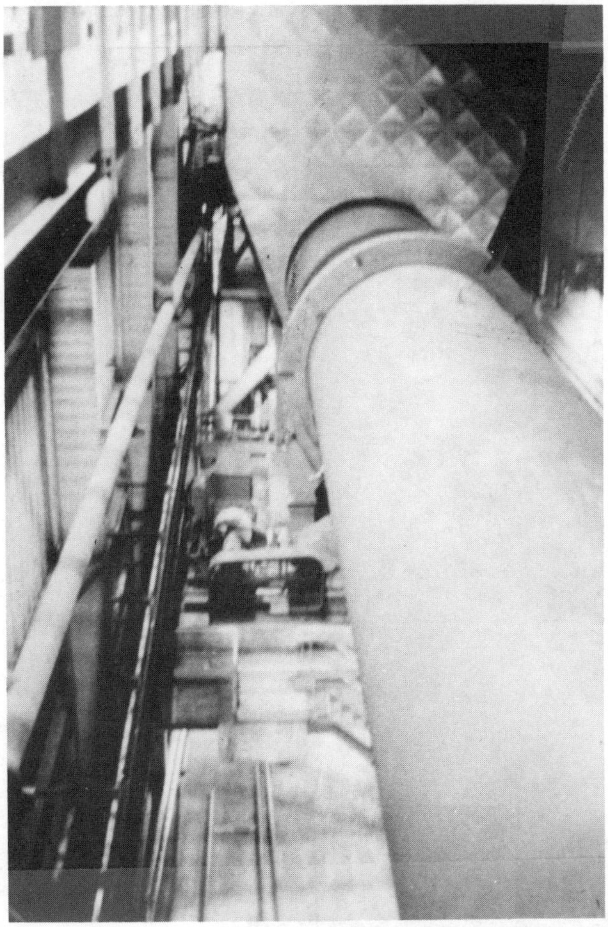

Fig 3 : Dryer :
View to outlet and drive

Fig 4 : Dryer :
View to feed bin

Fig. 5: Feed device and hot
Gas generator

Fig 6: Dried product outlet

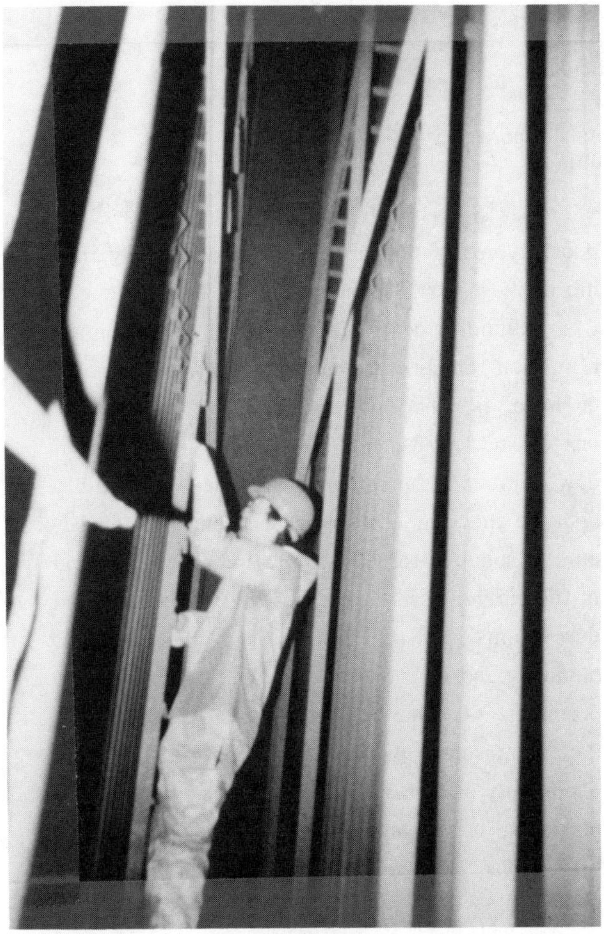

Fig 7 : Electrostatic precipitator
Internal construction

Fig 9 : Chimney and stack gas

Fig. 8 Results of Drying Tests

Slurry No.	1	2	4.3
dried product, t/h	208	193	165
initial moisture, %	20.5	18.4	19.0
final moisture, %	7.0	7.5	6.0
evaporation, t/h	35.3	25.8	26.4
gas temperature			
hot gas generator, ^{0}C	1100	840	1150
dryer inlet, ^{0}C	865	708	780
dryer outlet, ^{0}C	115	126	105
power consumption			
dryer, kW	210	239	245
electrostatic precipitator, kW	180	125	176
spec. heat consumption, kJ/kg	3850	4330	4250

Fig.10 Heat distribution

heat entering the system	%
combustible	92
air	3
slurry	5
heat leaving the system	%
ash	0,01
product	11
radiation	2
evaporation	71
sensible heat of dry stack gas	16

RECENT ADVANCES IN CHILLED MIRROR HYGROMETRY

John C. Harding, Jr.

General Eastern Instruments Corporation
Watertown, MA 02172

ABSTRACT

Improved accuracy and precision requirements relating to Humidity Measurements are being imposed by certain applications such as energy management and conservation, nuclear power, industrial process control and combustion emissions measurements. Some industrial process control applications require water content or total enthalpy measurement in a gas over a broad range of temperature and water vapor pressure. Measurement of Dew Point or Frost Point has long been recognized as a superior method of humidity determination since this is a fundamental measurement. Some limitation of this method, e.g. constraint to display of humidity in terms of dew point and susceptibility to errors induced by contaminants are addressed. Recent developments overcome those problems which have prevented realization of the full advantage inherent to the dew point measurement method of humidity determination. A microprocessor based hygrometer system is described that provides continuous long term measurement to within 0.5% in the absence of any recalibration.

INTRODUCTION

Most methods of humidity determination involve a secondary measurement; i.e., some quantity or parameter of a humidity sensitive device is measured. The moisture dependent physical expansion/contraction of a strand of hair or the impedance of an electrical device, for example, have often been used for relative humidity determination. These types of measurement methods require calibration and typically do not hold the original calibration characteristic. In virtually all cases the humidity sensitive device is sensitive to or even attacked by other gaseous compounds, is non-linear, is sensitive to ambient temperature and exhibits hysterisis. The resulting accuracy is poor, typically 3% or even worse initially, and more like 5 to 10 percent in the long term.

Humidity determination by condensation/sublimation temperature measurement, on the other hand, is inherently superior in terms of accuracy and long term stability. Determination of dew point is a fundamental measure of humidity because the temperature where condensation occurs is specific to the water vapor concentration in the carrier gas. In the absence of error in the dew point and air temperature measurement, humidity and/or total enthalpy can be determined precisely. Another important advantage is that a dew point sensor can be fabricated entirely from chemically stable or inert materials such as gold, platinum, teflon and stainless steel. There are limitations, however, associated with this technique which have historically created difficulties since the inception of the method more than a generation ago. The most significant of these are:

1. Contaminant induced measurement errors.
2. Constraint to display of humidity in terms of dew point.
3. The required accuracy and stability of temperature measurement imposed by the method is difficult to achieve and maintain.

A recently completed research and development program, the results of which are herein summarized, addresses these limitations.

The major objectives underlying this effort were to (1) develop a method and instrumentation which provides continuous automatic display of absolute or relative humidity to within 0.5%, traceable to the National Bureau of Standards, (2) to eliminate the need for periodic recalibration which is difficult and expensive, and (3) to provide a high level of reliability in harsh industrial environments.

PRINCIPAL OF OPERATION

Optical Dew Point Hygrometers detect the onset of condensation on a reflecting surface and maintain control of the surface at the temperature of condensation, while measuring the surface temperature. The Hygrometer system is composed of two basic subsystems, one which controls the mirror temperature so that it is maintained at the dew point of the sampled gas, the other which measures the mirror surface temperature as well as the prevailing ambient or gas temperature and pressure. The basic configuration is illustrated in Figure 1.

The operation of the mirror temperature control loop is described as follows: An offset current is applied to the temperature control amplifier. This offset current causes the thermoelectric cooler which is in thermal contact with the illuminated mirror to pump heat at its maximum rate thereby causing the mirror to slew downward in temperature. When the mirror temperature crosses the dew point, condensation occurs which in turn causes a reduction in photo current due to scattering of the incident light. The reduction in photo current continues until the offset in the control amplifier is essentially cancelled and the closed loop configuration reaches an equilibrium condition where the mirror is maintained precisely at the point where no additional evaporation or condensation occurs. In actuality water molecules are continuously being trapped and are escaping from the surface of the mirror, but the net level of condensate density does not change once equilibrium is established.

FUNDAMENTAL RELATIONSHIPS

Humidity can be expressed in terms of the ratio of volumes, pressures or weights or percent saturation. Each is suitable in particular applications but also subject to confusion and limitations. In the strict sense, vapor concentration in a carrier gas is the ratio of the mass of the water vapor to the mass of the total air mixture as expressed by the following:

$$G = \frac{M_v}{M_v + M_a} \qquad (1)$$

M_v = the mass of water vapor

M_a = the mass of the dry air in which the water vapor is suspended

Expressing humidity or moisture content as the ratio, r of the mass of water vapor to the mass of just the dry air rather than the total mixture is equally explicit and has historically been more popular:

$$r = \frac{M_v}{M_a} \qquad (2)$$

where r is the mixing ratio.

In accordance with Dalton's law of partial pressures which states that the total pressure exerted by a gas mixture is equal to the sum of the pressures that each constituent gas would exert, were it to occupy the same total volume. The mixing ratio can be expressed as follows:

$$r = \frac{W_w P_w}{W_a(P-P_w)} \qquad (3)$$

where W_w = molecular weight of water vapor, 18.01

W_a = molecular weight of air 28.967

P_w = partial pressure of water vapor

P = total pressure

The saturation vapor pressure of pure water is a function of temperature only. Vapor pressure is always dependent on the substance and its temperature. The change of vapor pressure with temperature may be expressed by the Clausius-Clapeyron equation:

$$\frac{dP}{dT} = \frac{L_v}{T} \qquad (4)$$

Where L_v is the molar latent heat of vaporation and ΔV is the difference in molar volumes at absolute temperature T.

Except near the critical temperature, above which it is not possible to liquify a gas, the molar volume of the liquid is insignificant compared to that of the gas. Equation 4 can be rewritten in the following form:

$$\frac{dP}{dT} = \frac{L_V}{TV} \qquad (5)$$

where V is the molar volume of the gas.

In accordance with the ideal gas law: (PV=RT), the following form also holds:

$$\frac{1}{P}\,\frac{dP}{dT} = \frac{L_V}{RT^2} \qquad (6)$$

Assuming the latent heat of vaporation is constant, (over moderate temperature range) the above expression when rearranged and integrated, yields the following form:

$$\frac{P_2}{P_1} = \frac{L_V}{R}\,\mathrm{Exp}\,\left(\frac{1}{T_1} - \frac{1}{T_2}\right) \qquad (7)$$

From the above expression, if the vapor pressure at one temperature is known, the vapor pressure at any other temperature may be calculated. It should be pointed out however that the derivation of the expression assumes ideal gas behavior. The Goff-Gratch form and others (4,5) are substantially more complex based on integration of the Clausius-Clapeyron equation, but take into account the deviation from ideal gas behavior. The object here is to illustrate the principal which establishes the specificity of Dew Point as a measure of water vapor concentration.

For convenience, Equation 6 (six) may be written in the form:

$$p = A\,\mathrm{Exp}\,\frac{(L_V)}{RT} \qquad (8)$$

where A is a constant derived by inserting a temperature and corresponding saturation vapor pressure into equation six. Selecting a temperature in the meteorological measurement range minimizes the effects of non ideal gas behavior for that application.

The Dew Point is by definition that temperature where a gas with a given water vapor partial pressure (concentration) becomes saturated. The temperature of condensation occurs where the saturation vapor pressure of equation 8 equals the water vapor partial pressure of equation 3. Therein lies the specific relationship between Dew Point and Vapor concentration. It is this fundamental physical relationship which provides the performance advantage inherent to the condensation method.

CONTAMINANT INDUCED MEASUREMENT ERROR

The above argument holds as long as the gas mixture is in contact with pure water, since the whole technique is premised upon the saturation vapor pressure being dependent on temperature only. One of the aforementioned limitations inherent to dew point measurement is susceptibility to the effects of water soluble materials which often co-exist in the sampled atmosphere. The dew point method is premised on the specific relationship between saturation vapor pressure and temperature. As stated earlier, condensation occurs at the temperature where the saturation vapor pressure equals the partial pressure of the water vapor in the gas mixture. The specific relationship between dew point and vapor concentration breaks down when water soluble materials go into solution with the controlling layer of condensate on the chilled surface because the saturation vapor pressure at any given temperature is lowered in proportion to the mole fraction of solute in solution with the liquid condensate. This causes the mirror to be controlled at a temperature higher than the true dew point in order to elevate the saturation vapor pressure and maintain the required equilibrium with the partial pressure of the water vapor in the sampled gas.

The effects of soluble contaminants on the operation of the condensation Hygrometer are graphically illustrated in the family of curves in Figure 2. The upper curves represent saturation vapor pressure over temperature for pure water and solute contaminated water (of arbitrary concentration).

The effects of soluble contaminants are two fold:

1. The mirror reflectance is reduced which tends to diminish the density of the operating dew layer, and
2. The vapor pressure of water at any given temperature is lowered in proportion to the mole fraction, (concentration) of solute in solution.

The former effect has been historically addressed in conventional instrumentation by periodic (manual or automatic) interruption of the mirror temperature control loop. The interruption has been consistently accomplished by removal of the reference offset voltage. The offset in the control amplifier is necessary when operation commences so as to cause the mirror to slew down in temperature and establish the operating density of the dew layer. Removal of the offset voltage allows the mirror temperature to rise so as to evaporate the dew layer. Any remaining offset associated with the reduction of mirror reflectance and other secondary effects, would then be eliminated either by manually or automatically injecting an equal and opposite term so as to cause the output of the amplifier to be zero.

The control loop is then rendered operative by reinjecting the offset voltage so as to cause the heat pump to drive the mirror temperature down to the dew point.

This technique addresses the first effect (mirror reflectance) in that it tends to re-establish the operating dew layer. The operating dew density otherwise decreases as the contamination builds up since the total reduction in reflectance when in control is fixed, as set by the magnitude of the control amplifier offset term. The dew layer would eventually be reduced to zero by accumulation of foreign matter. Under these conditions, the system becomes inoperative in that it no longer controls on dew.

Balancing the control loop does not adequately address the problem of solute induced measurement error (effect #2) which occurs long before the system goes out of control, if the contaminating material is water soluble. The ideal reflectance versus temperature characteristics approximates that of curve No. 1. The horizontal dashed line at 60% reflectance represents the equilibrium reduction in photo current when in control at the dew point, as established by the amplifier offset voltage described earlier.

The effect on vapor pressure brought about by the presence of soluble material and the impact on the reflectance versus temperature characteristic is illustrated in curve 2. Balancing as described above creates the apparent shift back to the original dry reflectance but the sharp knee and infinite slope at the temperature of condensation are not recovered as illustrated in curve 3. As a result, the system equilibrates at the temperature where the modified saturation vapor pressure of the condensate equals the water vapor partial pressure of the gas mixture. The end result is that the mirror temperature measurement does not correspond to the true dew point as illustrated by the intersection of curve 3 with the horizontal dashed line representing the reflectance operating point.

ELIMINATION OF THE CONTAMINANT INDUCED MEASUREMENT ERROR

A technique has been developed* which addresses this interference problem as follows: A timing circuit periodically interrupts the control loop that maintains the mirror surface at the temperature of condensation. The surface is cooled well below the prevailing dew point at interruption for a time (1/2 minute) sufficient to allow a heavy growth of condensate and coalescence to occur. The excess solvent (water) encourages all of the soluble material to go into solution and the puddling (coalescence) effect creates a medium by which molecules of solute can migrate. In normal operation this does not occur because the controlling dew layer consists of individual tiny droplets separated by dry surface area. After cooling, the mirror is rapidly heated well above the prevailing dew point so as

*U. S. Patent No. 4,216,669

to cause total evaporation of the liquid. As evaporation occurs, the large puddles break up and shrink in size carrying the soluble material, holding higher and higher concentrations in solution, leaving bare, uncontaminated mirror surface behind. Eventually, each little puddle becomes saturated and the soluble material begins to precipitate out in aggregate clusters. The net result is that the redistribution of contaminating material provides surface area which is free of contaminant and available for undisturbed condensation. Experimentation reveals that the resulting isolated colonies of contaminant typically do not occupy more than 15 percent of the mirror surface even in severely contaminated situations.

After total evaporation of the solvent and precipitation of the solute, the system measures the resulting mirror reflectance, injects a reflectance compensating signal into the control amplifier, and then reverts back to normal operation. As the mirror temperature slews down to the dew point, deliquescence (water absorption) still occurs at temperatures above the true dew point, but only at the sites of redistributed contaminating material. Water sorption at the sites of clustered contaminant cannot cuase a reduction in photo current which is sufficient for control since the occupied surface area is small relative to the total reflective surface. The mirror temperature, therefore, continues to slew downward until condensation forms in the substantially larger bared regions of the mirror surface so as to cause further reduction of the photo current. An offset in the control loop is set so that only when this occurs can control equilibrium be established. The infinite slope at the dew point is essentially recovered as illustrated in curve 4. By forcing undisturbed growth of condensate on the reflecting surface, the surface temperature is controlled at the true dew point and the contaminant induced measurement error which would normally occur is eliminated.

OTHER DIFFICULTIES INHERENT TO THE METHOD

The second limitation inherent to the condensation temperature measurement technique is constraint to display of humidity or water vapor concentration in terms of dew point. In many applications this is a serious problem because the actual water vapor concentration is the information that is required and conversion from dew point to water vapor concentration requires a fairly complex mathematical solution or the tedious employment of look-up tables. To address this limitation an on board microcomputing system which is described later was developed that continuously and automatically converts the measured dew point to more convenient operator selectable units of concentration, such as percents by weight or by volume or percent relative humidity.

The third item addressed is the requirement on the accuracy of the temperature measurement (both dew point and gas temperature) which must be imposed if the potential accuracy and long term repeatability inherent to the basic technique are to be fully realized. The design objective is to provide humidity information to within 0.5%. This represents a substantial but realizable improvement in the state-of-the-art over commercially available instrumentation. A 0.1 degree measurement error (in opposite directions) in each of the measured temperatures creates a .5% relative humidity error at 40% relative humidity. It is thus apparent that not only a highly accurate initial calibration of the temperature measurement system is required, but an extremely high precision and long term repeatability is necessary in order to maintain this accuracy in the absence of any recalibration.

The method and hardware described addresses this requirement. Measurement data has demonstrated that 0.1°C accuracy is realized and maintained in the long term by automatic control of the transducer measurement process along with automatic data reduction.

The scheme is premised upon the calibration and stability of high quality industrial grade platinum resistance temperature transducers but imposes no requirements on these devices which are inconsistent with the present state-of-the-art.

TEMPERATURE TRANSDUCER REQUIREMENTS

100 ohm platinum temperature transducers can be obtained commercially which exhibit repeatability and long term stability specifications in the order of .02 to .03°C. These devices are assembled with four leads (2 at each electrical end) and are subjected to high accuracy calibration of resistance versus temperature. Calibration to within .05C° is achieved by performing a 4-Wire resistance measurement using the four wires provided and subjecting them to four or five different temperatures that are determined by fundamental physical constants such as the triple point of water (.01°C), the freezing point of tin (231.968°C), and the boiling point of oxygen (-182.962°C). The capability exists, with these devices to perform measurements with long term total accuracy of .07 to .08°C. Given this transducer performance level, the problem reduces to one of high accuracy transducer resistance measurement (Figure 3) and resistance to temperature conversion (Figure 4) by the supporting instrumentation.

TRANSDUCER MEASUREMENT AND DATA PROCESSING

A microprocessor based hygrometer system performs a 4 wire resistance measurement on each of the transducers and in addition, on two reference resistance elements for offset suppression and normalization in the computing subsystem. This eliminates both offset and slope (gain) errors associated with the analog resistance measurement process. Constant current excitation of each transducer is provided with one pair of wires and the resulting voltage across the transducer is sensed by a chopper stabilized instrumentation amplifier through the other pair of wires. The sensed voltage is therefore linearly proportional to the resistance at the terminals of the transducer. By using a four wire resistance measurement technique, not only are measurement errors associated with lead and contact resistances eliminated, but the integrity of the original transducer calibration which utilizes the same four wires is maintained. The resulting accuracy of the resistance measurement is .01 ohms which corresponds to .025°C for a (100 ohm) platinum element which exhibits a temperature coefficient of .00391 ohms per ohm per °C. Conversion of the transducer resistance to temperature on the International Practical Temperature Scale of 1968 (IPTS-68) is accomplished by solving the characteristic equation in the microprocessor using the calibration coefficients stored in a programmable read only memory (microelectronic) device which is plugged in to the rear of the instrument chassis. Interpolation on the IPTS 68 to within 1 millikelvin is accomplished by this method. As stated earlier, the overall temperature measurement accuracy is typically better than 0.1°C.

Each of the resistance elements to be measured, two temperature transducers and two reference resistors, are sequentially provided excitation current and the resulting voltage sensed. (Refer to Figure 3). The sequential sampling is accomplished by a set of solid state switches under control from the computer. The polarity of the transducer excitation current can be alternately reversed back and forth and the resulting pairs of data samples algebraically summed in the computer. This eliminates the parasitic thermal E.M.F.'s and resulting errors which are in series (summed) with the RTD voltage that is resistance dependent. The instrumentation amplifier which senses the voltage provides an output to four sample and hold capacitors which are addressed synchronously with the four resistance elements so as to store the magnitude of each measured parameter for the duration of the sequence in order that the settling time be reduced each time a given element is addressed. The sample and hold capacitors are sequentially interrogated by an Analog to Digital (A/D) convertor (Figure 4) which in turn provides binary coded decimal data to the microprocessor on command. The computer system executes a series of instruction which are stored in an electrically programmable read only memory device (EPROM). Each instruction location is sequentially addressed by the program counter which is clocked by the Ready function available from the microprocessor. The microprocessor is capable of executing input and output instructions, memory store and recall functions and arithmetic operations. The microprocessor is also capable of conditional branching operations. A 1,024 word random access memory (RAM) is employed for intermediate storage of computed numbers. In addition to the RAM and A/D convertor inputs to the microprocessor, there is also employed a plug-in microelectronic programmable read-only memory (CALPROM) located on the rear panel of the instrument which stores calibration coefficients of the transducers and other operational parameters such as atmospheric pressure and the molecular weight of the gas medium. The front panel switches are also interrogated by the computer for command/control purposes in order that the users selected system of units be computed. Processed data is provided to the front panel displays as well as a latch for binary coded decimal output and analog voltage and current outputs.

Humidity data can be selected or scanned for display in terms of dew point, parts-per-million by weight or by volume or % Relative Humidity. Temperature and dew point can be displayed on either the Fahrenheit or Centigrade scale. Determination of water vapor concentration by weight or by volume requires computation of the saturation vapor pressure at the measured dew point along with total pressure measurement. In addition, the saturation vapor pressure at the measured ambient or dry bulb temperature is required for relative humidity determination. Optionally, the equivalent adiabatic saturation temperature or total enthalpy can be computed. An approximate solution to vapor pressure is accomplished in the on board computing system using a formulation[5] developed by the National Center for Atmospheric Research which provides very close agreement (less than 0.1% over the meteorological range) to the NBS vapor pressure tables.[6,7,8]

DEW/FROST POINT SENSOR REQUIREMENTS

The sensor materials and geometry (Figure 5) substantially impact on the systems performance in terms of response time, accuracy, repeatability and measurement range. The sensor cavity volume must be kept to a minimum to insure rapid equilibration for step changes in vapor pressure or dew point. This to some extent is in conflict with maximizing the thermal resistance from the condensing surface to the sensor body which is usually at a substantially higher temperature. The surface material that surrounds the mirror must serve as a vapor barrier but cannot significantly increase heat conduction to the condensing surface as this would cause a surface temperature gradient and a reduction in the temperature depression capability. A number of materials have been tried as vapor barriers that exhibit low thermal conductivity. In certain applications this surface must also be able to withstand high pressure or vacuum. Generally speaking, both structural integrity and resistance to water vapor transmission are improved with material density but the thermal conductivity also increases. For most applications, the optimum solution is to employ a very thin (0.5 mils) sheet of 316 stainless steel. The area under the barrier is encapsulated with polyurethene foam to minimize unnecessary heat loading of the thermoelectric cooler. The surface of the stainless steel is electropolished to reduce the dimensional amplitude of the microscopic peaks and valley that normally occur. This is to reduce the surface area and potential water entrapment on the surface. An unpolished surface also collects contaminating materials over a period of time, some of which are water soluble or hydrophyllic.

This condition would increase the response time. In high pressure applications, the stainless shim can be backed up with rigid material to provide support. This is necessary to prevent the foam from collapsing and destruction of the seal around the mirror. When the support ring is added, there is some loss in depression but normally under high pressure situations low frost point measurement is not a requirement. The sensor body (except the base) is fabricated entirely of 316 stainless steel when the required measurement range includes very low frost points. Aluminum can be used down to approximately -30°C. The inside geometry of the sensor body is designed to maximize the incidence of water molecules at the mirror surface. In addition to directing the gas flow by the mirror; there is a parallel path. A small fraction of the gas flow is diverted upward to the optics assembly region to eliminate the possibility of stagnation or water entrapment which would slow down the response. Natural diffusion due to partial pressure gradient would occur for a long time if this region were not purged.

The optics assembly consists of matched pairs of light emitting diodes (LED) and phototransistors. One LED illuminates the mirror while one of the phototransistors is located so as to collect the reflected light. Specular reflection occurs when the mirror is dry. It is reduced due to scattering losses when condensation or sublimation occurs. The photodetector signal is used to control the heat pumping power and maintain the mirror surface temperature precisely at the point where no additional evaporation of condensation occurs once control equilibrium is reached, i.e., once the controlling dew or frost layer is established. The additional phototransistor and LED is used for temperature compensating the signal optoelectronic pair. Both LEDS and phototransistors exhibit a very large temperature coefficient. This would cause the operating dew density to shift substantially over the specified range of sensor operating temperatures (-40 to +100°C), if this differential measurement circuit or some other form of temperature compensation were not employed.

The mirror assembly is designed to maximize the thermal coupling from the condensing surface to the top surface of the thermoelectric heat pump and to the temperature transducer. An attempt has also been made to minimize the thermal coupling to the surrounding vapor barrier surface so as to maximize the mirror temperature depression capability and minimize thermal gradients which cause measurement error. The thermoelectric heat pump is a 3 stage device which has been optimized in terms of design parameters for the heat load slope associated with the assembly materials and geometry. Each stage operates at a coefficient of performance (COP) of approximately .35. The cascaded coefficient of performance is approximately .03. This means that the heat pump requires an electrical power input that is 30 times the heat power being pumped. The heat pumped from the mirror surface is of the order of 1 watt at -70°C. The heat discharged at the base is the sum of the two (power input plus heat pumped) and is therefore approximately 30 watts. The sensor base is designed to provide maximum transfer from the heat pump to liquid coolant in the heat exchanger which is attached to the base.

The fins in the coolant cavity provide a substantial increase in the surface area and therefore improves the thermal coupling from the coolant to the base of the heat pump. The material between the thermoelectric device and the heat exchanger is copper which has been flashed with gold to avoid corrosion. Below the heat exchanger there is a cavity that is only .060 in height which has an inside surface of stainless steel. The gas passes through this cavity for precooling prior to entering the dew point sensor cavity when very low frost points are to be measured. This reduces the heat load on the mirror surface which extends the lower end of the measurement range and improves the response time. Precooling is particularly effective when sampling gases with very high thermal conductivity and/or specific heat such as helium or hydrogen.

The above described sensor configuration when cooled to -20°C by means of a recirculating chiller package can measure frost points as low as -75°C (one part per million by volume at atmospheric pressure). The sensor and associated plumbing can also be heated to a maximum temperature of 100°C for condensation temperature measurement approaching 100°C. The sensor can accommodate a microscope for purposes of viewing the condensing surface while in control. This serves as a useful tool particularly while operating in the range between 0 and -20°C where the sensor can control on liquid (super cooled) water or frost. This is important when a high degree of accuracy is required because the saturation vapor pressure over ice differs from that over water. The phase of the operating layer must be known to properly convert to units of absolute humidity, such as parts per million by weight or volume.

CONCLUSION

The resulting instrumentation has demonstrated the accuracy and precision which are in principle achievable by dew point measurement can in fact be realized. The NBS calibration[9] data in Figure 6 illustrates a typical initial accuracy. In addition, long term performance testing was done in the engineering laboratory of General Eastern Instruments Corporation. These tests conclude that the long term stability of both the reference resistance elements and the Platinum Temperature Transducers is well within their manufacturers specifications, thus insuring that the initial accuracy of the instrumentation is maintained.

Bibliography

1. List, R.J. Smithsonian Meteorological Tables, 1979 Ed. 6 350

2. Reid, R.C., Prausnity, J.M. Sherwood T.K., The Properties of Gases and Liquids, pp 181-186, Third Edition, McGraw Hill Book Co., 1977.

3. Riddle, J.L., Furkawa, G.T. Plumb, H.A. Platinum Resistance Thermometery, NBS Monograph 126, April 1973.

4. Curtis, D.I. Temperature Calibration and Interpolation Methods for Platinum Resistance Thermometers, RMT Report No. 6802-3F, Rosemount Inc., July, 1980.

5. Buck, A.L., Short Equation for Computing Vapor Pressure and Enhancement Factors, National Center for Atmospheric Research, February, 1981.

6. Wexler, A., Vapor Pressure Formulation for Water in Range 0 to 100°C. A Revision, J. Res. National Bureau Stand. (U.S.), 80A, Nos. 5 and 6, 775-785 (Sept. - Dec. 1976).

7. Wexler, A., Vapor Pressure Formulation for Ice, J. Res. Nat. Bur. Stand. (U.S.) 81A No. 1, 5-20 (Jan. - Feb. 1977).

8. Greenspan, L., Functional Equations for the Enhancement Factors for CO_2 - Free Moist Air, J. Res. Nat. Bur. Stand. (U.S.) 80A No. 1, 41-44 (Jan. - Feb. 1976).

9. Hasegawa, S. and Little, J.W., the NBS Two-Pressure Humidity Generator, Mark 2 J. Res. Nat. Bur. Stand. (U.S.), 81A No. 1, 81-88 (Jan. - Feb. 1977).

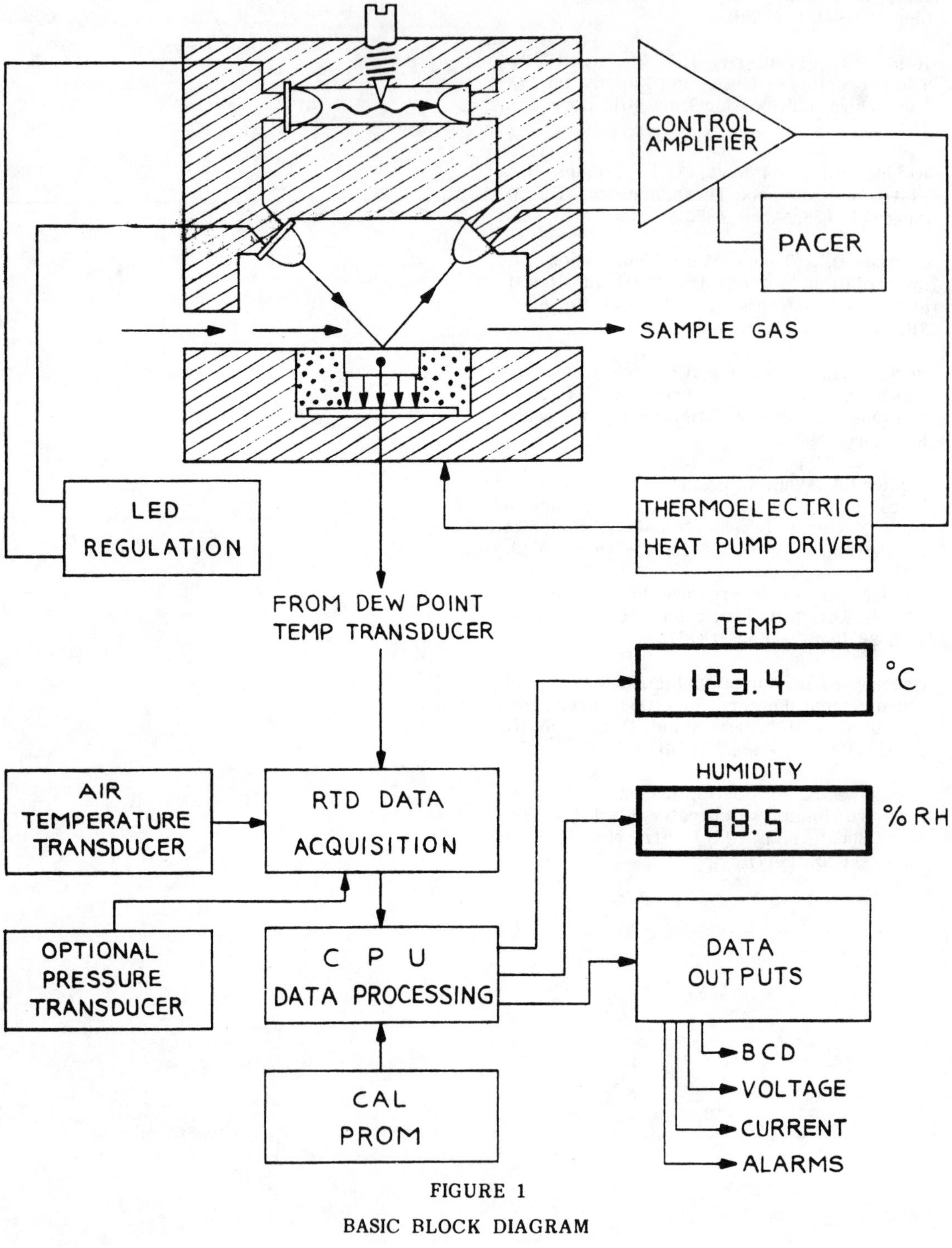

FIGURE 1

BASIC BLOCK DIAGRAM

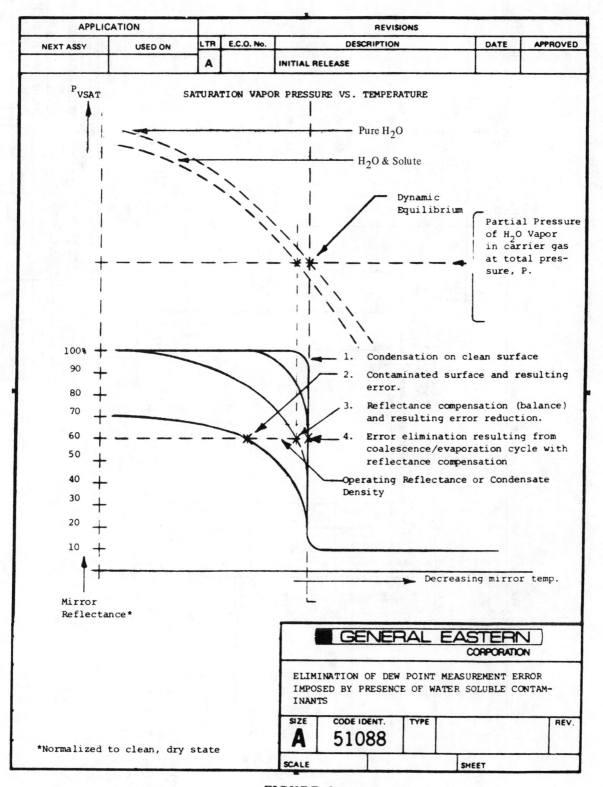

APPLICATION		REVISIONS				
NEXT ASSY	USED ON	LTR	E.C.O. No.	DESCRIPTION	DATE	APPROVED
		A		INITIAL RELEASE		

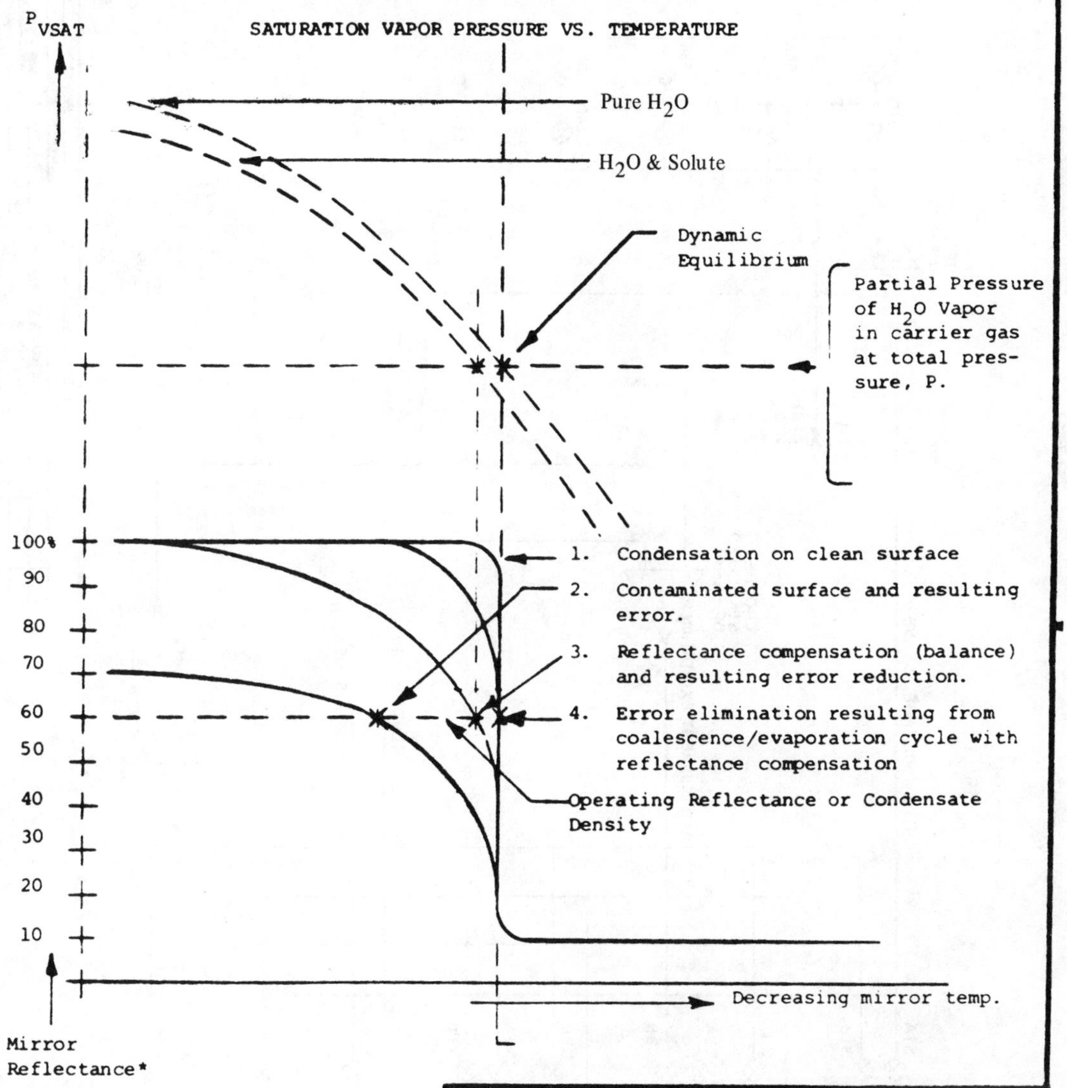

*Normalized to clean, dry state

GENERAL EASTERN CORPORATION				
ELIMINATION OF DEW POINT MEASUREMENT ERROR IMPOSED BY PRESENCE OF WATER SOLUBLE CONTAMINANTS				
SIZE A	CODE IDENT. 51088	TYPE		REV.
SCALE		SHEET		

FIGURE 2

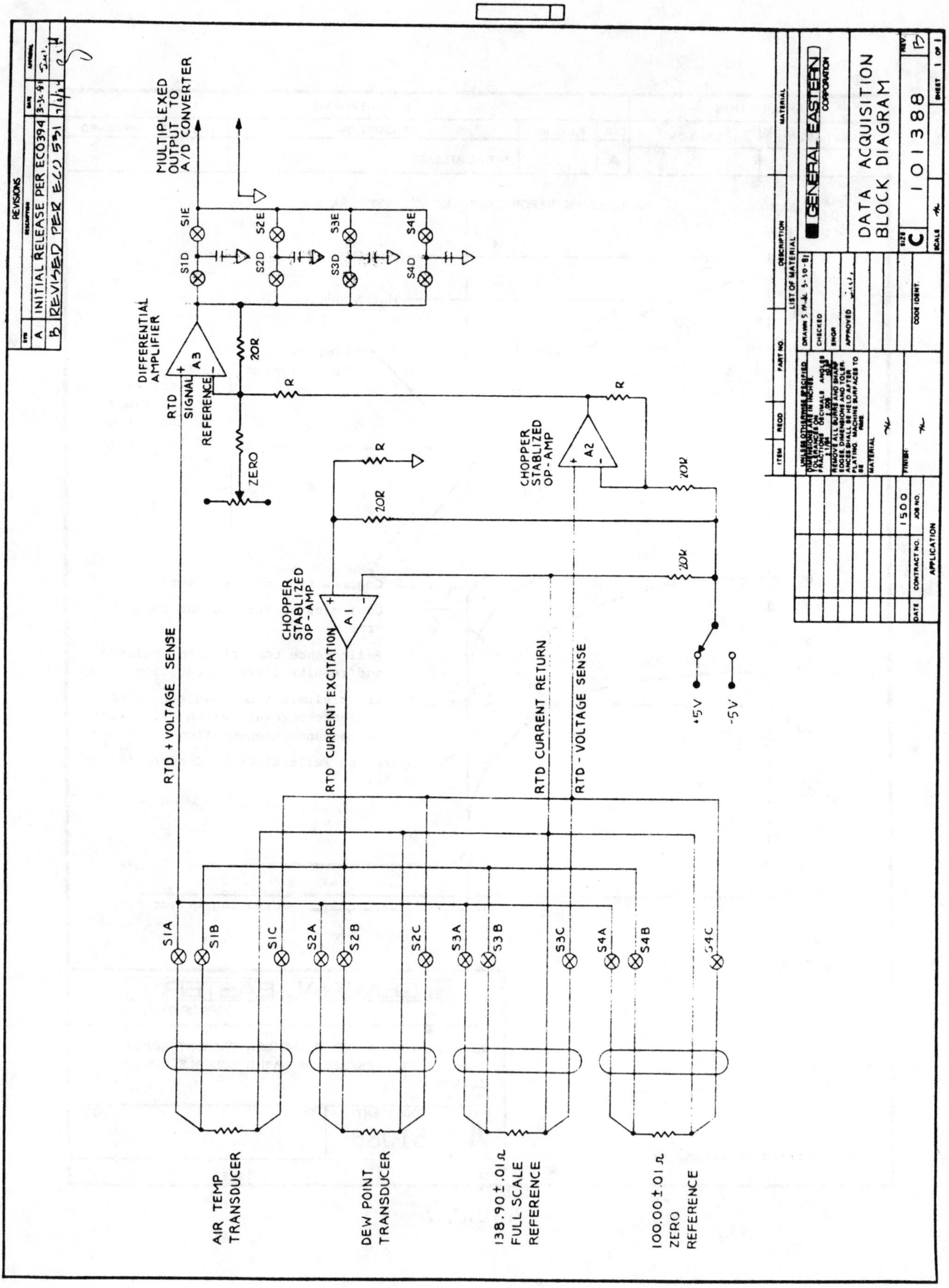
REVISIONS
A INITIAL RELEASE PER ECO394
B REVISED PER ECO 531
MULTIPLEXED OUTPUT TO A/D CONVERTER
DIFFERENTIAL AMPLIFIER
A3
RTD SIGNAL
REFERENCE
20R
R
ZERO
CHOPPER STABLIZED OP-AMP
A1
A2
RTD + VOLTAGE SENSE
RTD CURRENT EXCITATION
RTD CURRENT RETURN
RTD - VOLTAGE SENSE
+5V
-5V
S1A
S1B
S1C
S1D
S1E
S2A
S2B
S2C
S2D
S2E
S3A
S3B
S3C
S3D
S3E
S4A
S4B
S4C
S4D
S4E
AIR TEMP TRANSDUCER
DEW POINT TRANSDUCER
138.90±.01Ω FULL SCALE REFERENCE
100.00±.01Ω ZERO REFERENCE
GENERAL EASTERN CORPORATION
DATA ACQUISITION BLOCK DIAGRAM
C
101388
1500

FIGURE 3

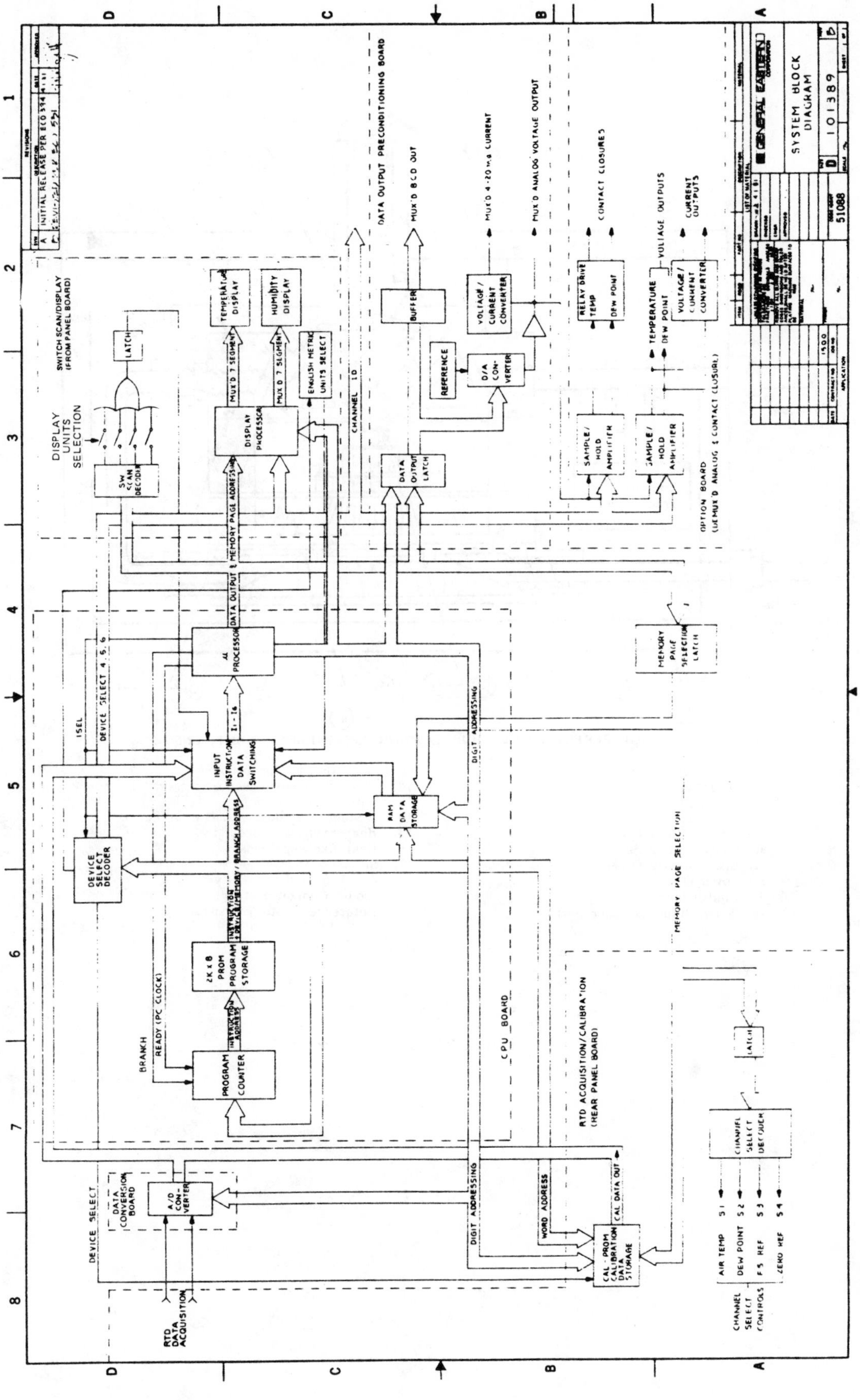

SYSTEM BLOCK DIAGRAM
GENERAL EASTERN CORPORATION
101389
51088
INITIAL RELEASE PER ECO 594
SWITCH SCAN/DISPLAY (FROM PANEL BOARD)
DISPLAY UNITS SELECTION
SW SCAN DECODER
LATCH
TEMPERATURE DISPLAY
HUMIDITY DISPLAY
MUX'D 7 SEGMENT
ENGLISH METRIC UNITS SELECT
DISPLAY PROCESSOR
CHANNEL I.D.
DATA OUTPUT PRECONDITIONING BOARD
MUX'D BCD OUT
BUFFER
DATA OUTPUT LATCH
REFERENCE
D/A CONVERTER
VOLTAGE/CURRENT CONVERTER
MUX'D 4-20 mA CURRENT
MUX'D ANALOG VOLTAGE OUTPUT
CONTACT CLOSURES
RELAY DRIVE TEMP
DEW POINT
TEMPERATURE
VOLTAGE OUTPUTS
CURRENT OUTPUTS
SAMPLE/HOLD AMPLIFIER
OPTION BOARD (DEMUX'D ANALOG & CONTACT CLOSURE)
DATA OUTPUT & MEMORY PAGE ADDRESSING
μ PROCESSOR
MEMORY PAGE SELECTION LATCH
ISEL
DEVICE SELECT 4, 5, 6
INPUT INSTRUCTION DATA SWITCHING
I1 - I6
DIGIT ADDRESSING
RAM DATA STORAGE
DEVICE SELECT DECODER
MEMORY PAGE SELECTION
2K x 8 PROM PROGRAM STORAGE
INSTRUCTION ADDRESS
BRANCH
READY (PC CLOCK)
PROGRAM COUNTER
CPU BOARD
RTD ACQUISITION/CALIBRATION (REAR PANEL BOARD)
LATCH
CHANNEL SELECT DECODER
DEVICE SELECT
DATA CONVERSION BOARD
A/D CONVERTER
WORD ADDRESS
CAL DATA OUT
CAL-PROM CALIBRATION DATA STORAGE
AIR TEMP S1
DEW POINT S2
FS REF S3
ZERO REF S4
CHANNEL SELECT CONTROLS
RTD DATA ACQUISITION

FIGURE 4

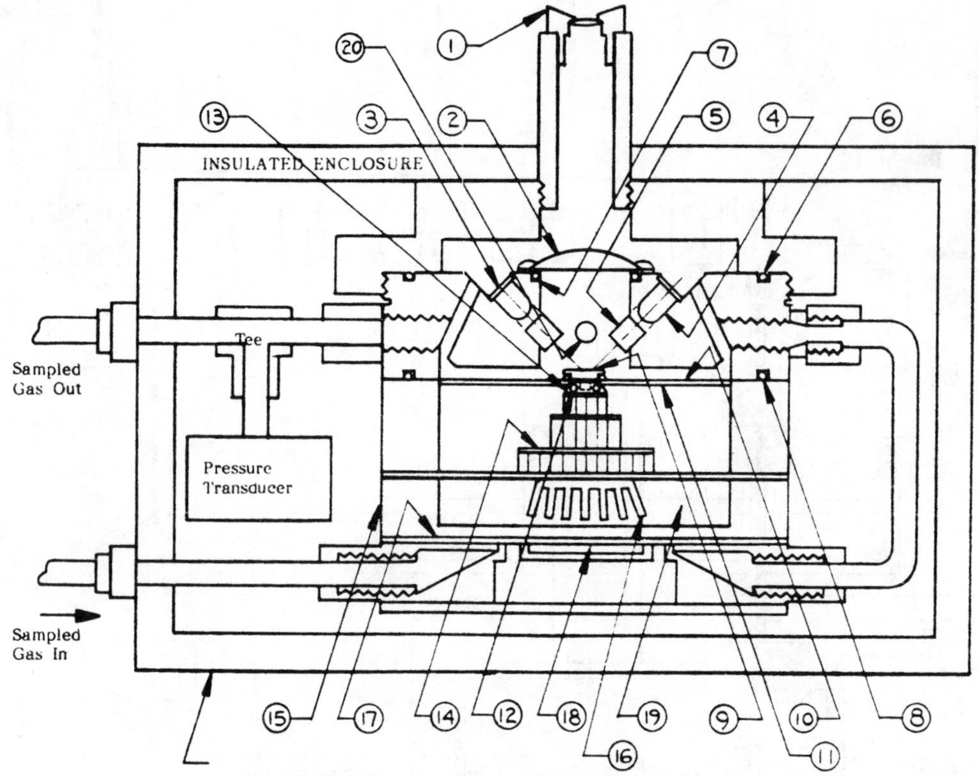

HIGH PERFORMANCE CHILLED MIRROR DEW/FROST POINT SENSOR

1. Microscope Eyepiece
2. Objective Lense
3. Signal Photo Detector
4. LED
5. Ground Glass Rod
6. Outer O-Ring Seal
7. Upper Cavity O-Seal
8. Lower O-Seal
9. Chilled Mirror
10. .5 mil Stainless Vapor Barrier
11. Vapor Barrier Support Ring
12. Platinum Resistance Temperature Transducer
13. Mirror RTD Block
14. 3 Stage Thermo-Electric Heat Pump
15. Heat Exchanger Body
16. Heat Exchange Fins
17. Stainless Shim
18. Gas Pre-Cooling Cavity
19. Liquid Coolant Fitting
20. Reference Photo Transistor and LED

FIGURE 5

Figure 6

Dew Point Hygrometer Calibration

NBS Identification No. H-3847

Dew Point Temperature (a) (°C)	Displayed Dew Point (°C)
24.81	24.7
9.98	9.94
4.46	-4.51
-20.30	-20.4
-30.35	-30.4
-57.23	-57.2
-69.97	-69.9

Dew Point Hygrometer Calibration S/N 2961

NBS Identification No. H-3874

Dew Point	%RH	Displayed Dew Point	%RH
25.07	99.4	24.9	99.0
20.78	76.8	20.6	76.6
15.24	54.2	15.1	54.3
10.08	38.6	10.0	38.8
5.26	27.8	5.3	28.0
-0.46	18.4	-0.4	18.6

RECENT DEVELOPMENTS IN NATURAL AIR DRYING OF CEREALS WITH RESPECT TO SYSTEM DESIGN

Digvir Singh Shahab Sokhansanj

Department of Agricultural Engineering
University of Saskatchewan, SASKATOON, Canada S7N 0W0

1. BACKGROUND

The present trend in cereal grain drying is to dry grain with natural air instead of artificially heated air. In natural air drying the ambient air is forced through the grain. As the air passes over the fan motor its temperature rises slightly (1.1°C), therefore the process is also called near ambient temperature drying. In near ambient aeration and drying the main objective is to move air through grain such that the temperature and moisture contents of grain are brought sufficiently low to prevent spoilage.

A typical storage bin equipped to dry grain with natural air is shown in Fig. 1. The main components are the fan, transition duct and fully or partially perforated floor. The purpose of these components is to force and distribute ambient air into the grain. The air lowers the grain temperature and dries grain slowly. The recommended airflow rates for natural drying are from 3.0 (L/s)/m3 (1/4 CFM/Bu) to about 30.0 (L/s)/m3 (2 CFM/Bu). Lower air flow rates are used for aeration purposes and higher flow rates are used in high temperature drying systems.

The amount of energy to dry grain by near ambient temperature drying technique is less than by high temperature drying. However in specific cases of wet and hot weather conditions the natural air drying may not be more efficient than hot air drying. For London (Ontario) area natural air drying was not energy efficient for 62 percent of the years during the years 1965-1977 (Mittal and Otten 1982). Fraser and Muir (1980a, 1980b) showed that natural drying is feasible for most regions of prairies and is economically superior to high temperature drying or solar assisted drying. In near ambient drying wasted heat is minimal and the initial capital expenditure is less than the high temperature drying installations. However the fans must operate for a longer time and the danger of spoilage is more probable than in high temperature drying.

Experimental and simulated results have indicated that adding 1 to 2 °C temperature rise to the drying air with electric or fossil fuel sources significantly increases the energy requirement per unit of water removed compared to ambient air drying (Morey et al 1981).

The purpose of this article is to review the information available for near ambient drying system design. Specifically resistance to air flow through grains, low temperature simulation models, thin layer drying and rewetting equations and EMC-ERH models has been reviewed. These models can be used in a computer program to design and select an optimum system. Example of such a computer program was developed by Sokhansanj et al (1983).

2. PRESENT STATUS

2.1. Air Flow Resistance

The resistance to air flow in a natural grain drying bin is from several barriers. The most significant is from the grain bed itself. The perforated floor, underfloor ducting and the transition between the floor and the under floor duct each are responsible for an increase in pressure drop.

2.1.1. Grain Beds

The energy necessary to force air through a mass of grain is dependent upon the magnitude of the resistance to the air flow. Shedd (1953) presented graphs relating airflow, (L/s.m2) to pressure drop (Pa per meter of grain depth) for 22 types of grains. The data for air flow ranged from 5 L/s.m2 to 1000 L/s.m2 (1 to 100 ft3/min.ft2) and for static pressures from 1.0 to 5000 Pa per meter of grain depth. Shedd's data has been adopted by American Society of Agricultural Engineers (ASAE D272.1). The note on the graph states that the data are for loose fill of clean and dry grain. The note recommends using only 80 percent of the graph values for high moisture grain, and using twice of the graph values for packed grain. No recommendation for fines and foreign materials is given. Shedd's data has been fitted to the following equation by Hukill and Ives (1955).

$$dp = \frac{a\, Q^2}{\ln(1+b\, Q)} \qquad \ldots(1)$$

where:
dp = pressure drop, Pa/m
Q = airflow rate, $m^3/s.m^2$
a and b are constants for particular grain. Values of a and b for several grains are given in Table 1.

Friesen (1982) measured static pressure in several grain bins in Manitoba. He used a multiplier to the equation describing Shedd's data to correct for field measurements.

Grain bulk density influences the pressure drop across the bed depths. Bern and Cherity (1975) studied the effect of grain bulk density on air flow resistance of corn. An increase in bulk density resulted an increase in the air flow resistance per unit bed depth of corn. They suggested following empirical equation:

$$\frac{dp}{L} = X1 + X2\,\frac{\left(\frac{pb}{pk}\right)^2 Q}{\left(1-\frac{pb}{pk}\right)} + X3\,\frac{\left(\frac{pb}{pk}\right)^2 Q}{\left(1-\frac{pb}{pk}\right)^3} \qquad \ldots(2)$$

where pb = corn bulk density, kg/m^3

pk = corn kernel density, kg/m^3
X1, X2, X3 are constants and are given in Table 2.

Bulk density and its changes in natural grain drying are function of grain's moisture content and the method of filling. Nelson (1978) gave the following empirical equations derived from the experimental data for hard red winter wheat:

$$pk = 1.3981 + 0.0068\, M - 0.0006086\, M^2 + 0.00000747\, M^3 \qquad \ldots(3a)$$

$$pb = 0.7744 - 0.00703\, M + 0.001851\, M^2 - 0.00014896\, M^3 + 0.000003116\, M^4 \qquad \ldots(3b)$$

M is the moisture content of the grain in percent; pk and pb are the kernel density and bulk density in g/cm3. Nelson gave similar data and empirical equations for yellow dent corn.

Method of filling affects the grain bulk density and hence the resistance against airflow. Chang et al (1982) observed an increase in the bulk density from 5 to 9 percent for wheat, 6 to 10 percent for corn and 11 to 12.5 percent for sorghum when spreaders were used. These increases in bulk density increased resistance to airflow by about 55 to 67 percent in wheat, and more than doubled the resistance in corn at airflow rate of 80 L/s-m2.

As the grain dries there will be a reduction in volume due to shrinkage. This shrinkage in volume of grain results in increased pressure drop. Bowden et al (1983) reported a 10 percent reduction in grain volume due to drying. The change in density was inversely related to grain's moisture content as follows:

$$\frac{1}{pb} = a\, M + b \qquad \ldots(4)$$

where a and b are constants and M is the moisture content of grain.

Bowden et al (1983) also reported that the change in thickness of the grain layer due to shrinkage can be estimated from the following equation:

$$dx = Wd\ /\ pb \qquad \ldots(5)$$

where dx is the amount of shrinkage, Wd is weight of grain per unit area and pb is the bulk density.

Haque et al (1978) studied the effect of fines in shelled corn on air flow resistance. They modified Hukill and Ives's (1955) empirical equation to include fines as follows:

$$(dp)'' = (dp)' \; (1+(14.5566 - 26.418Q)fm) \qquad \ldots(6)$$

where fm = weight fraction of fines, decimal.
dp′ = pressure drop across clean grain bed.
dp″ = pressure across grain bed with fines.

A natural flow of air is caused due to temperature gradients in the bin. These flows may be at the opposite direction to the forced ventilation. Gunasekaran et al (1983) reported pressure drop across rough rice shallow bed depths (5 to 30 cm) under low air flow rate conditions caused by natural convection. Temperature gradients up to 40°C in the bin were used to create air flow rates in the range 1.0 to 12.0 L/s.m2. The pressure drop created by temperature gradients was described by:

$$Q = a \, (dp)^{b} \qquad \ldots(7)$$

where values of a and b obtained by regression analysis were 0.0008 and 0.87 respectively. Q is the air flow rate in $(L/s)/m^3$, and dp is pressure drop in Pa.

2.1.2. Perforated Floors

Various floor systems to introduce air at the bottom of the grain mass are being used. These include fully perforated floors, partially perforated floors, circular, rectangular, square or semi-circular on floor ducts and flush in-floor ducts. Henderson (1943) found that the relationship between airflow and pressure drop in perforated ducts could be expressed as follows:

$$Q = 3000 \, Of \, dp^{0.52} \qquad \ldots (8)$$

Brooker et al (1974) included the effect of grain on the perforated floor by introducing the porosity of the grain and rearranged the equation to the following form:

$$dp = \frac{10^{-6}}{9}\left(\frac{Qa}{e \; Of}\right)^{2} \qquad \ldots(9)$$

where:
dp = pressure drop, inches of water
e = void fraction in grain, decimal
Qa = air flow rate, CFM/ft2
Of = percent perforation.

It should be noted that the original work by Henderson used slotted type holes and cleaned corn.

If the hole area is 10 percent or more of the total surface area the pressure drop through perforated floor has been regarded negligible as compared to pressure drop in grain. For circular ducts laid on the floor about 80 percent of the duct surface area is taken as the effective area. The air velocity entering the bin from the perforated duct should be about 10 m/min or less (Mckenzie and Foster 1980). Therefore to determine the minimum perforated floor area the total air flow is divided by 10.

Airflow patterns within the bin has been termed linear or non-linear for perforated floors and ducts respectively. Linearity and non-linearity refers to the variations in isopressure lines within the bed and variability in traverse time of the air within the bin. Air flow pressure patterns for ducts are nonlinear and for fully perforated floors is linear as was shown experimentally by Brooker (1969). Brooker (1969) has shown that traverse time in a bin equipped with a false floor is almost half of the time in a bin having the perforated ducts. This difference is due to the unevenness in air flow distribution in the latter system.

Brooker (1961) developed the mathematical model for air flow and pressure patterns in the bin. His finite difference numerical solution did not predict non linearity of pressures in grain with regard to pressure in the duct. However, Segerlind (1982) and Mikentinac and Sokhansanj (1983) used finite element method to model air flow and pressure patterns in the bin and demonstrated the fact that non linearity existed.

2.1.3. Transition Ducts

To connect the outlet of the fan to under floor duct often a piece of transition is needed. Calculating pressure drop in transitions is well established by analytical models such as:

$$dp = C V^2 \qquad \ldots (10)$$

The value of constant C depends on shape and geometry of transition and its value must be found experimentally.

Gebhardt (1981) experimentally demonstrated that transition ducts connecting fan housing with floor duct created pressure drops from 0 to 572.3 Pa (0 to 2.3 inches of water). He found that for transition ducts of same inlet area and length the static pressure drop across transition is much dependent on the outlet area. For a transition duct of 2918 cm2 inlet area and 101 cm length the static pressure increased from 99.5 Pa to 447.9 Pa when the outlet area was reduced from 2129 cm2 to 1641 cm2. In other words smoother the transition is lesser the pressure drop would be. Specific recommendation for design angles in transitions are lacking.

2.1.4. Fans

Many types and fan configurations are used in natural drying. One significant requirement is that they must overcome the resistance of various parts of the bin and the grain and deliver the required airflow rate. Other features such as noise level, ease of installation, power requirement durability and the initial cost are important. Metzger et al (1980b) tested 11 axial fans and compared their performance with the manufacturer's data. Test conditions were similar to a low temperature in bin drying system. They found that the manufacturer's airflow rate vs static pressure values were consistently 15 percent more than the tested values. They also found that fans of the same nominal size and rated power had significant differences in performance. The published performance by manufacturers were often 10 to 20 percent more than those found by Metzger et al (1980b). VanEe and Kline (1979) also observed that the fan performance data supplied by assorted manufacturers of different fans typically varies in kilowatt consumption from 0.75 to 1.25 times the nominal horse power rating while fans are operated in the range of 500-1000 Pa (2-4 inches of water) static head. These studies indicate that the selection of fan based on manufacturers' recommendation would result in less or more airflow rates than required for successful system operation.

2.2. Drying, Cooling and Rewetting Rates

2.2.1. General

Accurate drying prediction requires that the drying rate of a particular grain under a range of relative humidity and temperatures be known. In addition, equilibrium moisture-relative humidity relationships are needed to estimate the extent of drying and the final moisture content.

2.2.2. Simulation Studies

A comprehensive review of low temperature simulation models to date has been presented by Sharp (1982). In near ambient drying and aeration cooling airflow rates from 0.7 (L/s)/m3 to 67.0 (L/s)/M3 (1/20 cfm per bushel to 5 cfm per bushel) has been recommended (Noyes 1967, Brooker et al 1974). These recommendations are for shelled corn and soybean. Experimentally based airflow rate recommendations for grains such as wheat, barley, rapeseed and pulse crops are scarce or non existent.

Sharma and Muir (1974), Metzger et al (1980a), and Fraser and Muir (1980a, 1980b) have simulated aeration cooling and low/ solar temperature drying using an equilibrium model. The studies are useful in terms of evaluating the mechanics of the propagation of drying and / or cooling fronts. The value of predicted moisture contents and temperatures are questionable because they utilized sparse data or regression models developed or collected from Europian sources i. e. Pichler (1956), Kreyger (1972). Recently Muir and Friesen (1982) has reported the experimental measurements on the rate of drying and deterioration of damp wheat. In the same publication they report disagreement between the simulated and experimental results. They used an equilibrium model to predict the drying of wheat with near ambient temperature conditions. The major assumption in equilibrium model is that air leaves in equilibrium with grain which may not always be true.

2.2.3. Thin Layer Drying and Rewetting Equations

Thin layer drying and rewetting equations could be used to model the natural air drying systems, provided these equations are determined for the temperature and humidity ranges normally encountered in natural air drying conditions. In this modeling technique system under consideration is divided into finite number of thin layers. Depending on the air conditions in the vicinity of the thin layer, the amount of moisture removed or absorbed is determined by a mass balance equation.

The models of thin layer drying reported in the literature may in general be classified in three subgroups based on theoretical approach used to derive them.

(1) Diffusion models using the moisture and/or vapor concentration gradient as the driving force.

(2) Semi-theoretical models.

(3) Empirical models.

The models under first category take advantage of the fact that thermal diffusivity is large compared to moisture diffusivity and hence these models are solved only based on moisture diffusion equation. This method has been applied to drying of peanuts by Whitaker and Young (1972); to drying of corn by Henderson (1974) and to drying of rough rice by Wang and Singh (1978).

Semi-theoretical and empirical models are derived based on experimental data and the inherent disadvantage of the models is that these can not be applied over wide range of the moisture contents. More than one model may be required to fully cover and describe the drying phenomena. The most commonly used equation used to describe the thin layer drying of cereals is that of Page (1949). The modified form of this equation is given below:

$$MR = \mathrm{Exp}\,(-K\,t^{N}) \qquad \ldots (11)$$

Values of K and N are determined by best fit to experimental data. Most of the thin layer drying models from literature are given in Table 3. Wang and Singh (1978) have suggested to use a quadratic equation instead of Page's equation because of simplicity for rough rice. Most of the work on thin layer drying and rewetting is related to corn. Thin layer drying and rewetting equations for air temperatures and relative humidity ranges encountered in near ambient drying are scarce or non existent for most of the grains.

Thin layer drying and rewetting models with their range of applicability from various sources are summerized in Tables 3 and 4 respectively.

In the region of non-availability of data extrapolation of high temperature drying results is usually done for low temperature drying studies using an Arrhenius type relationship between drying constant and drying air temperature. As pointed out by Smith and Bailey (1983) the extrapolation of these results does not agree to the measured results. The main reason for disagreement is the short time used for low temperature drying studies and thus the drying constants only reflects the initial period of whole process. According to these researchers the test period for low temperature drying must be extended sufficiently long.

Work must be undertaken to understand thin layer drying of other cereals, legumes and oilseeds. A need to arrive at a theoretical model based on basic scientific principles to describe drying of single kernels under low temperature drying conditions also exits.

2.3. EMC-ERH models

Potential for natural air drying systems is very much dependent on the difference between the equilibrium moisture content (Me), and the moisture of grain.

Equilibrium moisture is dependent on the approach of reaching a point of equilibrium i.e. sorption or desorption. In drying applications equilibrium is attained by desorption. If during drying simulations and calculations an EMC model based on sorption results is to be used it must be corrected for hysteresis effect.

Timbers and Hocking (1974) presented the equilibrium moisture content of rapeseed for temperatures in the range of 25 °C to 60 °C. Roa and Pfost (1980) reported equilibrium moisture content of twenty seed grains grown in United States. Their tests were performed in an environment where relative humidity and temperature were maintained by adjusting the concentration of acid solutions. The temperature ranged from 5 to 40 °C and relative humidity ranged from 30 to 90 percent. The method yields a static equilibrium moisture and the results may not conform to a dynamic situation that often prevails in drying. Further the grain is treated with chemicals to prevent mold growth during extended storage in the controlled environment which in turn may effect the final equilibrium moisture content.

Equilibrium moisture content of a particular grain is also dependent on the temperature. As an example EMC for wheat at 70 percent relative humidity and -1.1 °C temperature is 15.7% whereas at 48.9 °C temperature and 70 percent relative humidity is 10.3% (Haynes 1961). Most of the EMC-ERH data for agricultural seeds have been summerized in ASAE 1982. The data in low temperature drying range is scarce. As suggested by Brooker et al (1974) none of the theoretical or semi-empirical model is capable of predicting the EMC of cereals over the entire range of temperature and relative humidity encountered in low temperature drying. Empirical models based on the experiments are the most reliable models because they predict the EMC of the cereals more closely over the studied range of temperature and relative humidity. EMC data at relative humidities above 90 percent is also non-existent because of mold development during the period required to reach equilibrium. EMC studies above 90 percent relative humidity are required to simulate rewetting during aeration and low temperature drying studies. Further research work to determine desorption EMC models at low temperatures and high relative humidities must be undertaken. During development of these models rapid method of reaching equilibrium must be devised to avoid long times required to reach equilibrium at low temperature drying conditions and to prevent the grain from onset of mold growth.

2.4. Storage structure stresses

It has long been stipulated that some of the structural collapses of grain bins have happened immediately or after the bin has been aerated. Risch and Herum (1982) measured hoop stresses in a laboratory grain storage bin containing shelled corn and ventilated by forcing air through the grain. The hoop stress increased in all cases except when air with 55 °C and zero percent relative humidity (continuous drying) was used. The greatest stress was introduced when air temperature was 10 to 25 °C and relative humidity was more than 95 percent (wetting). Much of their remainder work is by simulation which is doubtful and must be validated.

2.5 Grain quality

2.5.1 General

As such natural grain drying is slow process and sometimes spoilage of grain may begin before it is dried. Farmers usually follow the extension bulletins such as published by Saskatchewan Department of Agriculture and Manitoba Department of Agriculture (Anon 1982a, Anon 1982b). Extension publications on natural air drying often lack specific recommendation on minimum air flow rates and on duration of safe storage of the grains. The reason is the lack of reliable data. A graph showing safe storage time as a function of moisture and grain temperature for cereals, commonly used in technical literature has been traced back to Danish data of some 20 years ago (Moysey 1982, Personal Communications). This graph is reproduced in Fig. 2. Moysey believes that no one in North America has seen the original experimental results on which the graph is based. Reliability and use of the graph is questionable. Smith and Bailey (1983) stated that a ventilated bin was less susceptible to spoilage than a non-ventilated bin containing the grain with same moisture content and held at the same temperature.

The main source of recommendation for safe storage is that of Kreyger (1972) who compiled and reported the available data to date (1972). Bailey and Smith (1968) as reported by Bowden et al (1983) developed an empirical equation from data of Kreyger in order to predict storage time for barley:

$$ts = 67 + \exp(5.124 + (39.6 - 0.8107\ Tg))\ [1/(M-12) - 0.0315 \exp(0.0579\ Tg)] \quad \ldots (12)$$

The above equation is valid for grain stored in the temperature range of 5 - 25 °C and moisture content of 16 - 26 percent wet basis. Most reported safe storage time data are based with respect to germination.

Storage time and conditions has effects on viability of grains. Curves of seed survival, defined as the percentage viability plotted against time have a sigmoid shape. The slope of the seed survival curve is:

$$v = vi - (1 / var)\ tv \qquad \ldots (13)$$

where v is the probit viability and is defined as (tv′ - tv)/var, tv′ and tv are mean and instantaneous values of time variable, vi is the initial value of v and var is variance defined by:

$$\log(var) = 9.983 - 5.896 \log M - 0.040\ Tg - 0.00428\ Tg^2 \qquad \ldots (14)$$

M and Tg are the grain moisture (d.b. decimal) and temperature °C respectively. Bowden et al (1983) used a value of 3.35 for var which corresponded to a viability of 99.96 for barley.

2.5.2 Mold growth

Micro-organism, baking quality and germination of naturally dried wheat were reported by Muhlbauer et al (1982). They compared five alternatives in near ambient drying of wheat and found that a 4 m deep bed of wheat could be safely dried from initial moisture contents up to 23 percent at airflow rates of 20 (L/s)/m3 without grain deterioration. The five alternatives compared were as follows:

1. Continuous fan operation with air blowing in upward direction.

2. Continuous fan operation with air blowing in downward direction.

3. Intermittent fan operation. Fan shut off when relative humidity of drying air exceeds 70 percent.

4. Continuous fan operation with air conditioned to 70 percent relative humidity.

5. Continuous fan operation with 5 K temperature rise above ambient.

The best alternative from the stand point of energy was test number 3 and from mould growth point was test number 5.

Burrell et al (1980) reported mold growth and germination for aerobically stored rapeseed in England at 5 temperatures ranging from 5 to 25 °C and moisture levels ranging from 6% to 17% to facilitate the determination of the time available for natural air drying. For combinations of the temperatures and moistures they reported the time available before clumping, before appearance of visible fungal colonies and before decline of seed germination. For example, seeds at 25 °C and 10.6% moisture content clumped together after 11 days and visible fungal colonies appeared after 21 days but germination was still unaffected after 40 days. Depending on the investigation criterion safe storage times available for natural air drying could be determined using such studies.

Muir and Friesen (1982) tested and examined germination of wheat in a test bin at Winnipeg under near ambient temperature drying conditions. They found that continuous airflows of 22.2 and 9.5 (L/s)/m3 dried 19.6% moisture content wheat and an air flow rate of 5.7 (L/s)/m3 dried 16.1% wheat with no evident deterioration in less than 44 days. They also reported that wheat at 19.6% moisture content and ventilated at 4.1 and 4.4 (L/s)/m3 for 86 days was not dry and its rate of germination was 63 to 68% while the rate of germination of dried wheat was 95 to 98%. Halderson and Sandvol (1982) monitored aerated stored wheat in Idaho and attributed 5 percent of grain loss to germination, mold, insect damage and sprouting.

2.5.3 Chemicals in natural drying

To increase the storage life of grain without attack from moulds for natural drying a combination of chemical treatment and ambient drying has been reported. Eckhoff et al (1982) fumigated high moisture corn (29 %) by injecting 0.2 percent weight / weight sulfur dioxide gas into the natural air blowing at the rate of 10 to 13 (L/s)/m3. Over a five month period no microbial deterioration or effect to grain quality was detected. Vancauwenberge et al (1982) compared ammonia and sulfur dioxide for control of mold deterioration in high moisture corn. They concluded that sulfur dioxide was a better mold inhibitor than ammonia. However ammonia is sufficient for the purpose and is less expensive and easily handled.

Shove et al (1974) investigated the possibility of using propionic acid to extend the allowable drying time by inhibiting the mold growth. They found that propionic acid applied at the rate of 2 g/kg of corn at 26 percent moisture content doubled the allowable drying time. For corn at 30 percent moisture content they recommended an application rate of 3 g/kg of corn.

3. Conclusions

We reviewed the recent literature available on natural grain drying as related to system design. For an optimized system design we need to know consistent and reliable data to be able to develop computer model. It is evident that the data available is discontinuous and often is old and unreliable. For better understanding of natural drying, research work on safe storage times available for drying, thin layer drying and rewetting equations, static pressure drops across grains and different floor systems, EMC-ERH models and on alternate chemicals for safe storage must be undertaken.

REFERENCES

Agrawal, Y. C. and R. P. Singh. 1977. Thin layer drying studies on short grain rough rice. ASAE Paper No. 77-3531.

Anonymous. 1982a. Natural air grain drying. AGDEX 736.1. Saskatchewan Department of Agriculture, Regina, Canada.

Anonymous. 1982b. Movement of natural air through grain. AGDEX 732-1. Manitoba Department of Agriculture, Winnipeg, Canada.

ASAE. 1982. Agricultural Engineers Yearbook. Am. Soc. Of Agric. Engineers, St. Joseph, MI.

Bern, C. J. and L. F. Charity. 1975. Air flow resistance characteristics of corn as influenced by bulk density. ASAE Paper No. 75-3510. Am. Soc. Agric. Er., St. Joseph, MI.

Bowden, P. J., W. J. Lamond and E. A. Smith. 1983. Simulation of near-ambient grain drying. 1. Comparison of simulation with experimental results. J. Agric. Eng. Res. (In press).

Brooker, D. B. 1961. Pressure patterns in grain drying systems established by numerical methods. TRANSACTIONS of the ASAE 4(1):72-74, 77.

Brooker, D. B. 1969. Computing air pressures and velocity distribution when air flows through a porous medium and nonlinear velocity pressure relationship exist. TRANSACTIONS of the ASAE 12(1):118.

Brooker, D. B., F. W. Bakker-Arkema and C. W. Hall. 1974. Drying cereal grains. AVI Pub. Co., Inc., Westport, CI.

Burrell, N. J., G. P. Knight, D. M. Armitage and S. T. Hill. 1980. Determination of the time available for drying rapeseed before the appearance of surface moulds. J. Stored Prod. Res. 66:115-118.

Chang, C. S., H. H. Converse and C. R. Martin. 1982. Bulk properties of grains as affected by self-propelled rotational type grain spreader. ASAE Paper No. 82-3021.

del Gudice, P. M. 1959. Exposed layer wetting rates of shelled corn. Unpublished M. S. thesis, Purdue University.

Eckhoff, S. R., J. Tuite, G. H. Foster, M. R. Okos and R. A. Anderson. 1982. Sulfur dioxide as a mycological adjunct for low temperature grain drying. ASAE Paper No. 82-3516.

Fraser, B. M. and W. E. Muir. 1980a. Energy consumption predicted for drying grain with ambient and solar-heated air in Canada. J. Agric. Eng. Res. 25(3):325-331.

Fraser, B. M. and W. E. Muir. 1980b. Cost predictions for drying grain with ambient and solar heated air in Canada. Can. Agric. Eng. 22(1): 55-59.

Friesen, O. H. 1982. Report on field studies of the airflow resistance of grains and oilseeds. Unpublished Manitoba Department of Agriculture Report.

Gebhardt, P. D. 1981. Natural Air Grain Drying Project. Project report. Saskatchewan Agriculture, Regina, Canada.

Gunasekaran, S., V. K. Jindal and G. C. Shove. 1983. Resistance to airflow of paddy (rough rice) in shallow depths. TRANSACTIONS of the ASAE 26(2):601-605.

Halderson, J. L. and L. E. Sandvol. 1982. Results of a grain storage study in Idaho. ASAE Paper No. 82-3012.

Haque, E., G. H. Foster, D. S. Chung and F. S. Lai. 1978. Static pressure across a corn bed mixed with fines. TRANSACTIONS of the ASAE 21(5):997-1000.

Haynes, B. C. 1961. Vapor pressure determination of seed hygroscopicity. Tech. Bull. 1229 ARS. USDA, Washinton, D. C.

Henderson, S. M. 1943. Resistance of shelled corn and bin walls to airflow. Agric. Eng. 25:367-369, 374.

Henderson, S. M. 1974. Progress in developing the thin-layer drying equation. TRANSACTIONS of the ASAE 17(6):1167-1168,1172.

Hukill, W. V. and N. C. Ives. 1975. Radial air flow resistance of grain. Agric. Eng. 36(5):332-335.

Hutchinson, D. and L. Otten. 1982. Thin layer drying of soybeans and white beans. CSAE Paper No. 82-104.

Kreyger, J. 1972. Drying and storing grains, seeds and pulses in temperate climates. Inst. Stor. Proc. Agr. Prod., Wageningen, The Neth.

Marchant, J. A. 1976. The prediction of airflows in crop drying systems by finite element method. J. Agric. Eng. Res. 21:417-429.

Mckenzie, B. A. and G. H. Foster. 1982. Dry-aeration and bin cooling systems for grains. Pub. No. AE-107. Coop. Ext. Serv., Purdue University, West Lafayette, IN.

Metzger, J. F., W. E. Muir and P. D. Terry. 1980a. Aeration of wheat: Airflow rates and fan management for fall cooling. CSAE Paper NO. 80-104.

Metzger, J. F., P. D. Terry and W. E. Muir. 1980b. Performance of several axial-flow fans for grain bin ventilation. CSAE Paper No. 80-105.

Mikentinac, M. J. and S. Sokhansanj. 1983. Ventilation-pressure distribution in grain bins ... Brooker's model. Submitted for publication to the TRANSACTIONS of the ASAE.

Misra, M. K. and D. B. Brooker. 1979. Thin layer drying and rewetting equations for shelled yellow corn. ASAE Paper No. 79-3041.

Mittal, G. S. and L. Otten. 1982. Simulation of low-temperature corn drying. Can. Agric. Eng. 24(2):111-118.

Morey, R. V., R. J. Gustafson and H. A. Cloud. 1981. Combination high temperature ambient air drying. TRANSACTIONS of the ASAE 24(2):509-512.

Muhlbauer, W., T. Stahl and W. Hofacker. 1982. Comparison of low temperature wheat drying management procedures. ASAE Paper No. 82-3006.

Muir, W. E. and S. M. Friesen. 1982. Drying grain with near ambient temperature air. CSAE Paper NO. 82-102.

Nelson, S. O. 1978. Moisture dependent kernel and bulk density relationships for wheat and corn. ASAE Paper No. 78 - 3059.

Noyes, R. T. 1967. Aeration for safe grain storage. Pub. No. AE-71. Purdue Univ. Extension service.

Page, G. E. 1949. Factors influencing the maximum rates of air drying in thin layers. M. S. Thesis, Purdue Univ.

Pichler, H. J. 1957. Sorption isotherms for grain and rape. J. Agric. Eng. Res. 2:159-165.

Risch, E. and F. L. Herum. 1982. Bin wall stresses due to aeration of stored shelled corn. ASAE Paper No. 82-4072.

Roa, V. G. and H. B. Pfost. 1980. Physical properties related to drying of twenty food grains. ASAE Paper No. 80-3539.

Sabbah, M. A. H. M. Keener and G.

E. Meyer. 1979. Simulation of solar drying of shelled corn using the logarithmic model. TRANSACTIONS of the ASAE 22(3):637-643.

Segerlind, L. J. 1982. Solving the nonlinear airflow equation. ASAE Paper No. 82-3017.

Sharma, S. C. and W. E. Muir. 1974. Simulation of heat and mass transfer during ventilation of wheat and rapeseed bulks. Can. Agric. Eng. 6(1):41-44.

Sharp, J. R. 1982. A review of low temperature drying simulation models. J. Agric. Eng. Res. 27(3):169-190.

Sharp, J. R., P. J. Bowden and W. J. Lamond. 1982. Simulation of grain dryers using near ambient air. In: Proceedings of Third International Drying Symposium. Drying Research Limited, Walverhampton, England.

Shedd, C.K. 1953. Resistance of grains and seeds to airflow. Agric. Eng. 34(9):616-619.

Shove, G. C., M. F. Finner, G. P. Barrington and M. F. Walter. 1974. Grain preservative extends allowable drying time. ASAE Paper No. 74-3533.

Smith, E. A. and P. H. Bailey. 1983. Simulation of near-ambient grain drying 1. Control stratragies for drying barley in northern Britain. J. Agric. Eng. Res. (Submitted for publication).

Sokhansanj, S., Digvir Singh and P. Gebhardt. 1983. Natural grain drying simulation/animation on minicomputers. CSAE Paper No. 83-204.

Timbers, G. E. and R. P. Hocking. 1974. vapor pressure and moisture equilibria in rapeseed. Report No. 7142-2. Eng. Res. Branch, Agric. Canada, Ottawa, Ont. K1A 0C6.

VanCauwenberge, J. E., S. R. Eckhoff, R. J. Bothast and R. A. Anderson. 1982. A comparison of the trickle ammonia process with the trickle sulfur dioxide process for drying high moisture corn. TRANSACTIONS of the ASAE 25(5):1431-1434.

VanEe, G. R. and G. L. Kline. 1979. FALDRY - A model for low temperature corn drying systems. ASAE Paper No. 79-3524.

Wang, C. Y. and R. P. Singh. 1978. Use of variable equilibrium moisture content in modeling rice drying. ASAE Paper No. 78-6505.

Watson, E. L. and V. K. Bhargava. 1974. Thin layer drying studies on wheat. Can. Agric. Eng. 16(1):18-22.

Whitaker, T. B. and J. H. Young. 1972. Simulation of moisture movement in peanut kernels: evaluation of diffusion equation. TRANSACTIONS of the ASAE 15(1):163-166.

Table 1. Values for constants used in airflow resistance equation 1.

Grain	SI Units a	b
Barley	2.18×10	13.8
Oats	2.53×10^4	14.6
Shelled corn	2.06×10^4	30.7
soybean	1.14×10^4	18.1
Wheat	2.91×10^4	9.84

Note: Range of applicability 0.01 to 0.20 $m^3/s.m^2$.

Table 2. Values for constants (SI Units) for equation 3.

Airflow range ($m^3/s.m^2$)	X1	X2	X3
$0.027 < Q < 0.13$	-0.998	88.8	511.0
$0.13 < Q < 0.27$	-10.9	111.0	439.0
$0.27 < Q < 0.60$	-76.5	163.0	389.0

Note: Range of applicability 732 to 799 kg/m^3 corn bulk density.

Table 3. Thin layer drying equations.

Equation	Range Temp (C) Rel Humi (%) Air flow (m/s) Moisture (db)	Product	Reference
$MR = 1 + A\,t + B\,t^{2}$ $A = -0.001308\,T^{0.4687}\,RH^{-0.3187}$ $B = 0.00006625\,T^{0.03408}\,RH^{-0.4842}$	30-55 T 18-95 RH	Rough rice	Wang and Singh (1978)
$MR = \exp(-K\,t)$ $K = 18.4 \exp(-200/(T + 13)\ hr^{-1}$	< 30 T	Barley	Sharp et al (1982)
$MR = \exp(-K\,t^{N})$ $K = 0.0333 + 0.0003\,T$ $N = 0.3744 + 0.00916\,T\,RH$	32-49 T 34-65 RH	Soy-bean	Hutchinson and Otten (1982)
$MR = \exp(-K\,t^{N})$ $K = 0.0466 - 0.0104\,RH$ $N = 0.4002 + 0.00728\,T\,RH$	32-49 T 34-65 RH	White bean	Hutchinson and Otten (1982)
$MR = \exp(-K\,t^{N})$ $K = \exp(-7.1735 + 1.2793\,Ln(1.8\,T + 32) + 0.1378\,V$ $N = 0.0811\,Ln(RH) + 0.0078\,MO$	2-71 T 3-83 RH .025-2.33 V 18-60 MO	Shelled corn	Misra and Brooker (1979)
$MR = \exp(-K\,t^{N})$ $K = 0.02958 - 0.44565\,RH + 0.01215\,T$ $N = 0.13365 + 1.93653\,RH - 1.77431\,RH^{2} + 0.009468\,T$	30-55 C 25-55 RH		Agrawal and Singh (1977)
$MR = a \exp(-K\,t)$ $0.313 < a < 0.833$ $0.167 < K < 0.454$ a and K are dependent on temperature and relative humidity of drying air. Refer to original paper for exact values.	15-49 T 25-78 RH	Wheat	Watson and Bhargava (1974)

continued ....

Table 3. Continued.

$MR = \exp^{-K\,t}$ $K = b\,Ps^{m}$ $b = 6.272$ $m = 0.722$	10 T	Shelled corn	Sabbah et al (1979)

Table 4. Thin layer rewetting equations.

Equation	Range Temp (C) Rel Humi (%) Moisture (db)	Product	Reference
$MR = \mathrm{Exp}\,(-0.625(Ps)^{0.466R}\,(RH)^{3}\,t)$	16-41 T 11-83 RH	Shelled corn	del Gudice (1959)
$MR = \mathrm{Exp}\,(-\,K'\,t^{N'})$ $K' = \exp\,(-8.5122 + 1.2178\,\mathrm{Ln}(1.8\,T + 32) + 0.0864\,MO$ $N' = 2.1876 - 0.0167\,RH$	10-43 T >90 RH 18-60 MO	Shelled corn	Misra and Brooker (1979)

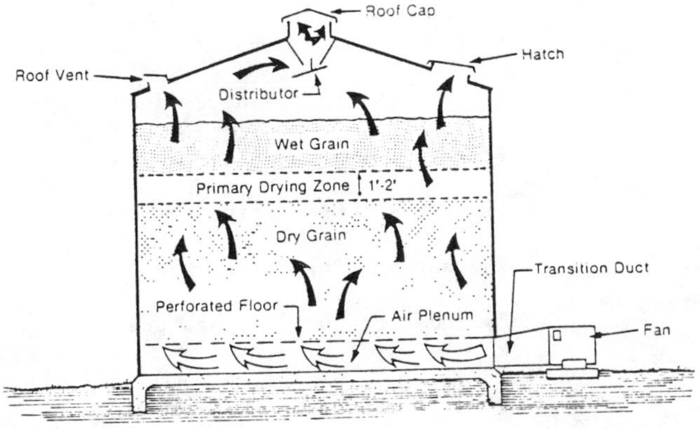

Figure 1. Three zones within grain during natural air drying in a typical bin.

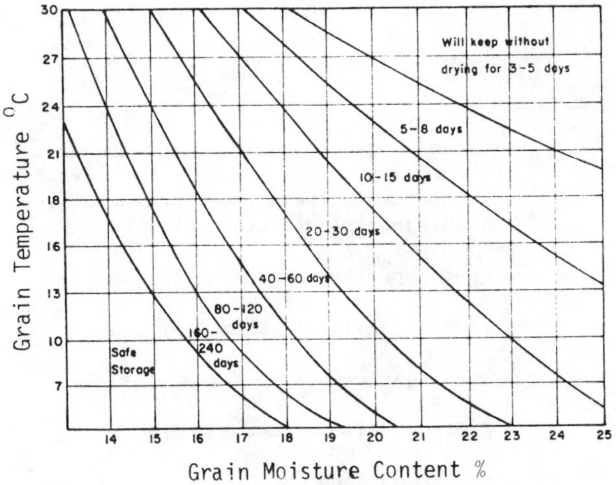

Figure 2. Safe storage time for cereal grains at various temperatures and moisture contents.

A NOVEL STEAM JET AND DOUBLE-EFFECT EVAPORATION DRYER

PART I - MATHEMATICAL MODEL

W.K. Cui and A.S. Mujumdar
Dept. of Chemical Engineering
McGill University
Montreal, Quebec
Canada.

ABSTRACT

A novel paper drying concept, combined pressurized superheated steam impingement and double-effect evaporation drying, is proposed. The proposed dryer is similar to the Yankee dryer. Pressurized superheated steam jets impinge on the moving web of paper. A part of the exhaust steam upon impingement is compressed with a thermocompressor to raise its pressure and used to heat the Yankee cylinder. The excess exhaust steam can be used directly in other thermal processes in the mill. A mathematical model is presented to estimate the drying rate, energy consumption, drying energy cost and the dryer thermal efficiency. The results indicate that, under certain operating conditions net heat consumption on the order of 700 kJ per kg of water evaporated and drying rates in excess of 260 kg/m^2 h, can be expected for drying of tissue paper. While the hot air Yankee dryers typically use 5020 to 7100 kJ/kg water (8) the drying rates are typically under 200 kg/m^2h. This new concept can also be expected to be used for drying packaging paper, board and other paper grades.

Part I of this two-part set of papers developes the mathematical model of the process while Part II presents representative results of the simulation runs.

NOMENCLATURE

A_{hoa}	Ratio of jet open area to drying area
B	Parameter in the correction factor for high evaporation rates
c_d	Heat capacity of paper fiber, kJ/kg.K
c_p	Heat capacity of steam, kJ/kg.K
c_v	Mean heat capacity of steam, kJ/kg.K
c_w	Heat capacity of water, kJ/kg.K
d	Nozzle opening, m
C	Energy costs, $/ton
Ew	Total energy consumption per each kg water evaporated, kJ/kg
Ey	Thermal efficiency of cylinder
h	Enthalpy of steam, kJ/kg
H	Nozzle-to-web spacing, m
Δh_A	Heat of adsorption of water on fiber, kJ/kg
Δh_0	Latent heat of vaporization of water at 0^0 C, kJ/kg
Δh_p	Latent heat of vaporization of water at the paper temperature, kJ/kg
I_c	Cross-flow interference parameter in Chance's correlation
L	Inter-nozzle spacing, m
$\dot{m}$	Mass flux of steam, kg/m^2 s
m_p	Dry fiber basis weight of paper, g/m^2
Nu	Nusselt number ($\alpha d/\lambda$)
n	Exponent for Reynolds number in Chance's correlation
P	System pressure, mpa
Py	Steam pressure inside cylinder, mpa
Pr	Prandtl number ($c_p\mu/\lambda$)
Q_i	Impingement heat transfer rate, kW/m^2
Q_r	Radiant heat transfer, kW/m^2
Q_y	Cylinder conduction heat transfer, kJ/m^2h
R	Total drying rate, kg/m^2h
Re	Reynolds number ($\rho_j V_j d/\mu_j$)
R_i	Impingement drying rate, kg/m^2h
R_y	Cylinder conduction drying rate, kg/m^2h

T	Temperature, K
T_r	$0.5(T_j + T_e)$, K
T_p	Temperature of paper, ^0C
t	Time, s
U	Overall cylinder heat transfer coefficient, kJ/m^2 h K
V_j	Jet velocity, m/s
V_{mach}	Machine speed, m/s
Y	Paper moisture content, kg water/kg dry fiber
Y_0	Initial paper moisture content, kg water/kg dry fiber
Z	Length of dryer, m
α	Heat transfer coefficient, $kW/m^2.K$
ε	Emissivity
λ	Thermal conductivity of steam, kW/m.K
μ	Viscosity of steam, kg/m. sec
ρ	Density, kg/m^3
σ	Stefan-Boltzman constant, $kW/m^2.K^4$
ϕ	Fraction of heat input after heat loss
ψ	Correction factor for high evaporation rates
$\omega_1 ... \omega_3$	Parameters in Chance's correlation
η_d	Dryer thermal efficiency, %
η_0	Overall process efficiency, %

Subscripts

j	Jet conditions
e,h	Hood exhaust conditions
p	Paper conditions
y	Yankee cylinder

1. INTRODUCTION

A major advantage of steam over air as the drying medium is that the net energy consumption can be reduced if the exhaust steam can be reused. There are several other advantages e.g. higher heat transfer coefficients which implies smaller heating surfaces and lower investment cost, no external diffusion-controlled drying rates; elimination of explosion and fire hazards usually associatiated with hot air drying etc. However, one operating limitation is that temperature of the steam must always be kept above saturation conditions in order to avoid condensation on the drying surface. Additional capital costs are associated with compression or reheating of exhaust steam. Furthermore some technological developments need to precede industrial applications.

In spite of the advantages application of steam drying in commercial processes is still very limited. One successful application of steam drying of pulp was reported by Svensson of MODO-Chemetics AB (1). In this process, the pulp is transported in superheated steam which doubles as the transport and drying medium. The water which evaporates from the pulp is recovered as process steam for use in other processes in the mill. As a consequence, the net steam consumption can be reduced to 0.3 tons per ton of water evaporated. In flash or pulp web dryers the energy consumption is on the order of 1.5~3 tons of steam per ton of evaporated water.

Mujumdar (3) proposed that significant improvement in drying rates and particularly the thermal efficiency may be obtained by using superheated steam rather than hot air, in impingement and through drying of paper. Loo and Mujumdar (3) have developed an approximate mathematical model for the combined impingement and through drying process using superheated steam as the drying medium (3), to calculate the drying rates, energy consumption, drying energy costs and dryer thermal efficiencies. Their simulated results, which incorporate a number of assumptions to be verified, show the drying rate is about 40% higher than the drying rate in the hot air Papridryer. If the total exhaust steam can be reused the dryer thermal effeciency is higher than in the conventional hot air drying process. Here the exhaust steam is at a pressure lower than the atmospheric pressure. It cannot be used directly in other processes except for air and water heating. The pressure of exhaust steam can be raised with a thermocompressor or mechanical steam recompressor or heat pump, which consumes high pressure steam or electrical power.

Svensson's study shows that superheated steam has higher heat transfer coefficient at higher pressures. (See Table 1.)

For directly reusing the exhaust steam to achieve higher drying rates, higher thermal efficiencies, lower energy consumption, we propose a novel concept - combined pressurized superheated steam impingement and double - effect evaporation dryer. Fig. 1 shows the flow diagram of this new design concept. In this system, pressurized superheated steam replaces hot air as the drying medium. The superheated steam comes from a reheat boiler and impinges on the moving wet paper wrapped around the Yankee cylinder. Most of exhaust steam is recycled as the impingement jet flow. Another part of exhaust steam is used to heat the cylinder. In order to raise the temperature differential between the heating steam and paper to be dried, a thermocompressor is used to compress the exhaust steam; the compressed exhaust steam is fed into the Yankee Cylinder. The excess pressurized exhaust steam can be used directly in other processes in the mill. This paper presents a mathematical model developed to predict the drying rate, energy consumption, drying energy costs and dryer thermal efficiencies for this new conceptual design. The model results will be used to assess the performance of such a dryer as compared to other conventional dryers. Several simplifying assumptions, not always explicitly stated, are built into this simplified first - generation model. The objective of this work was primarily to evaluate the feasibility of the concept. Experimental work is planned to support the model predictions and further refinement of the model. The reader is therefore cautioned to interpret the results in the light of this observation.

TABLE 1 - Heat transfer coefficient for air and superheated steam

Medium	Pressure (bar)	temp. (^{0}C)	α (kW/m^2 K)
Air	1	0~100	1.3
Steam	0.5	80	0.6
	1	100	1
	2	120	1.7
	5	150	3.4

2. MATHEMATICAL MODEL

When drying in pure superheated steam there is no diffusional resistance since water vapor does not have to diffuse through air. Heat and mass flows are countercurrent to each other, and resistances result only from the poor heat conductivity of the dried material near the surface and from the requirement of a pressure differential for moving the evaporated water from the inner regions of the material out to the surface via capillaries and pores.

After an initial warm-up period, the wet paper enters the hood and constant rate drying begins. It is assumed that when the moisture content of paper reaches 0.4 kg water/kg fibre, the falling rate drying period begins. This value of the critical moisture content for steam drying is an assumption which is probably conservative.

2.1 Constant Rate Drying

Water Balance

The overall water balance equates the inlet steam flow rate and the rate of water removal from the wet paper to the exhaust mass flow rate, as follows:

$$\dot{m}_j - \frac{m_p}{1000}\frac{dY}{dt} = \dot{m}_e$$

$$\frac{R}{3600} = -\frac{m_p}{1000}\frac{dY}{dt} = \dot{m}_e - \dot{m}_j$$

where

$\dot{m}_j$ is the mass flux of impinging jet flow, kg/m^{2}s

$\dot{m}_e$ is the mass flux of exhaust steam, kg/m^2 s;

m_p is the basis weight of paper, g/m^2;

Y is the moisture content of wet paper, kg water/kg fibre;

t is time, s, and

R is the drying rate, kg/m^{2}h

Heat Balance

The overall heat balance is derived by equating the heat input both by convective and radiant heat transfer from the hood and conduction heat transfer from cylinder, to the sensible heat in the exhaust streams and the change in the heat content (sensible heat) of the paper web, taking into account the heat of adsorption. Thus,

$$\dot{m}_j h_j \phi + Q_y + Q_r - \dot{m}_e h_e = \frac{m_p}{1000}\frac{d}{dt}((c_d + c_w Y)T_p - \Delta h_A Y)$$

where

ϕ is the fraction of heat input after heat loss (dimensionless)

h_e is the enthalpy of exhaust steam, kJ/kg

h_j is the enthalpy of steam jet flow, kJ/kg

The enthalpy for each stream is given by:

h_j = 1926.44 + 2.0033 T_j, kJ/kg (T_j in K, P=0.1mpa)

h_j = 1904.2+2.0333 T_j, kJ/kg (T_j in K, p=0.3 mpa)

h_j = 1890.24+2.053 T_j, kJ/kg (T_j in K, p=0.4 mpa)

h_j = 1878.09+2.07 T_j, kJ/kg (T_j in K, p=0.5 mpa)

c_d is the heat capacity of fibre, c_d=1.38 kJ/kg K

c_w is the heat capacity of water, c_w=4.187 kJ/kgK

Δh_A is the heat of adsorption of paper. It is as a function of paper moisture content and is given by (4):

$$\Delta h_A = \exp(7.2-17.3Y), \text{ kJ/kg}$$

Q_y is the conductive heat supplied by the cylinder and is given by

$$Q_y = E_y U(T_{st} - T_p)Y_i, \text{ kJ/m}^2\text{h}$$

Here, E_y is the thermal efficiency of cylinder.

(It is assumed that E_y=0.8~0.9)

U is the overall cylinder-to-web heat transfer coefficient. Depending upon the cylinder shell thickness, the grade and furnish of the sheet, and the condensate removal system, machine speed etc. U can vary over a very wide range (5), (6) and (7). A value of 2249 kJ/m^2 h K (=110 Btu/ h ft^2 ^0F) was selected for the simulation runs reported here.

T_{st}, T_p are the temperatures of steam and paper respectively, in ^0C.

Y_i is the ratio of cylinder surface area to the hood area

Q_r is the radiant heat flux to the paper, which approximated as follows

$$Q_r = \sigma f(T_r^4 - T_{pk}^4) \qquad kJ/m^2\ s$$

σ is the stefan-Boltzman constant, $\sigma=0.5669\times10^{-10}$ $kJ/m^2 s\ K^4$

f is the interchange factor which accounts for the emissivity of the paper (0.94), hood (0.74) and the geometry of the dryer. f is approximately 0.7.

T_{pk} is the paper temperature, K

T_r is the average absolute temperature of hood, K, i.e. $T_r=0.5(T_j + T_e)$

The overall heat balance, after expanding the differential term becomes:

$$\dot{m}_j h_j \phi + Q_y + Q_r - \dot{m}_e h_e = \frac{m_p}{1000}\left((c_d + c_w Y)\frac{dT_p}{dt} + c_w T_p\frac{dY}{dt} - \Delta h_A\frac{dY}{dt}\right)$$

For a fixed machine speed, the time differential can be replaced by the length differential via the machine speed V_{mach},

$$\frac{d}{dt} = V_{mach}\frac{d}{dZ}$$

Thus, the water balance equation becomes:

$$\frac{dY}{dZ} = \frac{1000(\dot{m}_j - \dot{m}_e)}{m_p V_{mach}} \qquad (1)$$

where

Z is the length of dryer, m

The heat balance equation can be rearranged as follows:

$$\frac{dT_p}{dZ} = \frac{1000(\dot{m}_j h_j \phi + Q_y + Q_r - \dot{m}_e h_e)}{m_p V_{mach} (c_d + c_w Y)} + \frac{(\Delta h_A - c_w T_p)\ dY}{(c_d + c_w Y)\ dZ} \qquad (2)$$

Finally, separate heat balance was made on the steam stream which is exhausted through the hood.

$$\dot{m}_j h_j \phi + (\Delta h_0 + c_v T_p)\ R/3600 - Q_i - \dot{m}_e h_e = 0$$

where

Δh_0 is the latent heat of vaporization of water at 273 K, kJ/kg, and

$\bar{c}_v$ is the mean heat capacity of steam, kJ/kg K, given by

$$c_v = \frac{1}{2}(3.55217+5.72196\times10^{-4} T_{pk})$$

Impingement Drying

In the absence of mass transfer resistance, pure impingement drying rate can be assumed to be controlled by the rate of heat transfer, viz.

$$R_i = 3600(Q_i + Q_r)/(\Delta h_p + \Delta h_A) \qquad (3)$$

where

Δh_p is the latent heat of vaporization of water at temperature T_p. An expression for Δh_p as a function of T_p is given by Huang and Mujumdar (9):

$$\Delta h_p = 2.26\times10^3((1-T_{pk}/647.3)/0.42353)^{0.38}, kJ/kg$$

Q_i is impingement heat transfer rate given by

$$Q_i = \alpha(T_j - T_{pk})\psi, \quad kJ/m^2 s$$

Here, ψ is a correction factor for high evaporation rates. For turbulent flow conditions, computed from Kast's (10) recent formulation, ψ is

$$\psi = \frac{1.5B}{\exp(1.5B)-1}, \text{ where } B = c_v(T_j - T_{pk})/\Delta h_p$$

α is the heat transfer coefficient, kJ/m^2K. For the case of a multiple round jet configuration, the correlation recommended by Chance (11) was used to estimate the heat transfer coefficient, viz.,

$$\alpha = Nu\ \lambda/d$$

$$Nu = \omega_1\ \omega_2\ \omega_3\ Re^n\ Pr^{1/3} A_{hoa}^{1.0146},$$

where

$$n = 0.561/A_{hoa}^{0.0835}$$

$$\omega_1 = 2.06$$

$$\omega_2 = 1.0-0.236\ I_c;$$

$$\omega_3 = 1.0-(H/d)(0.023+0.182\ A_{hoa}^{0.71})$$

$I_c = A_{hoa}L/d$, a cross-flow interference parameter;

A_{hoa} = ratio of jet open area to drying area

The ranges of validity are:

$2\le H/d \le 8$, $0.012 \le A_{hoa} \le 0.07$, $I_c \le 1.8$

Conduction through cylinder shell

Heat conduction from the cylinder augments the drying rate, it is evaluated as follows:

$$R_y = Q_y/(\Delta h_p + \Delta h_A), \quad kg/m^2 h$$

The total drying rate is then given by

$$R = R_i + R_y, \quad kg/m^2h \qquad (5)$$

2.2 Falling Rate Drying

In equations (3) and (4), Δh_A, the heat of adsorption term is included in order to account for the heat effect associated with removal of bound moisture in the paper below the critical moisture content. This term is used to calculate the drying rate in the falling rate drying period. Following Loo and Mujumdar (3) the falling drying rate is estimated as follows:

$$R_f = 680(EXP(0.544-14.5Y)-0.0052161)$$
(For $Y \leq 0.4$ kg water/kg fibre)

$$R_i = 3600(Q_i + Q_r)/(\Delta h_p + \Delta h_A + R_f)$$

$$R_y = Q_y/(\Delta h_p + \Delta h_A + R_f)$$

$$R = R_i + R_y$$

Here R_f is interpreted as the additional resistance due to depression of the bound water in solid.

2.3 Energy Consumption. Drying Energy Cost and Efficiencies

The following formulas are based on the flow diagram shown in Fig. 1. Because the system is closed and under pressure (0.2~0.5 mpa), the exhaust steam can be reused directly. A portion of this steam recycles and is reheated by the boiler for recycle as jets. Another portion of this steam is cleaned and saturated with the condensate before passage to the thermocompressor. The compressed steam is then fed to the cylinder as the heating steam. According to Blumberg's empirical data, (12) when the compression ratio is equal to 1.8. The thermocompressor requires 2 kg of motive steam for each kg of steam compressed. The pressure of the motive steam is 3.3 times the pressure of outlet steam. The remainder exhaust steam is saturated with condensate and used elsewhere in the mill. So the primary heat consumption is due to heat losses and equipment inefficiencies.

The circulation fan power consumption depends on the pressure required to drive the desired steam flux through the jet orifices. This pressure can be calculated from the well-known orifice equation. The energy costs used for the estimation of the fan cost was \$0.047/kW-hr and for the boiler it was taken as \$0.012/kW h.

Two different efficiencies were calculated in the program. These are defined as follows:

(a) Dryer thermal efficiency which is defined as the ratio of the total heat required for evaporation to the net heat input to the system. The latter is equal to the sum of the heat supplied by the boiler and motive steam and the enthalpy of water in the paper minus the overall enthalpy of the output steam and condensate, which can be reused directly in other processes.

(b) Overall process efficiency which is defined as the ratio of the energy required for evaporation to the total energy consumption which includes the fan power and the net heat input to the system.

The heat supplied by the boiler can be expressed as

$$Q_b = \frac{1}{\phi}(h_{eav}(\dot{m}_j - \phi\, \dot{m}_j + R_{av}/3600) - (Q_{yav} + 4.183\, T_{po} R_{av})/3600) + Q_r, \quad kJ/m^2s$$

The flow rate of the motive steam of the thermocompressor is given by

$$\dot{m}_m = \frac{2\, Q_{yav}}{3x3600\, E_y\, \Delta h_v}, \quad kg/m^2s$$

(Δh_v is the latent heat of motive steam, kJ/kg)
The overall enthalpy of output steam is given by

$$Q_{so} = h_{eav}\, R_{av}/3600 - \frac{Q_{yav}}{3xE_y x3600} + 4.183\, T_{st}\dot{m}_m, \quad kJ/m^2s$$

The heat consumption is thus given by

$$Q_h = Q_b + \dot{m}_m h_m - Q_{so} + 4.183\, T_{po}\, R_{av}/3600, \quad kJ/m^2s$$

$$Q_{hu} = 10^3 Z\, Q_h/(m_p\, V_{mach}) \text{ MJ/ton paper}$$

$$Q_{hw} = Q_{hu}/(0.95Y_0 - 0.05), \text{ kJ/kg water.}$$

The heat cost is given by

$$C_h = 277.77\, C_{oh}\, Q_h Z/(m_p\, V_{mach}), \text{ \$/ton paper}$$

The power consumption for circulating the steam is given by

$$P_f = \frac{1/2(\Delta P/\rho_{eav} + (\dot{m}_j/\rho_{eav})^2)\dot{m}_j}{1000}, \quad kW/m^2$$

$$P_{fu} = 10^3 P_f Z/(m_p\, V_{mach}), \text{ MJ/ton paper}$$

$$\Delta p = \rho_j V_j (1-A_{hoa}^2)/(2(0.62)^2)$$

The fan power cost is given by

$$C_p = 277.77\, C_{op}\, P_f Z/(m_p\, V_{mach}), \quad \text{\$/ton paper}$$

The total energy consumption is given by

$$E_t = Q_h + P_f, \quad kW/m^2$$

The total energy cost is given by

$$C_t = C_h + C_p, \quad \text{\$/ton paper}$$

The total energy consumption for drying one ton of paper is given by

$$E_u = Q_{hu} + P_{fu}, \quad \text{MJ/ton}$$

The total energy consumption per kg water evaporated is given by

$$E_w = E_u/(0.95Y_0 - 0.05), \quad kJ/kg$$

The heat required for evaporation is

$$E_e = R_{av}\, \Delta h_0/3600, \quad kJ/m^2s$$

Dryer thermal efficiency by definition is

$$\eta_d = \frac{100\, E_e}{Q_h}, \text{ and the}$$

$$\eta_0 = 100 \times E_e/E_t$$

Note that these efficiencies are defined in a simplistic manner since the objective here is only to evaluate the technical feasibility of the process. More precise calculations must be supported by laboratory scale tests.

Furthermore, the costs computed here do not include capital or maintenance costs. The latter must be included in a realistic technoeconomic evaluation of this novel concept after adequate experimentation to determine the best operating parameters.

3. CLOSURE

This paper presents a first order mathematical model to allow preliminary estimation of the drying rates, energy consumption and drying energy costs for the novel concept described. From a purely technical viewpoint the process is shown to be feasible and highly energy-efficient. Results of the simulation model will be presented in a companion paper which in turn will be followed by relevant experimental data.

4. ACKNOWLEDGEMENTS

The authors gratefully acknowledge the assistance provided by Purnima Mujumdar in the preparation of this final typescript. One of the authors (W.-K.C.) would also like to acknowledge the financial assistance of the Government of The People's Republic of China which permitted him to undertake this study at McGill University under the direction of the co-author (A.S.M.).

5. REFERENCES

1. Svensson, C., Drying' 80, Vol. 2, Ed. Mujumdar, A.S., pp 301-307, Hemisphere Publishing Corp., N.Y. 1980.

2. Crotogino, R.H. and Allenger, V., CPPA Trans., S (4), TR84 (1979).

3. Loo, E., and Mujumdar, A.S., Drying'84, Ed. A.S. Mujumdar, Hemisphere, N.Y. 1984.

4. Prahl, J.M., Thermodynamics of Paper Fiber and Water Minstures, Ph.D. Thesis, Harvard Univ., USA (1968).

5. Allander, C., Tappi, Vol. 44, No. 5, pp 332-337, 1961.

6. Cox, J.F., Tappi, Vol. 50, No. 7, pp 368-372, 1967.

7. Gardner, T.A., Tappi, Vol. 47, No. 4, pp 210-214, 1964.

8. Randall, K.R., Paper to be published in Drying'84.

9. Huang, P.-G. and Mujumdar, A.S., Drying'82, Ed. A.S. Mujumdar, Hemisphere-McGraw-Hill, pp 106-114, N.Y. (1982).

10. Kast, W., Heat Transfer Conf. (Munich), 2 pp 263-268 (FC47), F.R.G. (1982), Hemisphere, N.Y. (1982).

11. Chance, J.L., Tappi, 57 (6), 108 (1974).

12. Blumberg, K.M., Tappi, Vol. 66, No. 6, pp 69-70, 1983.

13. Pinter, R., and Holik, H., Drying'80, Vol. 2, Ed. A.S. Mujumdar, Hemisphere, 1980.

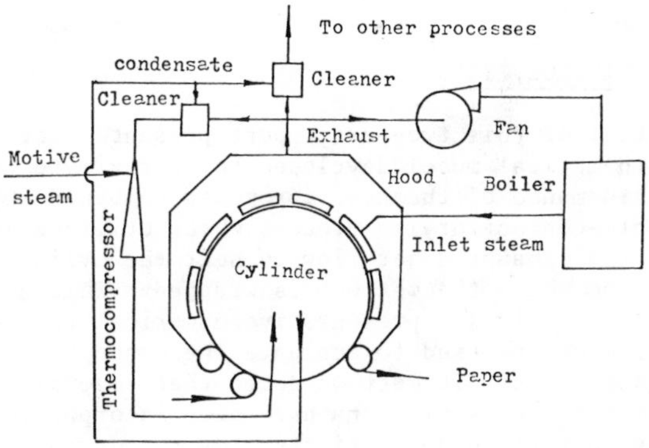

Fig.1 Flow diagram of combined steam jet and double-effect evaporation dryer

A NOVEL STEAM JET AND DOUBLE-EFFECT EVAPORATION DRYER
PART II: SIMULATION RESULTS AND DISCUSSION

W.K. Cui and A.S. Mujumdar

Dept. of Chemical Engineering
McGill University
Montreal, Canada

ABSTRACT

Part I of this two-part paper presented the differential model of the steam drying process, the definitions of the efficiencies and the cost figures. This paper presents representative results under plausible sets of operating parameters. It is shown that the proposed process yields a highly efficient process which also appears economically promising. Indeed, the loss of heat from the high velocity hood is shown to be the principal cause for process inefficiency since the drying medium (steam) as well as the steam produced during drying can be utilized to fullest possible extent. Experimental verification, optimization and a thorough technoeconomic evaluation needs to be accomplished in active collaboration with paper machine manufacturers and mills to exploit this highly promising novel concept.

1. INTRODUCTION

Part I of this two-part report presented the mathematical model developed to approximate the performance of the newly proposed combined impingement -contact drying process which utilizes a part of the exhausted jet flow to heat the cylinder around which the wet web is wrapped. This communication gives representative numerical results which can be used to evaluate the concept proposed. At the outset it must be noted that several simplifying assumptions have been incorporated to obtain first-order estimates for the drying rates, energy consumption and drying energy costs.

2. SIMULATION MODEL RESULTS

The two differential equations of water and heat balances along with the other related equations were integrated numerically using the fourth-oder Runge-Kutta technique. A step size of 2 mm (along machine direction) was chosen. The value of the convective heat transfer coefficient was calculated for each set of the given input conditions given by Chance (1). The program terminates when moisture content of the paper is lower or equal to 0.05 kg water/kg fibre.

This investigation includes an examination of the effects of various parameters (steam jet temperature and velocity, jet open area, heat losses, hood pressures etc) on the drying rate and energy consumption for the following illustrative case.

Jet Conditions:

d=0.005 m; h=0.016 m; L=0.022~0.036 m;

A_{hoa}=0.0144~0.0405; V_j=60~120 m/s;

T_j=520~640 K

P_j=0.3~0.5 mpa

Heat loss=0.5~3%

Yankee Cylinder Conditions:

Y_i=360/235=1.532; E_y=0.9; U =2249.4 kJ/m^{2}h K;

P_y=1.8 P_j; V_{mach}=6 m/s

Paper Conditions:

Initial paper moisture content,

Y_0=1.5kg water/kg fiber

Initial paper temperature T_{p0}=70~143^{0}C

Basis weight of paper m_p=24 g/m^2

2.1 Heat flow in the dryer

The heat available for evaporation of water and output steam enthalpy can be calculated from the equations presented earlier. The various possible losses of the primary energy in the process are taken into account. Figs. 1-3 show Sankey diagrams for the heat flow paths in the dryer under various conditions. We do not know precisely what is the temperature of the paper web during the constant rate drying period; this must be checked experimentally. However, the web temperature should be between 70^{0}C and the water saturation temperature at the hood pressure. For estimation pruposes, we consider the following three special cases:

Figure 1 assumes the paper web temperature to be T_p=70^{0}C. Other parameters are as follows:

Jet Conditions:

d=0.005 m; h=0.016 m; L=0.030 m; A_{hoa}=0.0207;

V_j=100 m/s; T_j=600K; heat loss from hood=0.5%$m_j h_j$
P_j=0.4 mpa;

Yankee Cylinder Conditions:

Y_i=360/235=1.532; E_y=0.9; U =2249.4 kJ/m^2 h K;

P_y=0.72 mpa; V_{mach}=6 m/s

If the total heat supplied by the boiler and the motive steam of the thermocompressor is taken as 100%,double-effect evaporation makes the heat available for evaporation increase 7.8%. 93.9% heat is available for evaporation; 91.9% of the heat output is in the form of superheated steam at a pressure of 0.4 mpa which can be used directly in other thermal processes in the mill. Heat losses from the hood and Yankee cylinder in total account for 17.2% i.e. the net heat consumption is only 17.2% of the input heat. Thus, a net heat consumption of 606 kJ per kg of evaporated water can be achieved. The effective dryer thermal efficiency is thus 545.9%. This figure is due simply to the definition used (it does not violate laws of thermodynamics). No commercial paper dryer boasts such low energy consumption.

Fig. 2 results are for a paper temperature, T_p=143^{0}C (i.e. water saturation temperature at the jet pressure). Other conditions are the same as those for Fig.1.

If the total of heat supplied by the boiler and the motive steam of thermocompressor is taken as 100%, double-effect evaporation concept makes the heat available for evaporation increase 3.0%; 98% of the heat is available for evaporation; 91.9% of the heat output is available in the form of superheated steam at a pressure of 0.4 mpa which can be used directly in other heating processes. Heat losses in total account for 22.0%, i.e. the net heat consumption is 722 kJ/kg water evaporated. The effective dryer thermal efficiency is thus 455% i.e. only 22% of input heat is effectively consumed.

Fig. 3 assumes the paper temperature T_p=60^{0}C. No thermocompressor is included in the process. After the exhaust steam is saturated and cleaned with condensate, it is fed to the Yankee cylinder directly. Further, P_j=0.5 mpa. A_{hoa}=0.0144 and T_j=580 K. Other conditions are the same as those in Fig. 1.

If the heat supplied by the boiler is taken as 100%, because of the double-effect evaporation concept effectively 111.5% input heat is available for evaporation; 81.7% of the heat output is in the form of superheated steam at a pressure of 0.5 mpa which can be used directly in other heating processes. Heat losses in total account for 19.7% i.e. the net heat consumption is only 19.7% of the input heat. Thus, a net heat consumption of only 476 kJ/kg water evaporated can be achieved in principle.

2.2 Drying rate, moisture content and paper temperature variations

Fig. 4 depicts the local drying rate, web moisture content and paper temperature along the length of the proposed dryer. As seen in this figure, there is a constant rate drying period over 1.4 m of the initial dryer length, and then a falling rate drying period of 0.6 m. The moisture content decreases linearly along the dryer. The critical moisture content is taken to be 0.45 kg water/ kg fibre. The paper temperature remains nearly constant at 70^{0}C in the constant rate period. However, in the falling rate zone, the paper temperature increases and reaches over 90^{0}C at the end of the dryer.

A breakdown of the drying rate contribution is also presented in Fig. 4. The Ri and Ry curves are nearly parallel to each other; both display constant and falling rate zones. The impingement drying rate is almost two times the cylinder conduction drying rate.

2.3 Effect of temperature of steam jet

The temperature of the impingement steam jet is one of the most important parameters affecting the drying rate as well as the energy consumption. Fig. 5 shows the computed heat consumption, total energy consumption and average drying rate as a fuction of the jet temperature. Both heat consumption and total energy consumption decrease with rise in temperature of the steam jet. The net heat consumption decreases from 676 to 560 kJ/kg water evaporated with rise in jet temperature from 267 to 347^{0}C. The average drying rate increases from 332 to 372 kg/m^2h with rise in jet temperature from 267 to 347^{0}C.

Fig. 6 shows the efficiencies and energy costs as a function of the jet temperature. The efficiencies increase with increase in jet temperature. The costs fall with rise in jet temperature. The effect of jet temperature is primarily on the temperature differential which is the driving force for heat transfer. It has the greatest effect on water removal by impingement.

In an air impingement dryer, air temperatures ranges from about 150 to 426^{0}C. Inlet air temperature of 315^{0}C is common with modern units on Yankee dryers. It is not known a priori what the right temperature range should be for steam drying.

2.4 Effect of steam jet velocity

The effect of the impingement velocity on the average drying rate and energy consumption is shown in Fig. 7. An increase in the impingement velocity produces a corresponding increase in the heat transfer coefficient and improves drying rate. However, the fan power increases with the cube of velocity, while the drying rate increases with approximately the first power of velocity. Thus the total energy consumption to evaporate 1 kg water increases roughly with the cube of the jet velocity. The heat required to evaporate 1 kg water increases slightly. The effect of impingement velocity on the efficiencies and costs is shown in Fig. 8. The dryer thermal efficiency decreases slightly, as the jet velocity increases. The overall process efficiency decreases with increase of the jet velocity. The heat cost increases slightly, but the power cost increases as the cube of velocity.

2.5 Effect of open area

The drying rate can be increased by increasing the dryer open area, however this is achieved at the expense of both increased fan power and increased heat consumption. At values above 2.0% open area, the required fan power and energy consumption become excessive as shown in Fig. 9. The effect of open area on the efficiencies and cost is shown in Fig. 10. The dryer thermal efficiency and overall process efficiency fall with an increase in the open area of the hood. The costs increase with an increase in open area. The optimum open area is a compromise between drying rate, energy consumption and the fan power.

2.6 Effect of hood heat loss

In the superheated steam yankee dryer system, the exhaust steam can be reused completely at least in principle. The effect of hood heat loss to the surroundings on the energy consumption is shown in Fig. 11. It can be seen that as the hood heat loss increases from 0.5% to 3%, the heat consumption increases over five-fold (from 583 to 3038 kJ/kg water evaporated). The effect of hood heat loss on the efficiencies and costs is shown in Fig. 12. It can be seen that as the hood heat loss increases from 0.5% to 3%, the thermal efficiency falls from 452% to 86.8%; the overall process efficiency falls from 273% to 77.7%. It is therefore very important to insulate the hood very well. Pinter and Holik (2) reported a value of 0.5% for the heat loss from the hood of a hot air yankee dryer. Here we present results over a wide range of values of hood heat losses. The lower range is probably a more realistic range.

3. CONCLUSIONS

A novel paper drying concept, combined superheated steam impingement and double-effect evaporation, has been proposed. A mathematical model which simulates the novel dryer process was developed to predict the feasibility of the novel concept. Because the temperature of the paper web during the constant rate drying period is not known precisely (in the absence of any data), several possible web temperature values have been considered. Even when the paper temperature attains the temperature of saturated water at jet pressure; i.e. the worst case for proposed concept, the simulation results indicate a net heat consumption of only 722 kJ per kg of evaporated water and an average drying rate of over 260 kg/m^2h for drying of tissue paper. If the paper temperature equals 60^0C no thermocompressor is needed i.e. this is the best process condition. The simulation results show that a net heat consumption of 476 kg per kg water evaporated and a average drying rate as high as 379 kg/m^2 h can be achieved. For comparison, the hot air Yankee dryer typically consumes 5020 to 7111 kJ/kg water evaporated (8) while the drying rates are typically under 200 kg/m^2 h. It is obvious that this novel concept is a viable alternative to other competing and conventional drying methods. The drying rates and specific energy consumptions are very attractive when compared to other drying alternatives.

Preliminary experiments are in progress to verify the simulation model. Further experimental tests concerning the effect of steam drying on the paper quality, optimization and detailed technoeconomic analyses need to be carried out in active collaboration with paper machine manufactures and mills to exploit this novel concept.

4. REFERENCES

1. Chance, J.L., Tappi, 57 (6), 108 (1974).

2. Pinter, R., and Holik, H., Drying'80, Vol. 2, Ed. A.S. Mujumdar, Hemisphere, 1980.

5. ACKNOWLEDGEMENT

The authors gratefully acknowledgement the assistance provided by Purnima Mujumdar in the preparation of this final typescript. One of the authors (W.K.C.) would also like to acknowledge the financial assistance of the Government of The People's Republic of China which permitted him to undertake this study at McGill University under the direction of the co-author (A.S.M.).

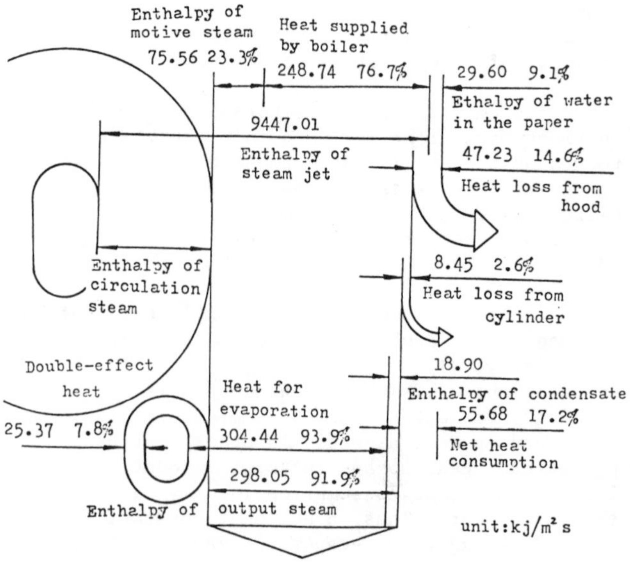

Fig. 1 Heat flux in the combined steam jet and double-effect evaporation dryer

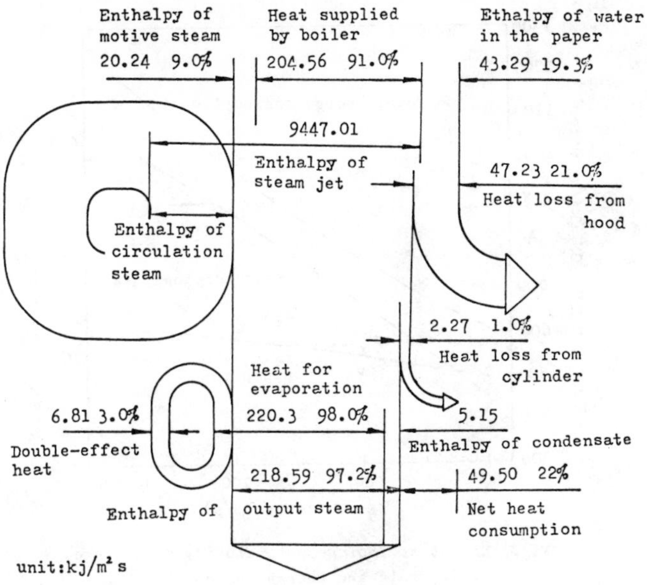

Fig.2 Heat flux in the combined steam jet and double-effect evaporation dryer

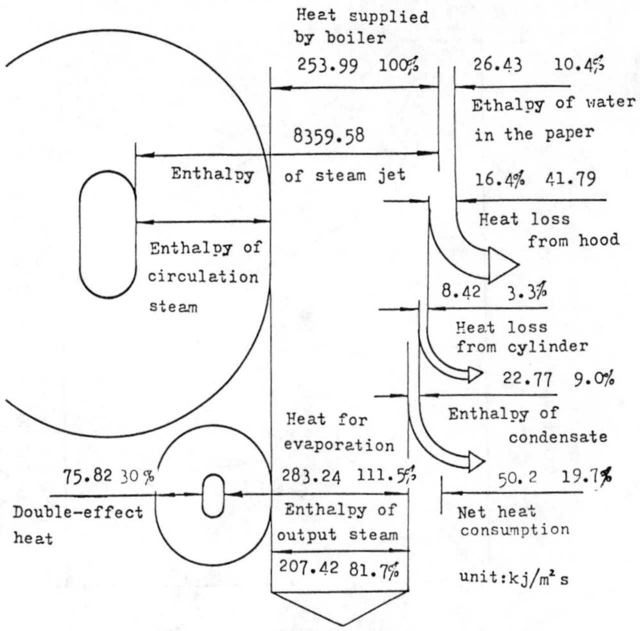

Fig.3 Heat flux in the combined steam jet and double-effect evaporation dryer

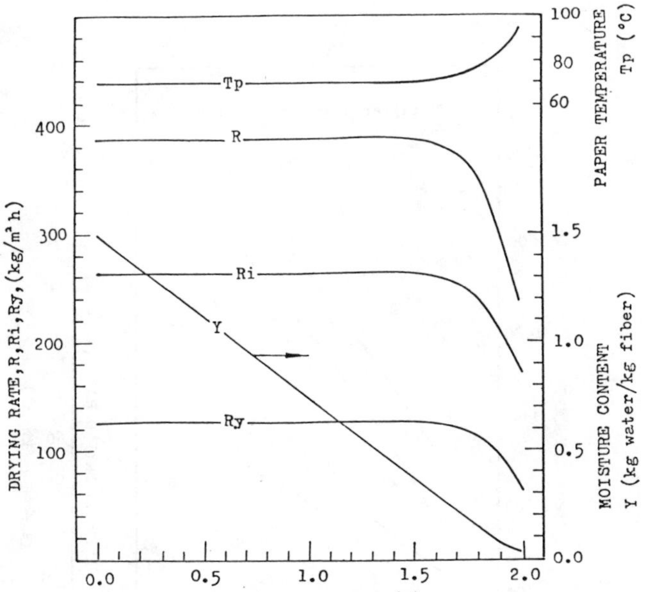

Fig.3 Drying rate and moisture content and paper temperature along the length of the novel dryer
(P_j=0.4 mPa,Tj=600 K,Vj=100 m/s,Ahoa=0.0207,Tpo=70°C)

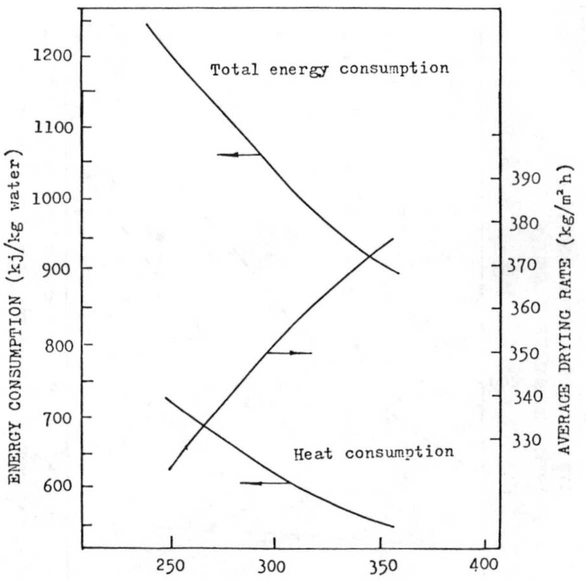

Fig.5 Effect of temperature of steam jet on average drying rate and energy consumption
(P_j=0.4 mPa,Vj=100 m/s,Ahoa=0.0207,Heat loss=0.5%,Tpo=70°C)

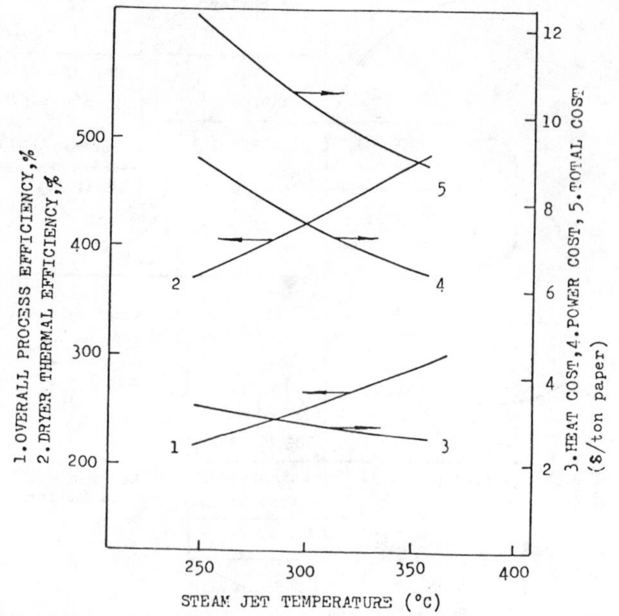

Fig.6 Effect of steam jet temperature on dryer thermal efficiency and overall process efficiency and costs
(P_j=0.4 mPa,Vj=100 m/s,Ahoa=0.0207,Heat loss=0.5%,Tpo=70°C)

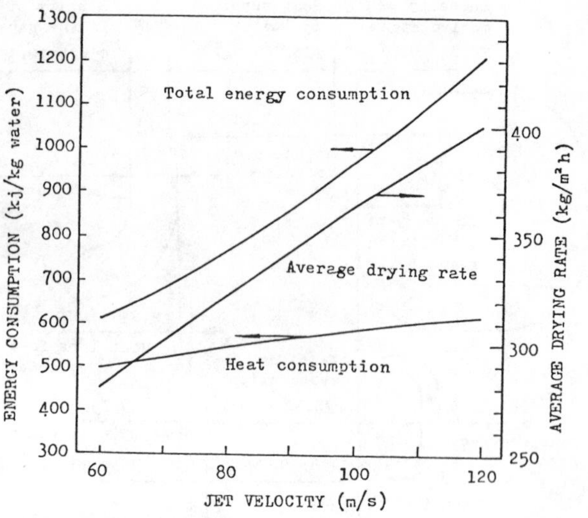

Fig.7 Effect of impingement velocity on average drying rate and energy consumption
(P_j=0.4 mPa,Tj=600 K,Ahoa=0.0207,heat loss=0.5%,Tpo=70°C)

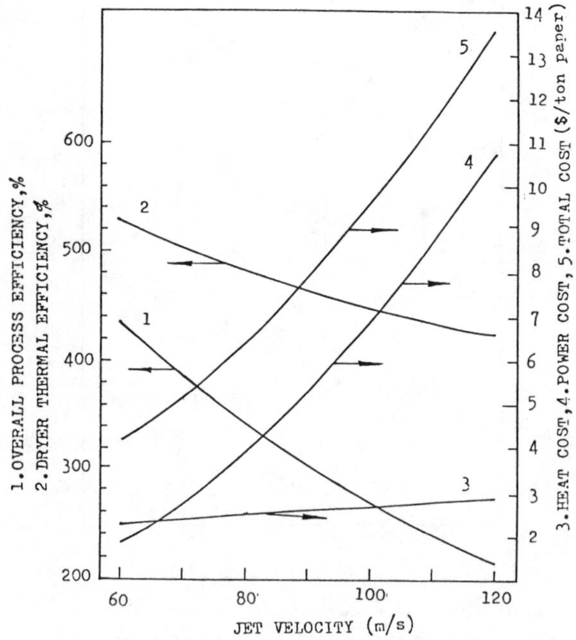

Fig.8 Effect of impingement velocity on dryer thermal efficiency and overall process efficiency and costs
(P_j=0.4 mPa,Tj=600 K,Ahoa=0.0207,heat loss=0.5%,Tpo=70°C)

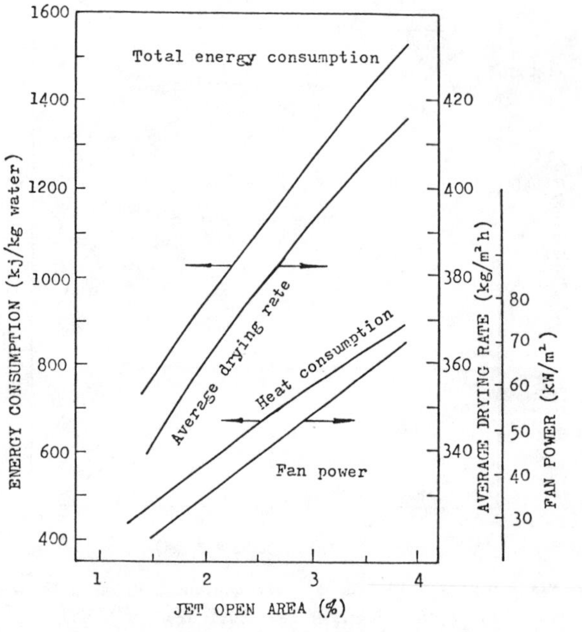

Fig.9 Effect of impingement jet open area on average drying rate and energy consumption
(P_j=0.4 mPa,Tj=600 K,Vj=100 m/s,Heat loss=0.5%,Tpo=70°C)

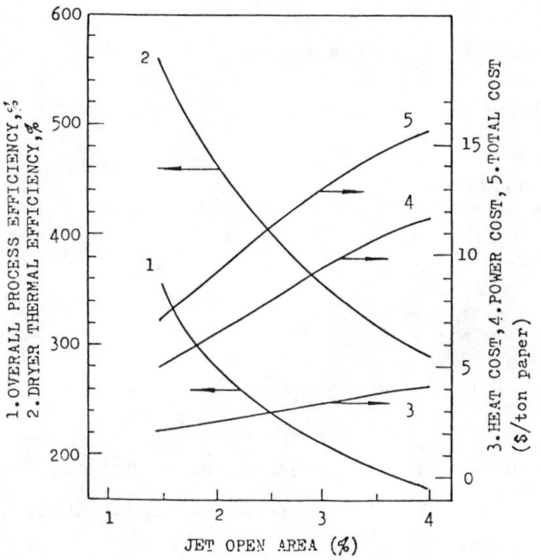

Fig.10 Effect of impingement jet open area on efficiencies and costs
(P_j=0.4 mPa, V_j=100 m/s, T_j=600 K, Heat loss=0.5%, Tpo=70°C)

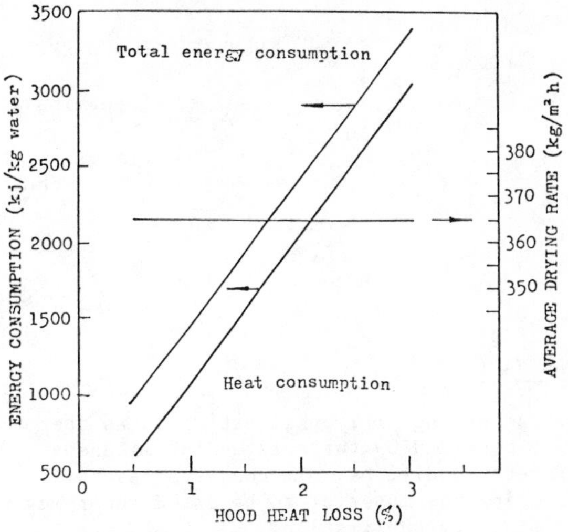

Fig. 11 Effect of heat loss on average drying rate and energy consumption
(P_j=0.4 mPa, T_j=600 K, V_j=100 m/s, Ahoa=0.0207, Tpo=70°C)

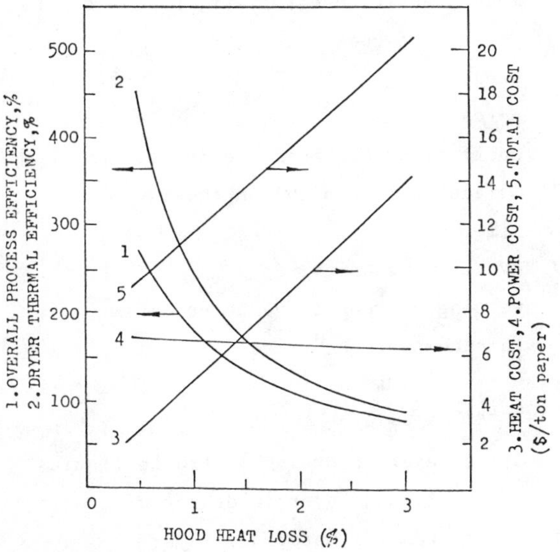

Fig. 12 Effect of hood heat loss on efficiencies and costs
(P_j=0.4 mPa, T_j=600 K, V_j=100 m/s, Ahoa=0.0207, Tpo=70°C)

THEORETICAL AND EXPERIMENTAL INVESTIGATION OF FREEZE DRYING

Determination of transport coefficients

L. Tomosy
Technical University Budapest
Budapest, Hungary

ABSTRACT

Freeze drying or lyophilization is an operation characterized by the outstanding mildness of the process conditions. It consists essentially of freezing the material to be dried and removing its moisture by sublimation.

Sublimation takes place at pressures below the triple point of the solvent (water, as a rule). Lyophilization is performed either in an evacuated chamber where only solvent vapours are present, or in the presence of an inert gas on the condition that the partial pressure of the solvent in the drying chamber is below the triple point pressure.

Lyophilized materials have the advantage of being readily and rapidly rewettable, recovering their initial properties completely. Freeze drying is utilized in the processing of solutions, suspensions and granular solids.

The paper presents a mathematical model describing the sublimation operation. An approach solution of the model allows to calculate the transport coefficients from mass decrease and temperature measurements carried out in dry-freezing experiments. The method is presented for the drying operation of frozen liquids and of granular material.

NOMENCLATURE

- a thermal diffusivity, $m^2\ s^{-1}$
- B intersection of the straight portion of curve
- c specific heat, $J\ kg^{-1}\ K^{-1}$
- D_k diffusion coefficient in free molecular flow, $m^2\ s^{-1}$
- F surface area. m^2
- g $g=\psi\rho_w$ and $g=\psi\rho_w\zeta$, dg m^{-3}
- j mass flow, drying rate, $kg\ m^{-2}\ s^{-1}$
- K_n Knudsen number
- l length, m
- L thickness of the material layer, m
- M slope of the straight portion of curve
- M_w molecular mass of the wetting medium, $kg\ mol^{-1}$
- $\Delta\bar{m}$ mass decrease, $kg\ m^{-2}$
- P pressure, Pa
- q heat flux density, $W\ m^{-2}$
- r_s heat of sublimation, $J\ kg^{-1}$
- R universal gas constant, $Pa\ m^3\ K^{-1}\ mol^{-1}$
- s slope of the secant of the vapour pressure curve, $Pa\ K^{-1}$
- t temperature, ^0C
- T temperature, K
- V volume, m^3
- W mass of moisture, kg
- z co-ordinate across the thickness of the layer, m

Greek letters

- β mass transfer coefficient, $kg\ m^{-1}\ s^{-1}\ Pa^{-1}$
- ε resistance coefficient, m^2
- ρ density, $kg\ m^{-3}$
- ψ porosity, $m^3\ m^{-3}$
- λ coefficient of thermal conductivity, $W\ m^{-1}\ K^{-1}$
- λ' molecular free path length, m
- τ time, s, h
- μ dynamic viscosity, Pa s
- ζ wetting degree of the pores, $m^3\ m^{-3}$
- ξ thickness of the dried layer, m

Subscripts

- I dried layer in drying of frozen liquids
- II moist, frozen layer in drying of frozen liquids
- III dried granular materials layer
- IV moist, frozen granular material layer
- 0 initial value
- p refers to pores
- K refers to the vapour condenser
- L value at the heating plate
- ξ value in the sublimation zone
- w refers to moisture

Drying of frozen liquids

Mathematical model

When a frozen liquid is being dried by conductive heat transfer, heat input takes place from the material-holding tray and sublimation of the solvent starts from the surface. The sublimation zone moves in the interior of the material towards the heated face. The dried part of the material forms a porous solid layer through which the vapour must pass (Fig. 1).

Let the drying layer be an infinite sheet. We assume that the sublimation front moves uniformly from the surface towards the heated face. In studying the operation we shall start from the enthalpy balance of the two layers.

For the dried, porous layer marded with the subscript I

$$c_I\rho_I\frac{\partial t_I}{\partial\tau}=\frac{-\partial}{\partial z}[q_I+j_wc_{pw}(t_\xi-t_I)].$$

Since in this layer heat spreads by conduction,

$$q_I=-\lambda_I\frac{\partial t_I}{\partial z}.$$

The drying rate, i.e. the sublimation mass flow is

$$j_w=-\frac{1}{F}\frac{dW}{d\tau} \tag{1}$$

Let us now express the moisture content W with the volume V_p of the pores formed in the drying process:

$$W=W_0-V_p\frac{\xi}{L}\rho_w$$

and

$$\frac{dW}{d\tau}=-\frac{V_p}{L}\rho_w\frac{d\xi}{d\tau}.$$

Hence

$$j_w=\rho_w\frac{V_p}{FL}\frac{d\xi}{d\tau}$$

where $\frac{V_p}{FL}=\psi$ is the porosity of the layer. Let us use the symbol g for $\psi\cdot\rho_w$.

We will then arrive to

$$j_w=g\frac{d\xi}{d\tau}$$

and hence

$$\frac{\partial t_I}{\partial\tau}=a_I\frac{\partial^2 t_I}{\partial z^2}-g\frac{c_{pw}}{c_I\rho_I}\frac{d\xi}{d\tau}\frac{\partial(t_\xi-t_I)}{\partial z} \tag{2}$$

Initial conditions: $t_I(z,0)=t_{II}(z,0)=t_0=\text{const.}$

Boundary conditions:
No heat exchange on the surface: at z = 0,

$$q|_{z=0}=0 \quad\text{and}\quad \left.\frac{\partial t_I}{\partial z}\right|_{z=0}=0$$

The amount of heat passing through the interface of the two layers, i.e. at z = ξ is

$$q_\xi=-\lambda_I\left.\frac{\partial t_I}{\partial z}\right|_{\xi-}$$

and

$$q_\xi=-\lambda_{II}\left.\frac{\partial t_{II}}{\partial z}\right|_{\xi+}+j_w\cdot r_s,\quad\text{resp.}$$

That is, on the basis of the above:

$$q_\xi=-\lambda_{II}\left.\frac{\partial t_{II}}{\partial z}\right|_{\xi+}+g\cdot r_s\frac{dz}{d\tau}.$$

For layer II, i.e. for the frozen liquid:

$$c_{II}\rho_{II}\frac{\partial t_{II}}{\partial\tau}=-\frac{\partial}{\partial z}(q_{II})$$

where

$$q_{II}=-\lambda_{II}\frac{\partial t_{II}}{\partial z}.$$

Hence

$$\frac{\partial t_{II}}{\partial\tau}=a_{II}\frac{\partial^2 t}{\partial z^2} \tag{3}$$

Boundary condition: at z = L, T_L = const.

In the above considerations it was assumed that layer I dried completely. In practice, however, about 3 to 4% bound moisture still remains in the layer. This can be removed at higher temperatures only.

The above mathematical model can be used for numerical computations with any desired accuracy by computer, or approached by some method. In the followings an approach will be discussed that is well suited to determine the transport coefficients.

Approach solution

The simplifying assumption for the two-layer model discussed in the foregoings is that sublimation proceeds at a steady "wet-bulb thermometer temperature" t_ξ.

Therefore

$$\frac{\partial t_I}{\partial z}=0,\quad\text{since}\quad\left.\frac{\partial t_I}{\partial z}\right|_{z=0}=0,$$

and

$$q_{II}=\frac{\lambda_{II}}{L-\xi}(t_L-t_\xi)=j_w\cdot r_s \tag{4}$$

The water vapour formed must pass through the porous layer, it must overcome the resistance of the layer. The nature of the flow in the layer, under the conditions of freeze-drying, can be decided on the basis of the Knudsen number (1, 2):

- in the case of $Kn>2\sim 10$ free molecular flow will take place,
- in the range $2\geq Kn>0.01$ the flow will be transient,
- in the case of $Kn<0.01$ constant flow of the medium will be established.

(The Knudsen number Kn is the ratio of the molecular free path and the diameter of the capillary:

$$Kn=\frac{\lambda'}{d}.)$$

Freeze drying, that is, sublimation of water is possible only below the triple point pressure of water, i.e. below 613 Pa. At such pressures the free path of the molecules is in the order of 0.01 mm.(3). At the pressures usually applied in lyophilization(0.1 to 1 Pa)the free path is in order of 0.5 to 1 cm. Hence the flow in the pores will take place in the range of free molecular flow. In the case of very thick layers the pressure in the layer will increase, so that the flow in the pores will be in the transient range.

In the range of molecular free flow the mass flow is

$$j_w=D_K\frac{M_w}{RT}\frac{\Delta P}{l}.$$

In the transient flow range, considering that flow takes place in a gas-free evacuated space, the simplified form of the mass flow is (1)

$$j_w=\frac{\rho\varepsilon\Delta P}{\mu l}$$

Let us combine all material constants contained in these expressions into the mass transport coefficient. This will then yield

$$j_w=\frac{\beta}{l}\Delta P,$$

and, using the symbols in Fig. 1,

$$j_w=\frac{\beta}{\xi}(P_{w\xi}-P_{wK}).$$

The value of the pressure P_{wK} established in the drying chamber is defined by the temperature of the condenser of the lyophilizing apparatus: its value is equal to the vapour pressure corresponding to the condenser temperature t_K.

The pressure difference in the above equation can therefore be expressed by a temperature difference, taking into account the water-ice vapour curve. The slope of the secant line substituting the portion in question of the vapour pressure curve is

$$s=\frac{P_{w\xi}-P_{wK}}{t_\xi-t_K}$$

and hence

$$j_w=\frac{\beta}{\xi}s(t_\xi-t_K) \qquad (5)$$

By substituting the value of t_ξ from Eq. (4)(the expression for q_{II})into Eq. (5), with Eq. (1) we obtain

$$j_w\left[\left(\frac{1}{\beta\cdot s\cdot r_s}-\frac{1}{\lambda_{II}}\right)\xi+\frac{L}{\lambda_{II}}\right]\frac{d\xi}{d\tau}=\frac{t_L-t_K}{g\cdot r_s}. \qquad (6)$$

Separation of the variables and integration will yield

$$\tau=\frac{g\cdot r_s}{t_L-t_K}\left[\frac{L}{\lambda_{II}}\xi+\frac{1}{2}\left(\frac{1}{\beta\cdot s\cdot r_s}-\frac{1}{\lambda_{II}}\right)\xi^2\right]. \qquad (7)$$

Equation (7) allows to determine drying time in the ideal case, i.e. when the sublimation front moves steadily from the surface to the heated face, no heat exchange takes place at the surface and the sublimation temperature is constant.

Our measurements demonstrated that these assumptions are justified for an about 5 mm thick layer up to removing 80% of the moisture content.

In order to be able to calculate drying times, the knowledge of the transport coefficients is necessary. The values of λ_{II} and β can be obtained from experimental drying data by means of Eq. (7).

For this purpose let us write Eq. (7) in the following form:

$$\frac{\tau}{g\cdot\xi}=\frac{r_s}{t_L-t_K}\left[\frac{L}{\lambda_{II}}+\frac{1}{2}\left(\frac{1}{\beta\cdot s\cdot r_s}-\frac{1}{\lambda_{II}}\right)\xi\right]. \qquad (8)$$

From Eq. (1):

$$\xi g=\int_0^\tau j_w\,d\tau=\Delta\bar{m} \qquad (9)$$

where $\Delta\bar{m}$ is the mass decrease of the material being dried per unit surface area during the period τ. In this manner the thickness of the dried layer which would be difficult to measure experimentally can be substituted by the readily measurable mass decrease. By combining Eqs (8) and (9) we arrive to

$$\frac{\tau}{\Delta\bar{m}}=\frac{r_s\cdot L}{(t_L-t_K)\lambda_{II}}+\frac{r_s}{2g(t_L-t_K)}\left(\frac{1}{\beta_s\cdot r_s}-\frac{1}{\lambda_{II}}\right)\Delta\bar{m} \qquad (10)$$

which is the equation of the straight line

$$\frac{\tau}{\Delta\bar{m}} = B + M \cdot \Delta\bar{m}$$

in a system with the coordinates $\tau/\Delta\bar{m}$ versus $\Delta\bar{m}$. This will allow to determine the transport coefficients λ_{II} and β from Eq. (10) by experimentally measuring the mass decrease and the temperatures as a function of time, analogously to the measurements performed with atmospheric macroporous drying (4). From the intersection B the coefficient of thermal conduction of the frozen liquid can be obtained as:

$$\lambda_{II} = \frac{r_s \cdot L}{(t_L - t_K)B},$$

and the slope M allows to obtain the value of :

$$\beta = \frac{1}{s(t_L - t_K)} \cdot \frac{1}{2M \cdot g + B/L}.$$

Experimental results

For the drying experiments we used a laboratory lyophilizing apparatus type OE-950 (manufacturer: Labor-MIM), supplemented - to comply with requirements - with a balance built into the drying chamber, and a system of thermocouples to be frozen into the material The thermocouples were fitted to the drying tray.

The lyophilization apparatus type OE-950 includes four electrically heated, temperature-controlled tray-supporting plates. In our experiments we only used the lowest plate.

The inbuilt balance consists of a force-measuring cell and an electrically driven lifting device fitted to the tray-holder frame(Fig. 2). The signal from the force-measuring cell is transformed by an tensometer bridge located outside the drying chamber.

For weight measurements, the drying tray, together with the force-measuring cell is lifted by means of the lifting device, the instrument reading is taken and the tray is lowered to its place. Weight measurements are taken every half hour, and since they last only about 1 minute, they interfere only negligibly with the drying process.

The thermocouples are made of 0.1 mm diam. copper and constant wire. They are fitted to a holder so as to measure both the tray temperature and the layer temperatures. The signals of the thermocouples are transmitted through the outlet channels built-in in the original apparatus to the recording compensograph Philips PM 9833.

Figure 3 is the mass decrease curve obtained in a lyophilization experiment performed with cow milk, and the corresponding function $\tau/\Delta\bar{m}$ versus $\Delta\bar{m}$ is plotted in Fig. 4. The required straight portion of this curve is clearly observable. The initial results deviating from the straight line are due to the formation of the porous layer, and the results towards the end of the experiment, also deviating from the straight line, indicate that- owing to the non-uniformity of the sublimation fromt - the front has in part already reached the bottom of the tray.

The experimental conditions were as follows:
-thickness of the layer: : = 4.4 mm
-pressure in the chamber: P_K= 3.9 Pa
-temperature at the bottom of the tray: t_L=-38.7^{0}C
-temperature of the condenser: t_K=-50^{0}C.

The values needed for the evaluation of the results were
-intersect: B = 1,41 m^2hkg^{-1}
-slope: M = 0.249 m^2hkg^{-2}

The coefficient of thermal conductivity computed from the experimental results is

$$\lambda_{II} = 0.214\ Wm^{-1}K^{-1}$$

and the coefficient of mass transfer is

$$\beta = 2.79 \cdot 10^{-8}\ kgm^{-1}s^{-1}Pa^{-1}.$$

Since in the experimental conditions the Knudsen number Kn 2, the diffusion coefficient D_K can be calculated from the value of β:

$$D_K = \beta\frac{RT}{M_w} = 3.19 \cdot 10^{-3}\ m^2s^{-1}.$$

In another experiment carried out under closely similar conditions we obtained

$$\lambda_{II} = 0.219\ Wm^{-1}K^{-1},$$
$$\beta = 3.79 \cdot 10^{-8}\ kgm^{-1}s^{-1}Pa^{-1},$$

and

$$D_K = 4.08 \cdot 10^{-3}\ m^2s^{-1}.$$

The difference in the value of β might be due to the differences in freezing.

Freeze drying of granular solids

The mass flow and heat flow developed in drying frozen granular solids is schematically shown in Fig. 5. Heat input is performed by thermal conduction from the material-holding tray. Let us assume that the size of the granules to be dried is negligible as compared to the thickness of the layer. The vapour evolved during the drying process is capable of passing between the particles through the frozen solid layer. Thus the drying process will start at the bottom of the layer and the sublimation front will move from the bottom towards the surface.

Let the drying layer be an infinite sheet, and the sublimation front move uniformly towards the surface.

Mathematical model

The enthalpy balance for layer III (cf.Fig. 5) consisting of dried particles is

$$c_{III}\cdot\rho_{III}\frac{\partial t_{III}}{\partial\tau}=-\frac{\partial}{\partial z}(q_{III})$$

assuming that in this layer the material will not be subjected to further changes in moisture content.

Heat spreads by conduction in the layer. In the present case

$$q_{III}=-\lambda_{III}\frac{\partial t_{III}}{\partial z}$$

and hence

$$\frac{\partial t_{III}}{\partial\tau}=\frac{\lambda_{III}}{\rho_{III}c_{III}}\frac{\partial^2 t_{III}}{\partial z^2}=a_{III}\frac{\partial^2 t_{III}}{\partial z^2}.$$

The enthalpy balance of layer IV is

$$c_{IV}\rho_{IV}\frac{\partial t_{IV}}{\partial\tau}=-\frac{\partial}{\partial z}[q_{IV}+j_w c_{pw}(t_\xi-t_{IV})].$$

Here too, heat spreads by conduction:

$$q_{IV}=-\lambda_{IV}\frac{\partial t_{IV}}{\partial z}$$

and hence

$$\frac{\partial t_{IV}}{\partial\tau}=\frac{\lambda_{IV}}{\rho_{IV}c_{IV}}\frac{\partial^2 t_{IV}}{\partial z^2}-\frac{c_{pw}}{c_{IV}\rho_{IV}}j_w\frac{\partial(t_\xi-t_{IV})}{\partial z},$$

i.e.

$$\frac{\partial t_{IV}}{\partial\tau}=a_{IV}\frac{\partial^2 t_{IV}}{\partial z^2}-\frac{c_{pw}}{c_{IV}\rho_{IV}}j_w\frac{\partial(t_\xi-t_{IV})}{\partial z}. \quad (11)$$

Initial conditions:

$$t_{III}(z,0)=t_{IV}(z,0)=t_0=\text{const.}$$

Boundary conditions:

$$\text{at}\quad z=0,\qquad t_L|_{z=0}=\text{const.}$$

The amount of heat passing at $z=\xi$ is

$$q_\xi=-\lambda_{III}\left.\frac{\partial t_{III}}{\partial z}\right|_{\xi-}$$

and

$$q_\xi=-\lambda_{IV}\left.\frac{\partial t_{IV}}{\partial z}\right|_{\xi+}+j_w\cdot r_s,\quad\text{resp.}$$

There is no heat exchange on the surface:

$$\text{at}\quad z=L,\quad q_{z=L}=0,\quad\text{and hence}\quad\left.\frac{\partial t_{IV}}{\partial z}\right|_L=0.$$

Let us express the mass flow j_w with the drying rate:

$$j_w=-\frac{1}{F}\frac{dW}{d\tau},$$

and the moisture content W with the pore volume emptied in the course of drying:

$$W=W_0-V_p\zeta\frac{\xi}{L}\rho_w$$

where ζ is the wetting degree of the pores. Then

$$\frac{dW}{d\tau}=-\frac{V_p\zeta}{L}\rho_w\frac{d\xi}{d\tau}$$

and hence

$$j_w=\rho_w\frac{V_p\zeta\,d\xi}{FL\,d\tau}=\rho_w\psi\zeta\frac{d\xi}{d\tau}.$$

Let us use the symbol g for $\psi\zeta\rho_w$:

$$j_w=g\frac{d\xi}{d\tau}$$

Thereby Eq. (11) will assume the form

$$\frac{\partial t_{IV}}{\partial\tau}=a_{IV}\frac{\partial^2 t_{IV}}{\partial z^2}-\frac{c_{pw}}{c_{IV}\rho_{IV}}g\frac{d\xi}{d\tau}\frac{\partial(t_\xi-t_{IV})}{\partial z}$$

Approach solution

The mathematical model describing the freeze-drying process of granular solids is analogous in form to that of lyophilization of liquids, therefore what was said above is valid in this case too. It is reasonable to find an approach solution. For this purpose we shall use the assumption that - as compared to the amount of heat spent for sublimation - the amount of heat required to warm layer IV to the sublimation temperature is negligible, that is, the total amount of heat passing by conduction through layer III is utilized for sublimation

$$q|_{z=\xi}=j_w r_s=\frac{\lambda_{III}}{\xi}(t_L-t_\xi).$$

The vapour flow j_w must pass through the granular layer:

$$j_w=\beta\frac{P_{w\xi}-P_{wK}}{L-\xi}.$$

The pressure difference can then be expressed in terms of the secant of the vapour pressure curve:

$$j_w=\frac{\beta s}{L-\xi}(t_\xi-t_k)$$

$$j_w\left[\frac{L}{\beta s}+\left(\frac{r_s}{\lambda_{III}}-\frac{1}{\beta_s}\right)\xi\right]=t_L-t_K$$

As shown earlier,

$$j_w = g\frac{d\xi}{d\tau}$$

and hence

$$g\left[\frac{L}{\beta_s} + \left(\frac{r_s}{\lambda_{III}} - \frac{1}{\beta_s}\right)\xi\right]d\xi = (t_L - t_K)\,d\tau$$

Integration of the above equation between the boundaries 0-τ and 0-ξ yields the drying time

$$\tau = \frac{g}{t_L - t_K}\left[\frac{L\xi}{\beta_s} + \frac{1}{2}\left(\frac{r_s}{\lambda_{III}} - \frac{1}{\beta_s}\right)\xi^2\right].$$

A suitable transformation yields

$$\frac{\tau}{g\xi} = \frac{r_s}{t_L - t_K}\left[\frac{L}{\beta_s r_s} + \frac{1}{2}\left(\frac{1}{\lambda_{III}} - \frac{1}{\beta r_s s}\right)\xi\right]$$

and by substituting $g\xi = \Delta m$ we obtain an equation formally similar to Eq. (10):

$$\frac{\tau}{\Delta\bar{m}} = \frac{r_s}{t_L - t_K}\left[\frac{L}{\beta s r_s} + \frac{1}{2g}\left(\frac{1}{\lambda_{III}} - \frac{1}{\beta r_s s}\right)\right]\Delta\bar{m} \qquad (12)$$

By means of Eq. (12) the transport coefficients β and λ_{III} can be computed from experimental data.

Experimental results

The lyophilization experiments with granular materials were carried out similarly to those of frozen liquids, in the same apparatus. However, reproducibility was more difficult to achieve, since a 4 to 6 mm uniform thickness layer must uniformly be compacted, and the thermocouple grid must be embedded into the layer, providing satisfactory compactness without injuring the sensitive thermocouples.

The materials used in our experiments were moisted perlite, and wheat grits roasted and stewed. Figs 6 and 7 represent the results obtained with wheat grits: mass decrease versus time and the function $\tau/\Delta\bar{m}$ versus $\bar{m}$, resp. The feature of interest in the latter is that the straight portion appears immediately at the start of the experiment, that is, the assumed state is established very rapidly in the granular layer. Towards the end of the experiment, when the sublimation front reaches the tray, the curve deviates from the straight portion.

The experimental ocnditions were as follows:
- -layer thickness: L = 4.6 mm
- -chamber pressure: P_K = 3.1 Pa
- -temperature at the bottom of the tray: $t_L = 27.2\,^0C$
- -temperature of the condenser: $t_K = -52\,^0C$.

The intersect in Fig. 7 is $B = 0.95\ m^2hkg^{-1}$, and the slope is $M = 1.187\ m^2hkg^{-2}$.

On the basis of Eq. (12) the mass transfer coefficient is

$$\beta = \frac{L}{Bs(t_L - t_K)} = 1.21 \cdot 10^{-8}\ kgm^{-1}s^{-1}Pa^{-1},$$

and the diffusion coefficient is

$$D_K = \beta\frac{RT}{M_w} = 1.31 \cdot 10^{-3}\ m^2\,s^{-1}.$$

The coefficient of thermal conductivity of the dry layer is (from Eq. 12)

$$\lambda_{III} = \frac{1}{t_L - t_K}\frac{Lr_s}{B + 2gML} = 9.76 \cdot 10^{-3}\ W\,m^{-1}\,K^{-1}$$

In another experiment performed in similar conditions the results were

$$\beta = 1.44 \cdot 10^{-8}\ kg\,m^{-1}\,s^{-1}\,Pa^{-1},$$

$$D_K = 1.45 \cdot 10^{-3}\ m^2\,s^{-1}$$

$$\lambda_{III} = 9.03 \cdot 10^{-3}\ W\,m^{-1}\,K^{-1}.$$

It is apparent that the coefficient of thermal conductivity of the loose layer consisting of dried particles is very poor. It is surprising that the diffusion coefficient is lower in this case thatn the value found for frozen liquids. This might be explained by the fact that the value obtained is an effective value, related to both intragranular and intergranular diffusion, with intragranular resistance being higher than the resistance of the pores in the frozen liquid.

ACKNOWLEDGEMENTS

The Editor is grateful to Purnima Mujumdar for the typing of this manuscript. Vibhakar Jariwalla assisted with the preparation of the camera-ready copy.

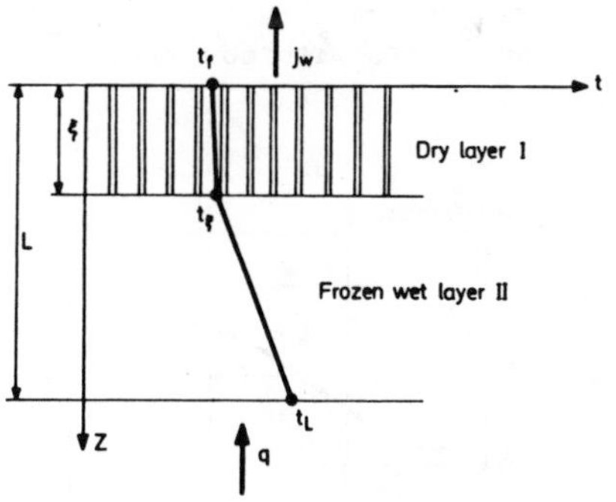

Fig. 1. Schematic diagram of the freeze drying of liquids

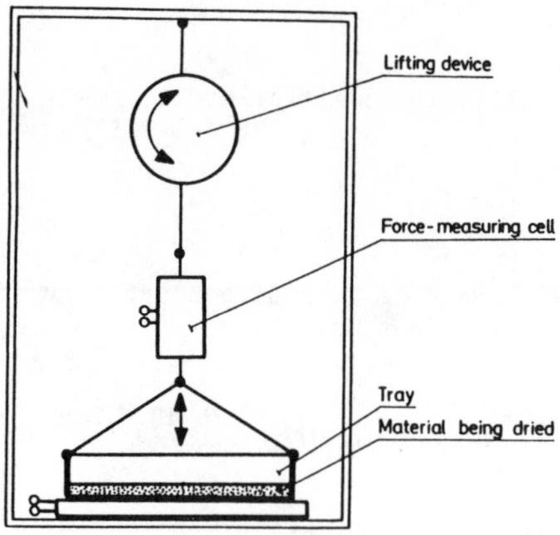

Fig. 2. Diagram of the apparatus used in the experiments.

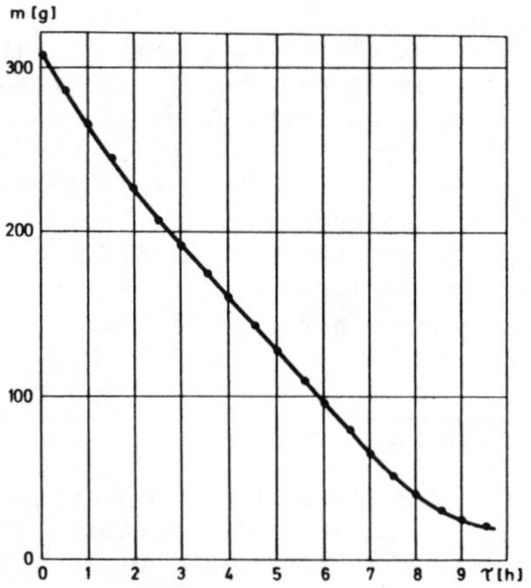

Fig. 3. Drying curve obtained in lyophilization experiments with cow milk

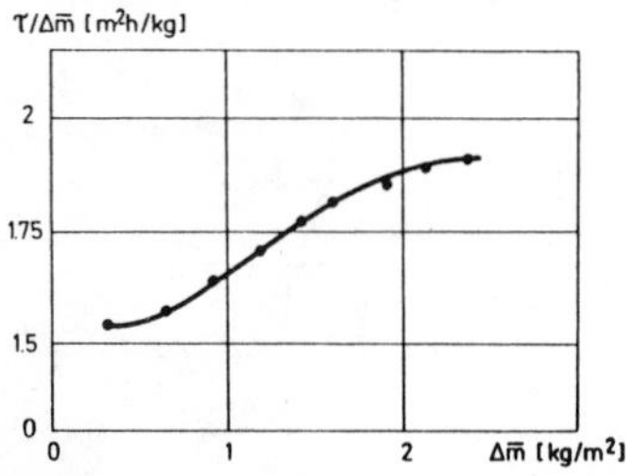

Fig. 4. The function $\tau/\Delta\bar{m}$ vs. $\Delta\bar{m}$ corresponding to the drying curve plotted in Fig. 3

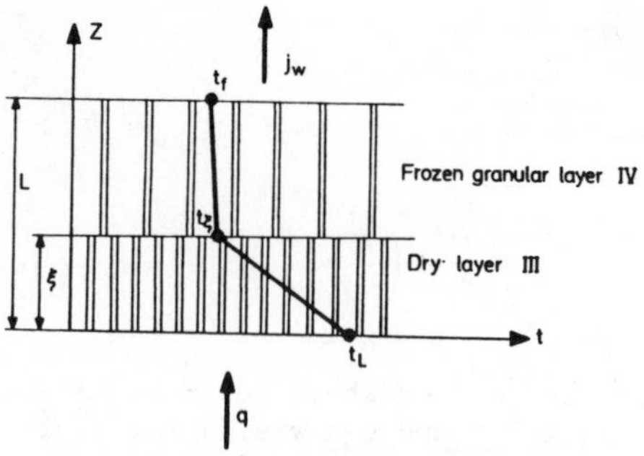

Fig. 5. Freeze drying of granular solids

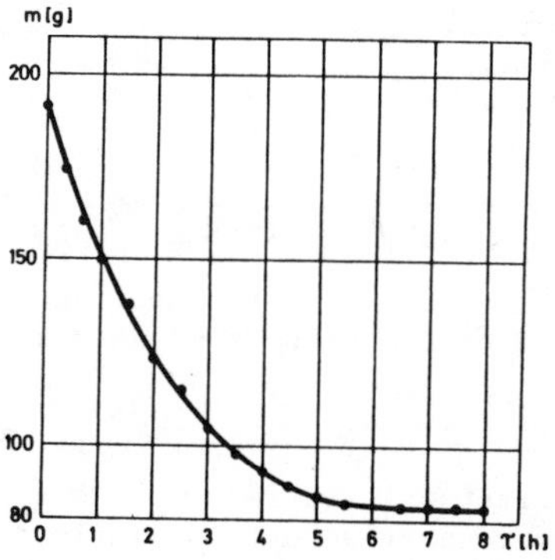

Fig. 6. Drying curve obtained in lyophilization experiments with wheat grits

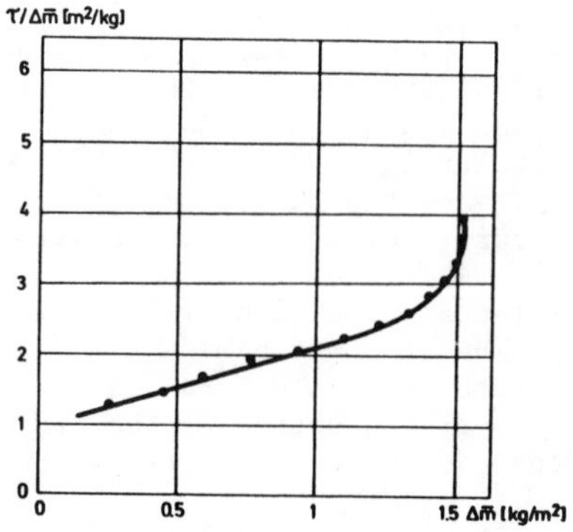

Fig. 7. The function $\tau/\Delta\bar{m}$ vs. $\Delta\bar{m}$ corresponding to the drying curve plotted in Fig. 6

AUTHOR INDEX

SUBJECT INDEX